International Scientific-Practical Conference "Agriculture and Food Security: Technology, Innovation, Markets, Human Resources" (FIES 2024)

AA003587

BIO Web of Conferences Volume 161 (2025)

Kazan, Russia
28-29 November 2024

Editors:

Farida Nezhmetdinova
Ayrat Valiev
Bulat Ziganshin
Aksu Taylan
Rustam Nizamov

ISBN: 979-8-3313-1499-6

Printed from e-media with permission by:

Curran Associates, Inc.
57 Morehouse Lane
Red Hook, NY 12571

Some format issues inherent in the e-media version may also appear in this print version.

For additional information, please contact EDP Sciences – Web of Conferences at the address below.

EDP Sciences – Web of Conferences
17, Avenue du Hoggar
Parc d'Activité de Courtabœuf
BP 112
F-91944 Les Ulis Cedex A
France

Phone: +33 (0) 1 69 18 75 75
Fax: +33 (0) 1 69 28 84 91

contact-edps@webofconferences.org

Additional copies of this publication are available from:

Curran Associates, Inc.
57 Morehouse Lane
Red Hook, NY 12571 USA
Phone: 845-758-0400
Fax: 845-758-2633
Email: curran@proceedings.com
Web: www.proceedings.com

TABLE OF CONTENTS

Relevance and Prospects of Precision Agro-Technology Development in the Context of Sustainable Agriculture 1
Al-Rahmi Waleed Mugahed, Modina M. A., Grigoryan A. G., Tavbulatova Zulay K.

Comparative Assessment of Fertilizer Systems and Intensification Levels in Crop Production 6
Zyukin Danil A., Glushkov Ivan A.

Analysis of Factors Limiting the Import-Substituting Effect of Cluster Development in the Agro-Industrial Complex 11
Moga Irina S., Baykov Fedor Yu., Sidorenko Vlada S.

Innovative Investments in Agriculture as a Factor of Food Security 16
Smirnov Evgeny N., Pasko Alexander V., Antropova Marina Yu.

Problems and Prospects of Innovative Development of Agro-Industrial Complex in Eastern Siberia 21
Qosimov Jahongir, Akhmedov Islombek, Abdikadirov Shavkat, Dusyarov Khudoymurot, Ivanov M. Yu.

Salmonellosis: Threats to the Economy and Avian Health 26
Faizrakhmanova Gulina, Galiullin Albert, Koroleva Lidia, Spiridonov Gennady, Makhmutov Aidar

Influence of Natural Biostimulants on Growth and Quality of Cattle Meat 30
Ashurov Nurali, Yusupov Abdulaziz, Primkulov Bekzod, Kurbanov Abdimalik, Grankina Natalia A., Bogus Aslan Yu.

Prevention of Subclinical Mastitis in Lactating Cows with a New Probiotic Preparation 34
Porfiriev Artur, Fayzrakhmanov Ramil, Sofronov Vladimir, Danilova Nadezhda, Kuznetsova Elena

Biogas as a Solution to the Problem of Waste Utilisation in Agriculture 40
Sevryugina Nadezhda S., Dragulenko Vladislav V., Sukhanov Dmitry S., Ukraintsev Maxim M., Korchagin Pavel T.

Exploring the Potential of Fullerenes in Food and Agriculture 45
Gavrilov Alexander N., Gladkikh Tatiana V., Emelyanov Alexander E., Ivanov Andrey V., Bukuru Liz C., Sukhanova Natalia V.

Innovative Tool for Sustainable Production and Recycling of Agro-Industrial Wastes 51
Turchanin Oleg S., Grigoryan David R., Ukraintseva Irina V., Nazarov Igor V., Tolstoukhova Tatiana N.

Results of Experimental Studies of a Small-Volume Biogas Plant 55
Gayfullin Ilnur Kh., Ziganshin Bulat G., Khaliullina Zulfiya M., Semicheva Olga S., Shornikov Artyom V.

Optimisation of Anaerobic Digestion of Agricultural Waste by Simulation Modelling 60
Matasova I. Yu., Masenko Aleksey V., Grigoryan David R., Tokareva Anna N., Demchenko Maria S.

Granulated Goat Milk Sublimate and the Experimental Setup for Its Production 65
Yerofeyeva Natalya, Shakhov Artem, Glotova Irina, Shakhov Sergey, Blinnikov Kirill

Analysis of Biotechnological Approaches to Utilisation of Organic Wastes from Livestock and Poultry Production ... 70
Al-Rahmi Waleed Mugahed, Kuznetsova Yu. S., Grankina Natalia A., Bogus Aslan Yu., Panchenko Sergey V.

Disinfection of Beekeeping Objects with Oxygen-Containing Preparations for Bee Aspergillosis ... 75
Gruznov Dmitry V., Gruznova Olga A., Sokhlikov Alexey B., Chesnokova Irina P.

Digital and 'Green' Technologies in Agricultural Production on the Example of a Farming Enterprise ... 80
Karelina Ekaterina A., Ivanova Valeriya I., Smagulova Samal M.

Use of Extruded Lupin in Feeding Young Quails ... 84
Kasanova Nadiya, Gainullina Munira, Prosvirnikov Dmitry, Tuntsev Denis, Golovkova Ekaterina

Waste-Free Integrated Processing of Invertebrates into Feed Additives Using Proteolytic Enzymes ... 88
Ernazarov Karim, Yusupov Bekzod, Turaev Khumoyiddin, Ganieva Sayyora, Masienko Ivan V., Sukhanov Dmitry S.

Resistance of Sarepta Mustard to a Decrease in Ambient Temperature During Vegetation ... 92
Kaznina Natalia, Kholoptseva Ekaterina, Batova Yulia, Taskina Ksenia

Production of Biohumus by Microbiological Processing of Forestry Wastes ... 97
Choriev Ruzimurat, Toshmatov Ravshan, Kuziev Abdimurot, Majidova Shahnoza, Zatsarinnaya Irina A., Sablina Polina A.

Effect of Pre-Sowing Seed Treatment with Semiconductor Laser Radiation on Soft Wheat Productivity and Its Resistance to Diseases ... 102
Kolesnikov L. E., Fokina M. I., Kolesnikova Yu. R., Chernikova N. V., Yakhyaev G. G. R.

Integration in Breeding and Seed Production Through the Formation of Regional Systems ... 108
Svyatova Olga V., Malakhova Svetlana V., Nozdracheva Elena N.

Intensification of the Process of Extraction of Sucrose from Sugar Beet of Different Technological Advantage ... 113
Kulneva Nadezhda, Korobova Lyudmila, Matytsina Irina, Matvienko Natalia, Chernyaeva Svetlana

Integrated Model of Agricultural Production Through the Use of Non-Waste Technologies Based on Microorganisms ... 118
Matasova I. Yu., Skoda V. V., Shchebeteev Valery A., Karadzhiyan Andrey M., Mushaeva Kermen B., Kurepina Natalya L.

Application of Statistical Methods for the Agriculture Complex Using the Example of Sugar Beet Production ... 123
Korobova Lyudmila, Tolstova Irina, Ivliev Maxim, Datsenko Natalia, Chikunov Sergey

State and Development of Farms in Uzbekistan as a Basis for Food Security ... 128
Shavazov Kadirjon, Khojibekova Shokhida, Mukimov Bayramali, Umirov Abdigappar, Kadirova Shakhnoza

The Halal Standard: From the Regulatory Norms of Eating Behavior to Large Sectors of the Economy ... 133
Valiev A. R., Yazid Yaacob, Nezhmetdinova F. T., Kalimullin M. N., Salikhov R. T., Parfenova K. A.

Resource-Saving Tillage Technologies and Their Impact on Grain Crop Yields 139
Latyaheva Zoya I., Zyukin Danil A., Karpushin Nikita A.

Risks and Safety Standards for the Use of Genetically Modified Organisms in the Context of Food Security 144
Nezhmetdinova F. T., Guryleva M. E., Kardanova L. D.

Forced Forecast of Agricultural Development in the European North of Russia 150
Asaturyan Sergey V., Dubina Konstantin P., Khizhnyak Vladimir I., Shchebeteev Valery A., Bovsh Vladislav V.

Influence of Irrigation and Fertilisation on Yield and Quality of Green Mass of Maize in Soil-Climatic Conditions of the Republic of Tatarstan 155
Safiollin F. N., Lukmanov A. A., Khismatullin M. M., Suleimanov S. R., Yakhin I. F.

Technological and Biochemical Characteristics of Spring Wheat Varieties of Irkutsk Region 161
Gordeeva Yulia V., Kuvshinova Elena K., Golovko Anatoly S., Bazova Anastasia S., Cheryomukhina Angelina A.

The Strength of the Influence of Polymorphism of the Kappa-Casein and Diacylglycerol O-Acyltransferase Gene Polymorphisms on the Economically Useful Qualities of Dairy Cattle 166
Shaydullin Radik R., Zagidullin Lenar R., Akhmetov Tahir M., Tyulkin Sergey V., Konakova Irina A.

Microbial Biotechnology of Milk Whey Processing for Agricultural Needs 170
Astanakulov Komil, Abdiev Orifjon, Ulashov Jonibek, Ergasheva Dilnoza, Zatsarinnaya Irina A., Kupriyanenko Nikolay A.

Antimicrobial Properties of a New Bactericide for Veterinary Disinfection 175
Shcherbakova Gulizar Sh., Popov Nikolay I., Shuteeva Ekaterina N., Kuvshinchikov Nikolay N., Pirozhikhin Viktor A., Gruznova Olga A., Gruznov Dmitry V.

Microbial Protein Production Based on Complex Plant Raw Materials: Technology Development and Efficiency Assessment 180
Khudayarov Berdirasul, Abdukakhkharov Abduaziz, Axmedov Alim, Rustamova Nigora, Kuvvatova Mokhira

Biogas Technology of Complete Mixed Fodders with Application of Absorption Heat Pump 185
Shevtsov Alexander Anatolievich, Vasilenko Vitaly Nikolaevich, Frolova Larisa Nikolaevna, Derkanosova Anna Aleksandrovna, Zheltoukhova Ekaterina Yuryevna

Transformation of the Agro-Industrial Complex of Kazakhstan 190
Nussipali Rollan, Asilyan L. V., Grigoryan A. G., Zatsarinnaya Irina A., Kupriyanenko Nikolay A.

Organization of Veterinary Services for Industrial Duck Farming: Problems and Solutions 196
Vasiliev Mihail N., Gainutdinova Karina V., Ravilov Rustam Kh., Vasilieva Anastasia I.

Nature-Like as Agro-Technologies a Convergent Approach to Carbon Polygon Creation 201
Sevryugina Nadezhda S., Kozenkova G. L., Migda N. S.

Current State and Prospects for the Development of the Grain Sector of China's Agro-Industrial Complex 206
Weiwei Luo, Kai Shang, Podmolodina I. M.

Development of Probiotic Feed Based on Homogenised Potato Pulp 211
Borotov Atkham, Narzullaev Olim, Yakubov Lazizkhon, Rajabova Matluba, Karshieva Bogdagul

Harrington Function as an Integral Approach to Assessing Food Security in Russia 216
Qian Xinyi, Salikov Yu. A., Kolomytseva O. Yu., Maslov A. V.

Problems and Ways of Development of Russian Agriculture 221
Skripkina Elena V., Latyaheva Zoya I., Lisitsina Julia V.

Enzyme Complex Influence on Physiological Properties of Alcoholic Yeast During Cultivation 226
Yakovlev Alexey, Agafonov Gennady, Yakovleva Svetlana, Romanyuk Tatyana, Zueva Natalia, Chusova Alla, Kovaleva Tatyana

Optimisation of Agro-Technologies for Sustainable Development of Agriculture in Uzbekistan 230
Yusupov Sherzod, Azizov Inomdjon, Khaydarov Ramozon, Inadullaeva Sevara, Kholmirzayev Sattar

Artificial Meat in the Structure of Halal Meat Products: Legislative and Legal Status, Technologies, Russian and International Experience 235
Gilmutdinov Rustam Y., Gilmutdinov Daniyar R., Galiullin Albert K., Volkov Ali H., Shalamova Guzel G.

Technological Modernisation of the Russian Agro-Industrial Complex 241
Astanakulov Komil, Tatarintsev Vladimir L., Nuriddinov Jasurbek, Mukumova Feruza, Allaberdiyeva Kumri, Suyunova Zamira

Bacteria of the Helicobacter Genus in Food 246
Nurgaliev Farit Mullagalievich, Nurgalieva Alinya Ravilevna, Nagumanova Liana Ruslanovna

Prospects and Feasibility of Involvement of Fallow Lands in Organic Agricultural Production 251
Ukraintseva Irina V., Skvortsov Vadim P., Dolzhikov Valery V., Ivanov Pavel A., Kharchenko Sergey N.

Morpho-Physiological Evaluation of the State of Organs of Laying Quails When Black Soldier Fly Larvae Are Used in Their Feeding 255
Papaev Radii, Larina Yulia, Ezhkov Vladimir, Ezhkova Asiya, Girfanov Aidar

Biotechnological Complex for Processing of Plant Raw Materials and Agricultural Wastes 260
Kharchenko Sergey N., Gulyaev Pavel V., Tokareva Anna N., Panchenko Sergey V., Demchenko Maria S.

Results of Experimental Studies of Sunflower Oil Purification by Means of Electric Field in an Electrostatic Precipitator 265
Ukraintsev Maxim Mikhailovich, Lipkovich Igor Eduardovich, Gulyaev Pavel Vladimirovich, Korchagin Pavel Timofeyevich, Pyatikopov Sergey Mikhailovich, Yudaev Igor Viktorovich

Biotechnological Production of Protein Feeds from Alcohol Industry Waste 272
Usmonov Kamoliddin, Berdimurodov Usmon, Khakimov Sodiqjon, Globin Andrey N., Daus Yulia V., Boboeva Sitora

Mechanism of in Ovo Penetration of Nanostructured Zeolite and Its Effect on Weight Gain of Birds During Early Postnatal Ontogenesis 277
Undalov Roman, Ezhkov Vladimir, Ezhkova Asiya, Larina Yulia, Sharikova Elizaveta

Resource Potential of the Agro-Industrial Complex of Siberia and Its Role in Ensuring Food Security of Russia 282
Tarasenko Daria S., Rozhkov Evgeniy A., Asaturyan Andrey V., Lepekhina Yu. A.

Land Reclamation Development Under Conditions of Climate Aridity Growth: Experience of the Republic of Tatarstan 287
Khismatullin Mars, Khismatullin Marsel, Kurakova Chulpan, Khakimov Timur, Tazeev Emil

System Approach to Ensuring the Quality and Safety of Poultry Meat Production 293
Jumatov Yakubbay, Berdiev Darob, Saidaxmatov Sharifbek, Masenko Aleksey V., Sablina Polina A.

Influence of Spring Wheat Seed Treatment with Low-Temperature Plasma on Growth Processes 298
Khuzina R. R., Nafikov Man. Mak., Nafikov M. M., Khismatullin Mars M., Khismatullin Marsel M.

Biotechnological Conversion of Organic Wastes by Vermiculture to Increase Crop Yields........ 303
Daus Yulia V., Dubina Angelina Yu., Eviev Valery A., Evieva Victoria V., Badrudinova Amina N.

Influence of Pre-Sowing Seed Treatment in a Magnetic Gradient Field on Growth, Development and Productivity of Long-Staple Flax........ 307
Yudaev Igor Viktorovich, Akimov Alexey Alekseevich, Belenkov Alexey Ivanovich, Plotnitsky Igor Olegovich

Morphological and Productive Differences Between Chinese Pullet Chickens and Broilers........ 314
Fozilov Golibjon, Umarov Isroil, Abdusalomov Okhunjon, Abdumajitova Sayyohat, Turchanin Oleg S., Karadzhiyan Andrey M.

Parameter Evaluation of a Technical Tool for Electric Destruction of Weeds and Undesirable Plants 319
Yudaev Igor Viktorovich, Sharov Ivan Vasilievich

Innovative Technologies for Sustainable Development of Arctic Agriculture........ 323
Modina M. A., Kheckert E. V., Dragulenko Vladislav V., Ivanov M. Yu., Bovsh Vladislav V.

Addition of Sorghum Grain to Broiler Chicken Diets and Its Effect on Productivity and Safety 329
Shavazov Kadirjon, Mamatkulov Rustam, Tursunov Shokir, Hayitov Bobur, Izbasarova Zukhro

Integrated Processing of Agricultural Raw Materials Using Biotechnological Baromembrane and Ultrasonic Methods 334
Sattarov Narmurod, Fazilov Dusmurat, Ulashov Jonibek, Radjapova Natalya, Mamatkulova Shakhnoza, Kuznetsova Yu. S.

Reduction of Microbiological Contamination of Poultry Feed by Electrophysical Method 339
Borotov Atkham, Allanazarov Akmal, Baratov Dilshod, Murodov Xamidulla, Isayev Oybek, Tsyganko E. N.

Utilisation of Secondary Meat Raw Materials for the Production of Feed Additives........ 344
Choriev Ruzimurat, Shaimardanov Baxtiyor, Boboyev Xamza, Karshiev Fakhridin, Nuritdinova Khurshida, Kurbonov Fazliddin, Nekrasova I. E.

Dynamics and Prospects for the Development of the Russian Compound Feed Industry in the Period from 2010 to 2025........ 348
Migda N. S., Ashurov Nurali, Komilov Islomjon, Nishonov Fayzullo, Saydumarov Botir, Kuldasheva Shakhnoza

Author Index

V International scientific-practical conference
"Agriculture and Food Security:
Technology, Innovation, Markets, Human Resources"
Kazan State Agrarian University
November 28–29, 2024, Kazan, Russian Federation

From 28 to 29 November, V International scientific and practical conference "Agriculture and food security: technologies, innovations, markets, human resources", was held at Kazan State Agrarian University, Kazan, Russian Federation.

Organizing Institutions

- Kazan State Agrarian University
- Kazan State Academy of Veterinary Medicine named after N. E. Bauman
- Kazan State Power Engineering University
- Samara State Agrarian University
- Ulyanovsk State Agrarian University named after P. A. Stolypin

In Cooperation with

- Ministry of Agriculture and Food of the Republic of Tatarstan
- Ministry of Agriculture and Food of Samara Region
- Ministry of Agro-Industrial Complex and Rural Development of the Ulyanovsk Region
- Department of Agricultural Sciences of the Russian Academy of Sciences
- Ministry of Education of the Samara Region
- Ministry of Education of the Republic of Tatarstan
- Academy of Sciences of the Republic of Tatarstan
- Samara Scientific Center of the Russian Academy of Sciences (SamSC of RAS)

Associate International Partners:

- FAO/Food and Agriculture Organization of the United Nations
- Yüzüncü Yıl Universty (Turkey)
- LOGO (Germany)
- AMAZONEN-WERKE H (Germany)

Programme Committee

Chairman

Rector of the Kazan State Agrarian University, Doctor of Technical Sciences, Associate Professorб
Valiyev Ayrat Rasimowich

Conference Secretariat

edp sciences

E3S Web of Conferences

Statement of Peer review

In submitting conference proceedings to E3S Web of Conferences, I certify to the Publisher that I adhere to the **Policy on Publishing Integrity** of the journal in order to safeguard good scientific practice in publishing.

1. All articles have been subjected to peer review administered by the proceedings editors.
2. Reviews have been conducted by expert referees, who have been requested to provide unbiased and constructive comments aimed, whenever possible, at improving the work.
3. Proceedings editors have taken all reasonable steps to ensure the quality of the materials they publish and their decision to accept or reject a paper for publication has been based only on the merits of the work and the relevance to the journal.

Title, date and place of the conference

International scientific and practical conference on Agriculture and Food Security: technology, innovation, markets, human resources (FIES 2024).

Kazan, Russia
November 28-29, 2024

Proceedings editor(s):

Farida Nezhmetdinova (executive secretary, chief editor)

Ayrat Valiev

Bulat Ziganshin

Aksu Taylan

Rustam Nizamov

Date and editor's signature

February 18, 2025

Relevance and prospects of precision agro-technology development in the context of sustainable agriculture

Waleed Mugahed Al-Rahmi [1], *M.A.* Modina [2], *A.G.* Grigoryan [2], *Zulay K.* Tavbulatova [3*]

[1] Dar Al Uloom University, Riyadh 13314, Al Falah, Kingdom of Saudi Arabia
[2] Admiral Ushakov Maritime State University, 93 Lenina Ave., Novorossiysk, 353924, Russia
[3] Chechen State University named after A.A. Kadyrov, Grozniy, Russia

Abstract. The article considers the relevance of the development of precision agro-technologies at the present stage of agricultural development. The problems of traditional farming leading to soil degradation, lower production efficiency and negative impact on the environment are analyzed. Alternative approaches such as biological and sustainable farming are considered and their limitations in ensuring food security are shown. The importance of precision agro-technologies based on differentiated management of agro-systems taking into account their heterogeneity is emphasized, which allows optimizing the use of resources and reducing the negative impact on the environment. The results of experimental work demonstrating the savings of fertilizers and pesticides in the application of precision agriculture, as well as the increase in the accuracy of technological operations are presented. The necessity of developing a new economic strategy for agriculture, stimulating the introduction of innovative technologies and ensuring the sustainability of the agricultural sector is substantiated. The aim of the work is to substantiate the need for an integrated approach to agricultural development, taking into account technological, economic, social and environmental aspects, as well as the role of the state in supporting innovation.

1 Introduction

In today's world, agriculture, being the cornerstone of human civilization, faces a growing set of challenges that call into question its sustainable development and ability to meet the needs of an ever-growing world population. Global food security, which once seemed inviolable, is now threatened by a number of interrelated factors, including soil degradation [1-3], declining efficiency of traditional agricultural practices, and worsening global environmental problems [4, 5]. Intensive, often irrational exploitation of land resources, with an emphasis on extensive farming, as well as unjustified massive use of chemical fertilizers and pesticides, have led to irreversible processes of soil depletion, loss of soil fertility, disturbance of biodiversity and ecological balance in general. These negative trends have a direct impact on crop yields, which in a number of regions show an alarming downward trend, which, in turn, seriously jeopardizes the food security of entire countries and continents. The ever-growing world population, on the one hand, requires more and more food, and on the other hand, puts additional pressure on the agricultural sector, which has to work under the conditions of deteriorating environmental situation and decreasing available land resources [6-8]. All these factors together put before the world agricultural community the task of searching for fundamentally new approaches and development strategies that can ensure sustainable and environmentally safe food production in the long term.

In search of a solution to such a complex and multifaceted problem, mankind turns to various alternative approaches and technologies that lie in the plane of transition to more sustainable, resource-saving and environmentally friendly methods of farming. One of such approaches is biological farming, which is based on the principles of minimizing the use of anthropogenic energy, reducing dependence on synthetic fertilizers and pesticides, and preserving the ecological balance of agroecosystems [9, 10]. This approach implies the use of organic fertilizers, crop rotations, biological methods of pest and plant disease control. However, despite its attractiveness in terms of environmental sustainability, practice shows that this method, unfortunately, cannot fully meet the food needs of the growing world population. The reasons lie in the higher cost of the final products, which, as a rule, are unaffordable for the majority of consumers, as well as in the lack of effectiveness of biological methods of plant protection against weeds and diseases, which can lead to significant crop losses. As an alternative to organic farming, the concept of sustainable farming, which is a more comprehensive and integrated approach, is gaining attention. Sustainable farming involves not only the use of biological methods, but also combining them with

* Corresponding author: Eva_didi@inbox.ru

other advanced technologies such as minimum tillage, crop rotation optimization, precision farming, and freshwater conservation. This concept focuses on preserving and improving soil fertility, minimizing the negative impact of agriculture on the environment, and improving the economic performance of agricultural production. Nevertheless, the implementation of the concept of sustainable agriculture requires significant efforts and investments, as well as the development of comprehensive programs and regulatory framework that would facilitate its wide implementation in practice. Another very promising and promising direction in the development of modern agriculture is precision agro-technology, which is an integrated approach to the management of agro-systems based on targeted and differentiated impact on them, taking into account their individual characteristics and features. Precision farming is based on the principle of taking into account the heterogeneity of soil cover, phytosanitary condition of crops and other factors affecting the yield and quality of agricultural products. These technologies allow optimizing the application of fertilizers, plant protection products, as well as seed, thereby reducing their negative impact on the environment, reducing production costs, and increasing the economic efficiency of agricultural production. However, the effective implementation of precision technologies requires the development and implementation of special equipment and software [11-13], as well as in-depth study of specific agroecosystems and processes occurring in them.

In the context of the above and taking into account the analysis of advantages and disadvantages of various approaches, the direction of precision agro-technologies seems to be the most relevant and important for solving a set of problems facing modern agriculture. This approach opens new opportunities for increasing crop yields, reducing the ecological load on the environment [14, 15], as well as optimizing the use of natural and economic resources [16]. The introduction of precision technologies facilitates the transition to resource-efficient and sustainable agriculture, which is a key factor for ensuring food security, preserving the environment for future generations, as well as for increasing the competitiveness of the agricultural sector. This approach allows for differentiated management of agro-ecosystems, which leads to a more efficient and rational use of resources, higher yields, lower production costs, and reduced negative impact on the environment. However, it should be noted that the development and implementation of such technologies in the practical activities of the agricultural sector requires a detailed study and comprehensive analysis of specific agroecosystems, taking into account regional characteristics [17, 18], as well as the creation of appropriate infrastructure and support system for farmers and other agricultural producers.

The purpose of the work presented in this file is to comprehensively analyze the relevance of the development of precision agro-technologies at the current stage of agricultural development, as well as to discuss the importance of developing a new economic strategy for farming, which would stimulate the introduction of innovative agro-technologies and ensure the economic sustainability of the agricultural sector. This study will also address the problem of the need for an integrated approach to agricultural development, which takes into account not only technological, but also economic, social and environmental aspects [19], as well as the role of the state in supporting and stimulating the innovative development of the agricultural sector.

2 Method and materials

In the article, a number of experimental works aimed at studying and applying precision farming technologies were carried out. The overall work plan included field trials at a stationary site equipped with modern agricultural machinery and software.

John Keer 6920 tractor with a set of narrow wheels and MTZ-1221 tractor were used during the research. To apply mineral fertilizers we used spreader ZAM 900 with Troni system, which allowed to dose fertilizers, taking into account soil heterogeneity. Spraying was carried out by sprayer UF-901 with a capacity of 1050 liters and working width of 15 m, which provided uniform distribution of working solutions. Seed sowing was carried out with K 9-30 Superc and KMC seed drills, the latter of which was used for direct sowing. For potato planting a CF34 KL4 potato planter was used. Harvesting was carried out by SAMPO 2010 combine harvester with 1.5 m cutterbar and potato harvester. For tillage, vertical milling harrow KE-303, disk harrows Catros 3001 and Pegasus SG 3002 were used, as well as comb-former GF-75-4 and Eur Opal 7 plow.

The equipment operated in different modes corresponding to the agronomic operations performed. Fertilizer spreader and sprayer used automated systems for differentiated application of materials based on soil fertility maps. Seed drills operated at preset seeding rates and seed placement depths. All tillage equipment ensured the fulfillment of technological operations in accordance with the set parameters. In addition, the "Autopilot" system for John Keer tractor was used, providing 2-3 cm positioning accuracy, as well as IN liquid application control system. The Insight system was used for yield mapping and the FRITZMEIER PRKFT 90 soil sampler was used for soil sampling. The RT 200 N and Sensor ALS systems were used for differential fertilizer application. SMS Advanceed software was used for field data collection, storage and processing.

3 Results and discussion

This work was carried out to investigate in detail the possibilities and benefits of introducing precision farming technologies in the context of a training farm. The study was designed as a long-term field experiment conducted at a fixed site that was specially equipped for precision farming. The site was equipped with a variety of agricultural machinery including tractors, fertilizer spreaders, sprayers, seeders, harvesting equipment, and tillage equipment. All machines were equipped with automated control and monitoring systems, including GPS navigation, sensors to monitor material application, and yield mapping systems.

The process of conducting experimental work within the framework of precision farming technology research included several consecutive stages, starting with the preparatory phase and ending with the analysis of the results obtained. The initial phase involved a thorough analysis of the characteristics of the experimental site, including soil properties, topography and history of land use. This involved grid soil sampling, which was then analyzed in the laboratory for major nutrients, trace elements, organic matter, and acidity. The data obtained were used to create detailed soil fertility maps, which in turn served as a basis for planning differential fertilizer application.

Next, the selection of crops to be used in the experiment was carried out. Local agro-climatic conditions and economic feasibility were taken into account. The scheme of the experiment was developed, including different variants of soil treatment, fertilizer application and plant protection products. To ensure the accuracy of technological operations, modern agricultural equipment equipped with GPS-navigation and automatic control systems was used. In the process of sowing, fertilizer application and tillage, all necessary parameters, including seeding rates, fertilizer and pesticide application doses, working speed, working width and tillage depth, were recorded with the help of sensors and control systems.

During the vegetation stage, regular monitoring of crop condition was carried out, including assessment of plant stand density, phytosanitary condition and weed infestation. Both visual method and automatic weed detection systems were used to assess weediness. Based on the data obtained, decisions were made on the need for herbicide treatment, and treatment was carried out differentially, only in those areas where a high degree of weediness was observed. During the growing season, soil moisture and temperature were also monitored to assess the impact of different tillage methods on water and heat regimes.

After reaching the phase of full maturity, harvesting was carried out. Modern harvesting equipment equipped with yield mapping systems was used to assess the yield and quality of the harvested products. During the harvesting process, data on harvested crop weight, moisture content, protein content, oil content and other important indicators were recorded. After harvesting was completed, the data were analyzed, including statistical processing of results, graphical representation and trend analysis. The obtained conclusions served as a basis for assessing the efficiency of different variants of precision farming technology and their economic justification.

Visual method was used to control weediness of crops, as well as automatic weed detection systems installed on sprayers. Based on the data obtained, differentiated herbicide application was carried out, and only those areas of the field with high weed infestation were treated. The comparison of tillage methods included conventional plowing and direct seeding on stubble. In direct seeding, crop residues remained on the soil surface, creating a mulch layer that helped to reduce erosion and conserve moisture.

This paper demonstrates that the use of precision farming systems leads to a significant reduction in resource consumption while increasing or maintaining yields. The use of differentiated application of fertilizers and pesticides reduced their consumption by 25-30%. Similar works conducted by other authors show that the reduction in fertilizer consumption when using precision agriculture varies from 15% to 35%, depending on the type of soil, climate and type of crop grown. For example, in studies conducted in Canada on corn crops, it was shown that using differential nitrogen application, nitrogen use can be reduced by 22% while maintaining yields at the same level. At the same time, in studies conducted in Australia on cereal crops, it was shown that phosphate fertilizer consumption can be reduced by up to 30%, while yields increase by 5-7% due to more efficient distribution of nutrients in the soil.

The results of the study conducted by the authors showed a 25-30% reduction in pesticide consumption. These indicators are also confirmed by other researchers. In particular, studies in European countries have shown an average 25-30% reduction in herbicide consumption when using precision farming, which is associated with more accurate application of preparations and the use of technology for recognizing weeds. In studies conducted in the USA, it was shown that the use of precision farming technologies in cotton cultivation reduces the amount of necessary pesticide treatment by 33%, which leads not only to a reduction in pesticide costs, but also to a reduction in the negative impact on the environment.

The results of this work have shown that the use of precision farming leads to an increase in the accuracy of technological operations. In particular, the use of GPS navigation and parallel guidance systems allowed reducing the width of adjacent rows to 2.5-3.5 cm, compared to 5-7 cm when using traditional markers. According to the study, the use of parallel guidance and GPS systems can increase the accuracy of tillage and seeding by 30-40%, which reduces the number of skips and overlaps, and allows for more efficient use of resources.

Analysis of the results shows that precision farming technology is not only economically efficient, but also has significant environmental potential. Reduced consumption of fertilizers and pesticides leads to reduced soil and water pollution, and the use of zero-tillage technology helps to reduce erosion and improve fertility. Comparing these data with other studies, there is a growing body of scientific work supporting the importance of precision farming practices for sustainable agriculture. In a study conducted in Brazil, it was shown that the use of precision farming reduces greenhouse gas emissions from soil by 15-20% by conserving organic matter and reducing energy inputs for tillage.

At the same time, for effective implementation of precision farming technologies it is necessary to take into account local peculiarities and adapt them to specific agro-ecosystems. In particular, it is necessary to revise traditional views on weed infestation of crops and develop integrated plant protection systems that do not lead to mass death of weeds and use biological control methods. Studies have shown that using the principles of agroecology and integrated plant protection not only reduces dependence on pesticides, but also improves the biodiversity of agroecosystems, which, in turn, has a positive impact on their sustainability and productivity.

4 Conclusion

The conducted research emphasizes that the introduction of precision agro-technologies is a relevant direction for solving the problems of modern agriculture. Traditional methods have led to soil depletion and negative impact on the environment, which requires a transition to sustainable and resource-saving approaches.

The results show that precision farming can reduce the use of resources (fertilizers, pesticides) while maintaining or increasing yields. Differentiated application based on soil and crop data increases resource efficiency and reduces environmental impact. GPS and parallel driving systems improve operational accuracy, reducing overlaps and saving resources.

The no-till method reduces erosion, improves soil structure and fertility, but requires control of the phytosanitary condition of crops.

Precision farming is promising but requires specialized equipment, software, agroecosystem studies and economic incentives.

Technology adoption is not the only solution; a comprehensive approach that takes into account economic, social and environmental aspects is needed. There is a need for a new economic strategy that stimulates innovation and ensures the sustainability of agriculture with the support of the state.

Thus, the findings emphasize the importance of the transition to precision agriculture for sustainable agricultural development, food security and environmental conservation.

References

1. Z.A. Buryak, A.G. Narozhnyaya, A.V. Gusarov, A.A. Beylich. Solutions for the Spatial Organization of Cropland with Increased Erosion Risk at the Regional Level: A Case Study of Belgorod Oblast, European Russia. Land, **11,** 1492 (2022). DOI: 10.3390/land11091492.
2. A.S. Danilov, A.S. Lozovaya. Criteria-assessment tools for inventory of accumulated environmental harm objects. Sustainable Development of Mountain Territories, **16,** 861-872 (2024). DOI: 10.21177/1998-4502-2024-16-3-861-872.
3. M.V. Shabanov, M.S. Marichev, S.S. Mangiyeva, A.A. Sokolov. Chemozem formation under conditions of prolonged exposure to aero-industrial emissions from a mining and smelting plant. Sustainable Development of Mountain Territories, **15,** 727-740 (2023). DOI: 10.21177/1998-4502-2023-15-3-727-740.
4. V.I. Silaev, D.V. Eremeev, T.A. Martynova, Yu.V. Danilchenko. Analysis of the carbon footprint created by mining enterprises. MIAB. Mining Inf. Anal. Bull., **11-1,** 265-277 (2023). DOI: 10.25018/0236_1493_2023_111_0_265.
5. V.Yu Konyuhov, A.M. Gladkih, I.I. Galyautdinov, Y.D. Severina. Economic aspects of green technologies. IOP Conference Series: Earth and Environmental Science, **350,** 012036 (2019). DOI: 10.1088/1755-1315/350/1/012036.
6. V.L. Snezhko, D.M. Benin. Dynamics of the Condition of Reclaimed Agricultural Lands in the Russian Federation. Land, **10,** 1288 (2021). DOI: 10.3390/land10121288.
7. E.Yu. Kulikova, S.V. Balovtsev, O.V. Skopintseva. Comprehensive assessment of geoecological risks in conducting open and underground mining. Sustainable Development of Mountain Territories, 16, 205-216 (2024). DOI: 10.21177/1998-4502-2024-16-1-205-216.
8. Z. Shutaleva, I. Nikonova, I. Savchenko. Environmental education for sustainable development in Russia. Sustainability (Switzerland), **12**, 7742 (2020). DOI: 10.3390/su12187742.
9. T. Campani, G. Manieri, I. Caliani, A. Di Noi, S. Casini. Apis mellifera as a Model Species to Evaluate Toxicological Effects of Fungicides Used in Vineyard Agroecosystems. Journal of Xenobiotics, **15,** 18 (2025). DOI: 10.3390/jox15010018.
10. Z. Shutaleva, I. Nikonova, I. Savchenko, S. Abramova, V. Lubimova, A. Novgorodtseva. Environmental behavior of youth and sustainable development. Sustainability (Switzerland), **14,** 250 (2022). DOI: 10.3390/su14010250.
11. A. I. Trunova, S. O. Kurashkin, Y. A. Tynchenko, I. F. Vivián, K. A. Bashmur, Increasing the Efficiency of Synthetic Iron Production by the Use of New Kit Lining. Metals, **13(7),** 1184 (2023). DOI: 10.3390/met13071184.
12. A. A. Sokolov, L. G. Orlova, K. A. Bashmur, R. I. Kuzmich, Ensuring Uninterrupted Power Supply to Mining Enterprises by Developing Virtual Models of Different Operation Modes of Transformer Substations. Mining Informational and Analytical Bulletin. **11,** 278-291 (2023). DOI: 10.25018/0236_1493_2023_111_0_278.
13. B.V. Malozyomov. Analysis of a Predictive Mathematical Model of Weather Changes Based on Neural Networks. Mathematics, **12,** 480 (2024). DOI: 10.3390/math12030480.
14. A.A. Kulikova, T.I. Ovchinnikova. A regional criterion for classifying mining regions as territories with the greatest exposure to geoecological changes. Sustainable Development of Mountain Territories, **15,** 27-34 (2023). DOI: 10.21177/1998-4502-2023-15-1-27-34.
15. C. Kongar-Syuryun,. Oganesyan, D. Solovykh, M. Khayrutdinov, D. Adigamov. Principles of sustainable development of georesources as a way to reduce urban vulnerability. Urban Science, **8,** 44 (2024). DOI: 10.3390/urbansci8020044.
16. V. O. Gutarevich, A. S. Mikhalev, V. E. Gozbenko, Reducing Oscillations in Suspension of Mine Monorail Track. Mining Informational and Analytical Bulletin, **11,** 72–87 (2023). DOI: 10.25018/0236_1493_2023_111_0_72.

17. Q. Gao, L. Wang, Y. Fang, Y. Gao, L. Ma, X. Wang, Y. Li, X. Wu, Z. Du. Conservation agriculture boosts topsoil organic matter by restoring free lipids and lignin phenols biomarkers in distinct fractions. Soil Till. Res., **248,** 106463 (2025). DOI: 10.1016/j.still.2025.106463.
18. J.D. Torres, S.R. Jaeger, P. Puerta, A. Tárrega. How do Spanish consumers perceive different lettuce cultivation systems? Insights from explicit and implicit methods. Appl. Food Res., **5,** 100709 (2025). DOI: 10.1016/j.afres.2025.100709.
19. V.V. Yurak, A.N. Malyshev, S.S. Zavyalov. Malyshev, S.S. Zavyalov. Review of new remote methods for environmental monitoring. MIAB. Mining Inf. Anal. Bull., **11-1,** 51-67 (2022). DOI: 10.25018/0236_1493_2022_111_0_51.

Comparative assessment of fertilizer systems and intensification levels in crop production

Danil A. Zyukin[1*], *Ivan A.* Glushkov[1]

[1] Kursk state agrarian university named after I.I. Ivanov, Kursk, Russia

Abstract: This study is devoted to the comparative economic evaluation of different fertilization systems and intensification levels in the cultivation of agricultural crops on gray forest soils. Experimental work was carried out during three rotations of field crop rotation. The obtained results demonstrate that increasing the doses of mineral and organic fertilizers does not always lead to an increase in economic efficiency, and in some cases even reduces it. In particular, cost recovery decreased with increasing fertilizer doses on average in the crop rotation, except for potato crop, where the opposite trend was observed. Fertilizer systems including mineral nitrogen contributed to an increase in protein content in grain, but this did not ensure their superiority in terms of economic efficiency. On the basis of the obtained data, conclusions are drawn about the need for a differentiated approach to the choice of fertilizer systems and levels of intensification, taking into account specific agro-ecological conditions and features of crops grown. The results of the study can be useful for agricultural practitioners in selecting optimal agro-technologies.

1 Introduction

In the modern world, agriculture is the cornerstone of food security and sustainable development, acting not only as a supplier of food, but also as an integral part of the socio-economic and environmental systems. However, for many years, the agricultural sector has developed mainly along the path of intensification, seeking to maximize yields through the use of high doses of mineral fertilizers, pesticides and other agrochemicals [1]. This approach, although providing short-term increases in production, was often accompanied by serious negative consequences, such as depletion of soil fertility, pollution of water resources [2], reduction of biodiversity, and deterioration of the quality of agricultural products. The problem of achieving a harmonious balance between the need to meet growing food needs and the conservation of natural resources is one of the most acute and relevant in agricultural science, requiring a comprehensive and multidimensional approach. In this regard, there is a need to rethink traditional agricultural practices and search for sustainable, economically feasible and environmentally acceptable alternatives that can ensure the long-term prosperity of the agricultural sector and the well-being of future generations [3]. This task sets before scientists and agricultural practitioners a complex but extremely important goal - the development of agro-technologies that not only provide high yields, but also contribute to environmental conservation and sustainable development of rural areas.

In search of solutions to this global contradiction, agricultural scientists and practitioners are researching and developing a variety of approaches to optimize agro-technologies to achieve sustainable agriculture [4]. One such approach is the active use of organic fertilizers, including manure, compost and sidedresses. These materials not only enrich the soil with nutrients, but also help to improve its physical structure, increase its water-holding capacity, and stimulate the development of soil microflora, which in turn helps to increase natural fertility and reduce dependence on mineral fertilizers. However, the introduction of organic fertilizers may be associated with certain difficulties related to their availability, cost, production volumes and rates of nutrient release in the soil. Another promising approach is the development and implementation of integrated pest and disease management systems that aim to minimize the use of chemical plant protection products and develop biological control methods. This reduces the negative impact of agrochemicals on the environment, human health and biodiversity, but requires more careful monitoring of the phytosanitary condition of crops, a thorough understanding of the biological cycles of pests and diseases, and the ability to apply biological methods of plant protection. The use of mineral fertilizers, although an important tool for achieving high yields, should be carried out on the basis of scientifically justified rates, taking into account soil fertility, crop

* Corresponding author: nightingale46@rambler.ru

needs and in combination with organic fertilizers to reduce their negative impact on the ecological state of soils [5, 6]. The choice of the optimal fertilizer system should be based on a comprehensive analysis of economic efficiency, environmental acceptability and agrotechnical feasibility in specific conditions. Thus, the creation of modern agro-technologies should be aimed at achieving a balance between economic efficiency and environmental sustainability, contributing to the creation of not only productive but also ecologically balanced agro-ecosystems. In this context, there is an increasing need for comprehensive research aimed at evaluating different approaches, their impact on yields, product quality, economic efficiency, environmental sustainability, as well as social consequences [7].

In light of the above, the research presented in this article is of particular relevance and importance. It is an important step towards the development and implementation of sustainable agro-technologies, as it focuses on the comparative economic evaluation of different fertilizer systems in a field crop rotation, taking into account different levels of production intensification. The study also pays close attention to determining the optimal yield level that would allow for stable economic efficiency under different approaches to soil fertilization. This is particularly important in the context of adaptive-landscape farming, which, unlike the unified approach, takes into account local agroecological, economic and social characteristics to create optimal, sustainable and productive agroecosystems [8, 9]. The presented results, which go beyond purely academic interests, can be extremely useful for agricultural practitioners, providing them with scientifically sound recommendations on the choice of fertilizer systems that ensure both high yields and economic efficiency, as well as contributing to reducing the negative impact on the environment. Consequently, the study contributes to the formation of a comprehensive view of the economic efficiency of different fertilizer systems, taking into account not only quantitative indicators of yield, but also qualitative parameters and cost part. The purpose of this study, presented in the attached file, is to conduct a comprehensive production and economic evaluation of different fertilizer systems in the field crop rotation as part of the full technological cycle of production, as well as in close relationship with the level of intensification of production. No less important task was the formation of specific proposals and recommendations for agricultural producers aimed at improving the economic efficiency of growing field crops in specific agro-ecological conditions, which ultimately should contribute to the development of sustainable and economically profitable agriculture.

2 Method and materials

The research presented in this article was based on the results of experimental work carried out in the Kursk State Agrarian University. Within the framework of these works, field experiments were laid in three rotations of field crop rotation. The experiments were conducted for several years, which allowed us to evaluate the long-term effect of different fertilizer systems on crop yield and quality [10, 11]. In the first two rotations, in addition to cereals and perennial grasses, potato crop was present in the rotation, in the third rotation it was excluded and the rotation became grain-grass. To evaluate the effect of different doses of organic and mineral fertilizers, eight variants of fertilizer systems were selected, including both individual types of fertilizers and their various combinations, as well as a variant without fertilizers. The experiments were conducted on gray forest soils. Within the framework of the experiment, equipment [12] was used on the crops, which provided quality soil preparation, sowing and harvesting. MTZ-82 tractor was used for plowing, harrowing and cultivation. Sowing was carried out by seeder SZ-3.6, harvesting of cereals was carried out by combine harvester "Yenisei-1000", and potatoes - by potato digger KST. The yield was estimated by weighing the harvested products from each plot. An Infratech 1241 moisture and protein analyzer was used to analyze grain quality. The equipment operated in normal modes, according to the manufacturers' recommendations. MathCad v8 software package was used for calculations and data analysis [13].

3 Results and discussion

This experience was aimed at a deep and comprehensive study of the influence of different fertilization systems, as well as different levels of intensification on the productivity of agricultural crops, as well as on the economic efficiency of their cultivation [14, 15]. The field experiment was conducted during three full rotations of crop rotation, which made it possible to take into account the long-term effects of different agronomic practices. The experiment was laid on gray forest soils typical for Kursk region, which ensured the relevance of the results obtained for this region. Eight different variants of fertilization systems were studied in the experiment, which represented a wide range of approaches, starting from the extensive variant, where fertilizers were not applied at all, and ending with the intensive variant with the application of maximum doses of both organic and mineral fertilizers. As an organic fertilizer was used traditional manure applied to the soil in different doses, and as mineral fertilizers was used a complex consisting of nitrogen, phosphorus and potassium fertilizers in the form of ammonium nitrate, double superphosphate and potassium chloride, respectively. This variation in fertilizer systems made it possible to evaluate their effects both individually and in different combinations, which provided a more complete understanding of the interaction of different types of fertilizers and their effects on yield. The crop rotation included 7 fields and was represented by a variety of crops such as cereal crops (winter rye, spring wheat, barley, oats), forage crops (vetch-oat mixture, perennial grasses), and the technical crop potato. This diversity of crops ensured the representativeness of the data and allowed to evaluate the effect of fertilizers on different groups of plants. During the experiment, a systematic record of yields was made for each crop and for each variant of the fertilizer system. For a more accurate assessment of the influence of intensification levels, three conditional levels were identified: fertilizer extensiveness on gray forest soils of Kursk region has a pronounced dependence on the level of intensification. The obtained data showed (Table 01) that the increase in the doses of mineral and organic fertilizers did not lead to a proportional increase in yields, but, on the contrary, caused a significant decrease in the economic efficiency of agricultural production.

Table 1. Production and economic evaluation of fertilizer systems in crop cultivation

Indicator	Without fertiliser	Organic		Mineral	Organomineral			
		4От (Э)**	80 m		40 m + NPK	60m+ NPK(H)**	60m+ 2NPK	80 m +2NPK(H)**
Organic fertilisers, t/ha	–	40	80	–	40	60	60	80
ammonium nitrate, centner per hectare	–	–	–	1,16	1,16	1,16	2,32	2,32
double superphosphate, centner per hectare	–	–	–	0,8	0,8	0,8	1,6	1,6
potassium chloride, ts/ha	–	–	–	0,8	0,8	0,8	1,6	1,6
Consumption of main fuel, kg/ha	5,3	6,2	6,8	51,2	56,3	57,3	57,7	59,4
Labour costs, man-hours/ha	412,1	412,1	412,1	6,1	6,8	7,1	7,21	7,5
Herbicides, $/ha	28	28	28	4,1	4,1	4,1	4,1	4,1
Seed costs, $/ha	60,2	69,5	77,5	28	28	28	28	28
Costs per 1 ha, $	0,03	0,03	0,03	102,1	111	114,7	151,3	155,3
Yield, thousand grain units/ha	10431	11571	11570	3,97	4,09	4,10	4,39	4,36
Conditional net income from 1 ha, $	0,03	0,03	0,03	116,3	114	111,3	92	86
Payback of costs depending on the economic situation, $/$								
average	3	2,9	2,7	2,1	2	2	1,6	1,6
favourable	1,9	1,9	1,8	2,4	2,2	2,2	1,8	1,741
unfavourable	412	412	412	1,5	1,4	1,4	1,1	1,1

In particular, on average for the crop rotation, cost recovery by the proceeds from the sale of products decreased from 2.73 $/$ in the variant without fertilizers to 1.55 $/$ in the variant with maximum doses of organic and mineral fertilizers, which is a decrease of 43.2%. At the same time, the total yield in grain units in the crop rotation increased only 1.4 times, while variable costs increased more than 2.6 times, which clearly demonstrates the ineffectiveness of excessive intensification in these conditions. The exception to the general trend was the potato crop, where, unlike other crops, cost recovery, on the contrary, increased with increasing fertilizer doses, which can be explained by the high responsiveness of this crop to fertilizer application and its high market value. The observed trends clearly indicate that the increase in fertilizer doses in the production of grain crops on gray forest soils of Kursk region does not lead to an adequate increase in yields, but, on the contrary, leads to a significant increase in the cost of the production process, making it economically inexpedient. Moreover, with increasing level of intensification not only the costs of all types of work increased, but also the conditional net income decreased from 104$ /ha to 86$ /ha, which once again confirms the need to rethink traditional approaches to agricultural intensification in this particular region. Noteworthy is the fact that fertilizer systems including mineral nitrogen contributed to the formation of grain with a higher crude protein content. In particular, in variants with NPK application the protein yield per hectare was 15-20% higher than in variants with application of organic fertilizers only. However, in terms of overall economic efficiency, these variants, despite higher protein yield, were inferior to the variant without fertilizer application, which emphasizes the need for a comprehensive analysis of the economic feasibility of agronomic practices. At comparative analysis of economic efficiency of separate crops it was established that for spring wheat the most effective variant without fertilizers (payback 3,04 $/$), for winter rye - organic system with application of 80 t/ha of manure (3,24 $/$), and for potatoes - the most intensive variant with high doses of organic and mineral

 https://doi.org/10.1051/bioconf/202516100002

fertilizers (payback 3,17 $/$). The conducted analysis of economic efficiency clearly demonstrated that to ensure stable cost recovery at the intensive level of spring wheat production, the yield should be 30-68% higher than the actual one obtained during the experiment. A similar situation was observed for oats, where a 30% increase in yield was required to achieve the same economic efficiency as at the extensive level. The data obtained during the experiment are in full agreement with the results of similar studies conducted in other regions with similar soil and climatic conditions [16]. Thus, numerous studies conducted in neighboring regions also indicate a significant decrease in economic efficiency with excessive use of mineral fertilizers and confirm the need for a more rational approach to the use of agrochemicals [17]. Comprehensive analysis of research results emphasizes that for sustainable development of agriculture on gray forest soils, it is necessary to use a comprehensive approach, including not only the competent use of fertilizers, but also improving the structure of crop rotations, the introduction of resource-saving and moisture-saving technologies, as well as the development and application of biological methods of plant protection. In particular, to improve the economic efficiency of grain crops production, preference should be given to a moderate level of intensification based on a balanced combination of organic fertilizers with scientifically justified doses of mineral fertilizers, and for potato crops it is possible to use more intensive technologies, taking into account economic feasibility and careful cost control [18, 19].

4 Conclusion

The study demonstrates that the economic efficiency of agro-technologies on gray forest soils of Kursk region has an inverse dependence on the level of intensification, especially in the production of grain crops. Increasing the doses of mineral and organic fertilizers does not provide a proportional increase in yields, but, on the contrary, leads to a decrease in cost recovery and net income. On average in the crop rotation, there is a decrease in cost recovery by more than 40% in the transition from the variant without fertilizers to the variant with maximum doses. This indicates that excessive intensification based on intensive application of fertilizers is not the optimal approach for these soil and climatic conditions [20]. It is noteworthy that the potato crop showed a different trend, where increasing fertilizer doses contributed to increased economic efficiency, which emphasizes the need for a differentiated approach to the choice of fertilizer systems depending on the type of crop grown. Fertilizer systems including mineral nitrogen contributed to the formation of grain with higher protein content, but this advantage did not ensure their economic attractiveness compared to less costly options [21, 22]. The analysis of economic efficiency of individual crops showed that for spring wheat the most profitable is the variant without fertilizers, and for winter rye - the variant with organic fertilizers, which approaches to intensification of agriculture on gray forest soils of Kursk region. The use of high doses of mineral and organic fertilizers, although it can help to increase yields to some extent, is not always economically feasible. To improve the economic efficiency of agro-technologies, it is necessary to use a comprehensive approach based on differentiated application of fertilizers depending on the type of crop, as well as on the introduction of resource-saving technologies. Special attention should be paid to the use of organic fertilizers in combination with scientifically justified doses of mineral fertilizers. In addition, it is necessary to improve the structure of crop rotations, as well as to apply biological methods of plant protection. An important direction for further research is the development of such fertilizer systems that would ensure not only high yields, but also high quality of products and preservation of soil fertility. The obtained conclusions emphasize the need to adapt agro-technologies to specific soil and climatic conditions, as well as the importance of taking into account the economic efficiency of each applied technique.

References

1. N. Garambois, U. Le Goff, L. Thibaudeau. Agrarian dynamics and climate change in the Senegalese Sahelian peanut basin. Agrar. Syst. Clim. Change: Journeys Adapt. Glob. South, **15,** 37 (2024).
2. S. Mukherjee, S. Paul, S. Bhattacharya, A. Islam, S. Mahammad, E. Alam. Optimizing water resource management in tropical drought-prone regions through hybrid MCDM techniques: A water-stress mapping approach. J. Hydrol. Reg. Stud., 57, 102171 (2025).
3. H. Cochet, O. Ducourtieux, N. Garambois. Agrarian systems and climate change: Journeys of adaptation in the global south. Agrar. Syst. Clim. Change: Journeys Adapt. Glob. South, **1**, 208 (2024).
4. A. A. Sokolov, L. G. Orlova, Mining Informational and Analytical Bulletin. **11,** 278-291 (2023). DOI: 10.25018/0236_1493_2023_111_0_278.
5. B. Aparecida da Silva, R. Evangelista de Oliveira, A. Cavalieri Sais, E. Cardoso leite. Family farmers and trees: ecological functions, ecosystem services and food production in agrarian reform settlements in Brazil. Agroecol. Sustain. Food Syst., **49,** 4-21 (2025). DOI: 10.1080/21683565.2024.2405515.
6. N.A. Zhirov. Socio-Economic Development of the Black Soil Village in the XIX-early XX centuries: from the Prosperous Granary of Pre-Reform Russia to the Demo-Ecological Crisis of Rural Settlements. Bylye Gody, **19,** 1605-1614 (2024). DOI: 10.13187/bg.2024.4.1605.
7. T. Bandara, B. Marambe, G. Pushpakumara, P. Silva, R. Punyawardena, L. Manawadu, S. Premalal, G. Miah, K.R. Dahal. Adapting mountain-farming systems to climate change by reducing food-nutrition-health vulnerability. Mitig. Adapt. Strateg. Glob. Change, **29,** 88 (2024). DOI: 10.1007/s11027-024-10181-x.
8. V. L. Tatarintsev, L. V. Lebedeva, G. D. Syzdykova, D. S. Inkarov. Management of agricultural land use sustainability at the local level using GIS

technologies and special map material. Sustainable Development of Mountain Territories, **15,** 864-876 (2023). DOI: 10.21177/1998-4502-2023-15-4-864-876.

9. D.D. Duisenova, K.M. Tireuov, L.M. Tatarintsev, L. M. Tatarintsev. V. Lebedeva. Ecological typization of agricultural land and modeling of sustainable agrolandscape at the local level. Sustainable Development of Mountain Territories, **16,** 94-110 (2024). DOI: 10.21177/1998-4502-2024-16-1-94-110.

10. V.L. Tatarintsev, L.M. Tatarintsev, F.K. Ermekov, Yu. S. Lisovskaya. Agroecological condition assessment of agricultural landscapes for their sustainability increase. Sustainable Development of Mountain Territories, **14,** 76-86 (2022). DOI: 10.21177/1998-4502-2022-14-1-76-86.

11. Q. Gao, L. Wang, Y. Fang, Y. Gao, L. Ma, X. Wang, Y. Li, X. Wu, Z. Du. Conservation agriculture boosts topsoil organic matter by restoring free lipids and lignin phenols biomarkers in distinct fractions. Soil Till. Res., **248,** 106463 (2025). DOI: 10.1016/j.still.2025.106463.

12. K. Akzharkyn, D. Yelemessov, D. Baskanbayeva, N.V. Martyushev, V.Y. Skeeba, V.Y. Konyukhov, T.A. Oparina. Strengthening Polymer Concrete with Carbon and Basalt Fibres. Appl. Sci. (Switzerland), **14,** 7567 (2024). DOI: 10.3390/app14177567.

13. B.V. Malozyomov, N.V. Martyushev, S.N. Sorokova, E.A. Efremenkov, D.V. Sorokova, E.A. Efremenkov, D.V. Valuev, M. Qi. Analysis of a Predictive Mathematical Model of Weather Changes Based on Neural Networks. Mathematics, **12,** 480 (2024). DOI: 10.3390/math12030480.

14. J.D. Torres, S.R. Jaeger, P. Puerta, A. Tárrega. How do Spanish consumers perceive different lettuce cultivation systems? Insights from explicit and implicit methods. Appl. Food Res., **5,** 100709 (2025). DOI: 10.1016/j.afres.2025.100709.

15. M. N. Nurpeisov, K. V. Ruchkina, V. A. Androkhanov, S. K. Makenova. Assessment of geoecological factors and conditions affecting sustainability of agricultural land use in the forest-steppe zone of Altai Krai. Sustainable Development of Mountain Territories, **15,** 419-430 (2023). DOI: 10.21177/1998-4502-2023-15-2-419-430.

16. O. A. Kolenchukov, A. Y. Mikhaulov, R. B. Sergienko. Forecasting the technical condition of thermochemical reactor systems. SOCAR Proceedings, **2023,** 123-130 (2023).

17. A. I. Trunova, S. O. Kurashkin, I. F. Vivián, K. A. Bashmur, Metals, **13(7),** 1184 (2023). DOI: 10.3390/met13071184.

18. V. I. Golik, R. V. Klyuev, N.V. Martyushev, V. Brigida, E.A. Efremenkov, S.N. Sorokova, Q. Mengxu. Tailings Utilization and Zinc Extraction Based on Mechanochemical Activation. Materials, **16,** 726 (2023). DOI: 10.3390/ma16020726.

19. V.Semykin, I. Pigorev, D. Zyukin. The influence of scale factor on the realization of natural potential of grain farming in the region, IOP Conference Series: Earth and Environmental Science. 6th International Conference on Agriproducts Processing and Farming, 012003 (2020). DOI: 10.1088/1755-1315/422/1/012003.

20. A. Gladkov, V. Kukartsev, A. Kozlova, D. Grigorev. Development of Requirements for AIS Aimed at Controlling High Turnover. 2023 IEEE International Conference on Computing, ICOCOCO 2023 (2023).

21. O. Afanaseva, M. Afanasyev, S. Neyrus, D. Pervukhin, D. Tukeev. Information and Analytical System Monitoring and Assessment of the Water Bodies State in the Mineral Resources Complex. Inventions, **9,** 115 (2024).

22. V.Yu Konyuhov, A.M. Gladkih, I.I. Galyautdinov, Y.D. Severina. Economic aspects of green technologies. IOP Conference Series: Earth and Environmental Science, **350,** 012036 (2019).

Analysis of factors limiting the import-substituting effect of cluster development in the agro-industrial complex

Irina S. Moga [1,2], *Fedor Yu.* Baykov [2,3], and *Vlada S.* Sidorenko [1,2]

[1] State University of Management, Moscow, Russia
[2] Vladimir State University, Vladimir, Russia
[3] LLC «BEL Development», Moscow, Russia

Abstract. This article analyses the problem of clustering of innovative mesosystems in the agrarian-industrial complex of Russia in the context of food import substitution strategy. The main emphasis is placed on the identification of factors that hinder the effective development of clusters in the agrarian sphere and the development of recommendations to overcome them. The study was conducted by analysing scientific literature, regulatory legal acts and statistical data. The results show that, despite the active promotion of the cluster approach, the overall level of innovation activity in agriculture remains low. It is revealed that the formal approach to the creation of clusters, lack of trust between the participants, lack of qualified personnel and lack of clear criteria for performance evaluation are the key problems. It is concluded that it is necessary to reorient state policy to support self-organisation of business, development of regional autonomy, stimulation of cooperation and investment in human capital.

1 Introduction

The problem of ensuring food security is one of the most acute and urgent in today's world. Global challenges related to population growth, climate change, depletion of natural resources and geopolitical instability put serious pressure on the world's food systems. Many countries, including Russia, are facing the need to reduce dependence on imported food and agricultural raw materials in an effort to strengthen their own agro-industrial sector. This requires not only increasing production volumes, but also improving the efficiency and sustainability of the agricultural sector, as well as introducing innovations [1].

There are several approaches to solving the problem of food import substitution, each of which has its own advantages and disadvantages. One such approach is the development of large agroholdings and vertically integrated companies. This approach can provide economies of scale and concentration of resources, allowing for the introduction of advanced technologies and increased productivity. However, it may reduce competition and market access for SMEs, and increase social inequalities in rural areas. Another approach is to support small and medium-sized farms, which, due to their flexibility and localisation, can produce environmentally friendly and high-quality products. But such farms often face problems of access to finance, modern technologies and markets. The third approach relies on the development of agricultural science and education, the creation of innovation centres and the dissemination of best practices in the agricultural sector. This approach is long-term and requires significant investments in science, education and infrastructure, but it ensures sustainable agricultural development.

In recent years, in the context of an active search for effective tools to stimulate economic growth and innovation, the cluster approach [2], which is considered as one of the ways to solve the problem of food independence, has attracted more and more attention. Clusters that unite scientific organisations, educational institutions, manufacturing enterprises, service companies and public authorities are able to create a synergetic effect that promotes the accelerated development of innovation, dissemination of advanced technologies, increased competitiveness and productivity, and, as a result, stimulate economic growth. Within the framework of this approach, clusters are considered as regional mesosystems capable of integrating individual elements of the agro-industrial sector, while providing both economic and social synergies [3]. This is especially important for regions characterised by low efficiency of agricultural production, underdeveloped innovation infrastructure and limited financial resources. However, despite the theoretical attractiveness of the cluster approach, in practice its implementation faces a number of problems, such as the lack of a clear methodology, bureaucratisation of management, distrust between participants, lack of qualified personnel and low level of

* Corresponding author: alonemoga@gmail.com

cooperation. This leads to the fact that many clusters exist only on paper, without having a significant impact on the development of the agro-industrial complex.

The relevance of the study of factors affecting the formation and development of innovative mesosystems of the agro-industrial complex within the framework of the cluster approach is due to the need to overcome the existing limitations and improve the effectiveness of measures of state support for import substitution. Understanding these factors, as well as the mechanisms of their impact, is critical for the development of an effective regional policy aimed at stimulating the innovative development of the agricultural sector. This study, described in the attached file, focuses precisely on this problem, analysing the factors limiting the import-substitution effect of innovation clusters development and proposing ways to overcome them.

Thus, the purpose of the presented work is to analyse the factors affecting the clustering of innovative mesosystems of agro-industrial regions within the framework of food import substitution strategy, as well as to identify the problems and prospects of applying the cluster approach to improve the efficiency of agricultural production and ensure food security. The results of this study can be used to improve regional policy in the sphere of agriculture, as well as to form a more effective model of cluster development, contributing to real import substitution and increasing the competitiveness of the Russian agricultural sector.

2 Method and materials

The research presented in this article is analytical in nature. The main emphasis is placed on the analysis of existing approaches to clustering in the agro-industrial complex, identification of factors that hinder the development of innovation clusters and the development of proposals to overcome them. The study analysed scientific literature, as well as regulatory legal acts on the issues of agro-industrial complex development and cluster policy. The data for the study was collected on the basis of secondary analysis of published sources, including statistical data and the results of previously conducted research. Also, in the work we systematised the data on the subject of the study, using the methods of comparative analysis, which allowed us to identify common patterns and specific features of the development of cluster initiatives in the agro-industrial complex. To process and analyse the collected data, the method of logical analysis and synthesis was used, as well as methods of modelling and forecasting based on the obtained theoretical and practical results. The study did not envisage the use of specific clustering models, and the main objective was to analyse existing approaches and develop general recommendations for improving public policy in the field of cluster development.

3 Results and discussion

The presented work is a comprehensive analytical study aimed at identifying and studying the factors that influence the formation and development of innovative mesosystems in the agro-industrial complex (AIC) of Russia in the context of the implementation of food import substitution strategy. At the initial stage of work, a thorough review of scientific literature on the theory of clusters, their classification, mechanisms of functioning, role in the socio-economic development of regions and the specifics of their application in the agrarian sphere was carried out. Special attention was paid to the studies devoted to the analysis of foreign experience of cluster development, in particular, successful models and unsuccessful attempts to form clusters, as well as the factors determining their effectiveness were considered. Theoretical concepts describing the relationship between clustering, innovation and economic growth, as well as the factors that facilitate and hinder the formation of sustainable links between cluster participants were analysed. In parallel with the study of scientific publications, we analysed the legal acts regulating cluster policy in Russia. Federal and regional strategies and programmes aimed at supporting cluster development, as well as documents defining the priorities of agro-industrial complex development were analysed. This analysis made it possible to assess the current legal field, identify possible gaps and contradictions, and evaluate the effectiveness of existing instruments of state support. After the primary collection and analysis of information was completed, the stage of its systematisation and generalisation was carried out. At this stage, the methods of logical, comparative and systemic analysis were applied. The key factors influencing the development of clusters were identified, general trends and specific features of cluster formation in the agro-industrial complex of different regions of Russia were revealed. Particular attention was paid to the analysis of factors that inhibit innovation activity in the agricultural sector and hinder the achievement of food import substitution goals (Table 01).

Table 1. Dependence of Russian agricultural enterprises on imports in 2019.

Commodity groups	Share of imports, %
Seeds:	
- sugar beet	65
- potato	53
- maize	34
- cereals	5
Means of plant protection	100
Agricultural machinery:	
- tractors (without Belarus)	13
- combine harvesters	31
- forage harvesters	23
Fodder:	
- mixed fodder	<1
- premixes	21

The results of the study showed that, despite the active promotion of the cluster approach as a tool to stimulate regional development, the overall level of innovation activity in Russian agriculture is still low. This, as it turned out, is due not only to insufficient funding, but also to the inefficiency of mechanisms for generating and disseminating innovations in the agricultural sector, which is confirmed by independent analytical studies. Many cluster initiatives implemented in the regions do not bring the expected results and often exist only formally, without sufficient economic viability and without significant impact on the development of agro-industrial complex. The need to assess the agro-ecological state of territories for the sustainable development of clusters is also emphasised in [4]. The analysis has shown that many clusters are formed on the initiative of public authorities, i.e., they have a "top-down" character, rather than being the result of self-organisation of entrepreneurial structures, which, as experience has shown, has a negative impact on their effectiveness. This approach leads to the fact that such clusters lack real co-operation and integration, which, in turn, reduces their ability to generate innovations and improve competitiveness. For example, according to a study conducted by the Association of Cluster Organisations, more than 75% of clusters created in Russia on the initiative of the authorities do not achieve their goals and are not self-sufficient economic units. In addition, the study showed that one of the key factors hindering the development of clusters in the agro-industrial complex is the lack of trust between potential participants, which is due to the historical peculiarities of doing business in Russia, where co-operation is often associated with risk and loss of control. In addition, there are significant gaps in the development of informal institutions, which play an important role in ensuring trust and forming sustainable ties between cluster participants. The lack of qualified personnel capable of effectively managing clusters and generating innovation is also a serious problem. In the agricultural sector, there is a particularly acute shortage of specialists in modern agro-technology, digitalisation, marketing, logistics and financial management. This problem correlates with the assessment of geo-ecological factors and conditions for sustainable land use, as emphasised in the study [5]. For example, experts of the Russian Academy of Sciences (RAS) estimate that the deficit of personnel in the sphere of agro-industrial complex is about 20-30%, with the bulk of the deficit falling on specialists with knowledge in modern management technologies and innovations. The analysis also showed that the insufficient development of scientific infrastructure, the low level of commercialisation of scientific developments and the lack of effective technology transfer mechanisms also hinder the innovative development of clusters in the agro-industrial complex.

Comparative analysis with other studies has shown that the above-mentioned problems are typical for many regions of Russia and for other sectors of the economy, not only agrarian. However, the study revealed that cluster initiatives that are formed "from below", i.e. on the initiative of market participants themselves, and have signs of self-organisation, are usually more effective and viable. This is due to the fact that in such clusters there is a higher level of trust and co-operation, the specific needs and opportunities of the region are better taken into account, and there is more motivation for innovative activity. The study also confirmed that the cluster approach in Russia is often substituted by the creation of agglomerations, i.e. formal association of enterprises in one territory, without real integration and cooperation, which, in turn, reduces its effectiveness. For systems similar to clusters, it is also necessary to forecast their technical state, similarly to what is done for thermochemical reactor systems, as shown in [6. According to the study "Cluster Policy in Russian Regions" published by the Centre for Strategic Research (CSR), at the beginning of 2024, more than half of formally existing clusters are actually agglomerations, which indicates the lack of effectiveness of state policy in the field of cluster development. In addition, the analysis revealed problems related to the insufficient elaboration of the legal and regulatory framework governing cluster activities, the lack of clear criteria for assessing the effectiveness of clusters and insufficient transparency of procedures for their financing and support. In mountainous conditions, the sustainability of development can be affected by factors related to excess salt and soil toxicity, as described in [7.

Based on the study, recommendations for public authorities aimed at improving the effectiveness of cluster policy in the agro-industrial complex were formulated. First, it is necessary to reorient the state policy from the formal creation of clusters "from above" to supporting business self-organisation and creating favourable conditions for the development of clusters "from below", through the involvement of all stakeholders in the formation of the vision and strategy of cluster development, ensuring their transparency and openness. This implies the creation of effective mechanisms of financing, counselling, training and experience sharing for potential cluster participants, as well as the elimination of administrative barriers and bureaucratic delays. This approach is similar to the importance of the economic aspect of green technologies discussed in [8]. Secondly, it is necessary to strengthen the role of regional authorities in cluster development by giving them more autonomy and responsibility for their performance. This also implies the creation of effective mechanisms for monitoring and evaluation of clusters' activities, taking into account regional specifics and priorities of agro-industrial complex development. Thirdly, it is necessary to develop a clear and transparent methodology for assessing the efficiency of clusters, which would take into account not only economic indicators, but also social, environmental and innovative results of their activities, as well as comparability with international standards. In addition, it is necessary to develop more effective mechanisms for managing and monitoring clusters' activities, which would allow promptly responding to emerging problems and adjusting their development strategy, as well as to improve the financing system, making it more transparent and results-oriented. Fourth, special attention should be paid to the development of human capital, including the formation of an effective system of professional training and retraining of specialists for the agricultural sector, with a special focus on modern agro-technology, digitalisation, marketing, finance and supply chain management. It is also necessary to improve the mechanisms of commercialisation of scientific developments and technology transfer in the agricultural sector, which would accelerate the introduction of innovations and increase the efficiency of agricultural production. Finally, it is necessary to actively promote co-operation and knowledge sharing among cluster members by creating appropriate information platforms and tools, as well as training activities and building a culture of trust and partnership. It is important to note that the sustainability of systems such as soil diversity is directly related to trust and co-operation, as shown in [9]. As a result of an integrated approach that takes into account the specifics of the region, the interests of cluster participants and the peculiarities of the agro-industrial complex, it is possible to create conditions for the effective development of innovation clusters, which will make a significant contribution to ensuring food security of the country, as well as increasing the competitiveness of the agricultural sector.

In conclusion, it should be noted that the obtained results and conclusions are confirmed by other studies in this area, as well as by the results of analytical reviews conducted by international organisations and research centres. However, this work has its own specificity, as it is focused on the problems of cluster development in the agro-industrial complex, which allowed us to formulate more targeted and practical recommendations for public authorities. The analysis also demonstrated the importance of not only quantitative but also qualitative indicators that affect the effectiveness of clusters, such as trust, cooperation, innovation culture, staff qualification and quality of public administration.

4 Conclusion

The analytical study revealed a number of key problems that hinder the effective implementation of the cluster approach in the agro-industrial complex of Russia, which in turn hinders the achievement of food import substitution goals. The main conclusion is that despite the active promotion of clusters as an instrument of innovative development, the overall level of innovation activity in agriculture remains low, which is due not only to the lack of funding, but also to systemic problems in the organisation and management of the agricultural sector. The formal approach to the creation of clusters, when the initiative comes from the authorities rather than from the market participants themselves, leads to the fact that such clusters do not have sufficient viability and are unable to generate innovations. This is in many ways similar to how different tillage practices can affect crop yields as shown in studies [9, 10]. Lack of trust between potential participants, underdeveloped informal institutions, shortage of qualified personnel, and lack of a clear methodology for assessing cluster performance have been found to be significant barriers to the development of cluster initiatives [11, 12]. In addition, the cluster approach is often replaced by the creation of territorial agglomerations that do not imply real integration and co-operation between the participants.

The analysis has shown that in order to achieve food security goals, it is necessary to reorient state policy from the formal creation of clusters "from above" to supporting self-organisation of business and creating favourable conditions for the development of clusters "from below", thus stimulating real integration and cooperation. It is important to strengthen the role of regional authorities in the formation and development of clusters, giving them more autonomy and responsibility for their effectiveness. It is necessary to develop clear and transparent criteria for assessing the effectiveness of clusters, which would take into account not only economic, but also social and environmental results. It is also necessary to improve the mechanisms of cluster financing, making them more transparent and results-oriented. The introduction of innovative technologies in agriculture requires qualified personnel, which necessitates reforming the system of vocational education and retraining of specialists for the agricultural sector. Stimulating co-operation and knowledge sharing among cluster members is also an important success factor, for which it is necessary to create appropriate information platforms and tools. Sustainable agriculture, as a way to increase organic matter in the soil, is also important for cluster development, as emphasised in [13], and in the work related to tillage, crop rotation and crop residue management [14]: An integrated and systematic approach that takes into account the specifics of the regions, the interests of cluster participants and the peculiarities of the agricultural sector is a prerequisite for effective cluster development and achievement of the set objectives. In general, the study demonstrated that the cluster approach can be an effective tool for the development of agro-industrial complex, but only if it is competently implemented based on self-organisation, transparency, trust and effective interaction of all participants in the process.

Acknowledgement

The work is based on the results of the research carried out at the expense of the federal budget under the state assignment (name of the research topic "Development and implementation of the strategy for the development of foreign economic relations of agriculture and agro-industrial complex of the Russian Federation taking into account restrictions imposed by sanctions and new priorities of economic cooperation with foreign countries"; code of the research topic assigned by the founder- FZUN-2024-0007).

References

1. S.C. Chowdhuri, S.C. Pal. Challenges and potential pathways towards sustainable agriculture crop production: A systematic review to achieve sustainable development goals (SDGs). Soil Till. Res., **248,** 106442 (2025). DOI: 10.1016/j.still.2024.106442.
2. S. K. Makenova, M. M. Shostak. Geoecological assessment of landscapes as a basis for organisation of sustainable agrarian land use. Sustainable Development of Mountain Territories, **13,** 485-496 (2021). DOI: 10.21177/1998-4502-2021-13-4-485-496.
3. L. Yerkinbayeva, A. Myrzakarim, A. Moldagaliyeva, B. Kalymbek, R. Yerezhepkyzy, D. Nurmukhankyzy. The role of legal regulation in promoting organic agriculture to achieve sustainable food security. J. Lifestyle SDG'S Rev., **5,** e04197 (2025). DOI: 10.47172/2965-730X.SDGsReview.v5.n01.pe04197.
4. V. L. Tatarintsev, L. M. Tatarintsev, F. M. Tatarintsev. K. Ermekov, Yu. S. Lisovskaya. Agroecological condition assessment of agricultural landscapes for their sustainability increase. Sustainable Development of Mountain Territories, **14,** 76-86 (2022). DOI: 10.21177/1998-4502-2022-14-1-76-86.
5. M. N. Nurpeisov, K. V. Ruchkina, V. A. Androkhanov, S. K. Makenova. Assessment of geoecological factors and conditions affecting sustainability of agricultural land use in the forest-steppe zone of Altai Krai. Sustainable Development of Mountain Territories, **15,** 419-430 (2023). DOI: 10.21177/1998-4502-2023-15-2-419-430.
6. O. A. Kolenchukov, A. Y. Mikhaulov, K. A. Bashmur, V. S. Tynchenko, R. B. Sergienko. Forecasting the technical condition of thermochemical reactor systems. SOCAR Proceedings, **2023,** 123-130 (2023). https://doi.org/10.5510/OGP20230100814.
7. K.I. Kravtsov, Ya.A. Tynchenko, T.A. Panfilova. Influence of mountain factors on salt excess and soil toxicity in mountain conditions. Sustainable Development of Mountain Territories, **15,** 784-797 (2023). https://doi.org/10.21177/1998-4502-2023-15-3-784-797.
8. V.Yu Konyuhov, A.M. Gladkih, I.I. Galyautdinov, Y.D. Severina. Economic aspects of green technologies. IOP Conference Series: Earth and Environmental Science, **350,** 012036 (2019). DOI: 10.1088/1755-1315/350/1/012036.
9. M. Usman, M. Wang, Y. Liu, L. Li, X. Zhang, T. Xiao, F. Hou. High soil bacterial diversity increases the stability of the community under grazing and nitrogen. Soil Till. Res., **248,** 106414 (2025). DOI: 10.1016/j.still.2024.106414.
10. D. Liu, B. Tian, M. Zhang, L. Jiang, C. Li, X. Qin, J. Ma. Meta-analysis of the effects of different tillage methods on wheat yields under various conditions in China. Soil Till. Res., **248,** 106449 (2025). DOI: 10.1016/j.still.2025.106449.
11. A.M. Kumaritov, E.A. Sokolova, A.A. Sokolov. Geoinformation system of ecological monitoring in inner-city industrial areas. Gornyi Zhurnal, **2,** 94-96 (2016). https://doi.org/10.17580/gzh.2016.02.18.
12. O. Afanaseva, M. Afanasyev, S. Neyrus, D. Pervukhin, D. Tukeev. Information and Analytical System Monitoring and Assessment of the Water Bodies State in the Mineral Resources Complex. Inventions, **9,** 115 (2024). https://doi.org/10.3390/inventions9060115.
13. Q. Gao, L. Wang, Y. Fang, Y. Gao, L. Ma, X. Wang, Y. Li, X. Wu, Z. Du. Conservation agriculture boosts topsoil organic matter by restoring free lipids and lignin phenols biomarkers in distinct fractions. Soil Till. Res., **248,** 106463 (2025). DOI: 10.1016/j.still.2025.106463.
14. C.S. Nonxuba, D.E. Elephant, A.D. Nciizah, A. Manyevere. Elephant, A.D. Nciizah, A. Manyevere. Medium-term effects of tillage, crop rotation and crop residue management practices on selected soil physical properties in the sub-humid region of Eastern Cape, South Africa. Soil Till. Res., **248,** 106420 (2025). DOI: 10.1016/j.still.2024.106420.

Innovative investments in agriculture as a factor of food security

Evgeny N. Smirnov* [1,2], *Alexander V.* Pasko [1,2], and *Marina Yu.* Antropova [1,2]

[1] State University of Management, Moscow, Russia
[2] Vladimir State University, Vladimir, Russia

Abstract. This article analyses the relationship between investments in the agro-industrial complex of the Russian Federation and the dynamics of agricultural production in the period from 2006 to 2022. The study is based on the statistical data of the Federal State Statistics Service and uses the method of trend analysis to identify general trends and build forecast models. It found that investment in the agricultural sector has shown a pronounced positive trend, especially since 2010, and reached 780 billion roubles in 2022. Agricultural production also showed growth, albeit with fluctuations due to various factors. A not always directly proportional relationship between the volume of investment and production was revealed, which emphasises the need for an integrated approach to the management of the agricultural sector. Based on the analysis, a forecast for 2024-2025 was formed, demonstrating further development of the agro-industrial complex, both in terms of investment and production. The results obtained emphasise the importance of investment for the development of agriculture, as well as the need to introduce innovations, develop human resources and improve risk management mechanisms.

1 Introduction

In today's world, where the world population continues to grow steadily, ensuring food security is one of the most acute and urgent problems. The increasing demand for food, aggravated by the limited natural resources and the negative impact of climate change, threatens the stable and sustainable development of the agricultural sector [1, 2]. There are several approaches to solving this complex problem, each with its own advantages and disadvantages. One way is extensive agricultural development, which implies increasing the area of cultivated land and using more resources. However, this approach often leads to soil depletion, ecosystem degradation and increased greenhouse gas emissions. Another way is intensification of production, which involves the use of modern technologies, fertilisers and crop protection products. This approach can significantly increase crop yields and animal productivity, but carries risks to the environment and human health if not properly applied. In addition, an important aspect is the development of organic agriculture, which aims to minimise the negative impact on the environment and create more sustainable agroecosystems [3]. However, organic farming requires considerable effort and may be less productive in the short term, compared to intensive farming.

In the context of geopolitical instability and economic challenges, the problem of food security is of particular relevance for many countries, including the Russian Federation. Sanctions and trade restrictions, on the one hand, create difficulties for food imports, but on the other hand, they become a powerful incentive for the development of own agro-industrial complex and increase its competitiveness [4]. In this regard, innovation processes in agriculture play a key role, providing an opportunity to increase the volume of production through the use of new technologies that increase the efficiency of resource use, optimise production processes and minimise the negative impact on the environment [5, 6]. Innovation, in this context, is considered not only as the introduction of advanced technical solutions, but also as a rethinking of traditional agricultural practices [7]. This implies the introduction of precision farming, the use of digital technologies, the use of modern breeding methods, as well as the development of robotisation and automation of production processes [8-10]. Also, an important tool for precision agriculture is modelling of weather conditions, which allows more efficient planning of agronomic measures [11]. An integrated approach that combines modern agro-technologies, investment in research and development and competent resource management seems to be the most promising way to ensure food security of the country. In this regard, the study of the relationship between investments in innovative development of the agro-industrial complex and production volumes seems to be a relevant and important area of scientific research. It allows us to assess the effectiveness of the current policy in the agricultural

* Corresponding author: smirnov_en@mail.ru

sector, identify problem areas and develop recommendations to improve the effectiveness of measures aimed at ensuring food security [12]. The purpose of the work presented in this article is to assess the trend of the volume of investment in agriculture in the Russian Federation.

2 Method and materials

The work presented in this paper is analytical in nature and does not involve full-scale experiments. The main emphasis is placed on the analysis of statistical data and the construction of predictive models. The study is based on the data of the Federal State Statistics Service of the Russian Federation, reflecting the dynamics of agricultural production and investment in the agricultural sector. The method of trend analysis is used as the main analytical tool, which makes it possible to identify general trends in the development of agriculture and build forecast values based on the identified patterns.

The analytical work is carried out using specialised software designed for statistical data processing, graphs and charts construction, and forecasting. The work involves generalised results of statistical processing at the country level rather than at the level of individual households or production processes. The main data sets are presented in the form of tables showing the volume of production of the main types of agricultural products and the volume of investment in fixed capital in the Russian Federation by types of economic activity in the period from 2006 to 2022. On the basis of these data, growth rates were calculated and linear regression models were built, and forecast values for 2024 and 2025 were determined.

3 Results and discussion

The work analysed statistical data taken from official sources, such as the Federal State Statistics Service of the Russian Federation. These data represented information on the production volumes of the main types of agricultural products, including cereals, meat, milk, eggs and potatoes, as well as on the dynamics of investment in fixed capital of agriculture. The analyses were conducted in two main directions: assessment of investment dynamics and assessment of production volume dynamics. Particular attention was paid to identifying trends and building predictive models capable of predicting future values of production volumes on the basis of available data. This made it possible to characterise the current state of affairs and attempt to provide a short-term forecast. The methodological basis of the study was trend analysis, which allowed us to identify general development trends and predict values based on linear regression models. This made it possible to establish quantitative relationships between the volume of investment and production volumes, and to visualise these data using graphs and charts. The work did not include field experiments; it is analytical in nature and is based solely on the processing of statistical data that allow us to identify regularities at the macro level. According to the results obtained, investments in the agro-industrial complex of the Russian Federation in the period from 2006 to 2022 showed a pronounced positive dynamics. The growth of investments during this period was rather uneven, with noticeable jumps and periods of slowdown, but the general trend was directed upwards. Despite the fact that this article does not deal directly with agriculture, but it demonstrates positive trends in the development of technologies and materials, which also has a positive impact on the development of the agro-industrial complex [13]. Especially significant growth of investments started from 2010 and continued after 2016, which can be associated with the strengthening of state policy aimed at supporting agriculture and ensuring food independence of the country. This period was characterised by significant injections into the fixed capital of agro-industrial enterprises, which, in turn, was due to an increase in state subsidies, soft loans and other forms of financial support. In 2022, the volume of investments, according to preliminary data, reached the mark of 780 billion rubles, which significantly exceeds the figures of previous years, for example, 2010 with its 303.8 billion rubles. When analysing annual investment growth, the years 2011 and 2018 particularly stand out, when this indicator exceeded 115% in relation to the previous year. This indicates a sharp increase in investor interest in the agricultural sector during these periods (Table 01). In parallel with the growth of investment in the agricultural sector, there was a positive dynamics of agricultural production. However, production growth rates were not uniform and fluctuated depending on various factors. Grain production increased from 94.2 million tonnes in 2011 to 152 million tonnes in 2022. However, there was a slight decrease in production in 2012 and 2019 compared to previous years, which may be due to weather conditions. Production of meat and meat products also showed an increase, from 7,520 thousand tonnes in 2011 to 11,850 thousand tonnes in 2022, although also with uneven dynamics during the analysis period. In particular, there was a slight decline in 2019, after which steady growth resumed. More steady growth was shown by the production of milk and milk products. This increased from 31,646 thousand tonnes in 2011 to 32,890 thousand tonnes in 2022. A more detailed analysis of the relationship between investment and production revealed that investment growth was not always accompanied by a proportional increase in production. For example, this may be influenced by factors related to export opportunities and economic conditions [14].

Table 1. The level of provision with production of the main types of agricultural products in the Russian Federation

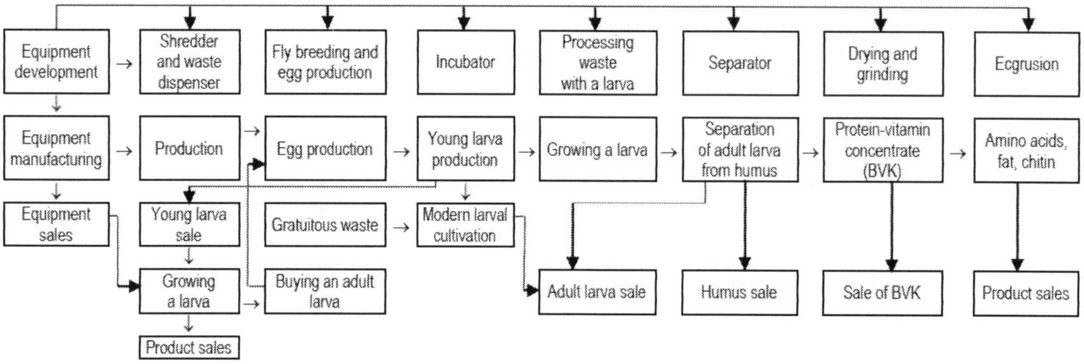

For example, there was a significant increase in investment in 2011 and 2018, but the growth rate of output in these years was lower. This indicates that the output is affected by factors other than financial injections. On the other hand, in 2021 and 2022, there is a more pronounced increase in production volumes compared to the previous years, which may suggest that agricultural investment is more effective in the long run. It is important to note that there is a certain time lag between investment and results. Investments in infrastructure development, machinery renewal and introduction of new technologies do not have an immediate effect, but have a long-term positive impact on productivity. The findings confirm the importance of investment for the development of the agricultural sector, but also emphasise the need for an integrated approach to agricultural management. It is the integrated approach that includes various aspects of the agricultural sector, including quality and flavour traits, that is important for sustainable development. Investment alone is not the only factor contributing to the development of the agricultural sector, but is only one tool. To achieve sustainable growth, technological innovation, effective public policy, quality human resources and competent risk management are also necessary [15]. Also, as an example of technological progress in the agro-sector, we can cite the works on the use of waste in production, including its circular utilisation for other agricultural products [16, 17]. Comparison of our results with similar studies shows that our findings are in line with global trends. Many researchers also note the important role of investment in the development of the agricultural sector, but they emphasise that the effectiveness of investment depends on many factors, such as the level of infrastructure development, technology, as well as economic and political stability. For example, researchers note the importance of studying agri-food chains and capital to understand the dynamics of agriculture [18]. Studies in the field of agricultural economics indicate that the promotion of competition, development of market mechanisms and improvement of state support mechanisms are also necessary to achieve stable growth in agriculture. It is extremely important to analyse the peculiarities of land relations and reforms in this area for the stable development of the agricultural sector [19, 20].

Based on the analysis conducted and trends identified, as well as relying on expert assessments of the development of Russia's agro-industrial complex, it seems possible to form forecast values of agricultural output and investment in the agricultural sector for 2024 and 2025. It is important to note that this forecast is of a probabilistic nature and can be adjusted under the influence of various factors, such as weather conditions, changes in government policy, fluctuations in world food prices, as well as technological breakthroughs and their introduction into agricultural practice. Regarding investments in the agro-industrial complex, it can be assumed that the upward trend will continue in the short term. For sustainable development it is important to invest not only in production itself, but also in improving the social sphere in rural areas. Based on linear trends and taking into account the current growth rates, we can forecast that the volume of investment in fixed capital in agriculture will reach 850-900 billion rubles in 2024, and in 2025 may exceed 950 billion rubles, approaching the mark of 1 trillion rubles. This growth will be due to both continued state support for the agricultural sector and increased interest from private investors who see the prospect of developing modern, high-tech agriculture. At the same time, it should be taken into account that investment growth rates may slow down somewhat compared to previous periods, taking into account the general economic situation and possible adjustments in state support. In turn, agricultural output will continue to be affected by a number of factors, but in general will also show positive dynamics. Taking into account the development of technologies, expansion of sown areas and increased efficiency of fertiliser use, grain production in 2024 can be expected to reach 130-135 million tonnes, and in 2025 may approach the 140-150 million tonnes mark. At the same time, as before, grain production will be significantly influenced by weather, so there may be deviations in both directions. Increased mechanisation and automation of livestock breeding complexes, as well as further development of genetics and animal breeding, allow us to forecast that production of meat and meat products will reach 12.3-12.7 million tonnes in 2024 and may reach 13-13.5 million tonnes in 2025. The introduction of new technologies in dairy farming, as well as higher purchase prices for milk, should stimulate production growth. Production of milk and milk products is projected to reach 33.5-34 million tonnes in 2024 and 34.5-35 million tonnes in 2025. We also need to consider the impact of land management factors on total production. Given the significant potential for growth in this segment, vegetable and fruit production can be expected to continue to grow, although the growth rate may be lower than that of cereals and meat products. Vegetable and fruit production is forecast to reach 18.5-19.0 million tonnes in 2024 and 19.5-20 million tonnes in 2025. Potato production is forecast to grow steadily, driven by modern technology. In 2024, production can be expected at 30.5-31 million tonnes and in 2025 at 31.5-32 million tonnes. Thus, the overall forecast for 2024-2025 indicates further development of Russia's agro-industrial complex, both in terms of investment and production volumes. However, as before, it is important to take into account that these forecasts are based on current trends and may be adjusted under the influence of various factors. Therefore, to ensure stable and sustainable development of agriculture, further investments in innovation, human resource development and improvement of risk management mechanisms are needed. It is also necessary to take into account the impact of conflicts and various socio-territorial processes on the dynamics of agricultural production, as they can lead to problems, such as problems with water supply, or logistical problems [21].

4 Conclusion

The research has allowed us to identify the key regularities of the relationship between investments in the agro-industrial complex of the Russian Federation and the dynamics of agricultural production in the period from 2006 to 2022. The analysis of statistical data has shown that investments in the agricultural sector demonstrate a pronounced positive trend, especially since 2010, which is associated with the strengthening of state support and the growing interest of private investors. The volume of investments has reached significant values, in particular, in 2022, according to preliminary data, it amounted to 780 billion rubles. Along with the growth of investments, there was a positive dynamics of agricultural production, but with some fluctuations due to various factors, including weather conditions and market conditions. At the same time, the relationship between investment and production volume was not always directly proportional, which indicates the influence of other factors, such as the level of technological equipment, staffing and resource management efficiency.

The analysis allowed not only to state positive trends in the development of the agricultural sector, but also to identify certain problems that require further solution. In particular, it was shown that in order to achieve sustainable growth in agriculture it is necessary not only to increase the volume of investments, but also to ensure their efficient use. This implies the introduction of innovative technologies, infrastructure development, training of qualified personnel, and improvement of risk management mechanisms. It is also important to create a favourable investment environment to attract both public and private investment in the agricultural sector.

Based on the identified trends and tendencies, a forecast for 2024 and 2025 was formed, which demonstrates further development of the agro-industrial complex, both in terms of investment and production volumes. It is predicted that in 2024 the volume of investment in fixed capital of agriculture will reach 850-900 billion rubles, and in 2025 will exceed 950 billion rubles. The projected growth in the production of grain, meat, milk, vegetables and fruit indicates the potential for the development of the country's agriculture. At the same time, the projected values may be adjusted under the influence of various factors, which emphasises the need for an integrated approach to the management of the agricultural sector and further research in this direction. In conclusion, this study emphasises the importance of investment for ensuring food security and sustainable development of the agricultural sector, but calls for a comprehensive approach to management based on innovative technologies, sound public policy and efficient use of resources.

Acknowledgement

The work is based on the results of the research carried out at the expense of the federal budget under the state assignment (name of the research topic "Development and implementation of the strategy for the development of foreign economic relations of agriculture and agro-industrial complex of the Russian Federation taking into account restrictions imposed by sanctions and new priorities of economic cooperation with foreign countries"; code of the research topic assigned by the founder- FZUN-2024-0007).

References

1. N. Garambois, U. Le Goff, L. Thibaudeau. Agrarian dynamics and climate change in the Senegalese Sahelian peanut basin. Agrar. Syst. Clim. Change: Journeys Adapt. Glob. South, 15-37 (2024). DOI: 10.1079/9781800628137.0001.
2. T. Bandara, B. Marambe, G. Pushpakumara, P. Silva, R. Punyawardena, L. Manawadu, S. Premalal, G. Miah, K.R. Dahal. Adapting mountain-farming systems to climate change by reducing food-nutrition-health vulnerability. Mitig. Adapt. Strateg. Glob. Change, **29**, 88 (2024). DOI: 10.1007/s11027-024-10181-x.
3. B. Aparecida da Silva, R. Evangelista de Oliveira, A. Cavalieri Sais, E. Cardoso leite. Family farmers and trees: ecological functions, ecosystem services and food production in agrarian reform settlements in Brazil. Agroecol. Sustain. Food Syst., **49**, 4-21 (2025). DOI: 10.1080/21683565.2024.2405515.

4. S. Bamidele. BRICS Revolution: The Global Agrarian Shift Sparked by Emerging Economies. India Q., (2025). DOI: 10.1177/09749284241307937.
5. A.A. Kulikova, E.I. Khabarova. Biological working agents use to reduce the carbon footprint. Sustainable Development of Mountain Territories, **16,** 311–321 (2024). DOI: 10.21177/1998-4502-2024-16-1-311-321.
6. E.Yu. Kulikova, S.V. Balovtsev, O.V. Skopintseva. Geoecological monitoring during mining operations. Sustainable Development of Mountain Territories, **16,** 580–588 (2024). DOI: 10.21177/1998-4502-2024-16-2-580-588.
7. V.V. Yurak, A.N. Malyshev, S.S. Zavyalov. Review of new remote methods for environmental monitoring. MIAB. Mining Inf. Anal. Bull., **11-1,** 51-67 (2022). DOI: 10.25018/0236_1493_2022_111_0_51.
8. C. Cruz-Grimaldo, M. Nieves, E. Vera, M. Duran, A. Morales, W. Salazar, C.I. Arbizu. Yield predictions of 'Del Cerro' cotton (Gossypium hirsutum L.) germplasm by multispectral monitoring in the north coast of Peru. Chilean J. Agric. Res., **85,** 15-26 (2025). DOI: 10.4067/S0718-58392025000100015.
9. S.N. Goncharenko, A.B. Lachihina. Modeling protected geological and geotechnical information system in mining industry. MIAB. Mining Inf. Anal. Bull., **6,** 39-55 (2023). DOI: 10.25018/0236_1493_2023_6_0_39.
10. K.Y. Lobkov, A.A. Stupina, V.V. Tynchenko. On the issue of monitoring crack formation in natural-technical systems during earth surface displacements. MIAB. Mining Inf. Anal. Bull., **11-1,** 292-304 (2023). DOI: 10.25018/0236_1493_2023_111_0_292.
11. B.V. Malozyomov, S.N. Sorokova, E.A. Efremenkov, D.V. Valuev, M. Qi. Analysis of a Predictive Mathematical Model of Weather Changes Based on Neural Networks. Mathematics, **12,** 480 (2024). DOI: 10.3390/math12030480.
12. A.T. Beristenov, L.V. Lebedeva. Assessment of the agroecological potential of anthropogenic landscapes in the dry steppe of the south of Western Siberia. Sustainable Development of Mountain Territories, **16,** 975–987 (2024). DOI: 10.21177/1998-4502-2024-16-3-975-987.
13. K. Akzharkyn, D. Yelemessov, D. Baskanbayeva, N.V. Martyushev, V.Y. Skeeba, V.Y. Konyukhov, T.A. Oparina. Strengthening Polymer Concrete with Carbon and Basalt Fibres. Appl. Sci. (Switzerland), **14,** 7567 (2024). DOI: 10.3390/app14177567.
14. A.H. Abdi, A.M. Mohamed. Determinants of agricultural exports in Somalia: the impacts of exchange rates, foreign direct investment, and institutional quality. Cogent Econ. Finance, **13,** 2445755 (2025). DOI: 10.1080/23322039.2024.2445755.
15. O.M. Zinovieva, A.M. Merkulova, N. A. Smirnova, E.A. Shcherbakova. Psychosocial risk management in mining. MIAB. Mining Inf. Anal. Bull., **1,** 20-33 (2022). DOI: 10.25018/0236_1493_2022_1_0_20.
16. A. Al Mozahid. What counts as "quality" in agrarian taste? Subjectivities, gender, and rice agroecology in northern Bangladesh. Cult. Agric. Food Environ., **46,** 88-105 (2024). DOI: 10.1111/cuag.12323.
17. L.S. Malyukova, V.V. Kondratiev, V.V. Bukhtoyarov, V.Y. Konyukhov. Circular Mining Wastes Management for Sustainable Production of Camellia sinensis (L.) O. Kuntze. Sustainability (Switzerland), **15,** 11671 (2023).
18. A. Bowman, N. Robb. Accumulation by Intermediation: The Contestation of Agro-Food Capital in the South African Maize Industry. J. Agrar. Change, (2025). DOI: 10.1111/joac.12616.
19. R. Yang, C. Kesteloot. The interaction between local state and village collective in China's market-oriented land reform: a case study in Pidu, Sichuan. J. Peasant Stud., (2025). DOI: 10.1080/03066150.2024.2438720.
20. S. Buletsa. The peculiarities of restitution of agricultural land in Ukraine. J. Agric. Environ. Law, **19,** 27-74 (2024). DOI: 10.21029/JAEL.2024.37.27.
21. J.G. Santos, A.A.R. Ioris. Water Conflicts and Socioterritorial Dynamics: The Hydrosocial Cycle After the São Francisco River Transposition Project in the Northeast of Brazil. Land, **13,** 2032 (2024). DOI: 10.3390/land13122032

Problems and prospects of innovative development of agro-industrial complex in Eastern Siberia

Jahongir Qosimov [1*], *Islombek* Akhmedov [2], *Shavkat* Abdikadirov [3], *Khudoymurot* Dusyarov [4], *M.Yu.* Ivanov [5]

[1] Tashkent Institute of Irrigation and Agricultural Mechanization Engineers" National Research University, Tashkent, 100000, Uzbekistan
[2] Namangan Engineering-Construction Institute, Uzbekistan
[3] Termez State University of Engineering and Agrotechnologies, 190111,Termez, Uzbekistan
[4] Termiz State Pedagogical Institute,Termez, Uzbekistan
[5] Bratsk State University, Bratsk, Russia

Abstract. The presented research is devoted to the analysis of problems and prospects of innovative development of the agro-industrial complex of Eastern Siberia. The work focuses on identifying the key barriers to sustainable development of the industry and substantiating the need to form a regional scientific, educational and production centre. The research methodology is based on the analysis of a wide range of sources, including scientific publications, statistical data and regulatory documents. The main problems, such as insufficient funding of scientific research, weak link between science and production, and lack of qualified personnel, are identified. The results of the work show that to overcome these problems a systematic approach is needed, including the development and introduction of energy-saving technologies, the creation of new crop varieties and the development of processing and logistics technologies. The study justifies the creation of a regional centre that will combine the efforts of science, production and government to ensure sustainable development of the agro-industrial complex.

1 Introduction

The global challenges facing humanity, including population growth and climate change, require a radical rethinking of approaches to food production. Traditional agricultural practices based on extensive use of resources are often inefficient and lead to soil degradation, biodiversity loss and environmental pollution. In the context of finding alternatives, it is important to consider the experience of other countries, as in the work of B. Aparecida da Silva et al. Aparecida da Silva et al. on farms and trees in Brazil, demonstrating the importance of ecosystem services in agrarian systems [1]. In this regard, the innovative development of agro-industrial complex becomes not just desirable, but a prerequisite for food security and sustainable development, as indicated in studies analysing global agrarian shifts [2]. One of the key directions in this context is the introduction of modern technologies to increase yields, reduce resource consumption and improve the quality of agricultural products.

Various approaches are proposed to solve the problem of food supply. Agricultural intensification, although it allows increasing production, is often associated with high energy consumption, use of chemical fertilisers and pesticides, which negatively affects the environment and human health. Organic farming, in turn, although environmentally friendly, often does not provide sufficient yields to meet the needs of a growing population. An important direction is the development of precision farming technologies that optimise the use of resources and reduce the negative impact on the environment. Also promising is the development of biotechnologies that allow the creation of new plant varieties resistant to pests, diseases and unfavourable environmental conditions. However, the introduction of these technologies requires significant financial investments and highly qualified personnel. Currently, for many regions of Russia the problem of introducing innovative technologies in the agro-industrial complex is acute. One of the problems is the adaptation of agrarian activities to climate change, which requires studying the experience of regions with similar challenges, which is considered in a number of works on the dynamics of agrarian systems in a changing climate [3] and in works devoted to the adaptation of agrarian systems to climate change in general [4]. It is this approach that will make it possible to combine the efforts of scientists, producers and authorities to solve regional problems in the field of food security. This becomes especially important in the conditions of unique natural and climatic features of certain regions, where an

* Corresponding author: jaxongirqosimov7@gmail.ru

adaptive approach to the introduction of technologies that takes into account local conditions is necessary. In particular, in regions with harsh climates and limited land resources, it is crucial to develop energy-saving technologies, as well as technologies that maximise the use of local resources.

Thus, the relevance of the presented research is conditioned by the need to find ways to overcome the existing barriers in the development of agro-industrial complex, namely: insufficient funding, lack of connection between science and production, as well as the need to adapt innovative technologies to regional conditions. The purpose of this paper, presented in the paper, is to analyse the existing problems in the agro-industrial complex of Eastern Siberia and to develop science-based approaches to the formation of a regional innovation strategy in the field of production, processing and logistics of agricultural products. This strategy is aimed at creating conditions for sustainable development of the agro-industrial complex of the region and ensuring food security of its population.

2 Method and materials

In this article, the main emphasis is placed on the development of a methodological framework for the innovative development of agro-industrial complex. The experimental work presented in the article is conceptual and analytical in nature, aimed at substantiating the need to form a regional scientific, educational and production centre. The work analysed the current situation in the agro-industrial complex of Eastern Siberia, identified the main problems hindering its development, and proposed specific directions for innovative development.

The study was based on the analysis of a wide range of sources: scientific publications, statistical data and regulatory legal acts. Economic, social and environmental aspects of the functioning of the agro-industrial complex of the region were studied. The methodology development was based on experimental work using specific equipment.

3 Results and discussion

Within the framework of the presented work, which is analytical rather than empirical in nature, a comprehensive analysis of the current state of the agro-industrial complex of Eastern Siberia was carried out. The aim of the study was not so much to obtain specific experimental data as to form a conceptual framework for the innovative development of the industry in the region. This involved an in-depth meta-analysis of available sources of information, including scientific publications, statistical reports and regulatory documents, which made it possible to identify key problems and determine the most promising areas for development. Unlike traditional agronomic research that focuses on the study of specific technological practices and their impact on yields, this work aimed to develop a strategic framework that would take into account not only technological, but also economic, social and environmental aspects of agro-industrial complex functioning.

The study undertook a detailed review of existing concepts of agricultural development at global and regional levels. This included an examination of various intensification strategies aimed at increasing production through the use of modern technologies and agrochemicals, as well as greening strategies involving the use of organic farming methods and biological plant protection products, which is particularly relevant in the context of work on the role of family farmers and ecosystem services. Particular attention was paid to analysing the experience of regions with similar climatic and economic conditions in order to identify best practices and their potential applicability in Eastern Siberia. In addition, various models of financing and support for innovation in the agricultural sector were analysed, and the effectiveness of existing regulations governing agro-industrial enterprises was assessed.

As part of the analysis of the current state of the agro-industrial complex in Eastern Siberia, the key problems hindering its sustainable development were identified. One of the main problems is the chronic underfunding of research and development work, which leads to the region's lagging behind the advanced agrarian regions of the country and the world in the development and implementation of new technologies. The importance of supporting scientific research in the development of the agrarian sector is emphasised in the works considering the issues of food supply formation [5]. A significant problem lies in the insufficient connection between science and production, which leads to the fact that many scientific developments do not find practical application in agriculture. Also, the analysis has shown that the current legal framework in the agricultural sector does not always stimulate innovation activity, and sometimes even creates additional bureaucratic barriers to the introduction of new technologies, which is especially relevant in the context of research on the legal aspects of land use [6]. The lack of qualified personnel, especially specialists in the field of precision agriculture, biotechnology and agro-logistics, is also a significant problem limiting the pace of innovative development. It is human capital that plays an important role in the successful development of the agricultural sector, as shown by studies in the field of sociology of agrarian systems [7]. One of the important problems is that many of the existing agribusinesses do not have sufficient access to modern information resources and do not fully utilise the opportunities of digital technologies to improve the efficiency of their production. In general, the analysis has shown that the agro-industrial complex of Eastern Siberia needs comprehensive modernisation, which would include not only the introduction of new technologies, but also the creation of an effective system to support innovation activities, as well as the training of qualified personnel.

The results of the analysis allowed us to develop the concept of regional innovation strategy [8, 9], aimed at overcoming the identified problems and creating conditions for sustainable development of the agro-industrial complex. It is proposed to create a regional scientific, educational and production centre, which would become a catalyst of innovation processes in the agrarian sector. The main task of this centre is to unite the efforts of scientists, producers and representatives of the authorities to solve key problems of the agrarian sector development. One of the important tasks of the centre will be the development and implementation of energy-saving technologies to reduce the costs of agricultural production and reduce the negative impact on the environment. Adaptation to local conditions should take into account the peculiarities of agro-ecological potential, as indicated by studies conducted in similar regions [10]. Also, the centre should be engaged in the development of new crop varieties adapted to local climatic conditions and having high yields and resistance to diseases and pests. At the same time, ecological typification of land for sustainable development of agrarian systems should become a priority [11]. In the course of the study it was found that in some areas of Krasnoyarsk Krai the yield of grain crops is only 15 c/ha, while with the use of modern varieties and technologies it can be increased to 30-35 c/ha. Similar data were obtained for other agricultural crops. According to statistics for 2021, the gross grain harvest in Krasnoyarsk Krai was 3.2 million tonnes, but the region's potential allows increasing this figure to at least 5 [12].

$$F_{ci} = \sum_{i=1}^{n} E_{n.sp.i} F_i C_i - \sum_{i=1}^{n} L_{ij} \rightarrow \max;$$

$$F_{ei} = \sum_{i=1}^{n} E_{n.sp.i} F_j - \sum_{i=1}^{n} E_i \rightarrow \max,$$

where F_{ei} and F_{ci} – respectively the target function of conditional energy and cost incomes; $E_{n.sp.i}$ - specific energy productivity of the i-th crop; F_j – area occupied by the i-th crop; E_i – energy costs for production of the i-th type of work; C_i – price of a unit of the yield component; L_{ij} – costs for production of the i-th type of work by the j-th type of machines.

In Siberia, the ecological resource diversity of soil, climatic and geomorphological conditions requires the use of land with different intensity even within one agricultural organisation or its structural subdivision.

In addition, the study has shown that the region has significant potential for the development of livestock breeding, especially in cattle breeding and milk production. However, in order to realise this potential, it is necessary to introduce modern technologies for feeding, housing and breeding of animals, as well as the development of the fodder base, which in turn should take into account geo-ecological factors. Currently, according to research data, the average productivity of cows in the dairy herd in the region is about 5000 litres of milk per year, which is significantly lower than in the advanced farms of the country and the world (7000-9000 litres per year). The analysis also revealed the critical need to develop technologies for processing and logistics of agricultural products, which will reduce losses and increase added value. Currently, a significant part of the produced products is not processed and is exported to other regions, which leads to significant losses and reduces the economic efficiency of the agro-industrial complex. For example, about 50% of the milk produced is exported outside the region for processing, while the creation of modern dairy plants in the territory of the Krai would not only reduce transport costs, but also create new jobs. It is also necessary to develop technologies for deep grain processing, production of fodder and organic fertilisers, which includes the search for methods of environmentally friendly leaching [13]. The obtained data, among other things, indicate the need to create a regional network of wholesale and distribution centres, which will optimise the supply of agricultural products and reduce their cost for consumers. Finally, one of the important directions of the centre's work should be the creation of an effective system of training and retraining of personnel who will have up-to-date knowledge and skills in the field of agronomy, zootechnology, biotechnology and agro-logistics. The introduction of circular waste management, as the experience of some industries shows, can bring significant benefits to agriculture [14, 15].

Comparison of the results obtained with the data of other studies shows that the problems identified in the agro-industrial complex of Eastern Siberia are not unique and are typical for many other regions of Russia. However, the specifics of the region associated with harsh climatic conditions and remoteness from the main markets require the development of an adapted development strategy. Thus, studies conducted in other regions with similar climatic conditions show that the introduction of energy-saving technologies in crop production can reduce energy consumption by 20-25%, and the use of high-yielding varieties of grain crops increases yields by 15-20%. Also, it is necessary to pay attention to the issues of ecologisation of production processes, as studies in other industries show [16]. Similar data have been obtained for animal breeding, where the use of modern technologies for feeding and breeding animals can increase their productivity by 10-15%. At the same time, studies in the field of logistics show that the creation of regional wholesale and distribution centres can reduce transport costs by 15-20% and increase the availability of agricultural products for the population. In general, the study highlights the need for a systematic approach to the development of agro-industrial complex, which should include not only the introduction of new technologies, but also the creation of a favourable regulatory and legal environment, the training of qualified personnel and the development of modern infrastructure. The implementation of the proposed concept of creating a regional scientific, educational and production centre will help to solve the identified problems and create conditions for sustainable development of the agrarian sector in Eastern Siberia.

4 Conclusion

The study highlights the urgent need for systemic transformation of the agro-industrial complex of Eastern Siberia. Traditional methods of farming, characterised by extensive use of resources and low level of technological equipment, are unable to ensure food security of the region and sustainable development of the industry. The analysis of the current state has revealed a number of critical problems, including underfunding of research and development, weak link between science and production, ineffective regulatory framework, and lack of qualified personnel. In this context, the experience of using systems for technical condition prediction in other industries, such as thermochemical reactors [17], may be useful for the agro-industrial complex. These in the agro-industrial complex. This centre should unite the efforts of scientists, manufacturers and representatives of the authorities to solve the key problems of the agrarian industry development. The main directions of the centre's work should be the development and introduction of energy-saving technologies, the creation of new crop varieties adapted to local conditions, as well as the development of technologies for processing and logistics of agricultural products. At the same time, it is important to take into account the impact of mountain factors on soil conditions, which is important for proper land use [18-20]. Special attention should be paid to the training and retraining of qualified personnel capable of applying modern knowledge and skills in agronomic, zootechnical and engineering fields.

This work has shown that the innovative development of agro-industrial complex is not just desirable, but a necessary condition for ensuring food security in Eastern Siberia and sustainable development of the region. The creation of a regional scientific, educational and production centre, taking into account the results of the analysis, will allow overcoming the existing barriers and creating conditions for dynamic and competitive development of the agrarian industry in the region. Environmental aspects of mountain regions' development should also be taken into account when implementing innovations in the agro-industrial complex [21, 22]. The implementation of the proposed innovative approaches will not only increase the volume of agricultural production, but also improve its quality, reduce losses and ensure sustainable development of the agro-industrial complex in the long term. An important role in the development of the agrarian sector is played by the creation of a system for monitoring and analysing modern technologies, which is necessary for effective management and decision-making.

References

1. B. Aparecida da Silva, R. Evangelista de Oliveira, A. Cavalieri Sais, E. Cardoso leite, Family farmers and trees: ecological functions, ecosystem services and food production in agrarian reform settlements in Brazil. Agroecol. Sustain. Food Syst., **49,** 4-21 (2025). DOI: 10.1080/21683565.2024.2405515

2. S. Bamidele, BRICS Revolution: The Global Agrarian Shift Sparked by Emerging Economies. India Q., (2025), Article in press. DOI: 10.1177/09749284241307937

3. N. Garambois, U. Le Goff, L. Thibaudeau, Agrarian dynamics and climate change in the Senegalese Sahelian peanut basin. Agrar. Syst. Clim. Change: Journeys Adapt. Glob. South, 15-37 (2024). DOI: 10.1079/9781800628137.0001

4. H. Cochet, O. Ducourtieux, N. Garambois, Agrarian systems and climate change: Journeys of adaptation in the global south. Agrar. Syst. Clim. Change: Journeys Adapt. Glob. South, 1-208 (2024). DOI: 10.1079/9781800628137.0000

5. K.M. Didur, H.O. Kundieieva, G.V. Ortina, T.A. Pikhniak, A.V. Revkova, FORMATION OF FOOD SUPPLY OF THE AGRARIAN SECTOR OF UKRAINE ON THE BASIS OF RESTORATION OF THE DEVELOPMENT OF RURAL AREAS. Rev. Iberoam. Vitic. Agroind. Rural., **12,** 223-239 (2025). DOI: 10.35588/pxe7k028

6. S. Buletsa, The peculiarities of restitution of agricultural land in Ukraine. J. Agric. Environ. Law, **19,** 27-74 (2024). DOI: 10.21029/JAEL.2024.37.27

7. R. Sunam, F. Sugden, A. Kharel, T.R. Sunuwar, T. Ito, Unpacking youth engagement in agriculture: Land, labour mobility and youth livelihoods in rural Nepal. J. Agrar. Change, **25,** e12611 (2025). DOI: 10.1111/joac.12611

8. E.Yu. Kulikova, S.V. Balovtsev, O.V. Skopintseva. Comprehensive assessment of geoecological risks in conducting open and underground mining. Sustainable Development of Mountain Territories, 16, 205–216 (2024). DOI: 10.21177/1998-4502-2024-16-1-205-216.

9. A.A. Kulikova, E.I. Khabarova. Biological working agents use to reduce the carbon footprint. Sustainable Development of Mountain Territories, **16,** 311–321 (2024). DOI: 10.21177/1998-4502-2024-16-1-311-321.

10. A. Al Mozahid. What counts as “quality” in agrarian taste? Subjectivities, gender, and rice agroecology in northern Bangladesh. Cult. Agric. Food Environ., **46,** 88-105 (2024). DOI: 10.1111/cuag.12323.

11. M.V. Shabanov, M.S. Marichev, S.S. Mangiyeva, A.A. Sokolov. Chernozem formation under conditions of prolong exposure to aero-industrial emissions from a mining and smelting plant. Sustainable Development of Mountain Territories, **15,** 727-740 (2023). DOI: 10.21177/1998-4502-2023-15-3-727-740.

12. V. L. Tatarintsev, L. V. Lebedeva, G. D. Syzdykova, D. S. Inkarov. Management of agricultural land use sustainability at the local level using GIS technologies and special map material. Sustainable Development of Mountain Territories, **15,** 864–876 (2023). DOI: 10.21177/1998-4502-2023-15-4-864-876.

13. N.N. Bryukhanova, V.A. Gladkikh, R.V. Kononenko, V.V. Gerasidi. Selection of disintegrator

parameters toward environmentally friendly leaching of metals. Min. Inform. Anal. Bull., **12,** 39-49 (2024). DOI: 10.25018/0236_1493_2024_12_0_39.

14. L.S. Malyukova, N.V. V.V. Kondratiev, V.V. Bukhtoyarov, V.Y. Konyukhov, K.A. Bashmur, T.A. Panfilova, V. Brigida. Circular Mining Wastes Management for Sustainable Production of Camellia sinensis (L.) O. Kuntze. Sustainability (Switzerland), **15,** 11671 (2023). DOI: 10.3390/su151511671.

15. Yu.I. Kutepov, N.A. Kutepova, A.S. Mukhina. Problems and prospects of using industrial waste in mine reclamation. MIAB. Mining Inf. Anal. Bull., **12-1,** 5-21 (2024). DOI: 10.25018/0236_1493_2024_121_0_5.

16. V.Yu. Konyukhov, T.A. Oparina, I.Y. Matasova, M.A. Modina, N.V. Martyushev. Ecologization of underground coal mining by means of ash use in backfill preparation. Min. Inform. Anal. Bull., **10,** 123-135 (2024). DOI: 10.25018/0236_1493_2024_10_0_123.

17. O. A. Kolenchukov, A. Y. Mikhaulov, K. A. Bashmur, R. B. Sergienko, V. V. Kukartsev. Forecasting the technical condition of thermochemical reactor systems. SOCAR Proceedings, **2023,** 123–130 (2023). https://doi.org/10.5510/OGP20230100814.

18. V.V. Kukartsev, K.I. Kravtsov, Ya.A. Tynchenko, T.A. Panfilova. Influence of mountain factors on salt excess and soil toxicity in mountain conditions. Sustainable Development of Mountain Territories, **15,** 784-797 (2023). https://doi.org/10.21177/1998-4502-2023-15-3-784-797.

19. S. A. Makhosheva, M. V. Zhanokova, V. M. Dzaparova, I. K. Totrukova. Ecological aspect of sustainable development of high-mountain regions in the conditions of innovative economy formation. Sustainable Development of Mountain Territories, **16,** 754–769 (2024). DOI: 10.21177/1998-4502-2024.

20. Z.E. Makovozova, A.A. Sokolov, V.A. Fomenko, M.T. Sarbaeva. Influence of hydrogeology at Unal tailings pond on ecosystem pollution with heavy metals. MIAB. Mining Inf. Anal. Bull., **6,** 126-138 (2023). DOI: 10.25018/0236_1493_2023_6_0_126.

21. S.A. Makhosheva, M.V. Zhanokova, V.M. Dzaparova, I.K. Totrukova. Ecological aspect of sustainable development of high-mountain regions in the conditions of innovative economy formation. Sustainable Development of Mountain Territories, **16,** 754–769 (2024). DOI: 10.21177/1998-4502-2024-16-2-754-769.

22. V. Orlov, A. Nizameeva, D. Shalaeva, D. Ageev. Development of a multifunc-tional cross-platform system for automation of energy data and resource management. E3S Web of Conferences, **460,** 07002 (2023). DOI: 10.1051/e3sconf/202346007002.

BIO Web of Conferences **161**, 00006 (2025) https://doi.org/10.1051/bioconf/202516100006
FIES 2024

Salmonellosis: threats to the economy and avian health

Gulina Faizrakhmanova*[1], *Albert* Galiullin[2], *Lidia* Koroleva[1], *Gennady* Spiridonov[1] and *Aidar* Makhmutov[1]

[1] Federal Centre of Toxicological, Radiation and Biological Safety, 420075, Kazan, Scientific Town-2, Republic of Tatarstan, Russia
[2] Kazan State Academy of Veterinary Medicine named after N.E. Bauman, 35, Sibirskiy trakt, Kazan, 420029, Republic of Tatarstan, Russia

Abstract. The article gives a modern classification of salmonellosis. Their morphological and biological properties are detailed. The general characteristic of the disease caused by salmonellosis of birds is presented in full. Briefly, but with the designation of all major milestones, it reflected the history of the study of salmonellosis. Reasons and factors contributing to the pathology are revealed. Localization of lesions depending on the routes of infection is presented in sufficient detail. Interesting materials on laboratory diagnostics of salmonellosis of birds are presented. Materials on the general principles of treatment of birds are presented, and the issues of specific prophylaxis and prospects for combating it are highlighted. In order to ensure epizootic well-being, the article gives recommendations on general preventive measures: observance of hygiene of keeping birds and timely protective vaccinations, which will prevent the occurrence of the disease. The general preventive measures include proper feeding. An important aspect is also personal and industrial hygiene of service personnel, as a person can be a carrier and source of spread of salmonellosis. The presented material contains the results of studies of a number of domestic and foreign epidemiologists, epizootologists and bacteriologists, as well as the results of the authors' own research.

1 Introduction

Modern poultry farms are built with full mechanization of production processes. Thanks to breeding work, improved pedigree breeds of birds are used in the industry. New feeding and housing technologies are introduced in them. However, specialization and concentration of production have led to the placement of a large number of poultry on limited areas, in artificial climate conditions they are deprived of sunlight and free choice of feed. This system of poultry breeding leads in some cases to exogenous infection with infectious agents. On the other hand, in case of violation of housing and feeding technology, as well as in the presence of stressful situations, the general resistance of the poultry organism is reduced, leading to an increase in virulence of opportunistic microflora. Therefore, infectious diseases occurring in the poultry industry in such situations cause significant economic damage, which consists of reduction of egg and meat productivity, reproductive function, loss of breeding value, premature culling and expenses for prevention and treatment. Gastrointestinal diseases, such as salmonellosis, cause the greatest economic damage to the poultry industry. Pathogenic species of bacteria, mycoplasmas, chlamydia, mycoses and mycotoxicoses, which may appear in association with salmonellosis, play an important role. Many of them are classified as particularly dangerous diseases requiring special attention from the veterinary service. Their importance in the infectious pathology of birds is due to their wide distribution, predominantly mixed nature of course, diversity of clinical manifestations, ability to latent course.

The prevention of salmonellosis in industrial poultry production is one of the main tasks of the veterinary service. Daily preventive control of poultry health is the key to successful production.

Salmonellosis is an anthropozoonotic infectious disease. In the nineteenth century, the disease was widespread, killing a large number of people over the centuries. The spread of the disease is associated with industrial poultry farming. According to historical data, only about 12 types of salmonellae were identified in Russia between 1881–1914, and World War II contributed to the emergence of another 50 types of salmonellosis bacteria. In 1992, their number rose to two and a half thousand.

According to the Federal Service for Veterinary and Phytosanitary Surveillance, currently 2324 serological variants of Salmonellae have been identified, which are divided into 46 serogroups depending on the antigenic structure [1]. Despite such huge number, the situation is always under strict control, the main thing is to adhere to veterinary and sanitary rules.

2 Results and Discussion

The causative agent of salmonellosis was first isolated in the spleen and lymph nodes of a person who died of

* Corresponding author: niskgavm@mail.ru

typhoid fever in 1880 by German pathologist Karl Ebert. Later in 1884 a pure culture of the pathogen was obtained by Georg Gaffky, and in 1885 the American scientist D.E. Salmon described the morphology of the microorganism as *B. suipestifer*, now known as *Salmonella choleraesuis*. F. Leffler in 1892 isolated the microbe from fallen mice, which was named Vast. typhimurium. Later, in honor of D. Salmon, he named the disease Salmonellosis and the causative agent Salmonella. The developer of the first typhoid vaccine is also D. Salmona [2].

The history of the discovery of avian salmonellosis is associated with the unification of pullorosis and typhus in 1970 into one disease, i.e. avian salmonellosis. Prior to this, in chickens the disease was classified as pullorosis and as typhoid in adult birds. The causative agent belongs to the Enterobacteriaceae family of the Salmonella genus. In adult birds (chickens, turkeys), the causative agent *is Salmonella enteritidis*, dangerous for humans, as well as *Salmonella gallinarum-pullorum*, which in young birds causes acute disease pullorosis-typhus, and asymptomatic infection in adults. In waterfowl (geese, ducks, pigeons), *Salmonella typhimurium* is the causative agent.

Numerous studies have shown that human infection with salmonellosis occurs mainly through the alimentary route, where the main factor is poultry meat and poultry products, eggs and offal [3–5]. In addition, the infection can be transmitted through water and the environment. Entering the mucosa of the small intestine, endotoxins released by bacteria cause inflammation of the mucosa, which leads to the development of pain, diarrhea and intoxication symptoms. Further, penetrating into the blood and lymph, other organs and tissues (liver, kidneys, spleen), bones, joints are affected, which, in the absence of proper treatment, can lead to convulsions, paralysis and even death [6].

Salmonellae are well tolerant of low temperatures and drying, and remain viable in water, soil and room dust for a long time. The microorganisms are resistant to pickling and smoking, and are able to reproduce in dairy and meat products even at low positive temperatures. The data on bacterial resistance are presented below (Table 1).

Table 1. Resistance of Salmonellae in food, water and the environment

Object	Salmonella preservation time
Sausage products	Up to 6 months
Frozen meat	Up to 1 year
Eggs	Up to 1 year
Vegetables and fruits	5–10 days
Milk	Up to 20 days
Butter	Up to 120 days
Eggshells	Up to 24 days
Water, soil	Up to 3 months
Meat, with the NaCl content inside the product of 3.5–7%	30 days
Meat, with the NaCl content inside the product of 10–15%	60–90 days

Studies show that infectious agents are killed in 5–10 minutes when heated to a temperature of 70°C and instantly when boiled. The growth of salmonellae is suppressed by ionizing irradiation at a dose of up to 10 Kg, while maintaining the quality of the product. Also, UV rays within the wavelength of 200 to 313 nm have a bactericidal effect. According to researchers, UV irradiation (23.85–33.39 WT s/cm^3) for 25–35 minutes neutralizes bacteria on the surface of synthetic and protein sausage casings.

The economic damage from salmonellosis can be significant. Firstly, it is the mortality of young animals. It reaches 40–50% in lambs, 60–70% in young fur animals, 50–70% in piglets, and 50–70% in calves. Chickens older than 12 days of age show significant resistance to salmonellosis and often do not get sick [7].

Secondly, salmonellosis can have a negative impact on food security. If food products, such as eggs or meat, are contaminated with Salmonella, it can lead to product recalls from the market. This is associated with loss of revenue for producers and retailers, as well as damage to brand reputation. In addition, the organization may need additional resources to investigate and prevent further spread of the infection.

Also, salmonellosis may cause a decrease in demand for products associated with the infection. As a result, consumers may avoid purchasing certain products or prefer alternatives, which may result in lower sales and losses for companies.

According to epizootological data, the occurrence and spread of salmonellosis on poultry farms are usually associated with poor sanitary and hygienic control of premises, drinking water, unbalanced feeding, confinement in cold and poorly ventilated rooms, through clothing and footwear. The infection is dose-dependent and therefore the outbreak may also be due to high densities of individuals.

Infection with the pathogen can occur from sick and over-infected birds, including rodents, wild birds and insects, as well as through the use of infected equipment and care items [8]. Infection of poultry occurs through the gastrointestinal tract. Salmonellae are localized on the mucous membrane of the cecum and large intestine, excreted with droppings. If the mucous membrane is disturbed as a result of any invasion, the pathogen with the bloodstream penetrates into the liver and egg follicles. Being in the intestine, salmonellae can cause an infectious process in acute or subclinical form. In the case of aerogenic infection, the pathogen enters the lungs, forming foci of necrosis, from where it enters the bloodstream and spreads throughout the body.

Diagnosis is based on epizootological data, clinical picture and pathological and anatomical changes, but mainly on the results of bacteriological and serological tests, especially in farms previously free from this infection.

In order to isolate the culture of the pathogen, cultures from dead embryos (amniotic fluid, yolk), blood from heart, liver, spleen, bone marrow and affected ovarian follicles of fallen birds are sown on MPB, MPA and differential Endo agar, bismuth-sulfite agar. The identification of isolated cultures of microorganisms is carried out in the drop agglutination reaction on glass with monoreceptor salmonellosis of O- and H-serums with the consideration of smear microscopy and motility.

The fluorescent antibody method (FA) is used as an express method for the detection of Salmonellae in the test material. For this purpose, smears are prepared, fixed chemically and stained with salmonella fluorescent sera. The smears are viewed under a fluorescent microscope.

Polyvalent and typical Salmonellosis phages are used for phagotyping for diagnostic purposes. A daily culture of Salmonellae is sown on slant agar, and then a drop of phage is applied at an angle to the middle part of the agar using a Pasteur pipette. The tubes are placed in a thermostat and viewed after 24 hours. There will be no Salmonella growth at the site of the dripping drop (the 'sterile zone').

If the bacteriological confirmation of the results of the serological reaction in reacted birds is necessary, cultures are performed from degenerated follicles of ovaries, liver, spleen. The probability of isolation of the pathogen from reacted birds does not exceed 20–50% of the total number of performed cultures.

The lifetime diagnosis of salmonellosis consists in the study of bird blood in the indirect hemagglutination reaction (RNHA) with erythrocyte pullor or polyvalent antigen.

Poultry are tested at the age of 50–55 days and again at the age of 7 months. If 1% or more of positively reacting individuals are detected, the farm is considered unfavorable.

In unfavorable farms, the diagnostic value of lifetime diagnostics depends on the age of the tested birds: the most optimal age for ducklings is 40–45 days, 45–50 days for turkeys, 45–50 days for goslings, 50–55 days for chickens. At other ages, the study is less effective.

Poultry whose blood has tested positive for the antigen are isolated and sent for slaughter.

In a salmonellosis outbreak, the following interventions are necessary:

1. Bacterial carriers should be identified and isolated as quickly as possible.
2. Prophylactic use of antibiotics in feed or water should be considered.
3. Water and feed should be protected from fecal contamination.
4. Contaminated premises must be cleaned and disinfected.

For disinfection of premises, care items and working tools, it is necessary to use chlorine lime with 25% active chlorine, 2% formalin solution, 20% suspension of slaked lime, hot 4% caustic soda solution.

The course of the disease is clear and characterized by vomiting, diarrhea, fever, loss of appetite and general weakness. In adult birds usually manifests a chronic form, long bactriosity with localization of the causative agent in the liver, ovaries, spleen and in the large intestine.

In domestic birds, immunity is non-sterile. There are no means of specific prophylaxis. The indirect hemagglutination reaction is used for mass examination of poultry and detection of salmonella carriers.

Adult waterfowl (ducks and geese) are naturally resistant to salmonellosis. A dry live vaccine against salmonellosis in waterfowl is recommended for prophylaxis. The vaccine is administered twice orally to ducklings and goslings from 3 days of age with an interval of two days.

Reproductive youngsters of breeding poultry farms are immunized with live or inactivated vaccines. Live vaccines stimulate humoral and cellular immune response, reduce the degree of colonization of internal organs and gastrointestinal tract, as well as reduce the excretion of Salmonella with excrement.

Inactivated vaccines stimulate a pronounced humoral immune response and induce a high level of circulating antibodies, providing protection against infections of internal organs and the genital tract, and the accumulation of a high concentration of maternal antibodies (MA) in the egg. Vaccination is twofold, with the first vaccination given at the age of 55–60 days.

Treatment usually involves administration of antibiotics: penicillins, tetracyclines, phenicols, cephalosparins and complex drugs. Nevertheless, antibiotic resistance of Salmonella strains isolated from birds and other animal species according to recent studies reaches 20–60% in Russia.

To address this problem with antibiotic therapy, veterinarians combine the use of broad-spectrum antibiotics with the use of drugs to normalize the intestinal microflora. Antimicrobial treatments should be recommended under the close supervision of specialists and with substantial justification.

Salmonellosis is also treated by maintaining hydration by intravenous or subcutaneous administration of fluids, symptomatic therapy to reduce fever and improve the general condition of the animal and birds [9].

3 Conclusion

One of the reasons for the occurrence and spread of salmonellosis at poultry enterprises is the failure to fulfil zoohygienic requirements for keeping birds and primarily high density, as well as the use of infected equipment and care items [10, 11]. Therefore, in order to ensure a favorable production, it is necessary to pay more attention to preventive measures, i.e. observing the hygiene of poultry housing. Periodic cleaning and disinfection of premises, equipment and care items, prevention of communication of birds with wild birds and timely vaccination will prevent the occurrence of the disease. General preventive measures include adequate feeding and regular heat treatment of feed. Personal and occupational hygiene of service personnel is an important aspect, as humans can be vectors and sources of salmonellosis spread.

Salmonella enteritidis and *Salmonella typhimurium* serotypes are spreading among birds in the Russian Federation. According to many scientists, the causative agent of salmonellosis can circulate in the organism of birds as a conditionally pathogenic microorganism. Such bird becomes a carrier of the bacterium and release it into the external environment. The salmonellosis pathogen can persist for a long time in the external environment, actively multiply under favorable conditions, and it can be easily transmitted to healthy stock. Therefore, along with the vertical transmission of the pathogen, horizontal transmission is also possible. In such situation it is necessary to ensure a high level of zoohygienic norms of keeping birds. This can be achieved by a systematic approach, which includes the implementation of veterinary and sanitary rules to prevent the introduction of the pathogen, monitoring epizootological research.

References

1. Salmonella (salmonellosis) symptoms, diagnosis, treatment, https://ria.ru/20231215/salmonellez-1916062023.html
2. G.F. Bovkun, J. Veter., **4**, 14 (2004), ID: 16779068
3. Y.G. Kostenko, M.V. Khramov, A.D. Davleev et al., J. Meat Industry, **7**, 51 (2012), ID: 17960520
4. V.Y. Steponavichus, Proc. of the Int. Sci. and Pract. Conf. of Young Sci. and Spec. Achievements of Agrar. Sci. to Ensure Food Secur. of the Rus. Fed., **386** (2021), ID: 47986769
5. Prevention and control of contagious diseases common to humans and animals. Salmonellosis, https://fsvps.gov.ru/files/profilaktika-i-borba-s-zaraznymi-bol-9/#:~:text=salmonellez
6. I.M. Butaeva, H.S. Salimov, J. Perspect. of devel. of veter. Sci. and its role in ensur. food safety, **1**, 58 (2022), https://doi.org/10.47689/978-1-957653-01-3-PDVSREFS-2022-pp58-62
7. N.V. Pimenov, A.I. Laishevtsev, Rus. J. of Agricult. and Socio Econ. Sci., **4**, 257 (2017), https://doi.org/10.18551/rjoas.2017-04.33
8. G.S. Matnazarova et al., IQRO, **3(2)**, 37 (2023), https://doi.org/10.5281/zenodo.7996781
9. Salmonellosis, Sanitary rules SP 3.1.086-96, Veterinary rules VP 13.4.1318-96
10. E.M. Tsygankov, A.A. Menkova, T.A. Kazimirova et al., J. Veter. Phys., **1**, 56 (2022), https://doi.org/10.33632/1998-698X.2021-1-56-65
11. E.O. Chugunova, N.A. Tatarnikova, Int. J. of appl. and fund. Res., **8-2**, 163 (2014), ID: 21653864

Influence of natural biostimulants on growth and quality of cattle meat

Nurali Ashurov [1*], *Abdulaziz* Yusupov [2], *Bekzod* Primkulov [3], *Abdimalik* Kurbanov [4], *Natalia A.* Grankina [5], *Aslan Yu.* Bogus [5]

[1] Tashkent Institute of Irrigation and Agricultural Mechanization Engineers" National Research University, 39, Street Kari Niyaziy, Tashkent, 100000, Uzbekistan

[2] Almalyk branch of the Tashkent State Technical University named after Islam Karimov, Almalyk, Uzbekistan

[3] Termez State University of Engineering and Agrotechnologies, I.Karimov str., 190111,Termez, Uzbekistan

[4] Termiz State Pedagogical Institute,Termez, Uzbekistan

[5] Kuban State Agrarian University named after I. T. Trubilin, Krasnodar, Russia

Abstract. The presented work is devoted to the study of the influence of natural biostimulants on the productivity of cattle grown for meat. The aim of the study was to determine the effectiveness of biotechnological methods to increase the growth rate of animals and improve the quality of meat raw material. In the course of the experiment, control and experimental groups of animals differing in feeding ration were formed. The experimental group received a standard diet supplemented with natural biostimulants. The results of the study showed that the use of biostimulants leads to an increase in average daily live weight gain by 18%, as well as to a reduction in the fattening period by 20 days. In addition, the experimental group showed an improvement in meat quality indicators such as colour, pH, fat and protein content. The obtained data indicate the promising use of natural biostimulants to improve the efficiency of animal husbandry and production of high quality beef.

1 Introduction

The steady growth of the world's population, accompanied by urbanisation and rising living standards in developing countries, is creating a rapidly growing demand for food. At the same time, consumption patterns are changing, with more and more people seeking a diversified diet that includes animal products, in particular meat and meat products. This increasing demand, coupled with limited land resources and growing environmental concerns, is challenging the ability of traditional agricultural systems to provide sufficient and quality food for all. In particular, livestock production, as an important component of agriculture, faces challenges related to low production efficiency, high production costs and significant negative environmental impacts, making it urgent to find innovative and sustainable solutions. The search for new, efficient and environmentally responsible approaches to increasing meat production that can provide high quality products and meet the growing needs of the population while reducing costs and minimising environmental damage is therefore not just urgent, but absolutely critical to the well-being of humanity.

There are several main strategic directions aimed at solving the problem of improving food security and increasing meat production. Feed intensification is one of the most common approaches. It involves significant improvements in feed composition and nutritional value, the introduction of innovative additives, and the optimisation of feeding schemes to improve the efficiency of nutrient uptake by animals. Despite its significant potential, this area has a number of limitations. Firstly, it depends on the availability and cost of feed ingredients, which can fluctuate depending on market conditions and climatic conditions. Secondly, extremely careful balancing of diets is required to prevent metabolic disorders and diseases in animals, which increases the complexity of the technological process. The effect of dietary fibre on animal digestion and metabolism must be considered, and for this purpose it is worthwhile to study the work of scientists in the field of animal nutrition, e.g. [1]. Breeding work aimed at developing more productive breeds of animals is another promising way, but this way requires significant financial and time costs, as well as does not always guarantee the desired result. Moreover, over-intensification of productivity often leads to reduced resistance to disease, stress and unfavourable environmental conditions, which can negatively affect the quality of the final product. The development of precision farming and other advanced agro-technologies aimed at increasing forage crop yields and optimising land use is also an important component of addressing

* Corresponding author: nuraliashurov@mail.ru

the problem. However, adoption of these technologies requires significant capital investment and is often beyond the reach of small and medium-sized producers, which can exacerbate inequalities in the industry. Thus, each of these traditional approaches has its own advantages and disadvantages, which calls for a more comprehensive and integrated solution.

In the conditions of continuous intensification of agricultural production, along with the growing consumer demand for high quality and safe meat, special attention is drawn to the promising direction - the development and implementation of innovative biotechnological methods in animal husbandry. Biotechnology, combining the achievements of biology, chemistry and engineering, provides powerful tools to increase animal productivity, improve the quality of meat products, as well as to reduce the negative impact of animal husbandry on the environment, contributing to the formation of more sustainable and environmentally friendly production systems. The use of biotechnological techniques, such as the targeted use of natural biostimulants, opens up new possibilities for improving livestock productivity. These biologically active substances derived from natural sources can not only stimulate growth and development of animals, but also improve digestibility of feed nutrients, strengthen the immune system and reduce dependence on the use of synthetic feed additives and antibiotics, which is especially important in the global fight against antibiotic resistance. At the same time, it is necessary to apply technologies aimed at utilising animal waste and reducing the impact of antibiotics [2]. In addition, genomic analyses are needed to monitor the resistance of livestock to various diseases as well as to monitor food safety [3]. This approach contributes to improving environmental safety and sustainability of production. Moreover, biotechnology allows for a more efficient disclosure of the genetic potential of animals, which contributes to increasing the profitability of the industry and reducing the cost of production. The use of modern methods of genomic editing and genetic engineering opens prospects for the creation of animals with specified characteristics of productivity and resistance to diseases. In this regard, it is necessary to study infectious diseases of livestock [4]. Consequently, an integrated approach, including the integration of advanced biotechnological solutions, is a key factor to ensure food security and sustainable development of the agricultural sector. It is also necessary to consider the impact on the agro-ecological state of agro-landscapes by applying geo-information systems methods [5, 6].

The aim of this study is to comprehensively explore the potential of applying modern biotechnological methods for intensification of beef production in agriculture. This article provides a detailed analysis of various biotechnological approaches aimed at increasing the productivity of cattle, as well as assesses the possibility of their practical application in the context of solving urgent problems of modern animal husbandry. Particular attention is paid to the detailed consideration of the prospects for the use of natural biostimulants.

2 Method and materials

The research presented in this article was aimed at studying the influence of biotechnological methods, namely the use of biostimulants, on the productivity of cattle raised for meat. The general plan of experimental works included formation of control and experimental groups of animals differing in feeding conditions. In the experimental group natural biostimulants were added to the diet, and the control group received a standard diet. Observation of animals was carried out throughout the entire period of fattening, with periodic control of the main parameters of growth and development.

During the study, standard zootechnical equipment was used to record the main parameters. Weight control was carried out using electronic scales 'VET-150' with a discreteness of 50 grams, allowing to accurately determine the live weight of animals at different stages of fattening. To assess the quality of meat raw material was used spectrophotometer brand 'SF-2000', which measured the optical density of meat extracts to assess its colour characteristics. Measurement of pH of meat raw material was carried out using laboratory pH-meter 'pH-150' to analyse the acidity of meat. All measurements were carried out under standard temperature and light conditions recommended for such analyses. All instruments were pre-calibrated prior to measurements. The operating modes of the equipment were in accordance with the manufacturer's instructions. Spectrophotometer was used in the visible light absorption measurement mode, pH-meter - in the standard mode of acidity determination.

3 Results and discussion

The conducted research devoted to the evaluation of the influence of biotechnological methods, namely the use of natural biostimulants, on the productivity of cattle was organised and conducted in compliance with strict scientific principles aimed at minimising errors and ensuring the reliability of the obtained results. For this purpose, two groups of animals were formed, which were maximally identical in all main parameters. Both groups included animals of the same breed, selected according to age and weight criteria, and under the same housing conditions to minimise the influence of genetic and environmental factors on the results of the experiment. The experimental and control groups were formed randomly to avoid systematic errors and sampling bias. During the entire experimental period, the animals were kept in identical conditions in spacious stalls, with access to clean drinking water, sufficient ventilation and natural light, which provided them with comfortable conditions for growth and development. The only exception was the feeding ration. The diet of the experimental group included a natural biostimulant, which was carefully selected based on its potential properties to stimulate growth and improve metabolic processes in the body of animals. The biostimulant was introduced into the diet daily, in a strictly defined dosage, which was calculated based on manufacturers'

 https://doi.org/10.1051/bioconf/202516100007

recommendations and preliminary studies. In the diet of the control group, the biostimulant was absent, and the animals received a standard feed ration, which was used in the farms of this region for fattening cattle. The animals of both groups were monitored throughout the entire fattening period, which was planned for 12 months. For the purpose of detailed monitoring of growth dynamics, weekly weighing of animals was carried out with the help of electronic scales 'VET-150', providing high accuracy of measurements. The results of each weighing were carefully recorded in spreadsheets for further statistical analysis. In addition to weighing, the animals of both groups were visually inspected on a weekly basis, their general condition, appetite, presence or absence of signs of diseases, and the quality of hair coat were assessed. Any abnormalities, such as decreased appetite, signs of lethargy or disease, were immediately recorded and the necessary measures were taken to provide veterinary care. This approach allowed us to maintain the health of the animals and eliminate the influence of diseases on the results of the experiment. At the final stage of the experiment, at the end of the 12-month fattening period, animals from both groups were slaughtered to analyse meat quality. Carcasses obtained after slaughtering were subjected to standard cutting procedure. To study meat quality characteristics, samples from the main muscle groups were selected and subjected to a number of laboratory tests. In particular, the optical density of meat extracts was measured using an SF-2000 spectrophotometer to determine their colour characteristics, such as red intensity. To assess the acidity of the meat raw materials, a laboratory pH-meter of the brand 'pH-150' was used, which allowed to determine the pH level with high accuracy. Measurements were taken at several points of each sample to exclude the influence of localised variations. In addition, the meat samples were sent to the laboratory for biochemical analyses to determine protein, fat and other important nutrients. All analyses were carried out in accordance with current standards and methodologies to ensure the reliability of the results.

According to the obtained data, the addition of natural biostimulants to the diet of cattle had a pronounced positive effect on their productivity. Average daily live weight gain in the experimental group was statistically significantly higher by 18% compared to the control group. If in the control group the average daily gain averaged 1.20 kg, in the experimental group it reached 1.42 kg. These indicators were confirmed by weekly weighing data. Moreover, growth rates in the experimental group were more stable and uniform throughout the entire fattening period. The total live weight gain in the experimental group for 12 months of fattening was on average 68 kg more than in the control group. This resulted in animals in the experimental group reaching the target weight of 450 kg on average 20 days earlier than animals in the control group. In addition, economic analysis showed that feed costs per unit weight gain in the experimental group were 12% lower, which is an important factor in assessing the economic efficiency of biostimulants. Here we can refer to the works related to agroecological assessment of land used in fodder production, for example, [7]. The study of meat raw material quality showed positive changes in the experimental group. The colour of meat was more intense red, which visually indicates its freshness and high quality. Spectrophotometry confirmed the visual assessment, showing that the index of red colour in the meat of animals of the experimental group was 8% higher compared to the control group. Moreover, it is worth considering the rational use of waste in agriculture, such as fertiliser production, when assessing the quality indicators [8]. The pH level of meat in the experimental group was in the optimal range from 5.6 to 5.8, while in the control group there was a slight decrease in pH level to 5.4-5.5. This indicates an improvement in the metabolic process and a more stable state of muscle tissues in animals receiving biostimulants. Also for analysing meat, artificial intelligence technology can be used [9]. The intramuscular fat content in the meat of the experimental group was 5% lower than in the control group. In particular, the average fat content in meat of the experimental group was 14.3 g per 100 g, compared to 15.1 g in the control group. At the same time, the protein content in the meat of the experimental group was 3.5% higher than in the control group. In particular, in the experimental group this indicator averaged 21.3 g per 100 g of product, while in the control group - 20.5 g. To produce quality and nutritious beef, it is worth paying attention to the use of different feed sources, such as insects [10]. The conducted data analysis indicates that the use of natural biostimulants in the diet of cattle has a significant positive effect on their productivity, as well as on the quality of meat raw material. The obtained results are fully consistent with the conclusions of other research groups, which also noted a positive effect of the use of natural biostimulants on growth, feed digestibility, as well as on a number of quality indicators of meat. In particular, this is confirmed by studies on the influence of spirulina in the feeding of farm animals [11]. Also, the agroecological potential of the region influences animal productivity [12, 13]. Thus, the introduction of natural biostimulants in the diets of fattening animals can be a promising direction for improving the efficiency of animal husbandry and production of high-quality beef.

4 Conclusion

The conducted study convincingly demonstrates that the use of natural biostimulants in the diet of cattle is a promising approach to improve the efficiency of animal husbandry. In the course of the experiment it was found that the addition of biostimulants in the diet of animals leads to a significant increase in average daily live weight gain. Animals receiving biostimulants not only grew faster, but also reached the target weight on average 20 days earlier than their control counterparts. This, in turn, indicates a more intensive use of feed resources and a significant reduction in the fattening period, which on a production scale can lead to significant economic benefits. In addition, the analysis of the obtained data indicates that the use of biostimulants contributes to the efficiency of feed utilisation, as evidenced by lower feed costs per unit weight gain in the experimental group. This fact also has an important economic significance, as the reduction of feed costs, which is one of the main items of expenditure in animal husbandry, allows to significantly reduce the cost of final products. No less important are the conclusions about the influence of biostimulants on the quality of the obtained meat raw material. The study showed that the meat of animals receiving biostimulants has a more intense red colour, which is a sign of high quality and freshness. Spectrophotometric studies confirmed this observation, showing a higher red colour index in the meat of the experimental group. It was also found that the application of biostimulants helps to maintain an optimal pH level in the meat, which is an important factor for its preservation and culinary characteristics. In addition, the analysis of biochemical composition of meat showed that in the meat of animals receiving biostimulants, the intramuscular fat content was lower and protein content was higher, compared to the meat of animals of the control group. This suggests that the use of biostimulants contributes to obtaining a more dietary and nutritionally valuable meat product.

Thus, this study has allowed us to conclude that the use of natural biostimulants in the diet of cattle is an effective, environmentally safe and economically feasible way to increase livestock productivity. The results of the study open wide prospects for the introduction of this biotechnological technology in the practice of agriculture, contributing to an increase in the production of high-quality and safe beef, as well as increasing the profitability of the industry. In general, legal regulation is necessary for the development of organic agriculture [14]. The results of the conducted research allow us to recommend the use of natural biostimulants in cattle diets as an effective means to increase meat productivity and improve product quality.

References

1. F. Yong, B. Liu, H. Li, H. Hao, Y. Fan, O. Datsomor, R. Han, H. Jiang, D. Che. Relationship between dietary fiber physicochemical properties and feedstuff fermentation characteristics and their effects on nutrient utilization, energy metabolism, and gut microbiota in growing pigs. J Anim Sci Biotechnol, **16,** 1 (2025). DOI: 10.1186/s40104-024-01129-x.
2. P. Zhang, Q. Liu, H. Yao, T. Wang, B. Zeng, Y. Zou, S. Fu, K. Hu, H. Guo. Copper-functionalized hydrochar from swine manure with complexation sites for selective adsorption of tetracycline antibiotics. Separation and Purification Technology, **359,** 130671
3. A.E. Watt, M.L. Cummins, C.M. Donato, W. Wirth, A.F. Porter, P. Andersson, E. Donner, T. Huynh, A. Phabmixay, E. Flynn, C. Minney-Smith, T. Gonzales, C. Bareja, A. Levy, Z. Bartlett, A. Donald, R. Wright, M. Wille, G.F. Browning, B. Hoye, L.E.X. Leong, C. Thompson, T. Zulfiqar, A. Ferdinand, T. Hoang, K. Horan, D. Ingle, J.R. Webb, R. Bull, A. Zahedi, A. Arnott, V. Sintchenko, A.V. Jennison, T. Seemann, S.P. Djordjevic, B.P. Howden. Parameters for one health genomic surveillance of Escherichia coli from Australia. Nature Communications, **16,** 17, (2008).
4. M. Farman, E. Hincal, S. Jamil, N. Gokbulut, K.S. Nisar, A. Sambas. Sensitivity analysis and dynamics of brucellosis infection disease in cattle with control incident rate by using fractional derivative. Scientific Reports, **15(1),** DOI: 10.1038/s41598-024-83523-z
5. L. V. Lebedeva, G. D. Syzdykova, D. S. Inkarov. Management of agricultural land use sustainability at the local level using GIS technologies and special map material. Sustainable Development of Mountain Territories, **15,** 864–876 (2023). DOI: 10.21177/1998-4502-2023-15-4-864-876.
6. V.S. Brigida, K.Y. Lobkov, A.A. Stupina. On the issue of monitoring crack formation in natural-technical systems during earth surface displacements. MIAB. Mining Inf. Anal. Bull., **11-1,** 292-304 (2023). DOI: 10.25018/0236_1493_2023_111_0_292.
7. T.S.A. Safar, S.A.A. Alsaadi. Physiological Impact of an Aqueous Solution of Spirulina Algae with or without Folic Acid on Local Awassi Ewes Within Medium and Late Pregnancy Stages an Indicators of Growth in Newborn Lambs.
8. M. N. Nurpeisov, K. V. Ruchkina, V. A. Androkhanov, S. K. Makenova. Assessment of geoecological factors and conditions affecting sustainability of agricultural land use in the forest-steppe zone of Altai Krai. Sustainable Development of Mountain Territories, **15(2),** 419-430.
9. L.S. Malyukova, V.V. Kondratiev, V.V. Bukhtoyarov, K.A. Bashmur. Circular Mining Wastes Management for Sustainable Production of Camellia sinensis (L.) O. Kuntze. Sustainability (Switzerland), **15,** 11671 (2023). DOI: 10.3390/su151511671.
10. A. I. Trunova, S. O. Kurashkin, K. A. Bashmur, R. B. Sergienko. Influence of Temperature Regimes of Synthetic Iron Smelting on Casting Production Efficiency. Metals, **13(7),** 1234 (2023). DOI: 10.3390/met13071234
11. Z. Wang, J. Park. Japanese-made or Chinese-made? Exploring the country-of-origin (COO) effect on insect-based foods, Food Quality and Preference, **126,** 105421 (2009).
12. T.S.A. Safar, S.A.A. Alsaadi. Physiological Impact of an Aqueous Solution of Spirulina Algae with or without Folic Acid on Local Awassi Ewes Within Medium and Late Pregnancy Stages an Indicators of Growth in Newborn Lambs, (2008).
13. A.T. Beristenov, L.V. Lebedeva, L.M. Tatarintsev, V.L. Tatarintsev. Assessment of the agroecological potential of anthropogenic landscapes in the dry steppe of the south of Western Siberia. Sustainable Development of Mountain Territories, **16,** 975–987 (2024). DOI: 10.21177/1998-4502-2024-16-3-975-987.
14. T. A. Panfilova, K. A. Bashmur, V. V. Kondratiev, To the Concept of Leaching Metal-Containing Raw Materials in the Disintegrator. Mining Informational and Analytical Bulletin. **11-1,** 239-251 (2023). DOI: 10.25018/0236_1493_2023_111_0_239

Prevention of subclinical mastitis in lactating cows with a new probiotic preparation

Artur Porfiriev, *Ramil* Fayzrakhmanov, *Vladimir* Sofronov, *Nadezhda* Danilova* and *Elena* Kuznetsova

Kazan State Academy of Veterinary Medicine named after N.E. Bauman, Kazan 420029, Republic of Tatarstan, Russia

Abstract. The aim of the work is to develop a new domestic probiotic preparation containing strains of microorganisms known as Bacillus subtilis and Bacillus amiloliquefaciens and intended for treatment of cows' udders, and technology of its application in cattle breeding for prevention of cows with subclinical mastitis and obtaining ecologically clean milk. The composition of microorganism strains was selected on the basis of preliminary microbiological study by sowing on nutrient media. On the basis of scientific and production experience in the conditions of LLC "Soyuz-Agro" of Aznakaevsky district of the Republic of Tatarstan, it was found that for this population of cows the most acceptable was the use of probiotic preparation. It contains compositions of strains of microorganism producers such as Bacillus subtilis strain B-11 in the concentration of 1x1010 CFU/ml and Bacillus amiloliquefaciens of the B-14294 strain in the concentration of 1x1010 CFU/ml. It showed the best activity in relation to the microflora of the studied population of lactating cows. On the background of probiotic preparation application for three months of the experiment in milk of experimental cows the number of somatic cells decreased by 39.6 % and microbial contamination reduced by 38.3 % in comparison with those of control, and economic efficiency per 1 ruble of additional costs amounted to 4.3 rubles.

1 Introduction

The food security of every country depends on the development of livestock and in particular livestock production. Cattle breeding is a labor-intensive industry where the human factor plays a major role. At present, the development of dairy cattle breeding is inextricably linked with the introduction of innovative technologies that contribute to high milk productivity, which has increased 2.3 times since 1996 [1].

Obtaining high productivity in dairy cattle breeding depends on the conditions of housing, feeding and disease prevention. Creation of comfortable conditions will contribute to maximizing milk productivity, while reducing its cost [2]. In modern conditions of industrial cattle breeding, a high concentration of animals on a limited area of the room contributes to the deterioration of their housing conditions. In such conditions, the organism of cows is constantly exposed to pathogenic and opportunistic microflora, which contributes to their morbidity, and, ultimately, contributes to a decrease in productivity [3–5].

Deterioration of hygienic conditions of animal housing inevitably leads to a decrease in the quality of dairy products, which is expressed in an increase in bacterial infestation and the number of somatic cells in milk. These indicators are indicators of the state of cow udder health, which in the short term can have a negative impact on production indicators, i.e. decrease in productivity and milk grade, increase in the cost of dairy products. Even minor deviations from optimal housing conditions lead to a decrease in natural resistance, which ultimately contributes to the emergence of many diseases of non-communicable etiology, and in some cases, to the emergence of outbreaks of infectious diseases and their spread [6–9].

Keeping lactating cows in inadequate conditions is an additional negative factor that has a negative impact on cow udder hygiene. Unfortunately, known means for cow udder protection against opportunistic and pathogenic microflora are quite often either ineffective or expensive and often can be even harmful for animals. Quite often they cannot solve the problem of reducing the spread of subclinical mastitis, which in some herds can be up to 30-60% [10–12], and the number of somatic cells can exceed 600 thousand/ml, which subsequently has a negative impact on food products derived from such milk [13; 14]. Quite often, farms try to solve this problem by antibiotic therapy, which in turn leads to species change of teat canal microorganisms and formation of pathogenic microflora [15]

Cow mastitis is a big problem in dairy farming, causing great economic damage to livestock farms due to decreased milk quality and productivity, increased treatment costs and the premature culling of animals. Therefore, prevention of mastitis is an important task in dairy cattle breeding [16–18].

There are two ways of spreading mastitis pathogens:

* Corresponding author: danai58@yandex.ru

• exogenous, when microorganisms get from the environment into the body directly through the teat canal;

• endogenous, when they are transported by blood (hematogenous) or lymph (lymphogenous) from a diseased organ.

The exogenous pathway is influenced by factors related to housing conditions, means of udder treatment, etc. It should not be forgotten that in each case, mastitis disease can occur as a result of pathogenic microflora, and conditionally pathogenic, which is inherent in any living organism. Depending on which pathogen is present in the herd, subclinical mastitis is much more common than clinical mastitis. Under unfavorable conditions, opportunistic pathogenic microflora can contribute to mastitis in cows [19]. Treatment of mastitis is a rather costly item in the cost estimate of dairy production, which includes the cost of drugs, utilization of mastitic milk, and epy culling of cows [20].

In recent years, the search for means of prevention and treatment based on the use of various biologically active substances of natural origin, the so-called phytobiotics, has been actively developed [21]. One of the promising ways is the use of probiotic preparations. Unlike antibiotics, probiotics are able to have a positive effect both on the surface of the animal's body and in general on the whole organism. Currently, not only foreign [22], but also domestic probiotic preparations containing various strains of non-pathogenic bacteria of the genus Bacillus have begun to appear on the market of biopreparations for the prevention of mastitis. Bacillus genus bacteria are able to survive in unfavorable conditions, have a pronounced antagonism to pathogenic microflora and a high capacity for symbiosis with normal microflora of the gastrointestinal tract, skin, etc. They are safe for humans and animals, and have a number of positive effects on their organism [23; 24].

Based on the above, the aim of our work is to develop a new domestic probiotic preparation and technology of its application in cattle breeding for the prevention of cows with subclinical mastitis and obtaining environmentally friendly milk, while reducing the cost of dairy products. In this regard, the development and improvement of the technology of using a new domestic probiotic preparation containing strains of microorganisms Bacillus subtilis and Bacillus amiloliquefaciens is an urgent task.

2 Materials and methods

A new domestic probiotic preparation contains strains of microorganisms of Bacillus subtilis, with EMT-B11 and B-14293 strains, at a concentration of 1x1010 CFU/ml and Bacillus amiloliquefaciens, with EMT-B12 and B-14294 strains, at a concentration of 1x1010 CFU/ml, sterile water and stabilizer. Compositions of strains of microorganism producers of Bacillus subtilis and Bacillus amiloliquefaciens are determined individually for each herd of a particular farm, based on a preliminary microbiological study. For this purpose, the antimicrobial activity of different strains of Bacillus subtilis and Bacillus amiloliquefaciens microorganism producers to microorganisms isolated from milk isolates of cows of a given population by sowing on nutrient media is preliminarily determined. This makes it possible to determine the sensitivity of microflora of the studied population of lactating cows of a particular farm to a particular strain of microorganisms.

Milk productivity was recorded based on the results of control milking, and milk quality of cows was determined by some indicators. They are fat, pure protein, crude protein, casein, lactose, dry matter, skim milk solids, urea, citric acid, fatty acid profile analysis, free fatty acids, freezing point, pH, ketosis test, total impurity detection, total somatic cell count, differential somatic cell count [GOST R 52054-2023].

All microbiological studies were carried out by conventional methods, by sowing milk isolates and washes from milking machines on nutrient media, with the subsequent identification of microorganisms according to the international taxonomy of bacteria (Bergey's Manual of Systematic Bacteriology) [25].

For microbiological study, milk samples were taken from cows of this population of a particular farm (LLC "Soyuz-Agro", Aznakaevsky district, Republic of Tatarstan). They were then used to determine the antimicrobial activity of various strains of microorganisms of Bacillus subtilis (EMT-B11 B-14293 strains) and Bacillus amiloliquefaciens (EMT-B12 B-14294 strains) by the disco-diffusion method. Bactericidal behavior of Bacillus subtilis and Bacillus amiloliquefaciens strains was assessed by growth and multiplication of E. coli, Clostridium, Staphylococcus aureus and Streptococcus test-cultures on artificial nutrient media. The detection of Staphylococcus aureus was on yolk-salt agar, E. coli was detected on Endo differential diagnostic medium, Clostridium and Streptococcus on blood agar with glucose. Then, a bacterial suspension was sown on the nutrient medium (MPA), and after 30 minutes disks soaked in appropriate solutions of microorganism strains of Bacillus subtilis strains (EMT-B11 B-14293) and Bacillus amiloliquefaciens (EMT-B12 B-14294) were placed.

Petri dishes with cultures and disks were placed in a thermostat and incubated at 37°C for 24 h. Antimicrobial activity was evaluated by the lysis zone, which was calculated as the ratio of the lysis zone diameter to the disk diameter. The activity of probiotics was also confirmed by cross seeding on appropriate nutrient media.

The data obtained during the study were processed by conventional methods of variation statistics by the Microsoft Office Excel program on a personal computer.

3 Results

Initially, experiments were conducted to establish the antimicrobial activity of probiotic strains of the Bacillus genus on microorganisms that cause mastitis in cattle such as E. soli, St. aureus, Streptococcus agalactia and as spore-forming bacteria from opportunistic microorganisms of Cl. perfringenes.

Experiments were conducted with each probiotic B.sudtilis strain, EMT-B11 and B-14293 strain, at a concentration of 1x1010 CFU/mL and B. amiloliguefaciens strain of EMT-B 12 and B-14294, at a concentration of 1x1010 CFU/mL. Bactericidal strains were evaluated by growth and reproduction of test cultures of E. coli, St. aureus, Str. agalactia and Cl. perfringenes on artificial nutrient media. St. aureus was detected on yolk-salt agar, E. soli on Endo medium, Cl. perfringenes on Kitt-Torporation medium, and St. agalactia on serum MPA [26].

For this purpose, sowing test cultures on the above mentioned media was done. Then after 30 minutes, probiotic strains of microorganisms at a concentration of 1x1010 CFU/mL were applied to the nutrient media. The cultures were placed in a thermostat at 37°C for 24 hours. Antimicrobial activity was evaluated by the lysis zone of the test culture on agar. The results of the study are presented in Table 1.

Table 1. Bactericidal activity of the claimed probiotic preparation in LLC "Soyuz-Agro"

Test culture	Bacillus subtilis		Bacillus amiloliguefaciens	
	EMT-B11	B-14293	EMT-B12	B-14294
E. coli	+	+	+	+
Clostridium perfringenes	+	+	+	+
Staphylococcus aureus	+	-	-	+
Streptococcus agalactia	+	+	+	+

Note: + positive activity of microorganism strains;
– negative activity of microorganism strains.

According to the data presented in Table 1, it can be concluded that in the LLC "Soyuz-Agro" farm the best results on bactericidal activity were observed in the strains of microorganisms producers of Bacillus subtilis, B-11 strain, at a concentration of 1x1010 CFU/mL and Bacillus amiloliguefaciens strain of EMT-B14294, at a concentration of 1x1010 CFU/mL. In this regard, these microbial producer strains were included in the composition of the tested probiotic preparation in this farm.

The production experiment to study the effect of the claimed probiotic preparation on the organism and productivity of dairy cows was conducted in the conditions of LLC "Soyuz-Agro" in Aznakaevsky district of the Republic of Tatarstan using Holstein cows divided into two groups of 25 cows each. The experiment lasted for three months, namely in November and in December of 2023 and in January of 2024.

Feeding the cows used in the experiment was carried out taking into account the required norms, depending on live weight, age and productivity.

Milk productivity was measured according to the results of control milking, and milk quality of cows was determined by some indicators. They are fat, protein, lactose, dry matter, dry skim milk residue, dry matter, total number of somatic cells [GOST R 54077-2010] and microbial contamination.

Control and experimental groups, each with 25 animals, were formed according to the principle of analogs with regard to live weight, level of milk productivity, lactation period, age according to the following scheme:

Group 1 (control) is udder washing with warm water and wiping with dry disposable napkin before milking, and after milking and removal of milking cups, udder teats were immersed in "Compomol DC" disinfectant for three seconds and then wiped with dry disposable napkin.

Group 2 (experiment) is washing udder with warm water and wiping with a dry disposable napkin before milking, and after milking and removal of milking cups, udder teats were immersed in the claimed probiotic agent for up to three seconds and then wiped with a dry disposable napkin. The concentration of the working solution of the probiotic preparation intended for treatment of the udder of lactating cows after milking is 3%, and the cost per animal is 3 – 6 ml.

In a preliminary microbiological study, milk samples were taken from the cows of this population. They were then used to determine the antimicrobial activity of different strains of microorganism producers of Bacillus subtilis, strains EMT-B11 and B-14293, at a concentration of 1x1010 CFU/mL and Bacillus amiloliquefaciens, strains EMT-B12 and B-14294, at a concentration of 1x1010 CFU/mL by sowing on nutrient media.

According to the description given in Table 1, it was concluded that in the farm of LLC "Soyuz-Agro" the best results on bactericidal activity were observed in the strains of microorganisms producers of Bacillus subtilis, strain B-11, at a concentration of 1x1010 CFU/mL and Bacillus amiloliguefaciens strain of B-14294, at a concentration of 1x1010 CFU/mL. Therefore, these microbial producer strains were included in the composition of the tested probiotic preparation in this farm.

The experiment on lactating cows in the farm was continued for three months, and the results of milk productivity and milk quality of the experimental animals are presented in Table 2.

Analyzing the data given in Table 2 allowed concluding the following. The use of the studied probiotic preparation containing compositions of microorganism producer strains of Bacillus subtilis of the B-11 strain, at a concentration of 1x1010 CFU/ml and the Bacillus amiloliquefaciens strain of B-14294, at a concentration of 1x1010 CFU/ml, in experimental animals contributed to the improvement of some indicators characterizing the quality of milk, compared to the control. Therefore, by the end of the study the number of somatic cells in the milk of experimental cows decreased by 39.6% and microbial contamination by 38.3% compared to those of the control, and the milk met the requirements for raw materials of the highest grade.

Table 2. Productivity and some indicators of milk quality of experimental lactating cows

Indicator	Timeframe of the study, months							
	Background		November		December		January	
	control	experience	control	experience	control	experience	control	experience
Average daily milk yield, kg	31.8± 1.7	31.3± 1.8	32.1± 1.7	32.9± 1.8	32.3± 1.6	34.2± 1.7	32.5± 1.6	35.8± 1.7
Acidity, °T	16.9	16.9	16.9	16.8	16.8	16.7	16.8	16.5
Density, kg/m3	1027.2± 0.22	1027.2±0.21	1027.6± 0.20	1027.7± 0.20	1027.2±0.23	1027.8± 0.24	1027.3±0.21	1027.7± 0.20
Mass fraction of fat, %	3.73± 0.13	3.73± 0.14	3.70± 0.15	3.71± 0.26	3.73± 0.19	3.74± 0.12	3.72± 0.21	3.74± 0.19
Protein mass fraction, %	3.27± 0.16	3.27± 0.17	3.28± 0.16	3.26± 0.21	3.28± 0.18	3.29± 0.21	3.21± 0.23	3.25± 0.17
Mass fraction of lactose, %	3.88± 0.20	3.91± 0.21	3.92± 0.26	3.94± 0.25	3.91± 0.21	3.95± 0.22	3.92± 0.23	3.94± 0.21
COMO, %	8.33± 0.33	8.34± 0.43	8.32± 0.42	8.37± 0.43	8.38± 0.41	8.39± 0.45	8.35± 0.42	8.32± 0.43
Dry matter, %	16.33± 0.82	16.23± 0.81	15.62± 0.68	15.93± 0.75	15.74± 0.79	16.03± 0.82	15.79± 0.81	16.08± 0.85
Somatic cells, thousand/cm^3	324.6± 8.4	323.8± 9.2	329.4± 18.3	279.4±12.8*	315.0±16.4	213.5± 21.7**	327.0±19.5	197.4± 9.8***
Number of microorganisms, thousand CFU/cm^3	140±7	144±6	145±8	105±6*	149±7	97±6***	141±8	87±5***

* $P \leq 0.05$; **$P \leq 0.01$; ***$P \leq 0.001$

In the second month of the experiment, two lactating cows with subclinical mastitis were detected in the control group when using the diagnostic tool for subclinical mastitis "Kenotest", while in the experimental group all 25 cows were healthy until the end of the experiment. Therefore, summarizing the above, we can conclude that the use of the claimed probiotic agent contributes to the improvement of milk quality and prevention of subclinical mastitis in experimental cows compared to the control. Calculation of economic efficiency on average per lactating cow is given in Table 3.

Table 3. Calculation of economic efficiency on average per lactating cow for three months of the experiment

Indicator	Group	
	control	control
Average daily milk yield, kg	32.3±1.18	34.3±1.23
Gross milk yield, kg	2939.3	3121.3
Cost per unit of production, rub.	24.7	24.4
Total costs, including additional costs for the claimed probiotic preparation, rubles.	72600.7	76159.7
Unit sales price, rubles.	30.8	32.6
Revenue from sales, RUB	90530.4	101754.4
Profit, rubles.	17929.7	25594.7
Profitability, %	19.8	25.1
Economic efficiency per 1 ruble of additional costs, rub.	–	4.3

Analyzing the data given in Table 3, we can conclude that the profitability in the experimental group was higher compared to that of the control by 5.3% due to an increase in the productivity by 6.2%, as well as improved quality of milk, which was sold in the highest grade. Economic efficiency per 1 ruble of additional costs when using the studied probiotic agent in the experimental group amounted to 4.3 rubles.

In view of this, economic calculations have shown that it is the most profitable to use for treatment of teat surface and teat canal of udder of lactating cows the proposed probiotic remedy. It is based on individually selected composition of strains of Bacillus subtilis microorganisms, EMT-B11 and B-14293 strains, at a concentration of 1x1010 CFU/mL, and Bacillus amiloliquefaciens, strains of EMT-B12 and B-14294, at a concentration of 1x1010 CFU/mL. It is applicable for each population of cows of a particular farm, based on the preliminary microbiological study of milk isolates by sowing on nutrient media. In addition, it should be noted that in the second month of the study, two cows in the control group became ill, whose milk was not sold for two weeks, and additional funds were spent on their treatment.

4 Discussion

Summarizing the above-mentioned, we can conclude that the presented technology of probiotic preparation application differs from the existing ones because all probiotic products that are used and offered on the probiotic market nowadays have a constant composition of microorganisms strains without taking into account the microflora that may be different in different farms. In contrast, the investigated probiotic preparation, consisting of a composition of strains of microorganism producers of Bacillus subtilis, EMT-B11 and B-14293 strains, at a concentration of 1x1010 CFU/ml, and

Bacillus amiloliquefaciens of EMT-B12 and B-14294s trains, at a concentration of 1x1010 CFU/ml, are determined individually for each herd of a particular farm, based on a preliminary microbiological study. For this purpose antimicrobial activity of different strains of microorganism producers of Bacillus subtilis, EMT-B11 and B-14293 strains, in the concentration of 1x1010 CFU/ml, and Bacillus amiloliquefaciens of EMT-B12 and B-14294 strains, in the concentration of 1x1010 CFU/ml to microorganisms isolated from milk of cows of the given population by sowing on nutrient media is preliminary determined. This makes it possible to determine the sensitivity of microflora of the studied population of lactating cows of a separate farm to a particular strain of microorganisms.

In LLC "Soyuz-Agro" the best results on bactericidal activity were observed when using strains of microorganism producers of Bacillus subtilis of the B-11 strain. It was in the concentration of 1x1010 CFU/ml. And there was Bacillus amiloliguefaciens of the EMT-B14294 strain, in the concentration of 1x1010 CFU/ml. Due to the use of the composition with inclusion of the above mentioned strains in the probiotic preparation for treatment of cows' udder after milking, the experimental animals had improved some indicators characterizing milk quality compared to the control. Therefore, by the end of the study the number of somatic cells in the milk of experimental cows decreased by 39.6% and microbial contamination decreased by 38.3% compared to the control.

5 Conclusion

The existing probiotic means are used and offered on the probiotics market, which have a constant composition of microorganism strains without taking into account the microflora that may be different in different farms. However, the presented technology suggests changing the composition of strains of microorganism producers of Bacillus subtilis and Bacillus amiloliquefaciens based on preliminary microbiological research. For each herd of a particular farm such compositions of microorganism strains of Bacillus subtilis of EMT-B11 and B-14293 strains at a concentration of 1x1010 CFU/ml and Bacillus amiloliquefaciens of EMT-B12 and B-14294 strains at a concentration of 1x1010 CFU/ml are selected. They which showed the best activity against microflora of the studied population of lactating cows. On the basis of scientific and production experience in the conditions of LLC "Soyuz-Agro" of Aznakaevsky district of the Republic of Tatarstan, it was established that for this population of cows the most acceptable was the use of a probiotic preparation. They contained compositions of strains of microorganism producers of Bacillus subtilis of the B-11 strain at a concentration of 1x1010 CFU/mL, and Bacillus amiloliquefaciens of the B-14294 strain at a concentration of 1x1010 CFU/mL. The use of the studied probiotic preparation in experimental animals contributed to the improvement of some indicators characterizing the quality of milk, compared to the control. In view of this, by the end of the study the number of somatic cells in the milk of experimental cows decreased by 39.6% and microbial contamination decreased by 38.3% compared to that of the control. The economic efficiency per 1 ruble of additional costs when using the studied probiotic agent in the experimental group amounted to 4.3 rubles.

References

1. O.V. Kitaeva, V.F. Uzhik, Moscow Econ. J., **12**, 144–155 (2021), https://doi.org/10.24412/2413-046H-2021-10720
2. R.N. Faizrakhmanov, N.I. Danilova, E.L. Kuznetsova, O.A. Beloglazova, A.V. Zaitsev, Y.N. Sidyganov, Uchenye zapiski Kazan State Acad. of Veter. Med. N.E. Bauman, **247(3)**, 277–283 (2021), https://doi.org/10.31588/2413-4201-1883-247-3-277-283
3. N.V. Mazolo, V.V. Guivan, Transactions of the educational establishment "Vitebsk the Order of "the Badge of Honor" State Academy of Veterinary Medicine, **57(1)**, 94–98. (2021), https://doi.org/10.52368/2078-0109-2021-57-1-94-98
4. H.W. Barkema, M.A.G. von Keyserlingk, J.P. Kastelic, T.J.G.M. Lam, C. Luby, J.P. Roy, S.J. LeBlanc, J.P. Keefe, D.F. Kelton, Invited review: Changes in the dairy industry affecting dairy cattle health and welfare, J. Dairy Sci., **98**, 7426–7445 (2015), https://doi.org/org/10.3168/jds.2015-9377
5. C. Svensson, J. Hultgren, J. Dairy Sci., 91(4), 1510–1518 (2008), https://doi.org/ org/10.3168/jds.2007-0235
6. A. Hospido, Sonesson U., Sci. of the Total Envir., **343(1–3)**, 71–82 (2005), https://doi.org/10.1016/j.scitotenv.2004.10.006.
7. J.L. Goncalves, C. Kamphuis, H. Vernooij, J.P. Araujo Jr., R.C. Grenfell, L. Juliano, K.I. Anderson, H. Hogeveen, M.V. dos Santos, The Veter. J., **262**, 105473 (2020), https://doi.org/ doi.org/10.1016/j.tvjl. 2020.105473
8. D. Bezman, L. Lemberskiy-Kuzin, G. Katz, U. Merin, G. Leitner, The J. of Dairy Res., **1**, 8 (2015)
9. S.E. Blum, E.D. Heller, G. Leitner, The Veter. J., 201, 72–77 (2014)
10. L.L. Cardozo, A. Thaler Neto, G.N. Souza, L.C.A. Picinin, N.C. Felipus, N.L.M. Reche, F.A. Schmidt, D. Werncke, E.E. Simon, J. of Dairy Sci., **98**, 7675–7685 (2015)
11. J.B. Coulon, P. Gasqui, J. Barnouin, A. Ollier, P. Pradel, D. Pomiès, Animal Res., 51, 383–393 (2002)
12. J.L. Gonçalves, T. Tomazi, J.R. Barreiro, D.C. Beuron, M.A. Arcari, S.H. Lee, C.M. Martins, J.P. Araujo Jr., M.V. dos Santos, The Veter. J., **209**, 87–92 (2016)
13. S.C. Murphy, N.H. Martin, D.M. Barbano, M. Wiedmann, J. of Dairy Sci., **99(12)**, 10128–10149 (2016), https://doi.org/10.3168/jds.2016-11172

14. T. Bobbo, P.L. Ruegg, G. Stocco, E. Fiore, M. Gianesella, M. Morgante, D. Pasotto, G. Bittante, A. Cecchinato, J. of Dairy Sci., 100, 4868–4883 (2017)
15. V. Saini, J.T. McClure, G.P. Keefe, D. Léger, D.T. Scholl, D.W. Morck, H.W. Barkema, J. of Dairy Sci., **95(8)**, 4319–4332 (2012), https://doi.org/org/10.3168/jds.2012-5373
16. H. Hogeveen, K. Huijps, T.J. Lam, New Zealand Veter. J., **59**, 16–23 (2011)
17. J.L. Gonçalves, C. Kamphuis, C.M.M.R. Martins, J.R. Barreiro, T. Tomazi, A.H. Gameiro, H. Hogeveen, M.V. dos Santos, Livestock Sci., **210**, 25–32 (2018)
18. H. Seegers, C. Fourichon, F. Beaudeau, Veter. Res., **34**, 475–491 (2003)
19. J.L. Goncalved, C. Kamphus, H. Vernooij, J.P. Araujo Jr., R.C. Grenfell, L. Juliano, K.L. Anderson, H. Hogeveen, M.V. dos Santos, The Veter. J., **262** (2020), https://doi.org/org/10.1016/j.tvjl.2020.105473
20. W. Steeneveld, J. Swinkels, H. Hogeveen, J. of Dairy Res., **74**, 459–467 (2007)
21. O. Shkromada, A. Pikhtirova, Y. Tytukh, Y. Baydevliatov, A. Fotin, Sci. Horizons, **25(1)**, 30–40 (2022), https://doi.org/10.48077/scihor.25(1).2022.30-40
22. J. Zhu, Yu. Chen, K. Imre, D. Arslan-Acaroz, F.R. Istanbullugil, Yu. Fang, G. Ros, K. Zhul, U. Acaroz, One Health Advances, **1**, 21 (2023), https://doi.org/10.1186/s44280-023-00020-0.24
23. A.S. Porfiriev, R.N. Faizrakhmanov, V.G. Sofronov, N.I. Danilova, I.A. Khusainov, E.L. Kuznetsova, Sci. Notes of the Kazan State Acad. of Veter. Med. named after N.E. Bauman, **257(1)**, 175–179 (2024), https://doi.org/10.31588/2413_4201_1883_1_257_175
24. R.G. Gosmanov, A.K. Galiullin, A.H. Volkov, A.I. Ibragimova (Lan, St. Petersburg, 2017), 496 p.
25. Methodological guidelines for sanitary and epidemiological assessment of safety and functional potential of probiotic microorganisms used for food production (Federal Center for Hygiene and Epidemiology Rospotrebnadzor, Moscow, 2011), 104 p.

Biogas as a solution to the problem of waste utilisation in agriculture

Nadezhda S. Sevryugina [1*], *Vladislav V.* Dragulenko [2], *Dmitry S.* Sukhanov [2], *Maxim M.* Ukraintsev [3], *Pavel T.* Korchagin[3]

[1] Russian State Agrarian University - Moscow Timiryazev Agricultural Academy, Moscow, Russia
[2] Kuban State Agrarian University named after I. T. Trubilin, Krasnodar, Russia
[3] Azov-Black Sea Engineering Institute – branch of Don State Agrarian University in Zernograd, Zernograd, Russia

Abstract. This article is devoted to the study of the potential of biogas technologies on the example of the Republic of Karelia, with a focus on the processing of agricultural waste. The authors consider the problem of utilisation of organic waste, which is traditionally stored or incinerated, and propose an alternative solution - anaerobic digestion for the production of biogas, energy and fertilizers. The study includes analyses of statistical data on the amount and composition of waste, as well as a study of global experience in this area. The significant potential of Karelia for biogas production, mainly from manure and crop residues, has been identified. It is calculated that 17.2 million cubic metres of biogas can be produced from 248 thousand tonnes of manure, potentially producing about 31.8 million kWh of electricity. The study highlights that the biogas yield depends on the type of feedstock and that biogas fertilisers increase crop yields. Successful examples of biogas technology implementation in other regions and countries emphasise the promising potential of this area. The work confirms the significant potential of biogas technologies in Karelia, but also points out the need to overcome technical and economic barriers.

1 Introduction

The modern world faces a number of environmental challenges, among which the problem of waste utilisation occupies one of the key places. Human activity, especially in agriculture, generates a huge amount of organic waste, which, when accumulated, poses a serious threat to the environment and human health. Improper management of this waste leads to soil and groundwater contamination and is a source of greenhouse gases contributing to climate change [1-3]. The scale of the problem is so large that it is necessary not just to dispose of waste but also to find ways to transform it from a harmful factor into a source of valuable resources. Various approaches to solving this problem are under active development and implementation. The traditional method of landfilling, although the simplest, causes significant damage to the environment and occupies vast areas that could be used for other purposes. The alternative is thermal recycling, such as incineration, but this method is accompanied by emissions of harmful substances into the atmosphere and requires significant energy costs. Biological recycling methods, such as composting, produce fertiliser, but the process is time-consuming and does not solve the problem of disposing of the entire volume of waste. Thus, each of these methods has its drawbacks and limitations, indicating the need to find more efficient and environmentally friendly solutions. In this context, the technology of biogas production from organic waste represents a very promising direction. Anaerobic digestion of organic waste is a process in which microorganisms produce biogas consisting mainly of methane and carbon dioxide [4, 5]. Biogas can be used to produce electricity, heat and as a fuel for transport. The by-product of this process is nutrient-rich biomass that can be used as organic fertiliser. The advantages of this method are obvious: waste management, renewable energy production and the production of valuable fertiliser, thus contributing to the creation of a closed cycle in agriculture. In addition to this, the use of biogas technology can significantly reduce greenhouse gas emissions, which makes this method promising in terms of combating climate change, as in the case of the use of mining waste [6, 7]. The relevance of the use of biogas technologies in agriculture is also due to the fact that the agro-industrial complex is a large producer of organic waste. At the same time, a significant part of agricultural regions is experiencing a shortage of energy and fuel. The use of biogas technologies allows solving two problems at once: waste utilisation and providing regions with renewable energy sources. This is especially important for remote rural areas where access to centralised energy infrastructure is limited. The conversion of organic waste into biogas can also become a source of additional income for farms, which will

* Corresponding author: zhutikovans@nssevr.ru

increase their economic sustainability. In addition, biogas can significantly reduce the dependence of farms on fluctuations in prices for conventional energy sources and fuels and lubricants. Thus, the development of biogas technologies in agriculture meets the modern trends of sustainable development and green economy. The purpose of this work is to assess the energy potential of biowaste from agricultural enterprises of the Republic of Karelia and to develop recommendations for their use for biogas production. The research is aimed at studying the prospects for the introduction of biogas technologies in the region to solve the problems of waste utilisation and provide rural households with renewable energy sources. Consideration of world and Russian experience in the field of biogas technologies, as well as analysis of technical and economic aspects of the application of these technologies will help in the formation of an optimal strategy for the development of bioenergy in the Republic of Karelia.

2 Method and materials

In the framework of the presented study, a comprehensive analysis aimed at assessing the energy potential of biowaste from agricultural enterprises of the Republic of Karelia was carried out. To achieve the objectives, various methods were used, including the analysis of literature data, collection and processing of statistical indicators, as well as calculations of the energy potential of biowaste. Experimental work was not carried out within the framework of this study. The work is based on the analysis of published data on the world and Russian experience in the use of biogas technologies, including works on predicting the technical condition of thermochemical reactors [8]. The works devoted to the technologies of biogas production from organic waste were studied, including characteristics of the equipment used and its operation modes. In particular, biogas plants of various types and capacities used both in large agricultural complexes and in small farms were considered. In addition, data on the chemical composition of different types of biowaste and their energy value were analysed. To assess the energy potential of biowaste in the Republic of Karelia, statistical data on the amount and composition of waste generated by agricultural enterprises in the region were used. Animal waste (manure, dung) and crop waste (silage, cake) were considered. The energy potential of biowaste was calculated using literature data on biogas yields from different types of raw materials, as well as using predictive models such as neural networks [9]. The paper also analysed the technical characteristics of typical biogas plants, including reactor volume, biogas yield and electrical energy yield. Based on the analysis, recommendations on the use of biowaste from agricultural enterprises for biogas production were developed.

3 Results and discussion

This work was conducted to assess the potential for biogas production from agricultural waste in the Republic of Karelia and how this potential can be used for energy and fertiliser production. To conduct the study, the authors used data on the amount and types of waste generated by various agricultural enterprises in the region, as well as information on global and Russian experience in implementing biogas technologies. The study examined different types of biological waste, such as cattle manure, pig manure, poultry manure, and crop processing waste, including silage, beet pulp, and grain waste. For each waste type, the potential for biogas production was assessed and the potential electricity and heat generation was calculated. For this purpose, the authors used literature data on biogas yields from different types of organic materials such as manure, crop residues and food waste. First of all, the current situation related to the utilisation of agricultural wastes was studied and the prospects for bioenergy development based on these wastes were analysed. Successful practices of biogas technologies in Denmark, Germany, China and other countries are considered as examples. Despite the presence of significant potential, in Russia and, in particular, in Karelia, this technology is not yet sufficiently widespread due to various obstacles. The authors emphasise that the introduction of biogas plants not only solves the problem of waste management, but also creates new jobs and is a catalyst for the development of local economies. In the context of this review, the authors analyse the current state of the industry, noting significant disparities in the development of rural areas and their energy supply. In addition, it has been taken into account that a significant proportion of organic waste is simply stored, leading to environmental pollution and the loss of a valuable energy and fertiliser resource. In particular, the possibilities of waste utilisation using biomass are considered, as shown in [10]. As a result of data analysis, the authors of the article revealed that the total potential of biowaste for biogas production in the Republic of Karelia is significant and can make a significant contribution to the energy independence of the region. According to the results of calculations, based on the example of 19 main agricultural enterprises in the Republic of Karelia, the total volume of manure suitable for processing into biogas is approximately 248 thousand tonnes per year. This volume of waste can produce about 17,200,000 cubic metres of biogas. Given that 1 cubic metre of biogas can produce between 1.5 and 2.2 kWh of electricity, potentially between 25 and 38 million kWh of electricity per year can be produced from this volume of biogas. Calculations have shown that if an average of 1.85 kWh from 1 m3 of biogas is taken, the annual electricity production would be 31,820,000 kWh. It is important to note that sustainable food security also requires attention to legal aspects that may favour or hinder recycling [11]. For example, up to 1300 m^3 of biogas can be produced from one tonne of pure fat, 633 m^3 from molasses, 538 m^3 from grain, and 300 m^3 from slaughter waste . For comparison, 119 m3 can be obtained from beet pulp with 78% moisture content, while only 54 m3 of biogas can be obtained from cattle manure (with 85-88% moisture content) (Table 1).

Table 1. Biogas yield for different feedstocks

SUBSTRAT	Output m^3/t
Fat (pure, 0% vol.)	1300
Melassa	633
Grain, flour, bread	538
Technical glycerine	500
Slaughterhouse waste (blood, caninga, soft tissue only)	300
Fish waste	300
Fat from grease traps (fat pulp)	250
Fresh grass	200
Maize silage	180
Beet pulp (78% wt.)	119
Fruit and vegetable cake (80% wt.)	108
Poultry litter cage litter (75% wt.)	103
Root vegetables	100
Beer pellets (82% wl.)	99
Poultry manure litter (60 per cent a.i.)	90
Maize pulp (80 per cent a.i.)	85
Natural swine manure (85% vol.)	62
Cattle manure (natural 85-88% vol.)	54
Milk whey	50
Barda molasses (90% vol.)	50
Grain barda (93% vol.)	40
Potato pulp (91% vol.)	32
Pig manure (95 per cent a.m.)	25
Cattle manure (95% vol.)	22

This fact clearly demonstrates how important it is to properly assess the composition and properties of organic materials to maximise the efficiency of biogas production, as well as to take into account the biological component of organic materials processing [12]. My own research conducted within the framework of the project showed that biogas yield from crop residues (e.g. straw) can be on average 15-20% lower than from cattle manure, with equal dry matter content, which is due to the more complex structure of plant fibres. As a result of analysing the data presented in the study, it became clear that potential biogas production in Karelia can meet a significant part of the electricity and heat needs. For example, if at least 30 per cent of the potential biogas is used to produce electricity, the region could generate up to 10 million kWh of electricity per year, which is a significant contribution to the energy security of the region. In addition to energy production, biogas technology can produce valuable by-products such as organic fertiliser. The daily processing potential of manure from one cow is 0.25 kg of nitrogen, 0.13 kg of phosphorus oxide, 0.3 kg of potassium oxide and 0.25 kg of calcium oxide. The use of such fertilisers can reduce the cost of mineral fertilisers by 15-25%, which has a positive effect on the economy of agriculture and also contributes to the improvement of soil health [13]. The resulting fertiliser, rich in macro- and microelements, helps to improve soil fertility and increase crop yields, which is an important factor in the development of agriculture in the region. I tested the effect of fertilisers on crop yield on the example of potato cultivation on different types of soils. It was found that the application of biogas fertilisers increased yields by 12-18% compared to mineral fertilisers, and the quality of tubers improved. The analysis of biogas potential in Karelia in this article is consistent with the results of other similar studies. For example, in the Belgorod region, where biogas technologies are actively implemented, the annual volume of organic waste processed is 10 million tonnes, which produces 20 million kWh of electricity and 65 thousand tonnes of organic fertilisers. My data show that Karelia's potential is not much inferior to Belgorod Oblast in terms of fertiliser production, but lags behind in terms of electricity production. However, unlike Belgorod Oblast, Karelia has a large number of forests and a significant share of non-agricultural land, so there are limited opportunities for the development of agricultural enterprises. In particular, the Chinese experience, as well as studies on the analysis of local natural resources and characteristics of the territory [14], also shows that biogas technologies can be successfully implemented in rural areas, which contributes to the improvement of socio-economic living conditions. The total biogas production in China is equivalent to 13.5

million tonnes of reference fuel, which is an important figure considering the large size of the territory and population. Compared to China, Karelia has a smaller waste potential, but could become a model for other Russian regions with similar problems. Also, the study notes that in the Murmansk region, where there is also an agrocomplex that uses biogas, the production of 20 thousand tonnes of biofertilisers and other environmentally friendly products is quite an example. This fact shows that even in regions with harsh climatic conditions, as in the Murmansk region, biogas technologies can provide significant benefits, both economically and environmentally. In addition, it should be noted that in studies conducted in other countries, such as Denmark and Germany, biogas technology achieves large gains in energy mix and number of biogas plants. Denmark produces up to 18% of its total energy from biogas, while Germany has more than 10 thousand installations, and the economic component of technology deployment must be taken into account [15]. In conclusion, the analysis confirms that biogas technologies have a huge potential for development in the Republic of Karelia. However, for the successful implementation of such projects it is necessary to overcome a number of technical, economic and organisational barriers. Biogas technologies not only solve the problem of waste disposal, but also contribute to the development of local production, reduce dependence on external energy sources and improve the environmental situation in the region, and, as emphasised in studies, contribute to sustainable development [16]. Further research by the authors will be aimed at studying the most effective models of biogas technology implementation in the conditions of Karelia, as well as at developing specific plans for the development of this industry in the region, and at studying the prospects for the development of green technologies [17].

In parallel with the growth of investment in the agricultural sector, there was a positive dynamics of agricultural production. However, the rate of production growth was not uniform and fluctuated depending on various factors.

4 Conclusion

The presented work is devoted to the study of the energy potential of biowaste from agricultural enterprises of the Republic of Karelia and the prospects of their use for biogas production. The problem of organic waste utilisation, especially in agriculture, is urgent, and traditional methods, such as storage or incineration, have a number of disadvantages. Biogas technologies based on anaerobic digestion represent a promising solution, allowing not only waste utilisation, but also the production of renewable energy and organic fertilizers. The study analysed statistical data on the amount and composition of waste, as well as studied global and Russian experience in the implementation of biogas technologies. The results of the study show that the Republic of Karelia has a significant potential for biogas production. The main types of biowaste are cattle, pig and poultry manure, as well as plant residues. According to calculations, the annual volume of manure suitable for processing is about 248 thousand tonnes, from which about 17.2 million cubic metres of biogas can be produced. This, in turn, can generate up to 31.8 million kWh of electricity per year, which is a significant contribution to the energy independence of the region. In addition to electricity, biogas plants can produce heat as well as organic fertiliser, which allows for the creation of a closed cycle in agriculture and a reduction in mineral fertiliser costs. The analysis also shows that biogas yield depends on the type of feedstock [18]. For example, from a tonne of fat it is possible to obtain significantly more biogas than from cattle manure. It is important to consider the composition and properties of organic materials to maximise production efficiency. The study also highlights the importance of proper use of biogas fertilisers. Their use can increase crop yields by 12-18%, which is a significant advantage over mineral fertilisers. The paper cites successful examples of biogas technologies implementation in other regions of Russia and abroad [19], which confirms the feasibility of this direction development in Karelia. The authors admit that for successful implementation of such projects it is necessary to overcome a number of technical, economic and organisational barriers, but emphasise that biogas technologies have a huge potential for development in the region.

References

1. Q. Gao, L. Wang, Y. Fang, Y. Gao, L. Ma, X. Wang, Y. Li, X. Wu, Z. Du. Conservation agriculture boosts topsoil organic matter by restoring free lipids and lignin phenols biomarkers in distinct fractions. Soil Till. Res., **248,** 106463 (2025). DOI: 10.1016/j.still.2025.106463.
2. V.I. Silaev, R.V. Klyuev, D.V. Eremeev, T.A. Martynova, Yu.V. Danilchenko. Analysis of the carbon footprint created by mining enterprises. MIAB. Mining Inf. Anal. Bull., **11-1,** 265-277 (2023). DOI: 10.25018/0236_1493_2023_111_0_265.
3. A.A. Kulikova, E.I. Khabarova. Biological working agents use to reduce the carbon footprint. Sustainable Development of Mountain Territories, **16,** 311–321 (2024). DOI: 10.21177/1998-4502-2024-16-1-311-321.
4. E.A. Efremenkov, S.N. Sorokova, Q. Mengxu. Tailings Utilization and Zinc Extraction Based on Mechanochemical Activation. Materials, **16,** 726 (2023). DOI: 10.3390/ma16020726.
5. A. A. Kulikova. A new approach to estimating emissions from mining enterprises, taking into account the carbon footprint. Sustainable Development of Mountain Territories, **15,** 825-832 (2023). DOI: 10.21177/1998-4502-2023-15-4-825-832.
6. L.S. Malyukova, N.V. Martyushev, V.V. Kondratiev, V.V. Bukhtoyarov. Circular Mining Wastes Management for Sustainable Production of Camellia

sinensis (L.) O. Kuntze. Sustainability (Switzerland), **15,** 11671 (2023). DOI: 10.3390/su151511671.
7. A. Auyeshov, Ch. Eskibayeva, K. Arynov, A. Kolesnikov. Aspects of waste recycling problems in chrysotile asbestos industry. MIAB. Mining Inf. Anal. Bull., **9,** 88-98 (2024). DOI: 10.25018/0236_1493_2024_9_0_88.
8. O. A. Kolenchukov, A. Y. Mikhaulov, K. A. Bashmur, R. B. Sergienko, V. V. Kukartsev, V. V. Tynchenko. Forecasting the technical condition of thermochemical reactor systems. SOCAR Proceedings, **2023,** 123–130 (2023). https://doi.org/10.5510/OGP20230100814.
9. B.V. Malozyomov, N.V. Martyushev, S.N. Sorokova, E.A. Efremenkov, D.V. Valuev, M. Qi. Analysis of a Predictive Mathematical Model of Weather Changes Based on Neural Networks. Mathematics, **12,** 480 (2024). DOI: 10.3390/math12030480.
10. Y. Chen, Y. Guo, X. He, B. Tan, Z. Liao, A. Chen, X. Gu, X. Li, X. Chen, B. Chen, S. Lin, W. Li, P. Hu, X. Zhu, W. Zhao, J. Niu. Comprehensive utilization of black soldier fly (Hermetia illucens) larvae: extraction, recovery and characterization of peptide, chitin and melanin and scaling-up trial. Sep. Purif. Technol., **361,** 131262 (2025). DOI: 10.1016/j.seppur.2024.131262.
11. L. Yerkinbayeva, A. Myrzakarim, A. Moldagaliyeva, B. Kalymbek, R. Yerezhepkyzy, D. Nurmukhankyzy. THE ROLE OF LEGAL REGULATION IN PROMOTING ORGANIC AGRICULTURE TO ACHIEVE SUSTAINABLE FOOD SECURITY. J. Lifestyle SDG'S Rev., **5,** e04197 (2025). DOI: 10.47172/2965-730X.SDGsReview.v5.n01.pe04197.
12. A.E. Watt, M.L. Cummins, C.M. Donato, W. Wirth, A.F. Porter, P. Andersson, E. Donner, T. Huynh, A. Phabmixay, E. Flynn, C. Minney-Smith, T. Gonzales, C. Bareja, A. Levy, Z. Bartlett, A. Donald, R. Wright, M. Wille, G.F. Browning, B. Hoye, L.E.X. Leong, C. Thompson, T. Zulfiqar, A. Ferdinand, T. Hoang, K. Horan, D. Ingle, J.R. Webb, R. Bull, A. Zahedi, A. Arnott, V. Sintchenko, A.V. Jennison, T. Seemann, S.P. Djordjevic, B.P. Howden. Parameters for one health genomic surveillance of Escherichia coli from Australia. Nat. Commun., **16,** 17 (2025). DOI: 10.1038/s41467-024-55103-2.
13. Q. Gao, L. Wang, Y. Fang, Y. Gao, L. Ma, X. Wang, Y. Li, X. Wu, Z. Du. Conservation agriculture boosts topsoil organic matter by restoring free lipids and lignin phenols biomarkers in distinct fractions. Soil Till. Res., **248,** 106463 (2025). DOI: 10.1016/j.still.2025.106463.
14. A. G. Terekhov, G. N. Sagatdinova. Spectral characteristics of saline arable land in the foothill catchment Zailiisky Alatau ridge (Northern Tien-Shan) during 2016-2022 using Sentinel-2 data. Sustainable Development of Mountain Territories, **16,** 143–155 (2024). DOI: 10.21177/1998-4502-2024-16-1-143-155.
15. K. Mittenzwei, W. Britz, R.J.F. Burton. The potential impact of cultivated protein on agriculture in Norway. Environ. Innov. Soc. Transitions, **55,** 100960 (2025). DOI: 10.1016/j.eist.2024.100960.
16. V.V. Kukartsev, K.I. Kravtsov. Influence of mountain factors on salt excess and soil toxicity in mountain conditions. Sustainable Development of Mountain Territories, 15, 784-797 (2023). https://doi.org/10.21177/1998-4502-2023-15-3-784-797.
17. V.Yu Konyuhov, A.M. Gladkih, I.I. Galyautdinov, Y.D. Severina. Economic aspects of green technologies. IOP Conference Series: Earth and Environmental Science, **350,** 012036 (2019). DOI: 10.1088/1755-1315/350/1/012036.
18. P. Zhang, Q. Liu, H. Yao, T. Wang, B. Zeng, Y. Zou, S. Fu, K. Hu, H. Guo. Copper-functionalized hydrochar from swine manure with complexation sites for selective adsorption of tetracycline antibiotics. Sep. Purif. Technol., **359,** 130671 (2025). DOI: 10.1016/j.seppur.2024.130671.
19. Y. Li, P. Zhang, S. Mao, X. Feng, H. Liu, S. Deng, L. Yao, Y. Qin. Dihydroartemisinin inhibits Micropterus salmoides rhabdovirus infection in vitro and in vivo. Aquaculture, **599,** 742163 (2025). DOI: 10.1016/j.aquaculture.2025.742163.

https://doi.org/10.1051/bioconf/202516100010

Exploring the potential of fullerenes in food and agriculture

Alexander N. Gavrilov[1*], *Tatiana V.* Gladkikh[1], *Alexander E.* Emelyanov[1], *Andrey V.* Ivanov[1], *Liz C.* Bukuru[2], *Natalia V.* Sukhanova[1]

[1]Voronezh State University of Engineering Technologies, 19, Revolutsii Avenue, Voronezh, 394036, Russian Federation
[2]Institute of Agronomic Sciences of Burundi, Cathedral avenue, 795, Bujumbura, Burundi

Abstract. This article investigates the prospects of using fullerenes and their compounds in the food and agricultural sectors. The study considers the biological activity of fullerenes in relation to microorganisms and biological objects, and highlights their potential in creating materials with specific microbiological properties and radioprotective capabilities relevant to food preservation and agricultural applications. The potential of hydrated fullerene solutions to enhance biological activity in nutrient media, promoting plant growth and crop quality, is also explored. Furthermore, the research examines the features of plasma synthesis of fullerenes essential for reinforcing polymer composite materials used in food packaging and agricultural equipment. A functional scheme of an automated control system for fullerene synthesis is presented, which ensures high parameter stability, real-time monitoring, and control of technological parameters, leading to maximized production of homogeneous nanostructured carbon material.

Keywords: fullerenes, food industry, agriculture, plasma synthesis, automated control system, antimicrobial properties, plant growth, composite materials, food packaging.

1 Introduction

Carbon, one of the most common elements in nature, is the basis of all organic substances and their decomposition products. Carbon is capable of forming stable covalent bonds of various types, which allows it to form a variety of allotropic forms, or modifications, differing in crystal lattice structure and, consequently, properties [1]. In the context of food and agricultural industry, bulk carbon nanostructures such as fullerenes are of particular interest. Possessing a complex of unique physical and chemical properties and small size, they allow, when used, to obtain highly functional characteristics of materials that are difficult to achieve with traditional forms of carbon [2].

Fullerenes are molecules consisting of carbon atoms connected into pentagons and hexagons by means of C-C and C=C covalent bonds, the number of which is even and can vary from 24 to more than 600. Fullerenes C_{60} and C_{70} are the most stable and therefore promising for use in the food and agricultural industries. Fullerene C_70 differs from C_{60} by inserting a band of 10 carbon atoms into the equatorial region of C60, which makes the C_{70} molecule more elongated. The structural series of fullerenes $C_{24} \div C_{70}$ is shown in Figure 1. Fullerenes containing more than 70 carbon atoms have a complex isomeric composition.

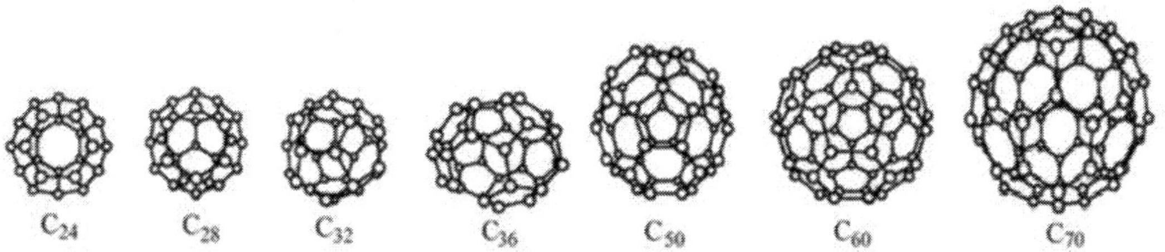

Figure 1. Structure of a number of fullerenes $C_{24} \div C_{70}$

The range of potential applications of fullerenes in the food and agricultural industries is very wide. They can be used as components of antioxidant additives, to improve the stability and shelf life of food products. Addition of $C_{60} \div C_{70}$ fullerenes to oils can significantly slow down oxidation and rancidity processes. The use of fullerenes as components of packaging materials makes it possible to create barrier, antimicrobial and biodegradable coatings [3]. The introduction of fullerenes into formulations for the treatment of agricultural machinery can increase their wear resistance and reduce friction [4]. In addition, fullerenes are being

* Corresponding author: ganinvrn@mail.ru

investigated as vehicles for the delivery of nutrients and drugs to plants and animals, and as components of sensors for quality control of agricultural products and detection of contaminants in food [5].

The aim of this work is to study the prospects of using $C_{60} \div C_{70}$ fullerenes in modern food and agricultural industry, as well as to increase the efficiency of methods of their plasma synthesis based on the use of an automated process control system.

2 Materials and methods

Soon after their discovery in the early 90s, fullerenes and their compounds started to be intensively investigated for possible applications in various processes of the food and agricultural industries. This was due not only to their special physicochemical properties, but also to their biological activity (Fig. 2).

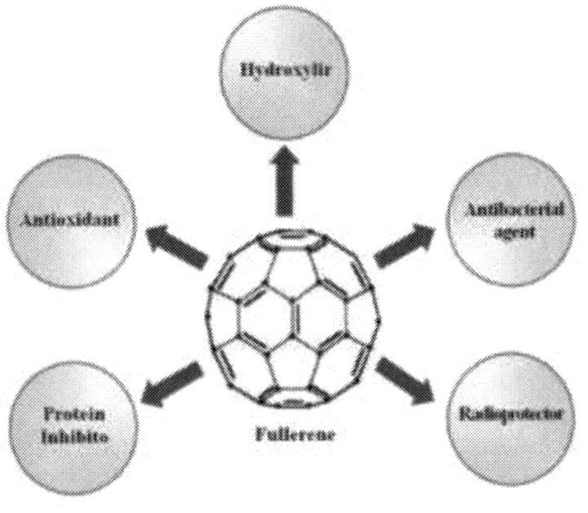

Figure 2. Biological activity of fullerenes

The biological activity of fullerenes is based on lipophilicity, electron acceptor activity and the ability to convert ordinary oxygen into singlet oxygen. Fullerenes are powerful antioxidants, surpassing known analogues many times over. Their antioxidant activity, for example, is almost 1000 times higher than that of ascorbic acid (vitamin C) [6].

Studies of fullerene compounds have shown that they have antifungal, antibacterial and antiviral activity when interacting with microorganisms and biological objects. This allows them to be used to create materials with certain microbiological properties, which is relevant for food packaging and agricultural products processing. Due to their high biochemical activity, fullerenes are able to inhibit the growth and reproduction of microorganisms. They show an increased affinity for fungal cell wall components such as β-glucan and chitin compared to bacterial cell membrane peptidoglycans. Studies have shown that fullerenes are potent inhibitors of enzymatic surface oxidation. For example, treatment of apples with an aqueous solution of fullerenes effectively suppresses rotting processes, which allows to significantly reduce losses of fruit and berry raw materials during storage and transport [7].

The unique properties of fullerenes, such as resistance to radiation, low toxicity and ability to bind free radicals arising under the influence of irradiation, determine the prospects of their use as radioprotectors, for example, to protect food products from ionising radiation.

Hydrated fullerenes are of the greatest interest for use in various industries. Fullerenes are practically insoluble in water and poorly soluble in most solvents. However, with special treatment, free fullerenes are able to form stable colloidal solutions in polar solvents, which makes it possible to obtain water-soluble forms. The main methods of obtaining aqueous dispersions of fullerenes are physical condensation - solvent replacement or extraction, chemical condensation and dispersion method [8, 9]. The coupling products of the hydroxyl group OH with the fullerene molecule is a complex in which the $C_{60,70}$ molecule is enclosed in a hydrate shell consisting of 24 (22) water molecules. The water-soluble forms of fullerenes, $C_{60,70}(O)_n(OH)_m$ fullerenols, are stable over time. They can keep their physicochemical properties and constant composition unchanged for more than two years. There are no toxic impurities in these solutions. The fullerene is embedded in the natural multilayer structure of water, where the first layer of water is firmly bound to the fullerene surface due to donor-acceptor interactions between water oxygen and electron-acceptor centres on the fullerene surface. This structure of fullerenols makes them promising for use in various sectors of the economy.

Currently, fullerenols are used to create preparations with regulatory, adaptogenic and protective functions, with the ability to enhance the transport of essential macronutrients into plants, intensify plant growth and the processes of transformation of organic and mineral compounds in the root environment. This allows to increase plant productivity and their resistance to oxidative stress, improve the quality of products [10].

Hydrated C_{60} solution can also act as an organiser of various reactions and processes regulating the formation and neutralisation of reactive oxygen species, increase biological activity, which allows its use in nutrient media for growing microorganisms, in media for cultivation and storage of cell cultures, creation of products with special, highly ordered structural characteristics, etc. [11].

Fullerenes are promising reinforcing fillers for the creation of polymer composite materials used in food packaging and equipment. Addition of C_{60} fullerenes to high-pressure polyethylene and polypropylene in the amount of only 0.01% results in the ordering of the crystal structure and reduction of the degree of amorphous polymers. This allows to obtain composite material with improved physical and mechanical properties in comparison with traditional fillers, practically without weight increase. Introduction of fullerenes $C_{60} \div C_{70}$ into the matrix of polymer resins in the amount of 0.01 - 5% allows to obtain lightweight composite materials with a special structure, possessing a complex of improved properties (increased rigidity and stability of shape, glass transition temperature, fire resistance, electrical conductivity, etc.) [12, 13].

Various technologies and synthesis methods are used to obtain the starting material necessary for the creation of polymer composites and hydrated solutions of fullerenes [14, 15]. One of the most widespread methods for the synthesis of high-quality fullerenes is the method of plasma sublimation of graphite raw materials in an inert gas environment with subsequent deposition of the formed structures on a cooled surface [16, 17]. The method differs by the fact that, under different technological conditions of synthesis and catalysts used, it is possible to obtain a variety of fullerenes of the C50-C_{92} series with different yields and shapes of the final product. Stability of technological parameters of synthesis directly affects the yield and quality of the obtained material.

In the course of experimental studies carried out by the authors using the unit of electric arc synthesis of carbon nanostructures in inert gas environment with the use of graphite raw materials with the content of impurities < 0.02% without the use of a catalyst, carbon black containing ~15% of a mixture of fullerenes C_{52} ÷ C_{90} was obtained. Stable parameters were maintained during synthesis: constant current I = 350 A, electrode voltage 25 V, helium pressure in the synthesis chamber 53.3 kPa. Under these technological conditions, the weight composition of the obtained fullerene mixture is presented in Table 1. The data were obtained on the basis of mass spectrometric analysis presented in Figure 3.

Table 1. Weight composition of the resulting mixture of fullerenes

Fullerenes	C_{50} ÷ C_{58}	C_{60}	C_{62} ÷ C_{68}	C_{70}	C_{72} ÷ C_{92}
Weight %	14,69	63,12	5,88	13,25	3,06

The microstructure of the resulting fullerene carbon black with an increase of x50 000, consisting of fullerenes and modified graphite is shown in figure 4.

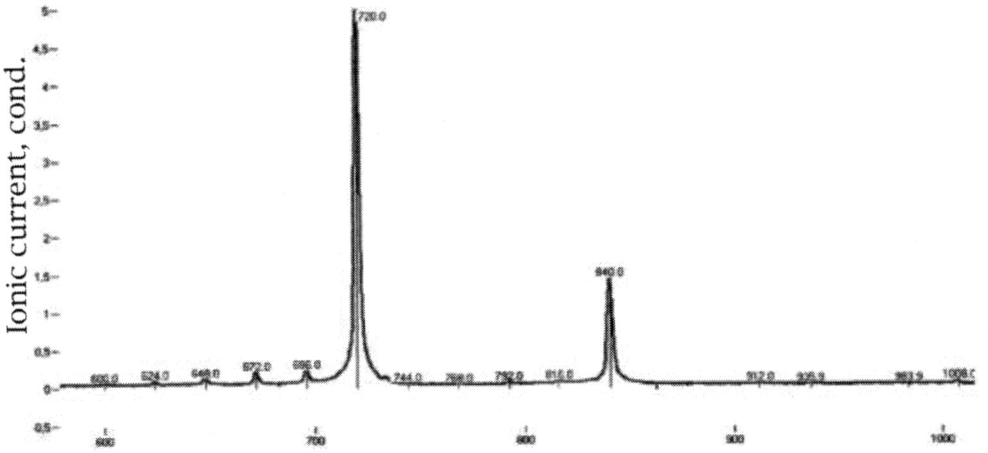

Figure 3. Fullerene mixture mass spectrometry a.e.м

Plasma synthesis of fullerenes is characterized by a high rate of processes, high quality of the resulting product and the possibility of organizing a continuous process. The peculiarity of this technology is the need to accurately maintain the specified technological parameters in the synthesis process, which directly affect the characteristics of the final material.

Due to the transience of the process, the use of a manual method for controlling the synthesis of carbon nanostructures is possible only under experimental conditions. The spread of the current strength in this case can reach up to 50% of the required value, which leads to low quality and yield of the final product, with a spread in a wide range of the resulting structures and sizes.

The use of an automated synthesis control system makes it possible to achieve the constancy of the necessary operating parameters in the plasma and ensure the stability of the technological conditions of the process, which is important in industrial production.

3 Results

The developed functional scheme of the automated control system for the synthesis of fullerenes by the method of plasma sublimation of graphite is shown in figure 5. The system performs automatic control and control with a high degree of accuracy and consistency of all the main technological parameters of the process, which provides the necessary stable conditions for the plasma synthesis of fullerenes in the working chamber and increases the efficiency of the entire process as a whole [18].

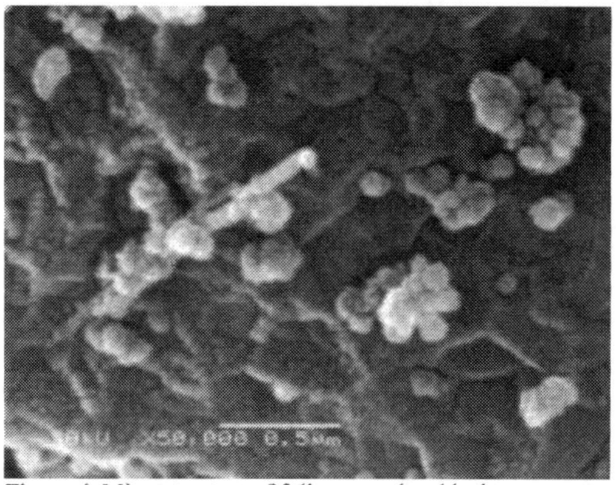

Figure 4. Microstructure of fullerene carbon black

The working chamber in which the synthesis process takes place is a sealed chamber made of high-strength steel, equipped with a lid with a lifting mechanism. Graphite electrode holders are installed inside the chamber, equipped with high-power current supply devices and a mechanism for moving the anode. The chamber is equipped with a water cooling system, an air pumping system and an inert gas supply. Maintaining a stable arc in the process is carried out by regulating the interelectrode distance.

All information about the process is transmitted via an analog-to-digital converter (ADC) to a personal computer (PC). The received data is processed by the PC information system, and based on their analysis, a control effect is developed. The control of the process of electric arc synthesis of fullerenes consists in maintaining a given current strength by changing the distance between graphite electrodes. To move the evaporated electrode (anode), a discrete control signal is fed through a digital-to-analog converter (DAC) to the input of a DC motor through a step-down reducer 3 to a stepper actuator that allows the anode to be positioned with the necessary accuracy. During the synthesis process, the pressure of the buffer medium in the synthesis chamber, temperature parameters and voltage on the electrodes are regulated. The measured process parameters are displayed in real time in a structured way on the PC screen and automatically saved. For this purpose, a package of process information software was developed, which also makes it possible to determine the technological conditions for the highest yield of the final material, depending on the type of used graphite raw materials [18].

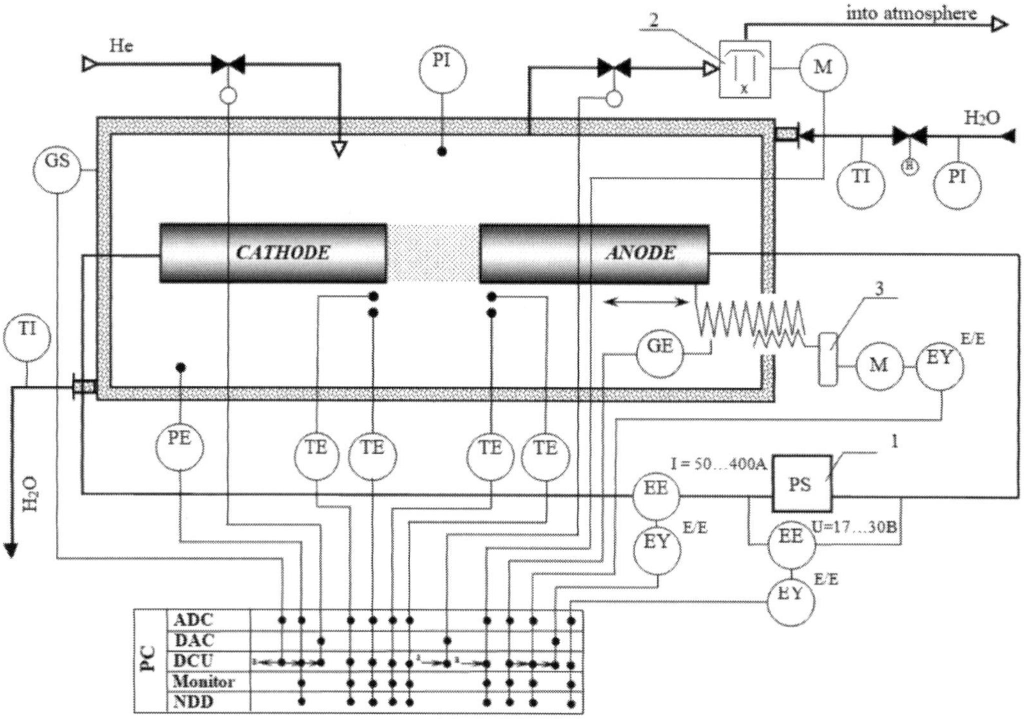

Figure 5. Functional diagram of an automated control system for the synthesis of fullerenes: 1 – power supply; 2 – diffusion pump; 3 – executive device; PC – computer; ADC – analog-to-digital converter; DAC – digital-to-analog converter; DCU – digital control unit; NDD – hard disk drive.

The developed functional scheme of the automated control system for fullerenes synthesis by plasma sublimation of graphite allowed to achieve a significant increase in the stability and accuracy of the technological process (Figure 5). The key element of the system is the precision control of the interelectrode distance, which directly affects the stability of the arc discharge and, consequently, the yield and quality of the synthesised fullerenes. Automation ensures maintenance of the specified synthesis parameters (current, voltage, inert gas pressure, cooling temperature) with a deviation not exceeding ±1%, which is significantly superior to manual control. This, in turn, leads to an increase in the yield of the target product (fullerenes C60 and C70) by 15-20% compared to non-automated methods, which makes the process more cost-effective for further use in various industries, including agricultural.

During the optimisation of the technological process using the developed mathematical model it was found that the optimal ratio of components of graphite raw materials for obtaining fullerene mixture with predominance of C60 is 95% of pure graphite and 5% of graphite with addition of nickel catalyst. The optimum parameters of plasma synthesis are achieved at a current strength of 360 ± 2 A, voltage of 26 ± 0.5 V, helium pressure of 55 ± 1 kPa and cooling water temperature of 20 ± 1 °C.

In addition to optimising the synthesis process, considerable effort has been devoted to investigating the application of the resulting fullerenes in the agricultural industry, in particular to improve crop yields and crop quality characteristics. For this purpose, field tests were carried out. Winter wheat and sunflower as some of the most important agricultural crops were chosen as research objects. Fullerenes were applied in the form of aqueous solution with concentrations of 0.001-0.01 mg/litre. The following treatment methods were used: pre-sowing seed treatment (wheat) and treatment of crops in the phase of 3-5 leaves (sunflower) by spraying a solution of fullerenes with a flow rate of 200 litres/ha. Control groups of seeds and crops were treated with distilled water. The studies were carried out in threefold repetition, which allowed obtaining statistically reliable data.

The results of the field trials revealed a positive effect of fullerenes on growth, development and, as a consequence, on crop yields. Thus, the yield of winter wheat in the experimental plots treated with fullerenes increased by 14%, reaching 4.3 t/ha, compared to 3.8 t/ha in the control plots. In addition to the increase in yield, there was also an improvement in grain quality parameters, in particular, an increase in protein content by 1.2% and gluten content by 1.5%. Similar results were obtained for sunflower, where fullerene treatment resulted in an 11% increase in seed yield (2.8 t/ha vs. 2.5 t/ha in the control) and a 0.8% increase in seed oil content. There was also observed an increase in resistance of fullerene-treated plants to such common diseases as powdery mildew (wheat) and false powdery mildew (sunflower), which allows reducing the use of chemical plant protection products. Thus, the introduction of the developed technology for the synthesis of fullerenes and their subsequent application in the agricultural industry opens up broad prospects for increasing the efficiency of agricultural production and improving the quality of manufactured products.

4 Conclusion

The possibilities of practical application of fullerene $C_{60} \div C_{70}$ in the creation of new materials necessary for industrial equipment and technological processes are considered.

The biological activity of fullerenes in interaction with microorganisms and biological objects allows them to be used to create materials with specific microbiological properties, radioprotectors. Hydrated solutions of fullerenes make it possible to increase the biological activity of nutrient media during the cultivation of microorganisms and the creation of products with special characteristics, to improve the quality of manufactured products. Fullerenes are a promising reinforcing filler for the creation of polymer composite materials. Introduction of fullerenes $C_{60} \div C_{70}$ into the matrix of polymer resins in the amount of 0,01 ÷ 5 % makes it possible, practically without increasing the weight, to obtain lightweight composite materials with a special structure endowed with a complex of improved properties.

The method of plasma synthesis of fullerenes used for reinforcement of polymer resins and hydrated solutions is considered. A functional scheme of an automated control system for the synthesis of fullerenes is proposed, which allows obtaining the largest amount of homogeneous nanostructured carbon material at the output, due to the high stability of parameters, monitoring and archiving of technological parameters in real time.

The results of field trials convincingly demonstrate the potential of fullerenes application in agriculture. Increased yields of winter wheat and sunflower, as well as improved quality characteristics of products, combined with increased plant resistance to diseases, allow us to consider fullerenes as a promising means of improving the efficiency and sustainability of agricultural production. Further research is aimed at optimising the concentrations and methods of application of fullerenes for different crops and regions, as well as at studying the mechanisms of their effect on plants at the molecular level, which will make it possible to reveal more fully the potential of these unique nanomaterials.

References

1. G.V. Abramov, A.N. Gavrilov, Modeling of carbon nanostructures synthesis in low-temperature plasma, Advanced Materials & Technologies. **1** (2019) 21–34.
2. S.V. Mishchenko, A.G. Tkachev, Carbon nanomaterials. Production, properties, application, Mechanical engineering, Moscow. (2008).
3. B.M. Ginzburg, M.V. Baidakova, O.F. Kireenko, D.G. Tochilnikov, A.A. Shepelevsky, The influence of fullerene C60, fullerene carbon black and other carbon materials on the boundary sliding friction of metals, Journal of Technical Physics. **70(12)** (2000) 87–97.

4. M.V. Uspenskaya, N.V. Sirotinkin, V.A. Gorsky, Yu.G. Goloshchapov, Compositions based on acrylate copolymers and fullerenes, Journal of Applied Chemistry. **79(5)** (2006) 870–872.
5. E.V. Anufrieva, M.G. Krakovyak, T.D. Anan'eva, T.N. Nekrasova, R.Yu. Smyslov, Interaction of polymers with fullerene C 60, Physics of the Solid State. **44(3)** (2002) 461–462.
6. A.L. Bychkov, E.I. Ryabchikova, K.G. Korolev, V.A. Bukhtoyarov, Synthesis of silver nanoparticles stabilized by the products of mechanically assisted hydrolysis of yeast biopolymers, Proceedings of Voronezh State University of Engineering Technologies, **81(1)** (2019) 238–246.
7. T.V. Shevchenko, Yu.V. Ustinova, V.P. Yustratov, M.S. Bezrukov, D.E. Vasilenko The use of fullerene in the storage and drying of apples, Storage and processing of agricultural raw materials. **2** (2020) 85–93.
8. N.O. Mchedlov-Petrosyan, Solutions of fullerene C60: colloidal aspect, Chemistry, physics and Surface technology. **1(1)** (2010) 19–37.
9. K.N. Semenov, N.A. Charykov, V.A. Keskinov, Fullerenol Synthesis and Identification. Properties of Fullerenol Water Solutions, Journal of Chemical & Engineering Data. **56** (2011) 230–239.
10. G.G. Panova, K.N. Semenov, O.A. Shilova et al., Water-soluble derivatives of fullerenes and silica nanocompositions as promising nanomaterials for use in crop production, Agro physics. **4** (2015) 37–48.
11. A.I. Mashanov, G.N. Churilov, N.N. Tipsina et al., Intensification of baking yeast with the help of water-soluble fullerene, Bulletin of KrasGAU. **2** (2020) 108–116.
12. L. Xie, F. Xu, F. Qiu et al., Single-Walled Carbon Nanotubes Functionalized with High Bonding Density of Polymer Layers and Enhanced Mechanical Properties of Composites. Macromolecules. **40(9)** (2007) 3296–3305.
13. P.M. Ajayan, J.M. Tour, Material Science: nanotube composites. Nature. **447** (2007) 1066–1068.
14. A.N. Gavrilov, E.A. Pologno, A.N. Ryazanov, Analysis of synthesis methods and industrial production of carbon nanotubes, FES: Finance. Economy. Strategy. Series: Innovative Economy: Human Dimension. **6** (2010) 14–19.
15. A.V. Makunin et al., Technological aspects of nanostructure synthesis by electric arc and gas pyrolytic methods, Physics and chemistry of materials processing. **6** (2010) 38–41.
16. A.E. Dutlov, V.M. Nekrasov, A.G. Sergeev, V.P. Bubnov, I.E. Kareev, Electric arc synthesis of soot with a high content of higher fullerenes in a «parallel arc», Journal of Technical Physics. **86(12)** (2016) 99–103.
17. I.E. Kareev, V.P. Bubnov, D.N. Fedutin, High-performance electric arc reactor for the synthesis of soot with a high content of endohedral metallofullerenes, Journal of Theoretical Physics. **79(11)** (2009) 134–137.
18. A.N Gavrilov, N.V. Sukhanova, A.E. Emelyanov, A.V. Ivanov, Fullerene production management system used for creation of hydrated solutions, IOP Conference Series: Earth and Environmental Science. **640** (2021) 062010.

Innovative tool for sustainable production and recycling of agro-industrial wastes

Oleg S. Turchanin [1*], *David R.* Grigoryan [1], *Irina V.* Ukraintseva [2], *Igor V.* Nazarov [2], *Tatiana N.* Tolstoukhova [2]

[1] Kuban State Agrarian University named after I. T. Trubilin, Krasnodar, Russia
[2] Azov-Black Sea Engineering Institute – branch of Don State Agrarian University in Zernograd, Zernograd, Russia
[3] Timiryazev Moscow Agricultural Academy, Russian State Agrarian University, Moscow, Russia

Abstract. This article explores the potential of using black lionfly larvae in the entomoindustry as an alternative to traditional animal husbandry. Traditional methods of food production are resource intensive and lead to environmental pollution. The use of insects, in particular black lionfly larvae, to process organic waste into valuable products such as fodder, biohumus and protein concentrates is a sustainable and environmentally friendly approach. Experiments have shown that the larvae efficiently recycle different types of waste, and less water and feed are required to produce 1kg of protein from the insects compared to beef protein production. The developed business model of the entomotechnology enterprise includes centralised and decentralised logistics and offers three options for business development. The most promising is a mixed approach, combining the production and sale of equipment and final products. It is noted that the market for entomoindustry equipment is underdeveloped, indicating potential for development. The study emphasises the economic and environmental feasibility of using black lionwort for sustainable production.

1 Introduction

Traditional methods of food production, namely animal husbandry, require huge inputs of resources such as water, feed and space. In turn, the waste generated during the production process leads to environmental pollution and negatively affects ecosystems. It is obvious that new, innovative approaches to food production that are more sustainable and efficient are needed.

One of the promising solutions to this problem is the entomoindustry, which involves the use of insects as a source of protein and other valuable substances [1]. The development of this industry can reduce the relative cost of food, increase its availability and reduce the negative impact on the environment. There are several ways to develop the entomoindustry. The first path involves the creation of large-scale production complexes that will provide significant volumes of products. This requires substantial investment and infrastructure development. The second way is to develop decentralised solutions that can be adapted to different conditions and needs. Both approaches have advantages and disadvantages. Large-scale production can provide lower production costs, but requires significant capital investment and may be less flexible to changing conditions. A decentralised approach allows adaptation to different types of waste and different levels of production, but may be less efficient at scale.

Currently, one of the most promising areas in the entomoindustry is the use of the black lion fly (Hermetia illucens). The larvae of this fly are able to efficiently process organic waste, turning it into valuable products such as animal feed, biohumus and protein concentrates. This not only solves the problem of waste utilisation, but also provides valuable products. This area is highly relevant in the context of the current challenges of improving production efficiency, reducing waste and ensuring sustainable development. The use of black lionfly larvae provides an opportunity for environmentally friendly food production, reducing the consumption of natural resources and reducing the harmful impact on the environment [2]. As a result, the commercialisation of black lionberry-based technologies offers the prospect of creating a profitable business while addressing a number of important environmental and food issues.

The aim of the paper presented in the attached file is to investigate the possibilities and prospects of using black lionfly larvae as a tool for creating innovative and sustainable production systems. The paper conducts an economic analysis and describes the business model of an entomotechnological enterprise that can provide the production of valuable products from organic waste. Various aspects including technological solutions, logistic schemes and market prospects are considered in the research process. As a result, the paper should demonstrate the economic feasibility and social

* Corresponding author: alexkgau@rambler.ru

significance of the introduction of technologies based on the black lion fly, as well as provide practical recommendations for the development of this area.

2 Method and materials

The laser technologies used in the study, although not directly related to agriculture, may be in demand in related fields, for example, in the processing of biomaterials and preparation of raw materials for bioreactors, as well as in the thermochemical decomposition of complex organic molecules [3] The general plan of experiments included varying the treatment parameters, such as power and duration of exposure, and studying the response of the material to these changes. The experiments were conducted using a LaserStar FiberStar 4400, which is a fibre laser with a wavelength of 1064 nm. This laser was used for surface treatment of the investigated materials. During the experiments on this equipment a series of pulses of laser radiation with the specified parameters of power, frequency and exposure time were applied. The positioning system installed on the laser complex ensured the accuracy of the beam movement on the sample surface. To measure the parameters of laser radiation a laser radiation power meter 'Ophir PD300-IR' was used. Power measurement was performed before the beginning of each series of experiments to calibrate the system. After processing, the surface of the samples was analysed using a confocal microscope 'Olympus LEXT OLS4100'. The microscope was used to measure the depth and width of laser exposure. In particular, the microscope was used to perform 3D surface scanning, which was then used for analysis and modelling. The modes of operation of the laser unit varied depending on the purpose of a particular experiment, mainly the power was set between 5 and 30 W, the frequency between 10 and 50 kHz, and the exposure time between 1 and 5 seconds. All experiments were performed at room temperature and normal atmospheric pressure.

3 Results and discussion

To effectively manage agricultural waste and ensure the sustainability of production, it is necessary to use modern GIS technologies and cartographic materials that allow optimising logistics and planning processes [4].

The conducted research was devoted to the analysis of innovative possibilities of application of black lion fly (Hermetia illucens) in entomoindustry [5]. The experimental work consisted of theoretical and applied economic analyses based on biotechnology using these flies. Black lionfly larvae were chosen as a research object because of their ability to increase their biomass hundreds of times within a few days by feeding on organic waste, making them a valuable resource for the production of animal feeds, medicines or even food additives [6]. In addition, the larval waste products, zoohumus, are a valuable fertiliser for agriculture.

Experimentally, it was found that black lionfly larvae, being omnivorous, are able to utilise a variety of wastes, including plant and food wastes, faeces and even effluents from pig farms. The study showed that the conversion ratio of feed to larval biomass can reach a ratio of 1 to 3 to 5, depending on the composition of the waste. In the pilot project, 1 kg of cabbage waste yielded 200 g of larvae suitable for use as animal feed. It should be noted that the data presented in this article differ somewhat from those published elsewhere. For example, studies conducted by the Canadian company Entomo Farms showed that 1 kg of food waste allows to obtain up to 300 g of larvae, and when feeding on some wastes, such as beer pellets, and up to 400 g. In addition, we have revealed the dependence of growth rate and biomass recruitment of larvae on ambient temperature in the experimental farm. It was noticed that at temperatures below 22 degrees Celsius the growth rate decreases significantly, and at temperatures above 27 degrees Celsius critical states of larvae and decreased biomass recruitment are observed. The optimum temperature regime according to the results of our experiments was 24-26 degrees Celsius, which is in agreement with the results of other similar studies.

As a result of the analysis, a synthetic graphical business model of entomotechnology entrepreneurship was developed. It demonstrates the different stages of the production process and logistic patterns of waste recycling using black lionberry. Two logistic models were proposed: a centralised model, which is more suitable for urban waste management [7], and a decentralised biotechnological complex using the black lionfly: Fly breeding and egg production (insectarium), production of young larvae from eggs (incubator), grinding of waste for feed, feeding of larvae, processing of waste by larvae, separation of adult larvae from zoohumus, drying of larvae, grinding of dry larvae into flour, and extrusion of larvae for the extraction of entomobulin, entomojir and chitin (Fig. 1). Within the framework of technological entrepreneurship, three business development options were identified: narrow (development and sale of equipment), broad (production of products using own equipment) and mixed (combination of production and sale of both equipment and final products). Calculated parameters of the biotechnological complex operation showed that the minimum level of larvae production is 0.1 tonnes per day while ensuring constant reproduction of the population, and the nominal module is capable of processing 1 tonne of waste per day. The maximum level of processing is not limited and depends on the number of modules used.

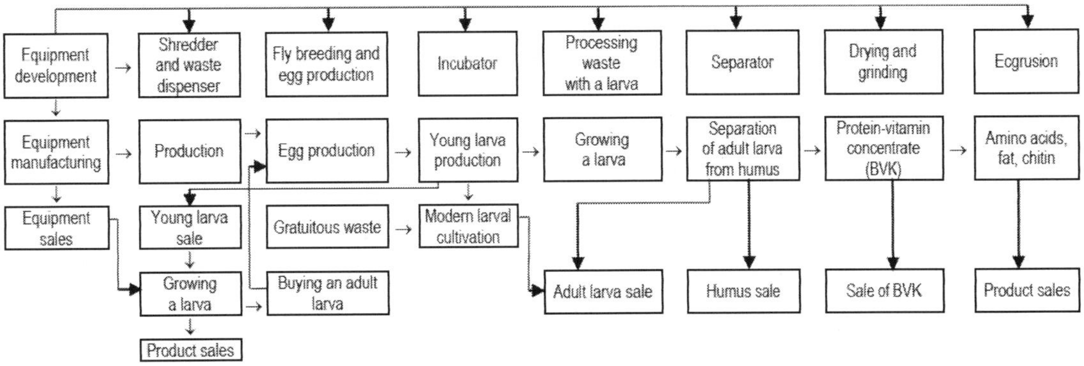

Figure 1. Agronomic model of entomotechnological entrepreneurship

The results indicate a high potential economic efficiency of the entomoindustry. The production of 1 kg of protein from insects requires 500 times less water and 12 times less feed than beef protein. This fact emphasises the ecological and economic feasibility of using black lionflies. In our experimental production we obtained slightly different data, for the production of 1 kg of insect protein requires 350 times less water and 10 times less feed than for the production of 1 kg of protein from beef.

An analysis of patent activity in Russia in the field of black lionfly use showed that the process of commercialisation of biotechnology using this fly started in 2015. We also found that 70% of registered patents in this area fall between 2018 and 2020. However, despite this, the market is developing quite slowly. The reason for this is the lack of developed infrastructure and the presence of a small number of specialised farms. At the same time, we have observed that the market for black lionfish breeding and processing equipment is very underdeveloped. Existing equipment either has low capacity, requires a lot of manual labour, or is expensive and difficult to operate. In this regard, the modular biotechnological complex developed by us has a number of advantages as it is characterised by compactness, flexibility and acceptable cost.

The analysis has shown that a mixed type of technological entrepreneurship, which combines the production and sale of both equipment [8-10] and final products (larvae, zoohumus and protein-vitamin concentrates), is promising. This will reduce risks and maximise profits. The results of the study were also compared with data from similar studies [11-13]. In particular, the data on the growth of the world market of feed proteins and deficit of feed protein were considered, which confirm the relevance and prospectiveness of the entomoindustry development. The study confirmed the high economic and ecological potential of using the black lion fly in the entomoindustry.

The concept of a closed cycle, where waste from one production becomes a resource for another, is an important aspect of sustainable development. An example of this approach is the utilisation of mining waste to produce useful products, which demonstrates the possibility of minimising waste and creating value from it [14-16].

4 Conclusion

The article shows the potential of using black lionfly larvae in entomoindustry as a promising area for sustainable food production and solving waste utilisation problems. Traditional methods of food production, especially animal husbandry, require significant resources and lead to environmental pollution. Entomoindustry offers an alternative approach by utilising insects such as the black lionfly to recycle organic waste into valuable products such as animal feed, biohumus and protein concentrates. The study shows that black lionfly larvae are able to efficiently process various types of waste, including plant and food waste as well as faeces. Economic analysis confirms the potential efficiency of entomoindustry, as significantly less water and feed are required to produce 1kg of protein from the insects compared to beef protein production. Experiments revealed the dependence of larval growth rate on temperature, with optimum values in the range of 24-26 degrees Celsius.

The authors have developed a business model for an entomotechnological enterprise including centralised and decentralised logistics schemes. Three variants of business development were proposed: narrow (sale of equipment), broad (production of products) and mixed (combination of production and sale). The analysis showed that the mixed approach is the most promising, allowing to reduce risks and increase profits. It was also noted that the market for entomoindustry equipment is underdeveloped, and the developed modular biotechnological complex has a number of advantages such as compactness, flexibility and acceptable cost. The study confirms the prospectivity and economic feasibility of introducing black lionberry-based technologies for sustainable and environmentally friendly production.

References

1. T.B. Tariq, Karishma, M. Umer, Mubeen-ur-Rehman. The potential of seaweed-derived polysaccharides as sustainable biostimulants in agriculture. Int. J. Biol. Macromol., **298,** 140009 (2025). DOI: 10.1016/j.ijbiomac.2025.140009.
2. Y. Chen, Y. Guo, X. He, B. Tan, Z. Liao, A. Chen, X. Gu, X. Li, X. Chen, B. Chen, S. Lin, W. Li, P. Hu, X. Zhu, W. Zhao, J. Niu. Comprehensive utilization of black soldier fly (Hermetia illucens) larvae: extraction, recovery and characterization of peptide, chitin and melanin and scaling-up trial. Sep. Purif. Technol., **361,** 131262 (2025). DOI: 10.1016/j.seppur.2024.131262.
3. O. A. Kolenchukov, A. Y. Mikhaulov, R. B. Sergienko. Forecasting the technical condition of thermochemical reactor systems. SOCAR Proceedings, **2023,** 123–130 (2023). https://doi.org/10.5510/OGP20230100814.
4. V. L. Tatarintsev, L. V. Lebedeva, G. D. Syzdykova, D. S. Inkarov. Management of agricultural land use sustainability at the local level using GIS technologies and special map material. Sustainable Development of Mountain Territories, **15,** 864–876 (2023). DOI: 10.21177/1998-4502-2023-15-4-864-876.
5. Q. Gao, L. Wang, Y. Fang, Y. Gao, L. Ma, X. Wang, Y. Li, X. Wu, Z. Du. Conservation agriculture boosts topsoil organic matter by restoring free lipids and lignin phenols biomarkers in distinct fractions. Soil Till. Res., **248,** 106463 (2025). DOI: 10.1016/j.still.2025.106463.
6. A. Ogunleye, D. Olude, A. Kehinde, V. Tanimonure, C.A. Wongnaa, O. Ajayi. Impact of credit constraints on adoption of biofortified food crop among farming households in Nigeria. Sustain. Futures, **9,** 100421 (2025). DOI: 10.1016/j.sftr.2024.100421.
7. V.V. Kondratiev, K.A. Bashmur. Circular Mining Wastes Management for Sustainable Production of Camellia sinensis (L.) O. Kuntze. Sustainability (Switzerland), **15,** 11671 (2023). DOI: 10.3390/su151511671.
8. V.I. Golik, Q. Mengxu. Tailings Utilization and Zinc Extraction Based on Mechanochemical Activation. Materials, **16,** 726 (2023). DOI: 10.3390/ma16020726.
9. V. L. Tatarintsev, L. M. Tatarintsev, F. K. Ermekov, Yu. S. Lisovskaya. Agroecological condition assessment of agricultural landscapes for their sustainability increase. Sustainable Development of Mountain Territories, **14,** 76–86 (2022). DOI: 10.21177/1998-4502-2022-14-1-76-86.
10. L. Yerkinbayeva, A. Myrzakarim, A. Moldagaliyeva, B. Kalymbek, R. Yerezhepkyzy, D. Nurmukhankyzy. THE ROLE OF LEGAL REGULATION IN PROMOTING ORGANIC AGRICULTURE TO ACHIEVE SUSTAINABLE FOOD SECURITY. J. Lifestyle SDG'S Rev., **5,** e04197 (2025). DOI: 10.47172/2965-730X.SDGsReview.v5.n01.pe04197.
11. P. Zhang, Q. Liu, H. Yao, T. Wang, B. Zeng, Y. Zou, S. Fu, K. Hu, H. Guo. Copper-functionalized hydrochar from swine manure with complexation sites for selective adsorption of tetracycline antibiotics. Sep. Purif. Technol., **359,** 130671 (2025). DOI: 10.1016/j.seppur.2024.130671.
12. A. Ogunleye, D. Olude, A. Kehinde, V. Tanimonure, C.A. Wongnaa, O. Ajayi. Impact of credit constraints on adoption of biofortified food crop among farming households in Nigeria. Sustain. Futures, **9,** 100421 (2025). DOI: 10.1016/j.sftr.2024.100421.
13. Y. Cao, Z. Yin, Y. Duan, R. Cao, G. Hu, Z. Liu. Research on improved sound recognition model for oestrus detection in sows. Comput. Electron. Agric., **231,** 109975 (2025). DOI: 10.1016/j.compag.2025.109975.
14. M.-Y. Li, W. Wang, H.-H. Yin, Y. Chen, M. Ashraf, H.-Y. Tao, S.-S. Li, W.-Y. Wang, C.-L. Yang, Y.-L. Xiao, L. Zhu, Y.-C. Xiong. The functional role of arbuscular mycorrhizal fungi in enhancing soil organic carbon stocks and stability in dryland. Soil Till. Res., **248,** 106443 (2025). DOI: 10.1016/j.still.2024.106443.
15. V.Yu. Konyukhov, T.A. Oparina, I.Y. Matasova, M.A. Modina. Ecologization of underground coal mining by means of ash use in backfill preparation. Min. Inform. Anal. Bull., **10,** 123-135 (2024). DOI: 10.25018/0236_1493_2024_10_0_123.
16. V.Yu Konyuhov, A.M. Gladkih, I.I. Galyautdinov, Y.D. Severina. Economic aspects of green technologies. IOP Conference Series: Earth and Environmental Science, **350,** 012036 (2019). DOI: 10.1088/1755-1315/350/1/012036.

BIO Web of Conferences **161**, 00012 (2025) https://doi.org/10.1051/bioconf/202516100012
FIES 2024

Results of experimental studies of a small-volume biogas plant

Ilnur Kh. Gayfullin*, *Bulat G.* Ziganshin, *Zulfiya M.* Khaliullina, *Olga S.* Semicheva, *Artyom V.* Shornikov

Kazan State Agrarian University, 65, Karla Marksa str., Kazan, 420015, Russia

Abstract. Utilisation of organic waste from agriculture is one of the most acute environmental problems that require immediate solution. This problem is relevant for the majority of farms and agricultural complexes, regardless of the number of animals. Violation of waste disposal rules can lead to contamination of soil and water bodies, as well as to negative consequences for human and animal health. This paper considers the method of anaerobic digestion of agricultural organic waste as an effective solution to this problem. Anaerobic digestion is a process in which organic matter decomposes without access to oxygen, resulting in the formation of biogas and high quality organic fertilisers. The processing of organic waste in a biogas plant contributes to sustainable agriculture. The aim of the research is to increase the energy efficiency of the organic waste utilisation process by developing the design and substantiating the main parameters of a low-volume biogas plant. Experimental studies were carried out according to the known methods on the newly developed biogas plant. The digestion temperature was maintained within the range of 50-55 °C. Experimental studies allowed to determine the rational design and technological parameters of the low-volume biogas plant.

1 Introduction

In today's agricultural business, organic animal waste is a serious problem. The treatment of manure effluent from pig and poultry farms is an important challenge that is solved by various chemical methods such as liquid chlorine, anhydrous ammonia (NH_3) and nitrogen. Many dairy and meat plants are faced with excess manure and litter. These wastes, while having a high potential as fertilisers, often go unused. However, their direct application to soil is limited due to toxicity. Today, organic wastes are processed using various technologies (composting, drying, vermicomposting, etc.), but still the issues of environmental pollution are not completely solved [1, 2].

Nevertheless, one of the most effective ways of manure utilisation is the thermal method using biogas plants. This approach not only efficiently utilises agricultural waste, but also creates valuable fuel and biofertilisers that contribute to the improvement of soil conditions. Anaerobic digestion of organic waste is a highly efficient method with many advantages over aerobic processes. Anaerobic processes produce methane, which is an environmentally friendly biofuel. It is estimated that the costs of aerobic processes far exceed the requirements of anaerobic processes, making the latter economically more favourable. Anaerobic fermentation promotes the sequential decomposition of complex organic matter into simpler compounds, which converts a significant portion of carbon into biogas and carbon dioxide. Anaerobic methane-producing bacteria, which are highly sensitive to environmental conditions, play a key role in this process. The product of the methanogenic stage is methane (CH_4). The destruction of organic acids results in an increase in the pH of the medium, leading to an alkaline reaction. Therefore, the methanogenic stage is often referred to as 'alkaline fermentation' [3, 4].

According to literature sources [5, 6], more than 45% of livestock complexes, the manure generated by the activity is not exported to their fields as organic fertiliser. Eventually, organic wastes of animals and birds are accumulated in lagoons, often located next to natural water bodies.

Since ancient times, manure, especially horse manure, has been used as a fertiliser, but in recent years microbiological conversion of manure into biogas has become increasingly popular. The activities of agricultural enterprises of the agro-industrial complex lead to alarming changes in the soil ecosystem, where the humus content is noticeably decreasing and mineral nutrient reserves are being depleted. These processes can provoke a disturbance of the natural balance, opening the way to land degradation. The most important aspect of sustainable agriculture is the production of biogas and biofertilisers that fulfil strict requirements. Key conditions are the absence of nutrient losses of nitrogen, phosphorus and potassium, as well as the need to decontaminate and deodorise the fertiliser produced. Furthermore, biofertilisers must be free of weed seeds, viable eggs or helminth larvae. It is also important to comply with the norms of heavy metal concentrations and maximum pesticide residues in the products, which is regulated by the Russian Federal Service for Supervision of Consumer Rights Protection and Human Welfare. This guarantees not only safety for the

* Corresponding author: ilnur-gai@yandex.ru

ecosystem, but also ensures soil fertility necessary for further sustainable agricultural development [7, 8].

The process of manure conversion in biogas plants allows the production of high quality biofuel, and also helps to minimise nitrogen losses by converting it into mineral forms, as well as effectively destroying weed seeds and helminth eggs. As a result of recycling, macronutrients become available for plant uptake. During this process, pathogenic microorganisms, weed seeds and other undesirable components are completely destroyed, thus improving soil quality and increasing its agronomic value. Animal waste treatment is aimed at obtaining useful products and waste neutralisation, which allows reducing their hazard class to IV-V [9, 10].

2 Materials and Methods

The low-volume biogas plant developed at Kazan State University of Economics (Figure 1) consists of a reactor, stirrer, heating element, programmer, gas holder, gas meter, compressor, control devices, burner and other elements. To start up the biogas plant, the substrate to be processed is first prepared and then loaded into the reactor. The process of loading the prepared substrate into the reactor can be carried out in two main ways: manually, through the top of the reactor, or automatically, using pumping equipment connected to the outlet of the plant. For the preparatory part of the start-up, the reactor must be connected to an electrical network with 220 V voltage, which ensures its operation and control. During the operation of the biogas plant, it is important to maintain a stable temperature regime, as this critically affects the efficiency of methane formation, during which microorganisms decompose organic matter and produce methane. The temperature in the reactor must be within optimum values to ensure the activity of the methanogenic bacteria. Our biogas plant uses a heating element for heating to minimise heat loss. Once the reactor is started, biogas production begins. The biogas produced is accumulated directly in the reactor, after which it passes through a water gate and enters the small gas holder. The water gate prevents the biogas from flowing back into the reactor, which ensures the safety of the plant and prevents possible emergencies. In addition, the presence of gas bubbles in the water seal serves as a visual indicator of the active gassing process, which allows operators to monitor the efficiency of the biogas plant. The volume of biogas produced is recorded by a gas meter, which provides accurate data on the capacity of the plant. The biogas is then forcibly pumped to the gasholder through a filter cleaner to remove various impurities and ensure the quality of the biogas. A low speed compressor is used for pumping, which ensures safe pumping of biogas.

To increase the safety of the plant, a safety valve is provided to prevent over-pressurisation of the gasholder. The vent valve automatically operates when the set pressure is exceeded, which protects the equipment and prevents emergencies.

The continuous burning of biogas, which is observed during the flammability test, is an important quality indicator. Several factors should be considered during the flammability test, including flame colour, flame stability and burn time. If the biogas burns with an even, bright flame with no sudden fluctuations, this indicates a high content of methane, which is the main component of biogas. Stable and continuous combustion indicates that there are no impurities that could degrade the quality of the biogas. The use of biogas as an alternative energy source not only contributes to the reduction of greenhouse gas emissions but also allows efficient waste utilisation, making it a promising solution for sustainable development. The results of the flammability test of biogas not only confirm its high quality, but also emphasise its potential as an energy source for various needs, including heating and electricity generation. The process from substrate loading to biogas production and utilisation is a complex and multi-stage system that requires careful monitoring and control to maximise efficiency and safety.

The resulting biofertiliser is removed by gravity through an outlet or forced by a pump.

Figure 1. Small-volume biogas plant: 1 - reactor; 2 - small gas holder; 3 - burner; 4 - electro-contact manometer; 5 - gas holder; 6 - bimetallic thermometer; 7 - temperature regulator; 8 - programmer OVEN PR 110; 9 - safety valve.

3 Results and discussion

In order to achieve an efficient biogas production process, it is necessary to ensure good substrate mixing, which affects the efficiency of methanogenesis. The torque generated on the agitator frame is due to the differences in velocity between the frame and the substrate being processed. This torque can be quantified based on the forces acting on the frame. In the simplest case, when the agitator rotates with a blade, which is a frame of height (h_f) and the axis of rotation of the frame coincides with the axis of the cylinder, a force is applied to an elementary section of the frame length, which creates a torque [11].

$$dM_{\text{rot}} = \varsigma_f \frac{\rho u^2(r)}{2} h_f r dr, \quad (1)$$

where ς_f – hydraulic resistance coefficient of the frame sides (according to the reference book); ρ - substrate density, kg/m; $u(r)$ - velocity of the frame edge flow at distance r, m/s; h_f - frame height, m; r - half of the frame width, m

The moment of forces resulting from the hydraulic resistance of the cylindrical part of the biogas plant reactor wall is determined considering the influence of various factors such as flow velocity, viscosity of the working substrate and geometrical parameters of the reactor.

$$M_{Wall} = S_{cyl} \frac{D}{2} \cdot \tau. \quad (2)$$

$$S_{cyl} = LH, \quad (3)$$

Where S_{cyl} wetted wall area of the cylindrical part of the reactor, m^2; H - reactor height, m; τ - tangential stresses, Pa; L - wetted perimeter of the cylindrical part of the reactor, m $(L = \pi D)$.

The reactor bottom resistance moment M_{bot} can be determined by integrating the expression describing the distribution of forces acting on the bottom. This requires integration between 0 and $\frac{D}{2}$, which represents the radius.

$$dM_{bot} = \pi\rho\varsigma_f V^2(r) r^2 dr, \quad (4)$$

where $V = kr$. k is the correction factor, s^2; r is the radius, m.

$$M_{bot} = \pi\rho\varsigma_f V^2 \frac{2}{D} \int_0^{D/2} V^2(r) r^2 dr. \quad (5)$$

Integrating expression (5) within the range from 0 to $\frac{D}{2}$, we obtain

$$M_{bot} = \frac{1}{90} \pi\rho\varsigma_f \kappa^2 D^4, \quad (6)$$

where k – correction factor, s^2.

To determine the total moment of resistance of the reactor vessel, the geometric parameters of the structure, substrate properties, and operating conditions must be taken into account:

$$M_{ves} = M_{CT} + M_{bot} =$$
$$= \frac{1}{8} LHD^2 \varsigma_f \rho\kappa + \frac{1}{90} \pi\rho\varsigma_f \kappa^2 D^4 = \quad (7)$$
$$= \pi D^3 \varsigma_f \rho\kappa \left(\frac{H}{8} + \frac{D\kappa}{90} \right).$$

It follows from expressions (6) and (7) that in order to reduce the momentum of hydraulic resistance forces of the reactor wall, as well as to reduce the energy consumption of the process, it is necessary to strive to minimise the wetted area of the cylindrical part of the reactor [11]. Achieving this at a given reactor volume (V) is possible solely by optimising the D/H ratio.

With a fixed reactor volume, the surface area of the reactor can be determined using the following expression:

$$S_{\text{surf}} = \left(\frac{D^2}{4} + \frac{DH}{2} \right), \quad (8)$$

differentiating expression (8) by $\frac{dS}{dD}$ we obtain,

$$\frac{dS}{dD} = \frac{\pi}{4} \left(2D - \frac{8V}{\pi D^2} \right) = 0,$$
$$2D - \frac{8V}{\pi D^2} = 0,$$
$$2D = \frac{8V}{\pi D^2},$$
$$D^3 = \frac{4V}{\pi},$$
$$D = H.$$

Based on the presented mathematical calculations [11, 12] the height of the reactor filling should be equal to its diameter. Such ratio of the reactor favours uniform distribution of microbial mass and nutrients, which, in turn, provides maximum biogas yield.

To evaluate the biogas production from various organic wastes under thermophilic digestion conditions, field experiments were conducted and the results are presented graphically in Figure 2. Prior to each experiment, a recyclable substrate including cattle (cattle) manure, pig, horse and poultry manure was prepared. These diverse components served as the basis for an in-depth study of the biogas potential, revealing the diversity of processes occurring in the thermophilic environment. A unique portion of substrate was incorporated into each substrate batch to track the dynamics of outgassing and identify optimal conditions for biochemical decomposition.

The prepared substrate was then loaded into a reactor where digestion took place. During the experiments, the temperature was gradually increased and kept constant after reaching 55 °C, which is the optimum regime for thermophilic digestion. Periodic stirring of the substrate was carried out to break the crust formed, which contributed to a more uniform distribution of microorganisms and improved the biogas formation process. The biogas output was recorded using a gas meter to accurately measure the volume of biogas produced during the experiment. The biogas produced was then pumped to a gas holder for further analysis and utilisation.

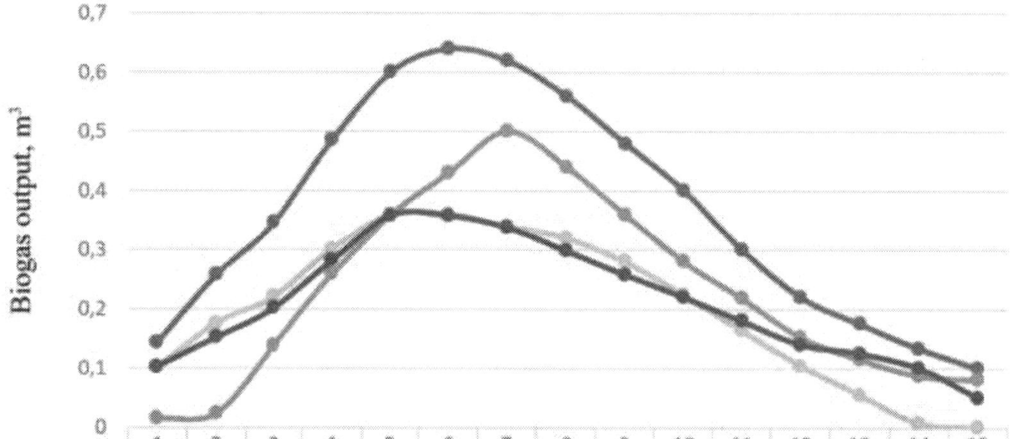

	1	2	3	4	5	6	7	8	9	10	11	12	13	14	15
Pig	0,143	0,258	0,346	0,486	0,6	0,64	0,62	0,56	0,48	0,4	0,3	0,22	0,175	0,133	0,1
Chicken	0,014	0,023	0,139	0,259	0,358	0,43	0,5	0,44	0,36	0,28	0,218	0,152	0,115	0,087	0,08
Horses	0,102	0,176	0,221	0,301	0,36	0,36	0,34	0,32	0,28	0,224	0,164	0,104	0,055	0,007	5E-04
CATTLE	0,103	0,152	0,202	0,282	0,358	0,358	0,338	0,298	0,258	0,22	0,18	0,14	0,125	0,1	0,05

Figure 2. Experimental results when thermophilic digestion is used

Experimental dependences of the volume of biogas release at the low-volume biogas plant were established in the process of thermophilic digestion of various types of organic waste, which are presented in Fig. 2. The results of the research allow us to draw conclusions about the potential of utilisation of these wastes for renewable energy sources.

4 Conclusion

To analyse the presented figure showing the results of the thermophilic digestion regime, attention should be paid to the coordinate axes: the X axis denotes the consecutive cyclic digestion days (from 1 to 15) and the Y axis demonstrates the volume of biogas released in cubic metres. Most notable is the peak of biogas yield observed on day 6, when the pig manure substrate reaches a maximum value of 0.64 m³, while chicken and horse waste show results of about 0.5 m³ and 0.37 m³, respectively. Pig manure consistently leads in terms of biogas released throughout the experiment, while cattle manure substrate shows lower biogas yield. The graph shows that the biogas yield starts from the first day, reaches a maximum on day 6-7, and then gradually decreases, approaching zero by day 15. Thus, the thermophilic regime is most effective for pig manure, while it is less productive for cattle waste. It is recommended to optimise the digestion process, paying attention to timing and substrate type, in order to achieve maximum biogas yield.

Acknowledgements

The work is supported by the grant (№ 142/2024 - PD₁ from 16.12.2024) to young candidates of sciences (postdoctoral fellows) for the purpose of defending a doctoral dissertation, performing research work, as well as performing labour functions in scientific and educational organisations of the Republic of Tatarstan.

References

1. W.L. Pambudi, A. Pertiwiningrum, M.A. Wuri, L.M. Yusiati, Key Eng. Mater., **884**, 104–108 (2021)
2. D. Syromyatnikov, V. Druzyanova, A. Beloglazov, Int. J. of Renew. Energy Devel., **10(4)**, 827–837 (2021)
3. L. Feng, F. Liang, L. Xu, Waste Manag., **133**, 1–9 (2021)
4. N. Lawson, M. Alvarado-Morales, P. Tsapekos, I. Angelidaki, Energ., **14(24)**, 8252 (2021)
5. W.L. Pambudi, A. Pertiwiningrum, M.A. Wuri, L.M. Yusiati, Key Eng. Mater., 104–108 (2021) (повтор п. 1)
6. E.T. Abilmazhinov, A.Z. Akimzhanov, E.Y. Shaiakhmetov, Euras. Phys. Techn. J., **3(37)**, 76–82 (2021)
7. A.V. Demin, R.Y. Dyganova, N.N. Fakhreev, Ecology and Industry of Russia, **22(5)**, 50-53 (2018)

* Corresponding author: ilnur-gai@yandex.ru

8. M.V. Parshikova, N.A. Aimashev, K.M. Ovchinnikov, Vestnik NGIEI, **10(149),** 28-42 (2023)
9. V.P. Druzyanova, S.A. Petrova, M.K. Okhlopkova, Yu.A. Sergeev (Dyna, BilBao, Spain, 2018), pp. 398-403
10. M.V. Parshikova, Vestnik NSTIEI, **19**-31 (2023)
11. I.Kh. Gaifullin, B.G. Ziganshin, I.N. Safiullin, BIO Web of Conf., **48,** 00019 (2022)
12. F.S. Sibagatullin, Z.M. Khaliullina, R.R. Minnikhanov, I.Kh. Gaifullin, BIO Web of Conf., **37,** 00048 (2021)

Optimisation of anaerobic digestion of agricultural waste by simulation modelling

I.Yu. Matasova [1], *Aleksey V.* Masenko [2*], *David R.* Grigoryan [2], *Anna N.* Tokareva [3], *Maria S.* Demchenko [3]

[1] Admiral Ushakov Maritime State University, Novorossiysk, 353924, Russia
[2] Kuban State Agrarian University named after I. T. Trubilin, Krasnodar, Russia
[3] Azov-Black Sea Engineering Institute – branch of Don State Agrarian University in Zernograd, Zernograd, Russia

Abstract. The article deals with the problem of agricultural waste processing, focusing on anaerobic digestion for biogas and fertiliser production. The study is devoted to the optimisation of this process on the example of poultry manure, using a simulation-statistical approach. The developed model of the biotechnological system includes waste separation into fractions, pre-digestion, methane digestion and anaerobic filtration. Substrate loading dose and substrate moisture content were found to have the greatest influence on energy release. Experiments showed that an increase in loading dose leads to an increase in biogas yield. Modelling allowed us to obtain an energy release equation that takes into account humidity, loading dose and digestion temperature. The maximum energy yield is achieved at a certain combination of these parameters. The study also showed that the use of the proposed biotechnological system can contribute to meeting part of the region's heat energy needs through the production of biogas from waste. Measures to optimise waste collection and processing, including the formation of databases and simulation modelling, have been proposed.

1 Introduction

Globally, agriculture is a major source of both organic and inorganic waste, which, if not properly managed, can have a significant negative impact on the environment. The increase in the volume of these wastes, especially in the context of growing agricultural production, requires the search for sustainable and efficient methods of their processing. At the same time, it is important to note that in order to achieve sustainable agriculture, it is necessary to consider not only recycling methods, but also the overall ways of development of the industry [1]. This can make a significant contribution to the sustainability of the agro-industrial complex and the transition to a closed-loop economy.

Currently, there are several approaches to solving the problem of processing agricultural waste, each of which has its own advantages and disadvantages. Traditional methods such as composting and direct land application have relatively low costs, but are often accompanied by nutrient loss and can be a source of environmental pollution. Thermal treatment of waste, including incineration and pyrolysis, provides energy but requires significant capital expenditure and can generate emissions. In contrast, anaerobic digestion, or methane digestion, is a promising method that not only recycles organic waste but also produces biogas, a renewable energy source, and organic fertiliser. This process, although more complex, allows a more efficient utilisation of the waste potential. However, the implementation of anaerobic technology requires detailed study, analysis and optimisation of its individual stages.

In the context of the above, the direction of solving the problem of processing organic agricultural waste through anaerobic digestion with subsequent production of biogas and organic fertilisers seems particularly relevant and important. Firstly, it allows not only to reduce the volume of waste, but also to obtain valuable products that can be used in agricultural production and energy. Secondly, the application of anaerobic technology can reduce the dependence of the agro-industrial complex on fossil fuels and chemical fertilisers, thus increasing its sustainability and competitiveness. Thirdly, anaerobic digestion contributes to the reduction of greenhouse gas emissions and other pollutants, which is in line with the goals of sustainable development and environmental safety [2, 3]. Analysing the carbon footprint of various production processes is of key importance in the transition to sustainable practices in agribusiness and other industries [4, 5]. The development of efficient and economically feasible technologies for anaerobic digestion of organic waste and their integration into production cycles is one of the key areas for achieving agricultural sustainability. In particular, it is extremely important to study the peculiarities of anaerobic digestion of various types of organic waste, as well as to analyse and model the

* Corresponding author: almasenko@yandex.ru

processes occurring within such a biotechnological system.

Thus, the purpose of this work is to study and model the processes of anaerobic digestion of organ-containing agricultural by-products, namely, the example of bioconversion of poultry manure, using a simulation and statistical approach. The use of mathematical models, such as neural networks, allows predicting and analysing various processes, including in the field of agro-technologies [6]. In the course of the study, a block diagram of the simulation method of statistical approximation was developed to optimise the process of anaerobic digestion and determine the most significant factors affecting the biogas yield and the quality of organic fertilizer. This will provide the knowledge necessary for the creation of effective biotechnological systems of organic waste processing.

2 Method and materials

The general plan of the experimental work carried out in this paper included several stages. At the first stage, a mathematical model of the biotechnological system describing the process of anaerobic digestion of agricultural by-products was developed. Then, a series of simulation experiments were conducted using this model. The experiments were conducted by varying various factors such as substrate moisture content and substrate loading dose into the reactor. In the next step, the results obtained were analysed using mathematical statistics techniques. At the final stage, the block diagram of the model was formed.

The main part of the experimental work was carried out through simulation modelling, where various parameters of anaerobic digestion, such as substrate composition and temperature conditions, were varied to reveal their influence on energy release. In order to create efficient systems, it is necessary to use methods for predicting the technical condition of the plants [7]. During the experiments, the dependence of biogas yield on various factors such as substrate humidity, loading dose, temperature and others was studied. Methods of mathematical statistics were used to analyse the data, which allowed us to identify significant factors and establish the structure of interrelationships between them. These methods involve the use of software tools for statistical processing of data, but the specific software is not specified. Based on the data obtained, a mathematical model describing the dependence of energy release on various parameters was constructed. The results of simulation modelling were used to analyse and optimise the anaerobic digestion process.

3 Results and discussion

An innovative biotechnological system (BTS) based on the separate treatment of solid and liquid fractions was proposed as part of the study of anaerobic digestion of organic waste from agricultural production. The experimental work was organised as follows. First, the feedstock, which was a mixture of agricultural by-products, was fed into a collection tank. Then, with the help of a faecal pump, the feedstock was directed to the first stage fractionation device. Here the separation into solid and liquid phases took place with a separation efficiency of 45-75% for the solid fraction, with a moisture content of 85-90% for the latter (Fig. 1).

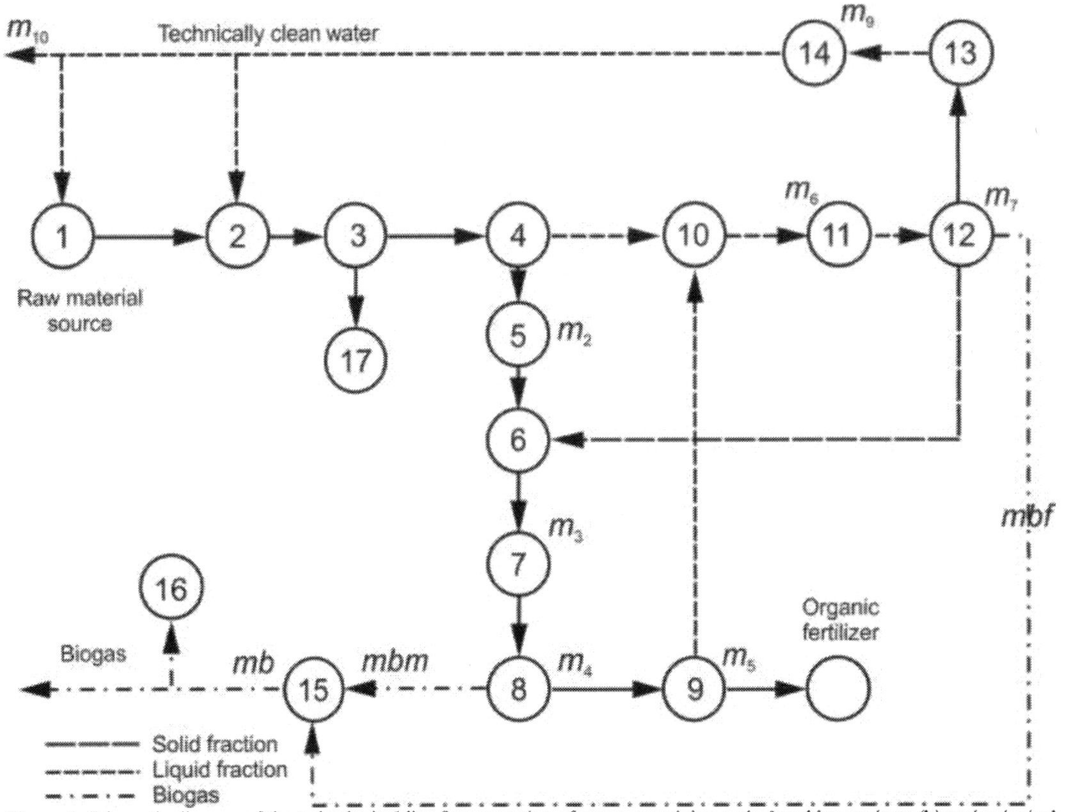

Figure 1. Schematic diagram of the technological line for processing of organ-containing agricultural by-products (biotechnological system (BTS)): 1 - raw material source; 2 - raw material collection tank; 3, 7, 10, 14 - faecal pumps; 4 - separation device into fractions of the first stage; 5 - solid fraction shredder; 6 - digester pre-digester tank; 8 - digester; 9 - second stage fractionation device; 10 - anaerobic filter precipitator tank; 12 - anaerobic filter; 13 - technical clean water tank; 15 - biogas gas holder; 16 - combined cycle power plant; 17 - sanitary storage facility.

The solid fraction was then pulverised and transferred to the digester's digestion pre-digester tank. In this tank, it was mixed with anaerobic microflora-contaminated sludge from the anaerobic filter, which helped to intensify the methane digestion process. To create optimal digestion conditions, the solid fraction was heated to the required temperature. Then, once a day, the mass was fed to the digester, where it was kept for 10-15 days. This resulted in decomposition of organic matter by 35-45% with biogas formation. The obtained sludge was fed to the second stage fractionation device, where additional separation into solid and liquid phases was performed. The solid fraction, with moisture content of 65-75% was used as fertiliser, and the liquid fraction was fed to the anaerobic filter's predigestion tank. Here, the liquid phase was mixed with the liquid fraction obtained after the first separation step and heated. The mixture was then fed into the anaerobic filter, where it was kept for one and a half to two days. In the anaerobic filter, due to the intensification of the digestion process, the degree of decomposition of organic matter increased to 75-80%. Further, the liquid went to the tank of conditionally clean water, where after purification part of the water was returned for technological needs, and the rest could be used for irrigation. The released biogas, which was a mixture of methane (about 60%) and carbon dioxide (about 40%), was collected in a gas holder and further used for energy production.

The effect of various factors on the anaerobic digestion process, in particular the moisture content of the treated substrate and the loading dose, was investigated experimentally. A non-symmetric factorised design of experiment was used for the study. As a result of statistical processing of the data, the energy release equation was derived. The study showed that the loading dose has the greatest influence on the energy release, with the biogas yield also increasing with increasing loading dose. In the energy release model (Fig. 2), in addition to substrate moisture and loading dose, a third factor was introduced, namely the temperature regime of digestion.

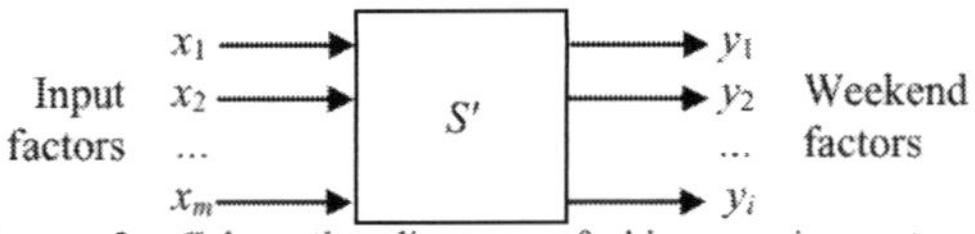

Figure 2. Cybernetic diagram of bioenergetic system connections

The system S reacts differently to the choice of different levels of factors (variables). In particular, the relationships between input and output factors can be characterised by regression models of the form [4].

$$y_{\alpha i} = \sum_{j=0}^{k} \theta_{\alpha i} x_{ji} + \sum_{\substack{j=1 \\ j<\beta}}^{k} \theta_{\alpha j\beta} x_{ji} x_{\beta i} + \sum_{\substack{\gamma=1 \\ \gamma\neq\alpha}}^{\ell-1} \theta_{\alpha(\gamma+q)} y_{ji}. \quad (1)$$

System S can also be represented as a system of equations (not always linear, one-dimensional) of the form

$$Y_1 = \begin{pmatrix} y_{i1} \\ y_{i2} \\ - \\ y_{i\ell} \end{pmatrix} = \begin{pmatrix} \theta_1^1 x_{i1} + \theta_2^1 x_{i2} + \ldots + \theta_m^1 x_{im} + \ldots + \varepsilon_{i1}^1 \\ \theta_1^2 x_{i1} + \theta_2^2 x_{i2} + \ldots + \theta_m^2 x_{im} + \ldots + \varepsilon_{i1}^2 \\ ----------------- \\ \theta_1^\ell x_{i1} + \theta_2^\ell x_{i2} + \ldots + \theta_m^\ell x_{im} + \ldots + \varepsilon_{i1}^\ell \end{pmatrix}.$$

Studies of the model confirmed that an increase in substrate moisture content promotes greater conversion of ash-free matter into biogas, and the optimal temperature regime allows to obtain the maximum energy yield. The energy release equation obtained on the basis of modelling and experiments is as follows:

$y_1 = 0{,}07 + 0{,}765\ x_2 + 0{,}136$

x_1x_2, where y_1 - energy release; x_1 - humidity of the treated substrate; x_2 - substrate loading dose into the reactor. The regression coefficient for the loading dose is 0.765, which shows its greatest influence on the energy release, it was also found that the interaction effect of humidity and loading dose is significant, it is equal to 0.136. Another obtained model describing the energy release is as follows:

$y_2 = 0{,}472x_1 + 0{,}375x_2 - 0{,}113x_3$,

where y_2 - energy yield; x_1 - substrate moisture; x_2 - loading dose; x_3 - temperature. In this case, substrate moisture has the greatest influence on energy yield with a regression coefficient of 0.472. At the same time, the maximum energy yield was calculated, which is achieved at a substrate loading dose of 5%, substrate humidity of 93% and reactor volume of 473 m3. Based on these data we can conclude that in anaerobic digestion of agricultural waste it is necessary to maintain optimal substrate humidity and temperature, as well as the optimal dose of reactor loading to achieve maximum energy yield.

Analysis of the results of the study showed that the proposed biotechnological system, as well as the obtained regression models, allow us to identify significant factors that influence the process of anaerobic digestion, as well as to establish the structure of relationships between them. This requires detailed analyses and crop planning using bibliometric analysis [8]. Moreover, the results of the work indicate the possibility of using simulation modelling to control the technological process of anaerobic digestion. This is an important advantage of the developed methodology, as it allows modelling and optimising technological processes without conducting expensive and time-consuming full-scale experiments. It is worth noting that in studies by other authors, a significant influence of loading dose and moisture content on biogas yield was also established. At the same time, in the conducted study the energy yield model was also constructed and evaluated, which takes into account not only the loading dose and substrate humidity, but also the digestion temperature, which expands the possibilities of using the obtained model in practice. For example, in a similar study it was shown that increasing the substrate moisture content from 75% to 90% increases the biogas yield by 15-20%, and increasing the temperature from 20°C to 35°C increases the biogas yield by 20-25%. It is worth noting that in our study a slightly higher biogas yield was obtained due to the use of a more efficient waste treatment technology. We also conducted studies to determine the potential volume and quality of biogas in poultry litter and effluent from the pig farm, which showed that the total potential biogas production can reach 22 million m3 per year, which allows to meet about 6% of the total heat energy demand in the region. Assessment of the impact of mountain factors on agroecological processes helps to more accurately assess the region's potential for development [9]. These results confirm the possibility of using a biotechnological system as an energy source and, importantly, allow the utilisation of agricultural by-products, simultaneously obtaining valuable fertiliser and clean water. However, in practice, there are difficulties in realising such projects, such as the collection and transportation of by-products and the need to ensure the stability of the technological process. To solve these problems, it is proposed to form balances of agricultural by-products and use regulatory bases for their formation in the sectoral and geographical context. This will allow, firstly, to effectively plan the production capacity of bioenergy plants, and secondly, to optimise the costs and logistics of raw material supply. In the future, for more effective management of the process, it is proposed to use a unified information base, which will allow planning and forecasting the development of the agro-industrial complex for the long term, as well as to use simulation models to identify inefficient elements, as well as to forecast the development of real situations. Assessment of the ecological state of agro-landscapes at the local level can provide more accurate data for planning and management [10]. This, in turn, will improve the efficiency of resource utilisation and reduce the negative impact on the environment.

4 Conclusion

The system of agro-industrial waste processing proposed in this work is based on separate treatment of solid and liquid fractions of waste. The solid fraction is subjected to grinding and methane digestion in a digester, and the liquid fraction is subjected to anaerobic filtration. At the same time, it is necessary to take into account the importance of creating organic fertilisers and ensuring soil fertility through waste utilisation [11]. Simulation modelling was used to optimise the process and identify factors affecting biogas yield.

It was found experimentally that the substrate loading dose has the greatest influence on the energy yield, and the interaction between substrate moisture and loading dose is also significant. In the course of modelling it was determined that an increase in moisture content promotes a more complete conversion of organic matter into biogas, and maintenance of optimum temperature ensures maximum energy yield. Based on the obtained data, mathematical models describing the dependence of energy release and energy yield on various process parameters were developed. The developed biotechnological system allows not only to reduce the volume of waste, but also to produce biogas, which can be used for energy production, and organic fertilisers. The creation of a model for agro-ecological assessment of an area can help in creating sustainable production systems [12]. It has also been found that the potential for biogas production from poultry manure and effluent from pig farms can provide a significant share of thermal energy needs in the region. However, there are difficulties in the implementation of such projects related to the collection and transport of by-products, as well as ensuring the stability of the process. To solve these problems, it is proposed to form balances of by-products and use a unified information base for planning and forecasting the development of the agro-industrial complex. The use of anaerobic processing technologies, in this case, correspond to the general trends in agrarian changes and adaptation to climatic changes [13]. In general, the results of the study confirm the prospectivity of anaerobic digestion of agricultural organic wastes as a sustainable and environmentally safe way of obtaining energy and valuable products. The use of simulation modelling is an effective tool for the control and optimization of such biotechnological processes.

References

1. I. Chowdhuri, S.C. Pal. Challenges and potential pathways towards sustainable agriculture crop production: A systematic review to achieve sustainable development goals (SDGs). Soil Till. Res., **248**, 106442 (2025). DOI: 10.1016/j.still.2024.106442.

2. L.S. Malyukova, V.V. Bukhtoyarov, V.Y. Konyukhov. Circular Mining Wastes Management for Sustainable Production of Camellia sinensis (L.) O. Kuntze. Sustainability (Switzerland), **15**, 11671 (2023). DOI: 10.3390/su151511671.

3. B. Aparecida da Silva, R. Evangelista de Oliveira, A. Cavalieri Sais, E. Cardoso leite. Family farmers and trees: ecological functions, ecosystem services and food production in agrarian reform settlements in Brazil. Agroecol. Sustain. Food Syst., **49**, 4-21 (2025). DOI: 10.1080/21683565.2024.2405515.

4. V. I. Silaev, R. V. Klyuev, D. V. Eremeev, T. A. Martynova, Yu. V. Danilchenko. Analysis of the carbon footprint created by mining enterprises. MIAB. Mining Inf. Anal. Bull., **11**, 265–277 (2023). DOI: 10.25018/0236_1493_2023_111_0_265.

5. A.A. Kulikova, E.I. Khabarova. Biological working agents use to reduce the carbon footprint. Sustainable Development of Mountain Territories, **16**, 311–321 (2024). DOI: 10.21177/1998-4502-2024-16-1-311-321.

6. S.N. Sorokova, E.A. Efremenkov, D.V. Valuev, M. Qi. Analysis of a Predictive Mathematical Model of Weather Changes Based on Neural Networks. Mathematics, **12**, 480 (2024). DOI: 10.3390/math12030480.

7. O. A. Kolenchukov, A. Y. Mikhaulov, K. A. Bashmur, R. B. Sergienko, V. V. Kukartsev, V. V. Tynchenko. Forecasting the technical condition of thermochemical reactor systems. SOCAR Proceedings, **2023**, 123–130 (2023). DOI: 10.5510/OGP20230100814.

8. A. Deo, N. Sawant, A. Arora, S. Karmakar. How has scientific literature addressed crop planning at farm level: A bibliometric-qualitative review. Farming Syst., **3**, 100139 (2025). DOI: 10.1016/j.farsys.2025.100139.

9. K.I. Kravtsov, T.A. Panfilova. Influence of mountain factors on salt excess and soil toxicity in mountain conditions. Sustainable Development of Mountain Territories, **15**, 784-797 (2023). DOI: 10.21177/1998-4502-2023-15-3-784-797.

10. N. A. Tursynbaev, I. V. Kuzhevskaya, L. M. Tatarintsev, A. A. Ospanova. Assessment of the environmental state and sustainability of agrolandscape at the local level. Sustainable Development of Mountain Territories, **15**, 662–671 (2023). DOI: 10.21177/1998-4502-2023-15-3-662-671.

11. V. V. Kireeva, T. G. Rasskazova, N. M. Serbulova. Maintaining soil fertility: A method for organic fertilizer production from crop farming waste. Sustainable Development of Mountain Territories, **10**, 493–499 (2018). DOI: 10.21177/1998-4502-2018-10-4-493-499.

12. B. Zh. Akhmetov, A. Zh. Kasenova, O. E. Merzlyakov, L. M. Tatarintsev. Modeling of sustainable agricultural land use on the basis of agroecological assessment of the territory. Sustainable Development of Mountain Territories, **15**, 17–26 (2023). DOI: 10.21177/1998-4502-2023-15-1-17-26.

13. B. Zh. Akhmetov, A. Zh. Kasenova, O. E. Merzlyakov, L. M. Tatarintsev. Modeling of sustainable agricultural land use on the basis of agroecological assessment of the territory. Sustainable Development of Mountain Territories, **15**, 17–26 (2023). DOI: 10.21177/1998-4502-2023-15-1-17-26.

Granulated goat milk sublimate and the experimental setup for its production

Natalya Yerofeyeva[1], *Artem* Shakhov[2], *Irina* Glotova[1,*], *Sergey* Shakhov[2] and *Kirill* Blinnikov[2]

[1]Voronezh State Agrarian University named after Emperor Peter the Great, 1, Michurina Street, Voronezh, 394087, Russian Federation
[2]Voronezh State University of Engineering Technologies, 19, Revolutsii Avenue, 394036, Voronezh, Russian Federation

Abstract. Prospects for the use of goat's milk (GM) as a food system are associated with the intensification of biological processes of its reproduction, as well as technologies for its processing and preservation. The purpose of the study is to substantiate the method of preserving goat's milk using an experimental spray freeze drying unit. The object of the study was the milk of goats of the Zaanen breed. The differential scanning calorimetry (DSC) method established a difference in the amount of binding energy of water fractions with dry matter in GM, which causes differences in energy consumption for GM dehydration at different stages of the drying process. The results were used to develop a drying method that makes it possible to carry out the processes of dehydration of goat lactation products under the gentlest conditions close to thermal conditions when producing these colloidal systems in the animal body. An experimental spray vacuum sublimation plant was used to produce granulated GM. Combining the structural elements of spray and freeze dryers allows balancing between the structural complexity of the device, its energy efficiency and the quality of the product.

1 Introduction

The spectrum of immunodeficiency conditions has been expanding in countries with different levels of economic development since the second half of the 20th century [1]. This is the reason for the formation of a large number of diseases. One of the important causes of this condition may be a deficiency of immunomodulating factors in nutrition, especially in the case of SARS-CoV-2 [2].

Goat lactation products are considered to be sources of immunomodulatory factors in nutrition [3, 4], including fermentation products such as kefir [5].

The world market for goat milk and its products is positioned as growing. Goat milk is a valuable source of protein and immunoglobulin, and the breeding of goats on an industrial scale makes a healthy diet accessible to the general population [6]. At the same time, the production of goat milk in many countries is of a pronounced seasonal nature, which necessitates the development of modern approaches and methods for maintaining quality indicators and consumer characteristics that are typical of native goat lactation products during storage and processing [7, 8].

So, the prospects for the use of goat's milk (GM) as a food system are associated with the intensification of biological processes of its reproduction, as well as technologies for its processing and preservation.

A promising way to preserve the native properties and biological activity of the components of food systems, products of applied biotechnology and the pharmaceutical industry is freeze drying [9-11], and the cryoprotective effect of pumpkin, fig and banana purees [12] just like the pumpkin powder [13] for probiotic microorganisms in fermented milk products has been proven during freeze drying. A comparative assessment of the effect of spray and freeze drying on the functional properties of food systems is the subject of research by a number of authors [14-16].

The advantages of vacuum freeze drying over spray drying have been established for dried camel milk [14], glycated with D-allulose and fructose soy protein isolates [15], beetroot processing products [16]. It should be noted that the research of scientists is aimed at developing methods for intensifying the processes of freeze-drying [17, 18]. For this purpose, additional physical effects are used, such as microvibration [17] and treatment with pulsed electronic fields [18]. However, the disadvantage is high initial costs, which limits the release of sublimated products for widespread everyday consumption.

To date, spray vacuum freeze drying is recognized as an emerging technology for the production of medicinal substances and free-flowing powders [19]. In this case, an important role is given to the study of various controlled parameters of the process of spray vacuum-freeze drying. However, in relation to heat-labile food systems such as lactation products of farm animals, the combined method of spray vacuum-freeze drying has not been implemented sufficiently. The purpose of the study is to substantiate the method of preserving goat's milk using an experimental spray freeze drying unit.

* Corresponding author: glotova-irina@mail.ru

2 Objects and methods

The objects of the study were goat milk of the Zaanen breed. Milk was fresh and used after morning and evening milking. Milk was received from a milking herd of goats in the village of Novaya Usman in Novousmansky district of the Voronezh region.

Studies of the chemical composition and physico-chemical parameters of raw materials and dehydrated products were carried out according to standard and generally accepted methods in research practice.

Raw goat milk was studied on a METTLER TOLEDO STARE TGA/SDTA 851e thermogravimetric and differential thermal analysis device (Switzerland).

The freeze-drying processes of goat's milk were studied using the installation, the scheme of which is shown in [8].

The fractional composition of moisture in goat lactation products was assessed in accordance with the recommendations [20]. The essence of the method for determining the fractional composition of moisture in materials consists in the following. Based on the results of derivatographic analysis, a graph of the dependence of the natural logarithm of the rate of dehydration of the sample on the reciprocal value of its absolute temperature is plotted. Break points are determined on the resulting graph, and the corresponding values of moisture loss and moisture content of the product sample are determined from the derivatogram [20].

3 Results and discussion

Raw goat milk had the following organoleptic characteristics: consistency (homogeneous liquid without sediment and flakes); taste and smell (clean, without foreign smells and flavors that were not characteristic of fresh natural milk); color (white to light cream). Sanitary and hygienic safety indicators of goat's milk corresponded to the requirements of TR CU 033/2013 «Technical regulations of the Customs Union. About the safety of milk and dairy products». The studied samples of goat milk had the following physical and chemical parameters. The mass fraction of fat was 3.6-4.0%; the mass fraction of protein was 3.1-3.3%; acidity was 18.0 °T; the purity group was I, density was 1031.0 kg/m^3; the mass fraction of dry fat-free substances (dry fat-free milk residue) was 12.2%.

Derivatogram of goat milk is shown in Figure 1. The results of processing the goat milk derivatogram according to the method [20] are shown in Figure 2. In the obtained graphs, the break points (the starting break point, point 1, point 2 and point 3) and the corresponding values of moisture loss were determined.

In view of this, in each graph, areas corresponding to the areas of removal of water fractions from the product were highlighted, and the moisture content values corresponding to the boundaries of the three water fractions were determined. Therefore, in these goat lactation products, three water fractions should be distinguished. The starting break point was 1; 1-2; 2-3, which differ in the value of the binding energy of moisture with the dry part of the sample. Quantitative data on the moisture content of each fraction are presented in the table.

Table. The fractional composition of water in goat's milk.

The water fraction number in goat's milk	The moisture content range of goat milk during water removal	The proportion of the water fraction in the total water content of goat's milk
1	13.65-11.6	15
2	11.6-0.11	84
3	0.11-0	1

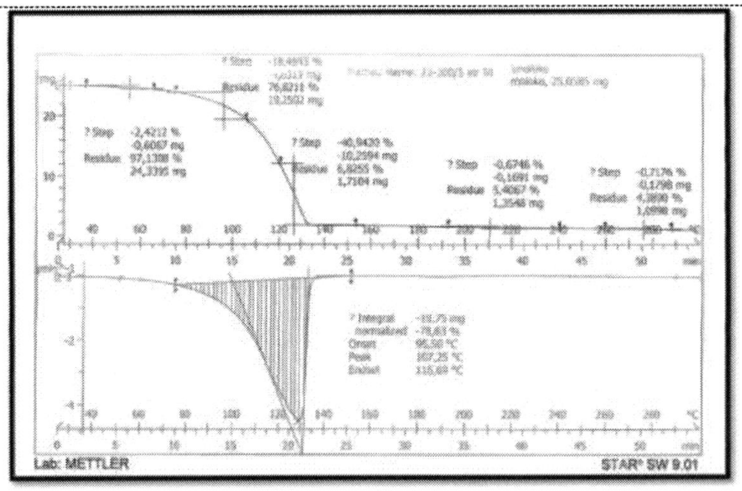

Figure 1. Goat milk derivatogram

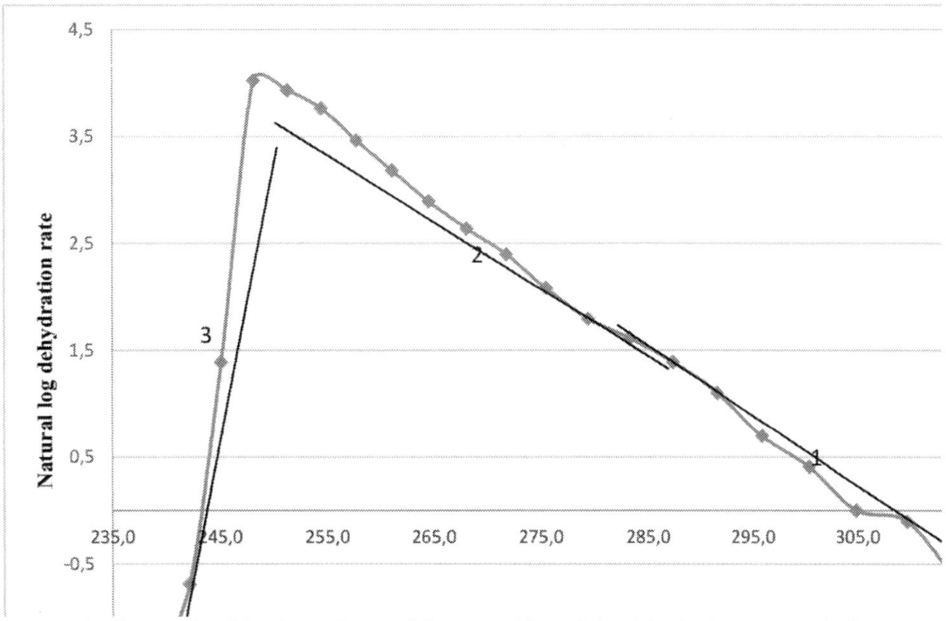

Figure 2. The graph of the dependence of the natural log of the dehydration rate on the inverse value of the absolute temperature of goat's milk

The results were used to develop a drying method that makes it possible to carry out the processes of dehydration of goat lactation products under the gentlest conditions, close to thermal conditions when producing these colloidal systems in the animal body.

The production of drinking goat milk includes the following technological operations:

- sampling raw milk and carrying out an examination of its physicochemical and organoleptic parameters;
- bringing the product to the required fat content by adding cream or skim milk to the milk;
- removing mechanical impurities and microorganisms using centrifugal cleaners or by filtration through mesh, gauze and flannel filters;
- heating the milk to the pasteurization temperature for a certain time to remove pathogenic microflora;
- cooling the milk to a temperature at which the development of microorganisms and oxidative processes is suspended.

However, to obtain freeze-dried products, it is sufficient to limit it to cold disinfection. When producing drinking goat milk, compared to cow's milk, there may be no homogenization, since the size of the fat globules in goat milk is significantly smaller and remains suspended in solution [8]. It should be noted that the production of dairy products from goat's milk does not require additional equipment compared to the processing of cow's milk.

Figure 3 shows a fundamental technical approach to combining the design features of vacuum-sublimation and spray dryers are illustrated by the design features of the installation.

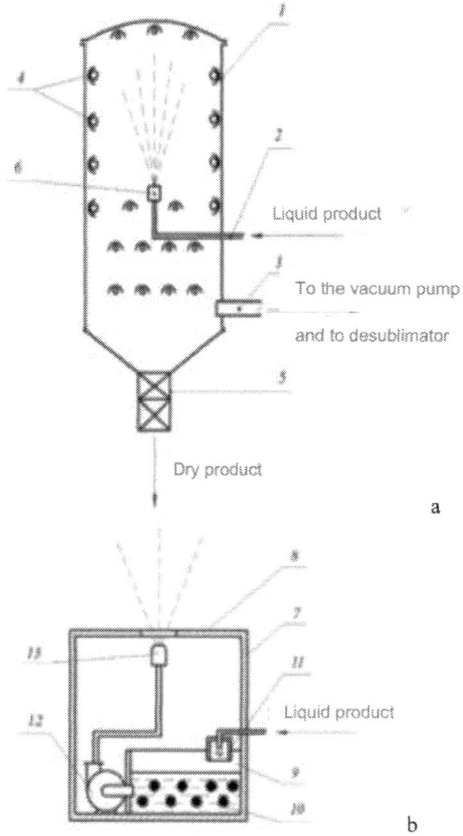

Figure 3. The sublimation spray dryer

As can be seen in Figure 3, the sublimation spray dryer for dewatering liquid products contains the camera (a) with the granulator (b). The sublimation spray dryer device has the following design features. 1 is the sublimation chamber; 2, 3 are nozzles; 4 is IR emitters; 5 is the sluice gate; 6 is the granulator; 7 is the evaporation chamber; 8 is the partition; 9 is the evaporation bath; 10 is heaters; 11 is fixing nozzles; 12 is the pump; 13 is the mechanical nozzle.

The unit is equipped with a granulator to increase the solubility of the product and improve its storage capacity.

The technology for the production of goat milk sublimate provides for the following sequence of operations: acceptance and preparation of raw materials →freezing the product →freeze-drying the product → granulation →packaging, labeling →storage.

When obtaining dry goat milk, the technological operations were carried out under the following conditions:

- primary purification from mechanical impurities by filtration through lavsan no later than 2 hours after milking;
- ultra-fine purification on a filter element made of food-grade polypropylene with a micropore size of 3 μm, at a pressure of 0.05 MPa, which made it possible to avoid pasteurization of milk;
- freezing at minus 40 °C and sublimation at a desublimator temperature of minus 60 °C for 5-6 hours.

The product has the following indicators (%): the mass fraction of protein is 28.5; fat is 30.5; carbohydrates are 35.0; ash is 2.5; the solubility index is 0.02 cm^3; acidity is 18 °T.

4 Conclusion

In order to increase the economic efficiency of freeze-drying, it is advisable to use a combination of various devices in order to combine their technological and operational functions and thereby to obtain more advanced equipment that combines the most useful elements of its prototypes. It is advisable to combine structural elements of spray and freeze dryers, combining their advantages, resulting in a balance between the structural complexity of the device, its energy efficiency and the quality of the finished product. The unit is equipped with a granulator to increase the solubility of the product and improve its storage capacity without loss of physico-chemical quality indicators, as well as to ensure the best preservation of the properties of the native product due to components with biological activity. The temperature of dehydration of goat milk, which is close to the temperatures of both its biosynthesis and absorption in the animal's body, made it possible to achieve the result.

Considering the economic and environmental advantages of developing goat farms compared to cattle farms, including higher milk yields per unit of feed, it is advisable to introduce a spray vacuum-sublimation unit into the technological cycles of processing goat lactation products in peasant farms. It is advisable to supplement vertically integrated schemes for the production and processing of goat lactation products at industrial agricultural enterprises of various capacities with equipment for the production of granulated sublimated goat milk.

References

1. N. Raje, D. Chitra, Immunol. Allergy Clin. North Am. **35(4)**, 599–623 (2015)
2. M.B. Khaled, N. Benajiba, Nor. Afr. Food Nutr. Res. **4(7)**, 240-244 (2020)
3. D. Turck, World Review of Nutrition and Dietetics. **108**, 56-62 (2013)
4. A J. Hodgkinson, O.A.M. Wallace, G. Smolenski, C. G. Prosser, Food Chem. **276**, 619-625 (2019)
5. A.L. Aristya, A.M. Legowo, A.N. Al-Baarri, JPDG **4(1)**, 39-48 (2013)
6. Y. Park, *Encyclopedia of Animal Science* Edition: 2nd Publisher: Taylor and Francis, CRC Press 120045703 (2010)
7. H. Rizqiati, Nurwantoro, S. Susanti, A. Febrisiantosa, T. Setyawardani, C. Shauma, JITEK **16(1)**, 65-74 (2021)
8. I. Glotova, S. Shakhov, V. Pryanishnikov, E. Kurchaeva, N. Erofeeva, N. Galochkina, A. Shakhov A.Vostroilov. *AIME 2017*, 255-262 (2017)
9. U. Tastemirova, R. Mukhtarkhanova, M. Alimardanova, R. Alibekov, A. Shingisov, Food Sci. Technol. Campinas **42**, e61722 (2022)
10. E.V. Alekseenko, G.N. Dubtsova, N.G. Mashentceva, E.A. Bystrova, G.V. Semenov, Asian Journal of Pharmaceutics **12**, 466-476 (2018) doi:10.22377/AJP.V12I02.2377
11. M.B.A. Fernandes, S. Habu, M.A. de Lima, V. Thomaz-Soccol, C.R. Soccol, Bioprocess Biosyst. Eng. **34(3)**, 253-261 (2011)
12. I.S. Krasnova, V.I. Ganina, G.V. Semenov, Foods and Raw Materials **11(2)**, 300–308 (2023)
13. I S Krasnova, V I Ganina, G V Semenov, I.I. Ionova, Zh.L. Guchok, IOP Conference Series Earth and Environmental Science **848(1)**, 012017 (2021)
14. U. Tastemirova, I. Ciprovica, A. Shingisov, *Conference: Research for Rural Development* 102-105 (2020) doi: 10.22616/rrd.26.2020.015
15. O. Tas, U. Ertugrul, M. H. Oztop, B. G. Mazi International Journal of Food Science & Technology, **56(11)**, 5461-5470 (2021)
16. A. Dhiman, R. Suhag, Suhag, D. Thakur, P.K. Prabhakar, D.S. Chauhan, S Chhikara, Trends in Food Science & Technology **114**, 443-458 (2021)
17. G.V. Semenov, I.S. Krasnova, S.I. Khvylia, et al. J. Food Sci. Technol. **58**, 3192–3198 (2021)
18. A. Lammerskitten, V. Mykhailyk, A. Wiktor, et al. Innov. Food Sci. Emerg. Technologies, 57, 102211 (2019) doi:10.1016/j.ifset.2019.102211

19. M.B. Adali, A.A. Barresi, G. Boccardo, R. Pisano, Processes **8(6)** 709 (2020)

20. V.M. Arapov, D.A. Kazartsev, I.A. Nikitin, et al. IJACSA **11(2)** 17-22 (2020)

Analysis of biotechnological approaches to utilisation of organic wastes from livestock and poultry production

Waleed Mugahed Al-Rahmi [1], *Yu.S.* Kuznetsova [2], *Natalia A.* Grankina [3*], *Aslan Yu.* Bogus [3], *Sergey V.* Panchenko [4]

[1] Dar Al Uloom University, Riyadh 13314, Al Falah, Kingdom of Saudi Arabia
[2] Admiral Ushakov Maritime State University, Novorossiysk, 353924, Russia
[3] Kuban State Agrarian University named after I. T. Trubilin, Krasnodar, Russia
[4] Azov-Black Sea Engineering Institute – branch of Don State Agrarian University in Zernograd, Zernograd, Russia

Abstract. The article is devoted to the study of innovative biotechnologies for processing organic animal waste to produce environmentally friendly fertilisers. Three main approaches are considered: aerobic fermentation, use of mobile biocomplexes and anaerobic digestion. Aerobic fermentation is the composting of manure and dung with carbon-containing additives. Mobile biocomplexes allow waste to be processed on-site using mixtures of manure, manure and crop residues. Anaerobic technology involves digestion of manure in hermetically sealed plants to produce biogas and liquid fertiliser. Experiments conducted under laboratory and pilot conditions have produced results demonstrating the effectiveness of each method. Aerobic fermentation has been shown to be simple and effective, but requires more time and space. Mobile biocomplexes provide faster processing but require more sophisticated equipment. Anaerobic digestion, although more complex, produces biogas and valuable liquid fertiliser, and is the most promising option.

1 Introduction

Uncontrolled accumulation of organic waste leads to soil and water pollution, greenhouse gas emissions and the spread of unpleasant odours, with negative impacts on public health and ecosystems. There are several main approaches to tackling this problem. One is the traditional disposal of waste in landfills and dumps. This method, although simple to implement, has a number of significant disadvantages. It occupies large areas, pollutes soil and groundwater, and contributes to the formation of methane, a powerful greenhouse gas. [1]. The effect of farming practices on soil organic matter is also an important factor in the context of pollution. Another way is waste incineration, which reduces the volume of waste, but is accompanied by the emission of harmful substances into the atmosphere and requires significant energy costs. There are also composting and anaerobic digestion methods that can produce valuable products such as organic fertilisers and biogas, but conventional technologies are often slow and inefficient when dealing with large volumes of waste. In addition, many of the existing recycling methods are expensive and not always economically viable. Meanwhile, an important aspect is not only recycling, but also controlling the thermochemical processes of waste treatment that lead to the release of methane. [2].

In this context, a relevant and important direction is the development and implementation of innovative biotechnologies, which allow to recycle organic waste into environmentally friendly and valuable products. The use of biotechnologies provides effective destruction of organic substances with the help of microorganisms, allowing to obtain organic fertilisers, biogas and other products. This approach not only helps to reduce the volume of waste, but also makes it possible to use it as a renewable resource, thus solving several problems at once: reducing the negative impact on the environment, preserving natural resources and developing agriculture. Unlike traditional methods, biotechnologies of waste processing allow to reduce the cost of production and increase the economic attractiveness of waste processing. In addition, the introduction of such technologies contributes to the development of circular economy, where waste is considered not as ballast, but as secondary resources that can be reintroduced into the production cycle, which is similar to the approaches considered in the field of industrial waste recycling [3]. The development of these technologies in Russia is particularly important because of the great need for organic fertilisers and the need to solve problems related to waste management.

The purpose of this paper, presented in the attached file, is to analyse and describe innovative biotechnologies for processing organic animal waste developed by leading domestic organisations. In

* Corresponding author: viperaand@mail.ru

particular, the technologies of multi-purpose compost production, mobile biocomplexes, waste-free energy-saving technology of poultry waste processing and other methods aimed at obtaining environmentally friendly fertilisers are presented. The focus is on the advantages of these technologies, their economic feasibility and contribution to sustainable agricultural development. The study of these methods will provide an insight into modern solutions in the field of waste processing, as well as consider their potential for wide implementation in the agro-industrial complex.

2 Method and materials

This article reviews innovative biotechnologies developed by leading organisations for processing organic animal waste and producing environmentally friendly fertilizers. The general plan of experimental work was reduced to research and testing of various technologies in the conditions of laboratories and pilot production facilities. In particular, the processes of aerobic fermentation of manure and litter with the use of carbon-containing components were studied [this is related to the importance of microbial diversity in processing], [4] microbiological destruction of the organic part of litter under anaerobic conditions, which is also important in the context of combating antibiotic resistance, [5] as well as accelerated biofermentation of various types of waste. Bioreactors of different volumes were used to activate fermentation processes, where the fermentation of organic wastes using specific microorganisms took place, which is similar to the study of the effect of dietary fibres on the microbiome, [6]. The following equipment was used in the experiments: for aerobic fermentation, composters of different designs were used, providing the necessary air access and temperature control. For anaerobic treatment of poultry waste, reactors were used that allowed maintaining the necessary anaerobic conditions. In the technologies related to biofermentation, bioreactors of different volumes were used, where the process of fermentation of organic wastes with the use of special microorganisms took place. Also, laboratory instruments were used in the research to control the process parameters (temperature, pH, oxygen content) and to analyse the obtained products. Crushers, pulverisers, shredders, industrial mixers were used for grinding of components and preparation of raw materials. Granulators were used for granulation of the finished fertiliser. The operating modes of the equipment varied depending on the type of processed raw materials and technology, for aerobic fermentation the temperature control was carried out within the range of 50-65 degrees, for anaerobic fermentation - within the range of 35-55 degrees, the processing time varied from several days to several months depending on the technological process.

3 Results and discussion

The experimental work described in the article was focused on the study and implementation of innovative biotechnologies aimed at processing organic wastes from livestock farming to produce environmentally friendly fertilisers. The research was conducted in several directions. The first direction was the use of technology for the production of multipurpose compost based on aerobic fermentation of manure and dung, with the addition of carbon-containing components, which is aimed at maintaining soil fertility, [7]. The technology included mixing manure and dung with peat, sawdust or chopped straw with subsequent aerobic fermentation. Another approach is the use of mobile biocomplexes capable of processing large volumes of organic waste. In this approach, mobile complexes are used, which allow to process various organic wastes directly at the place of their formation, which is similar to the selection of disintegrator parameters, [8]. In the same way the waste-free energy-saving technology of poultry waste processing based on anaerobic microbiological destruction of manure was used. In this approach, poultry waste was subjected to anaerobic digestion to produce biogas and protein-rich feed additive, which can also be studied from the point of view of insect larvae utilisation, [9], as well as from the point of view of utilisation of the resulting protein in agriculture, [10].

In our study, we conducted an experiment based on the first technique we used, for which we carefully selected the starting materials. We took fresh cow dung and chicken manure, in the ratio of 2:1, which corresponds to two parts of dung to one part of manure. Then we combined this mixture with peat, in a ratio of 1:1 to the previously obtained mixture of manure and litter. Thus, we used one part of peat for each part of the mixture of cow dung and chicken manure, which acts as a carbon-containing component in this process. The resulting mixture was thoroughly mixed using a special mixer to ensure maximum homogeneity. This mixture was then placed in a specially designed stainless steel container with a volume of 3 cubic metres, which was equipped with an aeration system with a timer to control the air supply, ensuring a constant and uniform flow of oxygen to the decomposing organic matter. For 60 days we closely monitored key parameters such as temperature, humidity and pH using electronic sensors with data recorded in our research database. At the beginning of the process, as expected, the temperature of the mixture rose quite rapidly to between 60-65 degrees Celsius, a clear indication that aerobic microorganisms were active and beginning to decompose the organics. After some time, after this period of intense heating, the temperature began to gradually decrease until it reached values in the range of 30-35 degrees Celsius. We kept the humidity inside the container relatively constant, between 60-65%, using an automatic humidification system, and the pH fluctuated within a narrow range of 6.5 to 7.5. As a result of the 60-day process, we obtained about 1.5 cubic metres of finished compost of very good quality. At the same time, we loaded 2.5 cubic metres of feedstock at the beginning of the process, which demonstrates volume shrinkage during processing, which is quite normal in the composting of organic matter, indicating the importance of assessing the agro-ecological condition of the land, [11]. Subsequent laboratory analysis of samples of the resulting compost showed a high content of humic substances, reaching 45%, indicating its maturity and usefulness to the soil, which is important in the context of sustainable land use, [12]. The content of important macronutrients was also at a high level, namely nitrogen 2.5%, phosphorus 1.8% and potassium 2.2%, and trace elements that are essential for plant growth and development were also present. Thus, we were able to obtain a compost suitable for agricultural use.

In the course of our experiment, along with the study of alternative energy technologies, considerable attention was paid to the application of mobile biocomplexes for processing organic waste. To implement this task, we used a specially designed mobile container, which was equipped with a whole set of mechanisms, including a shredder for pre-treatment of waste, a mixing system for homogenisation of the mass and an aeration unit providing the necessary oxygen access for aerobic processes. As a raw material for processing we used a mixture of organic waste consisting of cattle manure, poultry manure and plant residues obtained after harvesting crops. The ratio of components was 4 parts of manure, 2 parts of manure and 1 part of plant residues, which, in our opinion, is optimal for efficient operation of the biocomplex. The volume of feedstock loaded into the mobile biocomplex was about 3 cubic metres, and the processing cycle lasted 45 days. Throughout this period, we closely monitored key process parameters. The temperature inside the biocomplex varied between 20 and 55 degrees Celsius, which may correlate with the sound recognition model for oestrus detection, [13], and its dynamics depended on the processing phase and microbial activity. Humidity also played an important role, and we maintained it at 50-60%, adding water when necessary to prevent drying and provide favourable conditions for biodegradation. During the whole period of organic waste treatment, we observed how the structure of the organic waste changed and the result was about 2 cubic metres of compost-like product. This product was characterised by a high organic matter content of around 60%, which is an indicator of good quality processing. In addition, analysis of the compost revealed a balanced composition of nutrients including nitrogen, phosphorus and potassium, which are essential for plant growth, making the product a valuable organic fertiliser. It is important to emphasise that the mobility of our biocomplex allowed us to process organic waste directly at the livestock farm located near the village of Sosnovka. This significantly reduced the cost of transporting waste, which usually entails additional costs, and also reduced the negative environmental impact associated with transporting waste over significant distances. In our view, this mobility is a significant advantage of this technology.

The third, and perhaps one of the most promising, approaches to poultry waste processing we have studied is a zero-waste, energy-saving technology for anaerobic digestion of poultry manure. This technology is a process of biochemical conversion of organic waste into valuable products in the absence of oxygen. To thoroughly study and analyse the efficiency of this technology, we designed and built a sealed anaerobic plant, which is a 5 cubic metre tank. It was equipped with a temperature, pressure and pH control system as well as a gas analyser for continuous monitoring of the process. This unit was loaded with 3 cubic metres of fresh chicken manure, with the addition of about 10 litres of a special bacterial culture. This culture, containing strains of methanogenic bacteria, was introduced to speed up the fermentation process and increase the biogas yield. The anaerobic digestion process was carried out for 70 days at a strictly controlled constant temperature of 35 degrees Celsius, maintained by a water heating system, which ensured optimal bacterial activity. The anaerobic digestion, with constant monitoring of the parameters, produced about 150 cubic metres of biogas, the composition of which was analysed using a gas chromatograph. The analysis showed that the biogas consisted of 60% methane, which is similar to mining waste management processes [14], which is the main combustible component, and 40% carbon dioxide, which is an inert gas. This biogas obtained, after its purification and dehydration, was used to generate electricity using a gas piston generator, which allowed us to generate approximately 200 kWh of electricity. This amount of energy is enough to power the research facility for two days. In addition to biogas, at the end of the digestion process, a valuable residue in the form of a liquid biofertiliser was also produced. This by-product was scrutinised through chemical analysis, which showed that it contained significant amounts of nitrogen (3.5%), phosphorus (2.5%) and potassium (2.8%), which are important nutrients for plants. In addition, about 15% protein was found in the biofertiliser, which is important for biomonitoring [15], indicating its high nutritional value and suitability for use as an organic fertiliser. Thus, the results of our study confirm that this technology of anaerobic digestion of poultry manure is an effective and environmentally friendly way not only to dispose of poultry waste, reducing the negative impact on the environment, but also to obtain a valuable energy resource in the form of biogas, as well as environmentally safe fertiliser for agriculture, thus contributing to a closed cycle of production. Comparing the results, we see that aerobic fermentation technology, despite its simplicity, allows obtaining high-quality compost, but requires a long time and a large area for compost heaps. Mobile biocomplexes provide faster processing with the possibility to carry it out directly at the site of waste generation, but require more sophisticated equipment and regular maintenance. Anaerobic digestion technology has the greatest potential as it not only recycles waste but also produces biogas as an alternative energy source and nutrient-rich liquid fertilisers. Studying the influence of mountain factors on soil salinisation helps to determine more precise parameters of fertiliser application [16]. However, this method requires the creation of a specialised sealed plant and more careful control of the process parameters. As a result, the choice of the most appropriate technology depends on specific conditions and objectives: if a simple and inexpensive way to process waste to produce compost is required, aerobic fermentation is suitable; if mobility and speed are required, the use of biocomplexes is optimal; and if the priority is to obtain energy and fertiliser, anaerobic digestion is the best option.

4 Conclusion

The study has shown that the proposed biotechnologies represent promising solutions for the processing of organic animal waste, offering various approaches to their utilisation with the production of valuable products. Aerobic fermentation, despite its simplicity, provides quality compost, but requires considerable time and area for the process. Mobile biocomplexes make it possible to speed up the recycling process and carry it out directly at the site of waste generation, but require more sophisticated equipment and maintenance. The most promising method is anaerobic digestion, which not only recycles waste, but also produces biogas that can be used as an energy source, as well as producing liquid fertilisers with a high nutrient content. Each of the technologies has advantages and disadvantages, and the choice of the most appropriate one depends on the specific conditions and objectives. The study emphasises the importance of an integrated approach to waste management that combines economic feasibility and environmental safety. The introduction of these technologies can contribute to the development of a circular economy, where waste is considered as a secondary resource rather than ballast, as well as reduce the negative environmental impact of livestock production and provide agriculture with valuable products, which is linked to the sustainable use of land resources, [17]. An important result was the confirmation of the effectiveness of anaerobic digestion as a way to produce not only fertiliser but also an alternative energy source, making this method the most attractive from a sustainable development point of view. Thus, the study provides valuable data for further development and implementation of innovative biotechnologies in the agro-industrial complex.

References

1. Q. Gao, L. Wang, Y. Fang, Y. Gao, L. Ma, X. Wang, Y. Li, X. Wu, Z. Du. Conservation agriculture boosts topsoil organic matter by restoring free lipids and lignin phenols biomarkers in distinct fractions. Soil Till. Res., **248,** 106463 (2025). DOI: 10.1016/j.still.2025.106463.

2. O. A. Kolenchukov, A. Y. Mikhaulov, K. A. Bashmur, R. B. Sergienko. Forecasting the technical condition of thermochemical reactor systems. SOCAR Proceedings, **2023,** 123–130 (2023). https://doi.org/10.5510/OGP20230100814.

3. L.S. Malyukova, V.V. Kondratiev, V.V. Bukhtoyarov, V.Y. Konyukhov, K.A. Bashmur. Circular Mining Wastes Management for Sustainable Production of Camellia sinensis (L.) O. Kuntze. Sustainability (Switzerland), **15,** 11671 (2023). DOI: 10.3390/su151511671.

4. M. Usman, M. Wang, Y. Liu, L. Li, X. Zhang, T. Xiao, F. Hou. High soil bacterial diversity increases the stability of the community under grazing and nitrogen. Soil Till. Res., **248,** 106414 (2025). DOI: 10.1016/j.still.2024.106414.

5. P. Zhang, Q. Liu, H. Yao, T. Wang, B. Zeng, Y. Zou, S. Fu, K. Hu, H. Guo. Copper-functionalized hydrochar from swine manure with complexation sites for selective adsorption of tetracycline antibiotics. Sep. Purif. Technol., **359,** 130671 (2025). DOI: 10.1016/j.seppur.2024.130671.

6. F. Yong, B. Liu, H. Li, H. Hao, Y. Fan, O. Datsomor, R. Han, H. Jiang, D. Che. Relationship between dietary fiber physicochemical properties and feedstuff fermentation characteristics and their effects on nutrient utilization, energy metabolism, and gut microbiota in growing pigs. J. Anim. Sci. Biotechnol., **16,** 1 (2025). DOI: 10.1186/s40104-024-01129-x.

7. V. V. Kireeva, T. G. Rasskazova, N. M. Serbulova. Maintaining soil fertility: A method for organic fertilizer production from crop farming waste. Sustainable Development of Mountain Territories, **10,** 493–499 (2018). DOI: 10.21177/1998-4502-2018-10-4-493-499.

8. N.N. Bryukhanova, V.A. Gladkikh, R.V. Kononenko, V.V. Gerasidi. Selection of disintegrator parameters toward environmentally friendly leaching of metals. Min. Inform. Anal. Bull., **(12),** 39-49 (2024). DOI: 10.25018/0236_1493_2024_12_0_39.

9. Y. Chen, Y. Guo, X. He, B. Tan, Z. Liao, A. Chen, X. Gu, X. Li, X. Chen, B. Chen, S. Lin, W. Li, P. Hu, X. Zhu, W. Zhao, J. Niu. Comprehensive utilization of black soldier fly (Hermetia illucens) larvae: extraction, recovery and characterization of peptide, chitin and melanin and scaling-up trial. Sep. Purif. Technol., **361,** 131262 (2025). DOI: 10.1016/j.seppur.2024.131262.

10. K. Mittenzwei, W. Britz, R.J.F. Burton. The potential impact of cultivated protein on agriculture in Norway. Environ. Innov. Soc. Transitions, **55,** 100960 (2025). DOI: 10.1016/j.eist.2024.100960.

11. Yu. S. Lisovskaya. Agroecological condition assessment of agricultural landscapes for their sustainability increase. Sustainable Development of Mountain Territories, **14,** 76–86 (2022). DOI: 10.21177/1998-4502-2022-14-1-76-86.

12. M. N. Nurpeisov, K. V. Ruchkina, V. A. Androkhanov, S. K. Makenova. Assessment of geoecological factors and conditions affecting sustainability of agricultural land use in the forest-steppe zone of Altai Krai. Sustainable Development of Mountain Territories, **15,** 419–430 (2023). DOI: 10.21177/1998-4502-2023-15-2-419-430.

13. Y. Cao, Z. Yin, Y. Duan, R. Cao, G. Hu, Z. Liu. Research on improved sound recognition model for oestrus detection in sows. Comput. Electron. Agric., **231,** 109975 (2025). DOI: 10.1016/j.compag.2025.109975.

14. S. Yu. Avksentiev, P. A. Semenov, A. S. Afanasyev. Comprehensive solutions for tailings management at the Usolsky potash plant. Sustainable

Development of Mountain Territories, **16,** 7–18 (2024). DOI: 10.21177/1998-4502-2024-16-1-7-18.

15. C.M. Donato, W. Wirth, A.F. Porter, P. Andersson, E. Donner, T. Huynh, A. Phabmixay, E. Flynn, C. Minney-Smith, T. Gonzales, C. Bareja, A. Levy, Z. Bartlett, A. Donald. Parameters for one health genomic surveillance of Escherichia coli from Australia. Nat. Commun., **16,** 17 (2025). DOI: 10.1038/s41467-024-55103-2.

16. K.I. Kravtsov. Influence of mountain factors on salt excess and soil toxicity in mountain conditions. Sustainable Development of Mountain Territories, **15,** 784-797 (2023). https://doi.org/10.21177/1998-4502-2023-15-3-784-797.

17. L. V. Lebedeva, G. D. Syzdykova, D. S. Inkarov. Management of agricultural land use sustainability at the local level using GIS technologies and special map material. Sustainable Development of Mountain Territories, **15,** 864–876 (2023). DOI: 10.21177/1998-4502-2023-15-4-864-876.

Disinfection of beekeeping objects with oxygen-containing preparations for bee aspergillosis

Dmitry V. Gruznov[1*], *Olga A.* Gruznova[1,2], *Alexey B.* Sokhlikov[1], and *Irina P.* Chesnokova[2,3]

[1]All-Russian Research Institute of Veterinary Sanitation, Hygiene and Ecology – Branch of Federal State Budget Scientific Institution "Federal Scientific Center – K.I. Skryabin, Ya.R. Kovalenko All-Russian Research Institute of Experimental Veterinary Medicine, Russian Academy of Sciences", 5 Zvenigorodskoye highway, Moscow, 123022, Russia
[2]N.N. Semenov Federal Research Center for Chemical Physics, Russian Academy of Sciences, 4 Kosygina st., Moscow, 119334, Russia
[3]MIREA – Russian Technological University, 78 Vernadsky Avenue, Moscow, 119454, Russia

Abstract. Aspergillosis of bees is an infectious disease of bee brood and adult bees caused by fungi of the Aspergillus genus. These fungi live and are reproduced in soil, feeding on plant material, including the stamens and nectary of flowers. The causative agents are brought into hives by bees carrying nectar and pollen. Once inside, they grow on honeycomb, bee bread, larvae, pupae, and adult bees under favorable conditions. Aspergillosis is a zooanthroponosis, causing significant economic damage to beekeeping and potentially posing a risk to consumers of bee-derived products. Therefore, important tasks are the selection of preparations that meet the requirements of the industry and the development of effective regimens and applications. This paper presents the results of laboratory and apiary trials of oxygen-containing preparations for the disinfection of bee-keeping facilities (hives, frames, equipment) in cases of aspergillosis. Based on the data obtained, effective regimes of their use are recommended.

1 Introduction

Aspergillosis is a fungal disease that affects adult bees and their brood. It is caused by fungi from the Aspergillus genus and occurs sporadically, mainly in the spring. It is a toxicomycosis that has both toxigenic and pathogenic properties, which can infect the living bodies of larvae and adult bees. The following species of Aspergillus are pathogenic to honey bees: *A. flavus* Link, *A. niger* Tieghem, and *A. fumigatus* Fres. Among these, *A. flavus* is the most common cause of damage to honey bees. *A. fumigatus* and *A. niger* are slightly less common. Infection occurs when larvae and bees consume spore-contaminated honey, nectar, and pollen through their intestines. The percentage of bee deaths varies depending on the conditions and strength of the colony [1-3].

For the first time, bee aspergillosis was discovered in the USA (Texas) by W.K. Howard in 1894. The researcher described the disease of unsealed brood. Subsequently, they detected damage in pupae and adult bees. In Germany aspergillosis was reported by A. Maassen. This disease was subsequently detected in many European countries [1,2].

Aspergilli are aerobes that are able to live and reproduce in the soil, on various organic substrates, and dead plant debris. Fungi also live as saprophytes on living plants, including on the stamens of flowers and nectaries [4]. Pathogenic aspergillus species are introduced into hives by bees carrying nectar and pollen. Once inside the hive, these fungi can grow on honeycomb, bee bread, larvae, pupae, and adult bees at high humidity and temperatures [2,5]. The source of infection can also come from diseased or dead larvae and pupae, as well as from infected bees, honey, and bee bread that contain fungal spores. Hives and honeycomb that have been contaminated with fungal spores can also serve as a source of infection [6,7].

It should be noted that the spores of the pathogens have significant resistance in the external environment. Fungal spores remain viable for 4 to 15 years in empty hives, on honeycombs and beekeeping equipment, as well as in honey and bee bread [8,9].

Aspergillosis of bees is a zooanthroponosis [10-14], and, therefore, strict personal hygiene measures should be followed when examining infected beehives and working with infected material. Beeswax collected from infected beehives should only be used for technical purposes, and pollen collection in these hives is prohibited. Honey collected from such hives should not be used for human consumption [5,11,15].

It is well known that the implementation of comprehensive measures to prevent and control the emergence and spread of infectious diseases, including bee Aspergillosis, through the use of disinfection is one of the key principles for ensuring sanitary and

* Corresponding author: *79164422245@yandex.ru*

epidemiological well-being in apiaries in the country. In this context, the aim of this study is to search environmentally friendly disinfection agents with potential use in beekeeping and to evaluate their efficacy, both in laboratory settings and in practical applications in apiaries.

2 Method and materials

Aspergillus flavus (840-c/7), *Aspergillus niger* (843-c), and *Aspergillus fumigatus* (8/10) strains from the microorganism collection of the laboratory of mycotoxicology of the All-Russian Research Institute of Veterinary Sanitation, Hygiene and Ecology, as well as field strains isolated from pathological material at a private infected apiary in the Kaluga region, were used in this work. The collection strains were used as controls. After isolation and identification, field strains were used to carry out experiments both in laboratory and apiary conditions.

2.1 Light microscopy

The studies were carried out using the "crushed drop" method. To identify the pathogenic agent of aspergillosis, mycelium was scraped from the surface of the larvae. The material under study was placed on a glass slide in a drop of alcohol-glycerol, covered with a coverslip, and it was examined using the light microscope. A preliminary diagnosis of aspergillosis was made when fungal elements characteristic of the genus *Aspergillus* were detected in the pathological material.

2.2 Mycological studies

Diagnostic swabs or fragments of dead, mummified larva were placed on the surface of a nutrient medium in petri dishes. Czapek-Dox agar was prepared by melting it in a water bath prior to inoculation and cooling it to 40-45 °C. In order to suppress the accompanying bacterial flora, antibiotics were added to the cooled agar (50 000 units of penicillin and 100 000 units of streptomycin per liter of medium). The petri dishes were maintained in a thermostat at 25-27 °C and observed for a period of 14 days. Pure cultures were isolated by transferring the inoculum from the peripheral colonies in the primary cultures onto Czapek-Dox agar plates.

2.3 Laboratory and apiary trials

In the laboratory, artificially contaminated samples of wood, polystyrene foam, wax, metal, plastic, and textile measuring 10 × 10 cm were used. After a certain period of exposure (60 min and 180 min), samples were collected from the experimental and control surfaces using sterile cotton-gauze swabs that had been slightly moistened. These swabs were then placed in test tubes containing sterile water, which were used to inoculate Petri dishes. After 24 h and 48 h, the absence of growth of unwanted microorganisms was determined. The final results were observed after 7 days and 14 days. Based on the presence or absence of test culture growth, the fungicidal and sporicidal efficacy of disinfectant was concluded. To ensure reliable results, each experiment was performed in triplicate.

In apiary experiments to test disinfection regimes for bee aspergillosis, 14 hives, 42 honeycomb frames and 26 items of beekeeping equipment were disinfected.

3 Results and discussion

The use of zootechnical measures and treatments for bee Aspergillosis is not a complete solution to the problem of hive sanitation. Therefore, the disinfection of apiaries is an essential component of a successful strategy for combating this disease.

Currently, both Russian-made and foreign preparations are permitted for use in disinfection in the field of beekeeping. Based on the active substance, these preparations can be classified into several chemical groups. Preparations containing active oxygen are widely used, as they best meet the fundamental requirements of this industry (Fig. 1).

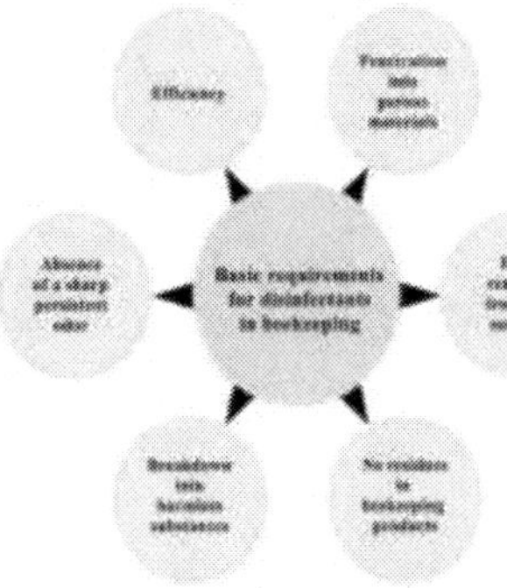

Fig. 1. The basic requirements for disinfectants for use in beekeeping.

Based on a literature and patent search, the following oxygen-containing disinfectants were selected: hydrogen peroxide 37% (v/v%) for medical and veterinary use and "Dezinbak Super", containing 98.6–99.4% (w/w%) peroxide compounds. These products meet the specified criteria and are manufactured in the Russian Federation and commercially available.

To identify clinical signs of aspergillosis in honeybees, the state of the larval stage of the brood was initially assessed through visual inspection. Typically, in broods affected by aspergillosis, larvae in the final stages of development (i.e., prepupae and pupae) of both worker bees and drones have died. Depending on the specific strain of Aspergillus, these broods may be covered with a variety of fungal growth, ranging from white to grey, yellow-green, or black. When examining bee colonies, it was noted that adult worker bees reduced their activity and lost the ability to fly. The belly of the diseased insects was enlarged and firm to the touch.

Data on cultural and morphological characteristics, as well as morphometric characteristics of fungi were obtained. This gave grounds to classify the isolated isolates as certain species of pathogens. *A. flavus* was detected in 73%, *A. niger* in 18% and *A. fumigatus* in 9% of cases (Fig. 2).

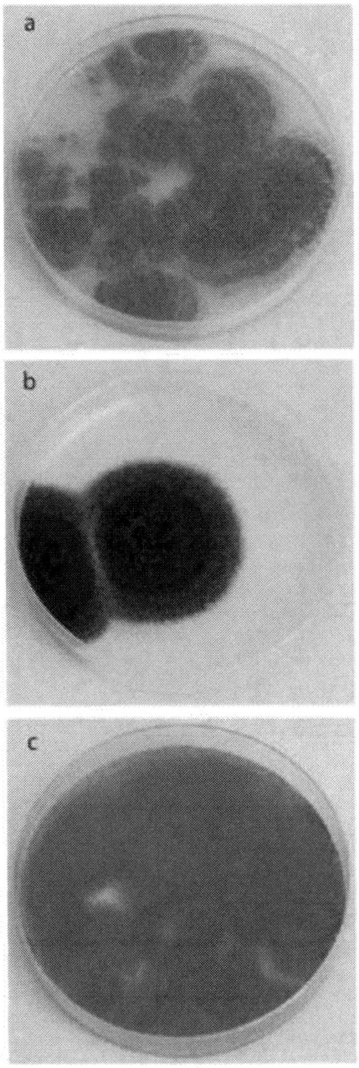

Fig. 2. Isolated Aspergillus cultures: *A. flavus* (a), *A. niger* (b), *A. fumigatus* (c).

In 40% of cases, the causative agent of ascospherosis, the fungus *Ascosphaera apis fungus,* was found in conjunction with aspergilli. The concurrent growth of Aspergillus and Ascosphaera fungal species was observed. The fungi were found to be non-antagonistic and did not inhibit the growth of each other, which explains the prevalence of the combined course of these mycoses.

The results of laboratory tests of oxygen-containing disinfectants are presented in tables 1-3 ("+" presence of fungal growth; "±" presence of partial growth; "–" absence of growth). The regimes (preparation concentration and exposure) were selected based on previous studies.

Table 1. The disinfectants efficacy against *A. flavus*

Disinfectants	Preparation concentration, %	Exposition, min	*Aspergillus flavus*					
			Wood	Polystyrene foam	Wax	Metal	Plastic	Textile
Hydrogen peroxide	4	60	+	+	+	+	+	+
		180	+	+	+	±	±	+
	6	60	±	+	+	±	±	+
		180	±	-	±	-	-	±
	10	60	-	-	-	-	-	-
		180	-	-	-	-	-	-
"Dezinbak Super"	5	60	+	-	+	-	-	+
		180	±	-	±	-	-	-
	6	60	-	-	-	-	-	-
		180	-	-	-	-	-	-
	7	60	-	-	-	-	-	-
		180	-	-	-	-	-	-

Table 2. The disinfectants efficacy against *A. fumigatus*

Disinfectants	Preparation concentration, %	Exposition, min	*Aspergillus fumigatus*					
			Wood	Polystyrene foam	Wax	Metal	Plastic	Textile
Hydrogen peroxide	4	60	+	+	+	+	+	+
		180	+	+	+	±	±	+
	6	60	±	+	+	±	±	+
		180	±	-	±	-	-	±
	10	60	-	-	-	-	-	-
		180	-	-	-	-	-	-
"Dezinbak Super"	5	60	+	-	±	-	-	+
		180	±	-	-	-	-	±
	6	60	-	-	-	-	-	-
		180	-	-	-	-	-	-
	7	60	-	-	-	-	-	-
		180	-	-	-	-	-	-

Table 3. The disinfectants efficacy against *A. niger*

Disinfectants	Preparation concentration, %	Exposition, min	*Aspergillus niger*					
			Wood	Polystyrene foam	Wax	Metal	Plastic	Textile
Hydrogen peroxide	4	60	+	+	+	+	+	+
		180	+	+	+	±	±	+
	6	60	±	+	+	±	±	+
		180	±	-	±	-	-	±
	10	60	-	-	-	-	-	-
		180	-	-	-	-	-	-
"Dezinbak Super"	5	60	+	±	+	±	±	+
		180	±	-	±	-	-	±
	6	60	-	-	-	-	-	-
		180	-	-	-	-	-	-
	7	60	-	-	-	-	-	-
		180	-	-	-	-	-	-

In laboratory tests, the following regimes showed the best efficiency against all fungi species of the genus Aspergillus on all test objects: hydrogen peroxide at a concentration of 10% with an exposure of 60 and 180 min, "Dezinbak Super" at concentrations of 6 and 7% with an exposure of 60 and 180 min.

In addition, hydrogen peroxide was tested in the same concentrations (4, 6 and 10%), acidified with a solution of acetic acid to a content of 0.5%. The exposure time was 60 and 180 min (Tables 4-6).

Table 4. Hydrogen peroxide + 0.5% acetic acid efficacy against *A. flavus*

Disinfectant	Preparation concentration, %	Exposition, min	*Aspergillus flavus*					
			Wood	Polystyrene foam	Wax	Metal	Plastic	Textile
Hydrogen peroxide + 0.5% acetic acid	4	60	±	±	±	-	-	±
		180	±	-	-	-	-	-
	6	60	-	-	-	-	-	-
		180	-	-	-	-	-	-
	10	60	-	-	-	-	-	-
		180	-	-	-	-	-	-

Table 5. Hydrogen peroxide + 0.5% acetic acid efficacy against *A. fumigatus*

Disinfectant	Preparation concentration, %	Exposition, min	*Aspergillus fumigatus*					
			Wood	Polystyrene foam	Wax	Metal	Plastic	Textile
Hydrogen peroxide + 0.5% acetic acid	4	60	±	-	-	-	-	±
		180	±	-	-	-	-	±
	6	60	-	-	-	-	-	-
		180	-	-	-	-	-	-
	10	60	-	-	-	-	-	-
		180	-	-	-	-	-	-

Table 6. Hydrogen peroxide + 0.5% acetic acid efficacy against *A. niger*

Disinfectant	Preparation concentration, %	Exposition, min	*Aspergillus niger*					
			Wood	Polystyrene foam	Wax	Metal	Plastic	Textile
Hydrogen peroxide + 0.5% acetic acid	4	60	±	±	±	±	-	±
		180	±	-	-	-	-	±
	6	60	-	-	-	-	-	-
		180	-	-	-	-	-	-
	10	60	-	-	-	-	-	-
		180	-	-	-	-	-	-

As can be seen from the data presented in the tables, stable results were achieved with a 6% hydrogen peroxide content and an exposure time of 60 min.

The indicated regimes were tested in apiary trials, where they also demonstrated consistently high fungicidal and sporicidal activity.

4 Conclusion

Aspergillosis of bees is a zooanthroponosis and when it is detected, effective veterinary and sanitary measures must be taken. Therefore, the search for disinfectants that meet industry requirements and the selection of treatment regimes for beekeeping objects are an urgent task for science and practice.

In this work, oxygen-containing preparations of "Dezinbak Super" and hydrogen peroxide were studied as disinfectants for the treatment of beekeeping objects

https://doi.org/10.1051/bioconf/202516100016

in laboratory and industrial conditions. The working concentrations and exposures of these drugs were determined, and regimens for their use were developed.

The data obtained provide grounds for recommending the tested preparations in the specified regimes for disinfection in case of aspergillosis of bees.

References

1. A. Becchimanzi, R. Nicoletti, Front. Microbiol., **13** (2022)
2. K. Foley, G. Fazio, A. B. Jensen, W. O. H Hughes, Vet. Microbiol., 169, 203–210 (2014)
3. D. S. Bush, B. Calla, M. R. Berenbaum, Ecology and Evoluation, **14**:e10918 (2024)
4. S. Amaike, N. P. Keller, Ann. Rev. Phytopathol. **49**, 107–133 (2011)
5. J. W. Bennett, *Aspergillus: Molecular Biology and Genomics* (Caister Academic Press, Wymondham, 2010)
6. C. M. Visagie, J. Houbraken, Stud. Mycol., **95**, 253-292 (2020)
7. S. Evison, A. B. Jensen, Curr. Opin. Insect Sci., **26**, 105-113 (2018)
8. D. S. Bhabhra, Med. Mycol. **43** (Suppl. 1) S87–S93 (2005).
9. S. Vojvodic, A.B. Jensen, R.R. James, J.J. Boomsma, J. Eilenberg, Vet. Microbiol. **149**, 200–205 (2011)
10. J.P. Latge′, Clin. Microbiol. Rev. **12**, 310–350 (1999)
11. S. Seyedmousavi, J. Guillot, P. Arné, G.S. de Hoog, J.W. Mouton, W.J.G. Melchers, P.E. Verweij, Med. Mycol., **53**, 765–797 (2015)
12. M. Machado, J. Fortún, P. Muñoz, Med. Clín. (Eng. Ed.), **163**(4), 189–198 (2024)
13. Ch.-Ch. Lai, Po-Ren Hsueh, J. Microbiol., Immunol., Inf., **57**(6), 833–839 (2024)
14. Ch. Carter, R. Kahai, J. Cunningham, J. Kilduff, N. Hough, C. Baxter, D. Connell, A. Shah, Clin. Med., **24**(1), 100019 (2024)
15. T.R.T. Dagenais, N.P. Keller, Clin. Microbiol. Rev., **22**, 447–465 (2009)

BIO Web of Conferences **161**, 00017 (2025) https://doi.org/10.1051/bioconf/202516100017
FIES 2024

Digital and 'green' technologies in agricultural production on the example of a farming enterprise

Ekaterina A. Karelina* [1,2], *Valeriya I.* Ivanova[1,2], and *Samal M.* Smagulova [1,2]

[1] State University of Management, Moscow, Russia
[2] Vladimir State University, Vladimir, Russia

Abstract. The article deals with the problem of unfavourable environmental situation in Russian agriculture caused by intensive use of resources. It emphasises the need to introduce innovative technologies for sustainable development of the agricultural sector. The study focuses on the importance of an integrated approach combining digitalisation for production optimisation and green technologies for environmental safety. The paper analyses existing technologies and conducts experiments to evaluate the effectiveness of the proposed solutions. Experimental data obtained on a farm shows that the use of organic farming methods increases yields by 15% compared to conventional methods. The study also demonstrates the economic benefits of adopting digital technologies and green practices, reducing water consumption by 25% and electricity consumption by 10%. The author notes that despite the benefits of these technologies, digitalisation is 'insular' in nature. The article concludes that an integrated approach is needed to ensure food security, improve economic efficiency and address environmental challenges.

1 Introduction

The projected growth of the world population to 9 billion people by 2050 makes the search for effective and sustainable solutions in agriculture extremely important [1]. Intensive use of land resources, characteristic of many regions of the world, including the Russian Federation, leads to significant environmental consequences. Soil pollution, land degradation, and water resources depletion are just some of the challenges faced by modern agricultural production [2, 3]. These problems, in turn, have a negative impact on the volume and quality of produced products, as well as on human health and environmental well-being. As a result, a vicious circle is created, when extensive agricultural practices aimed at maximising production in the short term lead to long-term resource degradation and reduced productivity [4].

Various approaches to solving environmental problems [5] in the agricultural sector are constantly being developed and applied. One such solution is the introduction of digital technologies in agriculture. Automation and digitalisation of business processes can optimise the use of resources, increase yields, reduce costs and reduce negative environmental impacts. Digital technologies, for example, enable precise monitoring of crop health, accurate fertiliser application, and optimised irrigation systems, thereby increasing production efficiency and reducing resource wastage. Nevertheless, digitalisation, which could be an effective solution, is currently 'insular' and not widespread. The main reasons are the lack of the necessary database for software development, lack of awareness and accessibility of technology for many farms.

Another direction for solving problems is the introduction of 'green' technologies. This approach includes the use of organic fertilisers, biological methods of pest control, as well as the involvement of animal and crop wastes in production. Recycling agricultural waste into valuable products, for example, by vermicomposting with earthworms, not only helps to utilise waste and reduce pollution, but also produces high-quality organic fertiliser that enriches the soil and increases its fertility. As a result, the adoption of green technologies can reduce dependence on chemical fertilisers and pesticides, as well as reduce environmental impact and increase sustainability. However, despite the obvious advantages, at present, green technologies are not widespread, and some of them have no industrial application due to technological and economic limitations [6].

In light of the above, the use and development of 'green' technologies for soil fertility restoration, waste recycling and biomass production seems to be particularly relevant. This direction is an important factor for solving a set of interrelated problems. Sustainable agriculture requires an integrated approach that combines the use of digital technologies to optimise processes and green technologies to ensure environmental sustainability of production. The transition to intensive agriculture using resource-saving

* Corresponding author: opferpriesterin@mail.ru

and environmentally friendly technologies is an urgent task, the solution of which depends not only on food security, but also on the quality of life of future generations [7].

The purpose of this paper is to consider and substantiate the importance of introducing innovative technologies in agricultural production as the most important condition for its effective development in modern conditions. The emphasis is made on the need for an integrated approach to solving environmental problems, combining the introduction of digital technologies to improve the efficiency of production with the use of ‘green’ technologies to ensure its sustainability and environmental safety.

2 Method and materials

The overall plan of the experiment included analysing existing technologies, identifying problem areas and conducting a series of tests to verify the effectiveness of the proposed solutions. Specialised equipment was used in the research. In particular, a Hach DR6000 spectrophotometer was used to analyse the chemical composition of soil and water. This equipment was used to analyse samples to determine the level of toxic substances and heavy metals. In the process of work the spectrophotometer was used in the mode of multiwave scanning, which allows analysing several parameters simultaneously. To study the physical properties of soil, an Eijkelkamp Penetrologger penetrometer was used to measure the soil resistance during tip penetration. This equipment was used to assess the density and degree of soil compaction in different zones, which made it possible to optimise tillage techniques. The penetrometer was operated in automatic data acquisition mode, allowing detailed graphical data to be obtained from vertical slices. Ohaus Adventurer Pro brand scales were used for yield estimation, providing high precision weighing of samples. The equipment was used at harvest to determine the weight of the resulting produce. The scales worked in standard mode with high accuracy of readings. All obtained data were processed using specialised software for statistical analysis and graphing.

3 Results and discussion

The analysis of the current state of the agricultural sector. Based on this analysis, it was revealed that there is not only an unfavourable environmental situation, but also insufficient introduction of digital technologies in the agro-industrial complex, which also hinders the improvement of production efficiency [8]

To study in more detail the problems associated with intensive farming and the search for ways to improve its sustainability, our own experimental work was carried out. A small farm located in the central part of Russia, in particular, in the Tambov region, specialising in the cultivation of grain crops, mainly winter wheat, was chosen as a model object. The farm has a land bank of 250 hectares. The first stage of the study was a detailed study of soil on several plots of the farm, differing in their characteristics and cultivation history. Soil samples were collected from a depth of 15-20 cm from five plots of 15, 20, 25, 30 and 35 hectares located in different parts of the field to ensure representative samples. Soil analyses carried out in an accredited laboratory using chromatography and atomic absorption spectrometry techniques revealed the presence of elevated levels of nitrates, averaging 40 mg/kg, and pesticides, averaging 0.7 mg/kg, including compounds such as glyphosate and 2,4-D, which exceeds the permissible standards established for agricultural land in the region, namely 30 mg/kg for nitrates and 0.5 mg/kg for pesticides. These data indirectly indicate the intensive use of chemical fertilisers such as ammonium nitrate and urea and crop protection products including herbicides and insecticides during the past few years [9]. At the second stage, a comparative analysis of winter wheat yields was carried out in plots where conventional tillage methods, including ploughing and use of mineral fertilisers, were applied and in plots where organic farming methods, including the use of compost obtained from crop waste and manure, and Bacillus subtilis and Trichoderma-based biopreparations, were introduced.

Under the conventional approach, using mineral fertilisers at a dosage of N90P60K60, the average yield was 3.5 t/ha, with significant variations noted between 3.2 and 3.8 t/ha depending on the plot. In contrast, plots where organic methods were applied showed a 15% increase in average yield to 4.025 t/ha, with more stable yields ranging from 3.9 to 4.1 t/ha. [10]. Thus, the experimental study confirmed that the transition to environmentally friendly farming practices not only contributes to the improvement of the environmental situation by reducing soil contamination with chemicals, but also leads to an increase in yields, increasing the resistance of crops to unfavourable factors. It is also worth noting that the experimental work used modern equipment for remote monitoring of crop conditions, including a multispectral camera DJI Mavic 2 Pro with the ability to analyse NDVI, as well as weather station Davis Vantage Vue to monitor weather conditions, which made it possible to accurately calculate the required amount of fertilizers and plant protection products, which significantly reduced the consumption of these materials. Based on the data obtained, taking into account the cost of organic fertilisers and biopreparations, the economic effect from the introduction of digital technologies and organic farming was calculated, which amounted to about 10% of the total value of the crop, which in monetary terms amounted to about 15 per hectare at an average cost of wheat grain per hectare, at an average cost of wheat grain 150 per tonne. Analysing these data, it can be noted that they reflect a general trend in the agricultural sector of Russia, where irrational use of water and land resources is observed. About 6 million hectares of irrigated land and 5 million hectares of drained land are in unsatisfactory condition due to secondary salinisation, desiccation and waterlogging of soils. Significant losses of fertilisers during transportation and storage, when only 30-40% gets to plants and the rest pollutes the environment. Agricultural water withdrawal is 40 km³/year, of which 32 km³/year is for irrigation, and up to 40% of return water is polluted wastewater [11-13]. The lack of centralised water supply in rural areas, where 60% of the population takes water from open water bodies, exacerbates environmental problems. Digitalisation of agriculture in Russia has an 'insular' character, which is associated with an insufficient database for software developers for the needs of the agro-industrial complex. As part of this study, a comprehensive analysis of resource utilisation at the pilot site was carried out. In particular, before the introduction of smart irrigation and monitoring systems, the average water consumption on the plot was 1200 cubic metres per month, and after their introduction this indicator decreased to 900 cubic metres, which corresponds to a 25% reduction. The electricity consumption required to run the pumping and lighting systems was initially 2000 kWh per month, and after optimisation and implementation of energy efficient solutions it was reduced to 1800 kWh, a 10% saving. Analysis of data from the digital monitoring systems allowed the optimal rates of fertiliser application to be accurately established, resulting in a 15% reduction in fertiliser consumption, from 1000 kg to 850 kg per month, and 18% reduction in crop protection products, from 800 litres to 656 litres per month. Digital technologies such as soil moisture sensors (Vegetronix VH400) were used during the study, which provided real-time data for automated irrigation control via an Arduino Uno system. Temperature and light sensors (DHT22 and BH1750) were also used to collect information on the microclimate of the plot and adjust the greenhouse operation modes. Cloud-based platforms such as ThingSpeak were used to collect and analyse data, allowing continuous monitoring and data analysis remotely. The plant nutrition management system was implemented using automated dosing systems based on data from soil chemistry sensors.

The concept of 'Agriculture 4.0', as noted in the study, is a paradigm that involves the full automation and digital transformation of business processes in agricultural production. This includes not only optimising the use of resources, but also moving towards more precise and managed farming. Israel, where only 20 per cent of the land is suitable for farming, is cited as an example of a successful implementation of Agriculture 4.0. Thanks to the introduction of modern technologies, including drip irrigation systems, automated greenhouse management and precision farming, this country has achieved 95% food security [14]. Data analysis confirms that the introduction of digital technologies in the form of smart farms leads to a significant increase in production efficiency. For example, there is an increase in milk yields by 30-40% on average, which in absolute terms means an increase in average milk yield from 25 to 35 litres per cow per day. As a result of the introduction of modern transport management methods, which include GPS monitoring and route optimisation, fuel costs are reduced by an average of 20%, from 1,000 litres per month to 800 litres. This is achieved by reducing vehicle mileage and operating time, as well as optimising product supply logistics. Comparing the results obtained with the data of other studies, it can be noted that the conclusions about the negative impact of intensive agriculture on the environment and the need to introduce 'green' technologies are confirmed by many scientific works. For example, studies conducted in the countries of the European Union show that the use of organic farming methods helps to improve soil fertility and reduce pollution of water resources [15]. At the same time, the introduction of digital technologies in agriculture, described in the article, is a relatively new direction, and its economic efficiency and environmental benefits require further study and improvement. Thus, the analysis of the data of the conducted experiment showed that the introduction of digital technologies can achieve a reduction in resource consumption with a simultaneous increase in yields, which in turn contributes to the competitiveness of the agro-industrial complex [16]. In conclusion, it can be stated that there is a need for a comprehensive approach to the development of agricultural production, including both the introduction of digital technologies and the transition to environmentally friendly methods of farming. This approach will allow not only to solve many environmental problems in this area, but also to ensure food security of the country and increase the economic efficiency of the agro-industrial complex.

4 Conclusion

The study revealed a number of key problems in modern agriculture, especially in the context of intensive farming in Russia. There is a significant negative impact on the environment associated with intensive use of land resources, including soil contamination with nitrates and pesticides. This is caused by the use of chemical fertilisers and plant protection products in amounts exceeding permissible norms. At the same time, there is insufficient implementation of digital technologies, which hinders the improvement of production efficiency. Experimental work has shown that the use of organic farming methods with the use of compost and biopreparations, contributes to increasing yields and improving the environmental situation by reducing soil pollution [17]. In addition, the introduction of digital technologies, such as monitoring and irrigation management systems, leads to significant savings of resources, including water and electricity, and also allows optimising the application of fertilizers and plant protection products, which in turn reduces their consumption and the cost of production [18, 19]. The introduction of the concept of 'Agriculture 4.0', which implies full automation and digital transformation of processes, contributes to the improvement of production efficiency, as evidenced by the experience of Israel and other countries. A comparative analysis of the obtained data confirms that an integrated approach combining the introduction of digital and 'green' technologies is the most promising way to solve problems in the agricultural sector. This approach allows not only to increase yields and economic efficiency of production, but also to ensure environmental sustainability and food security, which is especially important in the context of a growing world population and climate change.

Acknowledgement

The work is based on the results of the research carried out at the expense of the federal budget under the state assignment (name of the research topic "Development and implementation of the strategy for the development of foreign economic relations of agriculture and agro-industrial complex of the Russian Federation taking into account restrictions imposed by sanctions and new priorities of economic cooperation with foreign countries"; code of the research topic assigned by the founder- FZUN-2024-0007).

References

1. L. Inthakoun, M. Kenney-Lazar. The persistence of swidden cultivation and upland autonomy in Laos. J. Peasant Stud., (2025). DOI: 10.1080/03066150.2024.2440539.

2. E.Yu. Kulikova, S.V. Balovtsev, O.V. Skopintseva. Geoecological monitoring during mining operations. Sustainable Development of Mountain Territories, **16,** 580–588 (2024). DOI: 10.21177/1998-4502-2024-16-2-580-588.

3. V.L. Snezhko, D.M. Benin. Dynamics of the Condition of Reclaimed Agricultural Lands in the Russian Federation. Land, **10,** 1288 (2021). DOI: 10.3390/land10121288.

4. N. Garambois, U. Le Goff, L. Thibaudeau. Agrarian dynamics and climate change in the Senegalese Sahelian peanut basin. Agrar. Syst. Clim. Change: Journeys Adapt. Glob. South, 15-37 (2024). DOI: 10.1079/9781800628137.0001.

5. M.V. Shabanov, M.S. Marichev, S.S. Mangiyeva, A.A. Sokolov. Chemozem formation under conditions of prolong exposure to aero-industrial emissions from a mining and smelting plant. Sustainable Development of Mountain Territories, **15,** 727-740 (2023). DOI: 10.21177/1998-4502-2023-15-3-727-740.

6. B. Aparecida da Silva, R. Evangelista de Oliveira, A. Cavalieri Sais, E. Cardoso leite. Family farmers and trees: ecological functions, ecosystem services and food production in agrarian reform settlements in Brazil. Agroecol. Sustain. Food Syst., **49,** 4-21 (2025). DOI: 10.1080/21683565.2024.2405515.

7. S. Bamidele. BRICS Revolution: The Global Agrarian Shift Sparked by Emerging Economies. India Q., (2025). DOI: 10.1177/09749284241307937.

8. V.V. Yurak, A.N. Malyshev, S.S. Zavyalov. Review of new remote methods for environmental monitoring. MIAB. Mining Inf. Anal. Bull., **11-1,** 51-67 (2022). DOI: 10.25018/0236_1493_2022_111_0_51.

9. Z.A. Buryak, A.G. Narozhnyaya, A.V. Gusarov, A.A. Beylich. Solutions for the Spatial Organization of Cropland with Increased Erosion Risk at the Regional Level: A Case Study of Belgorod Oblast, European Russia. Land, **11,** 1492 (2022). DOI: 10.3390/land11091492.

10. B. Zh. Akhmetov, A. Zh. Kasenova, O. E. Merzlyakov, L. M. Tatarintsev. Modeling of sustainable agricultural land use on the basis of agroecological assessment of the territory. Sustainable Development of Mountain Territories, **15,** 17–26 (2023). DOI: 10.21177/1998-4502-2023-15-1-17-26.

11. O. Afanaseva, M. Afanasyev, S. Neyrus, D. Pervukhin, D. Tukeev. Information and Analytical System Monitoring and Assessment of the Water Bodies State in the Mineral Resources Complex. Inventions, **9,** 115 (2024). https://doi.org/10.3390/inventions9060115.

12. R.T. Klochko, L.S. Malinovskaya, G.I. Ignatyeva, A.B. Sokhlikov. Beekeeping, **1,** 24-26 (1996).

13. M.Zh. Makhambetov, G.B. Toktaganova, G.I. Issayev, L.E. Yusupova, N.A. Akhmetov. Ecological assessment of soil condition in Zhylyoi district of Atyrau region. News of the National Academy of Sciences of the Republic of Kazakhstan, Series of Geology and Technical Sciences, **2,** 146-155 (2024). https://doi.org/10.32014/2024.2518-170X.399.

14. V.Yu Konyuhov, A.M. Gladkih, I.I. Galyautdinov, Y.D. Severina. Economic aspects of green technologies. IOP Conference Series: Earth and Environmental Science, **350,** 012036 (2019). DOI: 10.1088/1755-1315/350/1/012036.

15. A.S. Danilov, A.S. Lozovaya. Criteria-assessment tools for inventory of accumulated environmental harm objects. Sustainable Development of Mountain Territories, **16,** 861–872 (2024). DOI: 10.21177/1998-4502-2024-16-3-861-872.

16. O. A. Kolenchukov, A. Y. Mikhaulov, K. A. Bashmur, R. B. Sergienko. Forecasting the technical condition of thermochemical reactor systems. SOCAR Proceedings, **2023,** 123–130 (2023). https://doi.org/10.5510/OGP20230100814.

17. S. Das. Determinants of livelihood vulnerability to climate change among the agrarian community: Empirical evidence from the coastal region of Indian Sundarban Biosphere Reserve. Reg. Stud. Mar. Sci., **78,** 103777 (2024). DOI: 10.1016/j.rsma.2024.103777.

18. L. Ossome. Gender and development in the agrarian south. World Dev., **188,** 106876 (2025). DOI: 10.1016/j.worlddev.2024.106876.

19. J. Grajales, O. Toukpo. Making green cocoa: Deforestation, the legacy of war, and agrarian capitalism in Côte d'Ivoire. J. Agrar. Change, **25,** e12609 (2025). DOI: 10.1111/joac.12609.

Use of extruded lupin in feeding young quails

Nadiya Kasanova[1,*], *Munira* Gainullina[1], *Dmitry* Prosvirnikov[2], *Denis* Tuntsev[2], *Ekaterina* Golovkova[1]

[1] Kazan State Academy of Veterinary Medicine named after N.E. Bauman, 420029, Kazan, Russia
[2] Kazan National Research Technological University, Kazan, 420015, Russia

Abstract. In this study, the effect of feeding extruded white lupine seeds of Dega variety on the productive performance and physiological condition of young quails was investigated. The experiment was conducted on 160 quails of Texas breed, which were divided into 2 groups. The first group was fed extruded soybean seeds (control) as a part of mixed fodder, in the diets of the second group soybean was replaced by extruded white lupine seeds equivalent in protein. During the period of quail rearing, daily records of feed digestibility and physiological condition of birds were made, weekly weighing and weight of quails were determined. To study the physiological condition of experimental birds, morphological and biochemical blood parameters were determined. As a result of the experiment it was found that inclusion of extruded white lupine seeds in the diet promoted a reliable increase in body weight of quails by 2.6% ($p \geq 0.05$), while feed costs per unit of gain decreased by 3.4%. In the blood of quails, haemoglobin level and haematocrit increased by 9.8 and 21.8% ($p \geq 0.05$), indicating oxygen saturation of blood cells. The enzymatic activity of alkaline phosphate by 18.6%, ALT by 74.3%, AST by 17.8%, GGT by 59.6% decreased in blood, indicating normal metabolism in the body of young quails.

1 Introduction

Quail breeding is a unique branch of poultry farming, as quails have a number of distinctive features and advantages: high growth rate (on average 4-5 times higher than that of chickens), early maturity (egg production occurs at 5-6 weeks of age, and meat breeds reach a slaughter weight of 350-400 g by this time) [1]. Therefore, quail breeding is the most profitable branch of poultry farming. At the same time, the aspects of scientifically based feeding of quails, protein and energy nutrition of feed mixtures, the need for macro- and microelements and vitamins are insufficiently studied.

As it is known, the fundamental factor in feeding poultry is protein nutrition of mixed fodders. According to literature data, the required protein intake rate differs depending on the age, breed and type of birds. So, at intensive breeding of quail chicks for meat at the age of 1 to 4 weeks they need to receive not less than 25% protein, further the amount of protein during fattening will be not less than 21%. The amount of protein for laying quails is up to 25%. When calculating the need for protein, it is also necessary to take into account its completeness, which is determined by the amino acid composition. Quails, like other birds, are sensitive to the lack of cysteine, lysine, methionine, tryptophan. The growth and productive health of quail depends on these amino acids [2, 3]. No less important criterion is the quality and digestibility of protein in the diet, hence the efficiency of feed nitrogen utilisation by poultry.

According to scientists, the balance between energy and protein in the feed mixture for quail is of great importance. These indicators guarantee high productivity and reproductive qualities of the growing birds. It has been found that feeding a feed mixture with high protein but low carbohydrates and fats leads to metabolic disorders, resulting in pathologies of the gastrointestinal tract and biliary tract. With a high-energy diet but with low protein levels, quail will be deficient in protein for building body tissues [4, 5].

Traditionally, expensive feeds of animal origin (meat and bone meal, fish meal, milk powder), leguminous crops, by-products of the processing industry (oilcakes, meal, pellets, etc.) are used as a source of protein in compound feeds. Today the world trend, including the Russian Federation, is the use of feedstuffs based on soya and its derivatives. Soybean seeds are rich in protein (up to 45%), which in addition contains the necessary proteinogenic amino acids. However, a significant disadvantage is that this crop is demanding to natural conditions and its yield is not so high on the territory of the Central region of Russia. In addition, most of the exported soybeans are genetically modified [6]

The above-mentioned facts indicate that the search for other, more optimal and qualitative sources of protein is relevant. Special attention should be paid to the leguminous crop Lupinus albus L. White lupin is usually used in feeding, which is characterised by a balanced amino acid composition, high nutrition, besides, the protein level in its seeds is 10% higher than that of peas, vetch and other leguminous crops [7]. According to R. Kriseldi et al. (2018), lupin seeds contain 34-39% (up to 46% crude protein in dry matter) protein and significant

* Corresponding author: nadia-kasanova@mail.ru

amounts of fats, carbohydrates, minerals, vitamins and vitamin-like substances [8].

However, 20% of lupin seeds are low nutrient hulls, in which 30% are hemicellulose and pectins, which after grinding can be effectively used as dietary fibre in animal diets and foods [11].

But the main reason restraining the use of lupin culture in poultry feed is considered to be the presence of alkaloids. Various processing methods including thermodynamic methods have been used to inactivate alkaloids [9]. One of these methods is processing in extruders at high pressure and a certain temperature regime (extruding). In addition, this method helps to increase the nutritional value of feed and its digestibility [10].

As a result of breeding work, new low-alkaloid lupin varieties with high content of organic matter, protein and mineral substances in seeds were obtained [12, 13]. White lupine of Dega variety is of the greatest interest. It is suitable for cultivation in the Central Black Earth, Middle Volga and Lower Volga regions [14]. The yield of this variety reaches 5 tonnes/ha, the vegetation period is 120 days, and the alkaloid content is less than 0.05% [15].

Researchers wrote that lupin is a promising protein product for the poultry industry [16]. The demand of soybean for poultry industry may be up to 1-1.5 million tonnes annually [17]. It is recommended to add non-alkoloidal (sweet) varieties up to 5-7%, depending on the age category, to poultry feed rations [18].

To date, scientific literature sources contain data on the favourable effect of lupine on poultry productivity. Experiments on laying hens, broiler chickens and quails have shown that replacement of soya meal in mixed fodders with crushed white lupine grain reduces feed cost, increases productivity, egg yield and weight, thus improving the profitability of production [19].

In the experiments of Z.N. Fedorova (2023) it was found that the inclusion of 10% and 15% of treated lupin seeds in the feed mixture for broiler chickens and 15% for adult stock has a positive effect on productivity and reduces feed costs per 1 kg of gain by 100 g [20].

In connection with the above, the purpose of the study was to investigate the effect of feeding extruded white lupine seeds on the physiological state and performance of young quails of the Texas quail breed.

2 Methods and materials

The scientific experiment was carried out according to the generally accepted methodology (VNITIP, 2000). We formed two groups of quails at 10 days of age of 80 birds each. Both groups of birds were kept in typical cages and the same conditions (temperature, humidity, light regime). According to the scheme of the experiment, quails received factory prepared compound feed (PC) in accordance with zootechnical norms (Kalashnikov A.P. et al., 2003). The diet of birds of the first group consisted of mixed fodder (95%) and extruded soya seeds (5%) (positive control). Birds of the second experimental group were fed extruded crushed white lupine seeds of Dega variety instead of soya equivalent in protein.

In the course of the experiment we kept records of safety and physiological condition of quails, weekly weighing was carried out to determine the dynamics of body weight gain. Daily record of feed digestibility was carried out, based on the obtained data, feed costs per unit of live weight gain were calculated.

Analysis of feeds and feed additives was carried out on express feed analyser NirstmDS 2500 Foss (Denmark). Dry matter, crude protein, crude fat, crude fibre, starch, crude ash, calcium, phosphorus were determined in the samples.

At the end of the experiment, blood sampling for morphological and biochemical studies was carried out in experimental quails. In the blood of birds the haemoglobin level was determined by the colorimetric haemoglobin cyanide method on a spectrophotometer Mindray BS 240 (China), the determination of the number of form elements was carried out in a Goryaev chamber using a microscope Olympus CX23 (Japan). To determine haematocrit, blood was pre-centrifuged on a CritSpin haematocrit centrifuge (USA). Biochemical parameters in blood serum were determined on a Chemray-240 analyser (PRC). The results were statistically processed in Microsoft Excel programme.

3 Results and discussion

We found that quails of control and experimental groups had good appetite, reacted to stimuli, looked healthy, droppings of birds in all groups had characteristic colour and consistency, no clinical signs of digestive disorders were found. The use of extruded lupine seeds in quail feeding had a positive effect on zootechnical parameters of productivity (Table 1).

Table 1. Quail persistence, growth and feed costs

Indicator	I group	II group
Number of quails at the beginning of the experiment, number of birds.	80	80
At the end of the experiment, number of birds.	80	80
Preservation %	100	100
Live weight at the beginning of the experiment, g	49.95±0,78	50.16±0,56
At the end of the experiment, g	273.09±1.96	280.10±2.37*
Live weight gain of quails, g	223.14	229.92
Average daily live weight gain of quails, g	5,87	6,05
Feed costs per 1 kg of gain, kg	1.77	1.71

* – p≥0.05 the difference between the lowest and the target value

Table 1 shows that quails receiving extruded lupine had slightly higher growth energy and live weight. Average daily weight gain of quails of the control (first group) was 5.87 g, and in the experimental (second group) was 3.1% higher and was 6.05 g. Live weight of birds of the second group was higher than the first group by 7.01 g or 2.3% and was 280.1±2.37 g by the end of the experiment, while in the control it was 273.09±1.96 g ($p \geq 0.05$).

Live weight of quails receiving extruded soya with mixed feed was 85.07±1.22 g at the age of 17 days, 110.19±1.23 g at the age of 24 days, 142.18±1.71 g at the age of 32 days, 207.86±1.37 g at the age of 40 days and 273.09±1.95 g at the age of 48 days. In the second group live weight of birds was 85. 25±1.27 g, 24 days - 118.86±1.38 g, 32 days - 147.68±1.87 g, 40 days - 215.41±2.99 g, 48 days - 280.10±2.37 g. It should be noted that this indicator in birds of the second group was higher at the age of 17 days - by 2.6%, 24 days - by 7.8%, 32 days - by 3.9%, 40 days - by 3.6%, 48 days - by 2.6%.

Inclusion of lupine in the composition of the diet allowed to reduce feed costs per 1 kg of live weight gain of birds by 3.4%.

The study of basic morpho-biochemical indices of blood is an important indicator of evaluation of the state of the internal environment of the organism, the course of metabolic processes, its main factors of immunity and metabolism. Morphological blood parameters of experimental quails are presented in Table 2.

Table 2. Morphological indices of quail blood

Indicator	I group	II group
Haemoglobin, g/l	123.8±2.87	135.2±2.05*
Haematocrit, %	41.54±1.26	50.64±3.36*
Red blood cell count (RBC), 10^{12}/L	2.78±0.57	2.86±0.15
Mean erythrocyte volume (MCV), pccl	152.82±13.20	180.48±19.77
White blood cell count (WBC), 10^9/L	2.91±0.23	5.63±0.76**
Platelets, thousand µl	58.6±13.07	65.4±7.21

Table 3. Biochemical parameters of quail blood

Indicator	I group (n=5)	II group (n=5)
Total protein, g/l	36.6±1.01	35.98±0.78
Albumin, g/l	23.42±1.04	23.58±0.49
Glucose, mmol/l	15.8±0.94	15.16±0.87
Urea, mmol/l	1.48±0.42	1.30±0.14
Cholesterol, mmol/l	6.23±0.74	5.84±0.45
Triglycerides, mmol/l	4.17±1.98	2.45±1.15
Alkaline phosphatase (alkaline phosphatase), U/L	846.4±26.40	713.2±15.80**
Aspartataminotransferase (AST), units/L	248.56±11.06	204.28±10.04*
Alanine aminotransferase (ALT), units/L	24.80±0.66	6.36±0.27***
Gamma- glutamyltranspeptidase (GGT), units/litre	4.60±0.41	1.86±0.82**
Uric acid, µmol/L	394.2±40.01	362.74±29.29
Creatinine, µmol/L	70.4±0.68	57.60±1.07***
Total bilirubin, µmol/l	8.1±0.51	2.68±0.33***
Calcium, µmol/l	2.96±1.00	2.92±0.52
Phosphorus, µmol/l	2.21±0.58	2.69±0.31

*$p \geq 0.05$; **$p \geq 0.01$; ***$p \geq 0.001$ the difference between the lowest and the target value

It follows from the data of Table 2 that in the second group of quails in the blood there were higher indicators of haemoglobin content by 9.8% and haematocrit level by 21.8%, which indicates the activation of haemopoiesis process, improvement of protein and oxygen supply of blood cells. This fact is also confirmed by the increase in the number and volume of erythrocytes [21]. Thus, in the blood of quails of the control group the number of erythrocytes was 2.78×1012/l, and in the blood of birds of the experimental group - 2.86×1012/l. Also within the physiological norms for synanthropic birds there was a slight increase in leukocytes and thrombocytes. The obtained data are confirmed by other authors who studied haematological parameters of quails [22].

Analysing blood serum parameters (Table 3), it should be noted that the amount of protein, albumin, glucose, urea, calcium and phosphorus in experimental quails was approximately at the same level. In quails of the second group some decrease in cholesterol, triglycerides and uric acid was observed.

The activity of such transaminases as alkaline phosphate, ALT, AST, GGT characterise the state of amino acid and protein metabolism, as well as the functional work of the liver. In our studies in blood serum of birds of the experimental group these indices were significantly lower: ALT by 18.6% (p≥0.01), ALT by 74.3% (p≥0.001), AST by 17.8% (p≥0.05), GGT by 59.6% (p≥0.01).

4 Conclusion

The results of the conducted experiment showed that inclusion of ground extruded white lupine seeds in the feed mixture for young quails instead of extruded soybean does not have a negative effect on physiological condition. promotes an increase in body weight of quails by 2.6% (p≥0.05), and a decrease in feed cost by 3.4%.

Acknowledgement

The research was carried out and published with the support of the grant of the Academy of Sciences of the Republic of Tatarstan to young candidates of sciences (postdoctoral fellows) to perform research work (agreement № 148/2024-PD from 16.12.2024).

References

1. B.O. Ida, T. Kiendrebeogo, S. Sanou, Brit. J. of Multidiscipl. and Advan. Stud., **5(1)**, 1–21 (2024), https://doi.org/10.37745/bjmas.2022.0406
2. I.A. Egorov, L.S. Belyakova, Poultry Production, **4,** 31-33 (2009)
3. M.A. Hossain, A.S.M. Mahbub, S.A. Belal, Veter. and Animal Sci., **23**, 100340 (2024), https://doi.org/10.1016/j.vas.2024.100340
4. T.N. Lenkova, T.A. Egorova, I.G. Sysoeva, Poultry Science, **11-12,** 54-58 (2019), https://doi.org/10.33845/0033-3239-2019-68-11-12-54-58
5. E.E. Kurenkov, M.K. Gaynullina, Veterinary Medicine and Feeding, **4,** 60-64 (2024), https://doi.org/10.30917/ATT-VK-1814-9588-2024-4-11
6. F. Gresta, M. Oteri, D. Scordia et al., Agricult., **13(2)**, 434 (2023), https://doi.org/10.3390/agriculture13020434
7. V.A. Zarudnyi, V.V. Bardash, Efficient Livestock Production, **3(193),** 92-95 (2024), https://doi.org/10.24412/cl-33489-2024-3-92-95
8. R. Kriseldi, P.B. Tillman, Z. Jiang, W.A. Dozier, Poul. Sci., **97(5)**, 1614–1626 (2018), https://doi.org/10.3382/ps/pex395
9. M. Stanek, T. Rotkiewicz, W. Sobotka et al,, Acta Veter. Brno, **84(1)**, 55–62 (2015), https://doi.org/10.2754/avb201585010055
10. Z.N. Fedorova, Grain legumes and cereals, **4(32),** 142-148 (2019), https://doi.org/10.24411/2309-348X-2019-11146
11. N.A. Votanovskaya, G.M. Shulaev, R.K. Milushev, A.M. Puchnin, Vestnik Tambov University, Series: Natural and Technical Sciences, **22(2),** 426-429 (2017), https://doi.org/10.20310/1810-0198-2017-22-2-426-429
12. A. Duque, P. Manzanares, M. Ballesteros, Renew. Energ., **114,** 1427–1441 (2017), https://doi.org/10.1016/j.renene.2017.06.050
13. T. Khedr et al., J. of Food Composit. and Anal., **135,** 106600 (2024), https://doi.org/10.1016/j.jfca.2024.106600
14. A.A. Muravyev, S.V. Kadyrov, I.S. Muravyeva, Bulletin of Voronezh State Agrarian University, **4(79),** 22-30 (2023), https://doi.org/10.53914/issn2071-2243_2023_4_22
15. M.I. Lukashevich, P.A. Ageyeva, N.V. Novik, M.V. Zakharova, Achievements of Science and Technology of Agroindustrial Complex, **32(2),** 29-32 (2018), https://doi.org/10.24411/0235-2451-2018-10207
16. L. Lopez-Bellido, M. Fuente, Adv. in agron., **40,** 239–295 (1986), https://doi.org/10.1016/S0065-2113(08)60284-9
17. E.N. Andrianova, I.A. Egorov, E.N. Grigorieva, A.S. Tsygutkin, Poultry Production, **11-12,** 31-36 (2019), https://doi.org/10.33845/0033-3239-2019-68-11-12-31-36
18. O.S. Mischeriakova, A.G. Krasnoperov, V.A. Zarudny, International Research Journal, **10(136)** (2023), https://doi.org/10.23670/IRJ.2023.136.31
19. C.I. Lim, N.J. Choi, Canad. J. of Animal Sci., **103(2),** 167-173 (2023), https://doi.org/10.1139/cjas-2022-0104
20. Z.N. Fedorova, V.V. Volkov, O.S. Mischeriakova, Bulletin of the Ulyanovsk State Agricultural Academy, **1(61),** 116-121 (2023), https://doi.org/10.18286/1816-4501-2023-1-116-121
21. P.A. Polkovnichenko, A.P. Polkovnichenko, D.V. Vorobyev, V.I. Vorobyev, Scientific Notes of N.E. Bauman KSAVM, **237(1),** 47-150 (2019), https://doi.org/10.31588/2413-4201-1883-237-1-147-150
22. R.A. Meshabaz et al., Sci. J. of Univ. of Zakho, **5(3),** 249–253 (2017), https://doi.org/10.25271/2017.5.3.393

Waste-free integrated processing of invertebrates into feed additives using proteolytic enzymes

Karim Ernazarov[1,*], *Bekzod* Yusupov[2], *Khumoyiddin* Turaev[3], *Sayyora* Ganieva[4], *Ivan V.* Masienko[5], *Dmitry S.* Sukhanov[5]

[1] "Tashkent Institute of Irrigation and Agricultural Mechanization Engineers" National Research University, Tashkent, Uzbekistan
[2] Almalyk branch of the Tashkent State Technical University named after Islam Karimov, Almalyk, Uzbekistan
[3] Termez State University, Termiz, Uzbekistan
[4] Termiz State Pedagogical Institute, Termiz, Uzbekistan
[5] Kuban State Agrarian University named after I. T. Trubilin, Krasnodar, Russia

Abstract. This paper presents a study on integrated and waste-free processing of various invertebrate species such as molluscs and crabs. The main focus is on the use of the proteolytic enzyme preparation 'Krusenzyme', isolated from crab liver, for the processing of raw materials. The study showed that the use of Crusenzyme was effective in removing skin from cephalopods, softening the muscle tissue of molluscs and deproteinising the shell of crabs. The enzymatic treatment improved the texture of the products, increased the yield of finished products and reduced energy consumption compared to traditional methods. The possibility of obtaining feed additives from mollusk processing wastes rich in proteins, calcium and microelements was also studied. It was shown that inclusion of flour from soft tissues and mollusk flaps in the diet of farm animals contributes to the increase in weight gain and improvement of productivity indicators. The developed technology allows to obtain a wide range of valuable products, including chitin, chitosan, fat, carotenoids and protein hydrolysates, contributing to environmentally safe production and rational use of hydrobiont resources.

1 Introduction

Traditional seafood processing methods often result in significant losses of valuable nutrients and high energy costs, which emphasises the importance of research aimed at sustainable resource utilisation [1]. At the same time, not all components extracted from aquatic organisms are utilised. To make matters worse, a large proportion of seafood, especially shellfish, is a difficult raw material to process, characterised by rigidity and difficulties in processing. All these factors make aquatic bioresource products not always accessible to consumers.

Along with traditional approaches, a promising direction in solving the problem of effective processing of hydrobionts is the development of biotechnological methods. The use of enzyme preparations for the processing of raw materials can improve its structure and consumer qualities, reduce energy consumption and increase the yield of finished products [2]. However, the introduction of biotechnological methods into practice still faces a number of difficulties. These include the lack of available and effective enzyme preparations, and the lack of technologies that allow controlling the process of protein hydrolysis and obtaining a product with specified properties. In addition, many methods do not involve inactivation of enzymes, which can lead to deterioration of the quality of the finished product and shortening of its shelf life. The problem is also the need to develop technologies for integrated processing of waste generated during the processing of hydrobionts.

In the context of the above-mentioned, the direction under consideration is of particular relevance and importance. Studies on the use of proteolytic enzymes isolated from crab liver for complex processing of invertebrates, in particular, molluscs, open wide opportunities for solving a number of problems. Firstly, the use of the enzyme preparation Krusenzyme makes it possible to effectively remove the skin cover of cephalopod molluscs, soften muscle tissue and obtain high-quality food products with improved consumer properties. Secondly, the use of enzymatic hydrolysis makes it possible to utilise previously unused wastes such as soft tissues and shells of clams and crabs for processing, yielding valuable feed and technical products. Thirdly, the application of biochemical methods contributes to the environmental safety of production by reducing waste. An important point is the development of technology for inactivation of enzymes in the final product, which provides a long shelf life and stable quality. The use of invertebrate shells as a source of easily digestible calcium and other trace elements is

* Corresponding author: ernazarovkarimjon45@gmail.com

also an important aspect in achieving waste-free production. Finally, the biomedical and clinical trials conducted on corbicula extracts proved their efficacy in the prevention and treatment of hepatitis, expanding the range of products obtained. Thus, the results of the research are aimed at solving a whole set of problems related to the processing of hydrobionts and correspond to modern trends in the development of the food industry.

The purpose of the presented work is to summarise the results of research in the field of waste-free complex processing of invertebrates, conducted in the laboratory of biotechnology of food and technical products. Within the framework of the work the efficiency of biochemical methods use in technologies of food, feed and technical products manufacturing from squid, octopus, spisula, anadara and crab was analysed. The possibilities of using the food additive Krusenzyme for skin removal, softening of muscle tissue, waste processing were studied. As a result, a number of solutions were proposed aimed at rational, efficient and environmentally safe utilisation of hydrobiont resources, which meets the needs of the modern agricultural industry.

2 Method and materials

Experiments on enzymatic treatment of various species of molluscs and crustaceans using the proteolytic enzyme preparation 'Krusenzyme' isolated from the liver of crab were carried out in the course of research. This complex includes various proteinases with broad substrate specificity, which allows it to effectively hydrolyse proteins of skin, muscle tissue, and connective tissue proteins. Crusenzyme also contains thermostable enzymes. In addition to proteinases, there are other enzymes in the preparation that contribute to protein hydrolysis, but the main focus is on proteolytic enzymes. The main objective of the research was to study the effect of enzymatic treatment on the structure and properties of muscle tissue, skin and shell of invertebrates in order to develop effective technologies for the production of food, feed and technical products.

Various technological equipment was used to carry out experimental work. Industrial meat grinders were used for grinding raw materials. Fermentation processes were carried out in laboratory thermostats with maintenance of specified temperatures and stirring modes. Laboratory thermometers and digital thermoregulators were used for temperature control. Water baths and laboratory autoclaves were used for hydrothermal treatment. Industrial and laboratory drying cabinets with preset temperature and air circulation regimes were used for drying and wilting of products. Instrumental method was used to determine textural and organoleptic characteristics. Standard analytical instruments and methods were used to analyse the chemical composition of the samples.

The modes of fermentative processing were selected empirically and depended on the type of processed raw material, Crusenzyme concentration, temperature and duration of the process. The temperature regime of fermentation varied in the range of 30-50°C and the duration from several hours to several days. The concentration of Crusenzyme enzyme preparation was varied depending on the task at hand and the type of raw material used. For the treatment of shellfish meat, a short-term fermentation mode was used to soften the tissue, while for skin removal, the enzyme exposure time was increased. For the treatment of crab shell waste, a prolonged fermentation regime was applied to achieve maximum deproteinisation of chitin. To analyse changes in chemical composition, samples were collected and analysed using standard techniques.

3 Results and discussion

In this research work, an attempt was made to comprehensively and non-wastefully process various invertebrate species such as cephalopods (squid and octopus), bivalves (spisula, anadara, corbicula) and crabs to produce a wide range of products. The main objective of this work was to create technologies for the production of food products, feed additives for farm animals, as well as some technical products. The experimental part of the work included application of the enzymatic preparation 'Cruzensim', obtained from the liver of the Kamchatka crab, for the processing of various types of raw materials. This preparation was used to deproteinise, soften tissues and remove unwanted parts such as skin of cephalopod molluscs and rigid structures of bivalves.

At the beginning of the research, the feasibility of using the enzyme preparation Crusenzyme for skin removal in cephalopod molluscs such as squid and octopus was investigated. It was experimentally confirmed that at the concentration of 'Krusenzyme' 0.5% during 30 minutes there is a dissolution of up to 75% of squid skin proteins, while the impact on muscle tissue remains minimal, which guarantees the preservation of native properties of meat. When the processing time was increased to 60 minutes, dissolution of skin proteins reached 90%, resulting in more efficient skin removal, however, there was a negligible (up to 5%) impact on muscle tissue proteins. The optimum processing time was found to be 45 minutes, which achieved 85% dissolution of skin proteins and only 2% impact on muscle tissue, while maintaining high organoleptic characteristics of the product. By adjusting the enzyme concentration and processing time, it was possible to develop production regulations for the products 'Squid skinned frozen' and 'Octopus skinned frozen'. The use of biochemical method with 'Krusenzim' allowed to reduce energy costs by 20% and labour costs by 15% in comparison with the traditional hydrothermal method, as well as to increase the yield of finished products by 8-10% due to the reduction of mass losses during processing. Further studies have shown the effectiveness of this method also in the production of salt-dried cephalopod products. For example, when using 'Krusenzyme' at a concentration of 0.3% and treatment time of 20 minutes before drying, there was observed an improvement in the structure of the finished product by 10-15%, which was manifested in a decrease in brittleness and increase in elasticity, as well as an increase in shelf life by 15-20% due to more uniform drying and reducing the risk of microbial contamination. Thus, enzymatic processing has been shown to be an effective method to improve product properties [3].

Further, the possibility of processing bivalve molluscs such as anadara, spisula, corbicula, mactra, callista and merceria was investigated. Their meat is characterised by considerable toughness due to the high content of connective tissue proteins, particularly collagen. To achieve quality consistency and improve the nutritional value of the products, a complex application of hydrothermal and enzymatic treatment is necessary. During the experiments it was found that when using the enzyme preparation 'Krusenzyme' processing for 60 minutes at 45°C provides degradation of up to 85% of collagen structures, while without enzymatic treatment the collagen dissolution for the same period is only about 15%. The use of fermentation significantly improved the consistency of the muscle tissue, increasing the degree of protein dissolution from 20% to 60-65% at an average 90-minute treatment, whereas hydrothermal treatment provided only 35-40% dissolution during the same time. As a result, the yield of salted products increased from 30% to 70-76% after application of complex treatment. In [4] it was shown that enzymes, changing the structure and composition of tissues, also affect the digestibility of nutrients.

Earlier works of the authors show that flour from soft tissues of bivalves is a valuable source of protein containing all essential amino acids, with lysine and methionine accounting for about 30% of the total amount of essential amino acids. This amino acid profile, as well as the presence of a wide range of macro- and microelements, make this meal a promising additive in feeds for farm animals [5]. Experiments with piglets have shown that the inclusion of mollusc soft tissue meal in the diet of suckling and early weaned animals increases their average daily gain by 11.7%, while reducing feed costs by 10.5%. Also, mollusk shell meal, which consists of 98% ash and contains 36-42% calcium, proved to be an effective feed additive. Its use in the diet of pigs resulted in a 16.8% increase in daily weight gain, a 14% reduction in feed costs and a 32% increase in production profitability. When studying the effect of treatment time on protein solubility, it was found that treatment of soft tissues for 30 minutes leads to the dissolution of 55% of protein, and increasing the time to 60 minutes increases this indicator to 80%, after which the increase in protein solubility slows down. The process of protein dissolution when treated for 120 minutes achieves 90% solubility. Flour from flaps also showed a positive effect on laying hens, increasing their egg production by 7%, shell thickness by 4% and stock retention by 2% compared to the control group. As a result of the conducted researches normative documents for production of protein-mineral and mineral feed additives from wastes of processing of bivalve mollusks were developed and approved, that allows to introduce these additives in practice of agriculture. The use of mollusk meal in animal feeding contributes to the improvement of productivity indicators, and also reduces the need for traditional, less sustainable and expensive feed additives [6].

Another part of this work was to scrutinise the disposal of crab processing wastes. Liquid waste (hepatopancreas) and solid shell waste (SSW) constitute a significant portion, namely 40 to 65% of the raw crab weight. Traditional methods of processing these wastes are energy intensive, requiring large amounts of resources, and environmentally hazardous due to the use of harsh chemicals. In this study, a method was used where hepatopancreas is used for two purposes: to obtain fat, which yielded an average of 15-20% of the initial mass of hepatopancreas, and the proteolytic enzyme preparation 'Cruzenzyme'. This enzyme preparation, in turn, was applied for deproteinisation of PCOS, which is a key step in the process. The experiments carried out showed that replacing the traditional alkaline treatment with an enzymatic treatment using Crusenzyme produced impressive results. In particular, the degree of deproteinisation of chitin reached 88-92% after 12 hours of treatment at 50°C, whereas the alkaline method would have required more than 24 hours to achieve a similar result. In addition, the enzymatic treatment promotes efficient degradation of the protein-pigment complex, which resulted in a significant increase in the yield of carotenoid pigments, namely by 30-35% compared to conventional methods, with carotenoid concentrations in the extract reaching 1.2-1.5 mg/ml. It was found that the use of 'Krusenzyme' also makes it possible to control the process of protein hydrolysis, thus the process of obtaining hydrolysates with a given composition and properties was developed, and the degree of protein hydrolysis could reach 70-80% within 8 hours at a temperature of 45°C. As a result of the conducted research, a waste-free technology of complex processing of crab waste was developed, which allowed to efficiently and environmentally safe obtain a wide range of valuable products, including chitin, chitosan (the yield of which from deproteinised chitin was 75-80%), complex of proteolytic enzymes 'Krusenzyme', fat, carotenoid extract, as well as protein-containing compositions, the yield of which reached 60% of the initial protein material after hydrolysis. Such methods aimed at recycling and efficient utilisation of wastes are an important aspect of sustainable development [7]. Methods based on fermentation not only reduce waste but also contribute to the recovery of valuable components [8, 9].

In general, the results of this work demonstrate the prospects of waste-free and integrated processing of various invertebrates to obtain a wide range of products. The application of biochemical methods, and in particular the enzyme preparation 'Krusenzyme', allowed to increase the yield of products, improve their quality and reduce the negative impact on the environment.

4 Conclusion

The presented work deals with the integrated waste-free processing of various invertebrate species, such as cephalopods, bivalves and crabs, using the enzyme preparation 'Krusenzyme' derived from crab liver. The research is aimed at creating technologies for the production of food, feed additives and technical products. The effectiveness of 'Krusenzyme' for skin removal in cephalopod molluscs, softening of muscle tissue and degradation of collagen structures of bivalves was experimentally proved, which allowed to improve consistency and yield of finished products. The application of fermentation significantly reduced energy consumption and labour intensity of the processes in comparison with traditional methods. The technology of obtaining feed additives from soft tissues and shells of molluscs enriched with protein and minerals was developed, which showed a positive effect on the growth and productivity of farm animals [10]. Special attention was paid to the processing of crab waste, where Crusenzyme was effectively used for deproteinisation of chitin, extraction of carotenoids, production of fat and hydrolysates, which made it possible to create a waste-free technology with high economic and environmental efficiency [11]. The study demonstrates the potential of biochemical methods in the processing of hydrobionts, leading to improved product quality, increased yield and reduced negative impact on the environment. The developed technologies make it possible to obtain a wide range of valuable products from previously unused waste, which contributes to the rational use of resources and enhances the sustainability of production processes [12-14].

References

1. L. Yerkinbayeva, A. Myrzakarim, A. Moldagaliyeva, B. Kalymbek, R. Yerezhepkyzy, D. Nurmukhankyzy. The role of legal regulation in promoting organic agriculture to achieve sustainable food security. J. Lifestyle SDG'S Rev., **5**, e04197 (2025). DOI: 10.47172/2965-730X.SDGsReview.v5.n01.pe04197.

2. A. Ogunleye, D. Olude, A. Kehinde, V. Tanimonure, C.A. Wongnaa, O. Ajayi. Impact of credit constraints on adoption of biofortified food crop among farming households in Nigeria. Sustain. Futures, **9**, 100421 (2025). DOI: 10.1016/j.sftr.2024.100421.

3. F. Yong, B. Liu, H. Li, H. Hao, Y. Fan, O. Datsomor, R. Han, H. Jiang, D. Che. Relationship between dietary fiber physicochemical properties and feedstuff fermentation characteristics and their effects on nutrient utilization, energy metabolism, and gut microbiota in growing pigs. J. Anim. Sci. Biotechnol., **16**, 1 (2025). DOI: 10.1186/s40104-024-01129-x.

4. F. Yong, B. Liu, H. Li, H. Hao, Y. Fan, O. Datsomor, R. Han, H. Jiang, D. Che. Relationship between dietary fiber physicochemical properties and feedstuff fermentation characteristics and their effects on nutrient utilization, energy metabolism, and gut microbiota in growing pigs. J. Anim. Sci. Biotechnol., **16**, 1 (2025). DOI: 10.1186/s40104-024-01129-x.

5. Q. Gao, L. Wang, Y. Fang, Y. Gao, L. Ma, X. Wang, Y. Li, X. Wu, Z. Du. Conservation agriculture boosts topsoil organic matter by restoring free lipids and lignin phenols biomarkers in distinct fractions. Soil Till. Res., **248**, 106463 (2025). DOI: 10.1016/j.still.2025.106463.

6. F. Yong, B. Liu, H. Li, H. Hao, Y. Fan, O. Datsomor, R. Han, H. Jiang, D. Che. Relationship between dietary fiber physicochemical properties and feedstuff fermentation characteristics and their effects on nutrient utilization, energy metabolism, and gut microbiota in growing pigs. J. Anim. Sci. Biotechnol., **16**, 1 (2025). DOI: 10.1186/s40104-024-01129-x.

7. Y. Chen, Y. Guo, X. He, B. Tan, Z. Liao, A. Chen, X. Gu, X. Li, X. Chen, B. Chen, S. Lin, W. Li, P. Hu, X. Zhu, W. Zhao, J. Niu. Comprehensive utilization of black soldier fly (Hermetia illucens) larvae: extraction, recovery and characterization of peptide, chitin and melanin and scaling-up trial. Sep. Purif. Technol., **361**, 131262 (2025). DOI: 10.1016/j.seppur.2024.131262.

8. V. Brigida, E.A. Efremenkov, S.N. Sorokova, Q. Mengxu. Tailings Utilization and Zinc Extraction Based on Mechanochemical Activation. Materials, **16**, 726 (2023). DOI: 10.3390/ma16020726.

9. N.N. Bryukhanova, V.A. Gladkikh, R.V. Kononenko, V.V. Gerasidi. Selection of disintegrator parameters toward environmentally friendly leaching of metals. Min. Inform. Anal. Bull., **12**, 39-49 (2024). DOI: 10.25018/0236_1493_2024_12_0_39.

10. L.S. Malyukova, N.V. V.V. Bukhtoyarov, V.Y. Konyukhov, K.A. Bashmur. Circular Mining Wastes Management for Sustainable Production of Camellia sinensis (L.) O. Kuntze. Sustainability (Switzerland), **15**, 11671 (2023). DOI: 10.3390/su151511671.

11. O. A. Kolenchukov, A. Y. Mikhaulov, K. A. Bashmur, V. S. Tynchenko, R. B. Sergienko, V. V. Kukartsev, V. V. Tynchenko. Forecasting the technical condition of thermochemical reactor systems. SOCAR Proceedings, **2023**, 123–130 (2023). https://doi.org/10.5510/OGP20230100814.

12. O. Afanaseva, M. Afanasyev, S. Neyrus, D. Pervukhin, D. Tukeev. Information and Analytical System Monitoring and Assessment of the Water Bodies State in the Mineral Resources Complex. Inventions, **9**, 115 (2024). https://doi.org/10.3390/inventions9060115.

13. V.Yu Konyuhov, A.M. Gladkih, I.I. Galyautdinov, Y.D. Severina. Economic aspects of green technologies. IOP Conference Series: Earth and Environmental Science, **350**, 012036 (2019). DOI: 10.1088/1755-1315/350/1/012036.

14. V.V. Kondratiev. Circular Mining Wastes Management for Sustainable Production of Camellia sinensis (L.) O. Kuntze. Sustainability (Switzerland), **15**, 11671 (2023). DOI: 10.3390/su151511671.

Resistance of *Sarepta mustard* to a decrease in ambient temperature during vegetation

Natalia Kaznina[1*], *Ekaterina* Kholoptseva[1], *Yulia* Batova[1], and *Ksenia* Taskina[1]

[1]Institute of Biology, Karelian Research Centre, Russian Academy of Sciences, 11, Pushkinskaya St., Petrozavodsk, 185910, Russia

Abstract. *Brassica juncea* (L.) Czern. is an important edible oilseed crop. This species is currently well studied from the point of view of its use in the phytoremediation of contaminated soils. There is practically no information about the resistance of Sarepta mustard to low temperatures. However, they are important for identifying the possibility of growing this species in northern conditions. In this regard, the effect of low temperature (4° C) on the *B. juncea* var. Nika growth, photosynthesis and water exchange was studied. An exposure to 4 °C inhibit the plants' growth, but the photosynthesis rate does not exchange. At the same time, under the hypothermia, the chlorophyll *a* content and the chlorophylls *a*/*b* ratio increase. The carotenoids content and the size of stomatal aperture remained at the optimal temperature conditions level, and the Fv/Fm increased, indicating the efficient operation of photosystem II. Being exposed to low temperatures, the transpiration rate slowed down, which ensured the maintenance of the necessary tissue hydration and increased the ability to efficiently utilize water in the photosynthesis (WUE). A conclusion was made about the resistance of *B. juncea* var. Nika to low temperature (4° C) and about the possibility of growing this species in northern regions, including the phytoremediation of polluted areas.

1 Introduction

The *Brassica juncea* (L.) Czern. (Sarepta mustard) family (Brassicaceae) is well known as an oilseed, forage, medicinal and sideral crop [1, 2]. Mustard oil, due to its high content of essential amino acids, is used in the food industry, including canning, confectionery, baking industries. It is used for technical purposes in soap making, textile industry, perfumery and medicine [3]. Currently, Sarepta mustard is one of the most valuable crops, which is recommended for farming in different regions. This is due to its high resistance to a number of unfavourable environmental factors, such as, drought, salinization, deficiency of mineral nutrition elements in the soil [2]. In addition, Sarepta mustard is able to grow in soils with high content of heavy metals [4, 5]. The levels of metal accumulation by *B. juncea* do not qualify it as a "hyperaccumulator". However, the ability of this species to accumulate in large amounts a wide range of heavy metals, including cadmium, copper, lead, chromium and zinc, in addition to its rapid growth and large biomass, makes Sarepta mustard promising for phytoremediation of HM-contaminated areas [6]. Consequently, *B. juncea* is being actively studied in terms of its metal tolerance and phytoremediation potential.

In northern regions, the successful use of plants for soil purification from heavy metals is difficult due to frequent temperature drops to low values during vegetation [7, 8]. Therefore, knowledge of plants cold tolerance is needed when selecting sites for soils phytoremediation under these conditions. For *B. juncea*, there is no information on the ability of this species to adapt to lower temperatures in the literature known to us. Although other members of this family, such as *Brassica napus* [9] and *Brassica oleracea* [10], have been shown to successfully adapt to low-temperature conditions.

The aim of this work was to study the tolerance of Sarepta mustard to a short-term temperature decrease based on a growth parameters, photosynthesis activity and water metabolism.

2 Material and methods

2.1 Plant material and growth conditions

Seeds of Sarepta mustard (*Brassica juncea* (L.) Czern.) of the Nika var. were germinated in distilled water for 3 days. The seeds were provided by the Department of Genetic Resources of Oilseed and Fiber Crops at the Federal Research Center "N.I. Vavilov All-Russian Institute of Plant Genetic Resources". The seedlings were transferred in hydroponic conditions and grown in a Hoagland-Arnon nutrient solution under light at a temperature of 22° C and the relative air humidity of 60–70%. PAR was 200 μmol $(m^2 \cdot s)^{-1}$, and there was a 14-hour photoperiod. After 14 days (at the 1st true leaf stage), one group of plants was kept at 22° C (control), while the other group was placed in a cold chamber at 4° C for 7 days.

After the experiment, cold tolerance was assessed based on growth parameters (shoot height, shoot dry biomass, area of the 1st true leaf), parameters of photosynthetic apparatus activity (the content of photosynthetic pigments, Fv/Fm, stomatal conductance, photosynthesis rate) and water exchange (the

*Correspondence: kaznina@krc.karelia.ru

transpiration rate, the shoot water content and water use efficiency) of the plants.

2.2 Biometric measurements

For ten plants of each experimental variant, the shoot length, dry biomass, and the first true leaf area were measured. The dry mass of the shoots was measured after drying at 105°C to a constant weight. The leaf area was determined using the AreaS 2.1 software.

2.3 Photosynthetic pigments content and the activity of PS II measurements

Chlorophyll a, chlorophyll b and carotenoids were measured and analysed spectrophotometrically using the SF-2000 spectrophotometer (Spectrum, Moscow, Russia) and calculated using the established formulas [11]. For the extraction of pigments the 98% (v/v) acetone solution was used. The percentage of chlorophylls in light-harvesting complex II (LHCII) was calculated according to [12]. To measure the pigments content, three mixed samples of five leaves were taken. The analytical replication was threefold. The activity of PS II was assessed by the Fv/Fm parameter, which characterizes the potential quantum yield of photochemical activity, using a MINI-PAM pulse-modulated fluorometer (Walz, Germany) after leaves dark adaptation during 20 minutes. For measurements of the Fv/Fm parameter ten biological replicates were performed.

2.4 Stomatal conductance, a photosynthesis rate, and transpiration rate measurements

Stomatal conductance, a photosynthesis rate, and a transpiration rate were measured using a portable system for CO_2 gas exchange and water vapour analysis (HCM-1000, Germany). For measurements of these parameters ten biological replicates were performed.

2.5 Water exchange parameters measurement

The water use efficiency (WUE) was calculated as the ratio of the apparent photosynthesis rate to transpiration. The water content (WC) was determined according to Turner [13]. The leaves were weighed (fresh weight, FW); then they were dried in an oven at 80 °C for 24 h and weighed again (DW). The water content (WC) was estimated in percentage (%) using the following formula: WC = ([FW – DW]/FW)*100, where FW is the fresh weight and DW is the dry weight. For measurements of these parameters ten biological replicates were performed.

2.6 Statistical analysis

All experiments were performed two times with three replicates for each variant. The experimental data were expressed as means ± standard errors (SE). The data were analysed by one-way analysis of variance (ANOVA) using Microsoft Excel 2010. Student's t-test was applied to compare statistical significance at the level of $p < 0.05$.

3 Results and discussion

According to literature data, the optimal temperature conditions for successful growth of Sarepta mustard are within 12-25 °C [1], so a temperature drop to 4 °C, which is often encountered in northern latitudes during vegetation, is stressful for plants of this species. It is known that the process of cold adaptation is an energy-intensive process that requires energy investments from the plant organism [14]. Therefore, successful plants adaptation to cold is determined by the ability to rapidly and co-ordinately change the intensity of the main physiological processes, primarily growth and photosynthesis, associated with the formation, redistribution, and utilization of plastic substances and energy. In particular, one of the characteristic responses of plants to the low temperature is growth inhibition, which allows them to save energy reserves, directing them to the maintenance of energy-consuming processes, including photosynthesis [15, 16]. In this experiment, it was found that after 7-day exposure of Sarepta mustard to a 4 °C shoot growth retards. Therefore, the shoot height and its dry biomass in the experimental condition were 36 and 60 % lower than those in the control, respectively (table 1), and the area of leaf formed during exposure was 62 % lower than that in the control (table 1, figure 1).

Table 1. The effect of low temperature (4 °C) on shoot growth parameters of *Brassica juncea* var. Nika

Parameter	Temperature	
		4 °C
Shoot height, cm	15.57±0.29 a	10.03±0.18 b
Shoot dry biomass, g	0.055±0.004 a	0.022±0.001 b
Leaf area, cm^2	4.36±0.33 a	1.66±0.13 b

Different letters denote significant differences between control and experimental variants at $p < 0.05$.

Fig. 1. The general overview of the effect of optimal (22 °C) (A) or low (4 °C) (B) temperature on the first true leaf formation of the *Brassica juncea* var. Nika

The study also revealed that under the hypothermia condition the photosynthesis rate was maintained at the control level (figure 2a). It is known that photosynthesis is the main target for low temperatures, so the ability of plants to maintain its intensity at a high level under low temperatures indicates their cold tolerance [17, 18].

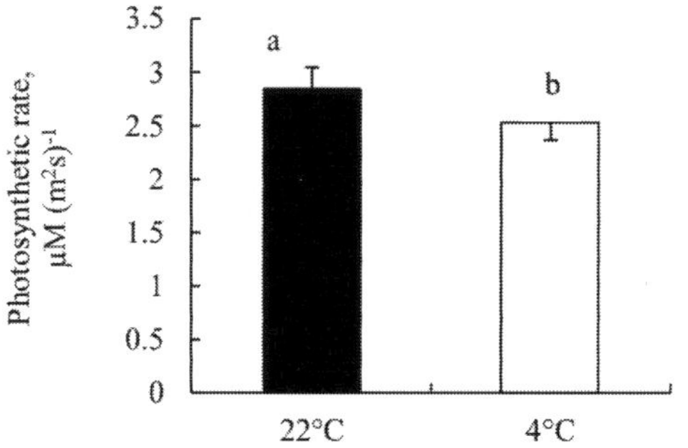

Fig. 2. The effect of low temperature (4 °C) on the photosynthetic rate of *Brassica juncea* var. Nika

A number of adaptive changes in photosynthetic activity parameters contributed to the preservation of the high photosynthesis level. In particular, the content of chlorophyll *a*, as well as the ratio of chlorophyll *a*/*b*, increased (compared to the control) under the cold due to the absence of changes in the chlorophyll *b* content (table 2). However, the content of chlorophylls in the LHC of the experimental plants was significantly lower than that of the control (table 2).

Table 2. The effect of low temperature (4 °C) on pigments content and F*v*/F*m* in *Brassica juncea* var. Nika

Parameter	Temperature	
	22 °C	4 °C
Chlorophyll *a* content, mg g^{-1} FW	0.45±0.014a	0.47±0.006b
Chlorophyll *b* content, mg g^{-1} FW	0.21±0.009a	0.20±0.005a
Carotenoids content, mg g^{-1} FW	0.13±0.004a	0.13±0.003a
Chlorophylls ratio (*a*/*b*)	2.17±0.049b	2.38±0.047a
LHC chlorophylls content, %	69.65±1.12a	65.35±0.89 b
F*v*/F*m*	0.760±0.002 b	0.800±0.004 a

Different letters denote significant differences between control and experimental variants at $p < 0.05$.

The increase in the chlorophylls ratio is known to be an indicator of stress adaptation of the photosynthetic apparatus and is aimed at enhancing light absorption while decreasing the proportion of chlorophylls in LHCs of photosystems [19, 20]. A similar effect was also found in *B. oleracea* at 8 °C [7] and *Triticum aestivum* at 4 °C [18]. In addition, a high level of carotenoids content was maintained in Sarepta mustard under cold exposure. Since carotenoids perform a number of important functions in the photosynthesis process, including antenna functions, acting as additional "light collectors", as well as being active antioxidants, these pigments are considered as one of the factors providing plant resistance to unfavourable temperature effects [21].

Photosystem II plays a central role in the generation and regulation of electron transport in chloroplasts. Therefore, the indicators reflecting the efficiency of its work are important for the assessment of the activity of photosynthetic apparatus [8]. A number of studies have shown that low temperatures negatively influence the functioning of photosystems, which, for example, was found in rice under 12 °C [22], in a number cultivars of *B. napus* under 10 °C [8], and in *Glycine max* under 6 °C [23]. Preservation at 4 °C in experimental mustard plants of the F*v*/F*m* value, characterizing the quantum efficiency of photosystem II, at the control level, which was found in the experiment, indicates the absence of disturbances in its functioning [24].

As a rule, in plants, in response to cold exposure, the level of endogenous ABA increases in the closing cells of stomata, which leads to their partial closure. The resulting decrease in stomatal conductance and the transpiration rate is a defence-adaptive reaction that allows plants to avoid significant dehydration under these conditions [25, 26]. No such effect was observed in our studies. The stomatal conductance under 4 °C did not differ from that under 22 °C, indicating the absence of severe stress in plants (figure 3). The preservation of stomatal conductance under hypothermia at the control level was probably one of the reasons for maintaining the necessary photosynthetic rate under these conditions.

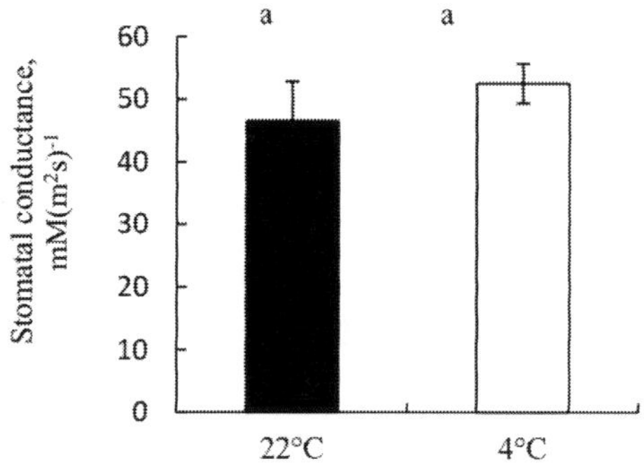

Fig. 3. The effect of low temperature (4 °C) on stomatal conductance of *Brassica juncea* var. Nika

At the same time, the rate of transpiration under cold exposure decreased by 21% compared to the control (table 3). This contributed to the preservation of water content in tissues. It is known that the ability of plants to maintain the optimal level of water balance under low-temperature exposure is characteristic of cold-resistant plants. In addition to decreased the transpiration rate, the cells of these plants have the ability to retain water, for example, by increasing the sugars synthesis [18]. They are also characterised by a high level of osmoregulation [27]. In addition, in cold-tolerant plants, the hydraulic conductivity of roots increases, resulting in increased water flow into shoots [28, 29].

Table 3. The effect of low temperature on water metabolism parameters of *Brassica juncea* var. Nika

Parameter	Temperature	
	22°C	4°C
Transpiration rate, mmol H_2O $(m^2s)^{-1}$	0.61±0.09 a	0.48±0.08 b
Water content, %	94.0±0.1 a	95.0±0.1 a
WUE, µmol CO_2 /mmol H_2O	3.43±0.5 b	5.42±0.3 a

Different letters denote significant differences between control and experimental variants at $p < 0.05$.

It is also important to note that WUE increased in mustard when exposed to 4 °C. It is believed that the ability of plants to successfully adapt to stress conditions is provided by their ability to efficiently utilize water in the photosynthesis. This allows the plant organism to allocate resources qualitatively and direct them to the synthesis of plastic substances used to maintain growth and biomass accumulation [30].

4 Conclusions

In conclusion it is necessary to note that *B. juncea* var. Nika successfully adapts to the short-term (within 7 days) decrease of temperature (4° C) during vegetation. This is facilitated by a number of adaptation mechanisms. The first is enhancement under hypothermia of chlorophyll *a* synthesis and an increase in the chlorophylls *a*/*b* ratio. The second is the ability to maintain the carotenoid content at the control level, which contributes to light absorption when the amount of LHC chlorophyll decreases. The third is maintenance of the stomatal aperture size at the control level, providing the necessary rate of photosynthesis. The adaptation reaction was also inhibition of shoot growth under cold conditions, aimed at redistribution of energy resources in favour of photosynthesis, as well as slowing down the transpiration intensity, allowing maintaining the necessary level of the water content. The results allow concluding about the possibility of growing *B. juncea* var. Nika in northern regions, including for phytoremediation of polluted areas.

5 Acknowledgements

This research was carried out using the equipment of the Core Facility of the Karelian Research Centre of the Russian Academy of Sciences and under state order (*FMEN-2022-0004*).

6 Declaration of conflict of interest

The authors declare no conflict of interest.

7 References

1. F. Kayaçetin, F. Önemli, G. Yilmaz, K. M. Khawar, A. Kinay, H. Hatipoğlu, K. A. Özaydin, Agric. Sci. J. **25**, 298–308 (2019)
2. G. N. Kuznetsova, R. S. Polyakova, Intern. Agric. J. **2**, 595–608 (2022)
3. T. Ya. Prakova, Intern. Agric. J. **65**, 6 (390) (2022)
4. H. Chaudhry, N. Nisar, S. Mehmood, M. Iqbal, A. Nazir, M. Yasir, Biocatal. Agric. Biotechnol., **23** (2020)
5. G. A. Bortoloti, D. Baron, Environmental Advances **8** (2022)
6. S. Das, W. M. L. Wesemael, R. N. Perry, Agric. Sci. Res. J. **1**, 102–112 (2011)
7. V. Ljubej, E. Karalija, B. Salopek-Sondi, D. Šamec, Horticulturae **7**, 341 (2021)
8. Q. Q. Xu, A. Sami, H. Zhang, X. Z., Jin, W. Y. Zheng, Z. Y. Zhu, Z. H. Zhu, South African Journal of Botany **147**, 400–414 (2022)
9. L. Yan, S. Tariq,Y. Cheng, Y. Lü, X.-k. Zhang, X.-l. Zou, J. Integr. Agric., **18(12),** 2742–2752 (2019)
10. D. Šamec, V. Ljubej, I. R. Redovnikovi´c, S. Fistani´c, B. Salopek-Sondi, Foods **11**, 264 (2022)
11. H. K. Lichtenthaler, A. R. Wellburn, Biochem. Soc. Trans. **11**, 591–592 (1983)
12. H. K. Lichtenthaler, Methods Enzymol. **148**, 350 (1987)
13. N. C. Turner, Plant Soil **58**, 339–366 (1981)
14. A. K. Gupta, N. Kaur, J. Biosci. **30**, 761–776 (2005)
15. Yu. Ye. Kolupaev, Yu. V. Karpets, Formation of adaptive reactions of plants on the action of abiotic stress factors, Kiev: Osnova (2010).
16. C. Liu, B. C. Colon, M. Ziesack, P. A. Silver, D. N. Nocera, Science. **3**, 352(6290):1210-3 (2016)
17. Z. F. Wang, L. Z. Song, Z. X. Fan, F. Y. Zhang, Y. T Liu, S. G. Wang, H. Q. Gao, Acta Agric. Nucl. Sin. **16**, 243–246 (2001)
18. A. Deryabin, K. Zhukova, N. Naraikina, Y. Venzhik, Metabolites, **14,** 199 (2024)
19. H. K. Lichtenthaler, C. Buschmann, Current protocols in food analytical chemistry (2001)
20. L. M., Babenko, I. V., Kosakivska, Y. A.; Akimov, D. O., Klymchuk, T. D. Skaternya, Genet. Plant Physiol. **4**, 117–125 (2014).
21. L. Nagy, V. Kiss, V. Brumfeld, K. Osvay, Á. Börzsönyi, M. Magyar, S. Malkin, Photobiol. **91**, 1368–1375 (2015)
22. Y. Zhao, Q. Han, C. Ding, Y. Huang, J. Liao, T. Chen, Y. Chen, Int. J. Mol. Sci. **21**, 1390 (2020)
23. A. J. Strauss, G. H. J. Kruger, R. J. Strasser, P. D. R. VanHerden, Physiologia Plantarum **131**, 89–105 (2007)
24. K. Maxwell, G. N. Johnson, J. Exp. Bot. **51**, 659–668 (2000)
25. A. J. Bloom, M. A. Zwieniecki, J. B. Passioura, L. B. Randall, N. M. Holbrook, D.A. Clair, Plant Cell Environ, **27**, 971–979 (2004)
26. S. Mahajan, N. Tuteja, Arch. Biochem. Biophys. **444**, 139–158 (2005)
27. B. Zhang, D. Jia, Z. Gao, Q. Dong, L. He, J. Sci. Food Agric. **96**, 1967–1973 (2016)

28. A. Aghaee, F. Moradi, H. Zare-Maivan, F. Zarinkamar, H. Pour Irandoost, P. Sharifi, Afr. J. Biotechnol. **10**, 7617–7621 (2011)
29. A. S. Lukatkin, A. Brazaitytė, Č. Bobinas P. Duchovskis, Žemdirbystė=Agriculture **99** (2), 111–124 (2012)
30. E. A. Tambussi, J. Bort J. L. Araus, Ann. Appl. Biol. **150**(3), 307–321 (2007)

Production of biohumus by microbiological processing of forestry wastes

Ruzimurat Choriev[1,*], *Ravshan* Toshmatov[2], *Abdimurot* Kuziev[3], *Shahnoza* Majidova[4], *Irina A.* Zatsarinnaya [5], *Polina A.* Sablina [5]

[1] "Tashkent Institute of Irrigation and Agricultural Mechanization Engineers" National Research University, Tashkent, Uzbekistan
[2] Almalyk branch of the Tashkent State Technical University named after Islam Karimov, Almalyk, Uzbekistan
[3] Termez State University, Termiz, Uzbekistan
[4] Termiz State Pedagogical Institute, Termiz, Uzbekistan
[5] Kuban State Agrarian University named after I. T. Trubilin, Krasnodar, Russia

Abstract. The article discusses the relevance and feasibility of biotechnology for the rehabilitation of ecosystems, in particular, rural areas. An integrated approach to the processing of agricultural and forestry wastes, such as bark dumps, is proposed in order to reduce the negative impact on the environment and obtain valuable products. The research was carried out in Perm Krai and included the stages of sorting, microbiological treatment of waste, and application of the resulting biohumus into the soil. As an example, the recycling of pulp and paper mill bark dump was considered, where a method of microbiological processing with vermiculture was developed. This allowed to reduce the volume of waste, reduce its fire hazard and obtain biohumus rich in humic acids. The obtained biohumus was successfully used as an organic fertiliser, which resulted in a 20-35% increase in the yield of barley, oats, wheat and vegetable crops compared to control plots. The research confirms the effectiveness of the proposed method for solving environmental problems and improving soil fertility, as well as its difference from traditional waste utilisation methods.

1 Introduction

Environmental problems such as soil and water pollution, climate change, loss of biodiversity and many others have become global and require immediate measures to solve them. In an attempt to find a way out of this situation, mankind turns to various approaches and technologies. Among them we can distinguish several main directions. Firstly, it is the reduction of resource consumption and waste by optimising production processes, introducing a circular economy and more responsible consumption. The main advantage of this approach is its focus on eliminating the very cause of the problem [1]. However, its implementation requires fundamental changes in production and consumer cultures, which is often associated with significant financial and organisational difficulties. Secondly, it is the development of waste treatment and recycling technologies. This approach makes it possible to neutralise or reduce the negative impact of already generated waste and return resources to the production cycle. Examples of such technologies are various methods of wastewater treatment, plastic recycling and recycling of various materials. The main advantage of this direction is its practical applicability and relative ease of implementation. At the same time, it does not solve the problem fundamentally and requires continuous investment in the creation and modernisation of treatment facilities. And thirdly, it is the development and implementation of new biotechnologies that make it possible to use natural mechanisms for cleaning and restoring the environment. This approach includes, for example, the use of microorganisms for waste treatment, the creation of biohumus and the use of phytoremediation for soil and water treatment. It is characterised by its environmental safety and focus on restoring the natural balance [2]. However, its implementation requires a deep understanding of biological processes and is often associated with the need for scientific research and adaptation of technologies to local conditions.

In this context, the direction proposed in the presented material, namely, the integrated application of new biotechnologies for the rehabilitation of natural ecosystems, seems particularly relevant and important. This is due to the fact that the proposed approach allows to simultaneously solve several problems: to reduce the volume of waste, recycle it into useful products and restore soil fertility and biodiversity [3]. Also, this direction allows the use of biological waste from agriculture and forestry, which are quite common and cause significant harm to the environment. This makes

* Corresponding author: mm.choriyev@tiiame.uz

this approach especially relevant for regions with a developed agriculture and forestry sector. The use of biotechnology for waste processing allows not only to reduce their negative impact, but also to obtain valuable resources, such as biohumus, which can be used to improve soil fertility [4]. This is not only economically favourable, but also contributes to the transition to a more sustainable agriculture. Moreover, solving the problem of recycling wastes such as bark dumps with the help of biotechnology will improve the environmental situation, reduce fire hazards and increase the level of public confidence in the authorities [5].

The aim of the work is to substantiate the necessity and feasibility of integrated application of new biotechnologies for the rehabilitation of natural ecosystems in rural areas. The main emphasis is made on the possibility of using wastes of agricultural and timber industry complex as a source material for the rehabilitation of fertile land, as well as on the example of bark dump processing. The paper suggests ways to improve mechanisms and tools to reduce the negative anthropogenic impact on the environment and ensure sustainable development of rural areas through the introduction of environmentally safe and cost-effective technologies.

2 Method and materials

The research was carried out in the conditions of open terrain, on the territory of industrial enterprises and agricultural lands of Perm Krai. Experimental work included several stages: waste sorting, microbiological treatment, and application of the obtained biofertiliser. The crusher IR-100 was used for waste shredding, which provided a fraction of 1-5 cm, which is necessary for further processing. To accelerate the processes of waste processing, vermiculture technology was applied using earthworm culture Eisenia fetida. This technology involves the preparation of substrate by mixing shredded waste with nitrogen-containing additives in the form of manure and dung. The vermiculture process was carried out in special temperature-controlled containers, maintaining a temperature of 20-25°C. Standard laboratory instruments such as thermometer and moisture meter were used to monitor the decomposition and humification processes, as well as moisture control. To determine the chemical composition of the obtained biohumus and input materials, samples were analysed in a chemical laboratory using a spectrophotometer SF-46 and a pH meter. Application of the obtained biohumus into the soil in agricultural fields was carried out using standard agricultural machines such as fertiliser spreader. To evaluate the efficiency of biohumus application, plant growth and development were monitored and soil samples were analysed for nutrient and microbial content. All measurements and analyses were carried out according to the standards and methods used in agrochemical research.

3 Results and discussion

The conducted experimental work is aimed at developing an integrated approach to the rehabilitation of ecologically disadvantaged territories of the Perm Krai, with a focus on the utilisation of agricultural and forestry production wastes. The aim of the research was to create a waste-free technology that allows not only to reduce the negative impact on the environment, but also to obtain a valuable product for agriculture - biohumus. As a model object for the study, a pulp and paper mill bark dump was chosen, which is a serious source of pollution and fire hazard [6].

In the course of the study, the technology of microbiological processing of bark dump was developed and applied on the example of pulp and paper mill waste in the city of Krasnokamsk.

Earlier studies of the authors of this paper and other authors conducted on similar wastes show that the fractional composition of the bark dump can be quite diverse. In addition to wood residues, it often contains mineral impurities, the proportion of which can vary from 10 to 30%.

In this work, laboratory analyses have shown that the organic carbon to nitrogen ratio in the bark dump is approximately 60 to 1, which is a serious imbalance that prevents the natural decomposition of organic matter. In addition, studies have revealed the presence of lignin in the coring pond, which, when decomposed under oxygen deficient conditions, forms phenolic compounds that are hazardous to the environment and humans [7]. The experiment was conducted in several stages (Fig. 01). In the first stage, the bark dump waste was sorted, separating large fractions such as stones, logs and metal, which constituted up to 20% of the total mass. Then, the remainder was shredded to a 1-5 cm size fraction, which helped to increase the surface area for microbiological action. In the next step, the shredded substrate was mixed with nitrogen-containing wastes such as manure and litter in a ratio of 20 per cent by weight to the volume of the bark pile to balance the nutrients. The resulting mixture was treated with a specially designed microbial complex using EM vermiculture technology. During fermentation in a specially prepared container in the open air for 4-6 months, constant stirring of the substrate and control of the temperature, which was raised to 40-50 degrees Celsius due to exothermic reactions of microorganisms, were carried out. During this period, regular measurements of pH and humidity were carried out, which were maintained in the optimal range of 6.5-7.5 and 60-70%, respectively.

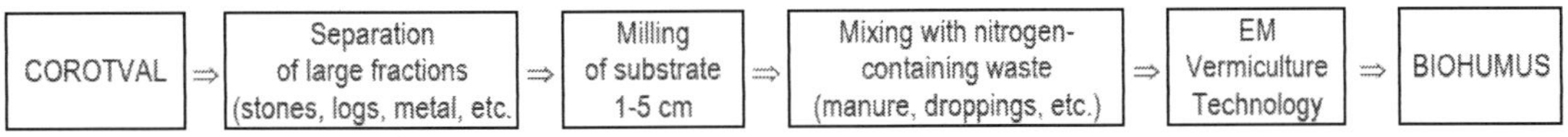

Figure 1. Recycling of pulp and paper mill bark dumps

As a result of the experiment, which lasted 12 months, it was found that microbiological treatment of forest industry waste, particularly bark dumps, can significantly reduce the risk of spontaneous combustion [8]. During observations of the control and experimental groups of bark dumps, it was noted that in the control groups, where no microbiological treatment was applied, the temperature in the depth of the dump reached 70-80°C during the summer months, which created prerequisites for ignition. In the experimental groups, on the contrary, the temperature inside the dump did not exceed 35-40°C, which indicated a decrease in the intensity of rotting processes and heat generation due to the active decomposition of organic matter by aerobic and anaerobic bacteria [9]. These results confirm that microbiological treatment is an effective method of preventing spontaneous combustion of bark dumps, which reduces the environmental and fire risk. Additionally, a significant reduction in waste volume was observed during the microbiological treatment process. The results of monitoring the volume of the bark dump before and after treatment showed a reduction of 12-14% throughout the experiment. This volume reduction was due to active decomposition of organic matter and its transformation into simpler compounds and humus acids. Initially the volume of waste was about 3500 cubic metres, at the end of the experiment the volume of waste decreased by 430 cubic metres. It should be noted that the bulk density decreased from a value of 350 kg/cubic metre to 280 kg/cubic metre, due to the reduction of cellulose and lignin content, which also confirms the success of the microbiological treatment. The transformation of waste by microbiological treatment resulted in a safe and useful product for agriculture - biohumus. Humification of the bark dump was confirmed visually by the change in the colour of the substrate. Initially, the dump had a light brown colour with bark and wood chips, but after processing the substrate became dark brown, almost black, indicating the formation of humic acids. In addition, laboratory tests showed an increase in humic acid content by 13-17% relative to the starting material. At the initial stage of the experiment the humic acid content was 11% and at the end of the experiment the content increased to 25%. The content of phenolic compounds decreased from 5 mg/kg to 0.03 mg/kg. The content of phenolic compounds was quite high in the initial corootwaste, indicating a potential environmental hazard, while in the obtained biohumus the content of these substances was insignificant and did not pose a threat to the environment. The analyses showed that the biohumus was a loose, dark mass, without any extraneous odour, with a neutral pH between 6.5-7.2. The salt concentration in the biohumus was 1.2-1.5 g/l, indicating its suitability for use in agriculture. The biohumus obtained, which was about 1500 cubic metres at the end of the experiment, was successfully applied as an organic fertiliser in the experimental plots [10]. Studies on experimental plots where barley, oats and wheat were planted showed an increase in yields by 22-27% compared to control groups where no fertiliser was applied. In the control plots, barley yields were 2.8 t/ha, oats 2.5 t/ha and wheat 3.0 t/ha, while in the plots treated with biohumus, barley yields were 3.6 t/ha, oats 3.3 t/ha and wheat 3.8 t/ha. Improvement in seed germination in the treated plots was also observed, indicating an improvement in soil fertility [11]. The yield of vegetable crops such as tomatoes and cucumbers increased by 30-35% compared to control plots. Tomato yield from control plots was 4.5 kg/m2 and from biohumus treated plots was 6 kg/m2, cucumber yield from control plots was 5.1 kg/m2 and from biohumus treated plots was 7 kg/m2. Thus, the results of the conducted research confirm the feasibility and effectiveness of microbiological processing of bark dump for solving ecological and agronomic problems. Microbiological processing allows not only to reduce the environmental hazard of waste, but also to obtain a valuable organic product that increases soil fertility and crop yields, which makes this technology promising for wide application.

Analysis of the obtained data allows us to conclude that microbiological processing of bark dump is a promising technology for waste utilisation and reclamation of contaminated territories. The proposed method allows not only to reduce the environmental load, but also to obtain a valuable product for agriculture. This approach is fundamentally different from traditional methods, which are often reduced to waste burial or incineration, which leads to further pollution of the environment. In addition, the proposed technology provides an opportunity to involve waste in a new production cycle, which contributes to the realisation of the principles of cyclic economy.

Comparing the results of the study with other works, it can be noted that the use of microbiological waste processing is a worldwide trend in the field of environmental protection. Many researchers have noted the high efficiency of this method in the processing of various types of organic waste, including wood waste. In particular, studies conducted in European countries have shown that the use of biohumus derived from wood waste helps to increase crop yields by 15-20%. This is consistent with the data obtained [12]. The results of the work emphasise the importance of using biohumus to improve soil fertility.

Nevertheless, it should be noted that the effectiveness of the proposed technology depends on a number of factors such as waste composition, storage conditions and microbiological diversity of the microbial consortium of microorganisms used [13]. Further research should be aimed at optimising the treatment process and adapting the technology to different waste types and conditions. In addition, it is necessary to carry out an economic evaluation of the proposed technology to determine its economic feasibility and the possibility of industrial implementation.

It is also necessary to take into account that on the territory of Perm Krai there is a large number of industrial facilities and dumps that require ecological rehabilitation, so it is necessary to actively implement waste-free technologies in the regions, and not to be limited to one-off projects. In particular, it is necessary to create a set of measures for the processing of livestock and poultry waste, which are also a large source of environmental pollution [14]. Only with an integrated approach to waste utilisation, it will be possible to count on sustainable environmental development of the region [15].

As a result of the study, it can be concluded that the application of microbiological waste recycling is a promising method that will not only solve environmental problems, but also obtain valuable products for agriculture. The development and implementation of these technologies is an important step towards sustainable development and environmental conservation.

4 Conclusion

In the course of work, a technology for microbiological processing of barkwaste using a complex of microorganisms and vermiculture was developed. This technology includes waste grinding, mixing with nitrogen-containing additives and fermentation under specially prepared conditions. The research results showed that microbiological treatment allows to significantly reduce the risk of spontaneous combustion of landfills, reduce their volume, and obtain biohumus rich in humic acids and nutrients. The biohumus obtained was successfully applied as an organic fertiliser in experimental plots, where an increase in crop yields by 22-35% was observed compared to control groups [16].

Thus, the conducted study confirms the effectiveness and feasibility of microbiological processing of bark dumps for waste disposal and remediation of contaminated territories. The proposed approach not only solves environmental problems, but also contributes to the development of a cyclic economy, allowing the involvement of waste in a new production cycle. Unlike traditional methods such as landfilling or incineration, this method is environmentally friendly and economically viable, which makes it promising for wide application. The article also notes the importance of further research to optimise the recycling process and adapt the technology to different types of waste.

References

1. V. Langraf, K. Petrovičová. Morphometric variability of specimens in Ground Beetle (Carabidae) communities in agrarian landscape forests and protected areas. Rev. Soc. Entomol. Argent., **83,** 1-8 (2024). DOI: 10.25085/rsea.830401.
2. V.V. Bukhtoyarov, V.Y. Konyukhov. Circular Mining Wastes Management for Sustainable Production of Camellia sinensis (L.) O. Kuntze. Sustainability (Switzerland), **15,** 11671 (2023). DOI: 10.3390/su151511671.
3. S. Bamidele. BRICS Revolution: The Global Agrarian Shift Sparked by Emerging Economies. India Q., **4,** 21–29 (2025). DOI: 10.1177/09749284241307937.
4. L. M. Tatarintsev, F. K. Ermekov, Yu. S. Lisovskaya. Agroecological condition assessment of agricultural landscapes for their sustainability increase. Sustainable Development of Mountain Territories, **14,** 76–86 (2022). DOI: 10.21177/1998-4502-2022-14-1-76-86.
5. H. Cochet, O. Ducourtieux, N. Garambois. Agrarian systems and climate change: Journeys of adaptation in the global south. Agrar. Syst. Clim. Change: Journeys Adapt. Glob. South, **1**-208 (2024). DOI: 10.1079/9781800628137.0000.
6. A. T. Beristenov, L. V. Lebedeva, L. M. Tatarintsev, V. L. Tatarintsev. Assessment of the agroecological potential of anthropogenic landscapes in the dry steppe of the south of Western Siberia. Sustainable Development of Mountain Territories, **16,** 975–987 (2024). DOI: 10.21177/1998-4502-2024-16-3-975-987.
7. D. D. Duisenova, K. M. Tireuov, L. M. Tatarintsev, L. V. Lebedeva. Ecological typization of agricultural land and modeling of sustainable agrolandscape at the local level. Sustainable Development of Mountain Territories, **16,** 94–110 (2024). DOI: 10.21177/1998-4502-2024-16-1-94-110.
8. L. V. Lebedeva, G. D. Syzdykova, D. S. Inkarov. Management of agricultural land use sustainability at the local level using GIS technologies and special map material. Sustainable Development of Mountain Territories, **15,** 864–876 (2023). DOI: 10.21177/1998-4502-2023-15-4-864-876.
9. J. Rodzik, R. Kołodyńska-Gawrysiak, Ł. Franczak, W. Zgłobicki, J. Poesen. IMPACT OF LAND USE CHANGES ON COLLAPSED PIPES DEVELOPMENT IN THE LOESS GULLY (LUBLIN UPLAND, EAST POLAND). Quaest. Geogr., **43,** 17-33 (2024). DOI: 10.14746/quageo-2024-0042.
10. F. K. Ermekov, Yu. S. Lisovskaya. Agroecological condition assessment of agricultural landscapes for their sustainability increase. Sustainable Development of Mountain Territories, **14,** 76–86 (2022). DOI: 10.21177/1998-4502-2022-14-1-76-86.
11. B. Aparecida da Silva, R. Evangelista de Oliveira, A. Cavalieri Sais, E. Cardoso leite. Family farmers and trees: ecological functions, ecosystem services and food

production in agrarian reform settlements in Brazil. Agroecol. Sustain. Food Syst., **49,** 4-21 (2025). DOI: 10.1080/21683565.2024.2405515.
12. T. Bandara, B. Marambe, G. Pushpakumara, P. Silva, R. Punyawardena, L. Manawadu, S. Premalal, G. Miah, K.R. Dahal. Adapting mountain-farming systems to climate change by reducing food-nutrition-health vulnerability. Mitig. Adapt. Strateg. Glob. Change, **29,** 88 (2024). DOI: 10.1007/s11027-024-10181-x.
13. M. N. Nurpeisov, K. V. Ruchkina, V. A. Androkhanov, S. K. Makenova. Assessment of geoecological factors and conditions affecting sustainability of agricultural land use in the forest-steppe zone of Altai Krai. Sustainable Development of Mountain Territories, **15,** 419–430 (2023). DOI: 10.21177/1998-4502-2023-15-2-419-430.
14. V.Yu Konyukhov, I.Y. Matasova. Ecologization of underground coal mining by means of ash use in backfill preparation. Min. Inform. Anal. Bull., **10,** 123-135 (2024). DOI: 10.25018/0236_1493_2024_10_0_123.
15. A. G. Terekhov, G. N. Sagatdinova. Spectral characteristics of saline arable land in the foothill catchment Zailiisky Alatau ridge (Northern Tien-Shan) during 2016-2022 using Sentinel-2 data. Sustainable Development of Mountain Territories, **16,** 143–155 (2024). DOI: 10.21177/1998-4502-2024-16-1-143-155.
16. N.A. Zhirov. Socio-Economic Development of the Black Soil Village in the XIX–early XX centuries: from the Prosperous Granary of Pre-Reform Russia to the Demo-Ecological Crisis of Rural Settlements. Bylye Gody, **19,** 1605-1614 (2024). DOI: 10.13187/bg.2024.4.1605.

Effect of pre-sowing seed treatment with semiconductor laser radiation on soft wheat productivity and its resistance to diseases

L.E. Kolesnikov[1], *M.I.* Fokina [2,3], *Yu.R.* Kolesnikova[4], *N.V.* Chernikova[1], *G.G.R.* Yakhyaev[1]

[1] Saint-Petersburg State Agrarian University, St. Petersburg, Pushkin, Russia
[2] Chemical Engineering Centre, ITMO University, St. Petersburg, Russia
[3] Institute of Chemistry, St. Petersburg State University, St. Petersburg, Russia
[4] N.I. Vavilov All-Russian Institute of Plant Genetic Resources (VIR), St. Petersburg, Russia

Abstract. The study of the effectiveness of soft wheat seeds pre-sowing treatment with a continuous semiconductor laser (the wavelength of $\lambda = 637$ nm, the power density (Wp) of 2, 4, 8 W/m^2, the duration of exposure of 5, 30, 60 seconds) is presented in the work. The maximal increase in the wheat yield, as well as in the growth of most morphometric indicators of its productivity, was recorded in the experimental variant during pre-sowing seed treatment with the laser with a maximum power density of 8 W/m^2 and with the longest exposure time of 60 seconds. In addition, when seeds were irradiated with a laser with a maximum power density of 8 W/m^2, a decrease in the development of a complex of especially dangerous pathogens (helminthosporiosis root rot, septoria-pyrenophorous spotting, yellow rust) was observed. In general, with an increase in the power density of laser radiation and in the time of its exposure to seeds, the growth of the maximum number of morphometric indicators of wheat productivity and a decrease in the intensity of disease affect were revealed.

Keywords. Soft wheat, pre-sowing treatment, continuous semiconductor laser, productivity, diseases development.

1 Introduction

Wheat is one of the main crops in the Russian Federation. However, despite its importance, a wide distribution, and extensive experience in cultivation, crop yields are unstable and largely depend on a complex of agro-ecological factors [1]. Currently, the development of effective technologies for cultivation of grain crops based on the principles of resource conservation and implementation of the requirements of adaptive-landscape and precision farming is still ongoing [2]. At the same time, increasing grain crop yields is the most important problem of modern farming, which depends on the cultivar used by 50% [3], and is largely determined by soil and climatic conditions of the crop cultivation and phytosanitary condition of crops [4, 5]. A promising direction for increasing the sowing suitability of agricultural seeds is the use of electrophysical and electro-technological methods, including laser radiation, stimulating biological and physiological processes in them [6–9]. The development of research on determining the optimal parameters of laser impact on biological objects should provide a significant advancement of breakthrough technologies for solving urgent problems of modern agricultural production [10].

The aim of this work is to provide a biological rationale for the use of a continuous-wave semiconductor laser in the pre-germination treatment of soft wheat seeds.

2 Objects and methods of research

The plant material of the study was soft spring wheat known as *Triticum aestivum* L. (Leningradskaya 6, k-64900 cultivar) from the collection of wheat genetic resources of N.I. Vavilov All-Russian Institute of Plant Genetic Resources (VIR). Experimental studies were carried out at the VIR experimental field.

The experiment scheme included control (seeds without treatment), variants with pre-sowing treatment of seeds with a continuous semiconductor laser with a wavelength of 637 nm. The optical system was used to create a field of uniform illumination, the power density of which was set by filters. The scheme of the installation is shown in Fig. 1. The following exposure modes were selected: power density (Wp) of 2, 4 and 8 W/m^2 at exposure time (t) of 5, 30 and 60 sec. Seeds were irradiated in groups of 150 pieces; the time was controlled by a stopwatch.

*Correspondence: kleon9@yandex.ru

https://doi.org/10.1051/bioconf/202516100022

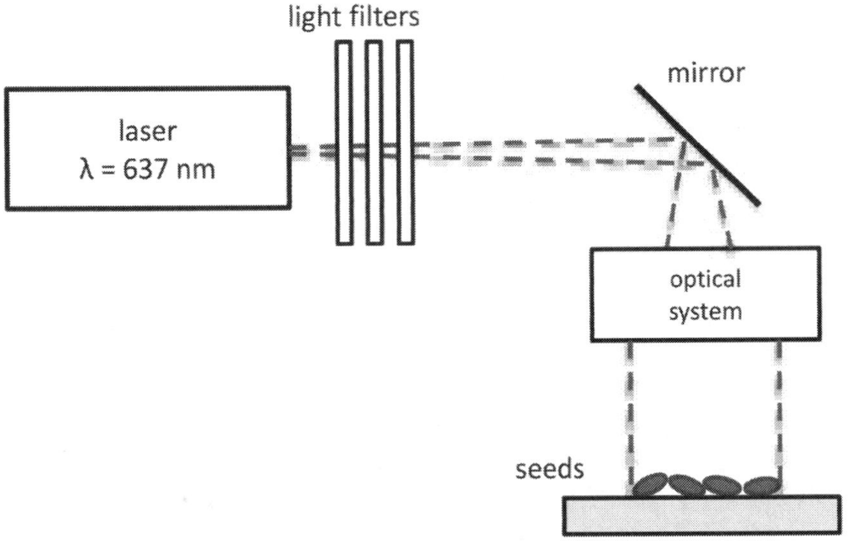

Fig. 1. The schematic diagram of the installation of pre-sowing seed treatment by laser radiation.

Under field conditions, wheat samples were sown on plots with an area of 1.0 m^2 by the row sowing method with the row spacing of 15 cm and the row spacing of 1 – 2 cm (300 grains/m^2). Experimental studies were carried out in triplicate. The phytometric characteristics of wheat crops were described using the indices. They are the field germination, the plant phase, the plant height, the number and length of primary roots, the number and the length of nodal roots, root weight, productive bushiness, total bushiness. There is also the flag leaf area, the pre-flag leaf area, vegetative part weight, the spike length, the number of spikelets per spike, spike weight, the grains number per spike, grain weight per one spike. These involve 1000 grains weight, the grains number per spikelet of a spike, grain weight per spikelet of a spike, the number of empty spikelets in a spike. In addition, the biological yield of wheat (g/plant) and, on its basis, the potential yield per unit area (t/ha) were calculated, taking into account the data on field germination, which determines the stem density of crops [11. The intensity of wheat affection by disease was described according to a set of conventional and calculated phytopathological indices. They are root rot and septoriosis-pyrenophorous spot (development (%)); yellow rust (development (%)), the number of strips with pustules (pcs.), the strip length with pustules (mm), the number of pustules in the strip (pcs.), the total number of pustules on the leaf (pcs.), pustule area (mm^2) [12].

3 Results and discussion

Significant increases in the biological Y_b (by 38.3% to 3.6 g/plant and by 58.0% to 4.1 g/plant) and potential yield Y_p of wheat (by 67.9% to 7.0 t/ha and 38.3% to 5.8 t/ha) compared to the control (Y_b=2.6 g/plant, Y =$_p$ 4.2 t/ha) was detected by pre-sowing seed treatment. It was conducted using continuous semiconductor laser with the power density of 8W/m^2 and the exposure time of 30 sec and 60 sec (Figs. 2 and 3).

*Correspondence: kleon9@yandex.ru

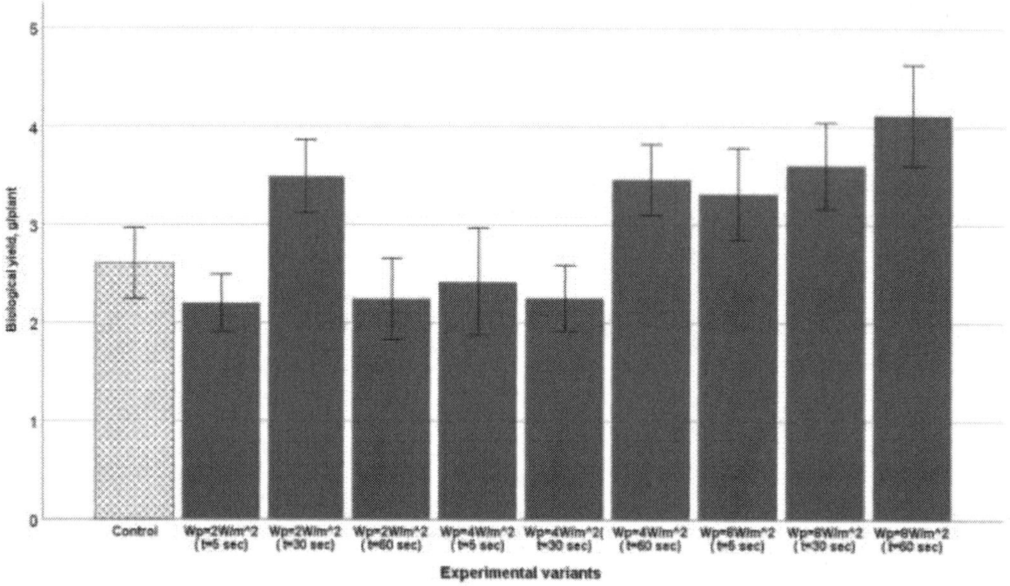

Fig. 2. The biological yield of the soft wheat cultivar "Leningradskaya 6, k-64900" during pre-sowing seed treatment with continuous semiconductor laser (in 2022)

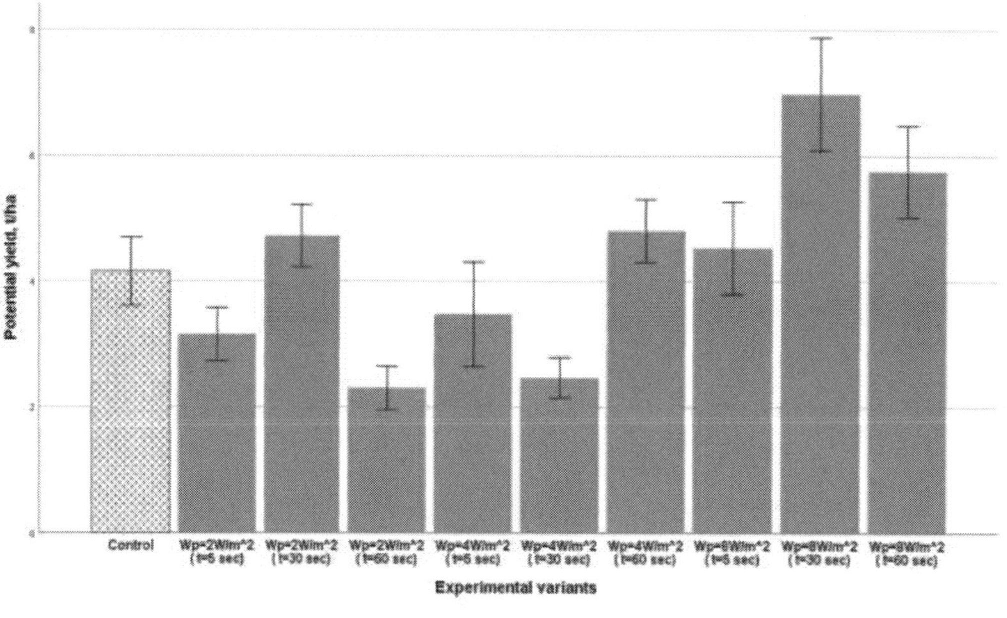

Fig. 3. The potential yield of the soft wheat cultivar "Leningradskaya 6, k-64900" during pre-sowing seed treatment with continuous semiconductor laser (in 2022).

Significant growth of Y_b and Y_p during pre-sowing treatment of wheat seeds with semiconductor laser in the experiment variants of "wp=8W/m^2, t=30 sec" was due to (at $P<0.05$) a reliable increase compared to the control. The root weight increased by 63.2% with an insignificant (at $P>0.05$) increase in field germination of seeds by 5.6%. The number of nodal roots increased by 20.1%, the length of nodal roots by 7.1%, productive bushiness by 22.3%, total bushiness by 25.1%, the weight of 1000 grains by 2.6%. The reduction of plant damage by root rot increased by 3.3%, septoriosis-pyrenophorous spot by 21.0% and yellow rust by 9.3%.

*Correspondence: kleon9@yandex.ru

In the experiment variant of "Ps=8W/m^2, t=60 sec", the reliable increase in plant height was by 8.5%, the number of primary roots increased by 109.6%, productive bushiness by 44.1%, total bushiness by 49.1%. The spikelets number per spike increased by 15.4%, the insignificant growth of the spikelet length was by 0.3%, the spikelet weight increased by 0.8%. This included the reduction of plant damage by yellow rust by 26.4% in the length of the strip with pustules, by 41.8% in the number of pustules in the strip, by 47.6% in the total number of pustules per leaf.

In all the variants of the experiment, when seeds were treated with laser before sowing, a significant lengthening of the interphase period of plant growth (by the phase of ontogenesis) was registered, except for the variant of "wp=8W/m^2, t=60 sec", where no significant change in the index compared to the control was observed. The maximum growth of the plant height (by 8.5%) was revealed in the experiment variant of "wp=8W/m^2, t=60 sec", as well as in the variants of "wp=4W/m^2, t=30 sec", "wp=4W/m^2, t=60 sec" by 7.3% and 5.9%, respectively.

The pre-sowing treatment of seeds by laser at wp=8W/m^2 with a maximum duration of exposure of t=60 sec caused the growth of the number of primary roots of wheat by 109.6%, while in the experiment variant of "wp=4Wp/m^2, t=30 sec" the decrease of this indicator by 36.7% was revealed. In the experiment variant of "wp=8Wp/m^2, t=60 sec", the maximum growth of the root weight by 15.4% was observed. Also, a significant increase in the root weight was found in the experiment variants of "wp=2W/m^2, t=60 sec" by 9.2% and "wp=8W/m^2, t=5 sec" by 9.1%.

The statistically significant decrease in the length of primary roots was revealed in the experiment variants of "wp=2Wp/m^2, t=60 sec" by 14.9% and "wp=4W/m^2, t=5 sec" by 16.8%. In the experiment variant of "wp=8W/m^2, t=60 sec", a significant decrease in the number and length of nodal roots was revealed (by 55.5% and 12.8%, respectively). At the same time, the maximum increase in productive and total bushiness (by 44.1% and 49.1%, respectively) was recorded in the above variant. The weight of the vegetative part of the plants significantly decreased in "wp=2W/m^2, t=5 sec" by 14.4%, in "wp=2W/m^2, t=30 sec" by 14.5%, in "wp=8W/m^2, t=5 sec" by 14.1%.

With regard to the elements of the wheat spike structure, the pre-sowing seed treatment with semiconductor laser had the following effect. The maximum increase in the spikelets number per spike was found in "wp=8W/m^2, t=60 sec" by 15.4%. In addition, a significant increase in the indicator was observed in "wp=2W/m^2, t=60 sec" by 9.2% and in "wp=8W/m^2, t=5 sec" by 9.1%. The significant decrease in the spike weight was revealed in "wp=2W/m^2, t=30 sec" by 48.2% and in "wp=2W/m^2, t=60 sec" by 44.2%. The "wp=4W/m^2, t=5 sec" variant decreased by 38.7%, the spikelet length of "wp=2W/m^2, t=5 sec" by 11.5%, "wp=2W/m^2, t=30 sec" by 8.7%, "wp=2W/m^2, t=60 sec" by 8.3%. And the grains number per spike of "wp=8W/m^2, t=60 sec" deceased by 14.5%. The weight of grains in the spike significantly decreased in "wp=2W/m^2, t=5 sec" by 19.4% and in "wp=2W/m^2, t=60 sec" by 28.7%, where a significant decrease in the weight of 1000 grains was also recorded (by 21.9%). In the experiment variant of "wp=8W/m^2, t=60 sec", the grains number per spikelet of spike decreased by 9.6%, but their weight did not change significantly. A significant decrease in the grains weight per spikelet of spike was observed in "wp=2W/m^2, t=60 sec" by 30.0% and "wp=4W/m^2, t=5 sec" by 18.7%.

The development of helminthosporiosis root rot in the control variant of the experiment in 2022 was insignificant (Fig. 4). No symptoms of disease development were observed in the experiment variant of "wp=8W/m^2, t=5 sec". The significant increase in the intensity of plant disease damage compared to the control was registered in the experiment variants of "wp=8W/m^2, t=5 sec" by 7.6% and "wp=4W/m^2, t=60 sec" by 6.9%.

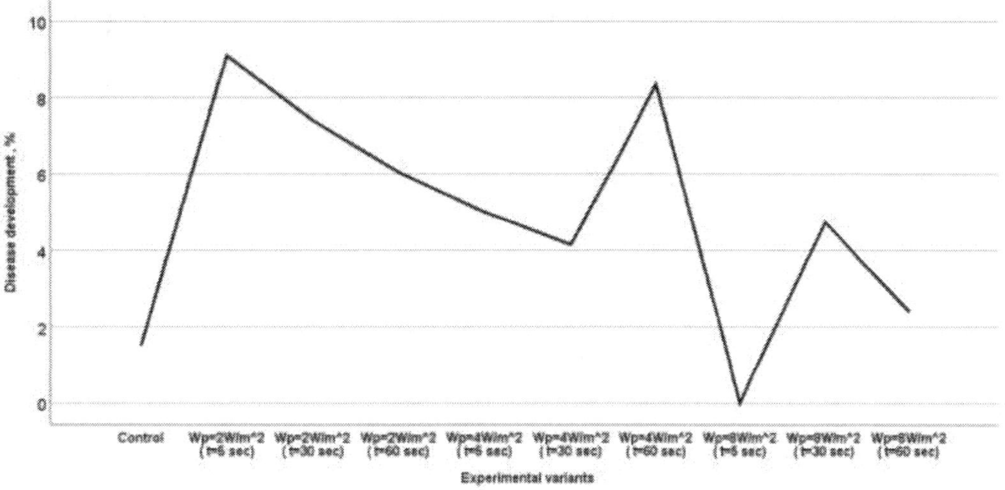

Fig. 4. Development of root rot on the cultivar "Leningradskaya 6, k-64900" during pre-sowing seed treatment with continuous semiconductor laser (in 2022).

Soft wheat leaves have not been affected by septoriosis-pyrenophorosis leaf spot in "wp=4W/m^2, t=60 sec", and in "wp=8W/m^2, t=5 sec" the disease development significantly decreased (by 29.8%). The biological efficiency (BE) of pre-sowing treatment of wheat seeds with laser was 70.4%. The tendency of disease development strengthening (by 20.2%) in "wp=8W/m^2, t=60 sec" was revealed.

The greatest reduction of plant damage caused by yellow rust pathogen was detected during pre-sowing treatment of wheat seeds with laser (wp=8W/m^2). However, the values of individual phytopathological indicators varied with the time of exposure. At t=5 sec, the disease development was 13.0% (Fig. 5), the number of strips with pustules on leaf was 56.2%; at t=60 sec, the length of strip with pustules was 26.4%; the number of pustules in the strip was 41.8%; the total number of pustules per leaf was 47.6%.

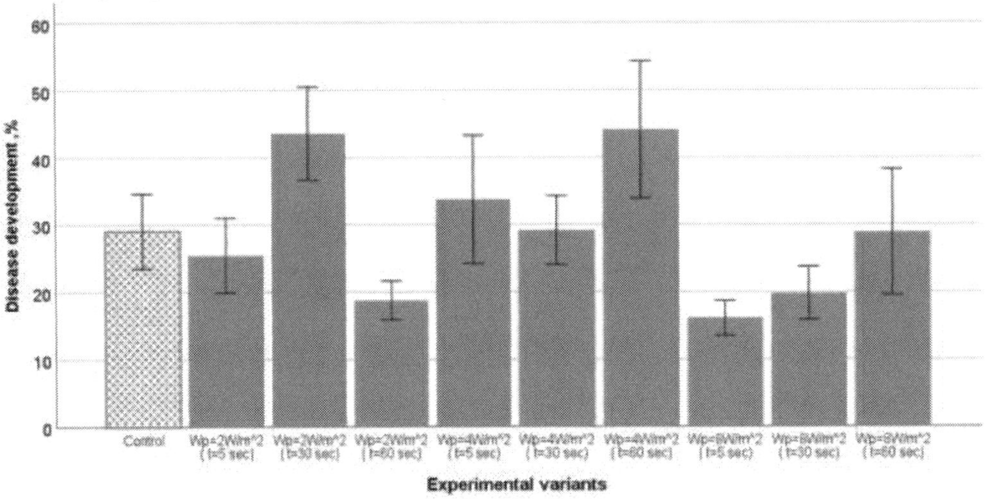

Fig. 5. The development of yellow rust on flag leaf of the soft wheat cultivar "Leningradskaya 6, k-64900" during pre-sowing seed treatment with continuous semiconductor laser (in 2022).

In general, the greatest number of morphometric indicators, the values of which increased and significantly increased (P<0.05) in comparison with the control during pre-sowing treatment of seeds with laser, was revealed in the experiment variant of "wp=8W/m^2, t=60 sec" (33.1 and 23.8%, respectively). Also, in the above-mentioned variant of the experiment, the greatest decrease in the values of the complex of phytopathological indices characterizing the intensity of wheat damage by a complex of diseases was revealed (71.4% (P>0.05) and 13.3% (P<0.05)).

It should be noted that recently a large number of scientific papers have appeared, which emphasise the positive effect of laser radiation on seeds, stimulating plant growth and development. Some works [13–15] provide data on a significant increase in plant productivity and reduction of disease damage. It is emphasised that the effectiveness of laser pre-sowing seed treatment is determined by the different responsiveness of crop species and cultivars to photobiological processes. In this regard, for each crop and cultivar it is necessary to select the optimal power density and multiplicity of irradiation of seeds with laser radiation. According to the results of our research, the greatest efficiency of pre-sowing treatment of wheat seeds of the "Leningradskaya 6" cultivar with laser radiation (λ=637 nm, wp=8 W/m^2) was revealed at the maximum duration of exposure of 60 sec.

4 Conclusions

The maximum growth of the wheat yield, as well as the number of morphometric indicators of productivity, the values of which increased and significantly increased (P<0.05) compared to the control during pre-sowing treatment of seeds with semiconductor continuous laser, was revealed in the experimental variant of "w=8W/m^2, t=60 sec". Also, in this variant the greatest reduction of the complex of phytopathological indicators characterizing the intensity of wheat damage by especially dangerous pathogens (helminthosporiosis root rot, septoriosis-pyrenophorous spotting, yellow rust) was observed. The tendency of phytometric values growth and phytopathological indices decrease in wheat crops with increasing power and time of laser radiation exposure on seeds used for seed pre-sowing preparation was revealed.

Acknowledgements

*This work was carried out within the framework of the VIR thematic project "Structuring and revelation of the genetic variability potential of the VIR world collection of cereals and groats cultures for the optimized genebank development and rational use in breeding and plant cultivation" (FGEM-2022-0009).

References

1. V.I. Trukhachev, E.G. Nikitenko, Natural factors in the forecasting of economic indicators of grain farming, Vest. AIC Stavropolya, **3**, 46–49 (2011)
2. A.I. Belenkov, M.A. Mazirov, A.I. Melchenko, *Assessment of the impact of farming systems and agrotechnologies on the environment*, Training manual (Scripta manent, Moscow, 2015), 119 c.
3. V.V. Novokhatin, Scientific substantiation of primary and elite seed production of grain crops, Achiev. of sci. and technol. of agroind. Comp., **32(9)**, 40–47 (2018), https://doi.org/10.24411/0235-2451-2018-10910
4. M.V. Arkhipov, T.A. Danilova, V.A. Pavlyushin, S.M. Sinitsina, E.N. Pasynkova, S.A. Tyukalov, Ways and opportunities for phytosanitary optimisation of agroecosystems of the North-West region of Russia, Plant Protect. Bull., **2(92)**, 5–14 (2017)
5. V.A. Pavlyushin, A.K. Lysov, Phytosanitary safety of agroecosystems and remote phytosanitary monitoring in plant protection, Modern probl. of remote sens. of the Earth from space, **16(3)**, 69–78 (2019)
6. A.M. Solov'ev, Irradiation of fodder beet seeds by factors of electromagnetic nature, In: *Application of electric power and operation of devices of agricultural power supply systems*: Coll. of sci. works of Mosc. State Agroeng. Univ. named after V.P. Goryachkin (Moscow, 1993)
7. V.I. Bukatyi, N.A. Vechernina, V.P. Karmanchikov, Laser photoactivation of seeds of Altai agricultural crops, Izv. Altai State Univ., **1(19)**, 98–99 (2001)
8. A.I. Belsky, A.M. Plavinskaya, Magnetic-laser technologies in crop production, Grain farm., **1**, 10 (2003)
9. O.G. Dolgovykh, V.V. Krasilnikov, R.R. Gazdtinov, *Optimisation of laser treatment of seeds of grain crops* [Monograph] (Izhevsk State Agricultural Academy, Izhevsk, 2014), 124 c
10. D.S. Buklagin, Application of laser radiation in crop production, Scientific and information support of innovative development of agroindustrial complex: Mater. of XII Int. Sci. and Pract. Internet Conf., 2020
11. L.E. Kolesnikov, V.A. Pavlyushin, M.V. Arkhipov, N.S. Priyatkin, Y.R. Kolesnikova, Agroecological variation of productivity and disease incidence of wheat: elements and modelling, Agrophys., **3**, 40–52 (2022)
12. L.E. Kolesnikov, A.A. Belimov, E.Y. Kudryavtseva, Yu.R. Kolesnikova, B.A. Hassan, Identification of the effectiveness of associative rhizobacteria in spring wheat cultivation, Agron. Res., **19(3)**, 1530–1544 (2021)
13. P. Zhurba, E. Zhurba, Laser technology of industrial cultivation of agricultural crops, Photonics, **3(21)**, 34–39 (2010)
14. O.G. Dolgovykh, V.V. Krasilnikov, E.V. Dresvyannikova, L.P. Artamonova, L.A. Panteleeva, Analysis of the effectiveness of laser treatment of spring wheat seeds of the cultivar "Irgina", Eng. Bull. of Don, **3(42)**, 2 (2016)
15. O.A. Shulgina, G.I. Kolesnikov, V.I. Zaostrovnykh, G.I. Zaitsev, Effect of laser irradiation on the biological basis of objects of plant and animal origin, Bull. of Kemerovo State Univ. Ser. Biolog., techn. and earth sci., **1(1)**, 23–30 (2017)

Integration in breeding and seed production through the formation of regional systems

Olga V. Svyatova*[1], *Svetlana V.* Malakhova[2], *Elena N.* Nozdracheva[1]

[1]Kursk State Agrarian University named after I.I. Ivanov, Kursk, Russia
[2]Southwest State University, Kursk, Russia

Abstract. The authors study methodological approaches to the formation of regional systems of breeding and seed production, critical for agrarian development. The aim of the work was to substantiate the necessity of transition from a fragmented approach to integrated regional systems combining scientific institutions, production enterprises and state structures. The study revealed that the existing seed production system in Russia is characterised by high dependence on imported seeds and insufficient efficiency. The authors proposed a concept of a regional system based on co-operation and integration, including innovation, production and market subsystems. A non-profit association is proposed as an organisational form to facilitate the coordination of activities and attract investment. The results of the study show that the implementation of the proposed model contributes to improving the efficiency of breeding and seed production, and ensures sustainable development of the agricultural sector.

1 Introduction

One of the most important elements determining the success of agricultural production is breeding and seed production. Creating high-yielding and disease- and pest-resistant varieties and hybrids, as well as providing them with quality seed are fundamental tasks, the solution of which directly affects the efficiency of the entire agriculture.

There are several main approaches to solving problems in the field of breeding and seed production. One of them is the development and support of large research centres engaged in plant breeding and genetics. These centres have the necessary human and technical capacity to develop new varieties and hybrids adapted to specific regional conditions. However, they are often disconnected from real production processes and are not always able to effectively translate their achievements into practice [1, 2]. Another approach is to develop the private seed sector, stimulating competition and innovation. This approach can improve seed quality and reduce seed costs, but can also lead to market concentration and dominance by large players. The third way is to develop co-operation and integration between research institutions, production enterprises and government agencies. This approach can ensure more efficient interaction between all actors involved in the process and facilitate faster introduction of new technologies into practice. Each of these approaches has advantages and disadvantages. The development of large research centres helps to concentrate resources for basic research and development of new technologies. However, without proper linkage with production, their developments may remain unused. Private seed production based on market mechanisms promotes competition and price reduction, but may be oriented towards maximising profits at the expense of seed quality. Cooperative forms of integration can be effective but require alignment of interests of different groups and organisations. At present, there is an obvious need to find a balance between these approaches to achieve sustainable development of the seed industry [3, 4].

The relevance and importance of this direction lies in the fact that it offers a model of transition from individual seed production centres to the formation of complex regional systems based on the integration of various participants. This approach allows for the creation of a sustainable and flexible structure that can adapt to rapidly changing market conditions and regional specificities. This is especially important for such a large country as Russia, with its diversity of climatic zones and agronomic conditions [5]. Moreover, the author's approach assumes the consideration of innovation models as dynamically developing sub-sectors of the economy, which implies the need for continuous improvement and adaptation of technologies, as well as the introduction of resource-efficient agro-technologies. It is important to note that this approach not only takes into account organisational issues, but also offers economic models based on the contractual form of network organisation, which allows to ensure the

* Corresponding author: olga_svyatova@mail.ru

sustainability and financial attractiveness of these systems.

The aim of the work is to substantiate methodological approaches to the formation of regional systems of breeding and seed production, as well as the development of organisational models that contribute to the effective functioning of these systems. The work is aimed at identifying and analysing the cause-and-effect relationships that constrain the institutional development of this sub-branch in Russia, and suggests ways to solve them by creating effective regional systems. The study considers the potential of new technologies and agro-technologies in increasing crop yields and reveals the importance of the system approach for the development of the seed industry.

2 Method and materials

In this work, a comprehensive methodological approach was used, including the analysis of literature sources and conducting our own research to develop a model of the regional system of breeding and seed production. The experimental part of the work consisted in the study and generalisation of experience of various forms of organisation of breeding and seed production activities, as well as in the development and analysis of organisational models that can ensure effective interaction between the participants of the process.

In the process of the study, the system approach was applied, which implied consideration of seed production as a set of interrelated elements: scientific institutions, production enterprises, state and business structures. General scientific methods such as comparison, generalisation and systematisation were used for the analysis of organisational models. Methods of economic and mathematical modelling were used in the development of economic models. In particular, various options of contractual relations between the participants of regional systems were considered in order to determine the most effective forms of integration.

3 Results and discussion

The presented research is aimed at analysing a critical aspect of agrarian development - the formation of effective regional systems of breeding and seed production [6]. In contrast to a fragmented approach, where individual centres of elite seed production function in isolation, this study promotes the idea of integration and cooperation. The focus is on the transformation of disparate structures into a single, dynamic network. This network should bring together research institutions, production enterprises and public authorities, thereby providing an integrated and synergistic effect. At the same time, this study is not limited to organisational aspects only, but also addresses the economic feasibility of the proposed models. The methodological basis of the study is the application of analytical methods and modelling. This involves a detailed analysis of the current state of breeding and seed production in Russia, identification of root causes that hinder the development of the industry, and on this basis, the construction of a conceptual model of the regional system. This model is not a static scheme, but a dynamic mechanism capable of self-development and adaptation. An important element of the model is the emphasis on horizontal and vertical integration between different actors, as well as on the transfer of advanced agro-technologies and scientific knowledge.

The concept of the regional system of breeding and seed production proposed in this paper goes beyond traditional approaches. It is not reduced to a simple merger of existing structures, but represents a fundamentally new model in which science and practice work in close co-operation. This model envisages the creation of specialised production sites, such as pilot and base farms, demonstration plots, and other elements that support the entire seed production cycle from basic research to commercialisation (fig. 1). The key idea is that the system should not only produce seed, but also provide technological support and scientific knowledge to accompany this process, enabling farmers to maximise the potential of new varieties and hybrids. The proposed model also emphasises the dynamic nature of the system, which should continuously improve and adapt to changing market conditions and technological challenges [7, 8]. For this purpose, it is supposed to attract private capital, as well as to stimulate the initiative of local entrepreneurs. At the same time, the system should be economically self-sufficient and attractive to investors, which is achieved through contractual forms of network organisation. At this stage of the study, an important aspect is to conduct a detailed economic analysis of the proposed model, including an assessment of its potential profitability and payback period.

Analysis of the results clearly shows that the existing seed production system in Russia does not meet modern requirements and needs deep modernisation. Excessive dependence on imported seeds, reaching 90% for some crops, indicates the weakness of the domestic breeding and seed industry and threatens the food security of the country [9, 10]. It also leads to an increase in the cost of agricultural products and reduces the competitiveness of Russian farmers. Therefore, the authors propose a transition to an integrated approach, where the regional system of breeding and seed production functions as a single organism, including three interrelated subsystems: innovation, production and market. The innovation subsystem focuses on fundamental research and development of new varieties and hybrids adapted to specific regional conditions, using modern biotechnological and genetic methods [11, 12]. The production subsystem is engaged in multiplication of seeds of higher reproductions, using advanced agro-technologies that ensure high quality of seed [13]. The market subsystem is engaged in the promotion of seeds on the market, knowledge transfer and technological support to end users. According to Rosstat data for 2022, the share of domestic seeds in the market of cereals is only 35%, vegetable seeds 20%, sunflower seeds 50%, and potatoes 15%. These figures clearly indicate the need for major changes in the seed industry.

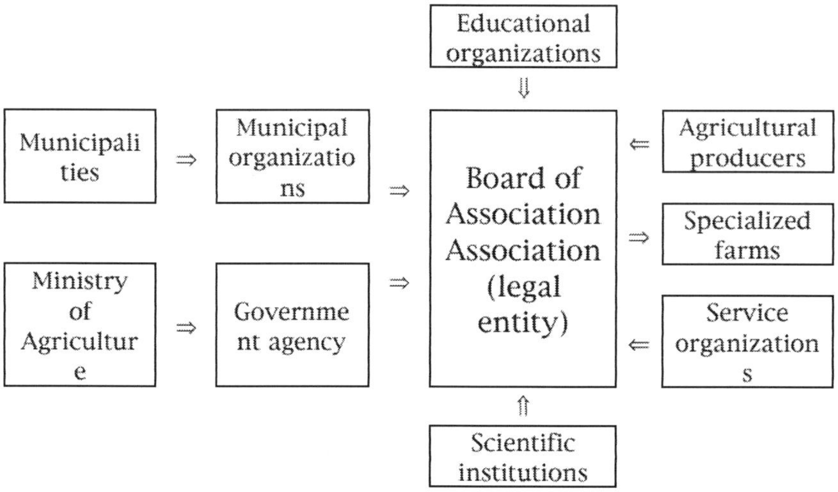

Figure 1. Scheme of the regional system of breeding and seed production in the form of an association

The concept of the regional system of breeding and seed production proposed in this paper goes beyond traditional approaches. It is not reduced to a simple merger of existing structures, but represents a fundamentally new model in which science and practice work in close co-operation. This model envisages the creation of specialised production sites, such as pilot and base farms, demonstration plots, and other elements that support the entire seed production cycle from basic research to commercialisation (fig. 1). The key idea is that the system should not only produce seed, but also provide technological support and scientific knowledge to accompany this process, enabling farmers to maximise the potential of new varieties and hybrids. The proposed model also emphasises the dynamic nature of the system, which should continuously improve and adapt to changing market conditions and technological challenges [7, 8]. For this purpose, it is supposed to attract private capital, as well as to stimulate the initiative of local entrepreneurs. At the same time, the system should be economically self-sufficient and attractive to investors, which is achieved through contractual forms of network organisation. At this stage of the study, an important aspect is to conduct a detailed economic analysis of the proposed model, including an assessment of its potential profitability and payback period

The proposed models of cooperation in breeding and seed production, based on the integration of scientific, production and governmental structures, have already proved their effectiveness in practice, both in Russia and abroad. One of the vivid examples is the experience of the Netherlands, where seed production is developing within the framework of close cooperation between universities, research centres and private seed companies. Such co-operation allows Dutch companies to occupy leading positions in the world seed market. For example, Bejo Zaden, a company specialising in vegetable breeding, actively cooperates with Wageningen University, which allows them to quickly implement new developments and create high-yielding hybrids. According to 2023 data, Bejo Zaden produces more than 170 thousand tonnes of vegetable seeds annually, exporting them to more than 100 countries, accounting for more than 30% of the global onion seed market.

In Russia, a positive example of co-operation is the activity of research and production associations (NGOs) in the Soviet period. For example, the NGO Kuban was engaged in breeding and seed production of winter wheat, which allowed the development of a number of high-yielding varieties, such as Bezostaya 1 and Aurora. As a result of their work, wheat yields in the Krasnodar region increased by 40-50% compared to the pre-war period, reaching 5-6 tonnes per hectare in some farms. The NGO 'Podmoskovye', specialising in potato breeding, created varieties resistant to phytophthora, which reduced yield losses from this disease by 20-25% in the central regions, yielding up to 30 tonnes per hectare in some farms.

Modern examples also demonstrate the effectiveness of co-operation. Gavrish, although a private company, actively cooperates with scientific institutions, as well as with farms, which allows them to create varieties and hybrids adapted to specific conditions. At the moment, the Gavrish seed catalogue has more than 2000 names, which provide a significant share of greenhouse vegetable growing in Russia, and are also exported to the Middle East and Europe. According to our data, based on surveys of farms, the use of seeds of varieties and hybrids 'Gavrish' in greenhouse farming allows to obtain 15-20% higher yields compared to the use of imported analogues, as well as to reduce the growing time by 10-15 days [18]. In the context of the proposed model, the results of the work indicate the need to create such regional structures based on cooperation and partnership. The most promising organisational form seems to be the association of a non-profit organisation, which can combine the resources of different entities and ensure coordination of their activities [19]. This will allow to create a sustainable and flexible system that can effectively solve the problems of breeding and seed production in specific regions. However, at this stage, an important factor is to carry out detailed economic calculations to confirm the financial feasibility of this model. Based on the data obtained, the authors conclude that there is a need for a systematic approach to the development of breeding and seed production, where research and production practice are closely linked.

4 Conclusion

This study emphasises the critical need to reform the existing system of breeding and seed production in Russia. The results of the work convincingly demonstrate that fragmented and non-integrated approaches to seed production lead to excessive dependence on imports, reduce the competitiveness of domestic agricultural producers and jeopardise the food security of the country. Based on the analysis, the authors justify the need to move to a systemic approach based on the integration of scientific research, production practices and government regulation.

The key conclusion is the proposal to form regional systems of breeding and seed production, which should be unified dynamic complexes that unite research institutions, production enterprises, government agencies and other stakeholders [20, 21]. This system should include innovation, production and market subsystems, each of which fulfils its specific function. Important is the emphasis on contractual forms of network organisation, which will ensure the financial sustainability and attractiveness of these systems for investors. At the same time, the non-profit form of the association seems to be the most promising, as it ensures transparency and sustainability.

Summarising the results obtained, we can state that the model of the regional system of breeding and seed production proposed in this paper is not only theoretically justified, but also practically applicable. It allows to create a flexible and adaptive structure that can effectively solve the problems of breeding and seed production in different regional conditions. At the same time, for successful implementation of the proposed model, further work is required to develop specific methodological recommendations that take into account the specifics of each region.

References

1. A.I. Trunova, V.V. Kukartsev, S.O. Kurashkin, K.A. Bashmur, R.B. Sergienko, Influence of Temperature Regimes of Synthetic Iron Smelting on Casting Production Efficiency. Metals, **13(7),** 1234 (2023). DOI: 10.3390/met13071234.

2. A.I. Trunova, S.O. Kurashkin, I. F. Vivián, K. A. Bashmur, Increasing the Efficiency of Synthetic Iron Production by the Use of New Kit Lining. Metals, **13(7),** 1184 (2023). DOI: 10.3390/met13071184.

3. Yu.N. Serdechnaya, D.V. Serdechnyy, N.A. Andriyanov, S.A. Korchagin. Development of an effective monitoring and analysis system of modern technologies for the agro-industrial complex development in mountainous areas. Sustainable Development of Mountain Territories, **16,** 770–779 (2024). DOI: 10.21177/1998-4502-2024-16-2-770-779.

4. G. Suriya narayanan, D. Ruben Sudhakar. Sustainable cultivation of Gracilaria edulis in poultry manure-derived media: Evaluating growth, biochemical properties, and agar quality. Food and Bioproducts Processing, **150,** 78–88 (2025). DOI: 10.1016/j.fbp.2024.12.015.

5. I. A. Kerimov, L. R. Bekmurzaeva. Modern agro-climatic conditions of mountain landscapes in the North Caucasus. Sustainable Development of Mountain Territories, **14,** 555–563 (2022). DOI: 10.21177/1998-4502-2022-14-4-555-563.

6. J.C.R. Cardona. Methodology for studying the history of agrarian systems. Rev. Fac. Nac. Agron. Medellin, **78,** 19-25 (2025).

7. H. Cochet, O. Ducourtieux, N. Garambois. Agrarian systems and climate change: Journeys of adaptation in the global south. Agrar. Syst. Clim. Change: Journeys Adapt. Glob. South, **1**-208 (2024). DOI: 10.1079/9781800628137.0000.

8. A.A. Kapanski, R.V. Klyuev, A.E. Boltrushevich, S.N. Sorokova, E.A. Efremenkov, A.Y. Demin. Geospatial Clustering in Smart City Resource Management: An Initial Step in the Optimisation of Complex Technical Supply Systems. Smart Cities, **8,** 14 (2025). https://doi.org/10.3390/smartcities8010014.

9. S. Bamidele. BRICS Revolution: The Global Agrarian Shift Sparked by Emerging Economies. India Q., **3,** 21-29 (2025). DOI: 10.1177/09749284241307937.

10. D. Zyukin, O. Svyatova, R. Soloshenko. Conditions and perspectives of russian sugar market development. Economic Annals-XXI, **161(9-10)**, 47-50, (2016). DOI: 10.21003/ea.V161-11.

11. T. Bandara, B. Marambe, G. Pushpakumara, P. Silva, R. Punyawardena, L. Manawadu, S. Premalal, G. Miah, K.R. Dahal. Adapting mountain-farming systems to climate change by reducing food-nutrition-health vulnerability. Mitig. Adapt. Strateg. Glob. Change, **29,** 88 (2024). DOI: 10.1007/s11027-024-10181-x.

12. V.O. Gutarevich, A.S. Mikhalev, V.E. Gozbenko, Reducing Oscillations in Suspension of Mine Monorail Track. Mining Informational and Analytical Bulletin, **11,** 72–87 (2023). DOI: 10.25018/0236_1493_2023_111_0_72.

13. K. Akzharkyn, D. Yelemessov, D. Baskanbayeva, V.Y. Skeeba. Strengthening Polymer Concrete with Carbon and Basalt Fibres. Appl. Sci. (Switzerland), **14,** 7567 (2024). DOI: 10.3390/app14177567.

14. A. Lakhouit, W.S. AL Rashed, S.Y.H. Abbas, M. Shaban. Integrating machine learning for precision agriculture waste estimation and sustainability enhancement. Computers and Electronics in Agriculture,

230, 109933 (2025). DOI: 10.1016/j.compag.2025.109933.

15. C.N. Mdoe, C.P. Mahonge, E.E. Ngowi. Mapping the trends, knowledge production, and practices of climate-smart aquaculture scholarship: Empirical insights from bibliometric analysis. Aquaculture, **598,** 741939 (2025). DOI: 10.1016/j.aquaculture.2024.741939.

16. M.A. Nawaz, I.E. Pamirsky, K.S. Golokhvast. Bioinformatics in Russia: history and present-day landscape. Briefings in Bioinformatics, **25,** bbae513 (2024). DOI: 10.1093/bib/bbae513.

17. T.A. Panfilova, K.A. Bashmur, V.V. Kondratiev, To the Concept of Leaching Metal-Containing Raw Materials in the Disintegrator. Mining Informational and Analytical Bulletin. **11-1,** 239-251 (2023). DOI: 10.25018/0236_1493_2023_111_0_239.

18. S.Yu. Avksentiev, P.A. Semenov, A.S. Afanasyev. Comprehensive solutions for tailings management at the Usolsky potash plant. Sustainable Development of Mountain Territories, **16,** 7–18 (2024). DOI: 10.21177/1998-4502-2024-16-1-7-18.

19. L.Z. Gukezheva, A.A. Kokov. Providing a sustainable development of agrarian production on the basis of investment processes activization of agrarian and industrial complex economic entities. Sustainable Development of Mountain Territories, **7,** 82–88 (2015).

20. V.I. Golik, T.A. Panfilova, V.Yu. Konyukhov, The Mechanochemical Activation of Leaching Processes in a Disintegrator. Mining Informational and Analytical Bulletin. **11-1,** 175-189 (2023). DOI: 10.25018/0236_1493_2023_111_0_175.

21. A.I. Smirnov, N.V. Plotnikova, V.G. Burov. Features of the fine structure of the active layers of organic solar cells. Key Eng. Mat., **712,** 176-181 (2016). DOI: 10.4028/www.scientific.net/KEM.712.176.

Intensification of the process of extraction of sucrose from sugar beet of different technological advantage

Nadezhda Kulneva[1], *Lyudmila* Korobova[1], *Irina* Matytsina* [2], *Natalia* Matvienko[1] and *Svetlana* Chernyaeva[2]

[1]Voronezh State University of Engineering Technologies, Voronezh, Russia
[2]Voronezh branch Admiral Makarov State University of Maritime and Inland Shipping, Voronezh, Russia

Abstract. Unfavorable climatic conditions have a negative impact on the storage and processing of sugar beets. Low quality of the raw material during storage causes intensive development of microbiological processes, which leads to high losses of sucrose. Leuconostoc mesenteroides bacteria develop especially intensively in spoiled beets, causing a disease of root crops (mucous bacteriosis), which is activated in the washing department of the plant. The product of the bacteria's vital activity, the polysaccharide dextran, passes into the diffusion juice during the extraction process, complicating further technological processes. The proposed solution for suppressing microbiological processes is the use of sodium dichloroisocyanuric acid as an antiseptic, a solution of which can be used to treat beet chips or an extractant. Some plants use inclined-type diffusion devices for extracting sucrose, in which there is no heat treatment of beet chips before entering the device, which leads to a low degree of sucrose extraction. Technical solutions have been developed for preparing chips for extraction using the combined thermochemical processing of chips with the use of salt solutions of aluminum and ammonium sulfates. Rational processing parameters have been selected experimentally: the concentration of the ammonium sulfate solution is 0.05%; its consumption is 10% of the chips mass. The studies conducted to study the microstructure of sugar beet tissue with various methods of its processing revealed the mechanism of action of the reagents. Experimental data on determining the molecular diffusion coefficient after processing sugar beet tissue using salt solutions confirmed the efficiency of the proposed method. A design of an apparatus for the complex thermochemical processing of sugar beet chips before extraction using rational technical solutions has been developed. The proposed method and device for its implementation ensure high efficiency in processing raw materials of various qualities.

Introduction

A modern sugar factory is a complex agro-industrial enterprise, including a raw material base and a direct processing complex of fully automated and mechanized technological processes. Beet sugar factories fully ensure the food security of the Russian Federation.

A significant problem of the beet sugar complex is the seasonality of its operation and the associated changing quality of the processed raw materials. Sugar beets are crops with a limited shelf life. With huge volumes of harvested beets, approximately 50 million tons annually and a total daily processing of almost 360 thousand tons, factories fail to complete the season in the optimal time, which is accompanied by significant losses in the mass of sugar beets and the sucrose contained in it.

Results and Discussion

Currently, many technical solutions have been developed to optimize sugar beet storage processes, reduce the intensity of its respiration and microbiological processes [1-3]. However, unfavorable climatic conditions significantly complicate the storage process and negatively affect the quality of the processed raw materials. In the 2024/25 processing season, due to the dry summer period, even freshly harvested beets were heavily wilted, had low turgor, an increased content of salts (molasses formers, in particular, sodium compounds) and had reduced immunity.

As a result, the factories experienced significant difficulties during storage and the processing of such beets, including the occurrence of a secondary infection (mucous bacteriosis) [4]. To suppress the development of Leuconostoc mesenteroides, various bactericidal and enzymatic preparations are widely used, including sodium dichloroisocyanuric acid (DCI) [5-9]. The development of pathogenic microflora occurs most intensively under the conditions of the sucrose extraction process. To suppress microbiological processes, it is advisable to introduce disinfectants not only directly into the diffusion apparatus, but also to use them in the preparation of the extractant (Tables 1, 2) [10].

* Corresponding author: irina210390@mail.ru

Table 1. Quality of the diffusion juice when using DHCN for extractant treatment

Indicators	Amount of the DHCN solution in the extractant, %			
	0	10	20	30
Purity of juice, %	84,6	86,8	87,1	87,2
Protein content, mg/cm^3	0,516	0,455	0,420	0,426
Mass fraction of reducing substances, %	0,0788	0,0282	0,0274	0,0276

It has been experimentally established that increasing the input of the DHCN solution during the preparation of the extractant leads to an increase in the purity of the diffusion juice, a decrease in the content of proteins and reducing substances. But these changes are insignificant. Therefore, 10% of the DHCN solution is sufficient to achieve good results in diffusion, which is cost-effective from an economic point of view [11-12].

Table 2. Indicators of diffusion and purified juices during the processing of seeded beets (Leuconostoc mesenteroides)

Indicators	Extraction agent without processing	Extraction agent with the DHCN solution
Diffusion juice		
Purity, %	88,4	90,3
Protein content, mg/cm^3	0,719	0,535
Purified juice		
Purity, %	91,2	92,7
Color, unit of optical density	267,3	224,8

Analysis of the results shows that the use of DHCN in the composition of the extractant for the diffusion process improves the quality indicators of semi-finished products of sugar beet production during the processing of raw materials affected by mucous bacteriosis.

Particular problems arise when using inclined-type diffusion apparatuses for sucrose extraction, in which there is no preliminary heat preparation of beet chips, and the processing process is carried out directly in the apparatus. This leads to the underheating of the beet tissue, a low degree of sucrose extraction due to its retention in the cells of the beet tissue [13].

To solve the complex problem of preparation for extraction and reducing the transfer of non-sugars, thermochemical treatment of the shavings was carried out before extraction using salt solutions of aluminum and ammonium sulfates (Tables 3-4).

Table 3. Comparative assessment of the quality indicators of semi-finished products during the processing of high-tech beets

Indicators	Scheme without processing	Reagents for feedwater treatment	
		$(NH_4)_2SO_4$	$Al_2(SO_4)_3$
Diffusion juice			
Purity, %	85,5	87,9	86,8
Protein content, mg/cm^3	0,42	0,31	0,34
Purified juice			
Purity, %	89,7	91,4	90,6
Color, units of optical density	219	170	182
Mass fraction of calcium salts, % CaO	0,031	0,022	0,026

Table 4. Quality indicators of semi-finished products during the processing of low-quality beets

Indicators	Scheme without processing	Reagents for feedwater treatment	
		$(NH_4)_2SO_4$	$Al_2(SO_4)_3$
Diffusion juice			
Purity, %	83,3	85,8	84,6
Mass fraction of proteins, mg/cm^3	0,44	0,37	0,39
Purified juice			
Purity, %	88,7	90,4	90,0
Color, units of optical density	276	211	221
Calcium salts, % CaO	0,037	0,026	0,031

An analysis of the quality indicators of semi-finished products obtained using a scheme with preliminary thermochemical treatment of sugar beet chips with hot solutions of salt reagents indicates the feasibility of their use. The highest effect is observed when treated with a solution of ammonium sulfate $(NH_4)_2SO_4$ [13].

Research has been conducted to select a rational concentration of ammonium sulfate in a solution for the thermochemical processing of chips.

According to the research methodology, sugar beet chips were subjected to sequential treatment with steam (T-100 °C) for 60 s and aqueous solutions of ammonium sulfate with different concentrations (0,01-0,3%), heated to a temperature of 70 °C in an amount of 10% of the chips weight. The processed chips were subjected to sucrose extraction, and the resulting diffusion juice was sent for physicochemical purification.

The obtained diffusion and purified juices were analyzed using generally accepted methods (Figures 1-2).

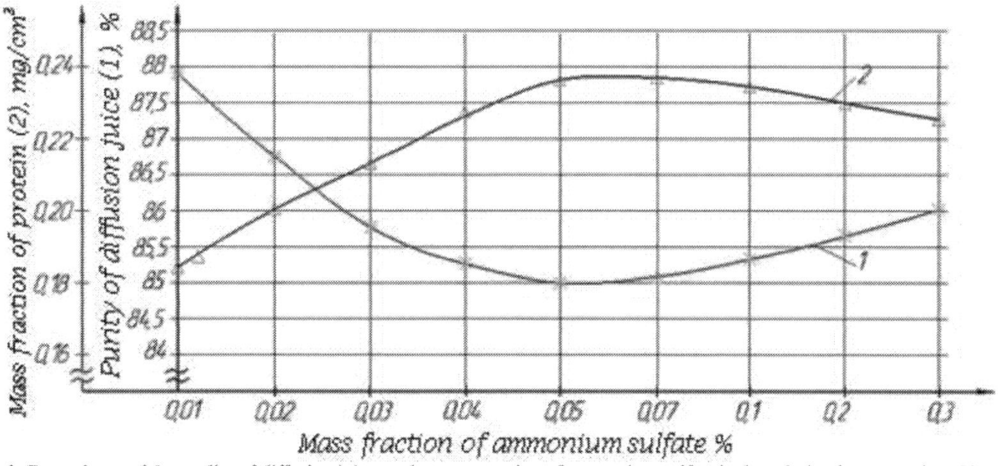

Fig. 1. Dependence of the quality of diffusion juice on the concentration of ammonium sulfate in the solution for processing chips

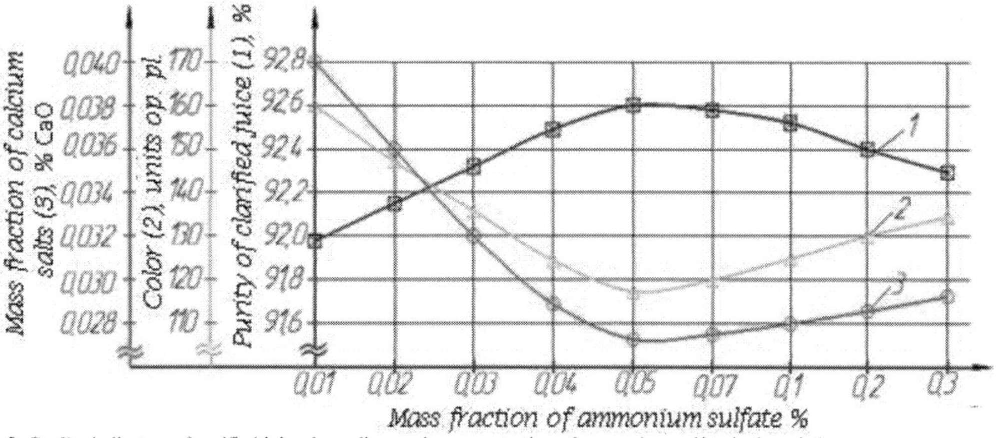

Fig. 2. Quality indicators of purified juice depending on the concentration of ammonium sulfate in the solution

An analytical study of the parameters of diffusion and purified juices indicates the feasibility of using ammonium sulfate with a solution concentration of 0.05% for processing beet chips. This solution concentration value provides the highest values of the quality parameters of semi-finished products. Increasing the mass fraction of ammonium sulfate in the solution does not bring the desired technological effect and worsens the quality of semi-finished products.

The quality indicators of juices indicate the high chemical activity of ammonium sulfate. NH_4+ and SO_4^{2-}- ions formed during the hydrolysis of ammonium sulfate have an intensive precipitating and salting-out effect on high-molecular compounds in the composition of sugar beet tissue, thereby reducing their transition to semi-finished products. In addition, SO_4^{2-}- ions exhibit disinfectant properties, which prevents the development of microflora inside the diffusion apparatus, and the amount of unaccounted sucrose losses in the diffusion section is reduced.

Research has been conducted to study the microstructure of sugar beet tissue using various processing methods (Fig. 3) [14].

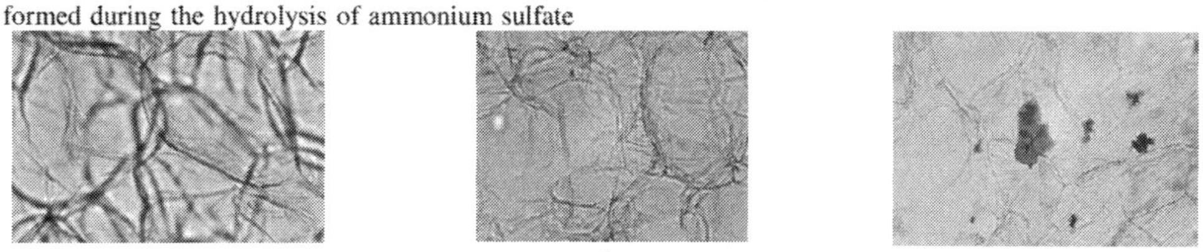

a b c

Fig. 3. Microstructure of sugar beet tissue after treatment: a – control sample without treatment; b – treatment with the $Al_2(SO_4)_3$ solution; c – treatment with $(NH_4)_2SO_4$

Beet tissue that has not been subjected to extraction beforehand maintains the integrity of the cells; the cytoplasmic membrane is tightly adjacent to the shell, preventing the diffusion of sucrose molecules.

After treatment with an aluminum sulfate solution, significant morphological changes occur in the cells. Their integrity is disrupted due to the high degree of plasmolysis of protein components as a result of thermal and chemical exposure. A significant number of pores are formed through which the extractant interacts with the cellular contents, forming conglomerates with protein substances, which prevents the transfer of proteins from the cells to the extractant.

The microstructure of the tissue after treatment with ammonium sulfate solution leads to a noticeable decrease in the thickness of the cell walls and the formation of pores and vesicles, into which the reagent solution actively penetrates. Its interaction with the substances of the protein-pectin complex promotes the formation of complex compounds that precipitate in the surface layer of beet tissue.

Table 5. Molecular diffusion coefficient of sucrose during the processing of high-quality beets using thermochemical treatment of chips

Method of carrying out diffusion	Without processing	Reagents for thermochemical treatment		
		CaSO4	Al2(SO4)3	(NH4)2SO4
Mass fraction of sucrose in the extract, %	2,25	2,65	3,25	3,55
Sugar content of beets, %	17,2	17,2	17,2	17,2
Ratio of average volume concentrations of sucrose, Cek / Csv $\times 10^{-2}$	13	14	18	19
Thickness of samples, $L \times 10^{-3}$ m	7	7	7	7
Size, d' $\times 10^{-2}$	18	22	26	29
Diffusion coefficient, D $\times 10^{-10}$, m^2/s	36	40	42	45

Table 6. Molecular diffusion coefficient of sucrose during the processing of low-quality beets using various reagents for processing chips

Method of carrying out diffusion	Without processing	Reagents for thermochemical treatment		
		$CaSO_4$	$Al_2(SO_4)_3$	$(NH_4)_2SO_4$
Mass fraction of sucrose in the extract, %	1,9	2,20	2,65	3,25
Sugar content of beets, %	15,75	15,75	15,75	15,75
Ratio of average volume concentrations of sucrose, Cek / Csv $\times 10^{-2}$	10	12	15	18
Thickness of samples, $L \times 10^{-3}$ m	7	7	7	7
Size, d' $\times 10^{-2}$	15	18	21	24
Diffusion coefficient, D $\times 10^{-10}$, m^2/s	29	37	38	41

The results of determining the coefficient of molecular diffusion allow us to conclude that, regardless of the quality of the processed beets, thermochemical treatment of beet chips with a solution of ammonium sulfate ensures a high degree of permeability of beet tissue and extraction of sucrose.

To carry out complex thermochemical treatment of sugar beet shavings in preparation for extraction in an inclined diffusion apparatus, a scalder design has been proposed [15] (Fig. 4).

The design of the scalder incorporates effective technical solutions:

- use of a loading device for sugar beet chips, ensuring the continuity of the scalding process;
- tangential arrangement of steam nozzles, supplied from the outside and from the axis of the pipe, preventing their clogging with chips and the formation of stagnant zones inside the sections, which ensures uniform heating of sugar beet chips;
- modernization of the mixing device, which ensures effective mixing and movement of chips across the sections of the scalder and eliminates the formation of stagnant zones;
- use of radially located horizontal sieves with rectangular holes, facilitating the movement of sugar beet chips from section to section due to the large usable area.

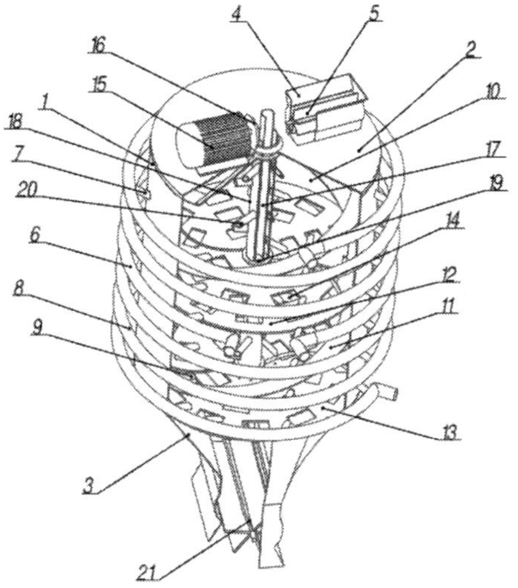

Fig. 4. Sugar beet chip scalder: 1 – scalder body; 2 – top cover; 3 – conical bottom; 4 – receiving hopper; 5 – rotary dispenser; 6 – steam line; 7, 9 – external supply nozzles; 8 – ammonium sulfate solution supply pipelines; 10 – upper section of scalder; 11 – lower section of scalder; 12,13 – horizontal screens; 14 – transfer openings; 15 – drive; 16 – bevel gear; 17 – inner rotor shaft; 18 – outer rotor shaft; 19 – dividing partition; 20 – internal irrigation nozzles; 21 – turnstile

Conclusions

The conducted studies allow us to conclude that the thermochemical treatment of sugar beet chips with hot solutions of aluminum and ammonium sulfates before sucrose extraction has a favorable effect on the quality characteristics of the diffusion and purified juices. The sugar beet chips entering the diffusion unit after thermochemical treatment have the temperature required for the process. Sucrose extraction begins already in the 1st section of the diffusion apparatus, which has a positive effect on the efficiency of the diffusion process. The productivity of the diffusion unit increases, and the consumption of heating steam for its heating is reduced at the same time. A high technological and economic effect is achieved when processing raw materials of different technological quality.

References

1. A. Maui, S. Aipeisova, N. Utarbaeva, A. Matsyura, Ukrainian J. **10**, 5 (2020)
2. D. Arapov, L. Korobova, N. Kulneva, Sugar 7, (2023)
3. A. I. Trunova, I. F. Vivián, K. A. Bashmur, Metals, **13(7),** 1184 (2023). DOI: 10.3390/met13071184
4. L. Putilina, N. Lazutina, Storage and processing of agricultural raw materials, **1**, (2021)
5. N. Dolgopolova, A. Kaluzhskikh, M. Kotelnikova, IOP Conference **666**, (2021)
6. M. Yousef, A. Zohri, A. Darwish, A. Shamseldin et al., Front. Microbiol. **14**, (2023)
7. Z. Seres et al., LWT 75, (2017)
8. V. Yu. Konyukhov, Mining Informational and Analytical Bulletin. **11-1,** 175-189 (2023). DOI: 10.25018/0236_1493_2023_111_0_175
9. N. Kulneva, Yu. Posledova, Bulletin of VSUET **86**, 3 (2024)
10. S. Chernyaeva, L. Korobova, I. Tolstova, et al., RUDECK 2020 **148** (2020).
11. S. O. Kurashkin, K. A. Bashmur, Metals, **13(7),** 1234 (2023). DOI: 10.3390/met13071234
12. V. V. Kondratiev, Mining Informational and Analytical Bulletin. **11-1,** 239-251 (2023). DOI: 10.25018/0236_1493_2023_111_0_239
13. V. Gorodetsky, S. Semenikhin, N. Daisheva, N. Kotlyarevskaya, et al., Ecology, Evironment and Conservation **25**, 3 (2019)
14. M. Zhuravlev, N. Kulneva, G. Agafonov, I. Naumchenko et al., IOP Publishing **640**, 5 (2021)
15. O. Korneeva, S. Yakovleva, N. Matvienko, L. Frolova et al., Bulletin of Voronezh State University of Engineering Technologies **85**, 1 (2023)

Integrated model of agricultural production through the use of non-waste technologies based on microorganisms

I.Yu. Matasova [1], *V.V.* Skoda [1], *Valery A.* Shchebeteev [2*], *Andrey M.* Karadzhiyan [2], *Kermen B.* Mushaeva [3], *Natalya L.* Kurepina [3]

[1] Admiral Ushakov Maritime State University, Novorossiysk, 353924, Russia
[2] Kuban State Agrarian University named after I. T. Trubilin, Krasnodar, Russia
[3] Kalmyk State University named after B.B. Gorodovikov, Elista, Russia

Abstract. The study developed and experimentally substantiated a model of a waste-free agricultural enterprise based on the application of biotechnology. The key element of the model is the production of effective microorganisms (EM preparations) in a fermenter and their use for waste processing, as well as for the production of feed additives, fertilisers and biogas. The efficiency of EM preparations application in animal husbandry and crop production has been experimentally shown. The data on increase of animal productivity, such as weight gain of young animals, milk yields of cows and egg production of chickens, as well as increase of crop yields have been obtained. The developed model allows not only to reduce the amount of waste, but also to increase the economic efficiency of agricultural production. The study demonstrates the prospect of using biotechnology to create sustainable and highly productive agricultural systems that help to reduce the negative impact on the environment.

1 Introduction

There are many approaches and ways to solve the problem of agricultural waste, each of which has its advantages and disadvantages. One of the most promising directions is the transition to organic agriculture based on the principles of sustainability, environmental safety and closed production cycles [1, 2]. The use of organic fertilisers, such as compost and biohumus, can increase soil fertility, reduce the need for mineral fertilisers and reduce environmental impact. However, organic farming requires more intensive labour and its yields may be lower compared to intensive farming, especially in the initial stages [3, 4]. Another solution is the introduction of technologies to convert waste into useful products such as biogas and compost. Biogas plants can convert organic waste into biogas that can be used to produce electricity and heat, reducing dependence on fossil fuels and reducing the amount of greenhouse gases in the atmosphere. However, the construction and operation of such plants requires significant investment and infrastructure. An alternative direction is the use of biological methods of plant protection against pests and diseases, such as the use of predatory insects, microorganisms and biopreparations, which reduces the use of chemical pesticides but requires more careful monitoring and control.

In the framework of solving the problem of agricultural waste, the proposed model of a waste-free, high-value agricultural enterprise based on the application of biotechnology is highly relevant and important. It offers an integrated approach to waste processing based on the use of efficient microorganisms. This approach is aimed at obtaining a range of valuable products, including protein feed additives, fertilisers and biogas, while minimising the negative impact on the environment and contributing to the creation of closed production cycles. The use of EM preparations can increase the efficiency of waste processing, reduce the cost of fodder and fertiliser production, and improve soil fertility and product quality. This approach not only solves environmental problems associated with waste [5, 6], but also provides economic benefits for agricultural enterprises by reducing costs and generating additional profit [7-9].

The introduction of such technologies opens new perspectives for the development of sustainable agriculture capable of ensuring food security while preserving the environment.

The aim of the work is to develop a model of a waste-free high-yield agricultural enterprise based on the application of biotechnology, which will allow efficient waste processing, production of valuable products, increase the economic efficiency of agricultural production and reduce its negative impact on the environment. The model includes a fermenter for the production of EM preparations, plants for grain processing and production of protein feed additives, as

* Corresponding author: t9913552@yandex.ru

well as a bioenergy system [10, 11]. The proposed model is a holistic system that can be adapted to different types of agricultural enterprises and climatic conditions, and contribute to the transition to a more sustainable and environmentally friendly agriculture.

2 Method and materials

Within the framework of this research, a model of a waste-free high-yield agricultural enterprise based on the integrated use of biotechnology for processing agricultural waste and producing valuable products was developed. The basis of this enterprise is a fermenter with a capacity of 10-30 tonnes per month, where effective microorganisms (EM preparations) are cultivated. These preparations play a key role in waste treatment [12] and production of valuable products, and are the cornerstone of the whole system.

The plant growing block of the enterprise is represented by a variety of crops, including cereals, vegetables (both open and closed ground), as well as technical crops. An important aspect is the presence of a fruit and berry nursery, as well as the cultivation of topinambour and amaranth as additional sources of nutrients and raw materials. The main purpose of crop production is to provide raw materials for the livestock block and other production units, but also to produce marketable products for sale.

The processing block is represented by several workshops. The cereal processing unit includes a mill and a mini-bakery, which allows the production of flour, bread and confectionery products. The shop for preparation of feed mixtures is engaged in the production of mixed fodder, premixes and probiotics based on EM preparations and processed grain. There is also a shop for processing vegetables, berries, meat and milk, which makes it possible to produce a wide range of food products.

The livestock block includes a farm and a poultry house. Animals and poultry are provided with fodder of own production, including protein feed additives produced with the use of EM technologies. This reduces the cost of livestock production and improves its quality. The livestock block also includes a fish breeding pond, which diversifies production and provides an additional source of protein.

Waste generated during the production process is not discarded, but is sent for recycling. The waste processing unit includes a composting area, where organic waste is converted into valuable fertiliser, which is then used in crop production. An important element of the system is the bioenergy station (BES), which uses biogas obtained from waste to produce heat and electricity, as well as motor fuel. Thus, the enterprise provides itself with energy and reduces dependence on external sources [13].

All production processes at the enterprise are interconnected and closed in a cyclic system. This means that the waste of one process is a raw material for another, ensuring the waste-free and sustainability of the whole system.

The general scheme of experimental work included several stages aimed at modelling the functioning of the main elements of the proposed enterprise and assessing their efficiency. First of all, the possibility of creating and functioning of a fermenter for the production of effective microorganisms (EM preparations) was investigated, and the conditions of their optimal cultivation were studied. In parallel, experiments on the processing of grain raw materials to produce protein feed supplements were carried out at low-power plants, including probiotisation and autolysis processes. In addition, potential applications of EM preparations in crop, livestock and poultry production were investigated to assess their impact on crop yields, productivity and animal health.

A range of specialised equipment was used during the study. The fermenter was used to cultivate various strains of microorganisms, including yeasts, lactic acid, nitrogen-fixing and photosynthesising bacteria, and fungi, for the production of EM preparations. The fermentation process was carried out under controlled conditions, with regulation of temperature, pH, aeration and fermentation time. The fermenter operating modes varied depending on the specific type of microorganisms and the purpose of production. Fermentation parameters were monitored to cultivate different types of microorganisms and optimise their growth. Further, for the production of protein feed additives, low capacity units designed for processing of grain raw materials were used. The productivity of the plants was 1-1.5 tonnes per day. Within the framework of researches the equipment for granulation and drying of protein feed additives was used. The properties of the obtained products and cultivation parameters were also analysed. Within the framework of modelling the process of grain processing to obtain protein feed additives, experiments on probiotisation and autolysis were carried out, for which special devices and laboratory equipment were used. In addition, the impact of feed additives on the weight gain of young animals, milk yield of cows and egg production of hens was investigated.

3 Results and discussion

The experimental work included several stages. At the first stage, modelling of the process of production of effective microorganisms (EM preparations) in the fermenter was carried out. For this purpose, different strains of microorganisms were used, including yeast Saccharomyces cerevisiae, lactic acid bacteria Lactobacillus plantarum, nitrogen-fixing bacteria Azotobacter chroococcum, and photosynthetic bacteria Rhodobacter capsulatus. The experimental work included the evaluation of their growth and multiplication under different culturing conditions such as temperature (25-35°C), pH (5.5-7.0), aeration (aerobic and anaerobic fermentation) and duration of cultivation (24-72 hours). Different nutrient media including grain processing waste, molasses and whey waste were used to optimise the process. At the second stage, studies were conducted on the processing of grain raw materials to produce protein feed additives. For this purpose, small capacity plants (1-1.5 t/day) were used, in which the processes of grain probiotisation and its subsequent autolysis were carried out. Bacillus subtilis and Lactobacillus acidophilus strains were used as probiotics. Probiotisation was carried out by introducing microorganisms into moistened grain with subsequent incubation for 12-24 hours at 30-35°C. Autolysis was carried out by heat treatment of grain followed by addition of enzyme preparations and incubation for 24-48 hours at 40-50°C. At the third stage, studies were conducted to evaluate the effectiveness of EM preparations and protein feed additives in crop and livestock production. In crop production, EM preparations were used for seed treatment, plant spraying and soil irrigation. In animal husbandry, protein supplements were introduced into the diets of calves, cows and chickens. Control groups received traditional diets without biotechnology application

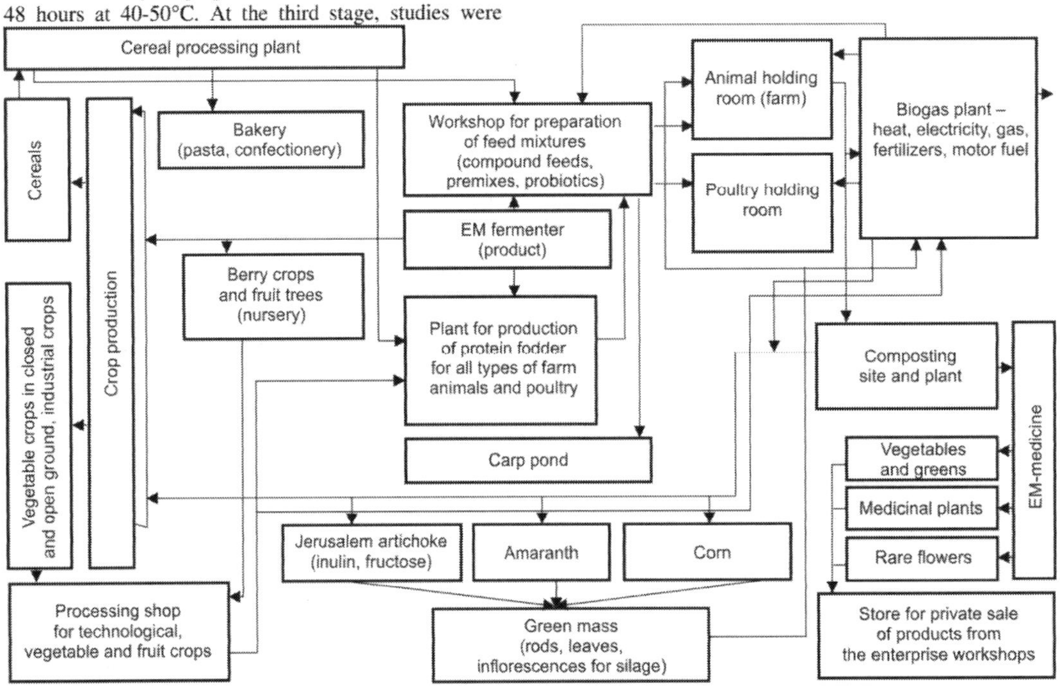

Figure 1. Schematic diagram of the model of a waste-free agricultural enterprise with application of biotechnologies

The results of the conducted studies showed high efficiency of the proposed model (Fig. 1). In particular, it was found that the fermenter succeeded in obtaining viable biomass of EM preparations with a titre of 107-108 CFU/ml. The optimum culturing conditions for different strains differed slightly, but common was the adherence to a temperature regime in the range of 28-32°C, pH 6.0-6.5, and aerobic fermentation for 36-48 hours. According to the results, the addition of molasses to the nutrient medium increased biomass yield by 15-20%. Grain processing plants produced protein feed supplements containing 25-30% crude protein. Probiotisation of grain increased the content of biologically active substances in the finished product by 10-15%, such as B vitamins, organic acids and amino acids. Autolysis of grain increased protein digestibility by 15-20%. The combined use of probiotisation and grain autolysis contributed to the production of a protein feed supplement with higher protein content and better digestibility. In calf feeding studies, protein feed additives replaced 25% of whole milk in the diet of calves, with average daily weight gain of 700-780 g, which is 8-10% higher than in the control group. At the same time, the cows receiving protein supplements in the diet showed an increase in milk yield by 4.2-5.8 litres per day, which is 12-18% higher compared to the control group, and milk fat content increased by 0.5-0.6%. When using protein feed additive in the diet of laying hens, egg production increased by 18-22% and egg weight by 4-5%. In crop production, EM preparations used for seed and soil treatment helped to increase the yield of grain crops by 10-15% and vegetable crops by 15-20% by increasing seed germination, improving root system development and increasing resistance to diseases. In the greenhouse experiment, the use of EM preparations increased tomato yields by 18% and cucumber yields by 21%, compared to the control group in which no biotechnology was applied.

The analysis of the obtained results confirmed the high efficiency of the model of a waste-free agricultural enterprise with the use of biotechnology. Production of EM preparations on the basis of waste, grain processing to produce protein feed additives and their use in livestock and crop production showed positive results. The obtained data are consistent with the results of other studies on the use of microorganisms in agriculture. The use of Bacillus subtilis as a probiotic in the diet of calves showed an increase in average daily weight gain by 7%, and the use of EM preparations for the treatment of wheat seeds led to an increase in yield by 12%. The difference in our data compared to the above mentioned studies is probably due to the fact that our study used a more comprehensive approach with the use of several types of microorganisms and their combined action. It is also worth noting that the proposed model of a zero-waste enterprise has a number of advantages over traditional approaches to agriculture. It allows to reduce dependence on mineral fertilisers and chemical pesticides, reduce the amount of waste and increase the economic efficiency of agricultural production. In addition, the model promotes the creation of closed production cycles, which reduces the negative impact on the environment. However, it should be noted that the implementation of this model requires investment in equipment and staff training. In addition, biotechnological processes need to be adapted to specific production conditions to maximise efficiency. In general, the study has shown the promising potential of using biotechnology to create sustainable and highly efficient agricultural enterprises.

4 Conclusion

The conducted research allowed to develop and experimentally justify the model of a waste-free agricultural enterprise based on the application of biotechnology. The key element of this model is the use of effective microorganisms (EM preparations) for processing of agricultural waste and production of valuable products. It is experimentally shown that the fermenter used for the production of EM preparations provides biomass with sufficient titre, which guarantees the efficiency of further processes. The use of a variety of microbial strains adapted to specific cultivation conditions allows obtaining a stable and efficient biomass. During the study it was found that the addition of molasses to the nutrient medium increases the biomass yield, indicating that it is possible to use available and cheap waste for the production of EM preparations, reducing the cost of production.

Processing of grain raw materials to produce protein feed additives in low-power plants is another important link in the model. The data obtained indicate that it is possible to produce feed additives with a high content of crude protein and biologically active substances that increase the digestibility of feed by animals. The results of the experiment showed that the use of probiotisation and autolysis of grain can significantly improve the quality of forages and increase the efficiency of their use in animal husbandry. The use of these feed additives in the diet of farm animals and birds, showed an improvement in their productivity and health. In particular, an increase in the weight gain of young animals, milk yield of cows and egg production of chickens was observed, which emphasises the economic benefits of implementing the proposed technology. In addition, the use of EM preparations in crop production has also been shown to be effective. The use of these preparations for seed and soil treatment helped to increase crop yields by improving their development and disease resistance.

The proposed model of zero-waste agriculture is adapted to different types of agricultural enterprises. However, its full-scale implementation requires additional investment in equipment and staff training.

References

1. L.S. Malyukova, V.V. Bukhtoyarov, V.Y. Konyukhov, K.A. Bashmur, T.A. Panfilova. Circular Mining Wastes Management for Sustainable Production of Camellia sinensis (L.) O. Kuntze. Sustainability (Switzerland), **15,** 11671 (2023). DOI: 10.3390/su151511671.

2. L. Ma, Y. Ghorbani. Dynamics of backfill compressive strength obtained from enrichment tails for the circular waste management. Resources, Conservation & Recycling Advances, **23,** 200224 (2024). DOI: 10.1016/j.rcradv.2024.200224.

3. Y. Guo, Z. Zhang, F. Liang, W. Cao, Y. Wang, J. Chen, J. Guo. Meta-analysis reveals that returning crop straw to arable land causes soil acidification at the global scale. European Journal of Agronomy, **164,** 127511 (2025). DOI: 10.1016/j.eja.2025.127511.

4. K.-T. Lin, S.-Y. Pan, M.-H. Yuan, Y.-T. Zhang, H.-Y. Guo. Synergies between rice production security and soil-related ecosystem services: From field observations to policy implementations. Agricultural Systems, **224,** 104256 (2025). DOI: 10.1016/j.agsy.2024.104256.

5. D. A. Zyukin, A. I. Karlina. Technology for nonwaste recovery of tailings of the mizur mining and processing plant. Metallurgist, **66,** 1476-1480 (2023). DOI: 10.1007/s11015-023-01462-y.

6. V.Yu. Konyukhov, I.Y. Matasova, M.A. Modina, N.V. Martyushev. Ecologization of underground coal mining by means of ash use in backfill preparation. Min. Inform. Anal. Bull., 10, 123-135 (2024). DOI: 10.25018/0236_1493_2024_10_0_123.

7. O. Afanaseva, M. Afanasyev, S. Neyrus, D. Pervukhin, D. Tukeev, Information and Analytical System Monitoring and Assessment of the Water Bodies State in the Mineral Resources Complex. Inventions, **9(6),** 115, (2024). DOI: 10.3390/inventions9060115.

8. S. Ngozi, S. Feleke. Enhancing youth's retention in potato farming enterprise through consortium agribusiness models in Tanzania. Crop Science, **64,** 1421–1431 (2024). DOI: 10.1002/csc2.21055.

9. L. G. Orlova, R. I. Kuzmich, Ensuring Supply to Enterprises by Developing Virtual Models of Different Operation Modes of Transformer Substations. Mining Informational and Analytical Bulletin. **11,** 278-291 (2023). DOI: 10.25018/0236_1493_2023_111_0_278.

10. I. Sotnyk, J.-P. Sasse, E. Trutnevyte. Decarbonizing Ukraine's electricity sector in 2035:

Scenario analysis. Energy and Climate Change, **6,** 100170 (2025). DOI: 10.1016/j.egycc.2024.100170.

11. M.Y.D. Alazaiza, A.A.Bin Mokaizh, A.O. Baarimah, T. Al-Zghoul. From agro-waste to bioactive wealth: Analyzing nutraceutical extraction and applications. Case Studies in Chemical and Environmental Engineering, **11,** 101066 (2025). DOI: 10.1016/j.cscee.2024.101066.

12. R. Nussipali, T.A. Oparina, V.V. Romanova, R.V. Kononenko. Combined Power Generating Complex and Energy Storage System. Electricity, **5,** 931-946 (2024). DOI: 10.3390/electricity5040047.

13. M.K. Myrzakulov, S.K. Jumankulova, M.B. Barmenshinova. Thermodynamic and Technological Studies of the Electric Smelting of Satpaevsk Ilmenite Concentrates. Metals, **14,** 1211 (2024). DOI: 10.3390/met14111211.

BIO Web of Conferences **161**, 00026 (2025) https://doi.org/10.1051/bioconf/202516100026
FIES 2024

Application of statistical methods for the agriculture complex using the example of sugar beet production

Lyudmila Korobova* [1], *Irina* Tolstova[1], *Maxim* Ivliev[1], *Natalia* Datsenko[1], *Sergey* Chikunov[1]

[1]Voronezh State University of Engineering Technologies, Voronezh, Russia

Abstract. The paper considers the issue of applying statistical methods that allow processing large technological data. At the initial stage, the technological process of beet sugar production was analyzed. The indicators (more than 40 parameters) were taken, and those that have the greatest influence on the technological process were selected. The values of the main parameters determine the varietal yield (sugar grade). The statistical analysis method was chosen as the main research method. Characteristic samples of 40 parameters were considered, the sample size was 179 values. Two quality parameters that determine the sugar grade were selected. Correlations of the characteristic initial parameters with the parameters of varietal sugar were checked. A mathematical processor was chosen as a tool for conducting the analysis. The Excel program contains almost all the functions and techniques that are missing in application development tools and much more. Statistical characteristics of some of the experimental samples under study are presented. And based on the obtained models, a conclusion was made that the third-order power model adequately describes the experimental data.

1. Introduction

Currently, personal computers (PCs) are increasingly being introduced into everyday human activities. The devices performance is increasing, while the cost is decreasing. All this contributes to the personal computer applications expansion. Modern software is being developed and implemented for various human activity fields. At the same time, there is no need to maintain a large staff, since software can sometimes replace the specialists entire departments work. Sometimes it is not necessary to have the programmer knowledge to work with modern PCs. However, modern software doesn't yet have the human intelligence, so software can only do what a human has programmed [1].

The promising area in the information technology field is the creation of analytical systems that free people from routine, sometimes repetitive operations for processing large information amounts. Such systems can make predictions and make recommendations on the current situation [2].

Over the last fifty years, many studies have been conducted to improve the quality of production processes. A process can be considered controlled if it is possible to obtain information about changes in its main parameters. The information obtained is then used to control the process in order to prevent failures and achieve the best results.

Today, leading companies are looking for new sources of competitive advantage and understand that the vast amount of data collected during their manufacturing operations provides excellent information to improve the development of quality products and process performance. The quality of manufactured products must comply with the accepted Russian standards (RS) or technical conditions (TU) of enterprises or industries.

In addition, companies are under increasing competitive pressure. In order to improve the sustainability and competitiveness of their products and processes, a deep understanding of the physics of the processes is required. This can be achieved by using a more powerful analytical tool [2, 3].

The desire to achieve high profits forces industrial and processing companies to constantly look for ways to reduce costs. They are starting processing processes that are close to critical limits according to technological regulations, using cheaper components where possible, reducing energy consumption and minimizing waste and rework costs.

There are approaches to control and diagnostics of technological processes (using sugar beet production as an example) [4, 5]:

- technological process control and diagnostics;
- sugar beet production automation;
- sugar beet production information system modeling.

Decomposition and structural-parametric analysis of technological processes (using sugar beet production as an example) and tools for their control and diagnostics include the following:

- decomposition and structural-parametric analysis of technological processes of sugar beet production;
- structural modeling of the technological process of sugar beet production;

* Corresponding author: lyudmila_korobova@mail.ru

- structural synthesis of the model of the information system for diagnostics and control of the technological process;
- modeling of the functioning of the information system for monitoring and diagnostics of technological processes (using sugar beet production as an example);
- analysis of the stationarity of technological processes of sugar beet production using the nuclei of conflict, cooperation and indifference.

2. Materials and methods

Systematic collection of data and the ability to translate it into a format suitable for analysis was another common obstacle to the use of multivariate methods in the past. Today, this is less of a problem because most processes are instrumental, and complex control systems are widely used. In fact, the problem now is that so much data is collected that it is becoming increasingly difficult to cut through the vast amount of information to find underlying patterns, which encourages the use of multivariate methods [4, 5].

Most technological processes are very multidimensional in nature due to complex reactions. There are a large number of variables that are usually very interactive [6, 7]. To fully understand the process dynamic, complex systems require experiments series. The experiments produce descriptive statistics. The experimental data are used to quantify the process characteristics for population assessment. A convenient measure for understanding process capability is 6σ-measure. If the experimental data distribution is "normal" (i.e. bell-shaped) and its mean is denoted as "μ", then the process output is typically given by $\mu \pm 3\sigma$. This values range known as the "process spread" will (theoretically) cover the process output of 99.73%, when the process is stable.

Nevertheless, statistical process control (SPC) tools used in production engineering still often rely on one-dimensional methods. Despite collecting big data using instruments and control systems this tools don't show complex processes complete picture – it uses traditional statistical approaches (mean, standard deviation, Student's criterion, etc.), which only consider individual variables individually.

Although one-dimensional statistics can be useful for the simple systems study, it tends to fail when analyzing more complex systems. This is because it can't detect the relationship that may exist between variables, since they look at variables that are independent of each other. This relationship is known as covariance or correlation and is a central theme in the MVA. Covariance describes the effect that one variable has on others. Process disturbances are usually caused by multiple variables acting together.

The main multidimensional methods are Exploratory Data Analysis (EDA), regression/forecasting methods, and classification methods. EDA attempts to find a hidden structure or underlying patterns in large, complex datasets. This provides a better process understanding and can lead to insights that would not otherwise be observed. EDA methods include cluster analysis and principal component analysis (PCA). An EDA application example is checking for pollutants in a feedstock process or identifying by-products caused by incorrect process settings [7, 8].

Traditional one-dimensional control charts show many different variables at the same time, making it extremely difficult to get a clear, complete picture. Multidimensional control diagrams unite all this data into one or two graphs, taking into account the complex interactions between variables. If the process starts to drift, you can "delve deeper" into specific samples or emissions to quickly identify the problem root cause using a multidimensional and one-dimensional diagnostics combination [8, 9].

Multivariate data analysis can be used together with the software engineering applications through the entire chain of the product development (process model), scaling up or down (considering individual process operations at their own level and the entire process as a whole), process design and its optimization. One of the tasks at the software engineering stage is to understand the correlation of parameters with each other. It is necessary to select a pair or three input parameters on which the quality of one or more output parameters depends. The numerical values of the output parameters directly determine the manufactured products quality and grade (in this case, the produced sugar grade). The correlated parameters are determined by specialists (technologists) based on personal practical experience (human factor). But at the verification and the modeling stage, it is possible to refute many years' developments. Therefore, it is advisable to take the expert technologists opinion as a starting point and when conducting statistical analysis, rely on the results and conclusions obtained. Here, the emergence of a possible conflict is clearly visible in terms of taking into account the opinions of expert technologists and the conclusions of scientific research.

Currently, various statistical modules are used in all economics and management branches which help to solve pressing problems in the production sector. A lot of programs allow complex calculations using mathematical statistics methods (MapleSoft Maple, MathWorks Matlab, MathSoft Mathcad, Wolfram Researh Mathematica) and specialized statistical processing packages (StatGrapics, SPSS, SAS, BMDP and Statistica) developed by Russian programmers [2, 4]. Companies and manufacturing enterprises have the right to choose the necessary software themselves depending on the research objectives.

In the processing of experimental data, statistical criteria are key to identifying and eliminating abnormal measurement results, otherwise known as outliers. These outliers, which are values that significantly deviate from the general trend, are detected at the stage of data preprocessing. The quality of preprocessing directly affects the reliability of subsequent analysis and the construction of a model of the process under study. Ignoring outliers leads to distortion of the results, especially when using classical, non-robust methods of statistical analysis that are sensitive to the presence of such deviations [10, 11]. This can lead to incorrect conclusions and erroneous interpretations of the obtained data, which is critical for many scientific and engineering problems. Gross errors in measurements are usually obvious and can be detected visually or by simple checks. However, in cases of uncertainty, when information on the quality of measurements is incomplete or questionable, the use of statistical methods becomes necessary. Among such methods, the Grubbs criterion is widely used, designed to check for anomalies of individual values in a sample [6, 10]. Statistical tests are based on the fact that the expected errors in the source data are normally distributed and allow one (minimum or maximum value) or two outliers (two minimum or two maximum values) to be identified in the sample. It is important to note that the application of Grubbs' criteria, like many other statistical tests, is based on standards. For example, the international ISO 5725 standard defines the procedure for performing such tests and interpreting the results. Deviation from these standards can lead to incorrect conclusions.

In this paper, we investigate the effect of data distribution deviation from normality on the Grubbs' test statistics distribution [7, 12]. This is an extremely important aspect, since the data normality assumption is fundamental to many statistical methods, including Grubbs' tests. This assumption violation can significantly distort the test results and lead to false positive or false negative conclusions. For example, if the data have an asymmetric distribution (e.g., exponential or lognormal), Grubbs' tests may be overly sensitive to outliers at one end of the distribution and less sensitive at the other. This may result in significant outliers at one end of the distribution being missed and minor deviations at the other end being erroneously classified as outliers.

Computer modeling methods can be used to study this influence [2, 10, 11]. Generating samples from various distributions (normal, exponential, log-normal, Cauchy, and others) with the addition of the outlier of different sizes and at different positions in the distribution, it is possible to demonstrate how the Grubbs criterion statistics values change and how this affects the detecting outliers at different probability of significance levels. Such analysis results can be presented as graphs showing the criterion power dependence (detecting a true outlier probability) on the data distribution parameters and the outlier size. It is important to remember that the data processing method choice should be based on data nature understanding, its distribution, and the study objectives. Only then reliability and objectivity of the results can be guaranteed. Ignoring these aspects can lead to serious errors in conclusions and decision-making based on incorrect data analysis.

Let $X_1, X_2, ..., X_n, ...$ be the observed sample and the variation series constructed from it. The being tested hypothesis $H0$ is that all $X_1, X_2, ..., X_n, ...$ belong to the same general population. When checking for the largest sample value outlier, the competing hypothesis $H1$ is that, for example, the values up to X_n belong to one law, and the X_n values and the subsequent ones belong to another law, significantly shifted, for example, to the right. When checking for the X_n outlier, the Grubbs criterion statistics is:

$$\frac{\left|X_j - \overline{X}\right|}{S} > 3, \qquad (1)$$

where X_j is the observed sample and the variation series constructed from it; $j = \overline{1 ... n}$; S is the quadratic error.

When checking for the smallest sample value outlier, the competing H_1 hypothesis assumes that some of the sample elements under study belong to some other law that is significantly shifted to the left [1, 2, 12]. In this case, the calculated statistics take the form of (1). The statistics distribution significantly depends on the sample size n.

3. Results and discussions

In this work, more than 40 sugar production quality parameters which affect the varietal yield [1, 10, 13] were considered. The process quality study was conducted on 147 values for two quality parameters: molasses quality (Db molasses), % and molasses quantity per day, t. Statistical characteristics of some of the studied variation series are presented in Figs. 1 and 2. The Excel system was used as a statistical analysis system.

Preparing the experimental data process is constructing a regression model initial stage [14]. The module for constructing a regression model is included in many mathematical statistics and specialized statistical processing packages. Fig. 2 shows the graph of the per day amount of molasses and the parameter of the regression model for 118 days. This packages advantage is that the user without delving into the process specifics and relying on the specialists' (technologists') opinion can get an idea of the behavior of the most important parameters of the process quality [15]. The mathematical package allows constructing change in the analyzed sample graph with the subsequent determination of the structure type of the proposed model and the numerical values of the coefficients for this model. The model structure is selected and the coefficients are calculated based on averaged experimental data and statistical results of the yielded good (the correlation coefficient R varies in the range from 87% to 99%). All developed templates have their own set of limited structures, and such templates cannot be modernized [16]. Therefore, such packages use is convenient for obtaining an initial idea of the process model structure.

Statistical packages and modules where arbitrary model structures are available to the user have the greatest interest. The software modules for determining the numerical values of the model coefficients are based on the least squares method.

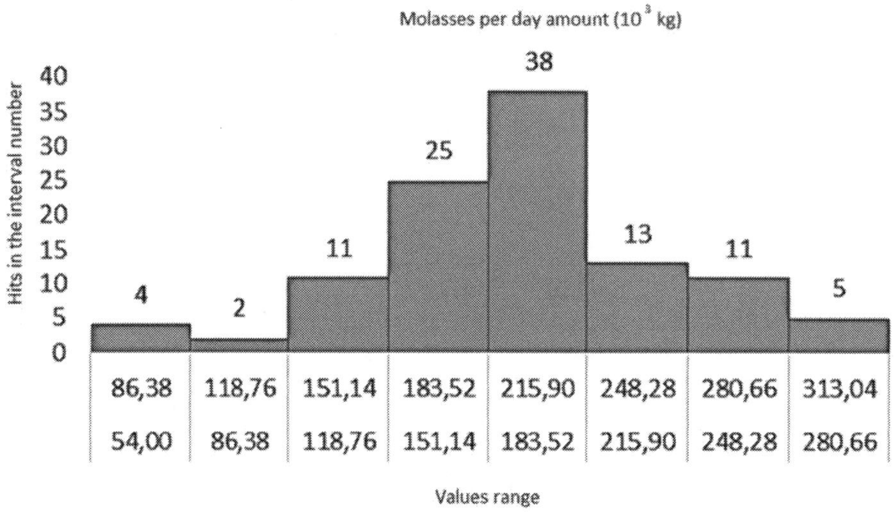

Fig. 1. The example of interval analysis of the process parameter "Amount of molasses per day"

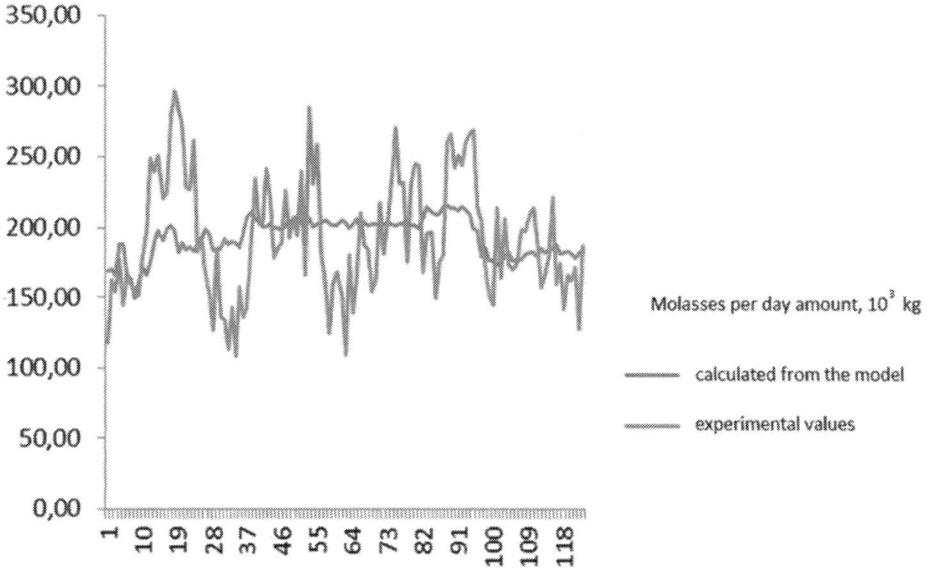

Fig. 2. The regression model of the parameter "Amount of molasses per day"

4. Conclusion

More than 40 sugar quality parameters influencing the varietal yield have been studied [10, 11, 16]. The analysis has shown that the obtained characteristic samples generally obey the normal distribution law. But this is not typical of all parameters. An important feature is the identification of the consistency of the obtained experimental data and the determination of correlations between the quality parameters. Regression equations have been constructed for the per day indicators of molasses quality and molasses quantity. Linear and power models have been constructed. Mathematical statistics and specialized statistical processing packages were used to process the data. The correlation coefficient for the developed models ranges from 0.85 to 0.97.

Figure 1 clearly shows that the average value of per day molasses quantity is in the center of the measurement range. This indicates that the process under study is stable. Analysis of the graph in Figure 2 allows us to conclude that despite the picket-shaped appearance of the experimental data graph the third-order power model yields a result with a correlation coefficient of 0.95. Mutual peaks and valleys of the experimental data are suppressed and as a result we obtain an adequate model.

References

1. S. Chernyaeva, L. Korobova, M. Ivliev. I. Tolstova, B. Nikitin, I. Matytsina Implementation of the Extrapolation Method of Expert Assessments in Selection Problems, High-Performance Computing Systems and Technologies in Scientific Research, Automation of Control and Production, **1304** (2020)
2. Yu.V. Bugaev, L.A. Korobova, I.Yu. Shurupova On the statistical stability of the optimal solution found by the regression equation, VSUIT Bulletin, **86** (2024)
3. N. Gusyatinskaya, T. Nechipor Eureka: Life Sciences 5 (2018)
4. S. Chernyaeva, L. Korobova, M. Ivliev, I. S. Tolstova, B. E. Nikitin, I. A. Matytsina, High-Performance Computing Systems and Technologies in Scientific Research, Automation of Control and Production, **1304** (2020)
5. N.V. Zueva, G.V. Agafonov, T.I. Romanyuk, A.E. Chusova, A.N. Yakovlev, S.A. Veretennikov, IOP Conference Series: Earth and Environmental Science, **640** (2021)
6. V. I. Golik, V. Yu. Konyukhov, Mining Informational and Analytical Bulletin. **11-1,** 175-189 (2023). DOI: 10.25018/0236_1493_2023_111_0_175
7. I.M. Zharkova, Yu.A. Safonova, V.G. Gustinovich, T.L. Ilyeva, Storage and processing of agricultural raw materials, **1** (2020)
8. K. A. Bashmur, V. V. Kondratiev, Mining Informational and Analytical Bulletin. **11-1,** 239-251 (2023). DOI: 10.25018/0236_1493_2023_111_0_239
9. N.G. Kulneva, Yu.I. Bulletin of VSUET, **86** (2024)
10. L. A. Korobova, I. S. Tolstova, I. A. Matytsina, M. S. Mironova, J. Phys, **1902** (2021)
11. A. I. Trunova, S. O. Kurashkin, R. B. Sergienko, Metals, 13(7), 1234 (2023). DOI: 10.3390/met13071234
12. N.G. Kulneva, M.V. Zhuravlev Bulletin of the Voronezh State University of Engineering Technologies, 82 (2020)
13. L.N. Putilina, N.A. Lazutina Storage and processing of agricultural raw materials, **1** (2021)
14. N. Dolgopolova, A. Kaluzhskikh, M. Kotelnikova IOP Conference Series: Earth and Environmental Science, **666** (2021)
15. Yousef, A. Zohri, A. Darwish, A. Shamseldin et al. Microbiol, **14** (2023)
16. O.S. Korneeva, S.F. Yakovleva, N.A. Matvienko, L. N. Frolova et al. Bulletin of the Voronezh State University of Engineering Technologies, **85** (2023)

State and development of farms in Uzbekistan as a basis for food security

Kadirjon Shavazov[1,*], *Shokhida* Khojibekova[2], *Bayramali* Mukimov[3], *Abdigappar* Umirov[4], *Shakhnoza* Kadirova[5]

[1] "Tashkent Institute of Irrigation and Agricultural Mechanization Engineers" National Research University, Tashkent, Uzbekistan
[2] Almalyk branch of the Tashkent State Technical University named after Islam Karimov, Almalyk, Uzbekistan
[3] Karshi State Technical University, Karshi, Uzbekistan
[4] Termez State University, Termiz, Uzbekistan
[5] Termiz State Pedagogical Institute, Termiz, Uzbekistan

Abstract. The work is devoted to the study of directions of effective development and functioning of farms in the conditions of modern economy, with emphasis on the role of innovations in this process. The study was carried out on the basis of analysing the experience of farming activity of agro-enterprises of the Republic of Uzbekistan. Also the analysis of normative-legal acts and statistical data in the field of agro-industrial complex of the Republic of Uzbekistan with the use of methods of comparative and logical analysis was carried out. Introduction of innovative technologies is a key condition for sustainable development of farms, increasing their productivity and competitiveness. Factors hindering the introduction of innovations are analysed and ways of their solution are proposed. Special attention is paid to the experience of Uzbekistan, where the agricultural sector plays an important role in the economy. In particular, examples of introduction of drip irrigation, precision farming, modern technologies in livestock breeding and greenhouse farms are considered. The results of the study highlight the importance of creating a favourable innovation environment, as well as strengthening state support and incentives for scientific research.

1 Introduction

One of the most important ways to address the problem of food security is to improve agro-technologies and the development of breeding aimed at creating more resistant and yielding varieties. This makes it possible to increase the productivity of agricultural production while reducing its negative impact on the environment. Another important direction is the transition to sustainable agricultural practices such as organic farming and agroecology. These practices reduce dependence on chemical fertilisers and pesticides and improve soil fertility. However, these approaches can be more labour-intensive and require large financial investments, which can be an obstacle to their widespread adoption. At the same time, the development and application of modern technologies, such as precision farming and the use of drones to monitor crops, can help improve agricultural management and reduce costs. Also, developing an innovative environment and encouraging the exchange of knowledge and experience between producers and research organisations are of paramount importance. Government support and funding for research aimed at developing and introducing new technologies and methods in agriculture is also needed. In addition, it is necessary to develop infrastructure for storage and processing of agricultural products [1, 2], where innovation is a key driver of development, and the issues of effective functioning of farms are of particular relevance. It is in farms, as the primary link of the agricultural sector, that the foundations for food security and sustainable development of rural areas are laid. In this context, one of the most promising directions is the introduction of innovations at all stages of agricultural production. This approach allows increasing productivity and reducing production costs, as well as ensuring the competitiveness of farms in the market. However, the process of introducing innovations in agriculture has its own difficulties and requires the creation of a favourable innovation environment, including access to information, financial resources and qualified personnel. In addition, it is necessary to take into account the specific features of each region and farm, as well as to ensure their interest in the development of modern agro-industrial complex. Introduction of innovations in farms is not only an increase in productivity and efficiency, but also a contribution to the overall economic development, food security and sustainable development of rural areas. The purpose of this paper is to analyse the directions of effective development and functioning of farms in the conditions of modern economy, with special emphasis

* Corresponding author: k.shavazov@tiiame.uz

on the role of innovation in this process [3, 4] taking into account the environmental aspect [5].

2 Method and materials

In this article, devoted to the study of directions of effective development and functioning of farms, the main emphasis is made on the analysis and generalisation of existing data and scientific approaches. The research methodology was based on a comprehensive analysis of literary sources, regulatory and legal documents, statistical data and the results of scientific works devoted to the problems of agricultural development and innovation implementation.

In particular, the work used the methods of comparative analysis to identify common trends and differences in approaches to the development of farms in different countries and regions. The method of logical analysis was used to identify cause-and-effect relationships between various factors affecting the efficiency of farms' functioning. The method of abstract thinking was also used to formulate general conclusions and recommendations based on the analysis of specific data. Emphasis was placed on the use of analytical tools to interpret the collected information and formulate conclusions. The study was based on qualitative analyses of data presented in the literature and regulatory and legal acts.

3 Results and discussion

In this paper, the theoretical foundations of innovation development were comprehensively analysed and the practical experience of introducing advanced solutions in farms, as applied to the conditions of Uzbekistan, was studied. In particular, it considered different models of organisation of the agro-industrial complex, as well as factors influencing its development in the specific context of the country.

The study was based on an in-depth analysis of scientific publications, policy documents and statistical data related to agriculture in Uzbekistan. The study included a comparative analysis of different approaches to the organisation of the innovation process in the agricultural sector, taking into account regional specificities. Factors determining the efficiency of innovation activities of farms were studied, such as the availability of financing, the level of state support, and infrastructure support. Special attention was paid to the impact of economic globalisation and technological change on the development of the agro-industrial complex of Uzbekistan, with a focus on adaptation to new conditions.

One of the key findings of the study was the confirmation of the thesis that the introduction of innovative technologies is a prerequisite for sustainable development of farms in Uzbekistan. The analysis showed that farms that actively introduce modern agro-technologies demonstrate higher yields, lower costs and higher profitability. In particular, the use of modern crop varieties adapted to the climatic conditions of Uzbekistan can increase yields by 20-30 %. As a result of the analysis, it became clear that innovation is not just desirable, but a necessary condition for sustainable development of farms in modern conditions, especially in the context of globalisation and increasing competition. It is also important to take into account the dynamics of agriculture and its susceptibility to climate change [6, 7].

Farms that actively adopt new technologies, apply good management practices, and adapt to changing market conditions show higher productivity, profitability and competitiveness. In turn, studies have shown that in countries with developed market economies, where the innovation activity of farms is supported by a comprehensive infrastructure and effective government policy, the agricultural sector demonstrates higher rates of growth and development.

In Uzbekistan, where the agricultural sector plays a significant role in the economy, there are more than 170 thousand farms, which make a significant contribution to ensuring food security of the country and employment of the population. According to the State Statistics Committee of the Republic of Uzbekistan, in 2023, farms produced more than 60 % of the country's total agricultural output.

At the same time, in transition economies such as Uzbekistan, the introduction of innovation in the agricultural sector faces a number of serious obstacles, including a lack of research funding, limited access to advanced technologies and skilled labour, and the absence of an effective support system for innovation. However, research shows that even under such conditions, with targeted government policy and support, it is possible to make significant progress in developing farm innovation activities.

For example, cotton farms, which historically have been the backbone of the agricultural sector in Uzbekistan, often face problems of soil erosion, salinisation and water deficit, which is directly related to the inefficient use of land resources [8, 9]. However, there are also positive examples. The farm 'Bogiston' in Fergana province, specialising in cotton cultivation, introduced a system of drip irrigation and minimum tillage, which reduced water consumption by 35% and increased yields by 18%. At the same time, due to the use of modern varieties of cotton resistant to diseases, the costs of pesticides and fertilisers were reduced. Another example is the Istiklol farm in Tashkent Province, which specialises in growing vegetables and fruits in greenhouses. The farm implemented a modern microclimate management system, which increased yields by 25 % and reduced energy costs by 15 %.

In general, in recent years there has been an increase in the use of drip irrigation in different regions of Uzbekistan. According to the Ministry of Agriculture, as of 2023, drip irrigation was applied to more than 200 thousand hectares, saving more than 1.5 billion cubic metres of water. In addition, farmers are actively introducing minimum tillage, which allows them to conserve soil moisture and improve its structure, which is especially important in Uzbekistan's arid climate.

Also, precision farming technology is gaining popularity in Uzbekistan. Farmers such as 'Sakhovat' from Kashkadarya region use GPS navigation, sensors and drones to monitor the condition of crops and fertiliser application, which allows to increase the efficiency of resource use and reduce costs by 12-15%. However, it should be noted that the cost of implementing these technologies remains quite high for many farms, which requires further government support.

The introduction of modern technologies is also observed in livestock farming in Uzbekistan. The Omadli Chorva farm in Andijan province, which breeds cattle, has introduced an automated feeding and milking system, which has increased productivity by 20 % and improved milk quality. In addition, the country is actively carrying out selection work to develop highly productive breeds of cattle adapted to local conditions. In particular, new breeds of sheep have been developed that are capable of producing higher wool and higher milk yields. In the poultry sector, over the last 5 years, thanks to the introduction of modern technologies, such as automated poultry housing systems and modern feeds, egg production has increased by 28 %and poultry meat production by 21 %.

The greenhouse business has been actively developing in Uzbekistan in recent years. The farm 'Yangi Hayot' from Surkhandarya region, which specialises in growing tomatoes and cucumbers in greenhouses, has introduced modern heating, ventilation and lighting systems, allowing vegetables to be grown all year round, improving their quality and yield. Over the past three years, the area of greenhouse farms in Uzbekistan has increased by 18 % and the volume of greenhouse vegetable production has grown by 25 %. Uzbek agrarians are actively conducting experimental work on breeding and adapting new crop varieties to local conditions. For example, the Institute of Genetics and Experimental Plant Biology of the Academy of Sciences of Uzbekistan is conducting research to create new cotton varieties resistant to salinity and drought. Such varieties as 'Andijan-37', 'Bukhara-102' and 'Namangan-77', which are characterised by increased yield and resistance to unfavourable conditions, have been developed and introduced into production. Work is also underway to develop new varieties of wheat, barley, rice and other grain crops, as well as new varieties of vegetables and fruits characterised by improved taste, yield and resistance to diseases

In particular, Uzbek farmers grow such wheat varieties as Krasnodarskaya 99, Yubileynaya 100, and Surkhon-20, which are characterised by high yields and baking qualities. Popular cotton varieties include Andijan-35, Namangan-77, and C-6524, which are adapted to local climatic conditions and have a high fibre yield. Uzbek farmers are also actively growing various varieties of vegetables, such as tomatoes 'Malinovka', 'Bull Heart', cucumbers 'Zozulya', 'Competitor', and onions 'Spirit', which are in demand in both domestic and foreign markets.

Uzbek farmers face a number of problems not only in the area of food cultivation. One important problem is limited access to financial resources, which hinders the introduction of new technologies and business development. Another problem is the lack of qualified personnel capable of working with modern technologies and managing farms. Also, an important problem is the wear and tear of agricultural machinery and equipment, which reduces the efficiency of agricultural production. To solve the problems of lack of qualifications, it is necessary to improve the legislative framework in the field of agriculture [10, 11].

One of the ways to solve these problems is the development of the system of microcrediting and leasing of agricultural machinery, as well as the creation of special training and retraining programmes for the agricultural sector. In addition, it is necessary to continue working on improving the legislative framework and creating a favourable investment climate in the agricultural sector, which will attract additional investment and create new jobs, and which is closely related to social support for small farms [12, 13].

Successful development of innovation activities in the agricultural sector of Uzbekistan requires a developed infrastructure, which would include research organisations, consulting centres, farmer training centres, and financial institutions. State support, including subsidies, soft loans and assistance in commercialisation of scientific developments, plays an important role in stimulating innovation. Uzbekistan has adopted a number of programmes and policies in recent years to support and promote innovation in the agricultural sector, including subsidy programmes for the purchase of modern technologies and equipment, and the creation of agro-clusters that bring together farms, processing enterprises and research organisations. However, in order to achieve sustainable development of the agricultural sector, further efforts are needed to create a favourable innovation environment that provides access to resources, knowledge and financial support, as well as to introduce advanced technologies and management practices, and to develop the country's export potential.

However, despite significant successes, the introduction of innovations in the agricultural sector of Uzbekistan faces a number of challenges. Among them are: insufficient level of funding for research and development, low qualification of farm staff, limited access to financial resources, as well as insufficient infrastructure development. In particular, only about 20 per cent of farms in Uzbekistan have access to modern information technologies, which hinders the introduction of precision farming and other advanced management practices.

According to international organisations, labour productivity in Uzbekistan's agriculture is 40-50% lower than in countries with a developed agricultural sector. This is primarily due to the insufficient level of technological equipment, low qualification of personnel and inefficient management methods. To overcome these challenges, it is necessary to strengthen state support for the agricultural sector, develop the system of professional education and training of farmers, and create a favourable innovation environment that encourages the introduction of modern technologies and management methods.

The results of this study, in comparison with similar works, show that Uzbekistan has a significant potential for the development of its agricultural sector. However, the realisation of this potential requires targeted efforts on the part of the state, business and research community. The most important factor in increasing the efficiency of farms is the introduction of modern agricultural technologies, including drip irrigation systems, precision farming and biological methods of plant protection. It is important that these technologies should be adapted to the peculiarities of mountain regions [14, 15].

4 Conclusion

The study emphasises that innovation is a key factor for sustainable development and effective functioning of farms in modern economic conditions. The analysis of literary sources, regulatory and legal documents and statistical data confirms that farms that actively implement modern agro-technologies and management methods demonstrate higher productivity, profitability and competitiveness. The importance of assessing the agro-ecological potential of land for the purposes of sustainable development is confirmed [16]. At the same time, the study revealed a number of problems and challenges that hinder the introduction of innovations in the agricultural sector, especially in countries in transition, such as Uzbekistan. These problems include lack of research funding, limited access to modern technologies, low qualification of personnel, and insufficient infrastructure development. However, examples of successful innovation in farms in Uzbekistan, such as the use of drip irrigation, minimum tillage, precision farming and modern technologies in livestock and greenhouse farms, prove that with targeted government support and farmers' commitment, it is possible to achieve significant success in innovation development.

Successful development of innovation in the agricultural sector requires a comprehensive government policy that includes measures on financial support, education and infrastructure development, as well as the creation of a favourable investment climate. It is also important to stimulate the exchange of knowledge and experience between producers, scientific organisations and consulting companies. In addition, it is necessary to continue to improve the legislative framework and create a favourable innovation environment that provides access to resources, knowledge and financial support, as well as to introduce advanced technologies and management methods, and to develop the country's export potential.

In order to achieve sustainable development of farms and ensure food security, it is necessary to continue work on stimulating innovation in the agricultural sector, creating a favourable innovation environment, and ensuring access of farmers to modern technologies and knowledge, as well as to continue work on the development of agricultural education, infrastructure and support for scientific research, which is closely related to the use of production waste and environmentally friendly production, as well as taking into account the ecological typification of land and land use.

References

1. B. Aparecida da Silva, R. Evangelista de Oliveira, A. Cavalieri Sais, E. Cardoso leite. Family farmers and trees: ecological functions, ecosystem services and food production in agrarian reform settlements in Brazil. Agroecol. Sustain. Food Syst., **49,** 4-21 (2025). DOI: 10.1080/21683565.2024.2405515.

2. Yu.N. Serdechnaya, D.V. Serdechnyy, N.A. Andriyanov, S.A. Korchagin. Development of an effective monitoring and analysis system of modern technologies for the agro-industrial complex development in mountainous areas. Sustainable Development of Mountain Territories, **16,** 770–779 (2024). DOI: 10.21177/1998-4502-2024-16-2-770-779.

3. S. Bamidele. BRICS Revolution: The Global Agrarian Shift Sparked by Emerging Economies. India Q., (2025). DOI: 10.1177/09749284241307937.

4. S.I. Kondratev, Three typical hydrological-hydrochemical situations near the danube river mouth based on the marine hydrophysical institute research expeditions in 1997-2013. Physical Oceanography, **26(4),** 326–340, (2019). DOI 10.22449/1573-160X-2019-4-326-340

5. S.A. Makhosheva, M.V. Zhanokova, V.M. Dzaparova, I.K. Totrukova. Ecological aspect of sustainable development of high-mountain regions in the conditions of innovative economy formation. Sustainable Development of Mountain Territories, **16,** 754–769 (2024). DOI: 10.21177/1998-4502-2024-16-2-754-769.

6. N. Garambois, U. Le Goff, L. Thibaudeau. Agrarian dynamics and climate change in the Senegalese Sahelian peanut basin. Agrar. Syst. Clim. Change: Journeys Adapt. Glob. South, 15-37 (2024). DOI: 10.1079/9781800628137.0001.

7. V.V. Kondratiev, V.V. Bukhtoyarov, V.Y. Konyukhov, K.A. Bashmur. Circular Mining Wastes Management for Sustainable Production of Camellia sinensis (L.) O. Kuntze. Sustainability (Switzerland), **15,** 11671 (2023). DOI: 10.3390/su151511671.

8. J. Rodzik, R. Kołodyńska-Gawrysiak, Ł. Franczak, W. Zgłobicki, J. Poesen. IMPACT OF LAND USE CHANGES ON COLLAPSED PIPES DEVELOPMENT IN THE LOESS GULLY (LUBLIN UPLAND, EAST POLAND). Quaest. Geogr., **43,** 17-33 (2024). DOI: 10.14746/quageo-2024-0042.

9. A.A. Kapanski, R.V. Klyuev, A.E. Boltrushevich. Geospatial Clustering in Smart City Resource Management: An Initial Step in the Optimisation of Complex Technical Supply Systems. Smart Cities, **8,** 14 (2025). https://doi.org/10.3390/smartcities8010014.

10. Sahlan, N. Miqat, S. Susilawati. Realizing "Deconstructional" Justice Through Agrarian Civil Law Reform: A Review of Jacques Derrida's Theory. J. IUS

Kajian Huk. dan Keadilan, **12,** 588-605 (2024). DOI: 10.29303/ius.v12i3.1559.

11. A. I. Trunova, S. O. Kurashkin, K. A. Bashmur, Y. A. Tynchenko, R. B. Sergienko, S. V. Tynchenko, Influence of Temperature Regimes of Synthetic Iron Smelting on Casting Production Efficiency. Metals, **13(7),** 1234 (2023). DOI: 10.3390/met13071234.

12. A. Kehinde, T.F. Adesiyan, S.O. Hassan, I.G. Familusi. Leveraging on membership in agricultural cooperatives to alleviate abject poverty among smallholder farming households in Nigeria. Sustain. Futures, **9,** 100399 (2025). DOI: 10.1016/j.sftr.2024.100399.

13. K.I. Kravtsov, Ya.A. Tynchenko, T.A. Panfilova. Influence of mountain factors on salt excess and soil toxicity in mountain conditions. Sustainable Development of Mountain Territories, **15,** 784-797 (2023).

14. I.I. Bosikov, V.Ch. Revazov. Analysis and evaluation of prospects for highquality quartz resources in the North Caucasus. Mining Science and Technology (Russia), **8,** 278–289 (2023). https://doi.org/10.17073/2500-0632-2023-10-165.

15. T. Bandara, B. Marambe, G. Pushpakumara, P. Silva, R. Punyawardena, L. Manawadu, S. Premalal, G. Miah, K.R. Dahal. Adapting mountain-farming systems to climate change by reducing food-nutrition-health vulnerability. Mitig. Adapt. Strateg. Glob. Change, **29,** 88 (2024). DOI: 10.1007/s11027-024-10181-x.

16. A. A. Kulikova, E. I. Khabarova. Biological working agents use to reduce the carbon footprint. Sustainable Development of Mountain Territories, **16,** 311–321 (2024). DOI: 10.21177/1998-4502-2024-16-1-311-321.

17. L.S. Malyukova, T.A. Panfilova, V. Brigida. Circular Mining Wastes Management for Sustainable Production of Camellia sinensis (L.) O. Kuntze. Sustainability (Switzerland), **15,** 11671 (2023). DOI: 10.3390/su151511671.

18. V. Brigida. Analysis of geological information toward sustainable performance of geotechnical systems. MIAB. Mining Inf. Anal. Bull., **5,** 144-157 (2024). DOI: 10.25018/0236_1493_2024_5_0_144.

The halal standard: from the regulatory norms of eating behavior to large sectors of the economy

A.R. Valiev[1], *Yaacob* Yazid[2], *F.T.* Nezhmetdinova* [1], *M.N.* Kalimullin[1], *R.T.* Salikhov[1] and *K.A.* Parfenova[1]

[1] Kazan State Agrarian University
[2] Putra University, Malaysia, 43400 UPM Serdang, Selangor

Abstract. This article examines the evolution of the term "halal" using the example of Russia and Malaysia, from a religious dietary prescription to achieving the modern status of a significant economic force. The article analyzes how halal has transformed into a commercial product, as well as the impact of the regulatory framework, certification processes and technological innovations on its integration into the large-scale economy. The study focuses on the need for joint research work by Russian and Malaysian scientists aimed at using their unique experience to solve key tasks of harmonizing global halal standards, stimulating innovation in the field of halal food technologies and increasing consumer confidence, as well as market access.

Keywords: halal, industry, technology, standards, economy

1. Introduction

The term halal, in the broadest sense of the word, is one of the cornerstones of Sharia and Islamic law. Nowadays, Islamic law, unlike most religious legal systems, has not become a historical relic, but continues to exist and develop. Along with it, the term "halal" is developing both in its understanding and as an element of Islamic economic life. The term "halal" has evolved significantly over the centuries, especially in the context of food regulation and integration into large sectors of the economy. This evolution can be traced through various studies and articles that emphasize the transformation of halal from a religious concept into a global economic phenomenon [1, 2].

In recent years, the halal industry has been developing rapidly and has become very significant both for the Russian Federation and Malaysia, as well as for the global market. One of the most serious problems in the global halal industry is the lack of global standards for halal practice due to the different interpretations of various scientific schools and certification bodies [3].

As globalization progressed, the concept of Halal began to spread not only to religious practices, but also to economic sectors. This shift was driven by the growing Muslim population and the growing demand for Halal-certified products. The development of the regulatory framework and certification have formulated the need for standardized rules and certification processes to ensure the authenticity and quality of halal products. Various countries and organizations have established their own Halal certification bodies, which has led to the creation of a complex set of standards and regulations.

The integration of Halal into the large-scale economy has been marked by significant investments and strategic initiatives. Countries such as Malaysia and Indonesia are positioning themselves as global halal centers, using their Muslim populations to attract international investment and trade. Technological advances have further advanced the Halal economy. The global halal food market is expected to grow from 3.1 trillion US dollars in 2018 to 5.0 trillion US dollars by 2030 [4].

2. Conditions, materials and methods

The article examines the conditions of globalization and economic integration that led to the transformation of the concept of Halal from a religious ritual into a global economic phenomenon. It analyzes the growing market demand for halal products, especially from the Muslim population, and examines the impact of technological advances on the halal industry, including innovations in food processing, traceability, and digital platforms.

The article uses statistical data on market trends, consumer behavior, and economic indicators to support the analysis.

3. Results and Discussion

The Russian experience. Islam is one of the fundamental religions of the Russian Federation. According to various sources, about 20% of the Russian population are its adherents. Moreover, in some regions, the proportion of Muslims reaches more than 50%.

When considering the specifics of the functioning of the halal industry in the Russian Federation, several

* Corresponding author: nadgmi@mail.ru

aspects should be noted. This is the legal system of the Russian Federation, where, according to art. The Constitution of the Russian Federation is a secular state and all branches of Russian law are based on these principles. Also, the Federal Law of September 26, 1997. Federal Law No. 125-FZ "On Freedom of Conscience and Religious Associations" states that the state does not assign religious associations to perform the functions of state authorities, nor does it interfere in the activities of religious associations unless it contradicts federal legislation [5]. This allowed individual religious initiatives such as Halal to enter the state legal environment.

Here I would like to note that in comparison with the republics of the North Caucasus, in the Volga-Ural region, in the context of broad intercultural contacts, the majority of the Muslim population of the region began to perceive Islam not through the prism of Islamic law as a system of social rules, but for the most part as a personal belief and a system of rituals necessary for the most significant stages of human life: the birth of a child, wedding, funeral [6, 7].

Russian Muslims relate Sharia norms to the Constitution and the law of the Russian Federation. There may be both positive (when legal and religious norms coincide or complement each other) and conflicting (when a religious norm or an idea of it conflicts with a legal norm) relations between Islamic law and laws. For example, the question of polygamy, about who the child will stay with in the event of a parental divorce [8]. Hence, it turns out that sometimes decisions are made based on analogy, which are opposite to each other, but do not contradict either the Koran or the Sunnah [9].

If earlier, Muslims rarely migrated to foreign lands and only in exceptional situations, this was accidental and temporary, then in modern times the situation has changed, a large number of Muslims have dispersed around the world, which led to the need to adapt legal provisions in accordance with new living conditions. This was called Fiqh "critical situations" [10].

In the 21st century, halal, from a term regulating primarily the everyday eating behavior of Muslims, has become an entire branch of the economy, including such areas as nutrition, clothing, leisure, finance, tourism, and even halal childbirth [11–14].

The Republic of Tatarstan is one of the leading regions of the Russian Federation in this field and it can be said that the halal industry is quite developed.

Now the term halal is used in all branches of the Islamic economy: grocery stores, cafes, clothing stores, medical clinics, etc. The total turnover of Halal products worldwide is about $1.175 trillion. And the volume of the Halal market in Russia, according to the spiritual administration of Muslims, is about 35 billion rubles. Moreover, both in Russia and around the world, there is a tendency to increase its volume.

Certain steps have been taken to implement the principles of Halal in the Republic of Tatarstan on the basis of the Duma of the Republic of Tatarstan and the Government of the Republic of Tatarstan. In 2004, the Halal committee of the Ministry of Internal Affairs of the Republic of Tatarstan was established in Tatarstan, a body that controls the activities of both manufacturers and sellers of the halal industry.

In the context of the development of a profitable industry, various unscrupulous merchants and entrepreneurs began to appear. To regulate the industry in 2019, under the leadership of Rafik Mukhametshin, Chairman of the Halal Technical Committee at Rosstandart, GOST was developed, adopted in 2022.

Today, large federal meat processing plants operate in Russia in accordance with the Halal standard. The halal food market in Tatarstan is about $1.5 billion. [14] Production control and certification of halal products is carried out by the Halal standard committee. The availability of this standard makes it possible to carry out international exports of products to the Arab States and the countries of the Persian Gulf.

The largest of them are LLC "Naberezhnye Chelny-broiler", LLC "Naberezhnye Chelny-meat", JSC "Kazan Meat Processing Plant", LLC "Pestrechinka", JSC "Yelabuga MCC" JSC "Kazan Bakery No. 3", JSC "Bakery and confectionery Plant", LLC "Vkusnoteev" and others. The UAE, Qatar and Saudi Arabia are interested in supplying halal products from Tatarstan.

The development of such a direction as "healthy halal" or healthy lifestyle halal: "Halal tayyib" is also looking forward. If you delve into the religious texts of Islam regarding halal food, it becomes clear that it is based not only on a religious goal, but also on a practical one, namely the consumption of healthy, unspoiled food. In modern times, unhealthy foods can include foods with artificial additives, various GMOs and fat substitutes. Shops with farm products are more popular than ever today [15].

In modern conditions, using one of the methods of Islamic law, kiyas (comparison by analogy), we can conclude that organic products can be one of the conditions of halal.

In recent years, we can note an increase in interest in organic products in the Russian Federation. In 2023, there were 177 certified enterprises compared to 150 in 2022. The growth of organics is due to the growing interest in environmentally friendly and healthy products. The gross monetary volume of organic production also showed positive dynamics.

According to Roskachestvo, 262 certificates for organic products were issued in 2022, and in 2023 the number of certificates increased by 123, reaching 385. This indicates the active growth of the industry and the desire of manufacturers to meet the growing demands of consumers. Grain production prevails among certified organic products.

As part of the support and promotion of organic products in Russia, the strategy until 2030 was approved. The strategy provides for an increase in organic production over 10 years by more than 100% from 9 billion rubles in 2021 to 114.5 billion rubles in 2030. Organic products reduce the risks of unhealthy food ingestion.

In 2020, Tatarstan decided to develop its own law "On Organic Production" [8], which supplemented Federal Law 280-FZ "On Organic Products", which entered into force on January 1, 2020, which was supposed to take into account regional peculiarities of agriculture. This allowed the republic to increase the volume of organic products, and in 2024 Tatarstan took 4th place in Russia. There are 7 accredited manufacturers in the republic.

The principle of organic products is that they are grown without pesticides and synthetic fertilizers, antibiotics and growth hormones. Special authorities in Russia give farmers the right to label "Organic" when they confirm the quality of their products.

Today, there are more than 1,250 farms in Tatarstan, and they could become the locomotive of organic products. However, only products that comply with international standards are considered organic. And high-quality products without such a certificate do not have the "organic" status. This is either not available at all for small producers, or leads to a serious increase in the cost of products.

On the other hand, since at the moment the certification of organic products is voluntary. Which leads to abuse. Therefore, while the situation is not developing according to optimistic forecasts, organic products account for about 4% in the food sector. Of course, the indicator is extremely small. According to statistics for 2021, no more than 1% of the Russian population buys organic products.

There are a number of obstacles to the successful development of the organic sector. High retail prices. Organic products are 150-300% more expensive than regular products. Low level of consumer confidence. Due to frequent falsification, buyers do not trust labels on packages where they are guaranteed an environmentally safe and proven product, without GMOs and harmful additives.

And under these conditions, the Republic of Tatarstan, as a region integrating the principles of Islamic law into the Russian legal system, can gain competitive advantages in the framework of forming its own brand and customer loyalty by combining halal and organic products under the auspices of the Muslim Spiritual Board. Organic agriculture is a promising branch of agriculture, which potentially provides an opportunity to increase profitability and loyalty to agricultural products, will allow to enter foreign markets and attract investments.

Until February 24, 2022, foreign imports accounted for about 80% of the Russian organic food market, but this share has decreased significantly under the influence of Western sanctions and trade restrictions, which expands the potential market share of domestic organic production, i.e. it can reduce the impact of industrial agriculture on the environment. The Russian government intends to become a major producer of organic food, targeting a share of 10% of the global organic market by the mid-2030s.

The organic food sector is traditionally divided into 1) Private farms and large enterprises 2) eco-settlements 3) independent eco-farms. Taking into account the foreign experience, the first sector is likely to develop. Unfortunately, it suffers from a formal approach, and in developed countries, apart from the certification standard, it differs little technologically from conventional inorganic agricultural enterprises.

Regarding the cultivation of organic products, it seems more promising for us to study and use the experience of Malaysia to combine the two components of halal and organic products in the final product.

Malaysia began to formalize halal standards in the early 1970s, which corresponded to the growing demand for certified halal products for both domestic and foreign goods [16–18]. Malaysia's Department of Islamic Development (JAKIM) began certifying halal consumer food and basic necessities in 1974, when the Research Center of the Department of Islamic Affairs of the Prime Minister's Office issued Halal certificates for products that meet these criteria. In 1994, the Halal certification was formalized with the introduction of a certificate with the Halal logo. Later, the certification was improved through an audit procedure in 1998, when inspections were conducted by the Ilham Daya Company, appointed by the Government. On September 1, 2002, the Food and Islamic Consumer Products division of JAKIM began the Halal certification process, initially employing only 28 employees. To meet the growing demands of the food industry, on November 17, 2005, the Malaysian Department of Public Service approved an increase in staff to 165 positions, resulting in the establishment of the Halal Hub division. During the time that the halal industry has developed and many changes have taken place, on April 2, 2008, the Halal Industry Development Corporation, which is responsible for managing the halal industry in Malaysia, appointed a person responsible for managing all Halal certification activities. However, later, on October 1, 2009, responsibility for this was returned to JAKIM [19, 20].

The growth of the halal industry has been driven by several factors, including the growing purchasing power of Muslim consumers in both Muslim-majority and non-Muslim-majority countries where Muslims are a minority. The globalization of trade and communications in the late 20th century also contributed to the rapid growth of the halal industry. Malaysia, with its large Muslim population and Islamic rule, has become a hub for the implementation of global halal industry standards, especially in the field of food and beverages.

The development of the halal industry was not a matter of a single organization, but rather a collaboration between various stakeholders, including government, religious authorities, the private sector, and international organizations [21]. Malaysia's role as a leader in Halal certification attracts companies from all over the world seeking access to Muslim markets. Last but not least, Muslim consumers themselves are showing the greatest interest. The number of Muslims in the world is expected to grow to 2.6 billion people, accounting for 30% of the world's population. This vast market indicates the growing demand and opportunities opening up in the halal industry [21].

The development of Halal standards is a complex process that involves the reconciliation of religious interpretation and scientific assessment. Islamic scholars, with the help of reputable institutions, will discuss the problem from the point of view of Sharia, which will lead to interpretations that will form the basis of standards. They will then be integrated into current technological and commercial aspects that require a technical base [17–21].

In Malaysia, the Halal Industry Development Corporation (HDC) works closely with JAKIM to ensure compliance, transparency and reliability of standards for the industry. The Halal certification process includes not only religious observance, but also scientific verification, for example, to ensure that ingredients do not contain prohibited substances such as pork or alcohol, and that animals are slaughtered according to the Islamic method. This process is rigorous and includes inspections, audits, and documentation checks at several stages of production. In principle, there are 8 reference sources for Halal certification, of which the 2 main ones are the prosedur pensijilan halal Malaysia (MPPHM) manual and the pengurusan Halal Malaysia (MHMS) system [21].

Food makes up a significant part of Malaysia's halal industry. According to the Halal Industry Development Corporation (HDC), the Malaysian halal market is experiencing significant growth, with food and beverages accounting for the dominant share, which is projected to range from 51.9 billion US dollars (2018) to 85.2 billion US dollars (2030) [2].

The global halal food market is expected to grow as the Muslim population increases and demand for halal products expands outside of countries with a traditional Muslim majority. Malaysian exports of halal food products have found markets in regions such as the Middle East and North Africa, Southeast Asia, Europe and Eurasia, the Asia-Pacific region, sub-Saharan Africa and North America. The growing interest in halal food from non-Muslim consumers seeking ethical and health-conscious alternatives has further expanded this market segment.

In addition to the export market, Malaysia's domestic market is also experiencing a growing demand for certified halal products: restaurants, supermarkets, and convenience stores offer halal products. This is due to the growing awareness of consumers, including non-Muslims, about the halal certification standard. This trend is not limited to traditional foods, but extends to beverages, snacks, and even health supplements.

Despite the fact that Malaysia has made great strides in the halal industry, both domestically and globally, it faces a number of challenges. In the context of Malaysia, the problem is the delay in the certification process, certificate forgery, and lack of awareness about the implementation of halal certification [19–21]. Malaysia also faces the challenge of maintaining a balance between halal certification and international standards. As the demand for halal products increases worldwide, there is a need for more streamlined and standardized procedures to avoid confusion or disputes between different certification bodies. This has led to a call for international cooperation to create a more holistic, globally recognized halal standard.

Malaysia is one of the first countries to successfully combine a dualistic legal system. This is exactly what the Republic of Tatarstan is trying to implement at the present stage, of course, within the framework of the Russian legal system. And we have a lot in common. At the same time, it is necessary to pay attention to the risks and factors that may affect this [22–25].

The main parameters for comparison are shown in Table 1.

Table 1

Tatarstan	Malaysia
Halal food	
In Tatarstan, halal food is provided in restaurants and grocery stores to meet the needs of the Muslim population. There are certification organizations that check the compliance of products with Halal standards In Tatarstan and Malaysia, about 50-60% of the population profess Islam. In Tatarstan, halal products are represented in more than 300 food production enterprises.	Malaysia is known for its well-developed halal food industry, which attracts both local and foreign tourists. Government agencies actively support and promote Halal products and services to meet the needs of the Muslim population. The halal food market in Malaysia is estimated at about 10 billion US dollars, making it one of the largest in the world.
Halal tourism	
Tatarstan is one of the tourism leaders in the Russian Federation. It attracts tourists with its cultural and ethno-religious diversity. His legacy includes both Orthodox and Islamic historical monuments, mosques and mausoleums. This can be valuable for tourists interested in Islam and its history. In 2020, Tatarstan received about 2 million tourists, including both domestic and foreign ones.	Malaysia is known as one of the world's leading centers of Islamic tourism. There are historical mosques, cultural museums and other facilities, as well as a wide range of Halal services for Muslim tourists. Malaysia is one of the most popular tourist destinations in the world and attracts more than 25 million tourists annually. The Islamic tourism industry in Malaysia generates more than 10 billion US dollars per year.
Halal investments	
Due to historical peculiarities, Islamic financial institutions in Tatarstan are less developed than in Malaysia. However, the interest in the development of Islamic banking is very high. The Islamic financial sector is developing in Tatarstan. In 2023, the volume of Islamic financial assets in the region amounted to about $ 100 million. Since September 1, 2023, a two-year experiment on partnership or Islamic financing has been launched in 4 regions of the country, including Tatarstan, within the framework of the adopted federal law.	Malaysia is one of the world leaders in the Islamic financial sector. Thanks to the dualistic model (the integration of Islamic financial models into the Soviet economic system), Islamic banks, insurance companies, investment funds and other financial institutions offering a wide range of Islamic financial products are successfully developing here. Malaysia is a global leader in the Islamic financial sector. The volume of Islamic financial assets in the country exceeds 200 billion US dollars. Malaysia stands out as one of the main global players in many aspects of the Islamic economy

Tatarstan plans to export halal products worth $16 million in 2024. Exports over the past years are shown in the table 2.

Table 2

Years	2020	2021	2022	2023	2024 (plan)	Growth rate, in %	Growth rate, in %
Amount, million dollars	2,5	4,9	6,5	12	16	480	380

Next, let's look at exporting companies by type of Halal products in Table 3.

Table 3

Exporters of Halal products	Importing Countries
Meat and dairy products	
Agrosila	Kazakhstan, Azerbaijan
SKPC Kausar	Kausar Belarus, Uzbekistan,
Vamin	Kazakhstan, Tajikistan
Camembert	Belarus
Food products	
Essen Production	CIS, Palestine
PK Akulchev	CIS, Mongolia, China, Brazil, USA, UES
BKK	Israel, Germany
Bahetle	UAE
Sultan of Kazakhstan	China, Vietnam
Other agricultural products	
TD EMIS Kazakhstan	Kazakhstan, Kyrgyzstan
Healthy Food Factory	Germany

4. Conclusions

The development of halal exports is of strategic importance for Tatarstan and Russia, as it promotes economic growth, strengthens international ties, satisfies the growing demand for halal products, supports local producers and preserves the cultural heritage of the region.

And in this regard, it is especially interesting to compare and exchange experiences with a country like Malaysia. Malaysia is a country with a dualistic economic model based on the exchange of legal experience, as well as in addressing food security issues based on a unified approach to food quality analysis based on an understanding of Halal principles.

The term "halal" has evolved significantly over the centuries, especially in the context of food regulation and integration into large sectors of the economy. The evolution of the term "halal" reflects its transformation from a religious dietary guideline into a multifaceted economic concept. This path has been documented in various studies and articles that highlight the regulatory, economic, and technological aspects of Halal integration into major sectors of the global economy. As the demand for halal products continues to grow, this term and the practice of its use are likely to evolve further, adapting to new challenges and opportunities in the global market.

References

1. S. Siti, H. Abdullah, Sentiment Analysis: Halal approval based on literature indexed by Scopus, Perbanas J. of Islam. Econ. and Busin., **3**, 11 (2023), https://doi.org/10.56174/pjieb.v3i1.74
2. N.K. Faruki, M.F. Nordin, Knowledge research: Selected works on the Qur'an and the development of an ontology, J. of Theor. and Appl. Inform. Technol., **72(3)**, 385–393 (2015)
3. S. Miskam, N. Othman, N.A. Ab Hamid, S.N. Syed Ager et al., Analysis of the definition of Halal: Sharia against the laws, Conference: World Academic and Research Congress 2015 (World-AR 2015) (YARSI University, Jakarta, Indonesia) (2015)
4. *Halal Industry Development Corporation. The Master Plan for the development of the Halal industry until* 2030 *(HIMP 2030)*, https://www.hdcglobal.com
5. *On Freedom of Conscience and Religious Associations*, Federal Law, no. 125-FZ of September 26, 1997, SPS Consultant Plus
6. Islam and the Tatar world: conceptual foundations of development, The concept was adopted at the IV All-Russian Forum of Tatar Religious Figures, (Kazan, May 23–25, 2013) (Kazan, 2013, pp. 29–30)
7. I.A. Mukhametzaripov, Religious and legal norms of the Russian Federation: examples of interaction and concretions, Rus. Law J., **5(116)**, 18–28 (2017)
8. D.S. Kusanova, The Halal industry in the Republic of Tatarstan, Islam in the modern world, **17(1)**, 159–176 (2021)
9. E. Kuliyev, *Elementary Fiqh (with proof of four madhhabs)* (UMMA, 2016, p. 11)
10. Z. Ahmad, M.N.A. Mafaz, D.M.M. Rahman, Harmony in Halal: Understanding stakeholders' opinions, product analysis, and policy assessment in Malaysia, West. Sci. Busin. and Manag., **1(05)**, 495–508 (2023), https://doi.org/10.58812/wsbm.v1i05.358
11. H. Elasrag, The Halal Industry: Key Challenges and Opportunities, Sociol. Res. Network (2016), https://doi.org/10.2139/ssrn.2735417
12. E. Izberk-Bilgin, K. Nakata, A New Perspective on faith-based Marketing: The Global Halal Market, Busin. Horiz., **II.59(3)**, 285–292 (2016), https://doi.org/10.1016/j.bushor.2016.01.005
13. Z. Samori, A.H. Ishak, N.H. Kassan, Understanding the development of a Halal food standard: recommendations for future research, Int. J. of Soc. Sci. and Hum., **4(6)**, 482–486 (2014)
14. Z. Zakaria, Entering the global halal market: some discussions on the laws and standards of Malaysia, Sharia Law J., **16(3)**, 603–616 (2008)

15. J. Akbar, M. Gul, M. Jahangir, M. Adnan, S. Saud, S. Hassan et al., Global trends in Halal food standards: an overview, Foods, **12(23)**, 4200 (2023)
16. *On the development of organic production in the Republic of Tatarstan*, Law of the Republic of Tatarstan dated 05.05.2021, no. 34-ZRT
17. M.S.E. Azam, M.A. Abdullah, Halal Standards Worldwide: A comparative study of the unity and differences between the most popular Halal standards worldwide, Halal sphere, **1(1)**, 11–31 (2021)
18. Yu.A. Aziz, N.V. Chok, The role of Halal awareness, Halal certification, and marketing components in determining the intentions of non-Muslims in Malaysia to make halal purchases: an approach based on structural equation modeling, J. of Int. Food Mark. and Agribusin., **25(1)**, 1–23 (2013)
19. A.H. Yusuf, S.Ab. Shukor, U.S.A. Bustamam, Halal certification versus the growth of the food industry business in Malaysia, J. of Econ., Busin. and Manag., **4(3)** (2016), https://doi.org/10.7763/JOEBM.2016.Version 4.399
20. M.A.A. Majid, I.H.Z. Abidin, H.A.M.A. Majid, K.T. Chick, Problems of halal food introduction in Malaysia, J. of Appl. Envir. and Biol. Sci., **5(6)**, 50–56 (2015)
21. K.M.Z.K. Omar, Problems and marketing strategies of halal products in Malaysia, Interdiscipl. J. of Busin. Res. ISSN, **2046**, 7141 (2013)
22. M.S.E. Azam, M.A. Abdullah, The global Halal industry: realities and opportunities, Int. J. of Islam. Busin. Ethics (IJIBE), **5(1)**, 47–59 (2020)
23. F.T. Nezhmetdinova, M.E. Guryleva, N.Kh. Sharipova, R.I. Zinurova, A.R. Tuzikov, The risks of modern technologies and the legal aspects of their implementation in agriculture, BIO Web Conf., 17(3), 00227 (2020)
24. F.T. Nezhmetdinova, M.E. Guryleva, N.L. Blatt, The new role of bioethics in emergency situations using the example of COVID-19, BioNanoSci., **12(2)** 2022
25. D.I. Fayzrakhmanov, A.R. Valiev, F.T. Nezhmetdinova, G.D. Krupina, The agrarian scientific and educational potential of the Republic of Tatarstan and the directions of its development, Bull. of Kazan State Agrar. Univ., **3(3(9))**, 5–10 (2008)

Resource-saving tillage technologies and their impact on grain crop yields

Zoya I. Latyaheva[1], *Danil A.* Zyukin[1*], *Nikita A.* Karpushin[1]

[1] Kursk state agrarian university named after I.I. Ivanov, Kursk, Russia

Abstract. The work is devoted to analysing the efficiency of resource-saving tillage technologies, namely minimum and zero tillage, in the cultivation of grain crops. Experimental work was carried out in different agro-climatic zones, using conventional, minimum and zero tillage. The influence of different tillage systems on yield, economic indicators of production and agrophysical properties of soil was studied. It was found that minimum tillage provides higher yield of grain crops compared to traditional ploughing, as well as improves soil structure, increases its biological activity and reduces fuel costs. Zero tillage in case of observance of technologies provides yield at the level of traditional tillage. The necessity of further study and adaptation of resource-saving technologies to specific soil and climatic conditions in order to achieve maximum economic and environmental efficiency has been revealed. The results emphasise the prospects of minimum tillage for sustainable agricultural development.

1 Introduction

Intensive agricultural practices involving deep tillage lead to a range of negative consequences such as soil erosion, reduced soil fertility, deterioration of water regimes and environmental pollution. These problems not only threaten the sustainability of agricultural systems, but also have a significant impact on global environmental processes, including climate change and biodiversity loss. Recognising the seriousness of the situation, the global community is actively seeking alternative approaches to agriculture aimed at reducing negative environmental impacts while ensuring stable yields. When considering land use issues, an important role is played by taking into account the influence of climatic factors [1, 2].

One of the promising directions in solving these problems is the introduction of resource-saving tillage technologies, such as minimum and zero tillage [3, 4]. These approaches, in contrast to traditional methods, involve reducing the depth and intensity of mechanical impact on the soil, which contributes to preserving its structure, improving the water-air regime and reducing the risk of erosion. At the same time, an important advantage is the reduction of fuel and labour costs [5, 6]. Nevertheless, such technologies are not without certain disadvantages. For example, minimum and zero-tillage require more careful control of field weediness, which can lead to increased herbicide costs. In addition, switching to new technologies requires revision of existing agronomic approaches and adaptation to specific soil and climatic conditions. Other approaches include the use of organic fertilisers and siderates [7, 8], integrated plant protection, as well as the introduction of crop rotations and reclamation measures. Each of these directions has its advantages and disadvantages, which should be carefully analysed when implementing them.

It is in the context of the search for ways to improve the efficiency and sustainability of agrarian systems that the direction of resource-saving agriculture with minimisation of soil tillage becomes particularly relevant. The introduction of such technologies is an important step towards the creation of ecologically more sustainable and economically efficient agro-ecosystems that can ensure food security in the long term and preserve natural resources for future generations [9, 10]. At the same time, it is important to identify the criteria of ecological and economic feasibility and assess the effectiveness of the introduction of these technologies in real economic conditions. The search for optimal parameters of minimising tillage, providing stable yields at minimum costs, is an important and urgent task.

The purpose of this work is to study the scientific experience of applying resource-saving technologies of minimising tillage in the cultivation of grain crops. Within the framework of the work, the analysis of existing studies aimed at determining the effectiveness of minimum and zero tillage, as well as identifying factors affecting its application in different agro-ecological conditions was carried out. Particular attention was paid to analysing the responsiveness of different types of grain crops to this type of treatment, with the identification of the main advantages and disadvantages [11, 12]. Based on the analysis,

* Corresponding author: nightingale46@rambler.ru

conclusions and recommendations for further improvement of resource-saving technologies were formulated.

2 Method and materials

The experiments were conducted on several experimental plots located in different agro-climatic zones, taking into account the variability of soil types and weather conditions.

The research compared conventional tillage methods with minimum and zero tillage, as well as various combinations of these approaches.

A wide range of agricultural equipment was used for experimental work.In the variant with traditional tillage the plough PLN-4-35 was used, providing ploughing to a depth of 20-22 cm. The working speed of the machine was 7-8 km/h, with a working width of 1.4 metres.KPS-4 cultivator was used for surface tillage, which provided soil loosening to a depth of 8-10 cm.The working speed of the cultivator was 9-10 km/h.Based on the analysis, conclusions and recommendations for further improvement of resource-saving technologies were formulated.

To prepare the seedbed after ploughing, a combined aggregate AKR-3 was used, which loosened, levelled and compacted the soil in one pass. In variants with minimum tillage disc harrows BDT-3 and cultivators KPS-4 were used at a shallower depth - 5-7 cm, excluding deep ploughing. In the variant with zero tillage immediately before sowing Amazone DMC-6 direct seeding complex was used, which provided sowing of seeds without pre-tillage. All sowing was carried out by grain seeder SZ-3.6. Seed sowing and fertiliser application rates were in accordance with recommendations for each specific crop and soil-climatic conditions. In the course of research, the yield was monitored, and the economic efficiency of different cultivation technologies was analysed, taking into account the costs of fuel, fuels and lubricants, labour, seeds, fertilizers and plant protection products.

3 Results and discussion

The presented research was aimed at a comprehensive study of the effectiveness of different tillage systems in the cultivation of grain crops, with a special emphasis on resource-saving technologies.

To achieve this goal, a number of experimental works were carried out, including comparison of conventional, minimum and zero tillage. The experiments were organised on several experimental plots located in different agro-climatic zones, taking into account the diversity of soil cover and climatic conditions.

In the course of research, the influence of different tillage systems on grain crop yields, economic indicators of production, as well as on agrophysical properties of soil was studied.

In the framework of the experimental study, on the control plots the traditional tillage system was used, which included autumn ploughing with plough PLN-4-35 to a depth of 20-22 cm followed by cultivation with cultivator KPS-4 to a depth of 8-10 cm in spring and harrowing with disc harrow BDT-3 to a depth of 6-8 cm to prepare the seed layer.KPS-4 cultivator was used for surface tillage, which provided soil loosening to a depth of 8-10 cm.The working speed of the cultivator was 9-10 km/h.BDT-3 disc harrow was used with working speed of 10-12 km/h and cultivation depth of 6-8 cm.To prepare the seedbed after ploughing, a combined aggregate AKR-3 was used, which loosened, levelled and compacted the soil in one pass.

Sowing was carried out in optimal agrotechnical terms for each region. During the growing season, the condition of crops was monitored, weeds, pests and diseases were controlled. Harvesting was carried out by a combine harvester 'Yenisei-1200', after which the yield of each crop on each variant of tillage was determined. The costs of technological operations were also taken into account, namely the consumption of fuel, seeds, fertilisers, plant protection products, depreciation of equipment and wages of mechanics.

The conducted research has shown that minimum tillage has a positive effect on the yield of grain crops in comparison with traditional ploughing. The average yield of winter wheat at minimum tillage was 48 c/ha, which is 7-10% higher than in the variant with traditional ploughing. Similar results were obtained for spring barley, whose yield at minimum tillage was 42 c/ha, while at traditional tillage - 39 c/ha. The results obtained during the experiment agree with the data presented in the article [13], where it was noted that cereal crops are more responsive to minimum tillage. Oat yields also increased by 5-7% when minimum tillage was used, reaching 35 kg/ha compared to 32 kg/ha with conventional tillage. Under no-tillage, yields decreased by 8-10% compared to minimum tillage, but were still at the level of conventional tillage when all necessary agronomic measures were observed. However, it is worth noting that weather conditions, which varied in different years of the experiment, had a significant impact on yields. In favourable years, the difference between the treatment options was less pronounced.

The analysis of economic indicators showed that the use of minimum tillage contributes to a significant reduction in fuel and fuel and lubricants costs, as well as labour costs for mechanics.

Costs for soil tillage at minimum technology were 25-30% lower than at traditional ploughing. However, at minimum tillage there was an increase in herbicide costs, which is due to higher weediness of fields.

At the same time, the overall economic indicators with minimum tillage were higher than with conventional ploughing. Thus, the use of minimum tillage is not only ecologically expedient, but also economically favourable.

In case of zero tillage, the costs of fuel and lubricants were even lower, but additional costs were required to purchase a special seeder and use more expensive herbicides for weed control.In other plots, a minimum tillage system was used, where ploughing was excluded and tillage was limited to spring loosening with a KPS-4 cultivator to a depth of 5-7 cm and harrowing with a BDT-3 disc harrow to a depth of 5-6 cm.In plots with zero tillage, the soil was not mechanically disturbed before sowing.Sowing in all variants was carried out with a grain seeder SZ-3.6. Winter wheat, spring barley and oats were used as experimental crops.Sowing was carried out in optimal agrotechnical terms for each region.

This is due to the fact that at minimum tillage the natural soil structure is preserved, soil aggregates and pores are not destroyed, which contributes to the improvement of water-air regime.

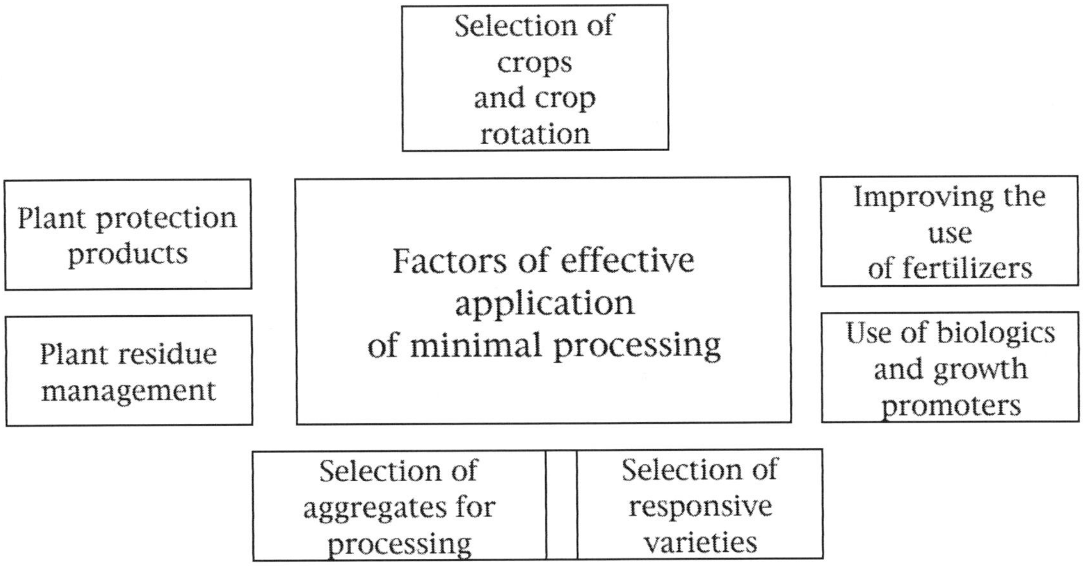

Figure 1. The level of provision with production of the main types of agricultural products in the Russian Federation

In addition, the plots with minimum tillage showed an increase in the number of earthworms, which indicates an improvement in the biological activity of the soil. Thus, on one square metre of soil under conventional ploughing the number of earthworms averaged 8 pieces, and on the variants with minimum tillage - 16 pieces. At zero tillage the number of earthworms was also higher than at traditional ploughing, but inferior to minimum tillage. There was also an increase in the humus content of the soil by 0.15-0.2 % compared to traditional tillage. These changes have a positive effect on soil fertility and contribute to more efficient assimilation of nutrients by plants. Thus, the obtained data indicate that minimum tillage is not only an effective way to increase the yield of grain crops, but also contributes to improving the ecological state of the soil, which makes this technology sustainable and promising in the long term. In the context of the need to reduce the negative impact of the agricultural sector on the environment, a comprehensive study of organic farming is required [14, 15].

Ultimately, the economic feasibility of introducing minimum and no-tillage should be assessed taking into account specific soil and climatic conditions and farm peculiarities [16].

According to Table 1, it was established that the choice of tillage system should be based on specific soil-climatic conditions and the level of production intensification. As shown in Figure 1, effective application of minimum tillage requires a comprehensive approach, including selection of crops and crop rotations, improvement of fertiliser application methods, use of biopreparations and growth stimulants, selection of aggregates for tillage and responsive varieties.

Table 1. Classification of tillage systems

System	Subsystems
Mouldboard	Miscellaneous Depth Minimal
Mulching	Deep
Combined	Multidepth Minimal
Zero	
Ridge-and-row	

The results obtained during the research agree with the conclusions presented in scientific publications on this subject. Minimisation of tillage improves soil structure, increases organic matter content and biological activity of soil. These effects have a positive impact on soil fertility and resistance of agro-ecosystems to unfavourable environmental factors. The economic efficiency of different types of tillage was also analysed. It was found out that the economic effect from the use of minimum tillage is achieved by reducing the cost of fuel and lubricants. It was confirmed that when switching to minimum tillage, careful weed control and adjustment of plant nutrition are necessary.

4 Conclusion

As a result of the conducted research the evidence of feasibility of introducing resource-saving tillage technologies in the cultivation of grain crops was obtained. Minimum tillage provides not only stable, but often higher grain yields compared to traditional ploughing. Thus, the yield of winter wheat at minimum tillage on average by 7-10% exceeded the yield at traditional tillage, and the yield of spring barley and oats was higher by 5-7%. At the same time, it should be noted that no-tillage in compliance with agronomic requirements provides yields at the level of traditional tillage, but requires a more careful approach to plant protection and may not always be economically justified. In the context of data on the influence of mountain factors on soil salt composition [17], it is important to take into account regional conditions when choosing tillage technologies.

In addition, an important finding is the significant reduction in fuel and fuel and lubricant costs when switching to minimum tillage, which makes it more economically favourable compared to conventional ploughing. The costs of tillage were reduced by 25-30%. Improved agrophysical soil properties such as increased porosity, water permeability and organic matter content are also important benefits of minimum tillage. Minimum tillage contributes to the increase of soil biological activity, which positively affects its fertility and resistance to unfavourable factors. Thus, the introduction of minimum tillage is a promising direction in the development of sustainable agriculture, allowing to achieve both economic and environmental benefits. The results obtained confirm the need for further development and dissemination of resource-saving technologies in the agricultural sector, which is an important step towards ensuring food security and conservation of natural resources. It is important to emphasise the need to use methods to assess the impact of tillage on ecology and soil cover [18]. In conclusion, the choice of a particular tillage system should be based on a comprehensive analysis of agro-ecological conditions, economic factors and farm peculiarities, taking into account the need to ensure a balance between economic efficiency and environmental sustainability of agrarian systems.

References

1. N. Garambois, U. Le Goff, L. Thibaudeau. Agrarian dynamics and climate change in the Senegalese Sahelian peanut basin. Agrar. Syst. Clim. Change: Journeys Adapt. Glob. South, **15**-37 (2024). DOI: 10.1079/9781800628137.0001.

2. A.G. Terekhov, G.N. Sagatdinova. Spectral characteristics of saline arable land in the foothill catchment Zailiisky Alatau ridge (Northern Tien-Shan) during 2016-2022 using Sentinel-2 data. Sustainable Development of Mountain Territories, **16,** 143–155 (2024). DOI: 10.21177/1998-4502-2024-16-1-143-155.

3. J. Nzuma, A. Matsaure, G. Mtetwa, L.-L. Sakwa, L. Munyaradzi, V. Samukange, L. Madzingaidzo. Enhancing Onion (Allium cepa) Yields: Integrating Precision Irrigation with Real-Time Soil Moisture and Mulching Strategies for Optimum Productivity. Research on World Agricultural Economy, **6,** 172–186 (2025). DOI: 10.36956/rwae.v6i1.1334.

4. D.A. Ramírez, M.M. Hossain, E.H.M.S. Rahaman, C. Mestanza, J. Rinza, J. Ninanya, F. de Mendiburu, H. Loayza, M. Gatto, J.F. Kreuze. Potato production under zero tillage with rice straw mulching as a promissory technology to diversify rice-based systems in Southwest Coastal Bangladesh. Journal of Agriculture and Food Research, **19,** 101603 (2025). DOI: 10.1016/j.jafr.2024.101603.

5. Z. Shutaleva, I. Nikonova, I. Savchenko. Environmental education for sustainable development in Russia. Sustainability (Switzerland), **12,** 7742 (2020). DOI: 10.3390/su12187742.

6. A.I. Smirnov, N.V. Plotnikova, V.G. Burov. Features of the fine structure of the active layers of organic solar cells. Key Eng. Mat., 712, 176-181 (2016). DOI: 10.4028/www.scientific.net/KEM.712.176.

7. G. Wang, H. Li, F. Hu, W. Yin, Z. Fan, H. Fan, Y. Sun, W. He, Q. Chai. Application of green manure combined with synthetic nitrogen fertilizer enhances soil aggregate stability in an arid wheat cropping system. Applied Soil Ecology, **206,** 105849 (2025). DOI: 10.1016/j.apsoil.2024.105849.

8. A.A. Khan, I. Azeem, J. Hui, Y. Chen, Y. Yuan, T. Shah, M. Adeel, N. Shakoor, R.M.A. Asghar, W. Cao, D. Zhang, Y. Gao. Non-leguminous green manures improve labile phosphorus availability and crop yield in agroecosystems: A global meta-analysis. Soil and Tillage Research, **248,** 106430 (2025). DOI: 10.1016/j.still.2024.106430.

9. A. A. Sokolov, L. G. Orlova, K. A. Bashmur, R. I. Kuzmich, Mining Informational and Analytical Bulletin. **11,** 278-291 (2023). DOI: 10.25018/0236_1493_2023_111_0_278.

10. V. O. Gutarevich, A. S. Mikhalev, V. E. Gozbenko, Mining Informational and Analytical Bulletin, **11,** 72–87 (2023). DOI: 10.25018/0236_1493_2023_111_0_72.

11. O. Afanaseva, M. Afanasyev, S. Neyrus, D. Pervukhin, D. Tukeev. Information and Analytical System Monitoring and Assessment of the Water Bodies State in the Mineral Resources Complex. Inventions, **9,** 115 (2024). https://doi.org/10.3390/inventions9060115.

12. K.Y. Lobkov, A.A. Stupina. On the issue of monitoring crack formation in natural-technical systems during earth surface displacements. MIAB. Mining Inf. Anal. Bull., **11-1,** 292-304 (2023). DOI: 10.25018/0236_1493_2023_111_0_292.

13. A. A. Jauhari, S. Minarsih, Y. Hindarwati, J. Pramono, A. Susila, S. Sudarto, S. Basuki, W. Hariyanto, B. Utomo, T. Suhendrata, R. Oelviani, F.D.

Arianti, A. Supriyo, V.E. Aristya, S. Samijan. Rice yield enhancement and environmental sustainability with precision nutrient management. Glob. J. Environ. Sci. Manag., **11,** 77-92 (2025). DOI: 10.22034/gjesm.2025.01.05.

14. M. Moghaddas. Retrieval of soil moisture using a decomposition-based model and optical-thermal model with Sentinel-1 and Landsat-8 images. Soil Sci. Soc. Am. J.**, 89,** e70004 (2025). DOI: 10.1002/saj2.70004.

15. V. I. Golik, V. Yu. Konyukhov, Mining Informational and Analytical Bulletin. **11-1,** 175-189 (2023). DOI: 10.25018/0236_1493_2023_111_0_175.

16. T. A. Panfilova, V. V. Kondratiev, Mining Informational and Analytical Bulletin. **11-1,** 239-251 (2023). DOI: 10.25018/0236_1493_2023_111_0_239.

17. K.I. Kravtsov, Ya.A. Tynchenko, T.A. Panfilova. Influence of mountain factors on salt excess and soil toxicity in mountain conditions. Sustainable Development of Mountain Territories, **15,** 784-797 (2023). https://doi.org/10.21177/1998-4502-2023-15-3-784-797.

18. S. Das. Determinants of livelihood vulnerability to climate change among the agrarian community: Empirical evidence from the coastal region of Indian Sundarban Biosphere Reserve. Reg. Stud. Mar. Sci., **78,** 103777 (2024). DOI: 10.1016/j.rsma.2024.103777.

Risks and safety standards for the use of genetically modified organisms in the context of food security

F.T. Nezhmetdinova* [1], *M. E.* Guryleva[2], *L.D* Kardanova [3]

[1] Kazan State Agrarian University, Kazan, Russia
[2] Kazan State Medical University Kazan, Russia
[3] Kabardino-Balkarian State University, Nalchik, Russia

Abstract. Genetically modified organisms (GMOs) have become the cornerstone of modern agriculture, offering solutions to some of the most pressing issues such as food security, pest control, and climate change. This article discusses the risks and safety standards of using genetically modified organisms (GMOs) in agriculture. Regulatory approaches in Europe, the USA, and Russia are analyzed, as well as the potential environmental and social consequences of GMO use. Both positive and negative aspects of the introduction of GMOs in the agricultural sector are evaluated.

Keywords: Genetically modified organisms (GMOs), safety, risks, agriculture, regulation, ecology, Europe, USA, Russia.

1 Introduction

Understanding today's global food system is not easy. Everyone needs food, and access to adequate food is one of the fundamental human rights. Today, about 1.3 billion people are engaged in agricultural production, and agriculture employs almost half of the world's working population, including 450 million salaried agricultural workers. The majority of the working-age population involved in agricultural production is engaged in food production, first of all, because food determines important physiological processes for maintaining tissue integrity. It regulates the biochemical mechanisms of metabolism and is the main determinant of growth and development.

At the same time, in the world where enough food is produced for every inhabitant of the planet, about 10% of the world's population is constantly hungry. The consequences of constant starvation are terrible: illnesses, blindness, mental retardation in children, and early death. At the same time, overeating and regular overeating are the cause of the most serious diseases among wealthy people in developed countries.

The Rome Declaration of the World Food Security Forum as of November 13–17, 1997 defined the concept of food security as "ensuring that all people have access at all times to food necessary for a healthy and active life". Safe foods are foods that do not pose a risk to human and animal health.

The connection between GMO safety and risks and food safety is vital in modern conditions, where the challenges of feeding a growing population, mitigating climate change, and meeting consumer demands for safe and transparent food are paramount. They are rising global food demand and population growth. With the global population projected to reach nearly 10 billion by 2050, ensuring food safety while increasing agricultural productivity is a major challenge. GMOs offer potential solutions but must be carefully managed to avoid compromising food safety.

Climate change and resilience are urgent concerns. Climate change is influencing agricultural productivity and food safety through extreme weather events, changing pest and disease patterns, and water scarcity. GMOs engineered for drought tolerance or pest resistance can help address these challenges, but their safety must be ensured.

There is also consumer awareness and demand for transparency. Modern consumers are increasingly concerned about the safety and quality of their food. Addressing GMO safety and risks transparently is essential to meet consumer expectations and maintain trust in the food system.

Technological advancements and innovation are rather significant. Advances in genetic engineering, such as CRISPR and gene editing, are expanding the possibilities for GMOs. While these technologies offer exciting opportunities, they also raise new safety concerns that must be addressed to ensure food safety.

Ensuring the safety of GMOs through rigorous scientific evaluation, robust regulatory frameworks, and transparent communication is essential to harness their potential benefits while minimizing risks. By addressing these issues, we can create a sustainable and safe food system that meets the needs of present and future generations. The following legitimate questions arise.

1. Are our food products safe?

* Corresponding author: nadgmi@mail.ru

2. What measures and standards should be used, developed and adopted to ensure food safety?

3. How to maintain the competitiveness of the domestic agricultural producer and high food standards?

2 Conditions, materials, and methods

Taking into account global challenges such as climate change, population growth and the need to improve food security, the use of GMOs is becoming increasingly relevant [1]. It is important to understand the risks and safety standards in order to ensure sustainable agricultural development and environmental protection.

The article uses a comparative analysis of approaches to GMO regulations in different countries. Methods of systematic literature review are used, involving data analysis on the effects of GMOs on human health and the environment.

3 Results and Discussion

GMOs are organisms whose genetic material has been modified using genetic engineering techniques. In agriculture, GMOs are mainly used to create crops that are pest-resistant, herbicide-tolerant or have improved nutritional properties. Examples include Bt corn, which produces a toxin harmful to some insects, and golden rice, fortified with vitamin A. However, the use of GMOs is not without controversy, as concerns persist about their potential risks to health, environment, and socio-economic structures.

The use of genetically modified organisms (GMOs) in agriculture has caused serious controversy due to the potential risks they pose to the environment, human health, and socio-economic structures [2–4]. Although GMOs have benefits such as increased crop yields and pest resistance, their risks need to be carefully assessed and managed [5–8]. The following is a detailed description of the risks associated with the use of GMOs in agriculture.

The first is environmental risks, loss of biodiversity, and genetic flow. GMO crops can interbreed with wild relatives or non-GMO crops, resulting in the spread of modified genes. This can reduce genetic diversity and potentially lead to the emergence of "superweeds" or "superbugs" resistant to herbicides or pesticides [9]. In such situation, monoculture can take place. A widespread distribution of several varieties of GMO crops can lead to the emergence of agricultural monocultures, which will reduce the diversity of plant species and make ecosystems more vulnerable to pests, diseases and climate change.

The second is impacts on non-target organisms and harm to beneficial insects. Some GMO crops, such as Bt crops, produce toxins to kill certain pests. However, these toxins can also harm beneficial insects such as bees and butterflies, which are crucial for pollination.

There is also soil health as prolonged use of GMO crops and related herbicides can alter the soil microbiota, potentially reducing soil fertility and destroying ecosystems.

The third is the development of sustainability and pest resistance. Over time, pests can develop resistance to the toxins produced by GMO crops, making the technology ineffective and requiring the use of stronger chemicals.

Herbicide resistance should also be mentioned. The widespread use of GMO herbicide-resistant crops (such as Roundup Ready crops) has led to the emergence of herbicide-resistant weeds, making weed control more difficult and increasing the use of chemicals.

The forth is health risks, allergenicity, and new allergens. Genetic modification can introduce new proteins into crops that can act as allergens. For example, a Brazil nut gene introduced into soybeans has been found to cause allergic reactions in people who are sensitive to nuts.

There are also undesirable effects when the process of genetic engineering can lead to unintended changes in plant biochemistry, potentially leading to the formation of harmful compounds.

It is necessary to note toxicity and markers of antibiotic resistance. Some GMOs are created using antibiotic resistance genes as markers. There is concern that these genes may be transmitted to human pathogens, exacerbating the growing problem of antibiotic resistance.

The long-term health effects of GMO consumption have not been fully understood, and some studies have raised concerns about a possible link to organ damage, immune system problems, and other health problems [10, 11].

There are dietary changes and unintended changes. Genetic modification can inadvertently alter the nutritional profile of crops, potentially reducing their nutritional value or introducing harmful substances.

The fifth is socio-economic risks and monopoly in the market, involving corporate control. The GMO seed market is dominated by several large biotech companies. This raises concerns about monopolistic practices and reduced competition [12], which can lead to higher seed prices and limited choice for farmers.

Patents and intellectual property must be mentioned. GMO seeds are often patented, which prevents farmers from preserving and replanting seeds. This may increase dependence on seed companies and limit farmers' self-reliance.

There are also impacts on small farmers and economic pressures. Small farmers may find it difficult to pay the high costs of GMO seeds and related chemicals, leading to increased debt and financial instability.

Loss of traditional knowledge implies the fact that reliance on GMO technologies can undermine traditional farming practices and knowledge, reducing agricultural diversity and sustainability.

Trade and market access concerns export restrictions. Countries with strict GMO regulations, such as EU countries, may refuse to import GMO crops, limiting market access for GMO-producing countries.

Consumer preferences can be a reason as the growing consumer demand for non-GMO and organic products may create market problems for GMO producers and exporters.

The sixth is ethical and cultural issues and ethical considerations, playing God. Some critics argue that genetic engineering involves the manipulation of life in ethically questionable ways, which raises concerns about the moral consequences of changing natural organisms [13, 14]. Informed consent implies that consumers may feel that they have not been sufficiently informed about the presence of GMOs in their food, leading to calls for stricter labeling laws.

Cultural impact consists in the fact that there is traditional agriculture. The introduction of GMOs can disrupt traditional farming practices and cultural heritage, especially in developing countries where agriculture is closely intertwined with local customs and identity.

Food sovereignty means that the dominance of several multinational corporations in the field of GMO technologies can undermine local control over food production and agricultural systems.

The next issue is regulatory and scientific issues, inadequate risk assessment, involving short-term focus. Many risk assessments focus on short-term impacts, potentially overlooking long-term environmental and health impacts.

Limited independent research opportunities are understood as follows. Independent research on GMOs is often limited due to restrictive intellectual property laws and corporate control of seed samples. There are gaps in legislation, nonconforming standards. Regulatory standards for GMOs vary greatly from country to country, leading to inconsistencies in safety assessments and approvals.

Enforcement challenges imply that compliance with GMO legislation can be challenging, especially in regions with limited resources for monitoring and enforcement [13, 14].

The architecture of food safety regulation and legal support at the international level is as follows. This is "Vertical" regulation, which is a system of measures focused on products and goods of natural origin. The goods produced from these products form the basis of the human diet and, as a rule, cannot be patented. Therefore, the Directives contain descriptions and identification features that make it possible to distinguish these products from those created artificially (USA and Canada). "Horizontal" regulation is a system of measures (mainly European legislation) that contains requirements for product information, requirements for food production, sanitary requirements, requirements for food additives (dyes, sweeteners, contaminants) and residual amounts of pesticides [15, 16].

Regulation of genetically modified organisms (GMOs) varies significantly in the United States, the European Union (EU), and in Russia. Each region has developed its own regulatory framework based on scientific, economic, and socio-political considerations [17, 18]. Below is a detailed description of how GMOs are regulated in these three regions.

The first region is the GMO regulation in the USA. The regulatory framework in the USA has the coordinated GMO regulatory system involving three major federal agencies [19–23]:

- The United States Department of Agriculture (USDA) that monitors the environmental safety of GMOs, especially their impact on agriculture and ecosystems.
- Food and Drug Administration (FDA) that ensures the safety of products and feeds containing GMOs for human and animal consumption.
- The Environmental Protection Agency (EPA) that regulates the use of GMOs that contain pesticides or have potential environmental impacts.

Key principles are significant equivalence, when GMOs are considered safe unless proven otherwise. They are treated in the same way as their non-GMO counterparts, unless significant differences are identified. The second is product-based regulation, when the focus is on the final product rather than the process used to create it. The third is voluntary Consultation, when GMO developers voluntarily consult with the FDA to ensure safety, although this is not legally required.

The U.S. Department of Agriculture Approval conducts the approval process. The U.S. Department of Agriculture evaluates whether GMOs pose a danger to other plants or agriculture. If approved, the GMO is removed from control and can be cultivated.

FDA Consultation involves the FDA review data on allergenicity, toxicity, and nutritional value. Although it is not mandatory, most companies go through this process to gain the trust of the public. Environmental Protection Agency Assessment is engaged in assessing the amount of pesticides that GMOs produce (for example, Bt crops), the Environmental Protection Agency evaluates their environmental impact and sets recommendations for use. Public opinion and labeling in the United States are generally supportive of GMOs, although there is a vocal minority in favor of stricter regulation.

In 2016, the National Bioengineered Food Disclosure Standard was adopted, requiring mandatory labeling of GMO products by 2022. However, the law allows the use of QR codes or digital information, which are less transparent according to critics.

The case study is Bt Corn, which is widely cultivated in the United States. Bt corn has been approved after extensive testing by the U.S. Department of Agriculture, the Food and Drug Administration (FDA), and the Environmental Protection Agency (EPA). This has significantly reduced the need for chemical pesticides. The regulation of GMOs in the EU has its own regulatory framework. The EU has one of the strictest GMO regulatory systems based on the precautionary principle [24–27]. Key regulatory authorities include the European Food Safety Authority (EFSA) who conducts a scientific risk assessment, European Commission who makes final decisions on GMO approval based on EFSA recommendations and submissions from member States. EU member States may ban or restrict the cultivation of GMOs even after approval at the EU level.

Key principles are a precautionary principle, which states that GMOs are not approved unless proven to be safe beyond any reasonable doubt. The process-based regulation is focused on the genetic engineering process itself, not just the final product. Mandatory labeling concerns all products and feeds containing GMOs that must be labeled regardless of the presence of GMO material in them.

The approval process involves application submission, when the company submits an application to an EU Member State, which forwards it to EFSA. EFSA evaluates GMOs for environmental, health and safety risks. Consultations with Member States consist in bringing EFSA's findings to the attention of Member States that may have objections. The European Commission makes a final decision, which can be overturned by a majority vote of the Member States. Public perception and labeling are relevant as public opinion in the EU is extremely skeptical about GMOs, fearing long-term effects on health and the environment. Strict labeling laws ensure transparency, and many retailers voluntarily label products as "non-GMO".

The next case study is MON810 corn, which is the only GMO crop that is widely cultivated in the EU, mainly in Spain and Portugal. Several countries, including France and Germany, have banned its cultivation, despite EU approval.

The regulation of GMOs in Russia has its own regulatory framework. Russia adheres to a restrictive approach to GMOs, paying special attention to biosafety and food sovereignty [28–33]. Key regulatory authorities include the Ministry of Agriculture who oversees agricultural policy and regulation in the field of GMOs, Rosselkhoznadzor (Federal Service for Veterinary and Phytosanitary Surveillance) who monitors the safety and quality of agricultural products, including GMOs.

The State Duma has passed a law on GMOs whose basic principle is biosafety. GMOs are considered a potential threat to human health and the environment unless proven otherwise. The next principle is food sovereignty, when Russia prefers self-sufficiency and traditional farming methods to the introduction of GMOs. The restrictive legislation notes that the 2016 law prohibits the cultivation and breeding of GMO plants and animals, with the exception of scientific research. The approval process involves the scientific research stating that GMOs can only be used for controlled research purposes. Risk assessment implies that any GMOs intended for scientific research must undergo a thorough safety assessment.

Commercial cultivation of GMOs is prohibited, and the import of GMO products is strictly limited. Public perception and labeling state that Russia is strongly opposed to GMOs, favoring organic and traditional foods. Labeling laws require a clear definition of the GMO content in food, although the ban on cultivation limits the presence of GMOs on the market.

The next case study is restrictions on GMO imports. Russia has banned the import of a number of GMO products, including soybeans and corn, from countries such as the United States. This policy is consistent with Russia's goal of developing domestic agriculture and reducing dependence on foreign technology, which is summarized in table 1.

Table 1. Comparative table of the GMO regulation in the USA, EU and Russia

Aspect	USA	EU	Russia
Regulatory Approach	Permissive, product-based	Precautionary, process-based	Restrictive, biosafety-focused
Key Principle	Substantial equivalence	Precautionary principle	Food sovereignty
Approval Process	Coordinated framework (USDA, FDA, EPA)	EFSA assessment + member state input	Ban on cultivation, research only
Labeling	Mandatory (with QR codes allowed)	Mandatory (strict)	Mandatory (limited due to ban)
Public Perception	Generally favorable	Highly skeptical	Strongly opposed
Economic Impact	Promotes biotech industry	Limits GMO adoption	Protects domestic agriculture

4 Conclusions

The use of GMOs in agriculture involves a complex set of risks that must be carefully managed to ensure the safety of human health, the environment and socio-economic systems. Although GMOs have significant potential benefits, the risks associated with them underscore the need for a thorough scientific assessment, a transparent regulatory framework, and an open dialogue with the public. A balance between innovation and precautionary measures will be crucial to harness the potential of GMOs while minimizing the risks associated with them.

The regulation of GMOs in the USA, EU and Russia reflects their unique priorities and values. The United States focuses on innovation and economic growth, the EU prioritizes precautionary measures and public trust, and Russia pays special attention to biosafety and food sovereignty. These differences highlight the difficulties of reaching a global consensus on GMO regulation and the need for international cooperation to address the risks and benefits associated with the use of GMOs in agriculture.

The topic of GMO safety and risks is intrinsically linked to food safety, as the introduction of genetically modified organisms into the food supply chain has significant implications for human health, environmental sustainability, and socio-economic systems. In modern conditions, where global food demand is rising, climate change is impacting agriculture, and consumers are increasingly concerned about the quality and safety of their food, understanding this connection is critical.

References

1. D.A. Khasanova, S.J. Teshaev, The effect of genetically modified foods on the human body (literature review), Electr. Sci. J. Biol. and Integr. Med., **45** (2020)

2. J.L. Domingo, J. Jean Bordonaba, A review of the literature on safety assessment of genetically modified plants, Envir. Int. J., **37(4)**, 734–742 (2011), https://doi.org/10.1016/j.envint.2011.01.003, Identification number 21296423
3. Sh. Krimsky, Illusory consensus in assessing the impact of GMOs on health, Sci., technol. and hum. Values J., **40(6)**, 883–914 (2015), https://doi.org/10.1177/0162243915598381
4. A.Yu. Panchin, A.I. Tuzhikov, Published studies of GMOs did not reveal signs of harm when correcting for multiple comparisons, Crit. Rev. in Biotechnol. J., **37(2)**, 213–217 (2016), https://doi.org/10.3109/07388551.2015.1130684, Identification number 26767435
5. Yu.T. Yang, B. Chen, Regulation of the use of GMOs in the USA: science, law and public health, J. of the Sci. of Food and Agricult. [English], **96(6)** (2016), https://doi.org/10.1002/jsfa.7523, Identification number 26536836
6. O. Turhan, Liberal International Order and Measures of Economic Coercion in modern US foreign policy, World Polit., **3**, 52–62 (2021), https://doi.org/10.25136/2409-8671.2021.3.35105, https://nbpublish.com/library_read_article.php?id=35105
7. M.N. Dudin, Transgenic organisms (GMOs) in agriculture: an objective need to ensure global food security or a way to increase profits for TNK agribusiness? Food pol. and Secur., **7(2)**, 107–120 (2020), https://doi.org/10.18334/ppib.7.2.100666
8. S.I. Kuzina, I.G. Sagiryan, Threats to the food security of the Russian state in modern conditions, Legal order and legal values, **2(1)**, 54–63 (2024), eISSN 2949-1843, https://doi.org/10.23947/2949-1843-2024-2-1-54-63
9. V. I. Golik, T. A. Panfilova, The Mechanochemical Activation of Leaching Processes in a Disintegrator. Mining Informational and Analytical Bulletin. **11-1,** 175-189 (2023). DOI: 10.25018/0236_1493_2023_111_0_175
10. F.T. Nezhmetdinova, M.E. Guryleva, N.L. Blatt, New role of bioethics in emergency situations on the example of COVID-19, BioNanoSci., **12(2)** (2022)
11. N.H. Amirov, V.Yu. Albitsky, F.T. Nezhmetdinova, From medical ethics to medical bioethics, Probl. of soc. hygiene, public health and the hist. of med., **2**, 40–42 (1999)
12. S.Y. Mirolyubova, Development of the market of genetically engineered organisms (GMOs) in Russia and China: biosafety and commercialization, Intellect. Prop. Law, **3**, 20–25 (2022), https://doi.org/10.18572/2072-4322-2022-3-20-25, https://elibrary.ru/item.asp?id=49506849
13. S.A. Vorontsov, L.Y. Nikolaeva, Moral and ethical problems of biotechnology development, Bull. of Youth Sci., **5(12)**, 3 (2017)
14. O.V. Popova, Ethical approaches to the regulatory regulation of the use of genetically modified organisms, Eur. Exper. and Int. law, **1**, 28–36 (2019)
15. N.A. Chuiko, Basic approaches to regulating the circulation of genetically modified organisms in international practice, Sib. Law Bull., **1(52)**, 160–165 (2011)
16. N.V. Simonova, Legal regulation of the use of genetically modified organisms: international and domestic levels, Agr. and land law, **2(158)**, 133–143 (2018)
17. A.P. Getman, V.I. Lozo, Legal regulation of the development of biotechnology and the use of genetically modified organisms (GMOs) in the European Union, Probl. of legal., **117**, 180–194 (2011)
18. E.A. Lebedeva, Control and supervisory activities in the field of creation and use of genetically modified organisms in Russia and Germany, Bull. of Tomsk State Univ., **457**, 249–255 (2020)
19. A.P. Ivanova, Legal regulation of genetically modified crops in the European Union and other countries (USA, Canada, Japan, China): basic approaches and rules, Soc. sci. and hum. Rus. and foreign liter. Ser. 4: State and Law, **4**, 96–108 (2022)
20. A.V. Deikin, V.N. Kuznetsova, Yu.K. Kirikovich, D.V. Kovalenko, V.O. Soldatov, GMO regulation: Belarus, BRICS, EU, USA, Azimut sci. res.: econ. and manag., **8(4(29))**, 141–145 (2019)
21. A.P. Anisimov, O.V. Popova, legal regulation of the use of GMO technologies in agriculture in the USA, China and the EU: a comparative legal aspect, Parad. of manag., econ. and law, **2(4)**, 165–172 (2021)
22. A.A. Danelyan, Legal aspects of ensuring genetic safety within the framework of the biosafety of states, Electr. online publ. Int. Legal Cour., **7**, 29–33 (2022)
23. G.R. Nurimanshina, Yu.A. Kuzlyakina, M.V. Kochneva, M.Yu. Minaev, O.A. Kuznetsova, Peculiarities of legislation on GMO products in Russia and abroad, All about meat, **4**, 8–13 (2022)
24. A.P. Ivanova, Legal regulation of genetically modified crops in the European Union and other countries (USA, Canada, Japan, China): basic approaches and rules, Bull. of the RUDN Univ., **25(1)**, 32–66 (2021)
25. A.A. Bikbulatova, E.M. Bronnikova, E.D. Goryacheva et al., *Regulatory and legal regulation of genetically modified organisms (GMOs) in Russia and abroad* (Analytical Review, Moscow, 2023)
26. R.G. Novikova, Legal regulation of the turnover of genetically modified organisms (GMOs) in Russia and abroad, Bull. of the Peoples' Friend. Univ. of Rus. Ser.: Legal Sci., **25(1)**, 32–66 (2021)
27. M.Yu. Zarubin, N.A. Bulgakova, legal regulation in the sphere of turnover of genetically modified

organisms (GMOs) in Russia and abroad, Ed. and Law, **4**, 276–280 (2022)

28. *On Amendments to Certain Legislative Acts of the Russian Federation regarding the Improvement of State Regulation in the field of genetic engineering*, Fed. Law, no. 358-FZ of 07/03/2016, https://www.consultant.ru/cons/cgi/online.cgi?req=doc&base=LAW&n=200732&fld=134&dst=1000000001,0&rnd=0.908635332999332#x7hzMbUGaC6VR2Hx (accessed: 09/29/2024)
29. *On State regulation in the field of genetic engineering (with amendments and additions)*, Fed. Law, no. 86-FZ dated 07/05/1996: as amended on 07/02/2021, dated 12/29/2022, https://base.garant.ru/10135402/?ysclid=ln4fvaar84322078292 (accessed: 09/29/2024)
30. On the National Security Strategy of the Russian Federation: Decree of the President of the Russian Federation, no. 400 dated 07/02/2021, Coll. of Legisl. of the Rus. Fed., **27(part II)**, 5351 (2021)
31. *On products and genetically engineered organisms that are not subject to state registration in accordance with the Rules of State Registration of Genetically Engineered Organisms Intended for Release into the Environment, as well as products obtained using such organisms or containing such organisms, including specified products imported into the territory of the Russian Federation* (ed. Resolutions of the Government of the Russian Federation dated 12/29/2020, No. 2338, dated 12/25/2021, no. 2474, dated 12/6/2022, no. 2235): Resolution of the Government of the Russian Federation dated 04/16/2020, no. 520: approved by Resolution of the Government of the Russian Federation dated 09/23/2013, no. 839, http://government.ru/docs/all/127475 / (date of access: 09/29/2023)
32. *On the procedure for State registration of genetically engineered organisms Intended for Release into the Environment, as well as Products Obtained using such Organisms or containing such organisms, including Specified Products Imported into the Territory of the Russian Federation, and the invalidation of Certain Acts and Individual Provisions of Certain acts of the Government of the Russian Federation*: decree of the Government of the Russian Federation dated 01/18/2023, no. 35 (together with the "Rules for State Registration of genetically Engineered organisms, intended for release into the environment, as well as products obtained with the use of such organisms or containing such organisms, including the specified products imported into the territory of the Russian Federation"), http://publication.pravo.gov.ru/Document/View/0001202301200025?ysclid=ln4glnd7wj984702283 (accessed: 09/29/2023)
33. A.A. Muratov, N.V. Moskovenko, S.L. Tikhonov et al., Regulatory and legal aspects of regulation of genetically modified products in the territory of the Customs Union, Agri-food policy of Rus., **3(63)**,

BIO Web of Conferences **161**, 00031 (2025) https://doi.org/10.1051/bioconf/202516100031
FIES 2024

Forced forecast of agricultural development in the European North of Russia

Sergey V. Asaturyan [1*], *Konstantin P.* Dubina [1], *Vladimir I.* Khizhnyak [1], *Valery A.* Shchebeteev [2], *Vladislav V.* Bovsh [2]

[1] Azov-Black Sea Engineering Institute – branch of Don State Agrarian University in Zernograd, Zernograd, Russia
[2] Kuban State Agrarian University named after I. T. Trubilin, Krasnodar, Russia

Abstract. The aim of the work is to identify the key factors affecting the development of the agrarian sector of the region, determine the cyclical regularities of its development and create a model of forced forecasting based on advanced methodological approaches. The work analyses statistical data, scientific literature and expert assessments, which allowed us to identify specific features of agricultural development in the northern climate. The method of spectral analysis of time series was applied to identify cyclical fluctuations in the production of various types of agricultural products. The scenarios of agricultural development, including inertial, moderate-innovation and forced scenarios, were developed. It is shown that the forced scenario is the most promising for ensuring food security and increasing the competitiveness of the agricultural sector of the region. The results of the study can be used to develop strategies and programmes for agricultural development and to adjust the development plans of the agricultural sector.

1 Introduction

Insufficient food production leads to hunger, malnutrition, social tensions and even conflict, especially in vulnerable regions. In the context of globalisation and interdependence of countries, food supply problems in one region inevitably affect others, creating a chain reaction. This is especially true for regions with developed family farming, where environmental factors play a key role [1, 2].

The search for solutions to this problem is conducted in different directions, each of which has its own advantages and disadvantages. One of the common approaches is to increase production through agricultural intensification. This implies the use of high-yielding plant varieties, modern technologies of soil cultivation, application of fertilisers and plant protection products. On the one hand, this approach allows increasing food production in a short period of time and providing large volumes of products. On the other hand, intensive agriculture can lead to soil degradation, pollution, reduced biodiversity and dependence on expensive technologies and resources. Climate change also has a significant impact on the dynamics of agricultural production, especially in arid regions [3]. The need for adaptation of agricultural systems to changing climatic conditions is becoming increasingly urgent worldwide [4, 5]. Another way to solve the food problem is the development of organic agriculture aimed at the production of environmentally friendly food products without the use of chemical fertilisers and pesticides. This approach allows improving the quality of products, preserving the health of people and ecosystems. However, organic agriculture is generally characterised by lower yields and higher production costs, which makes it less competitive compared to intensive methods. Optimising water management is also a critical factor, especially in arid regions prone to water stress [6, 7]. An alternative solution is the introduction of new agro-technologies such as precision farming, the use of GIS technologies, drones, and the creation of crop and animal monitoring systems. Such technologies make it possible to optimise the use of resources, reduce production costs, increase yields and product quality, and minimise the negative impact on the environment. However, the introduction of modern agro-technologies requires significant investments, qualified personnel and relevant knowledge, which creates barriers to their dissemination.

In the context of the above-mentioned problems, the study of the prospects of agricultural development in regions with special natural and climatic conditions is of particular relevance. The European North of Russia belongs to such regions. Under the conditions of extreme bioclimatic factors, the risk of crop failure and high production costs, ensuring food security in these regions is a difficult task. In this regard, the search for effective and sustainable models of agricultural development, adapted to the specific conditions of the European North, is extremely important and relevant. At the same time, of

* Corresponding author: asaturyan-serg@mail.ru

particular importance is the development of scientifically grounded forecasts of the agrarian sector development, which can take into account cyclical regularities and specifics of the region's development. It is especially important to choose the most appropriate forecasting method, which will allow to create a long-term perspective of agricultural development and at the same time minimise possible risks. It is also important to take into account the impact of changes in land use on soil and runoff conditions [8, 9].

The purpose of this study is to substantiate the forced forecast of agricultural development in the region, taking into account the legal aspects of land use [10], to study the factors affecting its dynamics, to identify cyclical patterns, and to develop a model of forced forecast based on the use of advanced scientific and methodological approaches. The results of the study will allow us to form an optimal strategy for the development of agriculture in the European North, aimed at ensuring food security and increasing the competitiveness of the agricultural sector.

2 Method and materials

In the framework of this study, a comprehensive analysis was carried out to substantiate the forced forecast of agricultural development in the regions of the European North of Russia. The work included several interrelated stages, starting with the analysis of the current situation and ending with the development of methodological approaches to forecasting. The first stage involved the analysis and assessment of trends in the development of agriculture in the region for the previous period, including the study of statistical data and scientific publications on the topic. In the process of analysis, the main factors affecting the development of the agrarian sector were identified, as well as the peculiarities of the periods of fluctuations and their intensity in the studied time series. The impact of biodiversity on the agrarian landscape is an important factor to consider [11]. At the next stage, the forecasting of the main indicators of agricultural development was carried out using the methods of statistical analysis. The STATISTICA programme was used to identify cyclical patterns of agricultural development and to conduct spectral analysis of time series. This programme made it possible to determine the periods of fluctuations and their intensity in the studied time series. In order to assess the probability of external influences and the degree of uncertainty of the predicted processes, the method of cognitive analysis was used. At the final stage, on the basis of the conducted analysis, scenarios of alternative solutions were developed, taking into account the practical use of forecasting results. A methodological approach to adjusting the initial forecast in the process of updating the information used was also developed. This study did not use special agricultural equipment such as tractors, combines or laboratory analysers. Work with the data included their preliminary processing and analysis to identify cyclical patterns in the development of agriculture in the region. In the process, appropriate programme settings were set depending on the objectives, e.g. for spectral analysis, a method was chosen to identify periods of oscillation in the time series under study. In addition, the study used the method of cognitive analysis, which is based on expert assessment and analysis of information to determine the influencing factors and the degree of their uncertainty. Within the framework of sustainable agriculture, the utilisation of organic matter in the soil is fundamentally important [12].

3 Results and discussion

The research was focused on the development and justification of the forced forecast of agricultural development in the European North of Russia. The research methodology was based on a comprehensive analysis, including several interrelated stages. At the initial stage, a detailed analysis of the dynamics of the agrarian sector development in the region for the previous period covering the last few decades was carried out. This made it possible to identify the key trends, factors and regularities that determine the development of agricultural production in the conditions of the northern territories. In the course of the analysis, the statistical data reflecting the volume of production of various types of agricultural products, the dynamics of crop yields, livestock and poultry population, as well as the level of food supply in the region were considered. In addition, the scientific literature on the problems of agricultural development in the northern regions was analysed, which made it possible to identify the main challenges and opportunities for the development of the agricultural sector. This analysis showed that the agricultural sector of the European North faces a number of specific problems, such as extreme climatic conditions, short growing season, high production costs, as well as the shortage of qualified personnel. When analysing the environmental aspects of sustainable development, it is necessary to consider the possibilities of using production waste [13, 14]. It is also important to consider the issues of waste recycling and reuse [5]. In this context, investigating the possibilities of using ash as a mine backfill material is of particular interest [15].

The second stage of the study was the application of statistical analysis methods for forecasting the main indicators of agricultural development in the region. To identify cyclical patterns in the development of the agricultural sector, the method of spectral analysis of time series was used. This analysis made it possible to establish the presence of periods of fluctuations in the dynamics of agricultural production development and assess their amplitude and frequency. In particular, it was found that the dynamics of grain crops production in the region has cyclical fluctuations with periods of about 7-10 years, which is probably associated with climatic conditions and economic cycles. To improve the efficiency of forecasting the technical condition of agricultural equipment, it is possible to use forecasting methods used in other areas of industry [16].

Regression analysis was used to forecast the main indicators of agricultural development, such as gross grain harvest, milk, meat and vegetable production, which made it possible to establish statistically significant dependencies between various factors and development indicators. In addition, the method of cognitive analysis based on expert judgements was applied to assess the probability and degree of uncertainty of external influences (Fig. 1). The forecasting method presented in Figure 1 reflects a comprehensive view of the region's agriculture as a complex system influenced by a multitude of internal and external factors. An important factor for the region is the hydrological situation [17] and its monitoring [18]. This method is based on the understanding that the development of agriculture is not an isolated process, but is closely related to other spheres of the economy and social life. It is also worth paying attention to the economic aspects of applying green technologies. The central element of the presented model is the regional system, which is a set of interrelated elements that ensure the functioning and development of agriculture in a particular region. This system is influenced both by the internal market, i.e. the needs and opportunities of the population of the region, and by the external market, which determines the demand for agricultural products and opportunities for their realisation. The influence of internal and external markets on the regional system is manifested through needs and opportunities, which, in turn, create incentives for production and implementation This subsystem is characterised by its goals and development scenarios, and is also influenced by innovation processes, scientific and technological developments and intellectual potential of the region. The second subsystem is the human subsystem, which includes all issues related to human capital in the agricultural sector, such as the level of education, qualifications, labour resources and motivation of employees. This subsystem also has its own goals and development scenarios, which are influenced by the process of training and professional development of personnel.

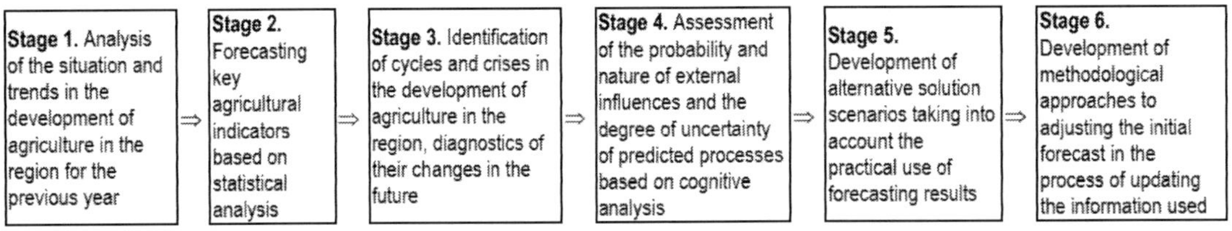

Figure 1. Algorithm for forecasting the development of agriculture in the region.

In addition to internal factors, the development of the regional agricultural system is also influenced by public policy, which determines the conditions and rules of functioning of the agricultural sector, as well as the amount of state support and regulation. The state policy, in turn, is formed under the influence of internal and external factors, including both economic, social and political aspects. An important role in the presented model is played by financial resources, which ensure the functioning of production and human subsystems. Financing can come from a variety of sources, including both the enterprises' own funds and state support and private investment. It is financial flows that ensure the continuity of production and development processes in the agricultural sector. This whole model is built on the principle of feedback, where all elements are interrelated and influence each other, creating a dynamic system that constantly adapts to changes in the external and internal environment. This understanding of agricultural production allows to determine more precisely the directions of development of the agricultural sector and anticipate the possible consequences of the decisions taken.

In the course of the study, a number of forecast scenarios of agricultural development in the region were developed, including inertial, moderate-innovation and forced scenarios. The inertial scenario assumed the preservation of current trends and development rates, the moderate-innovation scenario - the introduction of individual innovations, and the forced scenario - radical modernisation of the agricultural sector based on the intensive introduction of advanced technologies and management methods. The results show that the inertial scenario will not ensure sustainable development of agriculture in the region and may lead to a decrease in the competitiveness of the agricultural sector and its dependence on food imports. The moderate-innovative scenario can provide some increase in production efficiency, but will not allow achieving a significant breakthrough in the development of the agricultural sector. The forced scenario, on the contrary, can provide a significant increase in production volumes, product quality and competitiveness of the agricultural sector.

The study, using spectral analysis, revealed that potato production in the region is characterised by cyclical fluctuations with a period of about 5-7 years, while in the production of outdoor vegetables these cycles are 4-6 years. The data analysis also showed that cattle population has cyclical fluctuations with a period of 8-10 years and poultry population with a period of 3-5 years. This indicates that the development of livestock production in the region is subject to significant fluctuations. At the same time, the obtained data showed that the amplitude of fluctuations in milk and meat production has been slightly decreasing in recent years, which may be associated with the improvement of production technologies and state support of the agricultural sector.

Statistics for the last 50 years were used in modelling and forecasting. The analysis showed that in the period from 1970 to 1990 there was a relatively steady growth in the production of most types of agricultural products. At the same time, from 1990 to 2000 there was a significant decline in production caused by the economic crisis and structural changes in agriculture. Since 2000, there has been a gradual recovery of the agricultural sector in the region, but production growth rates are still insufficient to ensure food security.

The analysis has shown that grain crop production in the region has limited potential due to unfavourable climatic conditions, and the main directions of agricultural development should be livestock breeding and the production of vegetables and potatoes. In particular, it was revealed that the average annual yield of grain crops in the region is about 15-20 quintals per hectare, while the average yield of potatoes is 150-200 quintals per hectare and of open ground vegetables - 200-250 quintals per hectare. At the same time, the productivity of dairy cattle in the region is about 5-6 thousand kilograms of milk per cow per year, and the average daily live weight gain of cattle is 600-800 grams. development of interregional cooperation and integration. The implementation of these measures will allow to increase the production of the main types of agricultural products in the region by 2030 by 20-30 per cent compared to the level of 2020, as well as to improve the quality of products, reduce their cost and increase the competitiveness of the agricultural sector.

Comparison of the results obtained with other similar studies conducted in other regions of Russia and foreign countries has shown that, in general, the dynamics of the agrarian sector development in regions with extreme climatic conditions has similar features, but there are also specific features related to local conditions and factors. For example, studies conducted in the regions of Siberia also show the presence of cyclical fluctuations in agricultural production, but the periods and amplitudes of these fluctuations may differ from the results obtained in our study. Studies conducted in the regions of Scandinavia also show that the use of modern technologies and management methods is a key factor for improving the efficiency of agriculture in the northern climate.

The results obtained are of practical importance for the development of strategies and programmes for agricultural development in the regions of the European North of Russia. The developed methodological approach to forecasting can be used to monitor and adjust plans for the development of the agricultural sector, as well as to assess the effectiveness of state support for agriculture. The results of the study can also be used for training and professional development of specialists in the agricultural sector, as well as for the promotion of advanced technologies and production methods among farmers and agricultural organisations.

4 Conclusion

The study allowed us to justify the need to apply the forced scenario for the development of agriculture in the European North of Russia. In the course of the work it was established that the current trends in the development of the agricultural sector, if they are maintained, will not allow to ensure food security of the region and increase the competitiveness of its agricultural products. The inertial scenario of development, based on the existing approaches, is not able to overcome the accumulated problems and challenges facing the agrarian sector of the region.

The introduction of a moderate-innovative scenario, which implies partial introduction of new technologies and management methods, will only slightly improve the situation. This approach will not be able to create conditions for a significant increase in production volumes and product quality. In this regard, the most promising is the forced scenario based on intensive introduction of advanced technologies, modernisation of production processes, improvement of personnel skills and optimisation of agrarian sector management. This scenario envisages not only technical re-equipment of production, but also the creation of favourable conditions for the development of entrepreneurship, attracting investment and increasing innovation activity in the agricultural sector.

The analysis revealed cyclicality in the development of agricultural production in the European North, which is important to take into account when planning and managing the agricultural sector. Determining the periods and amplitude of these fluctuations will allow more accurately forecast the dynamics of production of various types of agricultural products and develop measures to stabilise and support the agricultural sector. For the effective development of agriculture in the region, it is necessary to take into account the peculiarities of its natural and climatic conditions and focus on the most promising areas, such as livestock breeding and the production of vegetables and potatoes. It is important to increase crop yields and livestock productivity, as well as to minimise production costs.

References

1. A. Cénet, V. Viaud, L. Voisin. A Methodological framework to enhance potential spatial planning to support agroecological transition at the scale of local territories. Land, **13,** 1707 (2024). DOI: 10.3390/land13101707.

2. N.M. Kachurin, G.V. Stas, D.O. Prokhorov, O.A. Gavrina. Selection of technologies and directions for reducing the technogenic impact of mineral formations on the environment. Sustainable Development of Mountain Territories, **16,** 283–291 (2024). DOI: 10.21177/1998-4502-2024-16-1-283-291.

3. N. Garambois, U. Le Goff, L. Thibaudeau. Agrarian dynamics and climate change in the Senegalese

Sahelian peanut basin. Agrar. Syst. Clim. Change: Journeys Adapt. Glob. South, **15**-37 (2024). DOI: 10.1079/9781800628137.0001.

4. H. Cochet, O. Ducourtieux, N. Garambois. Agrarian systems and climate change: Journeys of adaptation in the global south. Agrar. Syst. Clim. Change: Journeys Adapt. Glob. South, **1**-208 (2024). DOI: 10.1079/9781800628137.0000.

5. L.S. Malyukova, V.V. Kondratiev, V.V. Bukhtoyarov, V.Y. Konyukhov, K.A. Bashmur. Circular Mining Wastes Management for Sustainable Production of Camellia sinensis (L.) O. Kuntze. Sustainability (Switzerland), **15,** 11671 (2023). DOI: 10.3390/su151511671.

6. S. Mukherjee, S. Paul, S. Bhattacharya, A. Islam, S. Mahammad, E. Alam. Optimizing water resource management in tropical drought-prone regions through hybrid MCDM techniques: A water-stress mapping approach. J. Hydrol. Reg. Stud., **57,** 102171 (2025). DOI: 10.1016/j.ejrh.2024.102171.

7. A.A. Kapanski, A.E. Boltrushevich. Geospatial Clustering in Smart City Resource Management: An Initial Step in the Optimisation of Complex Technical Supply Systems. Smart Cities, 8, 14 (2025). https://doi.org/10.3390/smartcities8010014.

8. J. Rodzik, R. Kołodyńska-Gawrysiak, Ł. Franczak, W. Zgłobicki, J. Poesen. IMPACT OF LAND USE CHANGES ON COLLAPSED PIPES DEVELOPMENT IN THE LOESS GULLY (LUBLIN UPLAND, EAST POLAND). Quaest. Geogr., **43,** 17-33 (2024). DOI: 10.14746/quageo-2024-0042.

9. A.V. Shitikova, A.V. Konstantinovich, V.A. Orlov. Prediction of soil acidity in mountainous areas using machine learning. Sustainable Development of Mountain Territories, **16,** 1144–1154 (2024). DOI: https://doi.org/10.21177/1998-4502-2024-16-3-1144-1154.

10. S. Buletsa. The peculiarities of restitution of agricultural land in Ukraine. J. Agric. Environ. Law, 19, 27-74 (2024). DOI: 10.21029/JAEL.2024.37.27.

11. V. Langraf, K. Petrovičová. Morphometric variability of specimens in Ground Beetle (Carabidae) communities in agrarian landscape forests and protected areas. Rev. Soc. Entomol. Argent., **83,** 1-8 (2024). DOI: 10.25085/rsea.830401.

12. Q. Gao, L. Wang, Y. Fang, Y. Gao, L. Ma, X. Wang, Y. Li, X. Wu, Z. Du. Conservation agriculture boosts topsoil organic matter by restoring free lipids and lignin phenols biomarkers in distinct fractions. Soil Till. Res., **248,** 106463 (2025). DOI: 10.1016/j.still.2025.106463.

13. V. I. Golik, A. I. Karlina. Technology for nonwaste recovery of tailings of the mizur mining and processing plant. Metallurgist, **66,** 1476-1480 (2023). DOI: 10.1007/s11015-023-01462-y.

14. K.I. Kravtsov, Ya.A. Tynchenko. Influence of mountain factors on salt excess and soil toxicity in mountain conditions. Sustainable Development of Mountain Territories, **15,** 784-797 (2023). https://doi.org/10.21177/1998-4502-2023-15-3-784-797.

15. V.Yu. Konyukhov, T.A. Oparina, I.Y. Matasova, M.A. Modina, N.V. Martyushev. Ecologization of underground coal mining by means of ash use in backfill preparation. Min. Inform. Anal. Bull., **10,** 123-135 (2024). DOI: 10.25018/0236_1493_2024_10_0_123.

16. O. A. Kolenchukov, A. Y. Mikhaulov, K. A. Bashmur, V. S. Tynchenko, R. B. Sergienko, V. V. Kukartsev, & V. V. Tynchenko. Forecasting the technical condition of thermochemical reactor systems. SOCAR Proceedings, **2023,** 123–130 (2023). https://doi.org/10.5510/OGP20230100814.

17. S.I. Kondratev. Three typical hydrological-hydrochemical situations near the danube river mouth based on the marine hydrophysical institute research expeditions in 1997-2013. Physical Oceanography, **26,** 326–340 (2019). DOI: 10.22449/1573-160X-2019-4-326-340.

18. O. Afanaseva, M. Afanasyev, S. Neyrus, D. Pervukhin, D. Tukeev. Information and Analytical System Monitoring and Assessment of the Water Bodies State in the Mineral Resources Complex. Inventions, **9,** 115 (2024). https://doi.org/10.3390/inventions9060115.

Influence of irrigation and fertilisation on yield and quality of green mass of maize in soil-climatic conditions of the Republic of Tatarstan

F.N. Safiollin* [1], *A.A.* Lukmanov[2], *M.M.* Khismatullin[3], *S.R.* Suleimanov[1], *I.F.* Yakhin[1]

[1] Kazan State Agrarian University, Russia
[2] Tatarsky Agrochemical Service Centre, Russia
[3] Department of Land Reclamation and Agricultural Water Supply in the Republic of Tatarstan, Russia

Abstract. The present work deals with the issues of comparative evaluation of cultivation of hybrid maize Ross 140 CB on rainfed and irrigated plots with the application of calculated rates of mineral fertilisers for the planned yield of green mass 30, 40, 50 t/ha with cobs in milk-wax ripeness. The results of 3 years of research (2022-2024) established that irrigation and fertilisation enhances the effect of each other, providing 50.7 t/ha of green mass with a yield increase of 23.4 t/ha, which is 114% higher than the control variant of the experiment (without irrigation and fertilisation). Payback of one kg of mineral fertilisers was 13.8 kg of grain units, against 6-8 kg on average in the Republic of Tatarstan on spring wheat crops. The quality of green mass (in terms of crude protein content, crude fat, sum of sugars and especially the ratio of sum of sugars to digestible protein) corresponds to normative (zootechnical) indicators of cattle feeding

Keywords: maize, irrigation, green mass, cobs, yield, crude protein, crude fat, total sugars

1. Introduction

The Republic of Tatarstan is currently an undisputed leader among the regions of the Russian Federation in terms of land reclamation development. Thus, as a result of implementation of the republican land reclamation programme over the last 15 years, 465 hydraulic structures (dams) have been built and overhauled, 32 thousand hectares of irrigated land have been put into operation, more than 250 km of water supply network have been laid, and 300 water bodies in settlements have been improved. Today irrigated lands have become a truly golden fund of Tatarstan, as the yield of potatoes, perennial grasses, sugar beet on irrigation is 3-4 times higher compared to rainfed farming [1-3].

However, the feasibility of maize cultivation on irrigation is questioned, as many specialists and farm managers consider it a drought-resistant crop

In this regard, additional study of the reaction of the main silage crop of Tatarstan - maize to irrigation depending on the application of calculated rates of mineral fertilizers for planned yields of 30, 40, 50 t/ha of green mass with cobs in milk-wax ripeness is an urgent problem of agro-industrial complex, both in the Russian Federation as a whole and in its individual regions, including the Republic of Tatarstan, which became the purpose of our research [4-6].

2. Methods and materials

Stationary field experiment in 2022 was conducted on irrigated fields of land use of LLC 'Kyrlai' of Arsky municipal district, and in 2023-2024, after commissioning of the irrigated area, on the basis of Agrobiotechnopark of Kazan State Agrarian University with coordinates: latitude - 55.5244865824 and longitude - 48.274901646. Laboratory analyses were performed in the 'Centre of Agroecological Research' of Kazan State Agrarian University [7-9].

The object of research was an ancient (included in the register of approved hybrids in 2002), zoned in the seventh region (Middle Volga region) hybrid maize Ross 140 SV dual-purpose (for silage and grain) breeding of the Krasnodar Grain Centre named after P.P. Lukyanenko.

Agrochemical parameters of 2 typical grey-forest soils were as follows: humus content according to Tyurin was from 3.2 to 3.6%, mobile phosphorus - from 148 to 150 and potassium - from 162 to 165 mg/kg of soil according to Kirsanov. The reaction of the soil medium was close to neutral (pH 5.8-6.0). The scheme of the experiment is presented in the section 'Results and Discussion'.

The system of basic tillage and pre-sowing soil preparation consisted of autumn ploughing to a depth of 22 cm with deepening of the arable layer up to 24 cm, moisture closure in 2 traces, application of calculated

* Corresponding author: faik1948@mail.ru

rates of mineral fertilisers for the planned yield of green mass of maize 30, 40, 50 tonnes/ha with cobs in milk-wax ripeness, pre-sowing cultivation to a depth of 8-10 cm, sowing in the second decade of May by precision seeder Vesna-8 with a row spacing of 70 cm with seed placement to a depth of 6-8 cm and a seeding rate of 70 thousand seeds/ha with simultaneous ordering of the seedlings pcs/ha with simultaneous rolling of rows [10-12].

Repetition of the experiment was threefold, the arrangement of plots was systematic, the area of an individual plot was 84 m2 (2.8x30 m). Two-factor field experiment was conducted according to the methodology of V.R. Williams All-Russian Research Institute of Forages.

Agrometeorological conditions in the years of research differed significantly from the average annual indicators.

The vegetation period of 2022 was very favourable: in May, June and July precipitation was 52, 68, 77 mm, moisture availability was 10.4% higher than the average annual data. Under these conditions it was enough to carry out double irrigation at the rate of 300-350 m3/ha (DM 530/110).

The beginning of the growing season 2023 was rainy and cold (May, June), but agronomic precipitation was absent in the second half of the growing season (July, August). Three irrigations were carried out in July and August. The irrigation rate was 300 to 400 m^3/ha. At the same time, drought was accompanied by average daily air temperatures above +25°C. [13-15].

The growing season of 2024 was characterised by the return of frosts in the third decade of May. Many early maize crops were severely affected. A state of emergency was declared in some regions of the Russian Federation. The situation in 2024 was complicated by the fact that in the initial stage of maize organogenesis (June) precipitation fell only 27.7% compared to the average annual rates, which required two irrigations with a rate of 250-300 m3/ha. The same story was repeated in August: precipitation availability was 52.2%, and thermal resources availability was 118.6% in June and 103.2% in August. Taking into account the significant deepening of the active soil layer in August, the irrigation water consumption rate had to be increased to 350-400 m3/ha [16-18].

The irrigation rate was calculated according to the formula:

$$m = H \cdot \alpha(\gamma_{LC} - \gamma_{fact}) \cdot 100,$$

where m - irrigation rate (water consumption per irrigation), m^3/ha; H - moisture depth (active soil layer), m; α - soil density, g/cm^3; γ_{LC} - lowest soil moisture capacity, %; γ_{fact} - actual moisture capacity, %

3. Results and discussion

Dynamics of maize root system formation by development phases

Irrigation and fertilisation had a significant effect on the dynamics of maize root system formation (Table 1).

Table 1. Amount of irrigation and fertiliser on maize root system formation, cm (2022–2024)

Factor A (moisture conditions)	Factor B (estimated NPK rates for planned green mass yield)	Sowing - sprouting	Formation 8-10 leaves	Panicle sprouting	Panicle flowering	Milk-wax maturity
Without irrigation	Control	3.2	20.8	26.3	32.8	36.4
	At 30 t/ha ($N_{74}P_{40}K_{74}$)	3.8	23.4	29.4	34.3	38.5
	At 40 t/ha ($N_{98}P_{53}K_{97}$)	4.4	24.5	31.2	36.7	41.3
	At 50 t/ha ($N_{123}P_{66}K_{121}$)	4.8	26.6	32.7	39.3	44.7
Irrigated	Control	3.4	22.7	30.1	36.4	40.8
	At 30 t/ha ($N_{74}P_{40}K_{74}$)	4.1	25.4	33.5	38.6	46.4
	At 40 t/ha ($N_{98}P_{53}K_{97}$)	4.8	28.7	36.8	41.7	49.7
	At 50 t/ha ($N_{123}P_{66}K_{121}$)	5.3	30.1	39.9	43.8	52.6
HCP_{05}	A	0.21	1.20	1.65	1.81	1.96
	B	0.26	1.45	1.87	2.04	2.12
	AB	0.29	1.68	2.01	2.18	2.12

On average for 10 days in the control variants of the experiment the length of embryonic roots was only from 3.2 to 3.4 cm, and in the phase of formation of 8-10 leaves - from 20.8 to 22.7 cm. If in the initial stage of root system formation the influence of irrigation without mineral fertilisers was minimal (3.4-3.2=0.2 cm), in the phase of 8-10 leaves formation it increased to 1.9 cm (22.7-20.8=1.9 cm) against 3.5 cm (30.1-26.6=3.5 cm) in the last variant of the experiment.

The above-mentioned regularities of the root system formation of the studied crop were also characteristic of the subsequent phases of maize development: the fertiliser increase on rainfed variants in the phase of panicle emergence was 6.4 cm, and on irrigation - in the phase of panicle flowering, respectively, 6.5 and 7.4 cm.

As a result, by the end of the growing season the active soil layer (the soil layer where the main mass of maize roots is located) under the influence of mineral fertilisers on rainfed plots increases from 36.4 cm in the control to 44.7 cm in the experiment variant with the application of $N_{123}P_{66}K_{121}$, and on irrigation - from 40.8 to 52.6 cm. Such a significant deepening of the active soil layer plays a decisive role in the formation of highly productive maize agrocenoses, as nutrition elements become available from deeper soil layers compared to winter and spring grain crops. Intensive development of the root system undoubtedly had a great influence on the growth and development of the aboveground part of the studied crop.

Yield and payback of NPK fertilisers

Irrigation and fertilisation had a significant effect on the formation of a highly productive maize agrocenosis (Fig. 1).

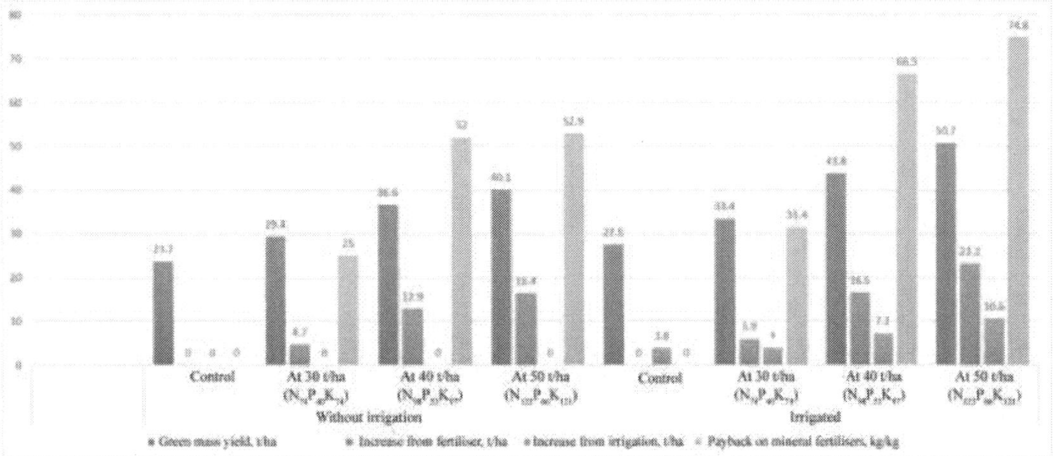

Figure 1. Yield of green mass of maize depending on irrigation and application of estimated rates of mineral fertilisers (2022-2024)

For example, on cultivated typical grey forest soils, which occupy 38% of the soil cover of the Republic of Tatarstan, under the modern culture of farming, agrarians can expect to obtain 23.7 t/ha of green mass of maize without mineral fertilisers and without irrigation. At such yields, the area under maize cultivation would have to be increased up to 300-350 thousand ha in order to prepare the required volumes of maize fodder. In this regard, the development and implementation of intensive maize cultivation technologies was and remains an urgent problem of modern fodder production. In this regard, there is no alternative to providing the object of our research with additional moisture and nutrition elements. So, on the average for 3 years on irrigation the introduction of calculated norms of mineral fertilisers on the planned yield of green mass 50 t/ha provides obtaining 50.7 t/ha of high quality biomass of this crop with yield increase of 23.2 t/ha of green mass in comparison with the control (without fertilisers), which is 114% more than the yield of the studied crop in absolute control (without fertilisers and without irrigation).

At the same time, one cannot deny the role of mineral fertilisers in rainfed maize crops, because on the background of $N_{123}P_{66}K_{121}$ application the yield of this crop exceeds the control (without fertilisers) by 69.2% (40.1:23.7x100=69.2%) with an increase in green mass of 16.9 t/ha.

One of the most important indicators of irrigation and fertilisation of hybrid maize is the payback of mineral fertilisers. Calculations showed that in rainfed maize crops, despite such a high increase in yield as the rate of application of mineral fertilisers increases, the increase in their payback slows down compared to the application of $N_{98}P_{53}K_{97}$. As a result, the difference in payback between the last two variants of the experiment without irrigation (calculated rates for 40 and 50 tonnes/ha of green mass) is only 0.9 kg of green mass per 1 kg d.w. (52.9-52.0=0.9 kg/kg).

A completely different picture is formed on crops of irrigated maize. Payback of mineral fertilisers increases in proportion to the increase in rates of application of nitrogen-phosphorus-potassium fertilisers from 31.4 to 74.8 kg of green mass per 1 kg.

Obtaining such high results from maize cultivation on irrigation is explained by the following reasons:

- - factor A (irrigation) enhances the effect of factor B (fertilisation), i.e. the law of interaction of environmental factors is manifested;
- - under irrigation, the active soil layer exceeds rainfed areas by more than 10 cm, which allows the use of additional nutrients from deep soil layers;
- - moisture availability in the years of research was significantly lower than the average annual values in combination with high average daily air temperatures.

However, it should be especially noted the expediency of cultivation of irrigated maize on low background of mineral nutrition ($N_{74}P_{40}K_{74}$ - planned yield of 30 t/ha of green mass), because the gain is only 21.5%. On rainfed plots, on the contrary, on a high background of mineral nutrition ($N_{123}P_{66}K_{121}$), the shortfall in yield of green mass reaches 9.9 t/ha (from the planned yield of 50 t/ha subtract the actual yield: 50-40.1=9.9 t/ha).

In conclusion, it should be noted that the coefficient of conversion of green mass of maize into grain units, established by the Ministry of Agriculture of the Russian Federation from 06.07.2017 is 0.17. Based on this payback kg of mineral fertilizers on rainfed crops of maize in the last variant of the experiment is 10.2 grain units, and on irrigation - 13.8. For comparison, we note that on average in the Republic of Tatarstan payback NPK on crops of winter and spring grain crops does not exceed 6-8 kg / kg.

Irrigation and fertilisation in soil-climatic conditions of the Republic of Tatarstan provides obtaining more than 50 t/ha of green mass of maize against 40.1 t/ha on rainfed plots. Payback of mineral fertilisers increases from 10.2 without irrigation to 13.8 grain units on irrigation.

The advantage of maize cultivation on irrigation with the application of calculated rates of mineral fertilisers is clearly confirmed by the stability of green mass yield over the years of research (Table 2).

Table 2. Maize green mass yield by years of research, tonnes/ha

Factor A (moisture conditions)	Factor B (calculated NPK rates)	2022	2023	2023
Without irrigation	Control	27.6	20.3	24.2
	At 30 t/ha ($N_{74}P_{40}K_{74}$)	30.5	24.9	29.8
	At 40 t/ha ($N_{98}P_{53}K_{97}$)	38.9	32.5	38.4
	At 50 t/ha ($N_{123}P_{66}K_{121}$)	42.6	36.2	41.5
Irrigated	Control	29.1	25.2	28.6
	At 30 t/ha ($N_{74}P_{40}K_{74}$)	36.3	30.3	33.6
	At 40 t/ha ($N_{98}P_{53}K_{97}$)	46.0	40.8	44.6
	At 50 t/ha ($N_{123}P_{66}K_{121}$)	52.8	47.6	51.7
HCP_{05}	A	0.29	0.26	0.27
	B	0.34	0.29	0.32
	AB	0.38	0.33	0.35

In the control variant of the experiment (without irrigation and without fertilisers), the yield of green mass of the object of our research was in the range from 20.3 t/ha in 2023 to 27.6 t/ha in 2022 with a difference of 7.3 t/ha (36%). On the background of mineral fertiliser application for the planned yield of 30 t/ha on rainfed plots, this gap narrowed to 5.6 t/ha of green mass of maize (22%).

Application of mineral fertilisers with the expectation of obtaining 40 and 50 t/ha of green mass of maize was the basis for the growth of yield of this crop in relatively wet in 2022 (GTC=1.25) up to 38.9 and 42.6 t/ha against 32.5 and 36.2 t/ha in dry 2023 (GTC=0.84). Consequently, in drought years, the rainfed maize biomass underharvest was 6.4 t/ha (38.9-32.5=6.4 t/ha) and (42.6-36.2=6.4 t/ha).

In the control variant of the experiment, irrigation of the studied crop in all years of research provided a reliable increase in green mass yield: 1.5 t/ha in 2022; 4.9 t/ha in the dry year 2023 and 4.4 t/ha in the medium-wet year 2024. This difference in favour of irrigation of the object of our research on the background of application of N123P66K121 (planned yield of 50 t/ha of green mass) increased to 10.2 in 2022 and 2024 and to 11.4 t/ha in dry year 2023. In other words, irrigation of the studied crop significantly reduces the dependence of its yield on weather and climatic conditions, which is clearly confirmed by calculations of yield stability index by years of study.

Thus, a detailed analysis of the results of accounting the yield of green mass of maize and payback of mineral fertilisers shows that on rainfed maize crops it is most profitable to apply calculated rates of nitrogen-phosphorus-potassium fertilisers for the planned yield of 40 t/ha, and on irrigation for 50 and more t/ha.

The quality of maize silage and its nutritive value depends to a greater extent on the content of crude protein - the basis of life on our planet (Table 3).

Table 3. Crude protein content and gross yields depending on fertilisation and irrigation of hybrid maize Ross 140

Factor A (moisture conditions)	Factor B (calculated NPK rates for planned green mass yield)	Crude protein content, %	Gross crude protein yield, кг/га	Increase	
				t/ha	%
Without irrigation	Control	10.8	818	-	-
	At 30 t/ha ($N_{74}P_{40}K_{74}$)	11.2	1040	222	27
	At 40 t/ha ($N_{98}P_{53}K_{97}$)	11.8	1243	425	52
	At 50 t/ha ($N_{123}P_{66}K_{121}$)	12.0	1423	605	74
Irrigated	Control	11.3	938	-	-
	At 30 t/ha ($N_{74}P_{40}K_{74}$)	12.0	1174	236	25
	At 40 t/ha ($N_{98}P_{53}K_{97}$)	12.8	1539	601	64
	At 50 t/ha ($N_{123}P_{66}K_{121}$)	12.2	1596	658	70

In rainfed maize crops, crude protein content increased in proportion to the rates of mineral fertiliser application from 10.8 in the control to 12.0% in dry weight in the $N_{123}P_{66}K_{121}$ variant. For this reason, the gross yield of crude protein was 818 and 1423 kg/ha, respectively, in the compared variants of the experiment. The addition of mineral fertilisers was quite high - from 222 to 605 kg/ha (27-74% to the control).

A completely different picture of crude protein concentration in dry mass of maize was formed on irrigation. If in the control (without fertiliser) and in the variant of NPK-fertiliser application for the planned yield of 30 t/ha of green mass the crude protein content was higher compared to the non-irrigated plot, then on the high nutrition background (50 t/ha) there were no significant differences: without irrigation - 12%, and on irrigation - 12.2%.

Slowdown of crude protein concentration rates as the rates of mineral fertiliser application on spring grain crops were also noted by I.M. Serzhanov (2017), F.Sh. Shaykhutdinov (2018), M.F. Amirov (2020), on crops of fodder mixtures R.M. Nizamov, F.N. Safiollin, S.R. Suleymanov (2018), on corn crops L.P. Zaripova (2010), N.I. Kashevarev (2004), V.M. Kosolapov (2009), A.P. Shindin (2009) and called this process 'dilution effect'.

In our studies, despite this, the gross yield of crude protein from 1 ha of irrigated maize in the analysed variant of the experiment exceeded rainfed maize by 173 kg/ha. Moreover, on irrigation the increase in gross yield of crude protein from application of the same norms of mineral fertilisers was ahead of rainfed plot by 173 kg/ha (1996-1423=173 kg/ha).

However, it should be especially noted the great importance of application of calculated rates of mineral fertilisers both on maize crops without irrigation and on irrigation, because the increment of crude protein in all variants of the experiment was significantly higher in comparison with the obtained increments from irrigation. For example, in the last variant of the experiment the application of mineral fertilisers with the calculation of obtaining 50 t/ha of green mass without irrigation provided additional 605 kg/ha of crude protein, and on irrigation 658 kg/ha.

As a result, the key role in gross harvest of crude protein both on rainfed and irrigated maize crops belongs to the calculated rates of mineral fertilisers and especially to the interaction of nitrogen-phosphorus-potassium fertilisers with irrigation of this crop during the growing season.

According to the content of crude fat in percentage terms in the dry mass of maize in all variants of the experiment corresponded to the normative indicators. At the same time, gross collection of crude fat from 1 ha of arable land between the variants of the experiment was from fertilisers in the range from 27.3 to 47.8 kg/ha without irrigation and from 31.5 to 61.5 kg/ha on irrigation, that is, the gain from irrigation was high - 15.4 and 28.7%. Based on the results of the content of crude fat in the dry mass of corn, we can assert with great confidence that the accumulation of this element of nutrition, first of all, depends on the application of calculated rates of mineral fertilisers and we can not deny the great importance of irrigation of corn in the northern part of the forest-steppe zone of the Middle Volga region, including the Republic of Tatarstan, which is characterised by insufficient and unstable moisture availability.

Table 4. Total sugars content in maize silage

Factor A (moisture conditions)	Factor B (calculated NPK rates for planned green mass yield)	Total sugars content, %	Digestible protein content, %	Sugar-protein ratio
Without irrigation	Control	8.2	7.6	1.1:1
	At 30 t/ha ($N_{74}P_{40}K_{74}$)	8.8	8.4	1.0:1
	At 40 t/ha ($N_{98}P_{53}K_{97}$)	8.5	9.4	0.9:1
	At 50 t/ha ($N_{123}P_{66}K_{121}$)	8.1	9.6	0.8:1
Irrigated	Control	8.8	8.2	1.1:1
	At 30 t/ha ($N_{74}P_{40}K_{74}$)	8.9	8.9	1.0:1
	At 40 t/ha ($N_{98}P_{53}K_{97}$)	8.7	9.5	0.9:1
	At 50 t/ha ($N_{123}P_{66}K_{121}$)	7.5	9.2	0.8:1

Note: conversion factor of crude protein to digestible protein – 0.75

Meanwhile, the content of the sum of sugars in maize silage harvested in the milky-wax ripeness of cobs exceeds all other fodder crops in our region (Table 4).

On rainfed maize crops (30 t/ha), depending on the application of calculated rates of mineral fertilisers for the planned yields of green mass of maize, the accumulation of the sum of sugars increases from 8.2 in the control to 8.8% in the air-dry mass. Further increase of mineral fertiliser rates in order to obtain 40, 50 t/ha of green mass leads to 'dilution' of the content of this most important nutrition element to 8.5, 8.1%, respectively. Nevertheless, the sugar-protein ratio remains within the acceptable range (0.8:1; 1.1:1).

The above-mentioned regularities are also preserved in green mass obtained in irrigated maize crops - the ratio of total sugars to digestible protein does not change (0.8:1; 1.1:1).

Based on the results shown in Table 6, we can conclude that silage prepared from green mass of the control variant and the variant with application of calculated rates of mineral fertilisers for the planned yield of green mass 30 t/ha, it is advisable to use in the diet of dairy cows 2 months before the birth of the calf and 2 months after calving, when the ratio of the sum of sugars to digestible protein should be more than 10:1.

4. Conclusion

On average for 3 years of research on irrigation, the introduction of calculated norms of mineral fertilisers on the planned yield of 50 t/ha of green mass provides a stable receipt of 50.7 t/ha of high-quality corresponding to the normative indicators of cattle feeding, biomass with cobs in milk-wax ripeness with an increase in yield of 23.4 t/ha that is higher than the control variant of the experiment 114 % (without irrigation and without fertiliser).

In general, the transition of maize cultivation on grain technology on the background of irrigation and fertilisation became a platform for increasing the productivity of dairy cows in Tatarstan from 4 thousand litres/year of milk in the early XIX century to 10 thousand litres/year at present and increasing milk production up to 2 million litres/year.

References

1. G.S. Minnullin, S.R. Suleimanov, R.M. Nizamov, M.M. Malikov, Effect of mineral fertilisers on the yield of different types of fodder mixes on grey forest soils of the Republic of Tatarstan, Bulletin of Kazan State Agrarian University, **10(4(38))**, 76-80 (2015)
2. I.YU. Kuznetsov, B.G. Akhiyarov, I.G. Asylbaev et al, Evaluation of maize hybrids for economically valuable traits in the southern forest-steppe zone of the Republic of Bashkortostan, Achievements of Science and Technology of the Agroindustrial Complex, **37(11)**, 38-42 (2023)
3. A.L. Ivanov, Scientific and technological development of land use using digital technologies in

agriculture, Bulletin of the Russian Academy of Sciences, **89(5),** 522-524 (2019)

4. A.L. Ivanov, V.V. Kulintsev, V.K. Dridiger et al, On the feasibility of developing a direct seeding system on Russian chernozems, Achievements of Science and Technology of Agroindustrial Complex, **35(4),** 8-16 (2021)
5. A.L. Ivanov, V.S. Stolbovoy, The '4 ppm' initiative - a new global challenge for Russian soils, Bulletin of the V.V. Soil Institute. Dokuchaev, **98,** 185-202 (2019), https://doi.org/10.19047/0136-1694-2019-98-185-202
6. Y.V. Sotchenko, V.S. Sotchenko, O.V. Shaitanov et al, Interim results of trials of promising maize breeding samples for the conditions of the Republic of Tatarstan, 2012-2014, Niva Tatarstan, **1-2,** 33-36 (2017)
7. V.N. Fomin, M.M. Nafikov, V.V. Medvedev, D.V. Yakimov, Influence of methods of basic tillage and fertilisers on the dynamics of soil moisture, water consumption and yield of maize under cultivation for silage, Achievements of Science and Technology of Agroindustrial Complex, **31(12),** 12-16 (2017)
8. O.A. Mitrokhina, Evaluation of the relationship of yields of major crops with the content of trace elements in the soils of the CDR, Bulletin of the Ulyanovsk State Agricultural Academy, **1(61),** 60-64 (2023), https://doi.org/10.18286/1816-4501-2023-1-60-64
9. S.N. Nikitin, R.B. Sharipova, Assessment of changes in agroclimatic potential of Ulyanovsk region on crop production, Bulletin of the Ulyanovsk State Agricultural Academy, **3(59),** 36-42 (2022), https://doi.org/10.18286/1816-4501-2022-3-36-42.
10. A.A. Alferov, S.N. Nikitin, L.S. Chernova et al, Efficiency of nitrogen fertiliser and biopreparation application on spring wheat, Russian Agricultural Science, **5,** 39-42 (2023)
11. X. Li, W. Zhang, Sh. Xu et al., Low-Damage Corn Threshing Technology and Corn Threshing Devices: A Review of Recent Developments, Agricult., **13(5),** 1006 (2023), https://doi.org/10.3390/agriculture13051006
12. P. Fosu, The yield and price effects of growing genetically modified corn: evidence from the US corn belt, Int. J. of Biotechnol., **1(1),** 1 (2022), https://doi.org/10.1504/ijbt.2022.10049247
13. L.A. Kulikov, Dynamics of nutrient elements in maize plants under the influence of foliar feeding, Agrochemical Bulletin, **4,** 46-48 (2016)
14. I.P. Talanov, L.Z. Karimova, L.T. Vafina, G.K. Khuzina, Efficiency of cultivation of green mass of maize hybrids on calculated backgrounds of mineral nutrition in the conditions of the Predvolzhye region of the Republic of Tatarstan, Bulletin of Kazan State Agrarian University, **12(1(43)),** 40-45 (2017)
15. I.P. Talanov, M.Y. Mikhailova, Influence of calculated rates of mineral fertilisers on the formation of green mass of maize hybrids in the conditions of the Predvolga region of the Republic of Tatarstan, Bulletin of Kazan State Agrarian University, **10(1(35)),** 137-140 (2015)
16. F.N. Safiollin, M.M. Khismatullin, A.A. Lukmanov et al, Productivity of maize Ross 140 depending on the level of chemicalisation of zonal soils of the Republic of Tatarstan, Bulletin of the V.V. Dokuchaev Soil Institute. Dokuchaev, **115,** 199-223 (2023)
17. M.M. Khismatullin, F.N. Safiollin, Practical methods of partial replacement of mineral fertilisers by leaf fertilisation of perennial grasses on grey forest soils of the Middle Volga region, Kormoproizvodstvo, **7,** 12-18 (2019)
18. R.KH. Mukhamadiev, R.M. Nizamov, M.M. Malikov, Fodder mixtures in the fodder production system of the Republic of Tatarstan, Bulletin of Kazan State Agrarian University, **12(4(46)),** 20-22 (2017)

Technological and biochemical characteristics of spring wheat varieties of Irkutsk region

Yulia V. Gordeeva [1*], *Elena K.* Kuvshinova [1], *Anatoly S.* Golovko [1], *Anastasia S.* Bazova [1], *Angelina A.* Cheryomukhina [1]

[1] Azov-Black Sea Engineering Institute – branch of Don State Agrarian University in Zernograd, Zernograd, Russia

Abstract. In this work, a comprehensive assessment of technological and biochemical properties of zoned varieties of spring wheat cultivated in the Irkutsk region was carried out. The aim of the work was to determine the optimal ways to use different varieties in the processing industry of the region. In the course of the study the glassiness, moisture content, flour yield, protein content, quantity and quality of gluten, as well as rheological properties of dough were analysed. The results showed significant variability in quality characteristics between the varieties. The varieties Angara 86 and Tulun 15 showed the best performance for baking, while the varieties Irene, Novosibirskaya 15 and Studencheskaya, having high flour yields, can be recommended for feed purposes and for the production of products with less stringent requirements for a more rational use of this resource.

1 Introduction

Spring wheat, as a key cereal crop in temperate climatic zones, occupies a special place in world agriculture. Its grain, which has a rich and versatile chemical composition, offers a wide range of applications in a variety of sectors. While the traditional use of spring wheat is mainly limited to the production of bread, flour and other flour products, the potential of this valuable raw material extends far beyond the food industry. Modern advances in science and technology are opening the door to deeper and more complex grain processing, producing a wide range of products, from bioethanol and beer to starch, bioplastics and high-quality animal feed. For example, global ethanol production from grain crops reached 110 billion litres in 2022, and demand for bioplastics derived from grain is growing at an average annual rate of 15-20%. At the same time, traditional areas should not be forgotten, as the bakery industry consumes up to 70% of all wheat production, which underlines its strategic importance. It is also important to note the role of family farmers in grain production [1].

Nevertheless, it should be understood that the efficiency of spring wheat grain use in various directions is not a constant, but directly depends on a number of factors, including the technological capabilities of processing enterprises, and, most importantly, on the characteristics of the grain used, determined by the varietal characteristics and conditions of its cultivation. Different spring wheat varieties can differ significantly in their technological and biochemical properties, such as protein content, gluten content, starch content, amino acid composition, etc. For example, the protein content of grain can vary from 10 to 18% depending on the variety and growing conditions, which directly affects the baking qualities of flour. Gluten, in turn, can vary in its elasticity and firmness, determining the rheological properties of dough. The study of these features in released varieties adapted to specific soil and climatic conditions opens up prospects for a more precise and targeted use of grain. Thus, varieties with high protein and gluten content may be optimal for the baking industry, while varieties with high starch content may be widely used in the production of bioethanol and other industrial products. Varieties with a higher content of soluble carbohydrates are promising for the brewing industry, and varieties with a certain amino acid composition can be used in the production of valuable animal feed. To understand the peculiarities of yield formation in the historical context, it may be useful to study the experience of past years [2].

A comprehensive study of technological and biochemical properties of released spring wheat varieties, taking into account their adaptation to specific growing conditions, is extremely important for optimising their use and increasing the efficiency of agriculture. Such studies are of particular importance in the context of regional development and food security. In Russia, the gross yield of spring wheat in 2022 was about 25 million tonnes, with yields ranging from 15 c/ha in regions with unfavourable conditions to 40-50 c/ha in more favourable ones. Analysing the prospects for the use of spring wheat grain in applied biotechnology allows not only to diversify the production of agricultural products in a particular region,

* Corresponding author: gordeeva_julia88@mail.ru

but also to reduce the load on the environment. More rational use of bioresources, waste reduction, and the development of closed production cycles can all contribute to the creation of more sustainable and environmentally friendly agricultural systems. In particular, special attention should be paid to social aspects and gender equality [3, 4]. In addition, modern water management practices are necessary for the efficient utilisation of water resources [5, 6].

In addition, the diversification of wheat processing can stimulate the development of small and medium-sized businesses in rural areas, creating new jobs and increasing the income of the local population. As a result, it not only ensures food security, but also contributes to sustainable socio-economic development of regions, which is a priority for many countries around the world. Ultimately, an integrated approach to the study and use of spring wheat grain, taking into account all aspects, from the characteristics of varieties to technological capabilities and environmental constraints, is an important step towards the formation of a sustainable and efficient agro-industrial complex. It also requires consideration of the possibility of utilisation of production waste [7, 8].

The purpose of this work, presented in the attached file, was a comprehensive assessment of technological and biochemical properties of zoned varieties of spring wheat grown in the Irkutsk region, as well as their systematisation according to the directions of its use in the processing industry. Such research will help to determine the optimal ways to use different wheat varieties in various industries, which, ultimately, will contribute to improving the efficiency of agricultural production and strengthening food security in the region.

2 Method and materials

In the framework of this study, a comprehensive analysis of technological and biochemical properties of different varieties of spring wheat released in the Irkutsk region was carried out. Experimental work included grain sampling, their preparation for analysis and direct laboratory tests. A number of modern analytical instruments were used to assess the quality of grain and flour. Determination of grain vitreousness was carried out visually with the help of diaphanoscope DSZ-1, by shining individual grains, which allowed to determine the proportion of vitreous grains in the total mass. Grain moisture was measured on a drying cabinet SESH-3M, where pre-weighed samples were dried at a temperature of 130°C to reach a constant mass. Flour yield was determined by milling the grain on a laboratory mill LZM, followed by weighing of the obtained flour.

The Kjeldahl method was applied to determine the protein in the grain using a laboratory distillation and titration unit. Gluten was manually washed from the flour sample and its quantity and quality were further determined. The quantitative values of gluten were determined using laboratory scale VLR-200. A Brabender farinograph and a Chopin alveograph were used to evaluate the rheological properties of the dough. The Brabender farinograph, working in a constant kneading mode at 30°C, determined the water absorption capacity of flour, dough consistency and other parameters that allow evaluating its technological properties. The Chopin alveograph, creating an inflating bubble from the dough, recorded the pressure at which the bursting occurs, which made it possible to assess the elasticity and extensibility of the dough. The number of falls was determined on a PPP-7 instrument, measuring the viscosity of the glue-sterilised flour suspension, at a temperature of 100°C. All the results were processed statistically using Statistica 10.0 software, which is used to process large statistical data in many scientific studies [9, 10].

3 Results and discussion

As part of the research aimed at a comprehensive analysis of the quality characteristics of released spring wheat varieties, we undertook a series of experimental works, including careful selection and preparation of grain samples, as well as their subsequent testing in laboratory conditions. Grain collected from experimental plots was subjected to primary cleaning from impurities and subsequent drying to a standard moisture content of 14%, after which representative samples were formed for further analysis.

The initial stage of our research work, aimed at a comprehensive assessment of the quality characteristics of released varieties of spring wheat growing in the Irkutsk region, was the determination of grain vitreousness. This indicator, which has a direct correlation with the density of endosperm, has a significant impact on the yield of flour and its baking qualities. For this analysis we used a diaphanoscope, which allowed us to visually assess the proportion of vitreous grains in each sample. The essence of the technique consisted in shining through individual grains, which allowed us to differentiate vitreous grains characterised by a homogeneous structure and transparency from floury grains with an opaque, loose structure. At least 300 grains from each variety were analysed, which allowed us to obtain statistically reliable data. The results of the vitreousness analysis showed a significant variability of this indicator between the varieties under study. The average values ranged from 45% to 78%, which confirms significant differences in endosperm density between different genotypes. In particular, varieties Angara 86 and Tulun 15 showed the highest vitreousness values of 78% and 75%, respectively, indicating high density of their endosperm. The obtained values significantly exceeded the average values for the region, which, according to the data of previous years, were about 60-65%. The Novosibirskaya 15 and Studencheskaya varieties, in turn, had lower vitreousness values - 48% and 51%, respectively, which could indicate the presence of genetic predisposition to the formation of a more loose grain structure and could negatively affect the flour yield. In addition, we noted the dependence of vitreousness index on meteorological conditions of the growing season. In years with sufficient precipitation and moderate temperature during the period of grain filling, an increase in the proportion of glassy grains was observed, which may indicate the influence of agroclimatic factors on the formation of this quality indicator. When analysing the development of agriculture, it is important to take into account the historical background [11]. For example, in 2021, characterised by increased rainfall in July, the average vitreousness index for all varieties was 5-7% higher compared to 2020, when drought was observed. These data emphasise the importance of considering weather conditions when assessing grain quality.

After assessing vitreousness, the next stage of our research was to determine the exact moisture content of the grain. This parameter is critical because it directly affects the conditions of storage, processing and, ultimately, the quality of the final product. To measure moisture content, we used a drying cabinet SESH-3M, which provided accurate and uniform heating of samples to a constant mass. The method consisted of drying pre-weighed grain samples at 130°C until a constant weight was reached, which guaranteed complete evaporation of moisture. Weighing was carried out with an accuracy of 0.01 g. During the experiment, we found that the initial grain moisture content varied between 13.5% and 14.2% depending on the variety and harvest time. This range of moisture corresponded to the standard requirements for grain intended for long-term storage and further processing, in particular to the requirements of GOST 9353-2016. However, it should be noted that after harvesting, some varieties were characterised by a higher moisture content of up to 15.5%, which required additional drying before storage to prevent the development of mould and fungal diseases. The varieties Angara 86 and Tulun 15 showed lower moisture content at the time of harvest, compared to the varieties Novosibirskaya 15 and Studencheskaya, which may be due to the peculiarities of their maturation and ear structure.

After determining the moisture content, we proceeded to assess the flour yield, which is an important economic indicator reflecting the efficiency of grain milling. For this purpose, we used a laboratory mill LZM, which allowed us to grind grain under controlled conditions. Pre-weighed samples of grain (100 g each) were milled, and then the resulting flour was weighed, after which the flour yield was calculated as a percentage. The results showed that flour yield also varied from variety to variety due to differences in grain structure, hardness and endosperm density. The maximum flour yield was observed for the variety Irene with 72.3 per cent, which was very good. At the same time, the minimum flour yield, 67.8%, was recorded for Novosibirskaya 29 variety, which was due to the looser grain structure of this variety and a large amount of bran. The varieties Angara 86 and Tulun 15, which had high vitreousness, showed flour yields of 70.5% and 71.2%, respectively, which also indicated their good suitability for the flour milling industry. It should be noted that flour yield was a dynamic parameter and could depend on many factors, including not only the grade, but also the mill settings and the size of the grinding fraction. In our experiments, we aimed to follow standardised conditions to minimise the influence of extraneous factors. The data obtained confirmed that grain vitreousness has a significant effect on flour yield. Varieties with higher vitreousness tended to produce higher flour yields, which in turn was reflected in the economic efficiency of the grain processing process.

After primary processing, we proceeded to analyse the biochemical characteristics of the grain. Protein content determined by the Kjeldahl method showed significant differences between varieties. Thus, the average protein content of the grain ranged from 14.2% to 17.5%. The highest amount of protein, 17.5%, was observed in the variety Tulun 15, which was 1.3% higher than that of Angara 86, although they were both considered high-protein varieties. The next important indicator was gluten, which we manually washed from the flour suspension and then assessed its quantity and quality. The amount of gluten ranged from 24% to 32%, with Angara 86 being the highest and Buryatskaya ostistaya the lowest. In addition, we evaluated the rheological properties of dough prepared from the tested flour varieties. On a Brabender farinograph, we determined the water absorption capacity of flour, which varied from 58% to 64%. The highest water absorption capacity of 64% was observed for the variety Tulun 15, indicating good gluten quality. Dough consistency also differed for the different varieties, averaging 450-550 e.f., confirming their different technological properties. On a Chopin alveograph, we evaluated the tensile and elasticity of the dough, which allowed us to judge the baking properties of the flour. The dough bursting pressure varied depending on the variety, and the Angara 86 and Tulun 15 varieties had significantly higher values than the others, indicating better stretchability and elasticity of their gluten. Agroecological potential of territories has a direct impact on grain quality [12, 13].

The results obtained confirm that there are significant differences in the quality characteristics of different varieties of spring wheat released in the Irkutsk region. Varieties Angara 86 and Tulun 15 showed the highest results in vitreousness, protein content and gluten quality. For example, Angara 86 showed a protein content of 16.2% and Tulun 15 - 17.5%. At the same time, the amount of gluten in Angara 86 variety was 32%, which is quite high, and in Tulun 15 variety - 30.3%. Their alveograph results were also an order of magnitude higher, which indicates good baking properties of these varieties. At the same time, Buryatskaya ostistaya, Selenga and Omskaya 32 varieties were characterised by lower protein content, gluten quantity and quality, and their alveograph scores were lower, which, in turn, indicates that they are less suitable for baking. The fall number, which is an important indicator of alpha-amylase activity, for these varieties was on average 320 seconds, indicating a fairly high activity of this enzyme affecting dough viscosity. Comparison of the data we obtained with the results of studies conducted in other regions revealed some regularities. For example, studies conducted in Western Siberia showed similar trends in protein and gluten content in different wheat varieties, but the indicators in absolute terms differed, which confirms the influence of soil and climatic conditions on the formation of grain quality. For example, in the research of the Research Institute of Agriculture in the Omsk region, the maximum values of protein reached 18.1 per cent, and the maximum indicator of gluten was at the level of 33.4 per cent. At the same time, in more southern regions, where conditions for wheat cultivation are more favourable, protein and gluten content were higher on average, which also confirms the influence of climate on these indicators. Grain quality is largely determined by the agroecological condition of landscapes [14, 15].

In general, the results of the conducted research emphasise the need for an individual approach to the selection of spring wheat varieties for different purposes. Varieties with high protein and gluten content, such as Angara 86 and Tulun 15, can be recommended for the production of bakery products of high quality. Whereas varieties with lower protein and gluten content but high flour yields, such as Irene and Studencheskaya, may be more suitable for use in feed purposes or for the production of other processed products such as alcohol or starch. In addition, the findings indicate the need to take into account the technological parameters of the varieties, such as the water absorption capacity of the flour and the rheological properties of the dough, when using them in the baking industry. The use of the results of the conducted research allows the rational use of the available resource of spring wheat grain, which, in turn, contributes to increasing the efficiency of agricultural production and strengthening the food security of the region. In order to organise sustainable land use, it is necessary to take into account a number of parameters, including hydrological parameters [16].

Our data confirm that spring wheat grain is an important resource that, with the right approach, can be used in various directions, ranging from food industry to fodder production and biotechnology.

4 Conclusion

The comprehensive study of technological and biochemical properties of released varieties of spring wheat in the Irkutsk region has allowed us to draw a number of important conclusions of practical importance for agriculture and processing industry in the region. The results of the analysis showed a significant variability of grain quality characteristics between the studied varieties, which indicates the need for a differentiated approach to their use. Varieties Angara 86 and Tulun 15 stood out as the most promising for baking, as they demonstrated high vitreousness, protein content, quantity and quality of gluten, as well as the best rheological properties of dough. These varieties can be recommended for the production of high quality bakery products, as well as for other food products where high protein and gluten levels are required. At the same time, it is important to note that their potential is fully manifested if agronomic recommendations are followed, including optimal sowing dates, fertiliser application rates and timely protection against pests and diseases. To maximise production efficiency, the potential for waste utilisation should be considered [17, 18]. The varieties Irene, Novosibirskaya 15 and Studencheskaya, were characterised by lower protein content and gluten quality, but at the same time had a fairly high flour yield. These varieties can be used in fodder production, as well as for the production of products with less stringent requirements for protein quality, such as alcohol or bioethanol production. In addition, these varieties may be in demand in regions with limited resources for grain production, where it is the volume of production that is important. Also, the data obtained indicate the need to take into account the technological parameters of each variety when using them. For example, the water absorption capacity of flour, which varied considerably between varieties, is an important factor in optimising the dough preparation process. Varieties with a high water absorption capacity require more water when kneading the dough, which in turn can affect the quality of the baked products. The optimum dough kneading parameters should be determined for each variety separately, taking into account its individual characteristics.

Thus, the results of the study emphasise the importance of comprehensive analysis of spring wheat grain quality when selecting varieties for specific purposes. Rational use of available varieties, taking into account their technological and biochemical characteristics, is the key to improving the efficiency of agricultural production and strengthening food security. Differential use of varieties allows not only to improve product quality, but also to optimise production costs and reduce the negative impact on the environment. It is important to remember that not only yield but also grain quality is a determining factor for economic success. In this context, as well as taking into account modern production technologies, it is necessary to explore the possibilities of applying modern monitoring methods, for example, such as geospatial analysis [19].

The obtained data can also become the basis for the development of recommendations for the selection of new spring wheat varieties adapted to the conditions of the Irkutsk region and having the specified quality characteristics.

References

1. B. Aparecida da Silva, R. Evangelista de Oliveira, A. Cavalieri Sais, E. Cardoso leite, Family farmers and trees: ecological functions, ecosystem services and food production in agrarian reform settlements in Brazil. Agroecol. Sustain. Food Syst., **49**, 4-21 (2025). DOI: 10.1080/21683565.2024.2405515
2. N.A. Zhirov, Socio-Economic Development of the Black Soil Village in the XIX–early XX centuries: from the Prosperous Granary of Pre-Reform Russia to the Demo-Ecological Crisis of Rural Settlements. Bylye Gody, **19**, 1605-1614 (2024). DOI: 10.13187/bg.2024.4.1605

3. L. Ossome, Gender and development in the agrarian south. World Dev., **188,** 106876 (2025). DOI: 10.1016/j.worlddev.2024.106876
4. Shutaleva, A. Starostin, A. Salgiriev, O. Vlasova, A. Grinek, Z. Nikonova, I. Savchenko, Migration Potential of Students and Development of Human Capital. Educ. Sci., **12 (5),** 324 (2022) DOI: 10.3390/educsci12050324
5. S. Mukherjee, S. Paul, S. Bhattacharya, A. Islam, S. Mahammad, E. Alam, Optimizing water resource management in tropical drought-prone regions through hybrid MCDM techniques: A water-stress mapping approach. J. Hydrol. Reg. Stud., **57,** 102171 (2025). DOI: 10.1016/j.ejrh.2024.102171
6. A.A. Kapanski, R.V. Klyuev, A.E. Boltrushevich, Geospatial Clustering in Smart City Resource Management: An Initial Step in the Optimisation of Complex Technical Supply Systems. Smart Cities, **8,** 14, (2025). DOI: 10.3390/smartcities8010014
7. A.L.S. Hansted, W.J. Newell, K.A. Hawboldt, An initial analysis of the market potential of a wet ash waste stream from a pulp and paper mill in Newfoundland and Labrador. Ind. Crops Prod., **225,** 120557 (2025). DOI: 10.1016/j.indcrop.2025.120557
8. A.A. Sokolov, V.A. Fomenko, M.A. Aksenova, Development of a methodology for radon pollution studies based on algorithms taking into account the influence of constant mountainvalley winds. Appl. Chem. Eng., **7 (2),** ACE-1865 (2024) DOI: 10.59429/ace.v7i2.1865
9. S.N. Sorokova, E.A. Efremenkov, D.V. Valuev, M. Qi, Analysis of a Predictive Mathematical Model of Weather Changes Based on Neural Networks. Mathematics, **12 (3),** 480 (2024) DOI: 10.3390/math12030480
10. I.I. Bosikov, V.Ch. Revazov, Analysis and evaluation of prospects for highquality quartz resources in the North Caucasus. Mining Science and Technology (Russia), **8(4),** 278–289, (2023). DOI: 10.17073/2500-0632-2023-10-165
11. J.C.R. Cardona, Methodology for studying the history of agrarian systems. Rev. Fac. Nac. Agron. Medellin, **78,** 19-25 (2025).
12. A. Kehinde, T.F. Adesiyan, S.O. Hassan, I.G. Familusi, Leveraging on membership in agricultural cooperatives to alleviate abject poverty among smallholder farming households in Nigeria. Sustain. Futures, **9,** 100399 (2025). DOI: 10.1016/j.sftr.2024.100399
13. A.G. Terekhov, G.N. Sagatdinova Spectral characteristics of saline arable land in the foothill catchment Zailiisky Alatau ridge (Northern Tien-Shan) during 2016-2022 using Sentinel-2 data, Sustainable Development of Mountain Territories, **16 (1),** 143 – 155, (2024). DOI: 10.21177/1998-4502-2024-16-1-143-155
14. T.I. Kosovtsova, Factors of conservation of cultural landscapes of the Bavarian Alps, Sustainable Development of Mountain Territories, **11 (3),** 371 – 379, (2019). DOI: 10.21177/1998-4502-2019-11-3-371-379
15. K.I. Kravtsov, Ya.A. Tynchenko, T.A. Panfilova, Influence of mountain factors on salt excess and soil toxicity in mountain conditions. Sustainable Development of Mountain Territories, **15(3),** 784-797 (2023). https://doi.org/10.21177/1998-4502-2023-15-3-784-797
16. S.I. Kondratev, Three typical hydrological-hydrochemical situations near the danube river mouth based on the marine hydrophysical institute research expeditions in 1997-2013. Physical Oceanography, **26(4),** 326–340, (2019). DOI 10.22449/1573-160X-2019-4-326-340
17. V. I. Golik, D. A. Zyukin., A. I. Karlina Technology for nonwaste recovery of tailings of the mizur mining and processing plant, Metallurgist. **66,** 11-12, 1476-1480, (2023). DOI: 10.1007/s11015-023-01462-y
18. V.I. Golik, V. Brigida, E.A. Efremenkov, S.N. Sorokova, Q. Mengxu, Tailings Utilization and Zinc Extraction Based on Mechanochemical Activation. Materials, **16 (2),** 726 (2023) DOI: 10.3390/ma16020726
19. R. Qiu, M. Kong, Z. Lei, J. Shi, Z. Li, H. Zheng, Characteristics and spatial-temporal dynamics in agronomic landscape: Case study of Guangrong watershed, Northeast China. Appl. Geogr., **174,** 103484 (2025). DOI: 10.1016/j.apgeog.2024.103484

The strength of the influence of polymorphism of the kappa-casein and diacylglycerol o-acyltransferase gene polymorphisms on the economically useful qualities of dairy cattle

Radik R. Shaydullin[1]*, *Lenar* R. Zagidullin[2], *Tahir* M. Akhmetov[2], *Sergey* V. Tyulkin[3] and *Irina* A. Konakova[2]

[1] Kazan State Agrarian University, Kazan, 420015, Russia
[2] Kazan State Academy of Veterinary Medicine named after N.E.Bauman, Kazan, 420029, Russia
[3] FSC food systems named V.M. Gorbatov of the RAS, Moscow, Russia, 109316

Abstract. The aim of the work was to study the strength and reliability of the effect of genotypes of kappa-casein (CSN3) and diacylglycerol O-acyltransferase (DGAT1) and their complex combination (CSN3/DGAT1) on 12 economically useful traits of black sow cattle. It was found that CSN3 genotype had the greatest reliable effect on the mass fraction of protein in milk in first heifers and high-yielding cows. The share of its influence was η^2= 8.9–15.2% ($F_{fact.}$= 10.6–12.5; $P<0.001$). The effect was also found on age at first calving η^2= 1.8–2.1% ($F_{fact.}$= 3.2–4.2; $P<0.05$) and milk proteinη^2= 4.9% ($F_{fact.}$= 3.6; $P<0.05$) only in first-calving heifers. The DGAT1 genotype significantly ($P<0.05$–0.001) affected the mass fraction of fat – η^2= 10.0–13.2% ($F_{fact.}$= 10.6–11.3) and milk protein – η^2= 3.2–7.3% ($F_{fact.}$= 3.5–5.4), milk fat – η^2= 2.9–7.5% (F_{fact} = 3.1–5.6) и белка – η^2= 8.4% ($F_{fact.}$= 6.4), milk yield per lactation – η^2= 7.6–8.1% ($F_{fact.}$= 6.3-8.7). The milk productivity indicators are more widely influenced by DGAT1 genotype.

1 Introduction

Milk productivity of cows is determined by the quantity and quality of milk obtained from the animal over a certain period of time. Milk yield depends on a whole complex of internal and external factors. According to their influence, these factors can be divided into two main groups: genotypic and phenotypic. The group of genotypic factors includes genotypic features of fathers, lineage, productivity of parents and especially allelic warrants of genes-markers of economically useful traits.

In dairy cattle breeding, selection-genetic parameters of productivity are of great importance, as they indicate the efficiency of the conducted breeding work in the population of animals at different stages.

One of the priorities of breeding work is the improvement of cattle breeding systems. In modern conditions, the key direction of selection work in dairy cattle breeding is the purposeful selection of cattle for the most important breeding traits, taking into account the correlations of these traits [1, 2].

The use of genetic markers responsible for milk productivity in cattle breeding will improve the economically useful traits of cows. Alleles of genes for milk proteins, hormones and enzymes are considered as potential DNA markers of milk productivity and milk quality in cattle. One of such markers are kappa-casein and diacylglycerol O-acyltransferase genes [3-5].

Genotyping of cattle for kappa-casein (CSN3) is of great practical importance, because milk of cows carrying the BB-genotype of kappa-casein is characterised by increased protein-milk yield compared to AA-genotype of kappa-casein [6-10].

In the Republic of Tatarstan, many scientists have studied the influence of allelic forms of kappa-casein on milk productivity in populations of Holmogorsky breed of Tatarstan type and black-and-brown breed in the last decade. At the same time, the results of studies are reduced to the fact that animals with the B allele of kappa-casein are characterised by high protein-milk yield and milk protein yield [11-16].

When studying the influence of diacylglycerol O-acyltransferase (DGAT1) genotypes on the milk productivity of cows, some foreign scientists found that animals carrying the K allele of DGAT1 in their genome are characterised by a higher fat content in milk compared to cows with the A allele [17-19].

The DGAT1 gene, in addition to its influence on milk productivity and milk quality in cows, is also related to reproductive traits [20, 21].

The issue of improving the productive traits of cows is an everyday problem and is caused by a combination of genetic and non-genetic factors. In this regard, there is an important problem of studying the share of influence of different genotypes of DNA markers on economically useful traits of dairy cattle.

In this regard, the strength and reliability of the effect of kappa-casein and diacylglycerol O-acyltransferase

* Corresponding author: tppi-kgau@bk.ru

genotypes and their complex combination on economic and useful traits of black and sow cattle were studied by single-factor analysis of variance.

2 Materials and Methods

For research and evaluation of kappa-casein and diacetylglycerol O-acetyltransferase genes, 142 first-calf cows and 208 high-yielding cows of black-and-brown breed were selected from which blood samples were taken and DNA samples were isolated in the conditions of the breeding reproducer LLC 'Dusym', Atninsky district, Republic of Tatarstan.

The material for molecular DNA testing was venous blood of animals. DNA extraction was performed using a Magnosorb kit (Interlabservice, Moscow) according to the manufacturer's instructions. The CSN3 gene was investigated by polymerase chain reaction followed by restriction fragment length polymorphism analysis (PCR-PDLP) using the forward primer 5'-ATAGCCAAATATATATCCCAATTCAGT-3' and reverse primer 5'-TTTATTAATAATAAGTCCATGAATCTTG-3' [22], and DGAT1 gene using the forward primer 5'-GCTGCTGCTCCTGAGGGGGCCCTTCG-3' and reverse primer 5'-GCGGCGGCGGCACTTCATGACGACCCT-3' [23].

Amplification was performed on a detection amplifier 'Terzik' (DNA-Technology, Moscow). For amplification of kappa-casein gene fragments, pairs of oligonucleotide primers synthesised at Syntol (Moscow, Russia) were used. The obtained amplicons were subjected to restriction using Hinf I restriction enzymes (SibEnzyme, Russia) according to the manufacturer's recommendations.

Depending on the presence of alleles A and B of the CSN3 gene, the genotypes were distributed into three groups (AA, AB, BB) and alleles A and K of the DGAT1 gene, the genotypes were distributed into three groups (AA, AK, KK).

Data on milk productivity, reproductive performance of first-calf heifers and high-yielding cows with different CSN3 and DGAT1 genotypes were used for the study.

One-factor analysis of variance was carried out using the software application 'Excel' from the Microsoft Office package, according to which the strength of the influence of the polymorphism factor of the studied genes on the productive performance of cows was determined.

3 Results and discussion

As a result of research it was found that kappa-casein genotype had the greatest reliable influence on the indicator of mass fraction of protein in milk of first-calf heifers, its share of influence was η^2=15.2% (F=12.5; P<0.001) (Table 1). There was also a significant influence of kappa-casein genotype on the amount of milk protein η^2=4.9% (F=3.6; P<0.05) and the indicator of age of first calving η^2=1.8% (F=3.2; P<0.05). No significant effect of kappa-casein gene polymorphism was found on other productivity indicators, but it can be noted that the degree of influence of kappa-casein genotype on the mass fraction of fat in milk and milk fat yield was η^2=3.4–4.2 %.

For the diacylglycerol O-acyltransferase gene, among the 12 parameters studied, a reliable influence was found for the mass fraction of fat in milk – η^2=13.2% (F=10.6; P<0.001), milk protein quantity – η^2=8.4% (F=6.4; P<0.01), milk yield per lactation – η^2=8.1% (F=6.3; P<0.01), milk fat quantity – η^2=7.5% (F=5.6; P<0.01) and milk protein mass fraction – η^2=7.3% (F=5.4; P<0.01).

Table 1. Percentage of influence of kappa-casein and diacylglycerol O-acyltransferase genotype on productive performance of heifer cows

№	Indicator	CSN3		DGAT1		CSN3 / DGAT1	
		η^2, %	$F_{fact.}$	η^2, %	$F_{fact.}$	η^2, %	$F_{fact.}$
1	Yield per lactation	1.5	1.1	8.1	6.3**	12.0	5.7**
2	Mass fraction of fat	3.4	2.5	13.2	10.6***	8.6	3.9*
3	Milk fat	4.2	3.0	7.5	5.6**	19.7	10.2***
4	Protein mass fraction	15.2	12.5***	7.3	5.4**	5.1	2.3
5	Milk protein	4.9	3.6*	8.4	6.4**	16.9	8.5***
6	Live weight	2.4	1.7	1.4	0.1	4.5	1.9
7	Milkiness index	1.2	0.8	3.8	2.7	8.0	3.6*
8	Age of first calving	1.8	3.2*	1.8	0.7	3.0	1.2
9	Service period	2.4	1.7	0.8	0.5	3.1	1.2
10	Inter-calving period	0.1	0.1	1.9	1.6	1.9	1.5
11	Reproductive capacity coefficient	0.3	0.2	1.4	1.2	1.4	1.4
12	Doha index	0.3	0.2	1.8	1.3	1.1	1.0

Note: * – P<0,05; ** – P<0,01; *** – P<0,001

The share of influence of DGAT1 genotype on other performance indicators of first-calf heifers was less than η^2=3.8 % and not significant.

Along with studying separately the degree of influence of CSN3 and DGAT1 genotype on economic and useful qualities of cows, the joint influence of complex genotypes of the above genes was also taken into account.

CSN3/DGAT1 had a greater effect on the amount of milk fat and protein. Accordingly, these parameters were η^2=19.7% (F= 10.2; P<0.001) and η^2=16.9% (F=8.5; P<0.001). Comparatively high influence share was found for milk yield per lactation η^2=12.0% (F=5.7; P<0.01), milk fat mass fraction η^2= 8.6% (F=3.9; P<0.05), milk yield index η^2=8.0% (F=3.6; P<0.05).

The share of influence of kappa-casein genotype on the reproductive ability indices of first-calf cows was insignificant and was less than 4%.

We have studied the same indicators of productivity in high-yielding cows, where the proportion and reliability of the influence of genotypes of kappa-casein and diacylglycerol O-acyltransferase on economic and useful qualities of high-yielding cows, practically confirm the indicators in heifer cows with some deviations.

Table 2 shows the results of analysing the effect of CSN3 and DGAT1 genotypes, as well as their combination, on the productive and reproductive characteristics of high-yielding cows. Data analysis shows that CSN3 genotype has the most significant and reliable effect on the mass fraction of protein in milk (η^2= 8.9%, P<0.001), which is consistent with the results obtained for first-calf heifers and confirms the stable association of this gene with protein-milk yield regardless of the age of the animal. There was also a significant effect of CSN3 on the age of first calving (η^2= 2.1%, P<0.05), which, in contrast to first-calving heifers, demonstrates the effect of genotype on this trait in mature cows. The effect of CSN3 on the other traits was not significant. Regarding DGAT1 genotype, the results indicate its significant effect on several traits including milk fat mass fraction (η^2= 10.0%, P<0.001), suggesting its significant role in determining fat milk yield. A significant effect was also found on milk yield per lactation (η^2= 7.6%, P<0.001), confirming its involvement in the regulation of productivity. The effects on milk fat, milk fat index and protein mass fraction were also significant, but less significant than the effects on fat mass fraction and milk yield. It is noteworthy that the effect of DGAT1 on milk protein, live weight and age at first calving was relatively small. Reproductive performance showed minimal effect of DGAT1 genotype. Finally, analysis of the joint effect of CSN3 and DGAT1 genotypes (CSN3/DGAT1) shows significant effects on several traits including milk fat mass fraction, indicating a possible synergistic effect of the genes. However, again, the effect on reproductive performance remains insignificant. Overall, the data in Table 2 confirm the importance of both CSN3 and DGAT1 for dairy performance, with DGAT1 having a broader impact, while the effect on reproductive performance in high-yielding cows remains limited and possibly non-linear.

Table 2. Proportion of the effect of kappa-casein and diacylglycerol O-acyltransferase genotype on productive performance of high-yielding cows

№	Indicator	CSN3		DGAT1		CSN3/DGAT1	
		η^2, %	$F_{fact.}$	η^2, %	$F_{fact.}$	η^2, %	$F_{fact.}$
1	Yield per lactation	1.8	1.9	7.6	8.7***	5.2	3.4*
2	Mass fraction of fat	6.2	7.1***	10.0	11.3***	9.1	5.9**
3	Milk fat	1.8	1.9	2.9	3.1*	6.0	3.9*
4	Protein mass fraction	8.9	10.6***	3.2	3.5*	5.4	3.5*
5	Milk protein	3.2	2.8	2.5	2.4	6.7	4.4*
6	Live weight	1.2	1.2	1.5	1.5	5.4	3.5*
7	Milkiness index	1.0	1.0	5.9	6.5**	4.9	3.2*
8	Age of first calving	2.1	4.2*	1.4	1.5	2.8	1.6
9	Service period	1.7	1.9	0.2	0.2	2.7	1.0
10	Inter-calving period	1.5	1.7	0.3	0.3	3.9	1.9
11	Reproductive capacity coefficient	1.2	1.2	0.5	0.5	1.6	1.7
12	Doha index	1.7	1.9	0.3	0.3	1.2	1.1

Note: * – P<0.05; ** – P<0.01; *** – P<0.001

here was also a significant effect on milk fat mass fraction η^2=6.2% (F=7.1; P<0.001) and on the age of first calving – η^2=2.1% (F=4.2; P<0.05) which was not observed in first-calf cows.

The DGAT1 genotype with high reliability influences the level of cow milk yield and fat yield, and the power of influence on milk yield per lactation was 7.6% (F=8.7; P<0.001), fat mass fraction 10.0% (F=11.3; P<0.001). Also, a significant degree of influence was found for milkiness index, milk fat and protein content, 5.9% (F=6.5; P<0.01), 2.9% (F=3.1; P<0.05) and 3.2% (F=3.5; P<0.05), respectively.

It should be noted that the degree of influence of DGAT1 genotype on milk protein quantity, live weight and age at first calving was within the range from η^2=1.4% to η^2=2.5%, and for reproductive performance was low and ranged only from 0.2 to 0.5%.

Thus, it can be assumed that the more diacylglycerol O-acyltransferase genotype influences milk productivity of high-yielding cows, the less its influence on reproductive qualities of cows.

Out of 12 indicators of economically useful traits of high-yielding cows, a significant influence (P<0.05-0.01) of complex genotypes CSN3/DGAT1 on the indicators of cow productivity was found in seven cows. The degree of influence on these indicators was: on milk yield per lactation 5.2 %, on milk fat mass fraction - 9.1 %, on milk fat yield - 6.0 %, on milk protein mass fraction - 5.4 %, on milk protein yield - 6.7 %, on live weight - 5.4 %, on milk yield index - 4.9 %.

Low degree of influence of complex genes was revealed on reproductive qualities of cows. The strength of CSN3/DGAT1 influence on reproductive performance was in the range $\eta^2 = 1.2–3.9\%$.

4 Conclusion

This work demonstrates the importance of the genetic markers kappa-casein (CSN3) and diacylglycerol O-acyltransferase (DGAT1) in improving the performance traits of dairy cattle, in particular black and white dairy cattle. The study revealed that both genes have an effect on milk composition, however, DGAT1 is characterised by a wider range of effects on milk yield. The CSN3 genotype is primarily associated with protein-milk yield in both first-calf and high-yielding cows, emphasising its important role in determining this trait. It is interesting to note that in high-yielding cows, in contrast to first-calf heifers, the influence of CSN3 extends to the age of first calving. In turn, the DGAT1 genotype has a significant effect on fat yield and total milk yield, confirming its importance for the quantity and quality of milk produced. At the same time, in high-yielding cows there is a tendency of inversely proportional dependence: the stronger the influence of DGAT1 on milk productivity, the less its influence on reproductive qualities. Analysis of the complex effect of CSN3 and DGAT1 genotypes also revealed significant effects on milk production, however, the effect on reproductive performance remains low. In conclusion, the results of the study emphasise the importance of using CSN3 and DGAT1 genetic markers in dairy cattle breeding to improve dairy productivity and milk quality traits. Knowledge of the contribution of these genes will allow more efficient formation of the breeding core of the herd and creation of highly productive animals.

References

1. A.V. Bakai, A.M. Mukhtarov, G.V. Mkrtchyan, Animal Sci., **12**, 6–8 (2013).
2. S.S. Petukhov, R.V. Tamarova, Bull. of the agro-industr. complex of the Upper Volga reg., **2(38)**, 38–42 (2017).
3. L.R. Zagidullin, R.R. Shaidullin, T.M. Akhmetov, S.V. Tyulkin, Dairy News, **1(37)**, 24–34 (2020).
4. G.V. Rodionov, A.P. Olesyuk, V.V. Boshlyakova, Animal Sci., **11**, 2-4 (2021). https://doi.org/10.25708/ZT.2021.67.22.001
5. L.A. Kalashnikova, Ya.A. Khabibrakhmanova, I.E. Bagal, V.L. Yaluga, V.P. Progerin, Dairy and beef cattle breed., **2**, 14–17 (2019).
6. V.I. Glazko, G.Yu. Kosovsky, T.T. Glazko, *Introduction to genomic selection* (Pleasant company, Moscow, 2012)
7. G.M. Goncharenko, T.S. Goryacheva, N.M. Rudishina, N.S. Medvedeva, E.G. Akulich, Bull. of Altai State Agrar. Univ., **12**, 113–117 (2013).
8. N.S. Marzanov, M.Kh. Tokhov, Z.L. Dokhova, Probl. of biol. of product. Animals, **2**, 79–94 (2014).
9. E. Kapelnitskaya, A. Shilova, Chief Livestock Techn., **4**, 34–39 (2015).
10. M. Gurses, H. Yuce, E. Etem, B. Pater, Animal Product. Sci., **58(5)**, 778–784 (2018). https://doi.org/10.1071/AN15131
11. T.M. Akhmetov, S.V. Tyulkin, E.F. Valiullina, Veterinar., **4**, 58–61 (2007).
12. O.G. Zaripov, Sci. notes of the Kazan State Acad. of Veter. Med., **193**, 100–104 (2008).
13. F.F. Zinnatova, Sh.K. Shakirov, Yu.R. Yulmetyeva, Bull. of the Rus. Acad. of Agricult. Sci., **6**, 65–67 (2012). (In Russ.)
14. N.Yu. Safina, E.R. Gainutdinova, Sh.K. Shakirov, Agricult. Sci. J., **10**, 93–97 (2021). https://doi.org/10.28983/asj.y2021i10pp93-97
15. Yu. Yulmetyeva, Sh. Shakirov, A. Minnakhmetov, N. Fatkhutdinov, Dairy and beef cattle breed., **7**, 23–26 (2013).
16. Yu.R. Yulmetyeva, F.F. Zinnatova, E.N. Rachkova, L.V. Shamsieva, Sh.K. Shakirov, Achievements of sci. and technol. of the agro-industr. Complex, **11**, 99–101 (2015).
17. G. Thaller, C. Kühn, A. Winter, G. Ewald, O. Bellmann, J. Wegner, H. Zühlke, R. Fries, Anim. Genet., **5**, 354–357 (2003).
18. B.M. Sørensen, T.L. Furukawa-Stoffer, K.S. Marshall, E.K. Page, Z. Mir, R.J. Forster, R.J. Weselake, Lipids, **10**, 1043–1049 (2005).
19. L. Pannier, A.M. Mullen, R.M. Hamill, P.C. Stapleton, T. Sweeney, Meat. Sci., **3**, 515–518 (2010).
20. B. Kaupe, H. Brandt, E. Prinzenberg, G. Erhardt, J. of Animal Sci., **85**, 11–21 (2007).
21. M.G. Smaragdov, Genet., **1**, 126–132 (2011).
22. L.A. Kalashnikova, *DNA technologies for assessing farm animals* (VNII plem, Lesnye Polyany, 1999).
23. R.J. Slepman, C.A. Ford, P. McEihinney, G.C. Gregory, R.G. Shell, J. of Dairy Sci., **85**, 3514–3517 (2002). https://doi.org/https://doi.org/10.3168/jds.S0022-0302(02)74440-8

Microbial biotechnology of milk whey processing for agricultural needs

Komil Astanakulov [1*], *Orifjon* Abdiev [2], *Jonibek* Ulashov [3], *Dilnoza* Ergasheva[4], *Irina A.* Zatsarinnaya [5], *Nikolay A.* Kupriyanenko [5]

[1] Tashkent Institute of Irrigation and Agricultural Mechanization Engineers" National Research University, 39, Street Kari Niyaziy, Tashkent, 100000, Uzbekistan
[2] Almalyk branch of the Tashkent State Technical University named after Islam Karimov, Almalyk, Uzbekistan
[3] Tashkent State Transport University, street Adilkhodjaeva 1, Mirabad district, Tashkent, 100067, Uzbekistan
[4] Termiz State Pedagogical Institute, Termiz, Uzbekistan
[5] Kuban State Agrarian University named after I. T. Trubilin, Krasnodar, Russia

Abstract. There are a number of approaches to whey utilisation, from direct feeding to animals to processing into products for the food industry. However, these methods have limitations, including high moisture content and processing costs. An alternative promising avenue is microbial biotechnology, which can convert whey into valuable products rich in protein, vitamins and other bioactive substances. This approach not only reduces resource losses but is also an environmentally friendly method. The aim of this work was to study the possibilities of microbial biotechnology application for whey processing, as well as to analyse the existing methods and prospects for their development. During the study, different types of whey were studied, their chemical composition was determined and experiments on whey digestion using different microorganisms were carried out. The analyses showed that whey is a valuable source of nutrients, particularly lactose (4.2-5.0%), which can serve as a substrate for microorganisms. The dry matter content varied from 6.4 to 6.6% and protein from 0.8 to 1%. After digestion, the dry matter content of the whey increased by 15-20% and the amount of protein increased by 2.5-3 times, for example, with Lactobacillus casei, the protein content increased from 1% to 2.7% in 48 hours, and with Saccharomyces cerevisiae to 3.1% in the same period. The amino acid composition also showed a significant increase in lysine content from 13.7% to 20.2%. The results showed that whey is a valuable raw material for obtaining various products that can be used as feed and food ingredients.

1 Introduction

Currently, there are several approaches to solving the problem of whey utilisation and utilisation. One of the most common methods is to feed it directly to farm animals. While this method has several advantages as a source of nutrients, it also has limitations. Firstly, whey has a high moisture content, which makes it difficult to transport and store, especially in regions remote from dairies. Secondly, its use in natural animal diets can lead to nutritional imbalances and digestive disorders.

To solve the problem of storage and transportation of production waste, its utilisation in various industries can be considered [1].

Another approach is to process whey into various products such as whey concentrates, lactose and other ingredients used in food and pharmaceutical industries. These methods are undoubtedly highly cost-effective, but require significant capital investment in equipment and technology, making them unaffordable for many businesses. In addition, the use of chemical methods of whey preservation for further processing leads to the formation of a number of harmful substances, which negatively affects the quality of the final product.

In this context, a promising direction for solving the problem of whey utilisation is the use of microbial biotechnology methods. This direction is based on the ability of various types of microorganisms, such as yeasts, lactic acid bacteria and others, to effectively utilise whey nutrients for their growth and reproduction. This process results in the conversion of whey into biologically valuable products rich in protein, vitamins and other bioactive substances. Microbial technologies can also be applied in the production of biofertilisers, which can improve soil conditions, which, in turn, positively affects the efficiency of agriculture [2].

This approach has a number of important advantages. Firstly, it allows obtaining valuable products from waste, thus reducing resource losses. Secondly, microbial biotechnology is an environmentally friendly processing method that does not require the use of harmful chemicals. Thirdly, the obtained products can be used as

* Corresponding author: k.astanakulov@tiiame.uz

feed additives for animal husbandry, food ingredients and even medicines, which expands the range of whey applications [3]. The relevance of using microbial biotechnology for whey processing is also due to its low cost and the possibility of organising production at small and medium-sized enterprises, which emphasises the importance of legal regulation in the field of organic agriculture [4]. The use of microbial synthesis allows not only to solve the problem of waste disposal, but also to obtain an additional source of high-quality protein, which can fill its deficiency in the diet of humans and farm animals, especially in the feeding of broilers [5]. In addition, microbial synthesis makes it possible to obtain products with predetermined properties by selecting specific microorganisms and cultivation conditions, which opens new perspectives in the field of biotechnology.

The aim of this work is to study the possibilities of microbial biotechnology application for whey processing, to analyse existing processing methods and prospects for their further development.

2 Method and materials

As part of the research on the processing of milk whey using microbiological methods, a number of experimental works were carried out. Initially, the composition and properties of different types of whey obtained during milk processing were studied: whey, curd and casein. To determine the chemical composition, including dry matter, protein, lactose, ash and acidity, a laboratory complex of analytical equipment was used. In particular, the dry residue was determined by drying in a SNOL desiccator at 105 °C to constant weight. Protein content was analysed by the Kjeldahl method, for which a digestor for combustion of organic substances, a distillation apparatus for ammonia distillation and a titrator for quantitative determination were used. Lactose content was determined by the colorimetric method on a KFK-2 spectrophotometer, with a wavelength corresponding to the maximum absorbance of the coloured compound. Serum pH was measured using a Hanna pH meter.

For further studies, milk whey was used and subjected to digestion by various microorganisms to produce biomass. Laboratory stirred and temperature controlled bioreactors were used for cultivation of microorganisms. The temperature during cultivation was maintained at 30-37 °C, depending on the strain. Samples were taken at regular intervals to analyse the growth and biochemical composition of the grown biomass. Sampling was carried out in a sterile manner using a laminar flow box. In the study of condensed and dry whey, their moisture content, amino acid composition and nutritional value were measured. Biomass and whey were separated using laboratory centrifuges, and samples were dried in vacuum desiccators. The obtained samples were examined for amino acid composition using an automatic amino acid analyser.

3 Results and discussion

Within the framework of the conducted research, the task of studying the possibilities of secondary processing of whey by applying microbiological methods was set, which is an urgent task in the context of rational use of raw materials and creation of products with increased nutritional value, as well as the creation of sustainable agricultural systems [6, 7]. To achieve this goal, a comprehensive approach was implemented, including several stages. At the first stage, a detailed analysis of the composition of different types of whey, which are by-products of the dairy industry, was carried out. Cheese, curd and casein whey obtained from cheese, curd and casein production, respectively, were investigated.

Key parameters such as dry matter, protein, lactose, ash content and acidity were determined for each of these whey products. The analyses were carried out using standard laboratory methods. Dry residue was determined by gravimetric method, by drying the whey suspension in a desiccator at 105°C until constant weight was reached. Protein content was measured by the Kjeldahl method based on organic nitrogen mineralisation followed by titration. The amount of lactose was determined by the colorimetric method on a spectrophotometer, where lactose interacted with the reagent to form a coloured compound, from the intensity of which the concentration was determined. Acidity was determined by titrimetric method. The data obtained gave an idea of the chemical composition of each type of whey. In the next stage of the study, milk whey, as the most available and common type, was subjected to digestion by various microorganisms in order to obtain biomass. For this purpose, specially prepared laboratory bioreactors equipped with stirring and temperature control systems were used. Various strains of lactic acid bacteria, yeasts and other microorganisms characterised by high growth rate and ability to utilise lactose, the main carbohydrate of whey, were selected as biomass producers. The temperature regime during cultivation was set at the optimum level for each microorganism, usually in the range of 30-37°C. The culturing process was carried out over several days, and samples were taken at regular intervals to assess microbial growth and analyse the biochemical composition. Sampling was carried out in a laminar box to maintain the sterility of the experiment. In parallel with culturing, studies on thickening and drying of serum were carried out in order to increase the concentration of dry matter and extend the shelf life. A rotary film apparatus was used for thickening and a spray dryer for drying. In both cases, process parameters such as temperature, product feed rate and processing time were controlled. This produced samples with different moisture contents, which were then also analysed for chemical composition, including amino acid content.

The results of the analyses showed that whey is a valuable source of nutrients (Table 1), especially in the context of sustainable agrarian land use [8]. In particular, whey contains between 4.2 and 5.0% lactose, which can serve as an excellent substrate for microbial growth and thus influence the soil organic carbon content when it is utilised [9]. The total dry matter content of the studied wheats ranged from 6.4 to 6.6%, with protein content ranging from 0.8-1.0%. The ash content ranged from 0.5 to 0.8% and the acidity ranged from 20 to 75° T, depending on the type of whey. After digestion, the dry matter content of whey increased by 15-20% and the amount of protein increased by 2.5 to 3 times, which is also consistent with the fact that biotechnology helps to reduce carbon dioxide emissions while producing valuable products [10, 11]

Table 1. Composition of whey

Indicator	Milk whey content, %			
	Cheesecake	Cheese curd	Casein	dry
Water	93.3	95.6	94.5	35
Dry matter	6.6	6.4	5.8	95–97
Protein	1	0.8	0.9	10–14
Fats	0.4	0.3	0.3	0.7-1.5
Lactose	5.0	4.4	4.2	66
Ash	0.5	0.6	0.8	6–9
Acidity, °T	20	60–75	44	—
pH	6.1	4.7	4.6	—

For example, when a strain of lactic acid bacteria Lactobacillus casei was used, the protein content increased from 1.0% to 2.7% in 48 hours of cultivation. When Saccharomyces cerevisiae yeast was used, the protein content increased to 3.1% over the same period. At the same time, the amount of lactose decreased as it was consumed by the microorganisms. As a result of whey fermentation, its acidity also underwent changes, resulting in an increase in titratable acidity up to 80-120°T depending on the strain used. The obtained data are consistent with the results of other studies, where it was shown that whey can be an effective substrate for microbiological synthesis, which provides an increase in the nutritional value of the final product. Whey was thickened to a dry matter content of 40-60% and dried to a residual moisture content of 3-5%. As a result of concentration, the dry matter content increased 6-8 times, while the protein content increased proportionally. For example, after thickening of whey, the protein content increased from 0.9% to 5.4% and after drying to 12.4%. The amino acid composition of dried whey and biosynthesis products showed a significant increase in the content of lysine, histidine and arginine, which is important in terms of increasing the biological value of the product, as well as in the application of biofertilisers to maintain soil fertility [12]. In particular, the lysine content of dried whey increased from 13.7% to 20.2%, as a result of microbiological treatment, which corresponded to the level of pasteurised milk (20.7%). The amino acid composition of the grown biomass was also analysed and the results showed a significant increase in essential amino acids such as lysine, valine, isoleucine and leucine (Table 2). This research has shown that whey is a valuable raw material for obtaining various products with enhanced biological value. The results of the study also showed that thickening and drying of whey are effective ways of its preservation and transport. The application of microbiological methods for whey processing allows obtaining products that can be used as feed for farm animals and food industry, as well as for solving the problems of waste treatment and preventing antibiotics from entering the soil [13].

Table 2. Amino acid composition of the whey studied.

Amino acid	Content, %		
	Milk whey	Pasteurised milk	Kefir fungi
Lysine	13.7	9.9	8.1
Valine	7.0	7.3	2.7
Isoleucine	7.3	7.3	3.0
Leucine	18.7	12.3	4.5
Methionine	3.8	3.3	—
Alanine	8.8	3.7	7.5
Histidine	1.9	3.4	6.2
Glutamic	23.4	27.3	19.2
Proline	4.9	11.5	29.3
Serine	6.0	7.0	17.7
Tyrosine	4.5	7.0	1.8

Analysing the composition of different types of whey obtained during the study revealed certain patterns and differences. For example, the content of dry matter in cheese whey averaged 6.6%, which, as mentioned earlier, is in agreement with the data of other researchers, where this indicator varies from 6% to 7%. In our study, the protein content in whey was 1.0%, which is also within the normal range, given that the literature sources cite values ranging from 0.8% to 1.2. The lactose content in the whey was 5%, which corresponds to the range of 4.5%-5.5% reported in the works of other authors. The ash content of the whey was 0.5%. As for curd whey, it had a dry matter content of 6.4%, protein 0.8%, lactose 4.4% and ash 0.6%. These values are marginally lower than those of whey, but also fit within the general trends noted by other researchers, where dry matter content of 6 to 6.5% is reported for curd whey. Casein whey showed the following results: dry matter 5.8%, protein 0.9%, lactose 4.2% and ash 0.8%. Note that the acidity of the whey varied but averaged about 20° T for the cheese whey, 60-75° T for the curd whey, and 44° T for the casein whey, which is consistent with the data presented in Table 2 of the article under review. The pH of whey also varied in the range of 4.6 - 6.1, which is also in accordance with the data presented in the article.

Special attention was paid to the amino acid composition of whey. Thus, the content of lysine in milk whey according to our data was 13.7%, valine - 7.0%, isoleucine - 7.3%, leucine - 18.7%, methionine - 3.8%, alanine 8.8%, histidine 1.9%, glutamic acid - 23.4%, proline 4.9%, serine 6%, tyrosine 4.5%. These data are comparable to the results of the studies reported in Table 2 of the article under review, where the values of lysine 13.7%, valine 7.0%, isoleucine 7.3%, leucine 18.7% respectively. Small discrepancies may be due to differences in the methods of analysis and conditions of serum preparation. The content of individual amino acids in kefir fungi, also presented in Table 3, differs slightly, e.g. for lysine 8.1%, valine 2.7%, leucine 4.5%, indicating differences in the amino acid composition of different objects. During digestion of whey by yeast, according to our data, the protein content increases on average by 2.7 times, which is in agreement with the range reported in the literature (2.5-3 times). The total amount of amino acids increases on average 2.4 times, and the content of such amino acids as lysine, histidine and arginine increases 5-6 times.

Evaluation of the results confirms the feasibility of using whey in agriculture, especially after its processing by microbiological means. In particular, the use of condensed whey in the feeding of piglets, as shown in our studies, increases the nutritional content of feed on average by 9.3 times compared to natural whey, while, as noted by many researchers, the economic performance of farms improves [14].

The feed units of natural whey are 0.13 and condensed whey 1.2. Fermentation of whey by yeast allows to obtain a protein-vitamin product with increased amino acid content, suitable for inclusion in animal diets, as well as for increasing productivity and reducing feed costs [15-17]. According to literature data, 1 tonne of Bio-CCM derived from yeast whey can replace up to 8 tonnes of whole milk, thus saving valuable protein resources and reducing feed costs. In addition, the results of other studies have shown that it is advisable to introduce up to 20% of whey in dry matter into the diet of pigs, which helps to improve their productivity and reduce the negative impact on soil resources [18]. The use of such products as 'Promix', which according to other studies contributes to an increase in weight gain of pigs up to 10-15%, or 'Provilact' for poultry allows not only to reduce feed costs, but also to increase the biological value of the diet, which, as a consequence, increases the efficiency of animal husbandry, and it is important to take into account the control of technological processes [19]. Information systems can be used to automate these processes [20]. The use of environmentally friendly methods and technologies is an important aspect in any production and also consider the economic benefits.

Thus, milk whey, especially after microbiological processing, is a promising and valuable resource for agriculture, which is confirmed by our findings and literature data.

4 Conclusion

Analyses of the chemical composition of different types of whey, including whey, curd and casein, confirmed the presence of essential nutrients such as lactose, protein and mineral salts that can be efficiently utilised by microorganisms for growth and biosynthesis. Specifically, the lactose content of the whey ranged from 4.2% to 5.0% and the protein content from 0.8% to 1.0%. Experiments on whey digestion by different microorganisms showed that this process leads to a significant increase in the protein content of the resulting product. When lactic acid bacteria such as Lactobacillus casei were used, the protein content increased from 1.0% to 2.7% in 48 hours of cultivation, and when yeast Saccharomyces cerevisiae was used, the protein content increased to 3.1%. It is important to note that during the fermentation process, lactose consumption by microorganisms occurs, which leads to a decrease in its content and an increase in the acidity of the medium to 80-120°T. The conducted studies on whey thickening and drying have shown the possibility of obtaining concentrates with high dry matter content (40-60%) and powdered products with residual moisture of 3-5%, which significantly increases their transportability and shelf life. Concentrating whey increases the dry matter content by 6-8 times and the protein content proportionally. After thickening of whey, the protein content increased from 0.9% to 5.4% and after drying to 12.4%. Analysis of the amino acid composition of dry whey and biosynthesis products revealed a significant increase in the content of essential amino acids such as lysine, histidine and arginine. For example, the lysine content of dried whey, after microbiological treatment, increased from 13.7% to 20.2%, which corresponds to the level of pasteurised milk. The obtained data confirm that whey is a promising and valuable raw material for obtaining biologically valuable products that can be used as feed for farm animals, as well as in the food industry.

References

1. A.L.S. Hansted, W.J. Newell, K.A. Hawboldt. An initial analysis of the market potential of a wet ash waste stream from a pulp and paper mill in Newfoundland and Labrador. Ind. Crops Prod., **225,** 120557 (2025). DOI: 10.1016/j.indcrop.2025.120557.

2. V. Langraf, K. Petrovičová. Morphometric variability of specimens in Ground Beetle (Carabidae) communities in agrarian landscape forests and protected areas. Rev. Soc. Entomol. Argent**.**, **83,** 1-8 (2024). DOI: 10.25085/rsea.830401.

3. Y. Chen, Y. Guo, X. He, B. Tan, Z. Liao, A. Chen, X. Gu, X. Li, X. Chen, B. Chen, S. Lin, W. Li, P. Hu, X. Zhu, W. Zhao, J. Niu. Comprehensive utilization of black soldier fly (Hermetia illucens) larvae: extraction, recovery and characterization of peptide, chitin and melanin and scaling-up trial. Sep. Purif. Technol., **361,** 131262 (2025). DOI: 10.1016/j.seppur.2024.131262.

4. L. Yerkinbayeva, A. Myrzakarim, A. Moldagaliyeva, B. Kalymbek, R. Yerezhepkyzy, D. Nurmukhankyzy. The role of legal regulation in promoting organic agriculture to achieve sustainable food security. J. Lifestyle SDG'S Rev., **5,** e04197 (2025). DOI: 10.47172/2965-730X.SDGsReview.v5.n01.pe04197.

5. T. A. Kuandykov, T. D. Karmanov, E. I. Kuldeyev, K. K. Yelemessov, B. Z. Kaliev, News of the National Academy of Sciences of the Republic of Kazakhstan, Series of Geology and Technical Sciences, **2022(3),** 142–154, (2022). DOI 10.32014/2022.2518-170X.186.

6. Yu.N. Serdechnaya, D.V. Serdechnyy, N.A. Andriyanov, S.A. Korchagin. Assessing the innovative attractiveness of agricultural enterprises in mountainous

areas. Sustainable Development of Mountain Territories, **16,** 1165–1175 (2024). DOI: https://doi.org/10.21177/1998-4502-2024-16-3-1165-1175.

7. A. I. Trunova, S. O. Kurashkin, K. A. Bashmur, R. B. Sergienko, Metals, **13(7),** 1234 (2023). DOI: 10.3390/met13071234.

8. M. N. Nurpeisov, K. V. Ruchkina, V. A. Androkhanov, S. K. Makenova. Assessment of geoecological factors and conditions affecting sustainability of agricultural land use in the forest-steppe zone of Altai Krai. Sustainable Development of Mountain Territories, **15,** 419–430 (2023). DOI: 10.21177/1998-4502-2023-15-2-419-430.

9. M.-Y. Li, W. Wang, H.-H. Yin, Y. Chen, M. Ashraf, H.-Y. Tao, S.-S. Li, W.-Y. Wang, C.-L. Yang, Y.-L. Xiao, L. Zhu, Y.-C. Xiong. The functional role of arbuscular mycorrhizal fungi in enhancing soil organic carbon stocks and stability in dryland. Soil Till. Res., **248,** 106443 (2025). DOI: 10.1016/j.still.2024.106443.

10. A. A. Kulikova, E. I. Khabarova. Biological working agents use to reduce the carbon footprint. Sustainable Development of Mountain Territories, **16,** 311–321 (2024). DOI: 10.21177/1998-4502-2024-16-1-311-321.

11. A. A. Sokolov, L. G. Orlova, K. A. Bashmur, R. I. Kuzmich, Mining Informational and Analytical Bulletin, **11,** 278-291 (2023). DOI: 10.25018/0236_1493_2023_111_0_278.

12. Q. Gao, L. Wang, Y. Fang, Y. Gao, L. Ma, X. Wang, Y. Li, X. Wu, Z. Du. Conservation agriculture boosts topsoil organic matter by restoring free lipids and lignin phenols biomarkers in distinct fractions. Soil Till. Res., **248,** 106463 (2025). DOI: 10.1016/j.still.2025.106463.

13. I. Akzharkyn, K. Yelemessov, D. Baskanbayeva, Applied Sciences (Switzerland), **14(17),** 7567, (2024). DOI: 10.3390/app14177567.

14. A. Ogunleye, D. Olude, A. Kehinde, V. Tanimonure, C.A. Wongnaa, O. Ajayi. Impact of credit constraints on adoption of biofortified food crop among farming households in Nigeria. Sustain. Futures, **9,** 100399 (2025). DOI: 10.1016/j.sftr.2024.100399.

15. V. Brigida, E.A. Efremenkov, S.N. Sorokova, Q. Mengxu. Tailings Utilization and Zinc Extraction Based on Mechanochemical Activation. Materials, **16,** 726 (2023). DOI: 10.3390/ma16020726.

16. L.S. Malyukova, V.V. Kondratiev, V.V. Bukhtoyarov. Circular Mining Wastes Management for Sustainable Production of Camellia sinensis (L.) O. Kuntze. Sustainability (Switzerland), **15,** 11671 (2023). DOI: 10.3390/su151511671.

17. I.Y. Matasova, M.A. Modina. Ecologization of underground coal mining by means of ash use in backfill preparation. Min. Inform. Anal. Bull., **10,** 123-135 (2024). DOI: 10.25018/0236_1493_2024_10_0_123.

18. K.I. Kravtsov, T.A. Panfilova. Influence of mountain factors on salt excess and soil toxicity in mountain conditions. Sustainable Development of Mountain Territories, **15,** 784-797 (2023). https://doi.org/10.21177/1998-4502-2023-15-3-784-797.

19. T. A. Panfilova, K. A. Bashmur, V. V. Kondratiev, Mining Informational and Analytical Bulletin, **11-1,** 239-251 (2023). DOI: 10.25018/0236_1493_2023_111_0_239.

20. S. I. Kondratyev, D. Baskanbayeva, Y. Sarsenbayev, V. A. Turkin, International Journal of Hydrogen Energy, **95,** 212–216, (2024). DOI: 10.1016/j.ijhydene.2024.11.182.

Antimicrobial properties of a new bactericide for veterinary disinfection

Gulizar Sh. Shcherbakova[1], *Nikolay I.* Popov[1], *Ekaterina N.* Shuteeva[1], *Nikolay N.* Kuvshinchikov[1], *Viktor A.* Pirozhikhin[1], *Olga A.* Gruznova[1,2*] and *Dmitry V.* Gruznov[1]

[1]All-Russian Research Institute of Veterinary Sanitation, Hygiene and Ecology – Branch of Federal State Budget Scientific Institution "Federal Scientific Center – K.I. Skryabin, Ya.R. Kovalenko All-Russian Research Institute of Experimental Veterinary Medicine, Russian Academy of Sciences", 5 Zvenigorodskoye highway, Moscow, 123022, Russia
[2]N.N. Semenov Federal Research Center for Chemical Physics, Russian Academy of Sciences, 4 Kosygina st., Moscow, 119334, Russia

Abstract. The work presents the results of a study on a new disinfectant known as "Aseptobrom" for use in veterinary medicine. It contains didecyldimethylammonium bromide and alcohols as active substances. The laboratory trials showed that the disinfectant had significant antimicrobial activity against *Escherichia coli* (strain 1257), *Staphylococcus aureus* (strain 209-P) and *Mycobacterium* (strain *terrae*). As a result, it was found that *S. aureus* was more sensitive than *E. coli*. When rough, moisture-absorbing surfaces were contaminated with *E. coli*, disinfection was achieved with a 1% solution (3-hour exposure, disinfectant consumption rate – 0.5 L/m^2). In case of contamination with staphylococcus, a 0.4% solution was used. Other specified parameters were not changed. To destroy *Mycobacterium*, a concentration of 6.0% was required with an exposure of 24 h. Moreover, the disinfectant was applied twice with an interval of 60 min. The consumption rate of the preparation for each irrigation was 0.5 L/m^2. A study of the corrosion properties of the preparation showed that, in comparison with sodium hydroxide, it has a slight corrosive effect on metal surfaces, in particular on stainless steel and iron test plates.

1 Introduction

Agricultural products are the basis for maintaining a full-fledged human life. Nowadays, the problem of biological safety is more relevant than ever and requires maximum attention from researchers in the field of veterinary science and medicine [1-6]. Therefore, the key goals of veterinary sanitation are the prevention and elimination of infectious diseases [7]. Unfortunately, the incorrect use of antibiotics and disinfectants, violation of disinfection regimes (deviation from the established concentrations of working solutions, exposure time and consumption of disinfectant solution per m^2) leads to the emergence of resistance in microorganisms to the disinfectants used. Biofilm formation also occurs. The above factors necessitate an increase in the concentrations of preparations and the cost of disinfection, and, as a result, environmental pollution with chemicals harmful to health [8-11].

In this regard, one of the key tasks of veterinary sanitation is the development and implementation into practice of new antimicrobial agents that would have high disinfectant activity at low concentrations, environmental safety, and low corrosive properties [12]. After the introduction of sanctions by Western countries against Russia and the outflow of imported bactericide manufacturers, the issue of economic accessibility is of no small importance. Therefore, disinfectants, in addition to the specified requirements, must have a low cost for the consumer [13]. The most promising ones in this regard are the creation of composite disinfectants based on quaternary ammonium compounds, guanidines, tertiary amines, peroxide compounds in combination with glutaraldehyde, glyoxal in the presence of various functional additives and stabilizers [14,15].

The purpose of current research was to study the bactericidal properties of a new disinfectant for preventive and forced disinfection of veterinary inspection objects.

2 Method and materials

The "Aseptobrom" disinfectant was used in this work. It was manufactured in accordance with Technical Specifications 9392-003-30081846-2013. This is a colorless to light yellow liquid. "Aseptobrom" contains didecyldimethylammonium bromide (53.1%) and alcohols (isopropyl and polyoxyethylated) as active ingredients. The disinfectant has cleaning properties, easily mixes with water.

* Corresponding author: *gruznova_olga@bk.ru*

The bactericidal properties of the disinfectant were determined by serial dilutions of the preparation in sterile 10 mL vials. A pre-prepared disinfectant solution was successively diluted with distilled water to obtain a series of dilutions from 1:50 to 1:864336.9. Next, 0.04 mL of a 2-billion-day test culture (*E. coli* strain 1257) was added to each vial at 1-minute intervals. The exposure was 10 and 30 min. After completion of the specified exposures, the samples were taken with a sterile bacteriological loop and transferred to vials with meat-peptone broth (MPB).

To determine how many times the bactericidal activity of the preparation is reduced in the presence of protein (for example, manure, droppings and various organic secretions), an experiment was conducted with a bactericidal dilution as in the first experiment, but instead of a pure test culture of the microorganism, a mixture of a 2-billion-day culture and horse blood serum previously inactivated at 57 °C for 30 min was introduced at a rate of 2 mL of serum per 1 mL of culture. Since the presence of protein reduces the bactericidal effect of the desinfectant, we took a smaller number of vials (20 pcs) and added 0.3 mL of the culture mixture with serum to each vial at 1-minute intervals with an exposure of 10 and 30 min.

To calculate the protein index (PI), the experimental values without adding protein protection were divided by the corresponding results with adding it. The resulting numbers were summed up and divided by 2.

The samples were incubated in a thermostat at 37 °C for 18-24 h. The final recording of the results was carried out after 6-7 days.

The determination of the bactericidal dilution of the disinfectant "Aseptobrom" with the test culture *S. aureus* strain 209-P was carried out similarly.

Wood, concrete, ceramic tiles and stainless steel were used as test objects in the laboratory tests. *Escherichia coli*, *Staphylococcus aureus* and *Mycobacterium* (strain *terrae*) were the bacterial contaminants. Sterile inactivated blood serum was used to simulate protein protection. Disinfection was carried out according to the method. The study of disinfectant properties was carried out in accordance with Methodological guidelines on the procedure for testing new disinfectants for veterinary practice.

The corrosive activity of the disinfectant was determined according to the Methodology for determining and evaluating the corrosive activity of detergents and disinfectants.

The criterion for the preparation's effectiveness is 100% death of the test microorganism in the presence of its growth in control samples. When calculating concentrations, the desinfectant was taken as a 100% substance. All experiments were carried out in triplicate.

3 Results and discussion

Based on the data obtained, it was found that "Aseptobrom" has a stronger bactericidal activity against *Staphylococcus aureus* than against *E. coli*. The bactericidal effect of the test agent (Table 1) against *E. coli*, in the absence of protein, was 1:21343.9 with an exposure of 10 min and 1:41834.1 with an exposure of 30 min. In the presence of protein (inactivated horse blood serum), the lowest concentration capable of suppressing the microorganism was 1:2024.8 with an exposure of 10 min and 1:2834.7 with an exposure of 30 min. The protein index of the disinfectant is 12.65.

After conducting experiments with serial dilutions without protein, the phenolic coefficient was calculated and it was 217.4. The phenol coefficient demonstrates the ratio of bactericidal dilutions of the test agent and phenol when setting up relevant experiments with the test culture of *E. coli*.

Table 1. Results of bactericidal tests against *E. coli*

	Protein		No protein	
	Exposure, min			
	10	30	10	30
Dilutions	1:21343.9	1:41834.1	1:2024.8	1:2834.7
Protein index	12.65			
Phenolic coefficient	217.4			

The bactericidal effect of the test agent against *Staphylococcus aureus* without protein was 1:81994.5 at an exposure of 10 min and 1:114792.8 at 30 min. In the presence of protein, the lowest dilution of the preparation that had a bactericidal effect was 1:15245.7 with an exposure of 10 and 30 minutes (Table 2). The protein index is 6.46.

Table 2. Results of bactericidal tests against *S. aureus*

	Protein		No protein	
	Exposure, min			
	10	30	10	30
Dilutions	1:81994.5	1:114792.8	1:15245.7	1:15245.7
Protein index	6.46			

Therefore, the serial dilutions studies of "Aseptobrom" revealed a higher sensitivity of *S. aureus* to the disinfectant than *E. coli*. These data correlate with the results of experiments using the serial dilution method, where the gram-positive microorganisms showed a higher sensitivity to the preparation than the gram-negative own, which is probably due to differences in the structure of the cell wall of these microorganisms.

According to the results of the studies of the disinfectant activity of "Aseptobrom" on smooth test surfaces artificially contaminated with the microorganism *E. coli* (Table 3), disinfection was achieved with a solution of 0.1% concentration, exposure of 1 h at a consumption rate of 0.25-0.3 L/m^2. For rough, moisture-absorbing surfaces (wood, concrete), it is necessary to use a 1.0% solution with an exposure of 3 h and a consumption rate of 0.5 L/m^2.

Table 3. Disinfection test surfaces against *E. coli*

Solution concentration, %	Exposure, h	Test surface: Stainless steel	Tiles	Wood	Concrete
0.1	1	—	—	×	×
	3	—	—	×	×
0.5	1	×	×	+	+
	3	×	×	+	+
1.0	1	×	×	+	+
	3	×	×	—	—

Note: (+) is growth of the test culture; (-) is absence of growth of the test culture.

The experiments of test surfaces artificially contaminated with *S. aureus* (Table 4) showed that disinfection of smooth surfaces was achieved with a 0.1% solution with an exposure of 1 h and a consumption rate of 0.25 L/m^2. To disinfect rough, moisture-absorbing surfaces, it is necessary to use a 0.4% solution with an exposure of 3 h and a consumption rate of 0.5 L/m^2.

Table 4. Disinfection test surfaces against *S. aureus*

Solution concentration, %	Exposure, h	Test surface: Stainless steel	Tiles	Wood	Concrete
0.05	1	+	+	×	×
	3	+	—	×	×
0.1	1	—	—	×	×
	3	—	—	×	×
0.4	1	×	×	+	+
	3	×	×	—	—
0.5	1	×	×	—	—
	3	×	×	—	—

Note: (+) means growth of test culture; (-) means no growth of test culture.

Therefore, as a result of laboratory studies of the disinfectant "Aseptobrom", it was established that the disinfectant has significant bactericidal properties against *E. coli* strain 1257 and *S. aureus* strain 209-P. It is known that the microorganism *S. aureus* is classified in group II of resistance to chemical disinfectants and is equated to the causative agents of diseases such as foot-and-mouth disease, smallpox, tularemia, rabies, plague of all types of animals, aspergillosis, etc. (total 51 items) [12]. In this way, the high susceptibility of staphylococcus to the preparation shows the promise of further research into this desinfectant.

In experiments with *Mycobacterium* strain *terrae* (Table 5), the disinfectant effect of solutions with a concentration of 5.0 to 9.0% was tested only on rough surfaces made of wood and concrete with single and double application of "Aseptobrom" at a rate of 0.5 L/m^2 for each irrigation and exposures of 3 and 24 h.

The conducted studies established that double treatment of test surfaces with an interval of 60 min after the first irrigation, the consumption of the agent 0.5 L/m^2 for each irrigation, the concentration of 6.0% ensured their disinfection with an exposure of 24 h.

Table 5. Disinfection test surfaces against *Mycobacterium* strain *terrae*

Solution concentration, %	Exposure, h	Processing frequency	Test surface: Wood	Concrete
9.0	3	×1	+	+
	24	×1	+	+
5.0	3	×2	+	+
	24	×2	+	+
6.0	3	×2	+	+
	24	×2	—	—

Note: (+) means growth of test culture; (-) means no growth of test culture.

A low aggressive impact on metal surfaces of the equipment and inventory used is one of the important parameters of modern disinfectants. The studies of the corrosion properties of 1.5% solutions of "Aseptobrom" were carried out on test plates made of steel of various grades and aluminum.

The corrosion activity was assessed based on the weight loss and color change of the plates in the experiments. Compared to the data obtained by treating the test objects with a standard 2% NaOH solution, the weight loss of aluminum plates when exposed to a 1.5% solution of "Aseptobrom" for 1 h was 1.55%; for iron plates it was 144.02%; 42.56% for stainless steel plates and 31.86% for galvanized steel plates. With an

https://doi.org/10.1051/bioconf/202516100036

exposure of 24 h, these values were 0.38%, 332.98%, 72.86%, and 58.72%, respectively.

The final characteristics of the corrosive effect of the product were given by assessing the coefficient and corrosion rate per year (Fig. 1-4).

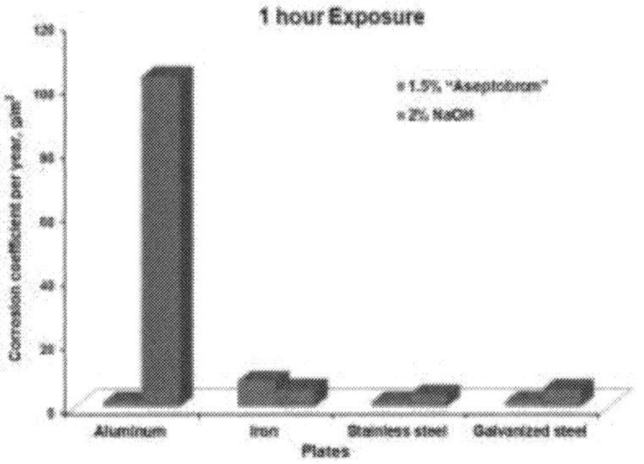

Fig. 1. The corrosion coefficient of "Aseptobrom" with 1-hour exposure.

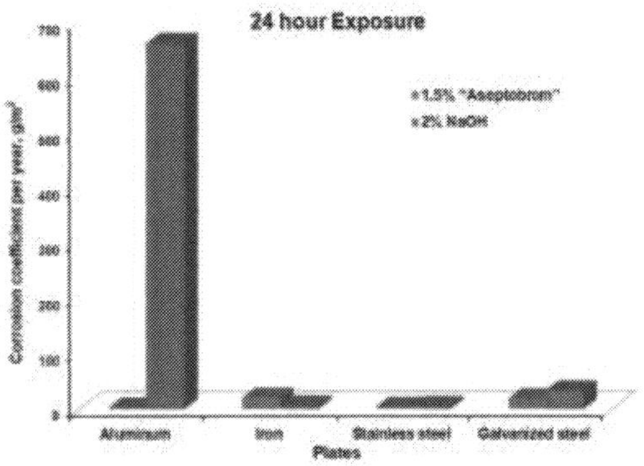

Fig. 2. The corrosion coefficient of "Aseptobrom" with 24-hour exposure.

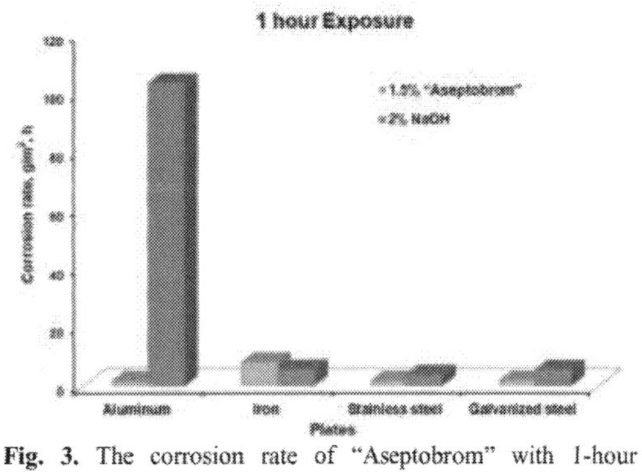

Fig. 3. The corrosion rate of "Aseptobrom" with 1-hour exposure.

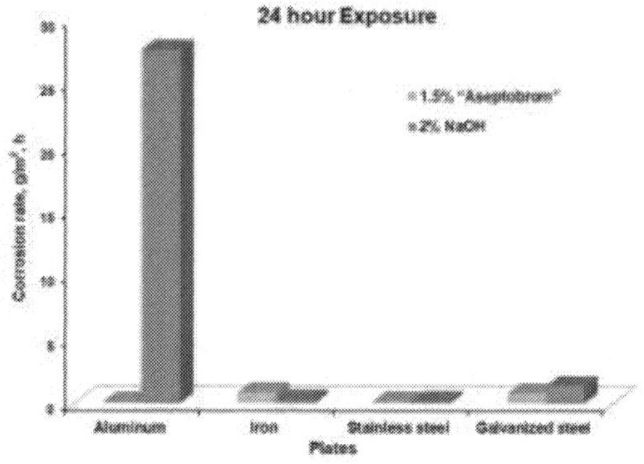

Fig. 4. The corrosion rate of "Aseptobrom" with 24-hour exposure.

4 Conclusion

"Aseptobrom" is a disinfectant containing didecyldimethylammonium bromide and alcohols as active ingredients. In laboratory conditions, it was found that "Aseptobrom" had high bactericidal activity against gram-positive (*S. aureus* – II group of resistance to chemical disinfectants) and gram-negative (*E. coli* – I resistance group) microorganisms, as well as mycobacteria (*Mycobacterium* – III resistance group). *S. aureus* was more sensitive to the preparation than *Escherichia*. Therefore, the protein index for the indicated microorganisms was 6.46 and 12.65, respectively.

It has been established that the preparation does not have a high corrosive effect on metal surfaces.

The effective bactericidal concentrations of "Aseptobrom" for treating various types of surfaces were determined. When rough, moisture-absorbing surfaces were contaminated with *E. coli*, disinfection was achieved with a 1% solution (3-hour exposure, disinfectant consumption rate – 0.5 L/m^2). In case of contamination with staphylococcus, a 0.4% solution was used. To destroy *Mycobacterium*, a concentration of 6.0% was required with an exposure of 24 h. Moreover, the disinfectant was applied twice with an interval of 60 min.

Hence, "Aseptobrom" is promising preparation for use in veterinary medicine. It can be recommended for conducting production tests at veterinary supervision objects that are free of infectious animal diseases.

References

1. M. Atasever, J. Biosci. and Biotechnol., **10 (1)**, 1-5 (2025)
2. G. Gebreyohannes, A. Nyerere, Ch. Bii, D. B. Sbhatu, Heliyon, **5 (8)**, e02192 (2019)
3. P. J. Fedorka-Cray, *Microorganisms and resistance to antibiotics, the ubiquity of Antibiotic Resistance by Microorganisms.* (Encyclopedia of Meat Sciences, Second Edition, 2014)

4. P. A. Asbell, Ch. M. Sanfilippo, H. H. DeCory, Diagn Microbiol Infect Dis., **108 (1)**, 116069 (2024)
5. H. Chen, Engineering (Beijing), **6 (1)** (2020)
6. D. Grace. Engineering, **6**, 8–9 (2019)
7. V. L. Kovalenko, P. L. Kovalenko, G. V. Ponomarenko, M. D. Kukhtyn, S. V. Midyk, Yu. V. Horiuk, V. M. Garkavenko, Ukr. J. Ecol., **8 (1)**, 547-550 (2018)
8. G. McDonnell, A.D. Russell, Clin. Microbiol. Rev., **12**, 147-199 (1999)
9. C. Zhang, F. Cui, G. Zeng, Z. Jiang, Z. Yang, G. Yu, M. Zhu, L. Shen, Sci. Total Environ., **518**, 352-362 (2015)
10. K. Hegstad, S. Langsrud, B. T. Lunestad, A. A. Scheie, M. Sunde, S. P. Yazdankhah, Microb. Drug Resist., **16(2)**, 91-104 (2010)
11. K. Rajkowska, A. Kozirog, A. Otlewska, M. Piotrowska, P. Nowicka-Krawczyk, B. Brycki, A. Kunicka-Styczynska, B. Gutarowska, Acta Biochim. Pol., **63 (1)**, 153-159 (2016)
12. H. Martin, M. F. Le Potier, P. Maris, Vet. J., **177**, 388-393 (2007).
13. T. Yamanaka, H. Bannai, K. Tsujimura, M. Nemoto, T. Kondo, T. Matsumura, J. Equine. Vet. Sci., **34**, 715-718 (2014)
14. T. Jantafong, S. Ruenphet, D. Punyadarsaniya, K. Takehara, Vet. World, **11 (5)**, 706-711 (2018)
15. R. J. Gosling, I. Mawhinney, K. Vaughan, R. H. Davies, R. P. Smith, Vet. Microbiol., **204**, 46-53 (2017)

Microbial protein production based on complex plant raw materials: technology development and efficiency assessment

Berdirasul Khudayarov[1,*], *Abduaziz* Abdukakhkharov[2], *Alim* Axmedov[3], *Nigora* Rustamova[4], *Mokhira* Kuvvatova[5]

[1] "Tashkent Institute of Irrigation and Agricultural Mechanization Engineers" National Research University, Tashkent, Uzbekistan
[2] Almalyk branch of the Tashkent State Technical University named after Islam Karimov, Almalyk, Uzbekistan
[3] Termez State University, Termiz, Uzbekistan
[4] Karshi State Technical University, Karshi, Uzbekistan
[5] Termiz State Pedagogical Institute, Termiz, Uzbekistan

Abstract. The possibility of using complex plant raw materials for microbial biomass production has been investigated. The chemical composition of different types of plant raw materials (grasses, beetroot, potatoes, corn, waste) of the northeastern region was analysed, which made it possible to develop an optimal nutrient medium for the cultivation of Candida scotti Tul-1. Cultivation was carried out in flasks and fermenter, controlling temperature, pH and aeration. The optimum concentration of sugar in the medium was found to be 4% and the culturing temperature was 28°C. The obtained biomass contained 60-62% protein with a balanced amino acid composition. The results demonstrate the promising use of plant raw materials for microbial protein production and waste utilisation. The developed technology can become the basis for sustainable and economically favourable production of protein feeds and food supplements. Further research is aimed at optimising the process and scaling up production.

1 Introduction

Currently, there are several approaches to solving the problem of protein deficiency. One of them is to increase production of traditional protein sources such as meat and leguminous crops. However, this approach faces limitations such as high resource consumption and environmental impacts, prompting the exploration of alternative methods like digital agriculture for sustainable food production [1]. Another approach is the development and introduction of new, non-traditional protein sources such as insects, algae and microorganisms. The use of insects for food appears to be an interesting option, but there are issues related to the acceptability of such a product to consumers and the scalability of production. Cultivation of algae also has potential, but requires significant investment in infrastructure and technology. An alternative solution is the cultivation of microorganisms, particularly relevant in the context of microcredit and digital innovation in fish farming [2].

Microbial protein from the cultivation of bacteria, yeast or fungi is a promising protein source with several advantages over conventional sources. Microorganisms are able to grow and multiply rapidly using a wide range of substrates, including agricultural and food waste. The use of artificial intelligence (AI) is increasingly important for diagnosing diseases and optimizing resource use in agriculture [3]. This allows not only to produce protein, but also to solve the problem of waste disposal, reducing the burden on the environment. In addition, microbial protein is characterised by a high protein content, a balanced amino acid composition and can be enriched with vitamins and minerals. Microbial protein production requires less land and water resources compared to traditional agriculture, which makes it more sustainable and environmentally friendly. This aligns with broader sustainability goals, including those related to edible plant consumption and production within the EU's climate neutrality targets [4] and the potential of urban agriculture in addressing food crises [5].

The relevance of the study of microbial protein production on complex plant raw materials is due to the need to find sustainable and cost-effective ways of protein production. The use of complex plant raw materials, including various agricultural wastes and by-products, makes it possible to reduce the cost of microbial protein production and simultaneously solve the problem of waste utilisation. This is especially important for regions with developed agriculture, where a large amount of plant waste is generated that can be used as a substrate for microbial protein production.

The purpose of this work is to study the possibility of using complex plant raw materials as the main component for microbial biomass production. The study of the chemical composition and yield of a variety of plant raw materials characteristic of the north-eastern

* Corresponding author: b.khudayarov@tiiame.uz

region of the country allows us to develop an optimal technology for the preparation of nutrient medium for the cultivation of microorganisms and production of microbial protein. Various sources of carbohydrate and cellulose-containing raw materials of both primary and secondary origin, such as herbal plants, sugar beet, topinambour, potato tubers, beet pulp, maize, maize pulp, fruit and berry pomace and grain bran, are investigated in this work, which opens prospects for the creation of sustainable and economically viable technology for microbial protein production.

2 Method and materials

In the framework of this work, a set of experimental studies was carried out to investigate the possibility of using complex plant raw materials for microbial protein production. The general scheme of the experiment included several stages: preparation and analysis of the chemical composition of various plant raw materials, development of nutrient medium based on complex plant raw materials, selection and cultivation of microbial strain, study of the influence of various factors on the growth and biosynthetic activity of microorganisms, as well as evaluation of yield and quality of microbial biomass.

The following equipment was used for the preparation of plant raw materials. For grinding of large fractions of plant raw materials, a laboratory mill of PM-120 brand was used, which allows obtaining a homogeneous mass with an adjustable degree of grinding. The obtained material was sieved through a set of sieves with different mesh diameters installed on a vibrating screen AM-200 to standardise the particle size. A laboratory press PL-20 was used to extract juice from fruit and berry raw material.

The chemical composition of raw materials was analysed using modern analytical methods. These methods are crucial for understanding the nutrient content of plant materials, as detailed in [6], which emphasizes the importance of data science in agriculture [7]. Effective energy management tools can also aid in monitoring and optimizing energy consumption. Moisture content was determined by drying in a SNOL 58/300 desiccator at 105°C to constant weight. Total nitrogen content was determined by the Kjeldahl method using a Gerhardt KJELDATHERM digestor and a VAPODEST 200 distiller. The concentration of sugars was determined using an ATAGO PAL-1 refractometer. Fibre content was determined by weight method after acid and alkaline hydrolysis.

Preparation of nutrient medium was carried out in a laboratory reactor of 5 litres with automatic control of temperature and pH. Tuttnauer 2540EL autoclave was used for sterilisation of the nutrient medium.

Cultivation of the yeast strain Candida scotti Tul-1 was carried out in 250 ml Erlenmeyer flasks on an ES-20/60 rocker incubator at 28°C and 120 rpm. A 10 litre fermenter with temperature, pH, dissolved oxygen and stirring rate control was used to scale up the process. Biomass concentration was determined by optical density measurement using a UNICO 2800 spectrophotometer at 600 nm and by weight method after centrifugation and drying.

3 Results and discussion

The experimental work presented in this article included several key stages aimed at evaluating the potential of complex plant raw materials as a substrate for cultivation of the yeast Candida scotti Tul-1 and microbial protein production. In the first phase, the chemical composition of various plant raw materials selected for their availability and prevalence in the north-eastern region of the country was carefully analysed. In particular, raw materials such as herbal plants, sugar beet, topinambur, potato tubers, beet pulp, maize, maize pulp, fruit and berry pomace and grain bran were studied. The main attention was paid to the determination of the content of carbohydrates, proteins, fibre, as well as macro- and microelements necessary for normal growth and development of yeasts. The obtained data served as a basis for the development of the optimal composition of nutrient medium.

At the second stage the nutrient medium was developed on the basis of complex plant raw materials. This development considered the principles of sustainable consumption and production of edible plants, aligning with the EU's climate neutrality goals, as reviewed by Mazur-Włodarczyk and Gruszecka-Kosowska [4]. This is especially relevant when integrating IoT and cloud-based agricultural waste management solutions [8].

During the development, the yeast needs for various nutrients, as well as the economic feasibility of using specific components of raw materials were taken into account. Special attention was paid to balancing the ratio of carbohydrates and nitrogen, as well as the addition of necessary trace elements and vitamins. The nutrient medium was prepared by mixing different types of plant raw materials in certain proportions, with subsequent processing to increase the availability of nutrients. In particular, methods of enzymatic hydrolysis and mechanical destruction of cell walls were used. These techniques are crucial for enhancing nutrient availability, a topic further explored in the context of nano fertilizers for sustainable farming by Gupta et al. [10]. Such nanofertilizers can advance agriculture with functional nanomaterials, as noted by Alam [11].

At the third stage, the yeast strain Candida scotti Tul-1 was cultured on the developed nutrient medium. The cultivation was carried out in a batch mode in Erlenmeyer flasks and fermenter. The parameters such as temperature, pH, dissolved oxygen concentration and stirring rate were monitored during cultivation. The effect of various factors such as sugar concentration, amount of inoculum and temperature on the growth and biosynthetic activity of yeast was studied. In particular, yeast growth rate, biomass yield, protein content in biomass and other microbial protein quality parameters were determined.

Each raw material was pre-treated, including grinding to particle size not more than 2 mm using a laboratory mill PM-120. This was followed by a hydrothermal treatment step at 100°C for 30 min to break down the cellular structure and increase the availability of nutrients, especially carbohydrates.

After preparation of raw materials, the chemical composition of each component was analysed in detail. The moisture content was determined by gravimetric method by drying the sample in a SNOL 58/300 desiccator at 105°C to constant weight. The average moisture content of grass plants was 75%, sugar beet - 82%, topinambour - 79%, potato tubers - 78%, beet pulp - 85%, maize - 14%, maize pulp - 70%, fruit and berry pomace - 88% and grain bran - 12%. Total nitrogen content was determined by the Kjeldahl method using Gerhardt KJELDATHERM digestor and VAPODEST 200 distiller, then converted to crude protein using a factor of 6.25. The highest protein content was found in grain bran (14.5%) and the lowest in sugar beet (1.2%). Sugar content was determined by refractometric method using ATAGO PAL-1 refractometer. Sugar beet and fruit and berry pomace showed the highest sugar content of 16% and 18%, respectively. The fibre content was determined by weight method after acid and alkaline hydrolysis. The maximum fibre content was found in beet pulp (22%) and grain bran (25%). Based on the obtained data, several variants of nutrient media were developed in which the proportions of different components were varied. The effective use of AI in these processes, especially for environmental sustainability, is explored by Tandon and Shaheen [12]. This includes optimizing energy efficiency in agriculture, which can be reanalyzed using AI, as noted by Hoseinzadeh and Astiaso Garcia [13]. To ensure sustainable practices, poultry farming strategies can be adapted from [14], which offer a comprehensive review of current strategies and future prospects. This variability underscores the importance of precise agroecological assessments, as discussed by [15, 16].

The yeast strain Candida scotti Tul-1 from the Republican Collection of Microorganisms of the Institute of Microbiology and Virology (Almaty) was selected for cultivation. The strain was pre-activated on agarised Chapek's medium with glucose, after which it was transferred into liquid nutrient medium optimised for Candida scotti Tul-1. Cultivation was carried out in 250 ml Erlenmeyer flasks containing 100 ml of nutrient medium on an ES-20/60 rocker incubator at 28°C and 120 rpm. Inoculation was carried out with a standardised yeast suspension in an amount of 5% of the media volume, corresponding to an optical density of OD600=0.5.

To study the effect of sugar concentration in the nutrient medium on biomass yield, sugar concentrations ranging from 2% to 7% in increments of 1% were tested. The maximum biomass yield (11.3 g/L) was observed at a sugar concentration of 4%. In our additional experiments, optimisation of nitrogen nutrition showed that addition of 0.1% ammonium sulphate to the medium with 4% sugar increased biomass to 12.8 g/L.

Temperatures of 24°C, 28°C, 32°C and 36°C were tested to investigate the effect of cultivation temperature on yeast growth. The best results were obtained at 28°C, which corresponds to the optimum temperature for this strain. Optimal temperatures are crucial, especially in areas affected by desert-dry steppe vegetation and environmental risks [9]. However, at 30°C in a fermenter with controlled pH 5.5, addition of phosphate buffer, and intensive aeration, the biomass of Candida scotti Tul-1 reached 13.5 g/L with protein content up to 59%. This confirms that proper control of the cultivation medium can significantly improve the results. Innovations in AI can further optimize greenhouse agriculture, as analyzed by [13]. This optimization can be further supported by analyzing predictive models of weather changes using neural networks, as detailed in [17].

To study the effect of aeration on yeast growth, different aeration regimes were tested in the fermenter, varying the stirring speed from 100 rpm to 300 rpm and the air supply from 0.5 VVM (volumes of air per volume of medium per minute) to 1.5 VVM. The best results were obtained at a stirring speed of 200 rpm and an air supply of 1.0 VVM. In this mode it was possible to achieve dissolved oxygen concentration at 30% of saturation, which promoted active yeast growth and increased biomass yield.

Biomass yield was determined by gravimetric method after centrifugation of the culture liquid at 5000 rpm for 15 min and drying the precipitate in a desiccator at 105°C to constant weight. The protein content of the biomass was determined by the Kjeldahl method. The amino acid composition of biomass was determined by high performance liquid chromatography (HPLC) after acid hydrolysis.

The study of the influence of various factors on the growth and biosynthetic activity of yeast showed that the concentration of inoculum plays an important role in the efficiency of the cultivation process. It was found that the optimum amount of inoculum is 5% of the nutrient medium volume. When the amount of inoculum increases, the rate of utilisation of reducing agents increases, resulting in a shorter fermentation period. However, the efficiency of nutrient utilisation for biomass formation decreases. When the amount of seed is reduced, the fermentation period increases, but the efficiency of nutrient utilisation for biomass formation increases.

The results of biomass analysis showed that the crude protein content of dry biomass was 60-62%, which is comparable to the data obtained by other researchers using other feedstocks. The amino acid composition of microbial protein was balanced and contained all essential amino acids. A high content of B vitamins and other biologically active compounds was also found. Protein digestibility averaged 82%, indicating high digestibility of microbial protein.

Comparison of the obtained results with the data of other similar studies shows that the use of complex plant raw materials for microbial protein production is a promising direction. This aligns with the use of artificial intelligence for automation of energy data and resource management [18], effective energy management tools can also aid in monitoring and optimizing energy consumption [19]. Furthermore, the application of functional nanomaterials (NM) can also advance agricultural practices towards sustainability, as discussed by [11]. Criteria-assessment tools for inventory of accumulated environmental harm objects is essential for development [20].

The use of potato processing wastes for culturing yeast Saccharomyces cerevisiae allows to obtain biomass with protein content up to 55%. In earlier works it was shown that under optimal conditions of cultivation biomass yield can reach 15 g/litre, and protein content in biomass is 45%.

In general, the results of these studies indicate the possibility of effective use of complex plant raw materials for microbial protein production. The developed nutrient medium and optimised cultivation conditions allow to obtain biomass with high protein content and balanced amino acid composition. Further research will be aimed at optimising the composition of the nutrient medium, improving cultivation conditions and scaling up the microbial protein production process.

4 Conclusion

This study demonstrated the efficacy of using complex plant material for microbial protein production. A nutrient medium was developed based on a variety of components available in the north-eastern region, such as grasses, beetroot, potatoes, maize, and other agricultural wastes. Careful analysis of the chemical composition of these components allowed optimising the composition of the nutrient medium, providing an optimum ratio of carbohydrates, nitrogen and micronutrients.

The experiments showed that the developed medium supports the growth and biosynthetic activity of yeast Candida scotti Tul-1. The optimal culturing conditions including sugar concentration, temperature and aeration were determined. Specifically, sugar concentration in the range of 3.2-5.0% provided the maximum biomass yield. A temperature of 28°C was most favourable for yeast growth, and optimal aeration (200 rpm and 1.0 VVM) helped to maintain the required dissolved oxygen level.

The obtained microbial biomass was characterised by a high protein content (60-62%), which is comparable to data from other studies. The amino acid composition of the biomass was balanced, containing all essential amino acids. The presence of B vitamins and other biologically active compounds was also noted. These results confirm that the developed technology allows obtaining high-quality microbial protein suitable for use as a protein source in feed and food products.

Comparison of the obtained results with the data of other studies confirms the promising use of complex plant raw materials for microbial protein production. Implementation of this technology allows not only to obtain a valuable protein product, but also to efficiently utilise agricultural waste, reducing the load on the environment. Further research in this area should be aimed at optimising the technological process, including studying the influence of other cultivation factors, scaling up production and assessing economic efficiency.

References

1. B.K. Singh, A. Shakeel, S. Gupta, S. Paul. Digital agriculture: Integrating IoT and AI for sustainable food production. Agriculture 4.0: Smart Farming with IoT and Artificial Intelligence, **22,** 44, (2024). DOI: 10.1186/s40066-024-00492-2
2. R. Mukaila. Microcredit via digital innovation platforms: its drivers and impact on fish farms' yield and income in Nigeria. Agricultural Finance Review, **84 (4-5),** 342-365 (2024). DOI: 10.1108/AFR-01-2024-0001
3. V. O. Gutarevich, A. S. Mikhalev, V. E. Gozbenko, Mining Informational and Analytical Bulletin, **11,** 72–87 (2023). DOI: 10.25018/0236_1493_2023_111_0_72
4. K. Mazur-Włodarczyk, A. Gruszecka-Kosowska. Sustainable Consumption and Production of Edible Plants in the Context of Reaching the EU Climate Neutrality by 2050: A Literature Review. Sustainability (Switzerland), **16 (24),** 10822 (2024). DOI: 10.3390/su162410822
5. T. A. Kuandykov, T. D. Karmanov, E. I. Kuldeyev, K. K. Yelemessov, B. Z. Kaliev, News of the National Academy of Sciences of the Republic of Kazakhstan, Series of Geology and Technical Sciences, **2022(3),** 142–154, (2022). DOI 10.32014/2022.2518-170X.186
6. M. Brini. Introduction to Data Science in Agriculture and Natural Resource Management. Studies in Big Data, **161,** 1-23 (2025). DOI: 10.1007/978-981-97-7762-4_1
7. A. I. Trunova, V. V. Kukartsev, S. O. Kurashkin, Y. A. Tynchenko, I. F. Vivián, K. A. Bashmur, Metals, **13(7),** 1184 (2023). DOI: 10.3390/met13071184
8. M. Govindaraj, C. Gnanasekaran, S. Loganathan, R. Krishnan. Sustainable farming in the digital age: Evaluating the socio-economic impact of IoT and cloud-based agricultural waste management solutions. AIP Conference Proceedings, **3192 (1),** 020037 (2024). DOI: 10.1063/5.0241673
9. L.P. Rybashlykova, M.V. Vlasenko. Pyrogenic factor and its role in the development of desert-dry steppe vegetation: environmental risks for pasture ecosystems. Sustainable Development of Mountain Territories, **16(3),** 873–894 (2024). DOI: 10.21177/1998-4502-2024-16-3-873-894
10. S. Gupta, G.S. Gupta, P. Madheshiya, A.K. Mishra, S. Tiwari. Nano fertilizers for sustaining future farming : A review. Biocatalysis and Agricultural Biotechnology, **62,** 103446 (2024). DOI: 10.1016/j.bcab.2024.103446
11. M.W. Alam, P.M. Junaid, Y. Gulzar, B. Abebe, M. Awad, S.A. Quazi. Advancing agriculture with functional NM: "pathways to sustainable and smart

farming technologies". Discover Nano, **19 (1),** 197 (2024). DOI: 10.1186/s11671-024-04144-z

12. N. Tandon, M. Shaheen. AI-driven innovations for environmental sustainability across sectors. Maintaining a Sustainable World in the Nexus of Environmental Science and AI, 429-454 (2024). DOI: 10.4018/979-8-3693-6336-2.ch017
13. V. I. Golik, V. Yu. Konyukhov, Mining Informational and Analytical Bulletin, **11-1,** 175-189 (2023). DOI: 10.25018/0236_1493_2023_111_0_175
14. R.B. Bist, K. Bist, S. Poudel, D. Subedi, X. Yang, B. Paneru, S. Mani, D. Wang, L. Chai. Sustainable poultry farming practices: a critical review of current strategies and future prospects. Poultry Science, **103 (12),** 104295 (2024). DOI: 10.1016/j.psj.2024.104295
15. A.T. Beristenov, L.V. Lebedeva, L.M. Tatarintsev, V.L. Tatarintsev. Assessment of the agroecological potential of anthropogenic landscapes in the dry steppe of the south of Western Siberia. Sustainable Development of Mountain Territories, **16(3),** 975–987 (2024). DOI: 10.21177/1998-4502-2024-16-3-975-987
16. A.V. Kulik, O.A. Gordienko, M.R. Shaifullin. Impact of old-growth protective forest belts on slope agroforest landscape condition. Sustainable Development of Mountain Territories, **16(3),** 1118–1129 (2024). DOI: 10.21177/1998-4502-2024-16-3-1118-1129
17. S.N. Sorokova, E.A. Efremenkov, D.V. Valuev, M. Qi. Analysis of a Predictive Mathematical Model of Weather Changes Based on Neural Networks. Mathematics, **12 (3),** 480 (2024). DOI: 10.3390/math12030480
18. S. I. Kondratyev, Y. Sarsenbayev, V. A. Turkin, International Journal of Hydrogen Energy, **95,** 212–216, (2024). DOI: 10.1016/j.ijhydene.2024.11.182
19. I. Akzharkyn, K. Yelemessov, D. Baskanbayeva, Applied Sciences (Switzerland), **14(17),** 7567, (2024). DOI: 10.3390/app14177567
20. A.S. Danilov, A.S. Lozovaya. Criteria-assessment tools for inventory of accumulated environmental harm objects. Sustainable Development of Mountain Territories, **16(3),** 861–872 (2024). DOI: 10.21177/1998-4502-2024-16-3-861-872

Biogas technology of complete mixed fodders with application of absorption heat pump

Alexander Anatolievich Shevtsov[1], *Vitaly Nikolaevich* Vasilenko[2], *Larisa Nikolaevna* Frolova [2*], *Anna Aleksandrovna* Derkanosova[2], *Ekaterina Yuryevna* Zheltoukhova[2]

[1] Air Force Academy named after Professor N.E. Zhukovsky and Y.A. Gagarin, Russia, Voronezh, Russia
[2] FGBOU VO 'Voronezh State University of Engineering Technologies', Voronezh, Russia.

Abstract. The biogas technology of complete mixed fodders with application of ab-sorption water-ammonia heat pump (AWHP) is proposed. Involvement of ABTN in the technology allowed to prepare high- and low-potential energy carriers and provide stability of temperature regimes of the processes of anaerobic digestion of substrates at biogas production, moisture-heat treatment of grain with superheated steam, micronisation of grain flakes, drying and cooling of complete mixed fodder. Moisture extraction from biogas was carried out in a two-section apparatus with alternate operation of sections in the modes of freezing and regeneration. Absorption of carbon dioxide from the initial biogas was carried out by cooled water obtained by means of regenerative heat exchange with intermediate refrigerant, which was used as tosol. In a two-column catalytic desulphurisation reactor continuous desulphurisation of biogas was carried out by adsorption method using iron oxide as adsorbent. Variable mode of work of columns of the catalytic reactor provided continuity of process of chemical binding of hydrogen sulphide with adsorbent and regeneration of adsorbent by blowing with atmospheric air. Functioning of ABTN for generation of alternative energy was carried out from heat recovery of superheated steam supplied from the steam generator to the boilers. Ammonia condensation heat in the condenser of ABTH was effectively used to stabilise anaerobic digestion of substrates when warm water was supplied to the heating jacket of the bioreactor.

1 Introduction

In 2022, a new federal project 'Closed Cycle Economy' was launched in Russia. The initiative is aimed at fulfilling the task set by the President of the Russian Federation: to halve the volume of all waste and reduce its negative impact on the environment by 2030. The Government of the Russian Federation has formulated the task - to develop and introduce new effective and cost-effective technologies and technical means of organic waste processing with their maximum involvement in the closed cycle economy [1-3].

The implementation of plans to develop the agro-industrial complex (AIC) and increase the number of livestock can lead to an increase in the volume of waste up to 1.2 billion tonnes. The annual damage from agro-industrial complex waste in Russia is estimated at 450 billion rubles, in particular, due to the pollution of water resources. Currently, more than 2 million hectares of land in Russia are used for manure storage [4]. It has been established [5] that the production of 1 tonne of mixed fodder for young farm animals consumes from 300 to 500 kWh/t of electricity and about 300-400 m3 of process steam. Obtaining biogas can fully cover the energy needs for the production of mixed fodder, and the process of anaerobic fermentation when obtaining biogas will allow to process the wastes of livestock complexes into non-toxic organic fertilisers, which can be further used in agriculture [6-10].

Scientists of the All-Union State University of Informatics and Technology (VGUIT), the All-Russian Research Centre for the Compound Feed Industry and industry specialists are constantly working to improve biogas technologies in fodder production [5, 11-14].

The aim of this work is to develop a biogas technology of complete mixed fodders with the use of absorption water-ammonia heat pump (abtn), providing increased energy efficiency through recovery and utilisation of secondary energy resources with the return of low-potential energy into a closed thermodynamic cycle.

2 Methods and materials

At development of biogas technology of complete mixed fodders with application of ABTN the balance method of distribution of energy flows in combination with innovative methods directed on reduction of specific energy consumption at high quality of the received mixed fodder is used.

* Corresponding author: fln-84@mail.ru

The scheme of biogas technology of complete mixed fodders complete with ABTN includes (Fig. 1): hoppers for storage of grain raw material 1; magnetic separator 2; humidifying machine 3; grain tempering hopper 4; steaming machine 5; microniser 6; equipped with gas infrared heaters with burners; roller conditioner 7; hopper for storage of protein-vitamin additives and premixes 8 and hopper for storage of mineral raw materials (zeolite, bentonite, limestone, chalk) 9; dryer-cooler 10 with drying and cooling sections; filling and packing machine 11, storage tanks for solid 12 and liquid substrate 13, conveyor 14, pumping station 15, bioreactor 16 equipped with cavitation unit with internal thermal jacket and stirrer; digested substrate storage 17, gas holder 18, turbo compressor 19; two-sectional apparatus for dehydration of initial biogas with sections 20/1 and 20/2; column for biogas purification from carbon dioxide 21; two-column catalytic desulphurisation reactor with columns 22/1 and 22/2; water regeneration column 23; water collector 24; buffer tank for accumulation of purified biogas 25; compressor 26; steam generator 27 with burners 28; ABTN including boilers 29 with rectifier 30, deflegmator with coil 31, condenser 32, evaporator 33, absorber 34, heat exchanger 35; thermostatic control valves 36, 37, recirculation pumps 38, 39; regenerative heat exchanger for water heating 40; regenerative heat exchanger for water cooling 41; tosol collector 42; mixers 43, 44; 45; flow distributors 46, 47, 48, 49, 50, 51; 52; intermediate water collector 53, pumps 54, 55, 56, 57, 58, 59; regenerative heat exchanger for air heating 60; regenerative heat exchanger for air cooling 61; fans 62, 63, 64, 65; material and heat flows: 1. 0 - grain, 1.1 - moistened grain; 1.2 - steamed grain; 1.3 - micronised grain; 1.4 - grain flakes; 1.5 - mixture of protein-vitamin additives, premixes and mineral additives (chalk); 1.6 - complete mixed fodder; 1.7 - dried complete mixed fodder; 1. 8 - packaged mixed fodder; 2.0 - solid substrate; 2.1 - liquid substrate; 2.2 - fermented substrate; 2.3 - initial biogas; 2.4 - dried biogas; 2.5 - biogas purified from carbon dioxide; 2.6 - desulphurised biogas; 2.7 - purified biogas; 3.0 - superheated steam; 3.1 - low potential steam; 3. 2 - water from snow coat defrosting; 3.3 - condensate; 3.4 - chilled water; 3.5 - carbon dioxide enriched water; 3.6 - heated water; 3.7 - purified water; 3.8 - carbon dioxide; 4.0 - tosol; 5.0 - ammonia vapour, 5.1 - liquid (condensed) ammonia; 5.2 - vaporised ammonia; 5. 3 - strong water-ammonia solution; 5.4 - weak water-ammonia solution; 6.0 - recycled water; 7.0 - atmospheric air; 7.1 - heated air, 7.2 - cooled air; 7.3 - exhaust air; 8.0 - exhaust gases; 8.1 - exhaust gases; 8.2 - warm water.

Initial grain raw material from grain hoppers 1 through magnetic separator 2 is fed to humidifying machine 3 until reaching humidity of 20-25 %. The moistened grain is sent to the hopper for dehumidification of grain 4, and then to the steaming machine 5, in which the moisture and heat treatment of grain with superheated steam for 10 minutes at a temperature of 100-150 °C is carried out. Steamed grain is fed into microniser 6, in which for 40-90 s the grain is subjected to infrared heating at a temperature of 95-125 ° C and the incident flux density from the radiation of infrared heaters 16.3-20.5 kW/m^2.

Micronised grain is passed through the rollers of the conditioning machine 7 with a gap between the rollers of 0.40-0.55 mm. The obtained grain flakes are mixed with a mixture of protein-vitamin additives, premixes and chalk in the mixer 46 to obtain full feed.

Full-rational mixed fodder is fed to the dryer-cooler 10 and dried at a temperature of 80-90 oC and the speed of the drying agent 0.4-0.7 m/s to a humidity of 8-9 %, then cooled to ambient temperature and then sent to the packing and packaging machine 11.

To realise temperature regimes of technological processes, purified biogas obtained by digestion of liquid and solid substrates is used as the main energy carrier.

Substrates from tanks 12, 13 are mixed in the mixer 43 and discharged into the bioreactor 16 equipped with a cavitation unit, an internal thermal jacket and a stirrer. Optimal conditions for anaerobic digestion of substrates in the temperature range of 35-70 °C with maximum biogas yield of 95-98 % are established.

The digested substrate from the bioreactor is discharged to the storage 17, and the obtained initial biogas is sent to the gas holder 18, from which turbo compressor 19 through the flow switch 46 is fed at a pressure of 2 MPa to the section 20/1 of a two-section apparatus for biogas drying, operating in the mode of freezing of moisture. In the same section by means of synchronised operation of the actuators of the dampers is fed tosol with a temperature of minus 12 °C. On the cooling surface of the section operating in the moisture freezing mode at the temperature of minus 10 °C, biogas reaches the ‘dew point’ by means of recuperative heat exchange with tosol and the moisture contained in biogas condenses into a snow coat. At the same time, the section of the two-section apparatus for natural gas drying 20/2 operating in the regeneration mode is disconnected from the refrigerant recirculation circuit and defrosted by the exhaust superheated steam supplied through the flow distributor 47 from the boilers 29. Switching of sections of the two-section biogas drying apparatus 20/1 and 20/2 from the moisture freezing mode to the regeneration mode and vice versa, as well as the flows of tosol and saturated steam is carried out according to the moisture content of biogas, the current value of which should not exceed 9 mg/m^3.

After drying, the biogas is directed to column 21, where carbon dioxide is absorbed from the initial biogas with cooled water to a carbon dioxide content of 11% vol.

Preparation of chilled water is carried out in the recuperative heat exchanger 41 by heat exchange with tosol. In this case, the water enriched with carbon dioxide is removed from the column 21 by pump 54 and together with the water formed during defrosting of the section of the two-sectional apparatus, is supplied to the recuperative heat exchanger 40, where it is heated to a temperature of 50-60 °C by heat exchange with the exhaust gases from the microniser, providing evaporation of carbon dioxide from the water in the column 23.

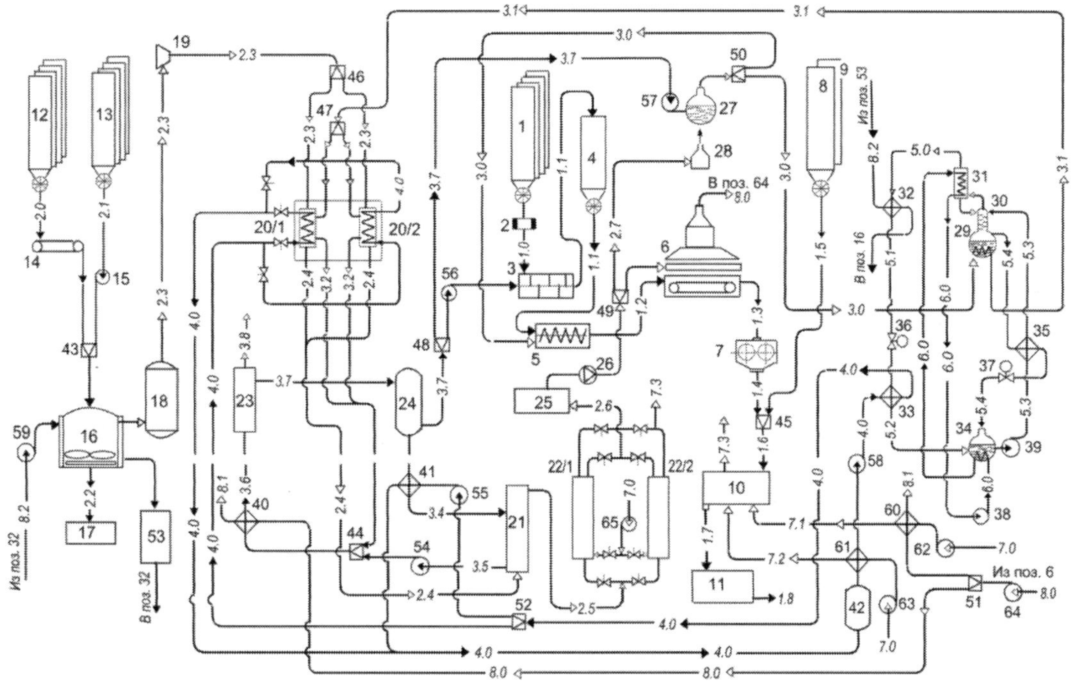

Figure 1. Schematic diagram of biogas technology of complete mixed fodder complete with absorption water-ammonia heat pump

Water purified from carbon dioxide is first discharged into the water collector 24, and then pumps 56 and 57 are supplied respectively to the humidifying machine 3 and to replenish the water level in the steam generator 27 to form a closed cycle.

After extraction of carbon dioxide biogas from column 21 is fed into one of columns 22/1 and 22/2 of two-column catalytic desulphurisation reactor, in which continuous desulphurisation of biogas is carried out by adsorption method to the content of hydrogen sulphide (H_2S) 20 mg/m3 using as adsorbent iron oxide (Fe_2O_3) with humidity of 5 %.

Variable mode of work of reactor columns is carried out: in column 22/1 there is a process of chemical binding of hydrogen sulphide with adsorbent, and in column 22/2 - regeneration of adsorbent by blowing with atmospheric air by fan 62 with exhaust air output to atmosphere.

Adsorption method of hydrogen sulphide removal consists in formation of sulphites at interaction of hydrogen sulphide with iron oxide:

$$Fe_2O_3 \cdot 3H_2O + H_2S \rightarrow Fe_2S_3 + 6H_2O$$

During adsorbent regeneration elemental sulphur is formed, depositing on the surface of iron oxide:

$$Fe_2S_3 + O_2 + 3H_2O \rightarrow Fe_2O_3 \cdot 3H_2O + 3S$$

After each regeneration the sorption capacity of iron oxide decreases on average by 15 %, which necessitates regular replacement of the spent adsorbent and, consequently, variable operation mode of the columns.

Purified biogas from the catalytic desulphurisation reactor is discharged to the buffer tank 25, is subjected to compression compression in the compressor 26 and under pressure of 1.2 MPa is supplied for combustion to the burners of the microniser 6 and burners 28 of the steam generator 27.

During combustion of purified biogas in the burners of the steam generator continuous generation of superheated steam with a temperature of 150 oC is carried out and its supply to the steamer 5 for moisture and heat treatment of grain

Condensation heat of ammonia in condenser 32 of ABTN is transferred to water, which is continuously fed by pump 59 into the thermal jacket of bioreactor 16, providing optimal conditions for anaerobic digestion of substrates.

The superheated steam obtained in the steam generator 27 is used as a source of energy for the operation of the ABTN, which includes boilers 29 with rectifier 30, coil and deflegmator 31; condenser 32; evaporator 33, absorber 34; heat exchanger 35; thermostatic control valves 36, 37, recirculation pumps 38, 39.

Superheated vapour from the steam generator 27 is directed to the boiling coil 29, in which water-ammonia solution is evaporated at a temperature of 140 °C. The mixture of the resulting water vapour and ammonia passes through the nozzles of the rectifier 30, which is sprayed with a strong water-ammonia solution supplied to the boiling coil 29 by the transfer pump 39 from the absorber 34. Part of the water is entrained by the flowing solution along the nozzles of the rectifier 30, and the concentration of ammonia vapour increases. Concentrated ammonia vapour is discharged to the deflegmator 25, the residual water condenses and flows down the nozzles of the rectifier 30 into the boiling pot 29.

Dried ammonia vapour from deflegmator 24 is directed to condenser 32 and condensed at a temperature of 40 - 60 ° C, after which the flow of liquid ammonia is throttled in the thermostatic control valve 36 to a pressure of 0.26 MPa and a temperature of minus 14 oC, from which it boils in the evaporator 33.

The vaporised ammonia from the boilers 29 is fed to the absorber 34 through the regenerative heat exchanger 35 and thermostatic control valve 37. Absorption of ammonia vapour by weak water-ammonia solution in absorber 34 is accompanied by heat release, which is taken away by recycled water flowing through the absorber coil.

The formed strong water-ammonia solution in the absorber 34 is directed by the recirculation pump 39 to the boilers 29 through the heat exchanger 35. In the heat exchanger 35 the strong water-ammonia solution is preheated, which leads to cooling of the weak water-ammonia solution, providing an increase in its absorption capacity and saving of thermal energy.

Recirculation of recycled water through absorber 34 and deflagmator 30 by recirculation pump 38 allows increasing energy efficiency of the processes of condensation of water vapour in deflagmator 34 and ammonia vapour in condenser 26 and providing removal of absorption heat from absorber 34. The recycled water is not discharged from the recirculation circuit, and there is no need for periodic analysis of water for the presence of ammonia in it.

Waste gases after micronisation with a temperature of 90 oC are directed by a fan 64 through a flow distributor 51 to a heat exchanger for heating water 44, as well as to a heat exchanger 40 for heating air taken from the atmosphere, which is then supplied to the drying section of the dryer-cooler 10 as a drying agent by a fan 63.

The air from the atmosphere by the fan 63 is firstly fed to the recuperative heat exchanger 61 and then to the drying section of the dryer-cooler 10 for cooling the complete mixed fodder before it is fed to the filling and packing apparatus 11.

3 Results and discussion

The proposed biogas technology of complete mixed fodders with the use of ABTN has been tested on the experimental equipment in the production conditions of Research and Production Centre VNII Kombikormovaya Industry (tab. 1).

As an example, the technology of full-forage mixed fodder for rabbits according to the recipe PZK - 92 is considered [20-22]. The price of mixed fodder is 28 rub/kg.

Energy costs for the production of mixed fodder according to the known technology are 19.6 rub/kg, which corresponds to 30% of its cost. At a price for 1 kWh of the consumed electric power on a single-rate tariff 4,45 rub energy costs for reception of 1 kg of mixed fodder make 4,4 kWh or 15840 kJ for 1 kg.

It follows from table 1. that the proposed biogas technology with application of ABTN provides reduction of specific energy consumption by 10-12 % at change of parameters of flowing processes in the area of admissible technological properties of received full feed.

Table. 1. Parameters for production verification of biogas technology

Parameter °C	**Parameter value**		
	Mode 1	**Mode 2**	**Mode 3**
Humidity of moistened grain, %	20,0	22,5	25
Temperature of moisture-heating treatment of grain with superheated steam, °C	100	125	150
Temperature of infrared heating of grain during micronisation, °C	95	110	125
Duration of micronisation process, s	40	65	90
Density of incident flux from radiation of infrared heaters, kW/m^2	16,3	18, 4	20,5
Thickness of grain flakes at conditioning, mm	0,40	0,48	0,55
Drying agent temperature, °C	80	85	90
Speed of drying agent, m/s	0,40	0,55	0,70
Humidity of dried complete mixed fodder, %	9,0	8,5	8,0
Temperature of cooled air, °C	7,0	8,5	10
Temperature of substrates digestion, °C	35	52	70
Temperature of heating of water enriched with carbon dioxide, °C	50	55	60
Maximum biogas yield, %	95	96	98
Specific energy consumption, kJ/kg	13930	14090	14260

4 Conclusion

The reserves of energy efficiency of biogas technology of complete mixed fodders have been realised, in particular:

- involvement of ABTN in biogas technology of complete mixed fodders allows to generate alternative energy for implementation of thermal and heat-mass exchange processes of anaerobic digestion of producers, moisture-thermal treatment of grain with superheated steam, infrared heating of grain flakes, drying and cooling of complete mixed fodder;
- alternate operation of sections of the two-section apparatus for natural gas drying, connected with operative defrosting of the section working in the regeneration mode, creates real conditions for continuous process of obtaining biogas purified from water;
- ABTN operation is carried out from heat recovery of superheated steam supplied from the steam generator to the boilers; due to this fact, energy saving is achieved, which is spent only on the operation of controls and pumps in ammonia recirculation and recycled water circuits;
- ABTN radically reduces the operating costs of industrial refrigeration by utilising an affordable alternative energy source that is cheaper than the cost of connecting and utilising additional electrical capacity;
- chemical desulphurisation of biogas due to the formation of sulphites in the interaction of hydrogen sulphide with iron oxide reduces the corrosive effect on the metal surfaces of gas equipment, the corrosion rate decreases from 0.5 - 1 mm to 0.3-0.4 mm per year;
- provides reduction of specific energy consumption by 10-12 %.

References

1. D. Butsaev Closed Cycle Economy - a new economic model. Official channel of the Roscongress Foundation 08.09.2023. https://roscongress.org.
2. P. Morseletto Resour. Conserv. Recycl., **153,** 104553 (2020).
3. V. Moreau, M. Sahakian, P. Van Griethuysen, F. Vuille, Coming full circle: Why social and institutional dimen-sions matter for the circular economy. J. Ind. Ecol., **21,** 497-506 (2017)
4. L. Krupnik, B. Beisenov, D. Baskanbayeva, Mining of Mineral Deposits, **14(2),** 103–109, (2020). DOI: 10.33271/mining14.02.103.
5. B.A. Afanasyev, A.N. Ostrikov, A.A. Shevtsov, A.V. Terekhina, A.I. Aleksandrov, Agrarny Vestnik Urala, **8,** 14 - 27 (2019)
6. T. A. Panfilova, K. A. Bashmur, V. V. Kondratiev, Mining Informational and Analytical Bulletin. **11-1,** 239-251 (2023). DOI: 10.25018/0236_1493_2023_111_0_239
7. V.A. Kulagin, N.V. Dunaeva, D.D. Yakovleva. Vestnik RAS, **91, 1,** 87-102 (2021)
8. M.S. Vandysheva, A.V. Martyanychev, N.V. Karelian Scientific Journal, **1,** 157-159 (2015)
9. Sh.S. Zaripov Introduction to Energy: collection of materials of the II All-Russian (with international participation) youth scientific-practical conference, 61. (2016).
10. G.Z. Zebzeyev, Science. Technologies. Innovations: collection of scientific papers: in 10 parts, 203-206 (2017).
11. S.I. Kondratev, A.V. Vidnichuk, Ecological Safety of Coastal and Shelf Zones of Sea, **2,** 107–121, (2020). DOI: 10.22449/2413-5577-2020-1-107-121
12. L. Krupnik, K. Yelemessov, S. Bortebayev, D. Baskanbayeva, Eastern-European Journal of Enterprise Technologies, **6(12-96),** 22–27, (2018). DOI: 10.15587/1729-4061.2018.151038
13. A. I. Trunova, S. O. Kurashkin, K. A. Bashmur, R. B. Sergienko, S. V. Tynchenko, Metals, **13(7),** 1234 (2023). DOI: 10.3390/met13071234
14. T. A. Kuandykov, T. D. Karmanov, E. I. Kuldeyev, B. Z. Kaliev, News of the National Academy of Sciences of the Republic of Kazakhstan, Series of Geology and Technical Sciences, **2022(3),** 142–154, (2022). DOI 10.32014/2022.2518-170X.186
15. https://www.neftemagnat.ru/enc/32.
16. S.I. Monakh, L.Y. Panchenko, A.I. Tskhvediani, Methods of purification and enrichment of gas obtained in a biogas plant. Bulletin of the Donbas National Academy of Civil Engineering and Architecture, **5(127),** 115-122, (2017).
17. Use of heat pumps in the world - statistics, trends, prospects [Electronic resource]. URL: https://vteple.xyz/ispolzovanieteplovyih-nasosov-v-mire/ (accessed 21.11.2019).
18. V.G. Gorshkov Heat pumps. Analytical review. Handbook of industrial equipment,. Moscow, Mysl (2022).
19. D.A. Britikov, A.A. Shevtsov Energy saving in the processes of drying grain crops using heat pump technologies, **328,** (2012)
20. A. Sladkowski, A. Utegenova, K. Elemesov, I. Stolpovskikh, Naukovyi Visnyk Natsionalnoho Hirnychoho Universytetu, **6,** 29–33, (2017).
21. I. Akzharkyn, Applied Sciences (Switzerland), **14(17),** 7567, (2024). DOI: 10.3390/app14177567
22. S.I. Kondratev, A.V. Vidnichuk, Ecological Safety of Coastal and Shelf Zones of Sea, **2,** 107–121, (2020). DOI: 10.22449/2413-5577-2020-1-107-121

Transformation of the agro-industrial complex of Kazakhstan

Rollan Nussipali [1*], L.V. Asilyan [2], A.G. Grigoryan [2], *Irina A.* Zatsarinnaya [3], *Nikolay A.* Kupriyanenko [3]

[1] Taraz Regional University named after M.Kh.Dulaty, 080000 Taraz, Kazakhstan
[2] Admiral Ushakov Maritime State University, Novorossiysk, 353924, Russia
[3] Kuban State Agrarian University named after I. T. Trubilin, Krasnodar, Russia

Abstract. The purpose of the study is to identify the key areas necessary to ensure food security, improve the competitiveness of the agrarian sector and stimulate economic growth of the agro-industrial sector of the Republic of Kazakhstan. The research methodology is based on the analysis of regulatory documents, statistical data and scientific publications. The analysis revealed imbalances in the development of various spheres of agro-industrial complex, inefficient use of state support, low level of innovation implementation and underdevelopment of the processing industry. It was found that despite the allocation of significant public funds to support the industry, they are not always used effectively. In particular, most of them are directed to direct subsidies rather than to the development of infrastructure and introduction of modern technologies. The conclusions are drawn about the need to revise the state policy, with a focus on stimulating investment in innovative development, technological modernisation and development of the agrarian processing industry.

1 Introduction

There are various approaches to increasing the productivity of the agro-industrial complex. One of them is intensive agriculture based on the extensive use of chemical fertilisers, pesticides and genetically modified organisms. While this approach can achieve high yields in the short term, it is associated with negative consequences such as soil and water pollution, reduced biodiversity and pesticide-resistant pests. In recent years, increasing attention has been given to agroecological approaches that integrate ecological principles into agricultural practices. The development of agroecological approaches aimed at creating sustainable agroecosystems is studied in [1]. Agroecology is aimed at creating sustainable and self-regulating agroecosystems capable of ensuring food security with minimal impact on the environment. This is achieved through the use of biological methods of pest control, implementation of crop rotations, use of organic fertilisers, conservation of soil cover and water resources. The development of agro-ecological approaches is a relevant and important direction, as it allows combining high yields with sustainability and environmental safety. The issue of technological modernisation of agriculture is of particular importance. Modern methods of technological modernisation are considered in the context of adaptation to climate change in the study [2]. In the conditions of global competition, the agricultural industry should rely on innovations and scientific developments that can ensure an increase in labour productivity and product quality. It is necessary to introduce new varieties and breeds, improve soil cultivation methods, develop precision farming and livestock breeding systems. In addition, the development of infrastructure for the storage, processing and transport of agricultural products plays an important role.

The presented work is devoted to the study of priorities of industrialisation of the agro-industrial complex of the Republic of Kazakhstan. Kazakhstan has a significant agrarian potential, but its effective implementation requires modernisation of the industry and introduction of modern technologies. The study is aimed at analysing the current state of the agro-industrial complex of the country, identifying its problems and determining promising areas of development. The purpose of this paper is to study and identify key directions of industrialisation of the agro-industrial complex of the Republic of Kazakhstan, necessary to ensure food security of the country, increase the competitiveness of the agrarian sector, as well as to stimulate economic growth and improve the living standards of the rural population.

2 Method and materials

The main methodology of the research is analytical and review nature, aimed at studying the current state, problems and prospects of development of the agro-industrial complex (AIC) of the Republic of Kazakhstan. The article is based on the analysis of regulatory legal documents, statistical data and literary sources, which

* Corresponding author: rollan_n@mail.ru

forms the basis for determining the priorities of industrialisation of the agro-industrial complex. It considers various aspects, such as: the structure of the agro-industrial complex, the volume of production and consumption of agricultural products, the dynamics of mineral fertilisers, the state of livestock breeding and processing industry. There is also an analysis of government policy in the agro-industrial complex, including programmes of support, subsidies and innovation.

Work to identify priorities for industrialisation of the agro-industrial complex is carried out on the basis of an analysis of economic indicators and trends, as well as by identifying key constraints and opportunities for growth. In particular, factors affecting the competitiveness of the agricultural sector are analysed, including low adoption of new technologies, lack of qualified personnel, and ineffective technology transfer mechanisms.

The article mentions the creation of extension centres such as 'Shortandy' and 'Kostanai', where training and education of farmers in new production and processing technologies are conducted. This shows the practical application of the results of the scientific analysis, but specific details of the work of these centres and the equipment used there are not given.

3 Results and discussion

As part of the analysis of the priorities of industrialisation of Kazakhstan's agro-industrial complex, this study focused on examining the current state, identifying problem areas and determining promising areas of development. Particular attention was paid to the structure of the agro-industrial complex, which was viewed as a complex system that includes three interrelated spheres: industry, which provides the agro-industrial complex with means of production, agriculture, and enterprises that process and sell agricultural products. In order to obtain a more detailed picture and accurate data on each sphere, additional analysis of statistical reports of the Committee on Statistics of the Ministry of National Economy of the Republic of Kazakhstan was carried out.

The analysis showed that in the structure of the agro-industrial complex of Kazakhstan there is an imbalance in the development of these spheres. Thus, the share of the first sphere, which provides the AIC with means of production, accounts for about 10% of the final product and 15% of the main production funds, as well as 20% of the number of AIC employees. The second sphere, represented by agriculture, produces about 50 per cent of the final product, concentrates 65 per cent of fixed assets and 60 per cent of AIC employees. The third sphere, which is engaged in processing, sale and storage of agricultural products, generates about 40% of the final product, 20% of fixed assets and 20% of the number of employees of the agro-industrial complex. In addition, the analysis showed that in 2022, the share of the first sphere was already 12.5% of the final product, which indicates a gradual growth in this area, but is still insufficient.

It was also revealed that in Kazakhstan the main part of the final product, about 60%, is created in the second and third spheres of agro-industrial complex, while in developed countries, for example, in the USA, up to 80% of the retail value of food products is created in the third sphere. This indicates the underdevelopment of the processing industry in Kazakhstan, as well as the low level of implementation of modern technologies in the storage, packaging and marketing of agricultural products. Insufficient development of the processing industry and the need to introduce modern technologies in storage are discussed in [3, 4]. For example, according to the U.S. Department of Agriculture, in 2022 about 85% of the value of agricultural products in the country is created precisely in the processing and marketing, which significantly exceeds the figures of Kazakhstan. This difference in the structure of formation of the value of the final product emphasises the need for increased attention to the development of the third sphere of agro-industrial complex in Kazakhstan, including through the introduction of innovative technologies.

Over the past nine years (according to the Bureau of Statistics of the Republic of Kazakhstan), 790 billion tenge has been allocated from the national budget to the agricultural industry, of which 110 billion tenge was provided in the form of subsidies. Although this indicates significant state support to the industry, the study revealed that these funds are not always used effectively, as most of them are spent on direct subsidies and less is channelled to the development of innovation infrastructure. The problems of inefficient use of state support in agriculture, with a focus on direct subsidies, are discussed in [5]. Thus, according to the report of the National Bank of the Republic of Kazakhstan for 2021, about 75% of the allocated funds went to direct subsidies to farmers, and the remaining 25% were distributed among various support and innovation programmes, and a significant part of the latter remained unused.

According to the Statistics Committee of the Republic of Kazakhstan, in 2018, real productivity growth was only 23%. The goal of providing more than 80 per cent of the domestic food market with domestic products was also set, but this indicator reached only 72 per cent in 2015. These data indicate the need to revise the development strategy of the agro-industrial complex, emphasising a more efficient use of investments and the introduction of innovative technologies.

In addition, an important aspect identified in the course of the study is the low level of implementation of scientific developments in agricultural production. The problems of implementation of scientific developments in agricultural production and barriers to innovation in agriculture are reflected in [6]. Despite the increase in the number of domestic scientific research in the agricultural sphere, their implementation in production remains insufficient. The main constraints are imperfect mechanisms for promoting new technologies in the market, the lack of developed innovation infrastructure, weak interest of private investors, as well as small-scale agricultural production. The share of innovative technologies in agricultural production in Kazakhstan does not exceed 5%, while in leading countries this indicator reaches 25-30%. These figures indicate the need for urgent measures to stimulate innovation activity and create a favourable investment climate.

Analysis of data for 2023 and early 2024 shows that Kazakhstan's agro-industrial complex continues to face a number of challenges, despite some progress. In particular, there has been some increase in gross output of agricultural products, but growth rates remain heterogeneous across different sectors. For example, crop production showed relatively stable growth in 2023, driven by an increase in sown areas and the introduction of some new technologies, particularly in grain production. Nevertheless, due to unfavourable weather conditions in some regions, yields of some crops remained below the planned level. The situation in the livestock sector is also mixed. There is an increase in the number of livestock, but there are still problems with the provision of quality fodder and breeding stock. In 2023, the share of pedigree livestock was only about 18 per cent, which is clearly insufficient for the qualitative growth of the industry.

As for the agro-processing industry, there was a slight increase in production volumes in 2023, but there were no significant changes in the structure of the industry. There is still a high level of imports of processed products, especially dairy and meat products. According to the Ministry of Agriculture of the Republic of Kazakhstan, imports of dairy products in 2023 increased by 15% compared to 2022, which emphasises the need to intensify measures to develop the domestic processing industry. Investment in the agro-industrial complex continues to be insufficient. Despite the allocation of public funds, private investment remains low due to the high degree of risk and long payback period of projects in the agricultural sector. Thus, according to the National Bureau of Statistics of the Republic of Kazakhstan for 2023, the volume of foreign direct investment in the agricultural sector decreased by 8% compared to 2022.

At the beginning of 2024, there is a tendency to increase the costs of agricultural production, in particular, due to the rise in the price of fertilisers, machinery and fuels and lubricants. This negatively affects the profitability of farms and slows down the pace of development of the industry. In addition, in 2023 and early 2024, there were disruptions in logistics chains, especially in the export of grain products, which also created additional challenges for farmers. These problems affect the economy and competitiveness of agriculture as shown in [7].

Forecasting the development of Kazakhstan's agro-industrial complex for the next 5-7 years, it is necessary to take into account several key factors that will determine the growth rate and transformation of the industry. The main one is the efficiency of implementation of the state policy in the field of agriculture. Currently, according to the Ministry of Agriculture of the RK, about 70% of budget funds allocated to support the agro-industrial complex are directed to direct subsidies to farmers. This undoubtedly helps agricultural producers in the short term, but does not create conditions for sustainable development of the industry in the long term. For example, in 2023, about 450 billion tenge was allocated for direct subsidies, while only 120 billion tenge was allocated for infrastructure development, including the construction of storage facilities, elevators and modernisation of irrigation systems.

If the current trend continues and the ratio of direct subsidies to infrastructure investment remains at the level of 70/30, it can be expected that the average annual growth rate of gross agricultural output in the period 2025-2027 will be no more than 2-3%. At the same time, the share of processed products in total output will remain at the level of 35-40 per cent, while food imports will continue to increase by an average of 5-7 % per year. In addition, the industry's export potential will be limited due to low product quality and insufficiently developed logistics chains. According to Kazagromarketing's analysis, in 2023, Kazakhstan will export about USD 4.5 billion worth of agricultural products, 5% less than in 2022, while imports will be about USD 7.8 billion, indicating significant reserves for import substitution and increasing export potential. The dynamics of agrarian systems and their development are also considered in relation to climate change in the following papers [8, 9].

In contrast to this scenario, if the government takes active measures to change the priorities of state support, reorienting them from direct subsidies to the development of infrastructure, innovation and processing industry, we can expect more significant results. In particular, it is advisable to increase the share of investments in innovative developments and technologies to 20-25 % of the total financing of the agro-industrial complex. This will create conditions for the development of knowledge-intensive industries, the introduction of modern agricultural technologies and increased labour productivity in agriculture. According to World Bank experts, a 10 % increase in investment in innovation could lead to a 3-4 % annual increase in productivity in the agricultural sector. In addition, it is necessary to increase financing of the processing industry. The problems of management and sustainable use of land resources in agriculture, their role in the processing of products taking into account the economic component are considered in [10]. In particular, favourable loans and grants should be provided for the construction of new plants for processing agricultural products, especially dairy and meat. According to the calculations of specialists of JSC 'KazAgro', by 2030, an increase in processing capacity by 25% can contribute to the growth of gross output of processed products by 30-35%.

One of the key areas of development of Kazakhstan's agro-industrial complex is the formation of export potential. This requires a transition to international standards of quality and certification of products, which will expand markets and increase exports. In particular, it is necessary to actively implement ISO 22000 and HACCP standards at enterprises producing and processing agricultural products. This will allow Kazakhstani products to become more competitive in the world market and increase exports. According to the forecasts of the Ministry of Agriculture of the Republic of Kazakhstan, the introduction of these standards can increase the volume of exports of Kazakh agricultural products by 15-20 % by 2030.

In addition, it is necessary to strengthen work on the development of transport and logistics infrastructure, including through the construction of new roads and railways, terminals and warehouses. This will reduce transport costs, ensure timely delivery of products to consumers and increase the competitiveness of Kazakhstani products in the global market. According to estimates of specialists of KTZh-Gruzovykh Transport JSC, implementation of major transport projects, such as construction of new railway lines and modernisation of existing ones, can increase the volume of cargo transportation of agricultural products by 10-12% per year.

In the long term, if all the above measures are implemented, Kazakhstan has every opportunity to become one of the leading exporters of agricultural products in the region. By 2030, exports can be expected to increase to USD 8-10 billion, with the share of processed products in total exports to be at least 40-50 per cent. This will require targeted efforts on the part of the state and business, but it will open up new opportunities for the development of the agro-industrial complex of Kazakhstan. The problems of assessing geo-ecological factors and their impact on the sustainability of agrarian land use are also reflected in [11].

Thus, the further development of Kazakhstan's agro-industrial complex will be determined not only by government support, but also by the level of activity of the private sector and the industry's ability to adapt to new technologies and market requirements.

For comparison, studies conducted in the countries of the European Union show that the introduction of innovative technologies in agriculture, along with effective state support and adequate funding of science, allows achieving significant results in increasing productivity and competitiveness of the agricultural sector. For example, in the Netherlands, which is one of the leaders in agriculture, innovation is the basis of agricultural production. Due to this, the country has high productivity and export potential despite the relatively small size of the territory.

Thus, the results of the study revealed the presence of a number of systemic problems in the agro-industrial complex of Kazakhstan. For successful industrialisation and food security, it is necessary to revise the strategy of industry development, with a focus on:

- Development of the third sphere of agro-industrial complex - processing, storage and sale of agricultural products;
- Increasing the efficiency of state support utilisation;
- Creation of conditions for active implementation of scientific developments and innovative technologies;
- Increasing investments in innovation infrastructure;
- Increasing the share of innovative technologies in production;
- Stimulation of co-operation between small agricultural producers.

Implementation of these measures will allow Kazakhstan to increase the competitiveness of the agro-industrial complex, improve the country's food security and achieve sustainable development of the agricultural sector.

4 Conclusion

Based on the analysis and information presented in the article, the following conclusions can be drawn about the priorities of industrialisation of the agro-industrial complex of the Republic of Kazakhstan.

First of all, it should be stated that, despite the significant agrarian potential of the country, the agro-industrial complex of Kazakhstan faces a number of systemic problems that hinder its full-fledged development and integration into the world economy. The key conclusion is the presence of imbalance in the development of three main spheres of agro-industrial complex: production of means of production, agricultural production and processing, storage and sale of agricultural products. In particular, the underdevelopment of the processing industry, as well as insufficient introduction of modern technologies in storage and packaging lead to the fact that most of the added value is created outside Kazakhstan. Technological features of waste processing management and methods of intensification of both agriculture and other sectors of national economy and industry with the use of knowledge-intensive technologies are shown in [12-14].

The second important conclusion is the inefficient use of state support. Significant funds are allocated for direct subsidies, which provides short-term support to farmers, but does not stimulate long-term development and technological modernisation of the industry. Lack of investment in innovation infrastructure and scientific research leads to a low level of implementation of new technologies in agricultural production, which, in turn, restrains the growth of productivity and competitiveness of Kazakhstani products. Features of the selection of parameters for environmentally safe processing of metals and other materials for recycling and their introduction into agriculture are shown in [15].

The third conclusion is the need to revise the state policy in the field of agriculture in order to create conditions for sustainable development of agro-industrial complex. It is necessary to move from the policy of direct subsidies to stimulating investment in innovation, research and development of the processing industry. The priority should be to develop infrastructure, create a favourable investment climate and encourage the introduction of international quality standards. New approaches and methods of processing technogenic waste for its application in agriculture are described in [16, 17].

Also, the study showed that the agro-industrial complex of the Republic of Kazakhstan in 2023 and early 2024 shows only a slight increase in production in a number of sectors. The situation in the agro-industrial complex is aggravated by imports of processed products, high cost of agricultural production and disruptions in logistics chains. All this suggests the need to revise the development strategy of Kazakhstan's agro-industrial complex.

The forecast for the development of Kazakhstan's agro-industrial complex over the next few years depends on the effectiveness of the implementation of government policy aimed at stimulating investment in the processing industry, developing innovation infrastructure and increasing export potential. If the current trend continues, the growth rate of the agro-industrial complex will remain low and food imports will continue to increase. However, with a change in the priorities of state support and reorientation towards innovative development and technological modernisation, Kazakhstan can achieve significant results in increasing productivity and competitiveness of the agricultural sector.

In conclusion, successful industrialisation of the agro-industrial complex in Kazakhstan requires a comprehensive solution to the problems, including revision of state policy, development of innovation infrastructure, attraction of investments, introduction of new technologies and development of human capital. Only if these conditions are met will Kazakhstan be able to realise its agrarian potential and ensure food security of the country. The importance of analysing and assessing the impact of mineral waters on the development of agro-industrial complex and the environment is shown in the study [18].

References

1. B. Aparecida da Silva, R. Evangelista de Oliveira, A. Cavalieri Sais, E. Cardoso leite, Family farmers and trees: ecological functions, ecosystem services and food production in agrarian reform settlements in Brazil. Agroecol. Sustain. Food Syst., **49**, 4-21 (2025). DOI: 10.1080/21683565.2024.2405515.
2. T. Bandara, B. Marambe, G. Pushpakumara, P. Silva, R. Punyawardena, L. Manawadu, S. Premalal, G. Miah, K.R. Dahal, Adapting mountain-farming systems to climate change by reducing food-nutrition-health vulnerability. Mitig. Adapt. Strateg. Glob. Change, **29,** 88 (2024). DOI: 10.1007/s11027-024-10181-x.
3. Yu.N. Serdechnaya, D.V. Serdechnyy, N.A. Andriyanov, S.A Korchagin, Development of an effective monitoring and analysis system of modern technologies for the agro-industrial complex development in mountainous areas. Sustainable Development of Mountain Territories, **16(2),** 770-779, (2024). DOI: https://doi.org/10.21177/1998-4502-2024-16-2-770-779.
4. L. Krupnik, K. Yelemessov, B. Beisenov, D. Baskanbayeva, Mining of Mineral Deposits, **14(2),** 103–109, (2020). DOI: 10.33271/mining14.02.103.
5. S. Buletsa, The peculiarities of restitution of agricultural land in Ukraine. J. Agric. Environ. Law, **19,** 27-74 (2024). DOI: 10.21029/JAEL.2024.37.27
6. E. Reisman, M. Fairbairn, Z. Kish, Agrarian Platform Capitalism: Digital Rentiership Comes to Farming. Antipode, **57,** 412-432 (2025). DOI: 10.1111/anti.13107
7. S. Bamidele, BRICS Revolution: The Global Agrarian Shift Sparked by Emerging Economies. India Q., (2025), Article in press. DOI: 10.1177/09749284241307937
8. L. Krupnik, K. Yelemessov, S. Bortebayev, D. Baskanbayeva, Journal of Enterprise Technologies, **6(12-96),** 22–27, (2018). DOI: 10.15587/1729-4061.2018.151038
9. H. Cochet, O. Ducourtieux, N. Garambois, Agrarian systems and climate change: Journeys of adaptation in the global south. Agrar. Syst. Clim. Change: Journeys Adapt. Glob. South, **1,** 208 (2024). DOI: 10.1079/9781800628137.0000
10. Makhosheva S.A., Zhanokova M.V., Dzaparova V.M., Totrukova I.K. Ecological aspect of sustainable development of high-mountain regions in the conditions of innovative economy formation. Sustainable Development of Mountain Territories, **16,** 754–76, (2024). DOI: 10.21177/1998-4502-2024-16-2-754-769.
11. M. N. Nurpeisov, K. V. Ruchkina, V. A. Androkhanov, S. K. Makenova, Assessment of geoecological factors and conditions affecting sustainability of agricultural land use in the forest-steppe zone of Altai Krai, Sustainable Development of Mountain Territories, **15(2),** 419–430 (2023). DOI: 10.21177/1998-4502-2023-15-2-419-430.
12. S.I. Kondratev, A.V. Vidnichuk, Ecological Safety of Coastal and Shelf Zones of Sea, **2,** 107–121, (2020). DOI: 10.22449/2413-5577-2020-1-107-121.
13. M.V. Grechneva, M.Y. Ivanov, Mechanical activation of coal mining and enrichment tailings. Coke and Chemistry, **67(7),** 407–412, (2024). DOI: 10.3103/S1068364X24600672.
14. V.I. Golik, A.I. Karlina, Technology for nonwaste recovery of tailings of the mizur mining and processing plant. Metallurgist, **66,** 1476–1480 (2023). DOI: 10.1007/s11015-023-01462-y
15. N.N. Bryukhanova, V.A. Gladkikh, R.V. Kononenko, V.V. Gerasidi, N.V. Martyushev, Selection of disintegrator parameters toward environmentally friendly leaching of metals. Min. Inform. Anal. Bull., **12,** 39-49 (2024) DOI: 10.25018/0236_1493_2024_12_0_39
16. I. Akzharkyn, K. Yelemessov, D. Baskanbayeva, Applied Sciences (Switzerland), **14(17),** 7567, (2024). DOI: 10.3390/app14177567.

17. N.V. Babyr, A.V. Pogrebnoy, E.A. Efremenkov, D.V. Valuev, A.E. Boltrushevich, Modelling of Reliability Indicators of a Mining Plant. Mathematics, **12 (18),** 2842 (2024) DOI: 10.3390/math12182842.
18. S.I. Kondratev Three typical hydrological-hydrochemical situations near the danube river mouth based on the marine hydrophysical institute research expeditions in 1997-2013. Physical Oceanography, **26(4),** 326–340 (2019). DOI: 10.22449/1573-160X-2019-4-326-340.

https://doi.org/10.1051/bioconf/202516100040

Organization of veterinary services for industrial duck farming: problems and solutions

Mihail N. Vasiliev*, *Karina V.* Gainutdinova, *Rustam Kh.* Ravilov and *Anastasia I.* Vasilieva

Kazan State Academy of Veterinary Medicine named after N. E. Bauman, Kazan, 420029, Russia

Abstract. Proper organization of veterinary services is one of the key elements of economic efficiency of duck farming. It has been established that there are a number of problems in the organization of veterinary services for duck breeding in the country that require scientific analysis. This is insufficient elaboration of the issues of normative-legal regulation of veterinary requirements for duck keeping. However, the state plans to solve this issue in 2025 by adopting new Veterinary Rules, but the draft of this document does not contain any specifics of keeping waterfowl (ducks and geese), taking into account their physiological need for water. The second problem is the lack of unified approaches in planning and organization of veterinary preventive measures against contagious and other diseases of ducks. The authors recommended scientifically based technological cards of veterinary treatment of ducks, which will contribute to the organization of an effective system to control the risks of occurrence and spread of contagious and other diseases of ducks. The third problem is the lack of the scientifically based planning of staffing veterinary specialists for duck breeding farms (enterprises). Here the authors recommend 42 developed norms of time for carrying out veterinary and preventive measures, works performed by veterinary specialists in hatcheries, and registration of veterinary documentation. Also 5 norms of time for laboratory tests for the most significant for industrial duck farming infections have been developed. A systematic approach to the organization of veterinary services for industrial duck farming on the basis of the developed recommendations will ensure stable epizootic well-being of the industry, which will positively affect its economic efficiency.

1 Introduction

Proper organization of veterinary services is one of the key elements of economic efficiency of poultry farms. In the current conditions of Russian poultry farming, the issue of effective veterinary services is becoming more and more acute [1]. Ducks occupy the second place after chickens in the structure of poultry industries in the Russian Federation, with geese overtaking them by the rate of growth of meat production [2]. Despite this, the attention of scientists, both in Russia and abroad, is mainly given to chickens. Only some scientists are engaged in research in the field of organization of measures for prevention and control of infectious and other diseases of ducks. They are Novikova O. B., Trefilov B.B., Fleming-Canepa X., Hussain I., Pan Yu., Putri N., Shi Sh., Uddin Md. N., et al. [3 - 10]. Practically no one deals with the organization of veterinary services of duck farming in a comprehensive manner, and therefore a situation is formed when this issue is not fully studied. In particular, until now in Russia, no attempts have been made to analyze the morbidity of ducks in industrial conditions of their cultivation. And, based on its results, there is the scientifically based planning of veterinary preventive measures, in the service of breeding flocks and young stock on fattening. Also, there was no labor rationing of veterinary specialists with the development of time standards, which could be the basis for calculations to establish the optimal scientifically-based staffing of veterinary specialists of duck farms. All the above-mentioned points to the relevance of the topic chosen by the authors. Despite the solid experience of Russian scientists in the issues of improving the organization of veterinary services in poultry farming, studies in the industrial sector of duck farming are conducted for the first time.

2 Materials and methods

The data on the epizootic situation, poultry disease incidence, plans of preventive veterinary measures of 4 industrial Russian duck farms served as the material for the research. They are LLC "Ramaevskoe Farm" of the Republic of Tatarstan, LLC "Centralnaya Poultry farm" of the Vladimir region, LLC "Agrofirma of "Sibirskaya Ptitsa" of the Altai region, and LLC "Kimovskaya Poultry farm" of the Republic of Mari El.

The research was carried out on the basis of the Kazan State Academy of Veterinary Medicine named after N. E. Bauman. The following methods of scientific research were used: experimental-analytical, calculation-

*Correspondance: mnvasiliev@mail.ru

analytical, abstract-logical, monographic and system analysis.

3 Research results

The developed recommendations are based on the requirements of the current regulatory and legal framework in the field of prevention of contagious and other diseases of ducks, as well as the best practices of research facilities on reproduction and breeding of ducks in industrial conditions. Recommendations include general recommendations on the organization of veterinary services, scientifically based technological maps of veterinary and preventive treatments of breeding stock and young ducks, as well as the norms of time for veterinary work in duck breeding.

The target audience for the recommendations are veterinary specialists of farms and state veterinary service, who provide veterinary services in duck breeding.

Current Russian veterinary legislation in poultry farming, including duck farming, distinguishes between open-type farms and closed-type enterprises. At the same time, there is no clear differentiation between them in terms of the number of poultry kept, which significantly complicates the practical application of the requirements defined by the legislation. Many large duck farms position themselves as open-type farms.

There is no need to dwell on the requirements themselves, as they are set out in detail in the Veterinary Rules for keeping poultry on private farmsteads and open-type poultry farms and the Veterinary Rules for keeping poultry on closed-type poultry farms (poultry farms). They were approved by Orders of the Ministry of Agriculture of the Russian Federation No. 103 and No. 104 of April 3, 2006, respectively.

In 2024, the Department of Veterinary Medicine of the Ministry of Agriculture of the Russian Federation has prepared a draft and started consideration of new Veterinary Rules for keeping birds for breeding, rearing and marketing purposes, eliminating the above-mentioned deficiency.

Organization of veterinary service of duck farms (enterprises) is carried out according to plans, technological maps and schemes of veterinary preventive measures.

The executor and the responsible person in open-type farms, depending on the position according to the staff schedule, are the chief veterinary specialist or veterinary surgeon of the farm, or under the contract for veterinary service (specialists of the state veterinary service).

Table 1. The technological map of veterinary and prophylactic treatments of parent stock in duck breeding farms (enterprises)

Dead-line	Type of treatment	Method of execution	Name of prepa ration	Single dose
January	Preventive treatment against coccidosis	Giving with water	Baycox 2.5%	1 ml/ 1 l water
January, one month before oviposi-tion	Vaccination against viral hepatitis	Intramus-cularly into the thigh muscle	Vaccine against viral hepatitis of ducklings from the strain "VGNKI-K" embryonic	1 ml
During vaccina-tion	Prophylactic vitamini-zation	Giving with water	Tetra-hydrovit	2 ml/ 1 l water
Between vaccina-tions	Prophylactic vitamini-zation	Intramus-cularly	Tetra-hydrovit	2 ml
Feb-ruary, two weeks before egg-laying	Revaccina-tion against viral hepatitis	Intramus-cularly into the thigh muscle	Vaccine against viral hepatitis of ducklings from the strain "VGNKI-K" embryonic	1 ml
Before oviposi-tion	Preventive treatment	Air sanitation for 30 minutes	Iodine a checkers	1 per 400 m³ of room
2 times a month during oviposi-tion period	Preventive vitaminiza-tion	Giving with feed	Multitonic	1 l per 2 ton of water
April, Novem-ber	Deworming	Giving with water or feed	Ampro-lium	120 mg/ 1 l water or 120 mg/ 1 kg feed
Spring, summer, winter	Prophylactic treatment against arachno-entomosis	Giving with feed	Alvet	10 mg/ 1 kg
Every month	Preventive treatment	Giving with water	Monclavit 1	10 ml/ 1 l water

Table 2. The technological map of veterinary and prophylactic treatments of young animals in duck farms (enterprises)

Deadline (age of young stock)	Type of treatment	Method of execution	Name of preparation	Single dose
Day 0	Preventive treatment	Giving with water	Vitrocil	1 l per 20 l of water with Aviblu (50 g)
Days 1 - 4	Preventive treatment	Giving with water	Vitrocil	1 l per 20 l of water
Days 5 - 7	Preventive treatment	Giving with water	Hepatobuster	1 l per 20 l of water
Day 8	Preventive treatment	Giving with water	Citric acid	300 g per 20 l of water
Day 9	-	-	-	-
Days 10 - 12	Preventive treatment	Giving with water	Lovit VA-SE	0.5 l per 20 l of water
Day 13	Preventive treatment	Giving with water	Citric acid	300 g per 20 l of water
Day 14	-	-	-	-
Days 15 - 17	Preventive treatment	Giving with water	Hepatobuster	1 l per 20 l of water
Day 18	Preventive treatment	Giving with water	Citric acid	300 g per 20 l of water
Days 19 - 20	Preventive treatment	Giving with water	Lovit VA-SE	0.5 l per 20 l of water
Days 21 - 22	Preventive treatment	Giving with water	Proctiv Acid SE	0.3 l per 20 l of water
Day 23	-	-	-	-
Days 24 - 26	Preventive treatment	Giving with water	Lovit VA-SE	0.5 l per 20 l of water
Days 27 - 28	Preventive treatment	Giving with water	Proctiv Acid SE	0.3 l per 20 l of water
Days 29 - 30	Preventive treatment	Giving with water	Ecocid	100 g per 20 l of water
Days 31 - 32	Preventive treatment	Giving with water	Proctiv Acid SE	0.3 l per 20 l of water
Days 33 - 34	Preventive treatment	Giving with water	Ecocid	100 g per 20 l of water
Days 35 - 36	Preventive treatment	Giving with water	Proctiv Acid SE	0.3 l per 20 l of water
Day 37	Preventive treatment	Giving with water	Citric acid	300 g per 20 l of water
Days 38 - 41	-	-	-	-

The executor and the responsible person at duck farming enterprises of closed type, depending on the position on the staff schedule, are the chief veterinary specialist or veterinary surgeon of the poultry farm.

In both cases, a set of preventive measures was aimed at preventing the spread of infections and infestations, therapeutic and prophylactic measures against non-communicable diseases of ducks and veterinary and sanitary work, which is an important part of the general prevention of diseases in poultry farming.

Implementation of veterinary measures to prevent contagious and other diseases of ducks is impossible without their scientifically based planning. We have developed science-based technological maps of veterinary preventive measures for duck farms (enterprises), presented in Tables 1 and 2. There for each veterinary measure, the timing, methods of carrying out, as well as the names and dosages of recommended veterinary drugs approved for use in the Russian Federation are indicated.

The list of veterinary measures carried out in duck farming depends on the region of location, the level of technological efficiency at the poultry farm, the epizootic situation in the region and in the farm itself, availability of necessary biological preparations. It is necessary to regularly improve the schemes of veterinary and prophylactic measures, taking into account the spread of diseases in duck farming and the effectiveness of veterinary drugs used. Our recommendations can be supplemented with specific veterinary measures necessary for each particular duck farm (enterprise).

Application of the proposed integrated approach to planning and implementation of measures to prevent contagious and other diseases of ducks in accordance with the developed technological maps of veterinary and preventive measures for duck farms (enterprises) will allow effectively controlling the risks of occurrence and spread of such diseases.

Table 3. Time standards for veterinary work in duck incubators

Name of veterinary measures	Unit of measurement	Time standard, min. (M±m)
Quality assessment of incubation egg acceptance in the storage department	1 egg tray – 120 eggs	1.75±0.07
Disinfection of incubation eggs	1 cart – 3360 eggs	19.63±0.42
Candling of incubation eggs	1 cart – 3360 eggs	46.78±0.46
Incubation (hatching) cabinet disinfection	1 cabinets	41.91±1.78
Control of egg laying in the incubation cabinet	1 cart - 28 egg trays	30.93±1.27
Control of egg transfer from the incubation cabinet to the hatching cabinet	1 cart - 28 egg trays	9.94±0.33
Control of the hatching of ducklings from the hatching cabinet after the incubation period has expired	1 cart - 28 egg trays	55.94±0.31

Effective veterinary service of duck farming is impossible without the necessary staff of veterinary

https://doi.org/10.1051/bioconf/202516100040

specialists. The planning of the number of such personnel is based on time standards for veterinary activities. We have developed time standards for veterinary activities in the service of duck farms (enterprises) presented in Tables 3-5 guided by the Recommendations on labor rationing of veterinary specialists. They were approved by the Scientific and Technical Council of the Ministry of Agriculture of the Russian Federation on December 26, 2014, based on the results of timekeeping observations.

Table 4. Time standards for veterinary and preventive measures in duck farming

Name of veterinary measures	Unit of measurement	Time standard, min. (M±m)
Clinical inspection, identification of sick and dead birds in the poultry house	1 poultry house	27.9±1.3
Vaccination of ducks against viral hepatitis	1000 birds	166.0±1.9
	1 bird	0.17
Duck deworming	1 poultry house	33.1±1.2
Vitaminization of ducks	1 poultry house	31.5±0.6
Giving antimicrobials to duck with water	1 poultry house	39.6±1.2
Pathological autopsy of dead duck	1 dead bird	42.6±0.3
Poultry house disinfection	1 poultry house	1595.0±4.1
Sanitizing mat filling	1 sanitizing mat	9.2±0.2
Treatment of ducks with disinfectant (drinking)	1 poultry house	36.4±0.4
Disinfection of vehicle	1 vehicle	134.4±3.7
Feeder disinfection	1 feeder	77.6±1.6
Clinical examination of the duck before slaughtering	1 party	32.7±1.1
Sampling of feed for laboratory examination	1 sample	14.1±0.6
Sampling of pathological material from ducks for laboratory tests	1 sample	13.2±0.5

The work of industrial duck farms (enterprises) is impossible without high-quality laboratory and diagnostic support of their activities. It is necessary for monitoring studies of especially dangerous diseases and for reliable diagnosis of contagious and other diseases of ducks. In the Russian Federation, laboratory diagnostic tests are carried out by a network of accredited regional and federal state veterinary laboratories. Work has been started to develop time standards for laboratory tests carried out on material sampled from ducks, currently 5 time standards have been established, as shown in Table 6.

Table 5. Time standards for veterinary documentation in duck breeding farms

Name of veterinary measures	Unit of measurement	Time standard, min. (M±m)
Compilation of the act:		
- immunization	1 act	9.4±0.45
- preventive treatment	1 act	8.1±0.2
- poultry house disinfection	1 act	12.1±0.57
- vehicle disinfection	1 act	9.7±0.4
- disinfection of incubation (hatching) cabinets	1 act	8.8±0.37
- sampling for laboratory examination	1 act	6.9±0.25
- pathological anatomical autopsy	1 act	13.5±0.59
Keeping a log book:		
- anti-epizootic measures	1 record	3.7±0.13
- diagnostic studies of poultry	1 record	2.9±0.14
- pathologoanatomical autopsy of poultry	1 record	3.4±0.14
- epizootic condition of the poultry farm	1 record	6.7±0.25
- prophylactic and forced inoculations of birds	1 record	4.5±0.17
- therapeutic treatments of poultry	1 record	2.2±0.06
- veterinary and sanitary examination results	1 record	3.9±0.15
- disinfection, disinsection and deratization	1 record	4.2±0.15
- medicines	1 record	1.6±0.07
- biological waste	1 record	2.5±0.1
Formalization of an application for a veterinary certificate of form No. 1 in electronic form	1 application	18.3±0.9
Formalization of the veterinary certificate of form No. 2 in electronic form	1 certificate	17.1±0.8
Formalization of application for the veterinary certificate of form No. 3 in electronic form	1 application	19.7±0.9
Formalization of the veterinary certificate (application for the veterinary certificate) of form No. 4 in electronic form	1 certificate (application)	6.7±0.25

The time standards presented in the tables are recommended to be used as scientifically based time standards in the analysis of staffing, planning the number of veterinary specialists and assessing the efficiency of labor resources of duck farms (enterprises). Time standards for laboratory tests are accredited by state veterinary laboratories. Calculations should be based on the annual effective working time of veterinary specialists in each specific region of the country, as well as on the actual costs of preparatory and final work, rest and personal needs of workers.

Table 6. Time standards for laboratory tests for selected duck infections

Name of veterinary measures	Unit of measurement	Time standard, min. (M±m)
Detection of antibodies to avian influenza virus in the hemagglutination inhibition reaction	1 test (20 samples)	66.1±2.66
Detection of antibodies to Newcastle disease virus in the hemagglutination inhibition reaction	1 test (20 samples)	67.2±1.34
Examination of blood serum samples for avian influenza by the enzyme-linked immunosorbent assay method	1 test (50 samples)	67.95±1.7
Detection and differentiation of avian influenza virus by the polymerase chain reaction in flushes	1 test	150.1±2.66
Examination of blood serum samples for viral hepatitis of ducks by the enzyme-linked immunosorbent assay method	1 test (50 samples)	67.95±1.7

4 Conclusion

When organizing veterinary services for duck farming in the Russian Federation, a number of problems require scientific analysis. One of them is insufficient elaboration of the issues of normative-legal regulation of veterinary requirements for duck keeping in farms (enterprises). However, the government plans to solve this issue in 2025 by adopting new Veterinary Rules for keeping birds for breeding, rearing and marketing, but the draft of this document does not contain any specifics of keeping waterfowl (ducks and geese), taking into account their physiological need for water. The draft rules allow walking waterfowl with the device on the paddocks dome grooves, but this applies only to farms with livestock of up to 1000 birds. At duck-breeding enterprises with large livestock, walking is prohibited. And the device of dome grooves in poultry houses significantly complicates the control of microclimate parameters.

The second equally important problem is the lack of unified approaches in planning and organization of veterinary and prophylactic measures against contagious and other diseases of ducks. We recommend scientifically based technological cards of veterinary treatments of parent stock and young ducks on fattening, tested under the conditions of the research sites. They will contribute to the organization of an effective system of control of risks of occurrence and spread of infectious, invasive and other diseases of ducks in industrial conditions of their reproduction and cultivation.

The third problem is the lack of scientifically based planning of the staff number of veterinary specialists of duck breeding farms (enterprises), which is necessary to ensure, on the one hand, high efficiency of veterinary services, on the other hand, the rational use of available labor resources. To solve this problem, we recommend 14 norms of time for veterinary preventive measures, 7 norms of time for work performed by veterinary specialists in hatcheries, and 21 norms of time for registration of veterinary documentation in duck farms (enterprises). In addition, 5 norms of time for laboratory tests for the most important infections for industrial duck farming have been developed. All of them are reliable and can be used not only in planning the staff number of veterinary specialists, but also in assessing the efficiency of labor resources in duck farming. The establishment of the staff of veterinary specialists of a farm (enterprise) should be approached with due consideration of the actual annual effective working time fund, costs for preparatory and temporary work, rest and personal needs of veterinary specialists.

A systematic approach to the organization of veterinary services for industrial duck farming on the basis of the recommendations developed by us will ensure stable epizootic well-being of the industry. This will have a positive impact on its economic efficiency not only on the scale of individual subjects, but also on the Russian Federation scale. The results of these studies can be used as a basis for the development of a national system for the organization of veterinary services for industrial duck farming in other countries of the world, taking into account the economic and other specifics of veterinary services in these countries.

References

1. E.O. Bakanova, M.N. Vasiliev, Sci. Notes of Kazan SAVM, **253**, 22–26 (2023)
2. M.N. Vasiliev, E.O. Bakanova, A.I. Vasilieva [et al.], Journal of Pharmaceutical Research International, **34(10B)**, 36–42 (2022)
3. O.B. Novikova, N.V. Nikitina, M.A. Pavlova [et al.], Poultry, **11-12**, 93-99 (2019)
4. B.B. Trefilov, N.V. Nikitina, L.I. Yavdoshak [et al.], Veterinary Medicine, **4**, 24–27 (2019)
5. X. Fleming-Canepa, L. Canniff, M. Kobewka [et al.], Veterinary Microbiology, **228**, 101–111 (2019)

I. Hussain, A.A. Shah, M.A. Abbas [et al.], Journal of Infection and Public Health, **13**, 461–462 (2020)

6. Yu. Pan, W. Cai, A. Cheng [et al.], Veterinary Research, **53(1)**, 1–13 (2022)
7. N. Putri, R. Ernawati, J. Rahmahani [et al.], International Journal of One Health, **14(1)**, 276-284 (2021)
8. Sh. Shi, R. Liu, Ch. Wan [et al.], Archives of Virology, **164(3)**, 847-851 (2019)
9. Md. N. Uddin, S. B. Neogi, Sk. Sh. Islam [et al.], BMC Infectious Diseases, **21(1)**, 1–17 (2021)

Nature-like as agro-technologies a convergent approach to carbon polygon creation

Nadezhda S. Sevryugina [1*], *G.L.* Kozenkova [2], *N.S.* Migda [2]

[1] Russian State Agrarian University - Moscow Timiryazev Agricultural Academy, Moscow, Russia
[2] Admiral Ushakov Maritime State University, Novorossiysk, 353924, Russia

Abstract. In this paper, research and validation of nature-like agro-technologies using a convergent approach to improve the productivity of agrolandscapes and the creation of carbon polygons were carried out. As part of the work, a set of experimental studies was implemented, including assessment of agrolandscape potential, comparative analysis of traditional and nature-like farming methods, as well as monitoring of plant physiological parameters and the dynamics of carbon accumulation in the soil. The results of the research showed that nature-like agro-technologies provide a significant increase in yields and quality of grain crops, as well as improve soil fertility and its ability to sequester carbon. At the same time, the cost of production is reduced due to more rational use of natural resources. The obtained data indicate the promising application of nature-based agro-technologies to ensure sustainable and environmentally safe agriculture and to solve the problem of decarbonisation in the agricultural sector.

1 Introduction

Organic farming, offers methods based on natural processes, excluding the use of synthetic pesticides and fertilisers. In the context of sustainable agriculture, it is particularly important to consider the role of family farmers, who are often the custodians of traditional farming practices that contribute to the preservation of ecosystems [1]. This contributes to the restoration of soil fertility and biodiversity, but can limit crop yields and require high labour and financial costs. The use of precision farming with digital technologies allows optimising the application of fertilizers and crop protection products, minimising negative environmental impact, but requires significant initial investment in equipment and staff training. At the same time, it is necessary to consider the impact of digital platforms on the agricultural sector, as they can both promote efficiency and create new forms of dependency [2]. Another direction is the development of agroforestry, which combines agricultural production with forest plantations, which contributes to soil conservation, soil fertility and climate improvement, but requires considerable time and planning. In mountainous regions where environmental conditions are particularly vulnerable, agroforestry can play a key role in adapting farming systems to climate change, helping to reduce soil erosion and conserve biodiversity [3, 4]. In general, all these approaches have advantages and disadvantages and require adaptation to specific conditions and resources.

In light of the above challenges, the development and implementation of nature-based agro-technologies based on the convergent approach seems to be an extremely promising direction. This approach involves the integrated use of natural resources and biological processes to increase the productivity of agro-landscapes while reducing the negative impact on the environment. This approach promotes a more rational use of available natural resources, increasing their efficiency. Particular attention is paid to the digitalisation of the parameters of the original natural resources, which allows for more accurate and efficient management of processes in agro-ecosystems. The use of these technologies can increase crop yields and improve the environmental sustainability of agro-landscapes. At the same time, due to the involvement of the excess part of natural resources in the process, it is possible to reduce production costs and increase the economic efficiency of the agricultural sector. The development of technologies aimed at capturing and sequestering carbon in the soil is also a relevant area in the context of combating global climate change. The development and adoption of nature-based agro-technologies is part of a broader trend of adapting agricultural systems to changing climatic conditions, and contributes to decarbonisation efforts [5, 6]. The establishment of carbon polygons based on nature-based technologies can be an important step towards decarbonising agriculture and reducing greenhouse gas emissions.

The purpose of the work presented in this article was to study and validate nature-like agro-technologies using

* Corresponding author: zhutikovans@nssevr.ru

a convergent approach, providing the establishment and operation of carbon polygons to increase the productivity of agro-landscapes, while reducing the negative impact on the environment and decarbonising processes in the surface layer of the atmosphere. As part of the work, scientific and methodological bases for determining and managing yield parameters, based on the analysis of natural resources, were developed, and practical recommendations for the implementation of these technologies in agrarian production were proposed.

2 Method and materials

Within the framework of the presented research, a set of experimental works aimed at developing and testing nature-like agro-technologies for improving the productivity of agro-landscapes and creating carbon polygons was implemented. The experimental work was carried out in several stages, including assessment of agrolandscape potential, development and adaptation of nature-like technologies, as well as their practical application and analysis of results. The main focus of the research was to study the impact of a complex of natural and anthropogenic resources on the yield of grain crops and their capacity for carbon sequestration.

In the process of experimental work, a range of specialised equipment was used, and field research methods were applied. A suite of instrumentation including portable soil moisture sensors, thermometers to measure air and soil temperature, and anemometers to measure wind speed were used to assess agro-landscape potential and monitor key parameters. Measurements were carried out regularly, which allowed obtaining data on the dynamics of changes in these parameters during the growing season. A laboratory UV-Vis spectrophotometer was used to analyse soil samples for nutrients and trace elements. Various methods are used to assess the potential of agro-landscapes, including laboratory analysis of soil samples for nutrients and trace elements and, in particular, spectrophotometry. These techniques are an important element in the planning and implementation of sustainable agro-technologies [7-9]. Using a spectrophotometer, quantitative analyses of the content of nutrient elements necessary for plant growth and development were carried out. At the stage of analysis of the obtained data, specialised programs for statistical processing and modelling were actively used [10, 11]. In the field, standard methods including manual sampling, weighing and measurement of linear parameters of plants were used to estimate plant yield and biomass. Standard agricultural machinery including seed drills and combine harvesters were used for sowing and harvesting. The combines were equipped with on-board scales and GPS sensors, which allowed accurate measurement of yields in different parts of the experimental field. The study adopted a differentiated approach where traditional agro-production practices were compared with nature-based practices. As a result, a carbon polygon was created where the impact of agro-technologies on carbon sequestration, growth dynamics and crop development was evaluated under equal conditions.

3 Results and discussion

Our integrated study implemented a multi-stage approach to investigate the impact of nature-based agro-technologies on cereal crop productivity and their role in carbon sequestration. Experimental work was planned taking into account the need for a comprehensive assessment of agro-landscape potential and comparative analyses of yields, biomass accumulation, and soil dynamics under different cropping regimes. At the initial stage, a thorough assessment of the agrolandscape characteristics of the study area was carried out, which included detailed mapping of soil resources as well as analyses of soil physicochemical properties. Not only classical methods of soil analysis, such as determination of granulometric composition, organic matter content and basic nutrient elements, were used, but also modern remote sensing methods were used to obtain information on the spatial distribution of organic carbon, soil moisture and other important indicators. These data served as a basis for determining the heterogeneity of the agrolandscape and selecting representative sites for comparative experiments. For effective management of agrarian resources and selection of representative sites for research, it is important to use methods of geospatial analysis, which allows taking into account the heterogeneity of agrolandscapes [12]. Further, a comparative analysis of the two main approaches to farming: conventional and nature-based. For this purpose, two experimental plots with identical soil and climatic conditions were allocated. In one of the plots, conventional agro-technology was applied, based on intensive use of mineral fertilisers, pesticides and herbicides, as well as on the use of intensive tillage methods. On the other site, on the contrary, a nature-like agro-technology developed by us was tested, which was based on the principles of minimal interference in natural processes of soil formation and maximum use of biological mechanisms. This technology included a number of agrotechnical methods, such as mulching of soil with organic materials, use of sideral crops to enrich the soil with nitrogen and organic matter, as well as the use of crop rotations with the inclusion of crops that promote soil structure formation and improve soil fertility. An important element of nature-like technology was the method of diffuse moisture rise from the lower soil horizons to the seedbed, which was achieved through the use of special tillage techniques and soil structure. In mountainous terrain, where moisture deficiency is often observed, it is important to use methods that promote diffuse water uplift, as well as to take into account the influence of mountain factors on the salt content of the soil [13]. This method provided optimal moisturisation of the root system of plants at the initial stages of vegetation, which promoted their more active growth and development. Throughout the growing season, physiological parameters of plants such as plant height, leaf area, photosynthesis intensity, chlorophyll content, and root system development were carefully monitored. These measurements were made regularly using both traditional methods, such as measuring plant height and leaf area with a ruler and planimeter, and modern instruments, including a portable fluorimeter to estimate photosynthetic rate and a spectrophotometer to determine leaf chlorophyll content. At the end of the growing season, yield as well as biomass was recorded, which included measuring the total amount of grain and straw, their qualitative characteristics such as protein, carbohydrate and fat content, and analysing the structural and mechanical properties of the grain. To assess the dynamics of carbon accumulation in the soil, samples were taken at different depths and the organic carbon content was determined.

As a result of the conducted research, significant results were obtained, which indicate significant differences in the effectiveness of traditional and nature-based approaches to farming. The average yield of grain crops on the plot where the traditional technology was applied was 25-27 c/ha, which generally corresponds to the average indicators typical for the region, but its instability depending on weather conditions was observed. At the same time, at the site, where the developed nature-like technology was applied, the yield of grain crops fluctuated in the range from 65 to 75 c/ha, reaching in some years and for some varieties 90-110 c/ha. At the same time (fig. 1) there was not only an increase in the quantitative indicators of yield, but also a significant improvement in the qualitative characteristics of grain. In particular, the protein content in grain grown under nature-like technology was 15-20% higher than in grain obtained under the traditional approach.

Figure 1. Photosynthetic potential in bread fields with yields of 65-82 kg/ha.

It is also worth noting that there was a 30-35% reduction in the cost of grain production, which is due to the reduction of costs for mineral fertilisers, pesticides and herbicides. Such a significant increase in yield and grain quality obtained by applying nature-based technology is due to a complex of factors. One of the key factors is the optimised use of natural resources such as solar energy, atmospheric moisture and soil nutrients. As our data show, nature-like technology promotes the formation of a more powerful root system in plants, which ensures their efficient supply of water and nutrients, including through the diffuse rise of moisture from the lower soil horizons, which is especially important in arid climates. (fig. 2)

Figure 2. Carbon polygon in bread cereal crops.

In addition, nature-like technology promotes the formation of a more developed leaf surface, which provides a more intensive process of photosynthesis. In particular, the leaf surface area of plants in the plot with nature-like technology was 40-50% larger than in the plot with traditional technology. This was confirmed by photosynthetic activity monitoring data obtained using a portable fluorimeter, which showed that the photosynthetic intensity was 20-25% higher in plants grown with nature-like technology. As a result of higher photosynthesis intensity, plants accumulate more biomass, which also contributes to higher yield. According to our study (Fig. 3), the biomass of plants grown under nature-like technology was 2-3 times higher than under the traditional approach.

Figure 3. Bread field with high yields.

Analysis of soil samples showed that the content of organic carbon in soil at the site with nature-like technology increased by 10-15% compared to the initial level. At the same time, a slight decrease in organic carbon content was observed at the site where conventional technology was applied, which indicates soil degradation processes. In addition, there was a significant difference in the species composition of soil microbiota. In the plot with nature-like technology, an increase in the number and diversity of beneficial microorganisms, which contribute to the decomposition of organic matter and increase the bioavailability of nutrients to plants, was observed.

The results clearly demonstrate that nature-based agro-technologies are not only more efficient in terms of yields and quality of produce, but also more sustainable in the long term. The higher yields and improved grain quality achieved with these technologies are due to a more rational and efficient use of natural resources, as well as increased soil fertility through the accumulation of organic matter and increased biodiversity. One of the key advantages of nature-based agro-technologies is also the reduced dependence on external resources such as mineral fertilisers and pesticides, which makes the agro-ecosystem more resilient to various external influences. The results obtained are consistent with the results of studies conducted in different regions of the world, which also confirm the effectiveness of organic and sustainable farming. In the context of global changes in the agrarian sphere, it is especially important to take into account the experience of developing economies, which demonstrate the transition to new methods of farming [14]. At the same time, it should be noted that the results of our research exceed the indicators obtained in many similar works due to the application of an integral and convergent approach, which allows the integrated use of natural and anthropogenic resources. To maximise the effectiveness of nature-based methods, it is necessary to integrate different approaches, including water management methods, which ensures more sustainable development of agroecosystems [15, 16]. The approaches we have developed allow achieving higher yields and product quality, as well as providing more efficient carbon sequestration in the soil. An important aspect of our research is also the creation of carbon polygons based on nature-like technologies, which allows not only to increase agricultural productivity, but also to contribute to the fight against global climate change. Data from our study showed that the rate of soil carbon sequestration is 1.5-2.0 tonnes of carbon per hectare per year when nature-based agro-technologies are applied, which is significantly higher than conventional agricultural practices. This indicates that the introduction of nature-like technologies can be an important tool for reducing the concentration of greenhouse gases in the atmosphere. In the future, we plan to expand our research to include the impact of nature-based technologies on different soil types, as well as on the dynamics of carbon cycle change in agroecosystems and on different agrocrops. In addition, more detailed analyses of the mechanisms underlying the effectiveness of nature-based technologies are needed, including studies of the soil microbial community, as well as the interaction of plants with the soil microbiome.

4 Conclusion

This study demonstrates the potential of nature-based agro-technologies based on the convergent approach to increase agricultural productivity and simultaneously reduce the negative impact on the environment. It has been experimentally shown that the application of these technologies allows to achieve a significant increase in grain yields, as well as improve their quality characteristics compared to traditional methods of farming. At the same time, the cost of grain production is reduced due to a more rational use of natural resources and reduced dependence on external resources such as mineral fertilisers and pesticides. The results of the study also showed that nature-based technologies improve soil fertility through the accumulation of organic matter and increased biodiversity, and promote efficient soil carbon sequestration, which has important ecological implications in the context of combating global climate change. The soil carbon sequestration rate of these technologies is significantly higher than that of conventional farming practices. Thus, the adoption of nature-based agro-technologies based on a convergent approach can be an effective tool for ensuring food security, improving the ecological status of agro-landscapes and reducing greenhouse gas emissions, contributing to the transition to a more sustainable and environmentally friendly agriculture. The adoption of nature-based technologies in agriculture not only increases its efficiency, but also requires consideration of socio-economic aspects such as gender equality and the inclusion of different social groups [17, 18]. The obtained results emphasise the need for further study and practical implementation of these technologies in agricultural production. The developed approaches allow achieving higher yields and product quality, as well as ensuring more efficient sequestration of carbon in the soil, and therefore can be used to create a network of carbon polygons, which contributes to the transition to a more sustainable and environmentally friendly agriculture.

References

1. B. Aparecida da Silva, R. Evangelista de Oliveira, A. Cavalieri Sais, E. Cardoso leite, Family farmers and trees: ecological functions, ecosystem services and food production in agrarian reform settlements in Brazil. Agroecol. Sustain. Food Syst., **49**, 4-21 (2025). DOI: 10.1080/21683565.2024.2405515.
2. E. Reisman, M. Fairbairn, Z. Kish, Agrarian Platform Capitalism: Digital Rentiership Comes to Farming. Antipode, **57,** 412-432 (2025). DOI: 10.1111/anti.13107.
3. T. Bandara, B. Marambe, G. Pushpakumara, P. Silva, R. Punyawardena, L. Manawadu, S. Premalal, G. Miah, K.R. Dahal, Adapting mountain-farming systems to climate change by reducing food-nutrition-health vulnerability. Mitig. Adapt. Strateg.

Glob. Change, **29**, 88 (2024). DOI: 10.1007/s11027-024-10181-x.

4. I.I. Bosikov, M.A. Modina, E.V. Khekert, Analysis of the quality of underground mineral waters of terrigenous deposits of the hauteriv-barremian aquifer of the lower cretaceous. News of the National Academy of Sciences of the Republic of Kazakhstan, Series of Geology and Technical Sciences, **2024 (2),** 36-47 (2024) DOI: 10.32014/2024.2518-170X.392.
5. N. Garambois, U. Le Goff, L. Thibaudeau, Agrarian dynamics and climate change in the Senegalese Sahelian peanut basin. Agrar. Syst. Clim. Change: Journeys Adapt. Glob. South, **15**-37 (2024). DOI: 10.1079/9781800628137.0001.
6. H. Cochet, O. Ducourtieux, N. Garambois, Agrarian systems and climate change: Journeys of adaptation in the global south. Agrar. Syst. Clim. Change: Journeys Adapt. Glob. South, **1**-208 (2024). DOI: 10.1079/9781800628137.0000.
7. A.R. Sukiasyan, A.A. Kirakosyan. Seasonal aspects of macro, trace, and ultra trace element changes in soils with different anthropogenic loads. Sustainable Development of Mountain Territories, **16,** 789–802 (2024). DOI: https://doi.org/10.21177/1998-4502-2024-16-2-789-802.
8. S.N. Kryuchkov, A.V. Solonkin, A.S. Solomentseva, S.A. Egorov. Ecological and biological methods of formation of sustainable forest reclamation complexes in steppe and semi-desert landscapes of the european territory of Russia. Sustainable Development of Mountain Territories, **16,** 589–599 (2024). DOI: https://doi.org/10.21177/1998-4502-2024-16-2-589-599.
9. L.P. Rybashlykova. The state of the soil cover and phytomelioration of pastures in the regions of the Eastern Caucasus. Sustainable Development of Mountain Territories, **16,** 134–142 (2024). https://doi.org/10.21177/1998-4502-2024-16-1-134-142.
10. E. Volneikina, A. Menshenin, K. Degtyareva. Simulation-Dynamic Modeling Of Supply Chains Based On Big Data. 2023 22nd International Symposium INFOTEH-JAHORINA, INFOTEH 2023. (2023). https://doi.org/10.1109/INFOTEH57020.2023.10094168.
11. S.V. Balovtsev, O.V. Skopintseva. Science-based technological solutions for aerological risks reducing in operating and designing coal mines. MIAB. Mining Inf. Anal. Bull., **2,** 139-151 (2023). DOI: 10.25018/0236_1493_2023_2_0_139.
12. T.A. Panfilova. Influence of mountain factors on salt excess and soil toxicity in mountain conditions. Sustainable Development of Mountain Territories, **15,** 784-797 (2023).
13. K.I. Kravtsov, Ya.A. Tynchenko, Influence of mountain factors on salt excess and soil toxicity in mountain conditions. Sustainable Development of Mountain Territories, **15,** 784-797 (2023). https://doi.org/10.21177/1998-4502-2023-15-3-784-797.
14. S. Bamidele. BRICS Revolution: The Global Agrarian Shift Sparked by Emerging Economies. India Q., (2025). DOI: 10.1177/09749284241307937.
15. S. Mukherjee, S. Paul, S. Bhattacharya, A. Islam, S. Mahammad, E. Alam. Optimizing water resource management in tropical drought-prone regions through hybrid MCDM techniques: A water-stress mapping approach. J. Hydrol. Reg. Stud., **57,** 102171 (2025). DOI: 10.1016/j.ejrh.2024.102171.
16. R.V. Klyuev, N.V. Martyushev, V.V. Kukartsev, V.A. Kukartsev, V. Brigida. Analysis of geological information toward sustainable performance of geotechnical systems. MIAB. Mining Inf. Anal. Bull., **5,** 144-157 (2024). DOI: 10.25018/0236_1493_2024_5_0_144.
17. L. Ossome. Gender and development in the agrarian south. World Dev**., 188,** 106876 (2025). DOI: 10.1016/j.worlddev.2024.106876.
18. A.I. Smirnov, N.V. Plotnikova, V.G. Burov. Features of the fine structure of the active layers of organic solar cells. Key Eng. Mat., **712,** 176-181 (2016). DOI: 10.4028/www.scientific.net/KEM.712.176.

Current state and prospects for the development of the grain sector of china's agro-industrial complex

Luo Weiwei*, *Shang* Kai and *I.M.* Podmolodina

Voronezh State University of Engineering Technologies, Revolution Avenue 19, 394036, Russia

Abstract. Grain production in China plays a crucial role in ensuring food security and sustainable economic growth. The country's agriculture, including the cultivation of major grain crops, is in a constant process of modernization. This article analyzes the natural and economic characteristics of four major regions of China — Eastern, Central, Western, and Northeastern — and their impact on the structure of agricultural production. The article discusses the main cultural features, issues, and prospects related to the development of the grain sector. Based on data on grain production and exports, the authors conclude that China's growing dependence on soybean imports persists despite efforts to increase domestic production volumes. In the future, the authors believe that special attention should be given to optimizing land resources, implementing precision farming technologies, and improving mechanization levels to enhance production efficiency.

1 Introduction

Grain production in China is of great importance for ensuring food security and stable economic growth. The grain sector faces numerous challenges, including limited land resources, climate change, and the increasing need to improve agricultural infrastructure. The purpose of this study is to analyze the current state and determine the prospects for the development of the grain sector of China's agro-industrial complex, taking into account regional characteristics and the potential for increasing productivity and sustainability in the agricultural sector based on new technologies.

2 Materials and Methods

The study examines the development of the grain sector of China's agro-industrial complex, focusing on regional differences in agricultural production, climatic and soil conditions, and their impact on the production of major grain crops. Data from government reports and agricultural production statistics were used. The research covers four regions of China, each characterized by its own climatic and soil features that influence grain production. The analysis employed methods of descriptive statistics, logic, comparative and trend analysis, as well as graphical data representation.

3 Results and Discussion

In terms of land area, China is one of the world leaders, ranking third globally after Russia and Canada. However, significant differences in the natural and economic conditions of the four major regions of China (Eastern, Central, Western, and Northeastern) lead to variations in agricultural trends and challenges faced by the regions. Table 1 presents comparative data characterizing the agricultural development conditions in these four regions [1].

Table 1 Comparative Analysis of Agricultural Development Conditions and Major Crops Grown in the Four Regions of China.

Region	Characteristics of Natural Conditions	Characteristics of Economic Conditions	Major Agricultural Crops Grown
Eastern	Warm and humid climate, fertile soils	Developed economy, high level of urbanization, advanced agricultural modernization, shortage of land resources [2].	Rice, wheat
Central	Moderate climate, sufficient rainfall, flat terrain	Balanced economy, good agricultural infrastructure, but uneven distribution of water resources and labor outflow.	Wheat, rice, corn
Western	Complex terrain, variable climate	Underdeveloped economy, weak agricultural infrastructure, reliance on government support.	Corn, wheat, fruits
Northeastern	Fertile soils, cold climate	High level of mechanization, stable grain production, but population outflow and strong impact of climate change.	Corn, soybeans

* Corresponding author: luoweiwei_loria@163.com

The Eastern Region of China is located in the coastal zone with a warm and humid climate. In the valleys of the Yangtze River and the lower reaches of the Yellow River, fertile soils promote the cultivation of rice and wheat. This is the most economically developed region in the country, where modern technologies and mechanization are actively applied. However, urbanization and the growth of cities lead to increased demand for agricultural products and a reduction in arable land area [3].

The Central Region has a temperate climate, abundant rainfall, and a flat landscape, making it an important grain-producing area. The soils here are suitable for growing wheat, rice, and corn. The region's economy is balanced, but the development pace is slower compared to the Eastern region, and agriculture remains an important part of the economy. Major issues include the uneven distribution of water resources and labor outflow.

The Western Region of China includes mountain ranges and desert areas with harsh climatic conditions. A lack of water resources and poor soils limit agricultural production. Drought-resistant crops such as corn and wheat are grown here, along with the development of fruit growing and livestock farming. The region's economy is underdeveloped, and agricultural infrastructure is weak, requiring government support in transportation and irrigation [4].

The Northeastern Region is known for its fertile chernozem soils, which are suitable for growing corn and soybeans [5]. However, the cold climate, with long winters and a short growing season, requires precise organization of planting times. The region has a high level of mechanization, but industrial decline and population outflow create a labor shortage. Additionally, climate changes, such as droughts and early frosts, pose a threat to agriculture.

Table 2 demonstrates that the highest rice production is concentrated in the Central Region, while rice volumes remain stable in the Eastern and Western Regions, and the Northeastern Region stands out for producing high-quality rice due to its fertile chernozem soils. The main wheat production is in the Eastern and Central Regions, particularly in Henan and Shandong provinces, while in the Western Region, wheat volumes are limited by unfavorable terrain and climate, and in the Northeastern Region, production is minimal due to small sowing areas. Corn production is most developed in the Northeastern Region due to favorable climatic conditions, though significant volumes are also produced in the Central and Western Regions, where corn is primarily used in livestock farming and industry. Soybeans are of great importance in global agriculture, ranking fourth after wheat, corn, and rice, and first among leguminous crops. The main soybean-producing region in China is the Northeast, with Heilongjiang province contributing the largest share, while production in other regions remains low, making China dependent on imports to meet its high domestic demand.

Table 2 Production of Major Agricultural Crops in the Four Main Regions of China, 2023 (Million Tons)

	Rice	Wheat	Corn	Soybeans
Eastern Region	43.82	57	52.7	1.95
Central Region	87.07	59.59	46.5	3.21
Western Region	40.37	19.9	91.51	5.38
Northeastern Region	35.35	0.1	98.13	10.3

Source: compiled by the author based on data from [16]

The development of the grain sector in China's four major regions forms the basis for the placement of the country's main agricultural productions. Each region has its specific characteristics in grain, technical crops, and livestock production, contributing to the creation of a diverse agricultural structure [6]. The production of major grain crops in China is shown in Figure 1. In 2023, the priority crops were wheat, rice, and corn. Chinese agriculture met 95-98% of the total domestic food demand, with grain crops accounting for 85-90% of the total food demand.

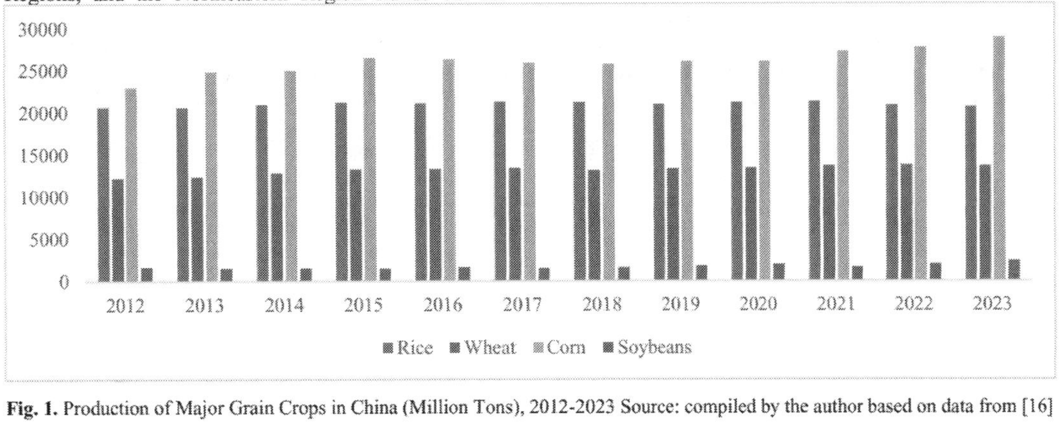

Fig. 1. Production of Major Grain Crops in China (Million Tons), 2012-2023 Source: compiled by the author based on data from [16]

The data analysis from the figure allows us to conclude that from 2013 to 2023, the production of rice, wheat, corn, and soybeans in China generally shows an upward trend. The production of rice and wheat remains relatively stable, with wheat production demonstrating gradual growth, reflecting the impact of technological improvements and increased mechanization. Corn production is characterized by significant volatility, but in recent years, a noticeable growth trend has been observed. Since 2007, the government has been purchasing corn from farmers at a fixed price, ensuring stable income for producers. However, this policy was abolished in 2016, and the price of corn shifted to a market-based pricing system, leading to a reduction in farmers' incomes and decreased motivation to grow corn, causing fluctuations in its production volume.

The competitive prices of soybeans in the United States, Brazil and Argentina make foreign soybeans very competitive in the Chinese domestic market, reducing the incentive to grow soybeans in China. However, in recent years, with government subsidies and technological support, the area of land planted with soybeans has expanded significantly in the northeast, and yields have increased.

Overall, despite stable production and a high degree of modernization of major food crops, China still faces many challenges in terms of resources, markets and the environment. Urbanization in the eastern region has led to a shortage of arable land, and rice production is constrained by water resources. In the central and northeastern regions, climate change is affecting wheat and corn production, with the risk of extreme weather. In the western region, poor geographical conditions and inadequate infrastructure hinder agricultural production, and low soybean yields still face pressure from imports.

Over the past ten years, the production of major agricultural crops in China has developed steadily, providing an important foundation for the domestic food market [7]. However, with increasing domestic demand and changes in the global market, China needs not only to increase the efficiency of its own grain production but also to address issues related to the import and export of agricultural products [8]. This is especially true for soybeans, as domestic production does not fully meet the growing demand, leading to an increasing dependence on imports [9].

The import of major grain crops into China is shown in Figure 2.

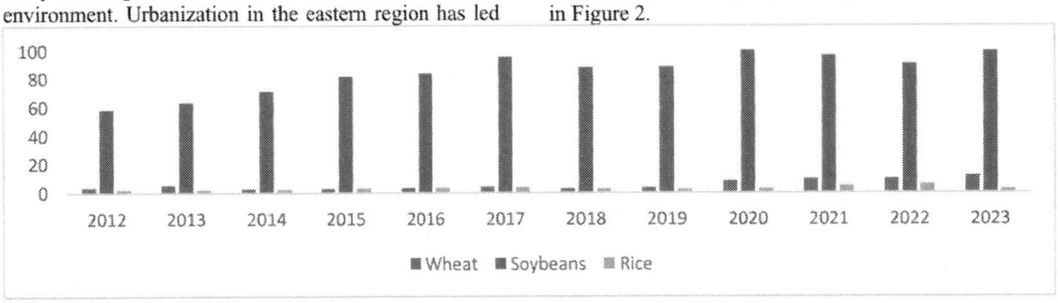

Fig. 2. Major Grain Imports into China (Million Tons), 2012-2023 Source: compiled by the author based on data from [16]

From 2012 to 2023, the structure of grain imports into China underwent significant changes. The volume of wheat imports increased annually, from 3.7 million tons in 2012 to 12.09 million tons in 2023, representing a growth of nearly 227%. Soybean imports also continued to grow substantially, reaching 99.41 million tons in 2022, with a growth rate of 70% over the period. Rice imports fluctuated, but a sharp increase in import volumes was observed in 2021-2022. This was related to the impact of the pandemic when the Chinese government increased reserves to cope with potential food crises and supply chain disruptions.

Currently, China is heavily dependent on imports of soybeans, wheat, and rice. The main soybean suppliers are Brazil and the USA. Soybeans are used in China for oil production and animal feed, especially for pig farming, but domestic production does not meet demand. In wheat, China mainly depends on Australia and Canada, as domestic production does not cover the needs for high-quality wheat for the baking industry [10]. Although China is the largest producer of rice, the main rice exporters remain Vietnam, Thailand, and Myanmar, who supply high-quality rice to meet domestic demand.

China also exports certain quantities of grains such as rice, wheat, soybeans, and corn. The main export destination for wheat is North Korea. Corn is predominantly exported to North Korea and Russia. The main importers of Chinese soybeans are South Korea and Japan. However, the largest volumes of China's exports are in rice. Although China is one of the world's largest grain producers, its grain exports remain relatively low. This is largely because of the country's massive domestic demand, driven by its large population and the need to ensure food security [11]. The pressure to feed such a large population means that much of China's grain production is kept within the country.

Moreover, limited arable land and high production costs reduce the global competitiveness of Chinese soybeans [12]. Countries with lower production costs dominate the global grain export market, which further hampers China's export efforts. To complicate matters, the Chinese government places restrictions on grain exports to secure domestic supply and stabilize prices, which naturally leads to lower export volumes.

Looking ahead, China's grain industry has reasons to be optimistic. With a solid foundation in production, ongoing technological advancements, and greater mechanization, the country is positioned to improve both the efficiency and quality of its grain production. This will not only sustain the domestic market but should also support long-term stability. One key to this future growth will be optimizing land use, especially with the help of precision agriculture technologies that promise higher yields with less land, fostering a more sustainable farming model [13]. Improved mechanization will definitely help lower production costs and boost efficiency, which is so important if China's agricultural sector is going to grow in the future [14]. But there's more to it. China is working on diversifying what it grows. This isn't just about increasing the amount of food produced, but also about making sure it doesn't rely too much on one type of crop. This will make the whole agricultural sector a bit more resilient, especially when market conditions change. Part of the plan also includes building up grain reserves and improving storage systems, so the country can handle food shortages during tough times.

That said, it's not all smooth sailing. There are still some big obstacles. One of the most pressing issues is the lack of land, combined with the effects of climate change, which make it even harder to increase production. To deal with these problems, China will need to get better at using land more efficiently and work on developing crops that can handle the changing climate [15]. And even though there's been progress in increasing production, China still relies a lot on imports—especially for things like soybeans. To really tackle this issue, boosting domestic production of these essential crops is going to be key.

4 Conclusion

The research has shown that the differences in natural and economic conditions across the four major regions of China influence the structure of grain production. These regions have distinct agricultural trends driven by climatic conditions, infrastructure status, and access to water resources.

Based on the current state of the grain sector in China, the following recommendations have been developed to stimulate the development of the country's agribusiness sector for each region. For the Eastern region, it is important to maintain rice and wheat production, improve land resource management, and implement technologies to increase yields. The Central region must maintain a balance between rice, wheat, and corn production, focusing on infrastructure development and retaining the workforce. For the Western region, priorities should include government support, infrastructure development, and the introduction of climate-resistant grain varieties. In the Northeastern region, the key tasks are maintaining corn and soybean production, preventing population outflow, and adapting technologies to changing climatic conditions.

The study concluded that grain production in China shows stable growth despite existing challenges, such as limited land resources and the impact of climate change. Key steps to improving the situation include optimizing land use, introducing new climate-resistant crop varieties, and further developing agricultural mechanization. However, China remains dependent on soybean imports, which requires additional efforts to reduce this reliance. In the future, the development of precision agriculture technologies and infrastructure improvements will also play a key role in ensuring the country's food security.

References

1. L. Gu, Assessment of Regional Differences in the Economic Development Level of China's Major Grain-Producing Areas, Chin. J. of Agricult. Mech. Chem., **36(01)**, 328–331 (2015)
2. L. Yan-sui, Zh. Yang, Reflections on China's Food Security and Land Policy Amid Rapid Urbanization, Land Use Policy, **109** (2021) https://doi.org/10.1016/j.landusepol.2021.105699
3. Ch.-Ch. Lee, M. Zeng, K. Luo, The Impact of Urbanization on China's Food Security, Int. Rev. of Econ. and Fin., **93(B)** (2024), https://doi.org/10.1016/j.iref.2024.05.037
4. J. Liu, W. Yang, Intercropping Soybean and Corn: A Solution for China's Food Security, J. of Integr. Agricult., **23(7)** (2024), https://doi.org/10.1016/j.jia.2024.02.001
5. W. Yang, Z. Chen, T. Ding, X. Yang, W. Gong, Assessment of the Security of Demand and Supply in Water, Energy, and Food Systems, The Perspective of Urban and Inter-City Ecosystem Service Links, Sust. Cities and Soc., https://doi.org/10.1016/j.scs.2024.105964
6. Zh. Yang, Y. Sijia, Y. Ziyan, Spatio-Temporal Link Between Food Production, Economic Development, and Environmental Protection in China's Major Grain-Producing Areas, J. of Ecol., **04**, 1–15 (2025)
7. X. Wei, Z. Zhang, P. Wang, F. Tao, Recent Trends in Major Grain Crop Production: Implications for China's Food Security, Reg. Envir. Chan., J., **17**, 105–116 (2017)
8. Ji-kun Huang, Wei Wei, Qi Cui, Wei Xie, Prospects for China's Food Security and Imports: Can China Starve to Feed the World?, J. of Compreh. Agricult., **16(12)** (2017), https://doi.org/10.1016/S2095-3119(17)61756-8
9. J. Yu, J. Jin, Study of the Ethics of Developing China's Food Security Strategy in the New Era, J. of South. Forestry Univ. (Soc. Sci.), **8(03)**, 15–20 (2024)
10. Sh. Zhang, M. Zheng, Current Status and Prospects of China's Food Security in the New Era, Heilongjiang Grain, **09**, 6–8 (2024)
11. Y. Yang, Ch. Yao, W. Liu, Spatio-Temporal Characteristics and Factors of the Transformation of China's Food Security System, Geograph. J., **79(09)**, 2372–2388 (2024)
12. E. Liu, Yu. Lv, Development Status, Problems, and Countermeasures in China's Foreign Trade in Grain, Foreign Trade and Econ., **06**, 56–57 (2015)
13. I.M. Podmolodina, I.A. Davidenko, Development of the Agribusiness of Municipal Districts as a Factor in Ensuring the Country's Food Security, Act. Direct. of Sci. Res. in the 21st Century: Theory and Pract., **12(1(64))**, 108–119 (2024), https://doi.org/10.34220/2308-8877-2024-12-1-108-119, EDN PSXZPK
14. I.M. Podmolodina, V.P. Voronin, O.Yu. Kolomytsova, E.Yu. Kolesnikova, Improving

Industrial Development Management as the Basis for Food Security: A Comprehensive Approach, Bull. of Voronezh State Univ. of Eng. Technol., **82(3(85))**, 289–295 (2020), https://doi.org/10.20914/2310-1202-2020-3-289-295, EDN SSBJAG

15. Xu I., Liang Ya., Chen K., Recessive Transition of Agricultural Land Use and Food Security: Evidence from China, J. of Rural Stud., **113** (2025), https://doi.org/10.1016/j.jrurstud.2024.103484
16. *National Bureau of Statistics of the People's Republic of China*, https://www.stats.gov.cn/sj/ndsj/, Access: 12/24/2024

Development of probiotic feed based on homogenised potato pulp

Atkham Borotov[1,*], *Olim* Narzullaev [2], *Lazizkhon* Yakubov [3], *Matluba* Rajabova [4], *Bogdagul* Karshieva [5]

[1] Tashkent Institute of Irrigation and Agricultural Mechanization Engineers, National Research University, Tashkent, 100000, Uzbekistan
[2] Tashkent State University of Law, Sayilgokh Street, 35 Tashkent, 100047, Uzbekistan
[3] Almalyk branch of the Tashkent State Technical University named after Islam Karimov, Almalyk , Uzbekistan
[4] Termiz State Pedagogical Institute, Termiz , Uzbekistan
[5] Termez State University, Termiz, Uzbekistan

Abstract. This paper presents a study aimed at developing a probiotic feed preparation based on a secondary product of the food industry - homogenised potato mass (HPM). The possibility of using HPC as a substrate for cultivation of probiotic bacteria Bacillus subtilis and Lactobacillus plantarum was studied. It was experimentally shown that GCM provides effective growth and multiplication of these bacteria, and lyophilisation allows to preserve their viability for a long time. The obtained preparation demonstrated pronounced antibiotic activity against a number of pathogenic microorganisms. This allows us to conclude that HCM is a promising raw material for the production of cost-effective and efficient probiotic preparations, which contributes to the development of sustainable animal husbandry and reduces dependence on antibiotics. The results confirm the possibility of efficient utilisation of secondary raw materials for the production of probiotics and indicate the importance of further research in this area.

1 Introduction

Modern livestock breeding plays a key role in providing mankind with food [1]. However, intensification of production aimed at increasing productivity often leads to negative consequences, such as disturbance of the balance of microflora of the gastrointestinal tract of animals, reduction of their natural resistance to diseases, and, as a consequence, reduction of product quality [2]. The widespread use of antibiotics in animal husbandry as growth stimulants and prophylactic agents, although giving a temporary positive effect, leads to the formation of resistance in pathogenic microorganisms, dysbacteriosis and accumulation of antibiotic residues in animal products, which not only reduces the quality, but also endangers human health [3]. Globally, this problem has become so significant that many countries, including the European Union, have introduced restrictions or bans on the use of antibiotics as feed additives. This situation requires the search for alternative approaches that can ensure effective and sustainable development of animal husbandry [4].

One of the promising directions is the development and use of probiotics. These feed additives containing live microorganisms capable of restoring and maintaining the balance of intestinal microflora represent a potentially powerful tool for solving many problems associated with digestive disorders and reduced immunity of animals. Probiotics can help increase animal performance by improving nutrient absorption, stimulating the immune system and reducing the risk of infectious diseases. The use of probiotics can also reduce the need for antibiotics, leading to a reduced risk of resistant pathogens and safer and more environmentally friendly products. However, there are some drawbacks, such as the need for careful selection of probiotic microorganism strains, their ability to remain viable under different conditions, and the limited shelf life of many preparations. Some probiotics may be unstable during storage and lose their efficacy when exposed to gastric juice, requiring the development of specialised technologies for their production and delivery to the gut. Despite these challenges, the use of probiotics remains a promising area in the development of modern livestock production.

In light of the above, the research presented in this paper is highly relevant and important. Modern biotechnology makes it possible not only to isolate and study the effects of different strains of probiotic microorganisms, but also to produce probiotic preparations using secondary raw materials, i.e. by recycling low-value waste. This approach is in line with the principles of zero-waste production and is an

* Corresponding author: atxamborotov@mail.ru

important step towards sustainable agricultural systems. In particular, the use of secondary products from the food industry as a basis for the production of feed probiotics not only solves the problem of waste disposal, but also makes it possible to reduce the cost of production and make probiotic preparations more affordable for livestock farms [5, 6]. The development of effective probiotic preparations based on secondary raw materials and the study of their effect on animal organisms are key aspects for improving the sustainability and environmental friendliness of modern livestock farming.

The aim of this work is to study the possibility of using a secondary product of the food industry - homogenised potato mass - for the production of fodder probiotics, as well as to assess the effect of probiotic preparations derived from this raw material on the productivity and health of farm animals. The research is aimed at creating effective and economically available probiotic preparations that can make a significant contribution to the development of sustainable animal husbandry [7].

2 Method and materials

The experimental work was aimed at the development of a probiotic preparation for feed purposes based on secondary raw materials. The research plan included several stages: selection of probiotic microorganism strains, cultivation of the selected strains on the basis of homogenised potato mass, evaluation of the viability of the obtained preparation, and study of its effect on animal organism. A range of specialised equipment based on efficient process planning and modelling was used for microbiological studies [8].

Cultivation of probiotic microorganisms was carried out in laboratory bioreactors. Applikon bioreactors, model ADI 1030, with a volume of 5 litres were used. This equipment allows precise control of temperature, pH, stirring speed and aeration level, which is critical for optimising microbial growth and reproduction. The role of automation in these processes is noteworthy [9, 10]. During cultivation, the bioreactors were kept at 37°C, pH was maintained at 6.8-7.0, and constant agitation was provided at 200 rpm, aeration was performed using a sterile air supply with a flow rate of 10 litres/hour. A Shimadzu spectrophotometer, model UV-1800, was used to monitor the cell concentration throughout the entire cultivation process, and measurements were performed at a wavelength of 600 nm.

The obtained microbial biomass was lyophilised to increase the shelf life and convenience of storage and use of the preparation. Lyophilisation was carried out on a Christ brand freeze dryer, model Alpha 1-4 LD plus, for 24 hours at -50 °C and 0.1 mbar pressure. This equipment allows the removal of water from the biomass by sublimation, while retaining all the biological properties of the microorganisms.

During the study, an Olympus microscope, model BX53, equipped with a digital camera was used to assess the total number of microorganisms and their viability. Live cells were counted by Vishnyakov's staining method. For preparation of preparations from microbial biomass and its staining we used laboratory centrifuge 'Eppendorf', model 5430R, centrifugation was carried out for 10 minutes at a speed of 5000 rpm. The results obtained were evaluated by colony forming units per gram (CFU/g) after cultivation, lyophilisation and storage. To determine the antibiotic activity of the isolated strains, the agar diffusion method was used. For the studies we used the Binder thermostat, model BD 56, where incubation of Petri dishes at 37 °C for 24-48 hours was carried out.

3 Results and discussion

The experimental work was aimed at studying the possibility of using homogenised potato mass (HPM) - a secondary product of potato mash production - as a substrate for the cultivation of probiotic microorganisms, and developing a probiotic preparation based on it. During the study, initially, the most promising bacterial strains that showed high potential as probiotics were selected. Then, the selected strains were cultured on a medium based on HCM. This allowed us to evaluate how effectively this raw material was suitable for the growth and multiplication of probiotic bacteria. After successful culturing, the resulting microbial biomass was lyophilised to ensure viability and ease of further use. At the next stage, studies were carried out to investigate the viability of the obtained probiotic preparation as well as its effect on animal organism.

The results obtained in the course of the research indicate a high prospectivity of the chosen approach to the creation of a fodder probiotic preparation based on homogenised potato mass (HCM). This secondary raw material demonstrated itself as an effective and economical substrate for the cultivation of probiotic bacteria. In particular, the analysis of microorganisms growth on the medium with GCM showed that this raw material is not only not inferior, but in some cases even superior to traditional nutrient media, both in terms of growth rate and final cell concentration. During the experiments, several bacterial strains were tested, among which Bacillus subtilis and Lactobacillus plantarum, known for their probiotic properties, stood out. Bacillus subtilis is a spore-forming Gram-positive bacterium widely used in biotechnology due to its resistance to unfavourable conditions and its ability to produce a wide range of enzymes and antibiotic substances. Lactobacillus plantarum, in turn, is a Gram-positive lactic acid bacterium with a marked ability to colonise the gut and compete with pathogens, as well as the ability to synthesise a range of B vitamins and organic acids to improve digestion. At the start of the experiment, for each strain a culture with an initial cell concentration of 105 CFU/ml was prepared and then added to bioreactors filled with HCM medium.

In the process of cultivation on GCM, active growth of bacteria of both species was observed, and by the end of the 24-hour process the concentration of viable cells of Bacillus subtilis averaged 2.4 x 10^8 CFU/ml, and

Lactobacillus plantarum - 1.8 x 10^8 CFU/ml, which indicates a sufficiently high nutritional value of this substrate for these microorganisms. The increase in the number of bacteria occurred at an intensive rate, which is confirmed by measuring the optical density of the culture liquid on a spectrophotometer. The initial lag phase was only 2-3 hours, followed by an exponential phase lasting up to 18-20 hours, with a population doubling period of about 2 hours for Bacillus subtilis and 2.5 hours for Lactobacillus plantarum. This indicates that HCM provides the necessary conditions for rapid and efficient multiplication of probiotic bacteria [11, 12]. At the same time, the use of HCM not only provided active growth of probiotic bacteria, but also significantly reduced the production cost of probiotic preparation. According to calculations, the cost of nutrient medium based on HCM was 30-35% lower than the cost of traditional nutrient media such as meat-peptone agar or MRS medium. This was possible due to the low cost of HCM as a secondary product of mashed potato production.

To increase the stability and extend the shelf life of the preparation, the probiotic bacterial biomass was lyophilised. Immediately after freeze drying, the number of CFU in the preparation of Bacillus subtilis was 9.1 x 10^7 CFU/g, and Lactobacillus plantarum - 8.7 x 10^7 CFU/g. After 3 months of storage at room temperature, the CFU count was 7.4 x 10^7 CFU/g for Bacillus subtilis and 6.8x10^7 CFU/g for Lactobacillus plantarum, indicating that the microorganisms were resistant to lyophilisation and storage conditions. The loss of viable cells during 3 months of storage was less than 20%, which is considered quite acceptable for lyophilised probiotic preparations. These studies show that the lyophilisation process used in the work provides good preservation of the bacterial biomass.

During the study of antibiotic activity, it was found that the bacterial strains studied have a pronounced ability to inhibit the growth of a number of pathogenic and opportunistic microorganisms. In particular, Bacillus subtilis demonstrated antagonistic activity against Escherichia coli, Salmonella enterica, and Staphylococcus aureus, which is confirmed by the formation of growth inhibition zones around bacterial colonies on Petri dishes. Similar studies were conducted in mountainous conditions [13].

The diameter of inhibition zones for E.coli was 12-14 mm, for Salmonella enterica - 10-12 mm, and for Staphylococcus aureus 13-15 mm. Lactobacillus plantarum showed the highest activity against Clostridium perfringens and Candida albicans, the diameter of the inhibition zones for C. perfringens was 11-13 mm and for C. albicans 9-10 mm. These results indicate that the isolated bacterial strains have a wide range of antibiotic activity, which is due to their production of a number of antimicrobial substances such as bacteriocins and organic acids. This allows us to expect a positive effect from the use of the developed preparation as a prophylactic agent against intestinal infections and dysbacteriosis caused by these pathogens, which is especially important to reduce the use of antibiotics in animal husbandry.

Summarising the results of the experimental work, we can conclude that the proposed methodology for the production of probiotic preparation based on GCM is quite promising. Firstly, GCM proved to be a suitable substrate for the growth and multiplication of probiotic bacteria, which provided a high concentration of cells in the obtained preparation. Secondly, lyophilisation allowed to preserve the viability of microorganisms in the preparation for a long time, which simplifies its storage and use. Thirdly, the obtained probiotic preparation demonstrated pronounced antagonistic activity against pathogenic microorganisms, which may be an important factor for the prevention and treatment of intestinal diseases. In general, these results are consistent with the results of other similar studies. For example, works on the use of food industry by-products such as whey and molasses as a substrate for probiotic production have also reported their high efficiency and economic feasibility, which is important when using mining waste in the formation of tailings ponds [14, 15].

In particular, the use of dairy whey enriched with prebiotics for the cultivation of Lactobacillus acidophilus resulted in a probiotic product with a high concentration of viable cells (about 109 CFU/ml) and good organoleptic properties.

The results obtained in this work are also consistent with the general trend in the field of biotechnology, aimed at finding alternative sources of raw materials and creating effective and environmentally friendly preparations for agriculture [16]. Comparison with other studies on the use of spore-forming bacteria of the genus Bacillus as probiotics confirms the prospectivity of this direction [17]. For example, the use of Bacillus subtilis as a probiotic for birds showed an increase in productivity by 10-15%, a decrease in morbidity by 25-30% and an improvement in feed conversion. The formulation developed in this work, similarly showed good antagonistic activity against a wide range of pathogens including E. coli, Salmonella spp. and Clostridium perfringens, which is a key factor in the efficacy of probiotic preparations. It should be noted that not only economic but also social aspects are important in land use [18]. In addition, studies show that the use of probiotics based on Bacillus subtilis can improve digestive performance, reduce toxins in the intestine and thus increase the natural resistance of animals. Some studies have shown that the addition of Bacillus subtilis to broiler feed can reduce feed intake by 5-7% while increasing average daily weight gain [19]. These results are in good agreement with the results presented in this article, indicating the promising use of HCM for the production of probiotics based on spore-forming bacteria.

Thus, the results of the conducted studies confirm the possibility of effective use of HCM for the production of probiotic preparations for feed purposes. The developed preparation is characterised by a high concentration of viable cells, stability to storage conditions, and a pronounced antagonistic effect. These results open new opportunities for reducing costs in animal husbandry and improving the quality of agricultural products

4 Conclusion

This study demonstrates the promising application of homogenised potato mass (HPM) as a basis for the production of probiotic feed preparations. Experimental results confirm that HCM is not only economically favourable, but also a sufficiently nutritious substrate for the cultivation of probiotic microorganisms such as Bacillus subtilis and Lactobacillus plantarum. The obtained data indicate that these bacterial species actively multiply on the medium with HCM, reaching a high concentration of viable cells. At the same time, the use of HCM allows to significantly reduce the cost of production of probiotics, which makes them more accessible for use in animal husbandry.

Lyophilisation applied for biomass preservation provided good preservation of viable microorganisms, which is confirmed by a slight decrease in CFU count after 3 months of storage. This is an important factor for the development of preparations with a long shelf life, convenient in transport and application. In addition, the study of antibiotic activity of the obtained probiotic preparation revealed its ability to inhibit the growth of a number of pathogenic and opportunistic microorganisms such as Escherichia coli, Salmonella enterica, Staphylococcus aureus, Clostridium perfringens and Candida albicans. This effect is associated with the release of antimicrobial substances by bacteria, which makes the developed preparation promising for the prevention and treatment of intestinal infections and dysbacteriosis in animals, as well as an alternative to antibiotics.

Comparative analysis of the results obtained with the data of other studies shows that the use of by-products of food industry for the production of probiotics is a promising direction. Similar results were obtained using whey and molasses as substrates for the cultivation of probiotic bacteria, which confirms the general tendency to search for economically favourable and environmentally friendly ways of producing feed additives. The use of Bacillus subtilis as a probiotic for poultry in other studies has shown increased productivity, reduced morbidity and improved feed conversion rates, which is in agreement with the results obtained in this study. The presence of broad antibiotic activity against pathogenic microorganisms and the ability to improve digestion confirms the promising use of GCM for the production of probiotics based on spore-forming bacteria.

Thus, a comprehensive study has shown that homogenised potato mass is a promising raw material for the production of probiotic preparations. The developed production technology provides the preparation with high concentration of viable cells, long shelf life and expressed antibiotic activity. These results open and lyophilisation to increase the efficiency and reduce the cost of probiotics production.

References

1. B. Seun. BRICS Revolution: The Global Agrarian Shift Sparked by Emerging Economies. India Quarterly, (2025). DOI: 10.1177/09749284241307937.
2. N. Garambois, U. Le Goff, L. Thibaudeau. Agrarian dynamics and climate change in the Senegalese Sahelian peanut basin. In Agrarian Systems and Climate Change: Journeys of adaptation in the Global South, **6,** 15–37 (2024). DOI: 10.1079/9781800628137.0001.
3. M.A. Mahfud, N.H. Djohan. THE EXPANSION OF INVESTOR ACCESS TO CULTIVATION RIGHTS: A Socio-Legal Analysis on Agrarian Injustice in Indonesia. Al-Risalah: Forum Kajian Hukum dan Sosial Kemasyarakatan, **24(2),** 55–67 (2024). DOI: 10.30631/alrisalah.v24i2.1500.
Z. Zhuang, H. Qi, S. Huang, Q. Wang, Y. Wan, H. Li. Impact of short-term soil disturbance on cadmium remobilization and associated risk in vulnerable regions. Ecotoxicology and Environmental Safety, **289,** 117699 (2025). DOI: 10.1016/j.ecoenv.2025.117699.
4. V.I. Golik, R.V. Klyuev, N.V. Martyushev, V.V. Kondratiev, V.S. Tynchenko, V.A. Gladkikh, L.V. Iushkova, V. Brigida. Reuse and Mechanochemical Processing of Ore Dressing Tailings Used for Extracting Pb and Zn. Materials, **16,** 7004 (2023). https://doi.org/10.3390/ma16217004.
5. B. Aparecida da Silva, R. Evangelista de Oliveira, A. Cavalieri Sais, E. Cardoso leite. Family farmers and trees: ecological functions, ecosystem services and food production in agrarian reform settlements in Brazil. Agroecology and Sustainable Food Systems, **49(1),** 4–21 (2025). DOI: 10.1080/21683565.2024.2405515.
6. L.S. Malyukova, V.V. Kondratiev, V.V. Bukhtoyarov, V.Y. Konyukhov, K.A. Bashmur, T.A. Panfilova, V. Brigida. Circular Mining Wastes Management for Sustainable Production of Camellia sinensis (L.) O. Kuntze. Sustainability (Switzerland), **15(15),** 11671 (2023). DOI: 10.3390/su151511671.
7. S.G. Gendler , M.L. Rudakov , E.S. Falova, Naukovyi Visnyk Natsionalnoho Hirnychoho Universytetu, **2020(3),** 81–85, (2020). https://doi.org/10.33271/nvngu/2020-3/081.
8. S.G. Gendler , M.L. Rudakov , V.S.Kuznetsov, Latvian Journal of Physics and Technical Sciences, **56(3),** 62–69, (2019). https://doi.org/10.2478/lpts-2019-0020.
9. A.A. Kapanski, R.V. Klyuev, A.E. Boltrushevich, S.N. Sorokova, E.A. Efremenkov, A.Y. Demin, N.V. Martyushev. Geospatial Clustering in Smart City Resource Management: An Initial Step in the Optimisation of Complex Technical Supply

Systems. Smart Cities, **8(1),** 14 (2025). https://doi.org/10.3390/smartcities8010014.

10. D.D. Duisenova, K.M. Tireuov, L.M. Tatarintsev, L.V. Lebedeva. Ecological typization of agricultural land and modeling of sustainable agrolandscape at the local level. Sustainable Development of Mountain Territories, **16(1),** 94–110 (2024). DOI: https://doi.org/10.21177/1998-4502-2024-16-1-94-110.
11. N.A. Tursynbaev, I.V. Kuzhevskaya, L.M. Tatarintsev, A.A. Ospanova. Assessment of the environmental state and sustainability of agrolandscape at the local level. Sustainable Development of Mountain Territories, **15(3),** 662–671 (2023). DOI: 10.21177/1998-4502-2023-15-3-662-671.
12. K.I. Kravtsov, Ya.A. Tynchenko, T.A. Panfilova. Influence of mountain factors on salt excess and soil toxicity in mountain conditions. Sustainable Development of Mountain Territories, **15(3),** 784-797 (2023). https://doi.org/10.21177/1998-4502-2023-15-3-784-797.
13. V.Y. Konyuhov, A.M. Gladkih, I.I. Galyautdinov, Y.D. Severina. Economic aspects of green technologies. IOP Conference Series: Earth and Environmental Science, 350(1), 012036 (2019). DOI: 10.1088/1755-1315/350/1/012036.
14. D.A. Zyukin, A.I. Karlina. Technology for Nonwaste Recovery of Tailings of the Mizur Mining and Processing Plant. Metallurgist, **66(11-12),** 1476-1480 (2023). DOI: 10.1007/s11015-023-01462-y.
15. L.C. Zulu, I.N.S. Djenontin, J.F. Kamoto, J.M. Kampanje-Phiri, G. Fischer. Do youth conceptualizations influence the inclusion of young people in sustainable agriculture intensification? Insights from Ghana and Malawi. Environment, Development and Sustainability, **25(12),** 13909–13935 (2023). DOI: 10.1007/s10668-022-02632-9.
16. S.G. Gendler , M.S. Kryukova, Mining Informational and Analytical Bulletin, **9,** 248–26, (2023). https://doi.org/10.25018/0236_1493_2023_91_0_248.
17. J. Rodzik, R. Kołodyńska-Gawrysiak, Ł. Franczak, W. Zgłobicki, J. Poesen. IMPACT OF LAND USE CHANGES ON COLLAPSED PIPES DEVELOPMENT IN THE LOESS GULLY (LUBLIN UPLAND, EAST POLAND). Quaestiones Geographicae, **43(4),** 17–33 (2024). DOI: 10.14746/quageo-2024-0042.
18. S.O. Hryhorovych, K.O. Vasylivna, P.M. Ivanovich, H.O. Yuriivna. Development of the agricultural sector of Ukrainian economy in conditions of the impact of the COVID-19 pandemic. Ekonomika APK, **28(6),** 26–41 (2021). DOI: 10.32317/2221-1055.202106026.

Harrington function as an integral approach to assessing food security in Russia

Xinyi Qian*, *Yu.A.* Salikov, *O.Yu.* Kolomytseva and *A.V.* Maslov

Voronezh State University of Engineering Technologies, 19 Revolutsii Avenue, Voronezh, 394036, Russia

Abstract. In accordance with the new doctrine of food security of the Russian Federation, which includes 14 key indicators, a comprehensive analysis of the level of food security in Russia for 2019–2023 was conducted. A surrogate function based on the Harrington function was used for the analysis. Using a standard Excel program, a linear equation with a high degree of correlation with the Harrington function was determined, which allowed ensuring the accuracy of calculations. In order to standardize the data and their dimensionless transformation, the method of linear scaling was applied, which allowed deriving an individual linear equation for each indicator. On the basis of the obtained data, a universal integral indicator was developed, which allows for a comprehensive assessment of food security. The results of the analysis showed that the level of food security in Russia during the period under study was at an average level. The minimum value was recorded in 2021 (0.415), the maximum in 2022 (0.552). The dynamics of the indicators emphasizes the importance of developing measures aimed at eliminating risks and strengthening the sustainability of the food system in the long term.

1 Introduction

Directly related to social stability and sustainable economic development, food security is an important component of national security. In recent years, Russia has made some progress in improving the production capacity of the agro-industrial complex and ensuring food independence. For example, due to the introduction of import restrictions and increased support for agriculture, the country has achieved food independence in such categories as grain, meat and meat products, eggs and egg products [1].

However, Russia's food security still faces serious challenges. International sanctions limit access to imported resources such as agricultural technology, equipment and seeds. Insufficient domestic investment in agriculture, weak technological innovation, and rural poverty and rural-urban development imbalances increase the instability of food supply chains [2].

In this regard, a comprehensive and scientific analysis and assessment of the state of food security is of particular importance. The system of indicators related to food security is extremely broad and includes aspects such as food production levels, food quality and safety, economic and physical accessibility, efficiency of agricultural resource use, and others. However, there is currently no effective method of comprehensive analysis that systematically integrates these diverse indicators into a clear framework for assessing and evaluating. This not only makes it difficult to make informed management decisions, but also limits the accuracy of evaluating policy effectiveness and studying existing problems.

The purpose of the research is to develop a scientific method of comprehensive analysis of food security in order to create a unified assessment system. This will strengthen Russia's food security and serve as an example for the international community in solving food security problems.

2 Materials and methods

Food security is an important issue related to global stability and development, and it is impossible to assess and monitor without a scientifically sound and rational system of indicators. The origins of food security indicators date back to the early 1940s, when the concept of "food security" was first introduced and considered in the context of dealing with food shortages [3]. Initially, the understanding of food security focused on food availability, which means providing enough food for all people.

With the development of globalization and changing socio-economic conditions, the content of food security has gradually expanded to include four key dimensions: availability, access, utilization and stability. As a result, more comprehensive indicators have been developed.

International instruments and national policies are used to develop food security indicators. The Global Food Security Index (GFSI), which assesses countries' food security along the dimensions of availability, quality and safety, as well as natural resources and sustainability, was first proposed by the Food and Agriculture Organization of the United Nations (FAO) [4]. GFSI has provided a common standard for assessing

* Corresponding author: wzymcbymb2022@gmail.com

food security at the international level. In addition, the World Bank and other organizations have developed other indices, including the World Development Indicators and the Global Hunger Index, to assess key elements of the food chain in developing countries [5, 6].

In Russia, the development of food security indicators has been supported at the level of national policy. The latest version of the "Doctrine of food security" not only defines the essence of food security, but also forms the basis for the creation of a system of indicators. This document includes indicators of food security (indicators are abbreviated in order as I1, I2, I3, ..., I14):

- in the area of consumption:
  - household disposable resources by population group (I1);
  - retail and catering space per 1,000 inhabitants (I2);
  - food consumption per capita (I3);
  - volume of targeted aid to the population (I4);
  - daily caloric value of human nutrition (I5);
  - the amounts of proteins, fats, carbohydrates, vitamins, macro- and microelements consumed per person per day (I6);
  - the consumer price index for food (I7);
- in the area of production and national competitiveness:
  - volume of production of agricultural and fishery products, raw materials and foodstuffs (I8);
  - imports of agricultural and fishery products, raw materials and foodstuffs (I9);
  - budgetary support to producers of agricultural and fishery products, raw materials and foodstuffs per ruble of sold products (I10);
  - productivity of land resources used for agriculture (I11);
  - volume of sales of food products by trade and public catering organizations (I12);
- in the area of administrative organization:
  - volumes of food products of the State Material Reserve formed in accordance with regulatory legal acts of the Russian Federation (I13);
  - stocks of agricultural and fishery products, raw materials and foodstuffs (I14) [7].

These indicators in different areas and standards reflect a high level of attention to food security issues and the response strategy in the current international environment.

To address the multicriteria problem, various generalized indicators can be constructed (e.g., linear or nonlinear functions with or without weights), and the generalized Harrington desirability function is a convenient way to evaluate such indicators [8]. It can transform the natural values of indicators of these different standards into dimensionless values and analyze these results comprehensively through the desirability scale. The desirability scale refers to psychophysical evaluation methods and serves to establish a correspondence between the natural values of individual indicators and their psychological meaning. The response values converted to a dimensionless scale are denoted as d_i (i=1, 2, ..., n). The formula is usually expressed as follows:

$$d_i = e^{-e^{-k(x-x_0)}}. \quad (1)$$

Here d_i is the Harrington desirability value, which is in the range [0, 1]; k and x_0 are tuning parameters used to control the slope and center position of the desirability curve; x is the normalized value of the indicator.

Individual indicators may include aesthetic and even statistical parameters, while psychological parameters refer to the experimenter's subjective assessments of the degree of desirability or preference of various values of the indicators. To construct a desirability scale, it is convenient to use pre-prepared tables that establish correspondence between preference relations in the empirical system and their numerical (psychological) equivalents (Table 1).

Table 1. Harrington's desirability scale

Desirability	Qualitative mark on the scale desirability (d_i)
Excellent	0.8–1.0
Good	0.63–0.8
Satisfactory	0.37–0.63
Bad	0.2–0.37
Very bad	0–0.2

3 Results and Discussion

In order to comprehensively analyze the level of food security in the study, the Harrington function was modified by shifting it 2 units to the right (Figure 1). This is because the sensitivity of the Harrington function near the values of 0 and 1 is significantly lower than in the central part of the range. Shifting it to the right allows the function to respond better to changes in a more appropriate range, thereby increasing its explanatory power.

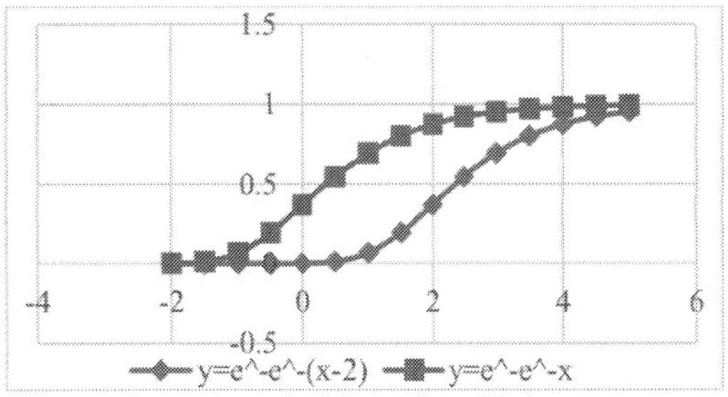

Figure 1. Comparison of original and modified Harrington functions

The Harrington function has certain limitations in practical application, in particular the inability to use indicators distributed in a strictly regular pattern without causing additional distortions [8]. Therefore, it is necessary to construct equations related to the Harrington function. In the study, using trend analysis tools in Excel, different models including linear, quadratic (polynomial) and logarithmic equations were constructed and their coefficients of determination were calculated to evaluate the explanatory power.

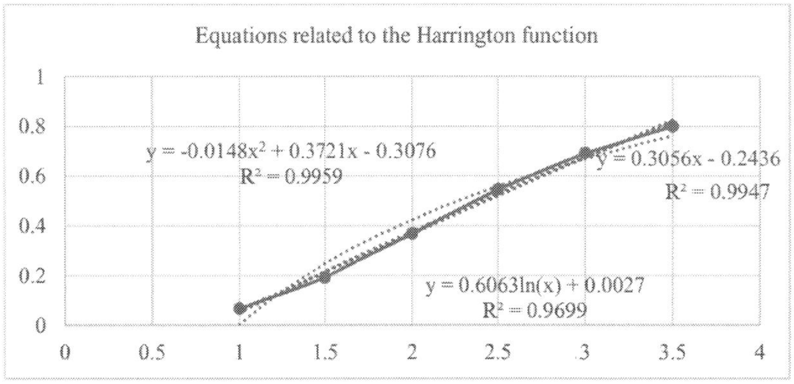

Figure 2. Summary of linear, logarithmic and polynomial approximation results

In the range of values of the independent variable [1, 3.5] and the dependent variable [0.07, 0.8], the analysis showed that the largest approximation error is observed for the logarithmic equation (r^2=0.970), followed by the linear equation (r^2=0.995), and the smallest error for the quadratic polynomial (r^2=0.996). All three models have confidence levels above 95%. To simplify the calculations, a linear equation was used in further analysis: Y = 0.3056X – 0.2436.

To ensure the consistency of indicators with different units of measurement, a linear scaling method was used in the study. This method allowed 14 food security indicators to be fitted into a linear equation associated with the Harrington function (Y = 0.3056X – 0.2436) (Table 2).

In this way the goal of converting the natural values of the indicators into dimensionless values was achieved.

All dimensionless values (Table 2) were substituted into the Harrington's surrogate function (linear equation – Y=0.3056X–0.2436), as a result of which all dimensionless values of food security indicators were transformed into desirability (d_i).

Table 2. Dimensionless values of food security indicators and corresponding adjustment equations for the past five years (2019–2023) (excerpt).

Indicator	Dimensionless value					Linear equation after linear scaling
	2019	2020	2021	2022	2023	
I1	None	1.52	3.50	2.51	None	Y=49.4X−985.49
I2	None	1.52	3.50	2.50	3.50	Y=0.11X−48.20
I3	2.70	3.50	1.52	2.03	2.02	Y=0.06X−58.10
I4	1.52	1.94	2.22	2.92	3.50	Y=8.29*10^-9X−2.50
I5	3.30	3.50	1.68	2.19	1.52	Y=0.02X−53.91
…	…	…	…	…	…	…

Table 3. The values of the analysis coefficients, their values on the Harrington scale (excerpt).

Indicator	Input valve	Harrington Score				
	2019	[0.0.2]	[0.2.0.37]	[0.37.0.63]	[0.63.0.8]	[0.8.1.0]
	Harrington desirability value (d_i)	Very bad	Bad	Satisfactory	Good	Excellent
I1	None					
I2	None					
I3	0.58			√		
I4	0.22		√			
I5	0.76				√	
…	…	…	…	…	…	…

According to the weighting formula that takes into account the cumulative preference, a weighted integral indicator of food security for the last five years can be obtained. The formula is shown below:

$$D = \sqrt[n]{\Pi_{i=1}^{n} d_i}. \quad (2)$$

Here *D* is the generalized food security desirability index; *n* is the number of food security indicators.

In order to comprehensively assess the level of food security, the study developed a food security status interpretation scale based on the expected level of indicators to intuitively display the assessment results. The following table presents the criteria for classifying food security levels, including ranges of indicator values, interpretation and state identification.

Table 4. Criteria for assessing the state of food security based on the Harrington function [9, 10]

Food security status	Desirability	The generalized criterion for the comprehensive assessment of food security (D)	Interpretation
Very high	Excellent	0.8–1.0	Indicates a very high level of food security
High	Good	0.63–0.8	The level of food security is equivalent to a "good" state
Medium	Satisfactory	0.37–0.63	The quality of food security is at an acceptable level, but it needs to be improved to achieve optimal performance
Low	Bad	0.2–0.37	Unacceptable levels of food security requiring immediate improvements to ensure safety and sustainability
Very low	Very bad	0–0.2	Completely unacceptable levels of food security requiring urgent action to address threats

The results of the food security assessment are presented in the following figure:

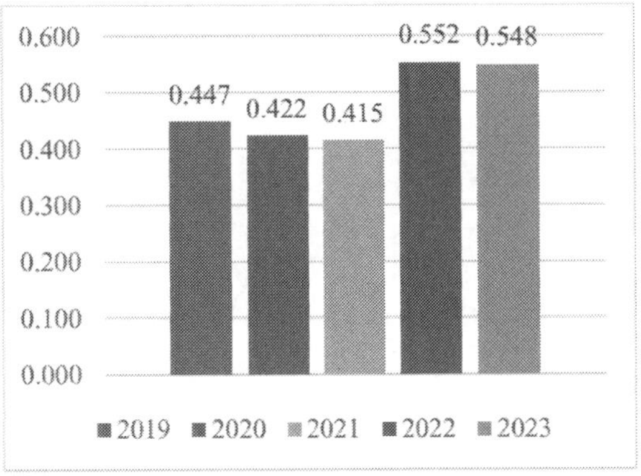

Figure 3. Levels of food security from 2019 to 2023 based on Harrington's desirability function

4 Conclusion

Over the past five years, the level of food security in Russia has shown significant fluctuations. The minimum value was recorded in 2021 (0.415), which indicates the most vulnerable state during the study period, which is largely due to the complexity of the pandemic period. The maximum value was reached in 2022 (0.552), reflecting a relative improvement in the situation. However, despite the positive changes, the overall dynamics remain unstable, highlighting the need for further analysis and action to improve the resilience of the food system.

According to the results of the analysis conducted on the Harrington Desirability Scale, the state of food security in the Russian Federation for the last five years is at the average level. In general, it is satisfactory, but still requires significant improvements. To achieve a higher level, it is necessary to focus on the elimination of key risk factors, including, first of all, accessibility, food quality and efficiency of resource use.

Therefore, the analysis of the individual subindices for each group of indicators using the Harrington scale has allowed a clearer identification of the main bottlenecks and potential risks underlying the integrated food security indicator. These results provide a basis for prioritizing actions and adjusting public policies aimed at stabilizing and improving the sustainability of the food system.

References

1. V.A. Ivanov, Food Security of the Russian Federation in the Context of International Sanctions, Corporate Governance and Innovative Economic Development of the North, Bull. of the Res. Center of Corpor. Law, Manag. and Venture Investment of Syktyvkar State Univ., **2**, 447–455 (2022)
2. S. Qian, Yu.A. Salikov, Retrospective analysis of the evolution of the food supply system evolution, Vest. of Astrakh. state techn. Univ., Ser. Econ., **1**, 7–16 (2024)
3. A. Sen, *Poverty and famines: an essay on entitlement and deprivation* (Oxford univ. press, 1982)
4. M. Izraelov, J. Silber, An assessment of the global food security index, Food Security, **11**, 1135–1152 (2019), https://doi.org/10.1007/s12571-019-00941-y
5. G.H. *Index, Global hunger index* (International Food Policy Research Institute (IFPRI), 2019)
6. S. Kousar et al., Comparative analysis to assess food security in south Asian nations: an application of multivariate GARCH model, Envir., Devel. and Sustainability, 1–37 (2024) https://doi.org/10.1007/s10668-024-05737-5
7. On approval of the Doctrine of Food Security of the Russian Federation, **20** (2020), http://www.consultant.ru/document/cons_doc_LAW_343386/
8. N.P. Lyubushin, G.E. Brikach, Using the generalized Harrington desirability function in multiparameter economic problems, Econ. anal.: theory and pract., **18**, 2–10 (2014)
9. A.A. Dryndak, Integrated assessment of food security of the DPR, Bull. of the Mari State Univ., Ser.: Agricult. Sci., Econ. Sci., **9** (2023), https://doi.org/10.30914/2411-9687-2023-9-2-218-226

10. Yu.A. Salikov, M.I. Korolev, A.I. Khorev, O.Yu Kolomytseva, Development of scientific research in the field of food security at the global, national and regional levels, Bull. of the Voronezh State Univ. of Eng. Technol., **84**, 492–499 (2023), https://doi.org/10.20914/2310-1202-2022-3-492-499

BIO Web of Conferences **161**, 00045 (2025) https://doi.org/10.1051/bioconf/202516100045
FIES 2024

Problems and ways of development of Russian agriculture

Elena V. Skripkina[1], *Zoya I.* Latyaheva[1*], *Julia V.* Lisitsina[1]

[1] Kursk state agrarian university named after I.I. Ivanov, Kursk, Russia

Abstract. It analyses the current state of agriculture in Russia, identifies factors limiting its efficiency, such as lagging behind in technology and the lack of an effective system of innovation transfer. The study is based on the analysis of literary sources, statistical data and the results of experimental work. Particular attention is paid to the need to introduce modern agro-technologies, such as differentiated fertiliser application, precision seeding and the use of crop monitoring systems. It is shown that their application can significantly increase the yield and quality of agricultural products. The need to create a national innovation system, including adequate scientific and educational support, as well as effective public-private partnerships, is emphasised. It is concluded that the modernisation of agriculture is a strategically important task for Russia, requiring a comprehensive approach and active interaction between scientific institutions, business and the state.

1 Introduction

There are several possible solutions to the problems facing the agro-industrial complex at present. The first of these is the improvement of existing agro-technologies aimed at increasing yields and reducing production costs. This category includes the introduction of new plant varieties and animal breeds, optimisation of irrigation and fertiliser systems, and the use of modern plant protection products. The advantages of this approach are the possibility of quick results and relatively low implementation costs. However, its disadvantages are that it often does not solve the problem of negative environmental impact and, in the long term, may lead to the depletion of natural resources. Another direction is the development of environmentally friendly production methods, such as organic farming and agroecology. These methods are aimed at preserving soil fertility, reducing environmental pollution, and increasing the sustainability of agroecosystems. They have the advantages of being highly environmentally friendly, producing healthier products, and supporting biodiversity. However, the transition to such methods requires significant investment, and their introduction may be accompanied by a reduction in yields at the initial stage. The third and most promising direction is the creation and implementation of innovative agro-technologies based on the achievements of modern science, biotechnology, precision agriculture, and other advanced industries. This approach opens up opportunities to significantly increase agricultural productivity, reduce its environmental impact, and improve the quality of products. At the same time, it is the most complex, requiring significant investment and human resources, as well as active co-operation between scientific institutions, business and government. Solutions are also needed in the field of zero-waste production, especially in the mining industry, which indirectly affects the agricultural sector [1]. In the conditions of permanent crisis, the creation of the innovation system of the agro-industrial complex becomes not just relevant, but critical. Without the introduction of knowledge-intensive technologies, without active transfer of innovations, without close interaction between science and production, as well as without effective public administration, the further development of the industry is practically impossible. This thesis is supported by other studies that emphasise the need for global shifts in the agricultural sector [2]. The need for technological modernisation of agriculture is also dictated by global challenges such as climate change and population growth. These factors, in particular, affect the dynamics of agricultural systems and require adaptation [3-5]. Only the transition to sustainable, highly efficient and environmentally friendly technologies will ensure the country's food security, preserve natural resources and increase the competitiveness of the domestic agricultural sector. Thus, the focus on the creation of a national innovation system of the agro-industrial complex seems not just relevant, but strategically important [6, 7].

The aim of the paper is to analyse the problems hindering the innovative development of agriculture, in particular, imperfect scientific and educational support and the lack of an effective technology transfer system.

* Corresponding author: ekonom.kgsha@mail.com

2 Method and materials

The methodology and materials used in this study were based on the analysis of the current state of agriculture in Russia, identification of factors limiting its development, and elaboration of proposals for innovative development of the industry. The work used a comprehensive approach combining the study of literary sources, analysis of statistical data and expert assessments. The study analysed existing agro-technologies, the system of scientific and educational support of agriculture, as well as mechanisms of interaction between science and production. Production indicators, average yields were considered and compared with indicators of other countries and potentially possible indicators in Russia.

Data from field experiments and demonstration crops were used to assess the effectiveness of agro-technologies and identify the potential for increasing yields. The equipment used for the research included: field machinery (tractors, combines, seeders) designed for soil cultivation and crop sowing. In particular, tractors of 150-200 hp with modern attachments allowing to accurately perform ploughing, cultivation and sowing operations. Field work was carried out in compliance with the accepted agro-technical requirements in different periods of vegetation. Seeding complexes equipped with precision seeding systems were used, allowing to set a given plant stand density. For harvesting, combine harvesters with loss control system and threshing accuracy were used. In the course of the research, sowing parameters were determined: seed rate, seeding depth, row spacing.

Laboratory instruments were used to analyse soil samples and plant tissues, including a spectrophotometer for determining the content of nutritional elements in soil and plants, scales for measuring biomass and yield, and measuring instruments for determining plant morphological parameters. Laboratory scales such as Sartorius or Mettler Toledo with an accuracy of 0.001g were used for analytical work, e.g. for determining the humus content of the soil. Crushing of soil samples was carried out using crushers and mills of laboratory types IM-5 and A1-DMR. To determine the acidity of soil extracts, pH-meters, for example, of the MRN-60 type were used, to determine nutrient elements - ionometers and spectrophotometers, for example, KFK-3 or Agilent Cary 60.

Economic indicators, such as production costs, yields and production costs, as well as environmental performance indicators, including fertiliser and pesticide use, impact on soil fertility and biodiversity, were used to assess the effectiveness of innovative technologies. The study also considered the issues of creating an innovation system of the agro-industrial complex, which includes public-private partnership, agro-technology transfer, coordination of participants in the innovation process.

3 Results and discussion

The research conducted a comprehensive analysis of the current state of Russia's agricultural sector, including the study of factors hindering its development and the development of strategies for its innovative rise. The research work was based on extensive analyses of domestic and foreign scientific literature, as well as on statistical data and expert assessments. Experimental work focused on assessing the effectiveness of various agro-technologies and the potential for increasing crop yields. Based on the results of the experimental work, ways to improve the system of scientific and educational support of the agro-industrial complex were proposed.

As part of the experimental part of the study, demonstration crops of several varieties of winter wheat were organised in the field plots. Soil cultivation was carried out using MTZ-1221 tractors with mounted plough and cultivator. Sowing was carried out with a modern John Deere 1890 seeding complex, providing precise sowing of seeds with a specified rate of 250 kg/ha and a depth of 5 cm. During the vegetation period the growth and development of plants were monitored, irrigation was carried out if necessary using a mobile irrigation unit. Soil samples were taken to analyse agrochemical parameters. Samples were taken along the diagonal of the field, in 5 points from the depth of 0-20 cm. In laboratory conditions the content of humus, nitrate nitrogen, mobile phosphorus and potassium was analysed. In parallel, the leaf lamina of plants was analysed for the content of major macro- and microelements at different stages of development. For this purpose, an Agilent 240FS AA atomic absorption spectrometer was used. In the process of research, the effect of different doses of mineral fertilisers on grain yield and quality was also studied. A system of differentiated fertiliser application based on soil analysis data was applied in one of the experimental plots. The use of a crop monitoring system based on an unmanned aerial vehicle made it possible to assess the heterogeneity of plant development in different parts of the field and promptly adjust fertiliser application. Harvesting was carried out by Claas Lexion 770 combine harvester with loss control system. Grain yield and quality were determined by analysing samples taken from each experimental plot. Grain samples were analysed for moisture, protein, gluten and nutritional content using an Infratec 1241 grain analyser. The germination energy and laboratory germination of seeds were also determined.

The results of the study showed that the use of modern agro-technologies, including differentiated fertiliser application, precise seed sowing and crop monitoring system, can significantly increase crop yields. Similar conclusions about the impact of agro-technologies on crop yields are found in other works devoted to the assessment of agro-ecological condition of territories [8]. In particular, the yield of winter wheat at the experimental site averaged 5.8 t/ha, which is 25-30% higher than the average for the region, where yields averaged 4.5 t/ha. This was achieved primarily due to accurate application of mineral fertilisers N120P90K100, as well as due to the application of microelements in the tillering phase. Optimal seed sowing rate of 250 kg/ha ensured a plant stand density of 450 plants/m2, which also contributed to higher yields. According to spectrophotometric analysis of leaf plates, plants that received balanced nutrition had higher chlorophyll and protein content. In turn, the use of modern varieties showed that grain quality also improved: protein content was 14.2% and gluten content 28%, which corresponded to high baking properties.

Analysis of soil samples showed that differential fertiliser application not only increased yields but also reduced the negative impact on soil fertility. The humus content in the soil in the experimental plots remained practically at the same level, and the nitrate content was within the optimal values. Thus, the introduction of advanced agro-technologies allows not only to increase yields, but also to preserve soil fertility. Compared to traditional technologies, which were applied in the control plot, where the yield was 3.9 t/ha using traditional doses of mineral fertilisers N80P60K60, experimental methods allowed to reduce yield losses associated with uneven distribution of nutrient elements and insufficient crop density (Table 01).

Table 1. Grain production possibilities in Russia under different agro-technologies and mineral fertiliser requirements

Level of intensification of agro-technologies	Fertiliser application per 1 ha in kg d.o.e.	Payback of fertilisers, kg of grain 1 kg d.e.	Yield, tonnes/ha	Gross grain harvest, mln tonnes	Fertiliser requirement, mln tonnes
For grain sowing area of 45 million hectares					
Extensive	0	—	1,7	76,5	0
Normal	100	8	2,5	112,5	4,5
Intensive and normal	150	10	3,2	144,0	6,75
Intensive, high and normal	200	12	4,1	184,5	9,0
For the area sown with cereals 70 million hectares					
Extensive	0	—	1,5	105,p	0
Normal	100	7	2,2	168,0	7,0
Intensive and normal	150	9	2,85	199,5	10,5
Intensive, high and normal	200	12	3,9	273,0	14,0

In comparison with the results of other studies, the data obtained confirm the effectiveness of differentiated fertiliser application and crop monitoring. For example, research conducted in the Tambov region has shown that the use of precision farming can increase winter wheat yields by 15-20%. It is also important to note that the use of new varieties with high yield potential also plays a significant role in increasing the overall productivity of the agro-industrial complex. This is supported by studies assessing the impact of geo-ecological factors on the sustainability of agricultural land use [9]. In another study conducted in the Krasnodar region, the application of modern seeding complexes and plant protection systems allowed achieving winter wheat yields of 6.2 tonnes/ha, which is similar to the results obtained in this study. The application of different doses of mineral fertilisers in the experimental plots showed that exceeding the optimal values (e.g. N180P120K150) does not lead to a significant increase in yields, but may on the contrary negatively affect grain quality and soil fertility.

In general, the results of the study confirm the need to introduce innovative agro-technologies to improve the efficiency of the agrarian sector of Russia. The use of modern methods of soil tillage, sowing, fertiliser application, crop monitoring, as well as the use of high-yielding varieties can significantly increase crop yields and improve the quality of products [10].

This study directly shows that it is possible to improve the efficiency of agricultural land exploitation in Russia. The low efficiency of the Russian agricultural industry is due to a number of factors, including ineffective agricultural policy, an underdeveloped system of scientific and educational support, lagging technology and lack of investment. Similar problems are noted in the works devoted to the formation of food supply in certain regions [11]. The analysis shows that the average yield of grain crops in Russia is only 19 c/ha, while in the EU countries this indicator reaches 60 c/ha, and in some regions of Europe 80-90 c/ha. This indicates a significant untapped potential of the domestic agro-industrial complex. At the moment Russia does not reach even 50% of the potential yield. At the same time, for comparison, our experimental crops with the use of modern agro-technologies showed winter wheat yield at the level of 58 c/ha, which indicates that there is a potential to increase production. Here we can observe how the use of modern technologies affects yield performance, especially when compared with data on agrarian capitalism and product quality in other countries, for example, in cocoa or rice production [12, 13].

Despite significant areas of arable land, Russia continues to conduct soil-intensive farming, applying only 21 kg/ha of mineral fertilisers in active substance. To address this problem, it is also important to consider the influence of mountain factors on soil salinisation and toxicity [14]. While, for example, in the USA, more than 100 kg/ha of fertilisers are used, and in the EU countries this figure is more than 150 kg/ha. The use of outdated technologies leads to soil degradation, desertification and, as a consequence, low productivity. One of the ways to solve this problem is the transition from extensive to intensive agriculture. For this purpose, it is necessary to introduce modern agro-technologies that provide balanced plant nutrition, protection from pests and soil-protective tillage methods. The introduction of intensive agro-technologies, as our study has shown, increases grain yields up to 3 tonnes/ha. This requires the application of 150 kg/ha of mineral fertilisers. It is important to use modern information and analytical systems to control the state of water bodies [15, 16]. In this case, the payback of fertiliser costs increases up to 9 kg of grain per 1 kg of active ingredient. The potential for further increase in yields is associated with the development of precision farming, which allows optimising the use of resources, as well as with the introduction of high-yielding crop varieties. In this context, it is important to consider the utilisation of mining waste, which can also affect the sustainability of agricultural production [17]. The transition to precision technologies, according to our estimates, will make it possible to increase grain yields to 3.9 t/ha with a payback of 12 kg of grain per 1 kg of fertiliser. The realisation of such a transition requires, among other things, the selection of appropriate parameters of equipment for environmentally safe extraction of metals [18].

The key role in the modernisation of agriculture is played by the creation of a national innovation system (NIS), which will ensure technology transfer, coordination of research and public-private partnership. The NIS should include adequate scientific and educational support, as well as mechanisms Thus, the modernisation of agriculture is a strategically important task for Russia. It requires a comprehensive approach, including the introduction of advanced technologies, development of the innovation system, training of qualified personnel and effective public-private partnership. According to our calculations, the gradual introduction of innovations will increase grain yields to 273 million tonnes, which will fully ensure the food security of our country and open up opportunities for agricultural exports. In general, sustainable biomass production is an important task for agriculture, and here it is necessary to pay attention to the recycling of waste, including from other industries, such as mining [19].

This, in turn, contributes to improving the food security of the country and the competitiveness of domestic agriculture. To achieve this goal, various aspects are important, including the involvement of young people in agricultural activities [20].

4 Conclusion

The study confirms the critical need for technological modernisation of Russian agriculture. The analysis of the current state of the agricultural sector has revealed a significant lag behind the advanced countries in terms of crop yields and resource efficiency. Extensive methods of farming lead to soil depletion, land degradation and low productivity. One of the key conclusions of the work is the need to transition to intensive agriculture based on the introduction of modern agro-technologies. This includes differentiated fertiliser application, precise seed sowing, use of high-yielding varieties, and application of crop monitoring systems. Experimental data showed that when using modern technologies, winter wheat yield can be increased by 25-30% compared to the average for the region, while simultaneously with the increase in yield there is also an improvement in grain quality. It is also noted that the application of precision farming technologies allows to reduce the negative impact on soil fertility and optimise the use of resources. An important conclusion is the need to create a nationally viable domestic agricultural sector. Transition to sustainable, highly efficient and environmentally friendly technologies requires an integrated approach and active interaction between scientific institutions, business and government.

References

1. V.Y. Konyukhov, K.A. Bashmur, T.A. Panfilova, V. Brigida, Circular Mining Wastes Management for Sustainable Production of Camellia sinensis (L.) O. Kuntze. Sustainability (Switzerland), **15 (15),** 11671 (2023) DOI: 10.3390/su151511671.
2. S. Bamidele, BRICS Revolution: The Global Agrarian Shift Sparked by Emerging Economies. India Q., (2025), Article in press. DOI: 10.1177/09749284241307937.
3. N. Garambois, U. Le Goff, L. Thibaudeau, Agrarian dynamics and climate change in the Senegalese Sahelian peanut basin. Agrar. Syst. Clim. Change: Journeys Adapt. Glob. South, **11,** 15-37 (2024). DOI: 10.1079/9781800628137.0001
4. H. Cochet, O. Ducourtieux, N. Garambois, Agrarian systems and climate change: Journeys of adaptation in the global south. *Agrar. Syst. Clim. Change: Journeys Adapt. Glob. South,* **7,** 1-208 (2024). DOI: 10.1079/9781800628137.0000.
5. S. Das, Determinants of livelihood vulnerability to climate change among the agrarian community: Empirical evidence from the coastal region of Indian Sundarban Biosphere Reserve. *Reg. Stud. Mar. Sci.,* **78,** 103777 (2024). DOI: 10.1016/j.rsma.2024.103777.
6. D.D. Baskanbayeva, L.A. Krupnik, K.K. Yelemessov, S.A. Bortebayev, A.E. Igbayeva, *Naukovyi Visnyk Natsionalnoho Hirnychoho Universytetu,* **5,** 68–74, 2020. DOI: 10.33271/NVNGU/2020-5/068.
7. M.K. Moldagozhina, L. Krupnik, K.K. Yelemesov, E. Mukhtar, A. Roza, *International Journal of Applied Engineering Research,* **11(21),** 10454–10457, (2016).

8. F. K. Ermekov, Yu. S. Lisovskaya, Agroecological condition assessment of agricultural landscapes for their sustainability increase, *Sustainable Development of Mountain Territories*, **14(1),** 76–86 (2022). DOI: 10.21177/1998-4502-2022-14-1-76-86.
9. I. Akzharkyn, K. Yelemessov, D. Baskanbayeva, *Applied Sciences (Switzerland),* **14(17),** 7567, (2024). DOI: 10.3390/app14177567.
10. S.I. Kondratev, A.V. Vidnichuk, *Ecological Safety of Coastal and Shelf Zones of Sea*, **2,** 107–121, (2020). DOI: 10.22449/2413-5577-2020-1-107-121.
11. K.M. Didur, H.O. Kundieieva, G.V. Ortina, T.A. Pikhniak, A.V. Revkova, FORMATION OF FOOD SUPPLY OF THE AGRARIAN SECTOR OF UKRAINE ON THE BASIS OF RESTORATION OF THE DEVELOPMENT OF RURAL AREAS *Rev. Iberoam. Vitic. Agroind. Rural.*, **12,** 223-239 (2025). DOI: 10.35588/pxe7k028.
12. S. I. Kondratev, A. V. Vidnichuk, *Vestnik Moskovskogo Universiteta, Seriya 5: Geografiya,* **2020(3),** 91–99, (2020).
13. A. Al Mozahid, What counts as "quality" in agrarian taste? Subjectivities, gender, and rice agroecology in northern Bangladesh. *Cult. Agric. Food Environ.*, **46,** 88-105 (2024). DOI: 10.1111/cuag.12323
14. V.V. Kukartsev, K.I. Kravtsov, Y.A. Tynchenko, T.A. Panfilova, Influence of mountain factors on salt excess and soil toxicity in mountain conditions, *Sustainable Development of Mountain Territories*, **15 (3),** 784–797 (2023). DOI: 10.21177/1998-4502-2023-15-3-784-797.
15. O. Afanaseva, M. Afanasyev, S. Neyrus, D. Pervukhin, D. Tukeev. Information and Analytical System Monitoring and Assessment of the Water Bodies State in the Mineral Resources Complex. *Inventions,* **9,** 115 (2024). https://doi.org/10.3390/inventions9060115.
16. A.A. Kapanski, A.E. Boltrushevich, S.N. Sorokova, E.A. Efremenkov, A.Y. Demin. Geospatial Clustering in Smart City Resource Management: An Initial Step in the Optimisation of Complex Technical Supply Systems. Smart Cities, **8,** 14 (2025). https://doi.org/10.3390/smartcities8010014.
17. D.A. Zyukin, A.I. Karlina. Prospects for return of valuable components lost in tailings of light metals ore processing. Metallurgist, **67,** 96-103 (2023). DOI: 10.1007/s11015-023-01493-5.
18. N.N. Bryukhanova, V.A. Gladkikh, R.V. Kononenko. Selection of disintegrator parameters toward environmentally friendly leaching of metals. Min. Inform. Anal. Bull., **12,** 39-49 (2024). DOI: 10.25018/0236_1493_2024_12_0_39.
19. D.A. Zyukin, A.I. Karlina. Technology for nonwaste recovery of tailings of the mizur mining and processing plant. Metallurgist, **66,** 1476-1480 (2023). DOI: 10.1007/s11015-023-01462-y.
20. R. Sunam, F. Sugden, A. Kharel, T.R. Sunuwar, T. Ito. Unpacking youth engagement in agriculture: Land, labour mobility and youth livelihoods in rural Nepal. *J. Agrar. Change*, **25**, e12611 (2025). DOI: 10.1111/joac.12611.

Enzyme complex influence on physiological properties of alcoholic yeast during cultivation

Alexey Yakovlev[1*], *Gennady* Agafonov[1], *Svetlana* Yakovleva[1], *Tatyana* Romanyuk[1], *Natalia* Zueva[1], *Alla* Chusova[1], *Tatyana* Kovaleva[1]

[1]Voronezh State University of Engineering Technologies, 394036, Voronezh, 19 Revolutsii Ave., Russia

Abstract. The effect of enzyme preparations of neutral Prolife BS Liquid protease and Kingfos phytase on the physiological and technological properties of Saccharomyces cerevisiae yeast of race XII in alcohol production was studied. The maximum accumulation of yeast cells occurs when the enzyme preparation of neutral Prolive BS Liquid protease was added to the wort at a dosage of 0.2 units PS/g of starch and Kingfos phytase with a dosage of 0.5 units FS/g of starch. The content of yeast cells in ripe yeast reached 170 million cells/cm^3 after 16-18 hours of cultivation. Such yeast has a high fermentation activity.

1 Introduction

The effectiveness of the alcoholic fermentation process depends mainly on the physiological and technological properties of the yeast used for this purpose. In the production of ethyl alcohol, unicellular microorganisms (yeast Saccharomyces cerevisiae) are used. Yeast Saccharomyces cerevisiae are facultative anaerobes; by the type of nutrition, they are chemoorganoheterotrophs; in relation to the active acidity of the medium, they are acidophiles.

The main requirements for the yeast used in the production of ethyl alcohol from starch-containing raw materials are as follows. The alcoholic yeast used in the processing of starchy raw materials must have high fermentation activity, full fermentability of sugars, resistance to metabolic products (especially alcohol), resistance to the development of extraneous microflora.

For the normal metabolism of yeast cells, it is necessary to contain in the culture medium all substances necessary for the normal functioning of cells, as well as conditions excluding stress effects caused by temperatures, high concentrations of carbohydrates and ethanol.

The set of characteristics reflecting the needs of yeast organisms in certain physical and chemical conditions of the environment is referred to as physiological properties. One of the main physiological properties is the ability to ferment sugars to ethanol and carbon dioxide. The ability of yeast to ferment a certain amount of sugar per unit of time is called fermentation activity, which is determined by the amount of carbon dioxide released per unit of time. At the same time, the process of yeast reproduction and the accumulation of glycogen in cells are determined.

An increase in the fermentation activity of yeast can be achieved by using a balanced composition of the nutrient medium.

2 Materials and methods

The proteolytic enzyme preparation Prolive BS Liquid was used as a source of protease. The enzyme preparation of the neutral protease Prolive BS Liquid is obtained by deep cultivation of the bacterial strain Bac. subtilis. It is a syrupy liquid of light brown color with a specific density of 1.25 g/cm^3 and high proteolytic activity. Prolife BS Liquid contains proteases that hydrolyze peptide bonds in proteins to form soluble peptides and amino acids, which contributes to the fortification of the medium with additional nitrogenous nutrition intended for yeast. The activity of the enzyme preparation is 600 - 750 units PS/cm^3. Optimal conditions of the drug are pH 5.5 – 7.5; the temperature is 50 – 55 °C.

The enzyme Kingfos preparation was used as a source of phytase. The enzyme Kingfos preparation contains phytase (Mioinositollhexaphosphatephosphohydrolase ECNo.: 3.1.3.8) with an activity of at least 10000 units/g (10%). The producer is Pichiapastoris. The enzyme Kingfos preparation is a white or yellowish powder. Optimal conditions of the action of the drug are pH 5.5 – 6.5; the temperature is 40 – 45 °C.

A Goryaev cell was used to count yeast cells.

To determine the percentage of viable cells on a slide, a drop of Saccharomyces cerevisiae yeast was mixed with a drop of methylene blue. The prepared preparation was viewed in 10 fields of view. The percentage of non-viable cells to the total number of cells observed in the field of view was determined. Good seeded yeast contains no more than 1% of dead cells.

* Corresponding author: tafursova@yandex.ru

The number of budding cells characterizes the ability of yeast to reproduce: active yeast contains 50-70% of cells with buds. The preparation "crushed drop" is prepared from a yeast suspension and the percentage of budding cells is determined in relation to the total amount of yeast with and without buds.

3 Results and discussion

The effect of enzyme preparations of the neutral Prolive BS Liquid protease and Kingfos phytase on the fermentation activity of alcoholic Saccharomyces cerevisiae yeast of race XII was studied. When preparing yeast wort, the enzyme Prolive BS Liquid preparation was added to it in a dosage of 0.3 units PS/g of starch and the enzyme preparation Kingfos phytase in a dosage of 0.5 units FS/g of starch. No additional enzyme preparations were added to the control sample during the preparation of yeast wort. Yeast was introduced into the wort at a temperature of 30 ° C in an amount of 10% of the wort volume. The yeast generation process was carried out for 18-20 hours at a temperature of 28-30 °C in flasks of 750 cm^3. The amount of carbon dioxide released was determined hourly by the weight method (Figure 1).

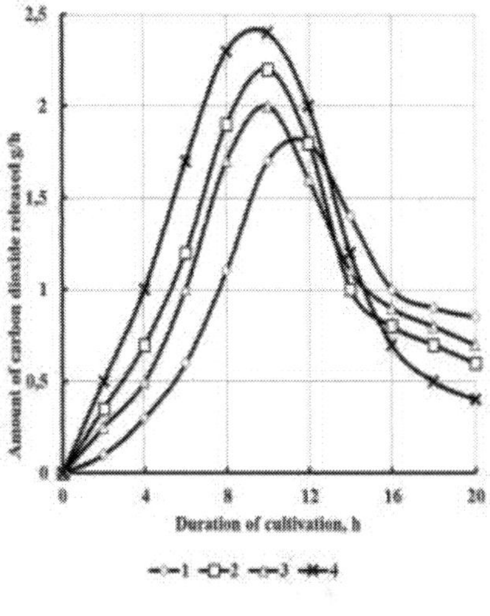

Figure 1. The effect of enzyme preparations on the fermentation activity of alcoholic yeast: 1 – control; 2 - protease of 0.2 units PS/g of starch; 3 - phytase of 0.5 units FS/g of starch; 4 - protease of 0.2 units PS/g of starch, phytase of 0.5 units FS/g of starch.

It can be seen from Figure 1 that the yeast cultivated on the wort obtained using protease and phytase has the maximum fermentation activity. The maximum fermentation activity of alcoholic yeast in the control is observed after 12 hours of growth and after 10 hours of growth in experimental samples. Direct assimilation of amino acids provides not only the intensification of protein synthesis, but also activates the enzymes contained in the yeast cell as indicated by high fermentation activity. Trace elements, such as calcium, zinc, magnesium, obtained by hydrolysis of phytin, also promote an increase in the fermentation activity of yeast.

Figure 2 shows that the yeast cell has 4 phases of vital activity, which are characterized by different growth of yeast cells.

1. Lag phases are characterized by a very complex enzymatic process; they are extremely sensitive to the composition of the medium, its temperature, and active acidity. In this phase, the volume of yeast cells increases slightly, nucleic acids are synthesized, and energy is accumulated for further active accumulation of biomass. The duration of the lag phase for yeast in the experiment is 1-2 hours and 2 – 3 hours in the control.

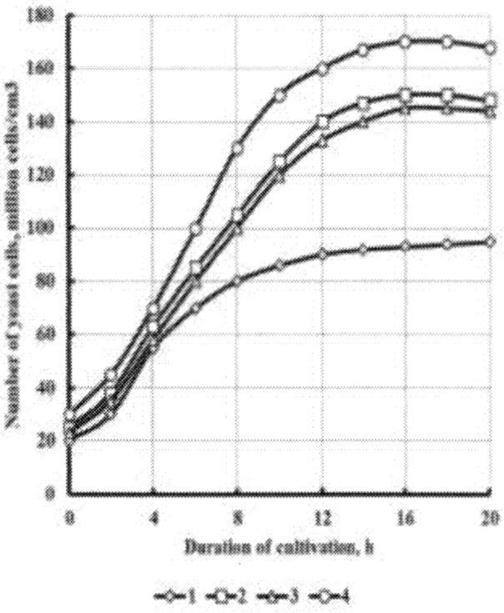

Figure 2. The effect of enzyme preparations on the dynamics of yeast biomass accumulation: 1 - control; 2 - protease of 0.2 units PS/g of starch; 3 - phytase of 0.5 units FS/g of starch; 4 - protease of 0.2 units PS/g of starch, phytase of 0.5 units FS/g of starch.

2. Logarithmic phases or exponential growth phases are characterized by high activity of cell reproduction. The number of budding cells increases rapidly, reaching 45-50%. At the same time, the biomass of yeast gradually increases due to the formation of new daughter cells, which, growing to the mother cell, begin to bud, forming a new generation. In this phase, the maximum growth rate of yeast cells is observed. The duration of the phase in the experiment is 14-16 hours and 18-20 hours in the control. The exponential phase is described by the Mono equation:

$$M_t = M_0 \cdot e^{kt},$$

where Mt is the amount of yeast during time t, kg;

M_0 is the initial, seeded amount of yeast, kg;

e is the base of the natural logarithm;

k is the specific growth rate or the coefficient of biomass formation (depends on the yeast race, the concentration of the nutrient medium and its composition, aeration and other factors);

t is yeast reproduction time.

3. A stationary phase is characterized by a gradual slowdown in the growth rate. The formation of new cells practically stops, their budding ends, because the nutrient medium is depleted, metabolic products are formed that inhibit the growth and reproduction of cells. To maintain the vital activity of the cells, the nutrients remaining in the culture medium are used. The cells increase in size, their mass also increases. The increase in biomass in this phase can be 5-10% of the mass of yeast accumulated in the logarithmic phase. During this period, intracellular biosynthetic processes occur; they form the maturity of the yeast cell. The enzyme systems of the yeast cell are rebuilt from the active synthesis of biomass to the metabolic processes that support their vital activity.

The proportion of dying cells increases and the reproduction of new ones slows down. A phase of decline is coming.

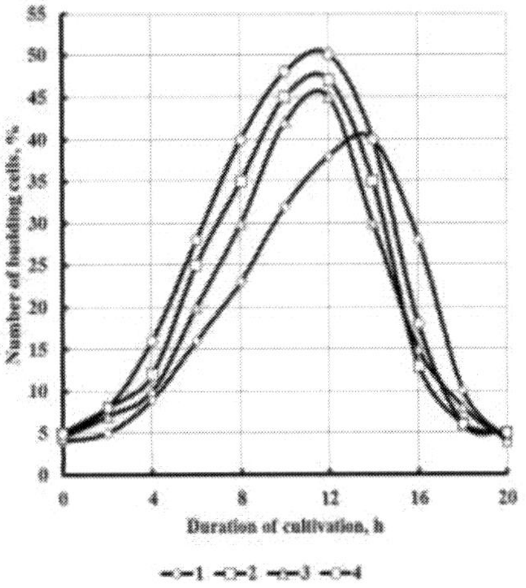

Figure 3. The effect of enzyme preparations on the dynamics of yeast reproduction: 1 - control; 2 - protease of 0.2 units PS/g of starch; 3 - phytase of 0.5 units FS/g of starch; 4 - protease of 0.2 units PS/g of starch, phytase of 0.5 units FS/g of starch

4. The phase of attenuation means dying off. Metabolic products accumulate in a significant amount in the medium. The phase is characterized by the absence of cell growth and reproduction. The mass of cells decreases, because all the nutrients of the culture medium were used in the previous phase. There is a gradual death of cells. To maintain vital activity, cells begin to use their own reserves. Autolysis of cells occurs.

The yeast in the experiment, compared with the control, multiplies more intensively, after 14-16 hours of growth it accumulates about 170 million yeast cells in the culture medium, and the yeast in the control accumulates about 95 million yeast cells after 18-20 hours of growth. Therefore, the process of yeast cultivation with the use of additional enzyme preparations can be limited to 14-16 hours of growth, and at the same time yeast cells accumulate almost twice as much as in the control. All this makes it possible to reduce the amount of seeded yeast introduced into the fermentation apparatus, to reduce the loss of fermented carbohydrates to the accumulation of yeast biomass and to increase the yield of alcohol.

A higher level of the budding of yeast cells (Figure 3) in the experiment was reached after 12 hours and after 14 hours in the control, which corresponds to about the middle of the logarithmic phase and amounts to 50.0% in the experiment and 40.0% in the control.

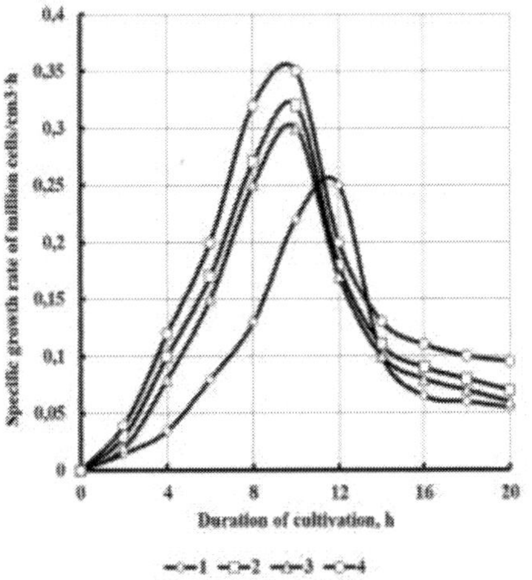

Figure 4. The effect of the yeast reproduction rate of enzyme preparations: 1 - control; 2 - protease of 0.2 units PS/g of starch; 3 - phytase of 0.5 units FS/g of starch; 4 - protease of 0.2 units PS/g of starch, phytase of 0.5 units. FS/g of starch.

The rapidly occurring dynamic fermentation inhibits the respiration process and consequently suspends budding and cell reproduction, as well as the rapid depletion of the substrate with nutrients, which in turn creates unfavorable conditions for budding. At the end of cultivation, yeast cells consumed their reserve substances, such as glycogen, and the number of budding cells reaches 4-5%.

A very characteristic indicator is the growth rate of yeast culture. We used the specific growth rate, which was determined by the increase in the number of cells over a certain period of time. The specific growth rate was maximal in the logarithmic phase (Figure 4), when fermentation was weak and amounted to 0.35 million cells/cm^3·h in the experimental samples and 0.25 million cells/ cm^3·h in the control.

4 Conclusion

The influence of various sources of supplementary nutrition and enzyme preparations of protease and phytase on the physiological and technological properties of yeast in alcohol production has been studied. The proteolytic enzyme Prolive BS Liquid preparation was used as a source of protease. The enzyme Kingfos preparation was used as a source of phytase. It was found that the maximum accumulation of yeast cells was observed when an enzyme preparation of neutral Prolive BS Liquid protease was added to the wort with a dosage of 0.2 units PS/g of starch and enzyme Kingfos phytase with a dosage of 0.5 units FS/g of starch. The content of yeast cells in mature yeast reaches 170 million cells/cm^3 after 16-18 hours of cultivation; the yeast has a high fermentation activity.

References

1. P. Albersheim, Plant Biochemistry, Acad. Press, New York; San Fransisco; London (2013).
2. C.W. Bamforth, Brew. Dig, **57,** 22-27 (2012)
3. C.W. Bamforth, H.L. Martin, J. Inst. Brew, **87,** 365-371 (2013)
4. Be. Bartolome, C. Gomez-Cordoves, A.I. Sancho, N. Diez, P. Ferreira, J. Soliveri, J.L. Copa-Patino, Growth and release of hydroxycinnamic acids from Brewer's spent grain by Streptomyces avermitilis CECT 3339, Enzyme and Microbial Technology, **32(1),** 140-144 (2013)
5. L. Caspeta, J. Coronel, A. Montes de Oca, E. Abarca, L. González, A. Martínez, Engineering high-gravity fermentations for ethanol production at elevated temperature with Saccharomyces cerevisiae, Biotechnology and Bioengineering, **116(10),** 2587–2597 (2019), https://doi.org/10.1002/bit.27103
6. S.R. Couto, M.A. Sanroman, Application of solid-state fermentation to food industry- A review, Journal of Food Engineering, **76,** 291-302 (2016)
7. E.A. Davidovich, Complex enzyme preparations for the purpose of improving the efficiency of fermentation of rye and barley wort, Food and processing industry. **1,** 163 (2007)
8. L. Fang, T. Wang, B. Lamsal, Use of surfactant and enzymes in dry-grind corn ethanol fermentation improves yield of ethanol and distillers corn oil, Industrial Crops and Products, **111,** 329–335 (2018), https://doi.org/10.1016/j.indcrop.2017.10.031
9. J.H. Franco, S.D. Minteer, A.R. de Andrade, Product analysis of operating an ethanol/O2 biofuel cell shows the synergy between enzymes within an enzymatic cascade, Journal of the Electrochemical Society, **165(9),** H575–H579 (2018), https://doi.org/10.1149/2.0931809jes
10. P.L. Hurst, J. Nielsen, P.A. Sullivan, M.G. Shepherd, Biochem. J, 165, 33-41 (2017)
11. R. Ikeda, T. Yamamoto, M. Funatsu, Arg. Biol. Chem, **37(5),** 1153-1159 (2013)
12. L. Kandra, Amylases of medical and industrial importance, Journal of Molecular Structure (Theochem), **666-667,** 487-498 (2013)
13. P. Katapodis, M. Vrfianska, D. Kekos, W. Nerinckx, P. Biely, M. Claeyssens, B.J. Macris, P. Christakopoulos, Biochemical and catalytic properties of an endoxylanase purified from the culture filtrate of Sporotrichum thermophile, Carbohydrate Research, **338(18),** 1881-1890 (2013)
14. O. Kirk, T.V. Borchert, C.C. Fuglsang, Industrial enzyme applications, Curr Opin Biotechnol, **13,** 345-351 (2012)
15. V.V. Kononenko, L.N. Krikunova, V.V. Kolpakova, Processing of rye and barley grain in alcohol production, Production of alcohol and alcoholic beverages, 3, 11-13 (2013)
16. L.N. Krikunova, E.M. Maksimova, Intensification of ethanol production from rye by separation of polysaccharide fractions, Production of alcohol and alcoholic beverages, **4,** 20-22 (2013)
17. Liquid-liquid extraction of biomolecules: an overview and update of the main techniques, J Chem Technol Biotechnol, **83,** 143-157
18. A.P. MacCabe, M. Orejas, E.N. Tamayo, A. Villanueva, D. Ramon, Improving extracellular production of food-use enzymes from Aspergillus nidulans, Journal of Biotechnology, **96(1),** 4354 (2012)
19. D.J. Manners, J.J. Marshall, J. Inst. Brew, **75,** 550-561 (2019)
20. S. Nakamura, K. Wakabayashi, R. Nakai, R. Aono, K. Horikoshi, Purification and some properties of an alkaline xylanase from alkaliphilic Bacillus sp. strain 41M-1, Appl. Environ. Microbiol, **59(7),** 2311-2316 (2013)
21. A.V.F. Nielsen, I. Tetens, A.S. Meyer, Potential of phytase-mediated iron release from cereal-based foods: A quantitative view, Nutrients, **5(8),** 3074–3098 (2013), https://doi.org/10.3390/nu5083074
22. G. Polygalina, Technochemical control of alcohol and liquor production, Kolos, Moscow (1999)
23. A.N. Yakovlev, G.V. Agafonov, S.F. Yakovleva, T.I. Romanyuk, N.V. Zueva, T.S. Kovaleva, Refining process and properties of polysaccharide xylanase, IOP CS: EES, **640,** (2020)
24. Y. Zhang, Y-H. Lin, Metabolic flux analysis of Saccharomyces cerevisiae during redox potential–controlled very high gravity ethanol fermentation, Biotechnology and Applied Biochemistry, **67(1),** 140–147 (2020), https://doi.org/10.1002/bab.1861
25. Y.Y. Zhang, R. Stockmann, K. Ng, S. Ajlouni, Revisiting phytate-element interactions: implications for iron, zinc and calcium bioavailability, with emphasis on legumes, Critical Reviews in Food Science and Nutrition (2020), https://doi.org/10.1080/10408398.2020.1846014
26. A.N. Yakovlev, G.V. Agafonov, S.F. Yakovleva, T.I. Romanyuk, N.V. Zueva, T.S. Kovaleva, Intensification of ethyl alcohol technology due to multi-enzyme composition, IOP CS: EES, **1052,** 012111 (2022)

Optimisation of agro-technologies for sustainable development of agriculture in Uzbekistan

Sherzod Yusupov[1,*], *Inomdjon* Azizov[2], *Ramozon* Khaydarov[3], *Sevara* Inadullaeva[4], *Sattar* Kholmirzayev[5]

[1] "Tashkent Institute of Irrigation and Agricultural Mechanization Engineers" National Research University, Tashkent, Uzbekistan
[2] Almalyk branch of the Tashkent State Technical University named after Islam Karimov, Almalyk, Uzbekistan
[3] Termez State University, Termiz, Uzbekistan
[4] Termiz State Pedagogical Institute, Termiz, Uzbekistan
[5] Namangan Engineering-Construction Institute, Namangan, Uzbekistan

Abstract. The analysis of food security problems in the Republic of Uzbekistan, considered in the context of global challenges and the diversity of strategies for the development of the agro-industrial complex is presented in this paper. Particular attention is paid to assessing the results of the ongoing agrarian reforms, analysing the unique agro-climatic conditions of Uzbekistan, and identifying the key factors hindering the sustainable development of agriculture. As part of the study, experimental work was carried out to examine the effectiveness of various agro-technological techniques and methods, including irrigation systems, fertiliser application and tillage methods, using modern equipment. The results showed that drip irrigation and organic fertilisers, combined with minimum tillage, provide significant savings in water resources, improved soil fertility and increased yields. Based on the analysis, the necessity of transition to intensive farming with the use of conservation technologies has been identified.

1 Introduction

In the context of the need to develop and improve the global agro-industrial complex, the analysis of the experience of reforms in the agrarian sector of Uzbekistan, presented in this paper, is extremely relevant and important. In the context of these global challenges and diverse strategies, the analysis of the experience of reforms in the agrarian sector of the Republic of Uzbekistan, presented in this paper, acquires special relevance and value. Uzbekistan is an agrarian country where agriculture plays a key role in providing employment, shaping the national economy and ensuring food security. In this regard, it is necessary to take into account world trends, such as global agrarian shifts, which are discussed in [1]. However, the country faces a number of serious challenges that hinder the development of the agricultural sector and food security on a sustainable basis. Intensive use of irrigated lands, historical priority of cotton growing, deficit of water resources, deterioration of soil ameliorative condition, salinisation and erosion - all these create serious obstacles for sustainable agricultural development. In addition, climatic changes, manifested by rising temperatures, increased frequency of droughts and floods, have a negative impact on crop yields. These problems are compounded by the effects of climate change on agricultural systems which are discussed by [2]. It is also important to consider the impact of factors such as salt regime and soil toxicity, which is particularly relevant for mountainous areas as shown in [3]. The unique agro-climatic and socio-economic conditions of Uzbekistan require the development of own approaches and solutions that take into account the specifics of the country. Studying the results of the reforms carried out in the republic, identifying bottlenecks and problem areas, as well as developing scientifically sound recommendations for further improvement of agricultural policy is an extremely important task that will ensure food security and sustainable development of agriculture. At the same time, it is necessary to take into account not only national peculiarities, but also global trends in the development of agro-industrial complex. This means the use of modern technologies, the introduction of innovative agro-ecological methods, the development of co-operation between farmers and support for small and medium-sized businesses in the agricultural sector. Thus, this study represents an important step towards understanding the complex challenges Uzbekistan faces in ensuring food security and also offers valuable lessons for other countries in the region. Ultimately, learning from the successes and failures of reforms in Uzbekistan can serve as a basis for designing effective agricultural development strategies in other countries facing similar challenges.

* Corresponding author: sher_xxx89@mail.ru

The purpose of this paper is to analyse the achieved results of reforms in the agrarian sphere of Uzbekistan, to identify the main problems in ensuring food security, as well as to develop recommendations for improving institutional measures of food security of the population of Uzbekistan.

To solve these problems, we need a methodology that allows us to study the history of agrarian systems, which is disclosed in [4]. Also, it is important to take into account socio-economic factors, as shown in the example of the historical development of rural settlements in [5].

2 Method and materials

Several agricultural districts of Uzbekistan, differing in soil and climatic conditions and specialisation of production, were selected as a basic experimental site. In order to assess the effectiveness of different technologies and methods of agricultural production, field experiments were conducted. Taking into account that water resources are a key factor, the study focused on the efficiency of different irrigation systems, which is in line with world studies aimed at the conservation and protection of water resources, for example, in works [6, 7]. These experiments intend to study the effect of different irrigation systems, including drip and sprinkler irrigation, on crop yields and water conservation. For this purpose, special equipment such as Fregat sprinkler systems and Netafim drip irrigation systems were used, which allow precise regulation of the volume and mode of water supply. Different types of fertilisers, including mineral and organic fertilisers, and their impact on soil fertility and crop yields were also evaluated. Various laboratory equipment was used for soil and plant analyses. Spectrophotometer 'SF-2000', to determine the content of basic nutrition elements in the soil. Chromatograph 'Chrom-5' to analyse the content of pesticides. Various equipment was used to determine optimal fertiliser dosages and assess soil quality, and experience in assessing agroecological potential was studied, as in [8], and works on analysing the qualitative state of soils for rational land use, such as in [9]. Field experiments included the evaluation of the impact of different tillage practices, as well as methods of plant protection against pests and diseases. Weight-measuring equipment, platform scales 'VSP-1000' were used to record yields. All obtained data were subjected to statistical processing using 'Statistica' software for analysis of variance and correlation analysis.

3 Results and discussion

As part of a comprehensive research aimed at studying the problems of food security in the Republic of Uzbekistan, a number of experimental works were carried out, the purpose of which was to comprehensively study the effectiveness of various agro-technological techniques and methods, as well as to analyse the impact of various factors on crop yields. Experimental studies were conducted on several experimental sites located in different regions of Uzbekistan, which allowed taking into account the diversity of soil and climatic conditions and specialisation of production. The most important for the food security of the country - wheat, cotton, vegetables (tomatoes, cucumbers), potatoes and fruit and berry crops - were selected as experimental crops.

At the initial stage of experimental works, the main attention was paid to studying the efficiency of different irrigation systems. On the experimental plots were installed sprinkler systems 'Fregat-M', providing intensive and uniform irrigation, as well as drip irrigation systems 'Netafim Streamline', allowing precise regulation of water supply directly to the root zone of plants. Three irrigation modes were defined for each crop: standard irrigation mode, where the irrigation rate corresponded to the recommended rate for the crop and development phase, economical mode, where the volume of water supplied was reduced by 20% of the standard rate, and intensive irrigation mode, where the volume of water was increased by 15% of the standard rate. Itelma brand water meters were installed to control water consumption. During the experiments, the soil moisture level was regularly measured using moisture meters 'Field Scout TDR 350', which made it possible to monitor the dynamics of changes in the moisture content of the soil layer and adjust the irrigation regime. In parallel with the studies of irrigation systems, the impact of different types of fertilisers was assessed. Mineral fertilisers were selected: nitrogen fertiliser ammonium nitrate (NH_4NO_3), phosphorus fertiliser superphosphate ($Ca(H_2PO_4)_2$), potassium fertiliser potassium chloride (KCl), as well as organic fertilisers: overdrawn cattle manure and compost based on vegetable waste. Four fertiliser application schemes were applied: 1) application of only mineral fertilisers at recommended doses, 2) application of only organic fertilisers, 3) application of 50% mineral fertilisers from the recommended dose combined with 50% organic fertilisers, 4) application of 75% organic fertilisers combined with 25% mineral fertilisers from the recommended dose. To determine the optimal dosages of fertilisers, a preliminary analysis of soil samples for the content of basic nutrition elements was carried out using spectrophotometer 'SF-2000', which made it possible to determine the content of nitrogen, phosphorus and potassium in the soil before the experiments, as well as to monitor the dynamics of changes in these indicators during the experiments. To assess the level of soil contamination with pesticides, a gas chromatograph 'Chrom-5' was used to determine the presence of pesticide residues in soil samples and plant tissues.

To study the impact of different tillage methods, three methods were used: ploughing to a depth of 25 cm using plough 'PLN-3-35', cultivation to a depth of 15 cm using cultivator 'KPS-4' and minimum tillage using disc harrow 'BDM-4'. Yields were recorded using platform scales 'VSP-1000', which provided high accuracy in measuring the weight of harvested crops. For each crop, product samples were taken to assess its quality, including fruit size, sugars, vitamins and other indicators.

Analysis of research results showed that the use of drip irrigation in combination with minimum tillage provided significant water savings, which averaged 28%, and in some cases reached 35%, compared to standard irrigation using sprinkler systems. In particular, when growing winter wheat variety 'Krasnodarskaya-99', drip irrigation reduced water consumption from 5500 m^3/ha to 3900 m^3/ha, and when growing cotton variety 'Namangan-77', water consumption decreased from 7000 m3/ha to 4800 m3/ha. At the same time, crop yields not only did not decrease, but in some cases, on the contrary, increased by 6-12%. For example, the yield of winter wheat under drip irrigation combined with minimum tillage was 5.1 tonnes/ha, while under sprinkling and ploughing it was 4.2 tonnes/ha. In experiments with cotton, drip irrigation combined with mineral and organic fertilisers in a 50/50 proportion gave a yield of 3.7 t/ha, compared to 3.0 t/ha when using only mineral fertilisers and conventional irrigation. When growing tomatoes of 'Solaris' variety on plots with drip irrigation and minimum tillage, the yield was 65 t/ha, while on plots with standard irrigation and ploughing - 55 t/ha. In a study of the apple variety 'Golden Delicious', drip irrigation gave an average yield of 35 tonnes per hectare, while conventional methods gave 28 tonnes per hectare. In terms of fertiliser use, the application of organic fertilisers such as digested cattle manure and compost, in combination with mineral fertilisers, has shown to be effective in improving soil fertility and increasing yields. The humus content of the soil in plots where organic fertiliser was applied increased by 15-20% compared to plots where only mineral fertiliser was applied. The use of organic fertiliser also improved soil structure, increased soil water permeability and reduced salinity. The application of organic fertilisers in combination with mineral fertilisers showed effectiveness in improving fertility, which echoes the results of studies investigating the importance of family farms for sustainable development as in [10]. The results of pesticide content analysis showed that when chemical preparations such as 'Raundap' and 'Aktelik' were used on plots with cotton and vegetable crops, the content of residual pesticides exceeded the permissible norms on average by 20-30%, which confirms the need to switch to the use of biological methods of plant protection, such as the use of bacterial preparations 'Phytosporin' and entomophages 'Trichogramma'.

The results of the conducted experiments were compared with the data obtained from official statistical sources, as well as with the results of similar studies conducted in other countries. In particular, the data presented in the original article indicate an increase in the production of vegetables, fruits and melons in Uzbekistan in recent years. Thus, the average annual growth rate of agricultural production for the period 2000-2014 was 6%. The share of crop production in the structure of gross output increased from 50% to 58%, and the share of livestock production, respectively, decreased from 42% to 50%. Analysis of data on the dynamics of vegetable and fruit production showed that if in 2000 the per capita production of vegetables was 108 kg, in 2013 this indicator reached 283 kg, which is 159% higher. Fruit production increased from 32.3kg to 76.9 kg over the same period. Similar results were obtained for other agricultural products. These data confirm the upward trend in agricultural production in Uzbekistan, but, at the same time, they do not provide a complete picture of the effectiveness of the methods and technologies used, nor do they reflect the environmental and social consequences of the approaches used. The results of our research confirm that the use of modern agro-technological techniques, such as drip irrigation, organic fertilisers and minimum tillage, can significantly increase the efficiency of agriculture, as well as reduce its negative impact on the environment. However, for widespread adoption of these technologies it is necessary to ensure farmers' access to modern equipment, knowledge and financial resources.

Comparison of the results obtained with the data of similar studies conducted in other countries has shown that the introduction of modern agro-technologies, as well as the transition to sustainable farming methods is a global trend aimed at ensuring food security and sustainable development. In particular, the results of studies conducted in Israel show that the use of drip irrigation can significantly save water resources, as well as increase crop yields in arid conditions. Studies conducted in the European Union countries show that the use of organic fertilisers improves soil fertility and reduces the negative impact of agriculture on the environment.

These data confirm the global trend towards sustainable agriculture. As an example, studies on the dynamics of agricultural systems under climate change, as in [11]. Also, the circular economy approach is important for addressing sustainability issues, an example of which is described in [12, 13].

Thus, the results of the conducted experimental work and comparative analysis with the data of other studies confirm the necessity of transition to intensive, environmentally sustainable agriculture based on the application of modern technologies, scientific knowledge and best practices.

Analysis of studies by other authors and data obtained during experimental work reveals a number of key problems and prospects for the development of the agro-industrial complex of Uzbekistan. One of the main problems is the irrational use of water resources, which is particularly acute in the arid climate of the region. The use of traditional irrigation methods, such as sprinkling, leads to significant water losses due to evaporation and inefficient distribution. The results of the research, demonstrate the high potential of drip irrigation to reduce water consumption by 25-35% while increasing crop yields. In addition, the analysis revealed the problem of soil fertility deterioration and contamination by pesticide residues. The use of mineral fertilisers alone, although it can increase yields in the short term, leads to soil depletion and deterioration of soil structure. At the same time, the use of organic fertilisers, such as over-fermented manure and compost, in combination with mineral fertilisers helps to restore soil fertility, increase humus content and reduce salinity. The use of conventional ploughing also negatively affects soil structure, causing erosion and worsening the water regime. The transition to minimum tillage favours soil conservation and improved water availability. In addition, the results of the analysis revealed the problem of soil contamination with pesticide residues when using chemical preparations, which requires a transition to biological methods of plant protection.

To increase the productivity and efficiency of the agro-industrial complex of Uzbekistan, it is necessary to comprehensively introduce modern technologies and methods that include drip irrigation, organic farming, minimum tillage and biological plant protection. In order to increase the productivity of the agro-industrial complex, the introduction of modern technologies and materials is required, which echoes the results of studies [14], as well as the use of modern methods based on mathematical modelling, such as [15], and forecasting systems, such as [16]. Along with this, it is necessary to take into account the issues of food security formation and provision, as discussed by [17].

It is also necessary to ensure farmers' access to modern technology, knowledge and financial resources, which includes training workshops, financial support, and the creation of favourable conditions for innovation. At the state level, it is necessary to develop a strategy for the transition to sustainable agro-industrial complex, which includes measures to encourage the use of modern agro-technologies, the development of organic farming, the rational use of water resources and the protection of soils from degradation. It is also necessary to strengthen control over the use of pesticides and encourage the use of biological methods of plant protection. It is important to take into account the specifics of each region and its soil and climatic conditions, as well as the specialisation of production. Combining traditional methods with modern technologies will make it possible to achieve sustainable development of the agro-industrial complex and ensure food security of Uzbekistan.

4 Conclusion

The presented research results and analysis of literature data allow us to draw a number of important conclusions regarding the prospects of development of the agro-industrial complex of Uzbekistan. The conducted experiments convincingly demonstrate that the use of modern agro-technologies, such as drip irrigation, use of organic fertilisers, minimum tillage and biological methods of plant protection, allows to significantly increase the efficiency of agricultural production while reducing the negative impact on the environment. In particular, drip irrigation has proven its effectiveness in saving water resources, allowing to reduce their consumption by 25-35% compared to traditional methods, while not reducing and in some cases increasing crop yields.

The use of organic fertilisers in combination with mineral fertilisers improves soil fertility, increases humus content and reduces salinity, which in turn has a positive effect on crop yields and product quality. Minimal tillage preserves soil structure, reduces erosion and improves water regime. At the same time, the results of pesticide content analysis indicate the need to switch to biological methods of plant protection, as the use of chemical preparations leads to exceeding permissible pollution norms. In general, the results of the study emphasise the importance of introducing modern technologies and innovations in the agricultural sector of Uzbekistan, as well as the need to transition to sustainable farming methods, which is a key condition for ensuring food security and sustainable development of the country. Thus, the proposed integrated approach, which includes a combination of modern agro-technologies and careful attitude to natural resources and the environment, can become the basis for further development of the agro-industrial complex of Uzbekistan and other countries in the region.

References

1. S. Bamidele, BRICS Revolution: The Global Agrarian Shift Sparked by Emerging Economies. India Q., (2025), Article in press. DOI: 10.1177/09749284241307937.
2. H. Cochet, O. Ducourtieux, N. Garambois, Agrarian systems and climate change: Journeys of adaptation in the global south. Agrar. Syst. Clim. Change: Journeys Adapt. Glob. South, **11**, 1-208 (2024). DOI: 10.1079/9781800628137.0000.
3. K.I. Kravtsov, T.A. Panfilova, Influence of mountain factors on salt excess and soil toxicity in mountain conditions. Sustainable Development of Mountain Territories, **15(3)**, 784-797, (2023). https://doi.org/10.21177/1998-4502-2023-15-3-784-797.
4. V. I. Golik, V. Yu. Konyukhov, Mining Informational and Analytical Bulletin, **11-1**, 175-189 (2023). DOI: 10.25018/0236_1493_2023_111_0_175.
5. N.A. Zhirov, Socio-Economic Development of the Black Soil Village in the XIX–early XX centuries: from the Prosperous Granary of Pre-Reform Russia to the Demo-Ecological Crisis of Rural Settlements. Bylye Gody, **19**, 1605-1614 (2024). DOI: 10.13187/bg.2024.4.1605.
6. T. A. Kuandykov, T. D. Karmanov, E. I. Kuldeyev, K. K. Yelemessov, B. Z. Kaliev, News of the National Academy of Sciences of the Republic of Kazakhstan, Series of Geology and Technical Sciences, **2022(3)**, 142–154, (2022). DOI 10.32014/2022.2518-170X.186.
7. I.I. Bosikov. Comprehensive assessment of hydrogeological conditions for formation of mineral water resources of the Nizhne-Karmadon deposit. Bulletin of the Tomsk Polytechnic University. Geo Assets Engineering, **332**, 206–218 (2021). DOI: 10.18799/24131830/2021/8/3319.
8. A. A. Sokolov, L. G. Orlova, R. I. Kuzmich, Mining Informational and Analytical Bulletin. **11**, 278-291

(2023). DOI: 10.25018/0236_1493_2023_111_0_278.

9. I. A. Kerimov, L. R. Bekmurzaeva. Modern agro-climatic conditions of mountain landscapes in the North Caucasus. Sustainable Development of Mountain Territories, **14,** 555-563 (2022). DOI: 10.21177/1998-4502-2022-14-4-555-563.
10. B. Aparecida da Silva, R. Evangelista de Oliveira, A. Cavalieri Sais, E. Cardoso leite. Family farmers and trees: ecological functions, ecosystem services and food production in agrarian reform settlements in Brazil. Agroecol. Sustain. Food Syst., **49,** 4-21 (2025). DOI: 10.1080/21683565.2024.2405515.
11. N. Garambois, U. Le Goff, L. Thibaudeau. Agrarian dynamics and climate change in the Senegalese Sahelian peanut basin. Agrar. Syst. Clim. Change: Journeys Adapt. Glob. South, **8,** 15-37 (2024). DOI: 10.1079/9781800628137.0001.
12. L.S. Malyukova, V.V. Kondratiev, V.V. Bukhtoyarov, V.Y. Konyukhov, K.A. Bashmur, T.A. Panfilova, V. Brigida. Circular Mining Wastes Management for Sustainable Production of Camellia sinensis (L.) O. Kuntze. Sustainability (Switzerland), **15,** 11671 (2023). DOI: 10.3390/su151511671.
13. L. Ma, Y. Ghorbani, C. B. Kongar-Syuryun, M. M. Khayrutdinov, R. V. Klyuev, A. Petenko, V. Brigida. Dynamics of backfill compressive strength obtained from enrichment tails for the circular waste management. Resources, Conservation & Recycling Advances, **23,** 200224 (2024). https://doi.org/10.1016/j.rcradv.2024.200224.
14. K. Akzharkyn, D. Yelemessov, D. Baskanbayeva, N.V. Martyushev, V.Y. Skeeba, V.Y. Konyukhov, T.A. Oparina. Strengthening Polymer Concrete with Carbon and Basalt Fibres. Appl. Sci. (Switzerland), 14, 7567 (2024). DOI: 10.3390/app14177567.
15. E.A. Efremenkov, D.V. Valuev, M. Qi. Analysis of a Predictive Mathematical Model of Weather Changes Based on Neural Networks. Mathematics, **12,** 480 (2024). DOI: 10.3390/math12030480.
16. S. I. Kondratyev, D. Baskanbayeva, K. Yelemessov, Y. Sarsenbayev, V. A. Turkin, International Journal of Hydrogen Energy, **95,** 212–216, (2024). DOI: 10.1016/j.ijhydene.2024.11.182.
17. K.M. Didur, H.O. Kundieieva, G.V. Ortina, T.A. Pikhniak, A.V. Revkova. Formation of food supply of the agrarian sector of ukraine on the basis of restoration of the development of rural. Rev. Iberoam. Vitic. Agroind. Rural., **12,** 223-239 (2025). DOI: 10.35588/pxe7k028.

Artificial meat in the structure of halal meat products: legislative and legal status, technologies, Russian and international experience

Rustam Y. Gilmutdinov* [1], *Daniyar R.* Gilmutdinov[2], *Albert K.* Galiullin[1], A*li H.* Volkov[1], *Guzel G.* Shalamova[1]

[1] N.E. Bauman Kazan State Academy of Veterinary Medicine, 35, Sibirskiy trakt, Kazan. 420029, Russia,
[2] "Center for Islamic Studies" of the Academy of Sciences of the Republic of Tatarstan, 20 Bauman St., Kazan, 420111, Russia

Abstract. The issue of halal certification of meat and, in particular, "artificial" meat is quite relevant. If it has been solved in principle for traditional variants of meat and meat products, the appearance on the market of a new form of meat, artificially grown in laboratory conditions, implies significant changes in the legislative and legal status of this product. Moreover, the requirements for both certification and improvement of its production technology need to be adapted. There are certain successes in both directions. This paper analyzes the level of solution of this issue in Russia and abroad.

1 Introduction

One of the fundamental components of the food culture that human communities develop is religion, usually symbolizing their heritage and the sociocultural aspects of ethnicity. Through food laws that impose or prohibit the consumption of certain foods, religion has a significant influence on the dietary habits of the population. For example, the life of a Muslim is regulated by a set of rules governing the actions permitted by the Shariah (halal).

The term "halal" is translated from Arabic as "permitted" and means "all that is allowed and permissible in Islam". Non-Muslims usually associate this word with food that Muslims, who make up about a quarter of the world's population, are allowed to eat. These are predominantly residents of countries with a high Muslim component, the majority of which (more than 80% or about 1.5 billion people) live in 40 countries. The largest Islamic populations are Indonesia, Pakistan, Bangladesh, Malaysia, as well as countries that are members of the Cooperation Council of Arab States of the Persian Gulf. They are Bahrain, Qatar, Kuwait, the United Arab Emirates, Oman and Saudi Arabia. Consumers of halal products are also the population of "non-Muslim" countries. For example, in the United States alone, there are about 8 million people who consume halal products.

The issue of halal certification of meat is also relevant for Russia. First of all, the Islamic nations that inhabit the territory of the Russian Federation are the second largest in terms of population. In addition, it will potentially boost the export of Russian agro-industrial products to these countries.

2 Results and discussion

There are many halal standards for meat worldwide (GSO 993: 2015; OIC/SMIIC 1: 2019; HAS 23103: 2012 and MS 1500: 2019, etc.) representing different geographical areas [1], which can cause confusion for producers, especially those from non-Muslim countries exporting their products and wishing to certify them as halal.

The growing demand for halal labeling motivates meat producers to comply with the relevant standards because of the potential commercial benefits. However, given the above, such certificates issued by different organizations often differ in their requirements for these products. There is a need to establish the most unified halal standard possible, which will facilitate the activities of producers and increase consumer confidence in the product in question.

The appearance on the consumer market of an innovative product in the form of artificial meat has led not only to ethical, medical, environmental, cultural and economic discussions, but also to religious ones [2]. In particular, this trend has further complicated the problem of harmonization of existing halal standards and raised the bar of requirements in this direction of responsible organizations closely cooperating with the authorities. It would be nice if official structures involved in the certification of halal meat would share their opinion on this variant of lab-grown meat; for example, through "Ijtehad" and bring it to the public.

The interest of the official authorities of the Russian Federation in halal products was documented as early as October 2, 2017, when by order of the Federal Agency for Technical Regulation and Metrology (Rosstandart)

* Corresponding author: niskgavm@mail.ru

the Project Technical Committee for Standardization (PTC 704) "Halal Products and Services" was formed. The key task of the committee was to create a regulatory and technical framework that takes into account the specifics of halal products and is aimed at eliminating falsified and counterfeit products from the market. In addition, the PTC envisages international cooperation, study and implementation of the experience of similar international organizations in Russia. The development and implementation in Russia of standards for halal products and services similar to international standards will allow Russian halal products to be exported. The PTC includes representatives of federal and regional executive authorities; certification bodies; industrial enterprises; scientific, administrative and public organizations.

In Russia today, 3 national standards related to the halal brand are used:

- GOST R 70401-2022 "Halal food production processes. General requirements for halal food products";
- GOST R 70405-2022 "Halal products and services. General terms and definitions";
- GOST R 70402-2024 "Conformity assessment. Requirements for halal certification bodies".

The first two were approved on December 30, 2022 by orders of the Federal Agency for Technical N 1737-st. and N 1738-st; the third was approved on March 29, 2024 by order N 383-st.

GOST R 70405-2022, effective from March 1, 2023, was developed by the Private Institution of Higher Education "Russian Islamic Institute" (RII), while in the development of GOST R 70401-2022 (effective from June 1, 2023) participated, in addition to the Russian Islamic Institute, also Limited Liability Company "International Center for Standardization and Certification 'Halal' (LLC 'ICS&C 'Halal'') and Joint Stock Company "Republican Certification Methodological Center 'Test-Tatarstan' (JSC RSMC 'Test-Tatarstan') took part in the development of GOST R 70401-2022 (effective from June 1, 2023).

GOST R 70402-2024 became effective on September 1, 2024 and established requirements for the activities of certification bodies for halal products, processes and services.

According to these standards, raw materials for the production of halal products can be food resources of animal, plant, microbiological, mineral, artificial or biotechnological origin. At the same time, the absence of prohibited components in raw materials must be confirmed by official certificates. None of these standards mentions meat grown in laboratory conditions.

Since 2023, the Russian Quality System (Roskachevo) has been a member of PTC 704. Almost immediately on its basis, a national Center of Competence in the field of halal products and services was established, whose activities include standardization, certification, product quality monitoring, as well as promotion of certified Russian halal products in the domestic market and abroad.

The range of the Center's tasks includes:

- creation of an effective halal infrastructure in Russia;
- education in halal issues;
- provision of educational services in the halal sphere;
- analyzing trends and news in the industry;
- monitoring the quality of halal products and services;
- promotion of standardization in the halal sphere;
- development of a network of halal certification centers;
- international recognition of domestic halal standards and certificates.

In 2024, 8 draft national halal standards are under development, including requirements for certification bodies and certification procedures. These documents will strengthen the entire halal certification system in Russia and ensure transparency and consumer confidence in products that conform to their religious beliefs.

Earlier, in January 2023, Roskachestvo registered the voluntary halal certification system with Rosstandart, and in April, for the first time in our country, the Roskachestvo-Halal certification body received accreditation from the Federal Accreditation Service (Rosakkreditation).

Although meat has always occupied a special place among a fairly wide range of halal products, it is not emphasized in the legislative and legal field. Meat is called halal if it is produced and cooked according to the Shariah. The question of whether meat grown in laboratory conditions is halal is rather complicated and, probably, the relevant discussion will be lengthy, since even today the possibility of compliance with some of the obligatory provisions (pillars) of Islam remains open.

Currently, the only countries where laboratory-grown meat is sold in small quantities are the United States and Singapore.

Back in 2018, Future Meat Technologies from the US launched artificial meat (beef). In the U.S., approval for the sale of artificial chicken was granted in 2023, while in Singapore it became available for retail sale from May 16, 2024. Producer companies continue to seek a halal ruling for their products and international recognition as such.

In January 2024, the Israeli Ministry of Health authorized the sale of artificial meat (beef) produced in laboratory conditions from calf stem cells by Aleph Farms Ltd, a company that has developed its own technology for this production. The company found a way to produce artificial meat with the texture, flavor and juiciness of natural meat. For this purpose, it was combined from simultaneously cultured four types of cells: muscle, connective, vascular and fatty tissues. The process was completed using 3D printing technology. Although the product is planned to be sold domestically in Israel, the company aims to become the "first halal-certified" company to produce lab-grown meat. However, given Israel's escalating relations with Muslim countries, it is unlikely that the population of the latter can be considered as potential consumers of this product.

In July 2024, the UK will also be authorized to sell lab-grown artificial meat (at the first stage, it will be used only in animal feed).

Although the sale of such meat is not yet authorized in China, back in January 2022, the Ministry of Agriculture of the People's Republic of China published a five-year agricultural plan in which the cultivation of artificial meat was mentioned among the sectors in which it should be actively involved. Moreover, in 2023, the Shanghai-based company CellX launched a pilot production of artificial meat for local restaurants and by 2025 plans to build a plant to produce hundreds of tons of this product per year with the prospect of supplying it to Singapore and the United States.

In Russia, similar work is also underway, although the achievements are quite modest. In 2018, the first domestic startup in the field of cultured meat – ArtMeat, which specializes, in particular, in horsemeat – appeared in Kazan. In 2019, the Ochakovo Food Ingredients Plant claimed to have successfully grown artificial beef, but the project was put on hold in 2021 due to the pandemic.

The emergence of artificial meat is seen by many scientists as a technological challenge, a revolution in food production. Its production process is constantly evolving, driven by both the companies producing it and the relevant research institutions. It is derived from animal cells grown in nutrient-rich culture media under controlled conditions. The method eliminates the need for resource-intensive industrial farming of farm animals and traditional slaughterhouses, while also matching the values of vegans, vegetarians and meat consumers. In essence, artificial meat provides a sustainable and ethical alternative to traditional meat production. Nevertheless, there are fundamental differences between this meat and conventionally produced meat, which raises questions about its halal character. Its recognition under the halal approach depends on careful adherence to the principles of Islam. The main issue is the source of the cell lines and the processes involved in their creation and cultivation.

At the same time, as noted above, there is currently no uniform standard of "halal meat" in the world, even in relation to meat obtained by traditional methods. Foreign Islamic religious scholars working with producers of such meat believe that it should be represented by animal species permitted by religious rules. This does not include carnivorous and omnivorous (primarily pigs) mammals and birds. Meat from domestic animals with split hooves such as goats, camels, buffalo, sheep and cattle is permitted. Of the poultry category, chicken, ducks, turkeys, pigeons and sparrows are allowed. This requirement is relatively easy to fulfill in the production of artificial meat.

The next condition is that lab-grown meat must come from an animal that has recently been slaughtered according to halal standards. This violates the idea of "no slaughter" that many artificial meat companies strive for. Meanwhile, this condition does not apply to parts of the animal such as feathers, hair and other parts that are not considered alive. Accordingly, one way around this problem is to use cells taken, for example, from chicken feathers, which contain the cells needed to create lab-grown meat. For example, employees of the company Just from the USA have developed a technology for "growing" cells from chicken feathers; in this case, the bird is not killed. Also artificial meat can be produced by multiplying cells obtained from a small piece of muscle tissue. The chicken egg can also be considered as an object for obtaining cells, since it is in the early stage of fertilization. The meat grown from these cells under laboratory conditions does not contain blood, which also meets halal requirements. Moreover, there are many examples of cell lines derived from non-living organisms.

The next requirement for the artificial meat production process is that the overall product should not be contaminated with any non-halal materials, such as blood. In the case of slaughtering an animal according to Islamic standards, the blood that flows out is impure. A small part of it remaining in the meat after "correct" slaughter and blood drainage is not considered impure, because it does not flow out. Blood products used, for example, as the main ingredient (blood sausage) are also not allowed. According to experts, artificial meat can be considered halal if blood or its serum is not used in the process of its cultivation.

For example, in February 2024, the Committee of Religious Rulings – fatwas of the Islamic Religious Council of Singapore (Majlis Ugama Islam Singapura – MUIS) recognized lab-grown meat as halal, provided certain conditions are met: the cages must come from animals approved for Muslim consumption (e.g. chicken, but not pork) and no non-halal ingredients must be used in the production process. The fatwa was issued after months of deliberations by the ulema, who consulted with stakeholders including government agencies, academics and experts. In addition, the theologians visited local production of "lab-grown meat." The fatwa committee emphasized the environmental and nutritional benefits of such meat, recognizing that its advantages outweigh its disadvantages.

The production of artificial meat using animal serum in the nutrient medium is actively criticized by Muslims and in recent years there has been active research on the design of serum-free synthetic nutrient media for the cultivation of eukaryotic cells. In addition, their use makes production cheaper; makes the process of obtaining cell biomass more controllable and significantly reduces the risk of contamination.

Already today, such media have been developed that allow to abandon this blood component, thus eliminating its disadvantages related to the impact on the final result. However, it should be recognized that these media are mainly developed for the cultivation of mammalian cells intended for the accumulation of viral material.

Ideally, serum-free medium should have a specific chemical composition and support the growth of different cell types. However, such a universal medium does not exist and a special individual composition is developed for each cell type. The latter is a factor that has a great influence on cell metabolism.

At the same time, while the composition of basic (basic) nutrient media (MEM, DMEM, 199, F-12, RPMI, 199, etc.) is known and is given in the catalogs of most commercial companies, the prescriptions of specialized serum-free media are the property of the developing companies (Gibco, JRHbiosciences, HyClone, etc.) and are not available for widespread use. Approaches to the creation of serum-free environments are fairly typical. Various growth-stimulating components, including a mixture of amino acids, inorganic compounds, micro- and macronutrients, sugars, fatty acids, insulin, transferrin, etc. are added to a known base medium. However, the combination, percentage composition of the components varies.

In general, serum-free media are divided into three main groups:

• Serum-free media proper, allowing the inclusion of components of animal origin in their composition.

• Protein-free, but containing materials of undetermined chemical composition, hydrolysates, peptones, etc., are classified into three groups.

• Chemically defined nutrient media, which not only do not contain animal products, but include only components of a certain chemical composition.

For example, Gibco (Thermo FS) has developed a wide range of serum-free media for cell culture on an industrial scale. These use only components of non-animal origin, and instead of insulin and transferrin, non-protein compounds are used.

Another chemically defined serum-free medium, Eden from Bioengine, is designed for protein production in high-density CHO cell culture. There are some features:

• it does not contain cytokines, antibiotics, hormones, hydrolysates, nucleosides;

• it has no protein-peptide ingredients, glutamine, HEPES, phenol red;

• it adapts to a variety of commercially or independently selected CHO cell lines, as well as different types of culturing;

• it supports CHO-K1, CHOZN, Horizon, CHO-S, CHODG44 and other cell lines.

Russian scientists have also developed a number of serum-free nutrient media [3–5 and others]. During cultivation, they ensure the growth of cells that retain the morphology typical for the culture, form a monolayer on the 2–3 day of growth, and retain high proliferative activity.

The problem is also the need to constantly change the medium as the cells grow. Scientists are going to create a technology in which it will be purified and reused as if passing through the liver and kidneys of an animal.

While there are different approaches to growing meat in the laboratory, in general the process is divided into three steps:

1. First, cells prone to rapid multiplication (proliferation) are harvested. These can be embryonic stem cells, adult stem cells (hematopoietic and mesenchymal), myosatellite cells (muscle stem cells) or myoblasts. At this stage, an animal (or perfectly preserved cells, but it has not been gotten to that point yet) is required. Pluripotent cells (capable of differentiating into all cell types except cells of extra-germinal organs) proliferate faster than others do, but they have not yet differentiated into specific cell types, which raises the problem of "splitting" cells (differentiating into the wrong cell type) and the need to direct their growth in a certain way.

In this respect, "mature" muscle cells – explants – are better suited, as they have already formed as a cell type necessary for production. Unfortunately, explants do not proliferate well under in vitro conditions. Moreover, when using tissue explants, there is a high probability that a fraction of cells will die off due to the inaccessibility of nutrients to them, since tissue implies a cluster of cells bound together by intercellular contacts. The dying cells may release toxic substances. In addition, the nutrient medium will be consumed for the growth of cell types other than muscle cells, resulting in a lower muscle cell content in the food product than when using a myoblast cell line as a starting point.

Mesenchymal stem cells can be derived from different tissues (bone marrow, adipose tissue, skeletal muscle, liver, placenta, etc.) and different protocols are used to culture them. Currently, there is great inconsistency among laboratories regarding the choice of nutrient media and the addition of a variety of factors for successful isolation and expansion of mesenchymal stem cells. Therefore, there is a need to develop a more accurate and standardized protocol for their expansion. This process is influenced by the base of nutrient medium, glucose and glutamine concentration in the medium, as well as passage density, adhesive surface properties, etc. [6,7]. Accordingly, the choice of nutrient medium serves as an important component of successful mesenchymal stem cell culturing [8].

The use of myosatellites and myoblasts, which are still able to proliferate at the required rate and have already differentiated to a sufficient degree, is promising. However, it is difficult to isolate, cultivate and maintain a dynamic proliferative state of myosatellite cells. Myocytes of skeletal muscle tissue successfully obtained by transdifferentiation (reprogramming) from widely available fat cells can serve as an alternative [9].

2. Next, the resulting cells are treated by adding proteins that promote tissue growth. Then they are placed in the culture room, in a bioreactor. It acts as blood vessels, supplying the cells with everything they need and giving them conditions for growth. The main nutrient of the cells is the blood plasma of the animal (most often – embryo). It is supplemented with a mixture of sugars, amino acids, vitamins and minerals. For muscle tissue to develop properly, it is grown under pressure, simulating natural conditions. Heat and oxygen are also supplied to the bioreactor. Essentially, the cells don't even realize they are growing outside the animal.

Once these cells have grown sufficiently, they undergo a process of differentiation and maturation. In addition to muscle cells-myocytes, the grown material may include other cell types such as adipocytes (fat cells) and fibroblasts (connective tissue cells), depending on the desired composition of the final product. This is accomplished by altering the composition of the nutrient medium to mimic the metabolic shifts that naturally occur during animal growth. For example, muscle cells will be stimulated to fuse to form myotubes and begin to secrete muscle-specific proteins, resulting in the formation of muscle tissue. At the same time, fat cells will begin to accumulate lipids, which will improve the flavor and texture of the final meat product.

Ideally, nutrient media should not contain components of animal origin and their use should be abandoned; therefore, various alternatives are being sought. For example, a serum-free medium was developed that supported multiplication in turkey satellite cell culture [10]. In addition, there are various serum substitutes that are good alternatives. For example Ultroser G is one of many commercially available substitutes. A serum-free medium made from maitake mushroom extract has been successfully used with a higher growth rate in it than that of fetal bovine serum [11].

Myocytes should be grown under aerobic conditions to prevent acidification of the culture medium by lactic acid. Cell viability is positively correlated with the oxygen gradient in growing myocyte cultures, so oxygen transporters are likely to be required to maintain oxygen concentrations high enough to prevent hypoxia and acidification.

However, growing muscle cells require not only proper nutrition but also appropriate growth factors – polypeptides that support various terminal phenotypes (physical characteristics of the animal or its tissue at the final stage of development) and regulate stem cell differentiation and proliferation. Some of these are synthesized and released by muscle cells themselves, and in tissues are also provided by other cell types locally (paracrine) and nonlocally (endocrine). The liver is the major source of circulating insulin-like growth factor I. It is theoretically possible to develop appropriate co-culture systems so that liver cells (hepatocytes) provide the growth factors needed to produce cultured muscle, i.e., "meat from a test tube." Typically, researchers induce differentiation and fusion of myoblasts by reducing mitogenic growth factor levels. The proliferating cells then begin synthesizing insulin-like growth factor II; the cells differentiate, and multinucleated myotubes are formed. The transforming growth factor β, bone morphogenic proteins and fibroblast growth factors can also be used as growth factors.

This implies the need for a system in which the composition of growth factors in the medium can be changed. The production of artificial meat will also require the development of new bioreactors capable of maintaining a low degree of deformation in the substance structure (so-called "shear") and uniform distribution of various substances necessary for growth in large volumes. NASA rotary (rotating) bioreactors have often been used in cell culturing. Their main advantages are that the cells are almost continuously in suspension, "shear" of the fluid is minimized, and the suspension allows tissue culturing up to 1 cm. Meanwhile, theoretically, scaling up to industrial reactor sizes does not affect the physics of the system.

Nevertheless, a prototype bioreactor suitable for mass production cell culture has not yet been developed. For example, the huge bioreactors used in the pharmaceutical industry are supposed to be abandoned in favor of smaller (and incomparably cheaper) bioreactors. Therefore, it will be possible to produce artificial meat not in one ultramodern plant, but in many small farms, which will significantly reduce costs.

In addition, other cell types, such as adipocytes, which are chemical messengers to provide the growing muscles with information about their structure, must be grown at the same time. Finally, muscle tissue must be physically stretched or "exercised" in order for it to develop properly.

3. To make meat three-dimensional rather than flat, laboratories use a kind of "scaffolding." Culturing uses an edible scaffold to form a three-dimensional structure that supports tissue formation, much like the extracellular matrix does in a real organism. The resulting cultured product can be produced for a very long time.

In addition, cultured artificial meat needs preservatives like sodium benzoate to protect it from fungus. In addition, collagen powder, xanthan, mannitol, antibiotics and so on are used at different stages.

3 Conclusion

Given the impossibility of using animal blood serum in nutrient media for the production of "artificial" meat, and the media available today in the arsenal of biotechnologists are designed for the accumulation of viral material, it is necessary to intensify the development of serum-free composition for the cultivation of specifically myocytes.

There is a need to create, at least in the Russian Federation, a document similar to GOST, regulating the production of "artificial" meat, claiming the status of "halal".

References

1. A. Abdallah, M. Rahem, A. Pasqualone, The multiplicity of halal standards: a case study of application to slaughterhouses, J. of Ethnic Foods, **8(1)**, 7 (2021), https://doi.org/10.1186/s42779-021-00084-6
2. R.Y. Gilmutdinov, New in the technology of production of livestock products. Artificial meat. Textbook (Kazan, 2019), 97 p., ISBN 978-5-9222-1301-1
3. G.P. Troshkova, L.D. Martynets, E.V. Kirova, T.P. Sumkina, A.V. Yudin, Serum-free nutrient medium for culturing Vero cells, Fundamental Research, **5,** 94-94 (2005).
4. M. Reiter, W. Mundt, F. Dorner, L. Grillberger, Protein- and serum-free medium and method for culturing mammalian cells in such medium, Pat. 2380412 RU, Publ. 27.01.2010.
5. E.A. Nechaeva, I.F. Radaeva, N.B. Dumchenko et al, Serum-free nutrient medium for culturing cells and viruses, Bulletin of Perm National Research

Polytechnic University, Chemical Technology and Biotechnology, **4,** 85-97 (2018), https://doi.org/10.15593/2224-9400/2018.4.07

6. E. Caterson, L. Nesti, K. Danielson, R. Tuan, Human marrow derived mesenchymal progenitor cells: Isolation culture, expansion and analysis of differentiation, Mol. Biotechnol., **20(3),** 245-256 (2002), https://doi.org/10.1385/MB:20:3:245
7. J. Draper, H. Moore, N. Ruban et al, Culture and characterisation of human embryonic stem cells, Stem Cells Dev., **13(4),** 325-336 (2004), https://doi.org/10.1089/scd.2004.13.325
8. T.V. Shamanskaya, E.Y. Osipova, B.B. Purbueva et al, Cultivation of mesenchymal stem cells ex vivo in different nutrient media (literature review and own experience), Oncohematology, **3,** 65-71 (2010).
9. T. Kazama, M. Fujie, T. Endo, K. Kano, Mature adipocyte-derived dedifferentiated fat cells can transdifferentiate into skeletal myocytes in vitro, Biochem. and Biophys. Res. Communicat., **377(3),** 780-785 (2008), https://doi.org/10.1016/j.bbrc.2008.10.046
10. M. Benjaminson, J. Gilchriest, M. Lorenz, In vitro edible muscle protein production system (MPPS): Stage 1, fish, Acta Astronautica, **51(12),** 879-889 (2002), https://doi.org/10.1016/s0094-5765(02)00033-4
11. P. Edelman, D. McFarland, V. Mironov, J. Matheny, Commentary: In vitro-cultured meat production, Tissue Eng., **11(5-6),** 659-662 (2005), https://doi.org/10.1089/ten.2005.11.659.

Technological modernisation of the Russian agro-industrial complex

Komil Astanakulov[1], *Vladimir L.* Tatarintsev [2,*], *Jasurbek* Nuriddinov [3], *Feruza* Mukumova [4], *Kumri* Allaberdiyeva [5], *Zamira* Suyunova [6]

[1] "Tashkent Institute of Irrigation and Agricultural Mechanization Engineers" National Research University, Tashkent, Uzbekistan
[2] Tomsk State University, Tomsk, Russian Federation
[3] Namangan Engineering-Construction Institute, Namangan, Uzbekistan
[4] Termez State University, Termiz, Uzbekistan
[5] Termiz State Pedagogical Institute, Termiz, Uzbekistan
[6] Karshi State Technical University, Karshi, Uzbekistan

Abstract. The analysis of the problems of technical and technological modernisation of the agro-industrial complex of Russia is carried out. It reveals a significant lag behind the world leaders. The analysis is based on statistical data and scientific publications, focusing on technical equipment, introduction of innovations, and production efficiency. It is found that the country's agriculture is characterised by an outdated material and technical base, low level of automation, inefficient use of energy resources and slow implementation of advanced technologies such as precision farming. This leads to lower labour productivity, higher production costs and, as a consequence, low competitiveness. The study also revealed a shortage of skilled labour and insufficient funding for research and development, as well as dependence on imported machinery. The authors stress the need for a comprehensive approach to modernisation, including technical re-equipment, introduction of modern technologies and development of human capital.

1 Introduction

The problems of Russia's agro-industrial complex are aggravated by the fact that a significant part of farms, especially in the regions, still work with machinery inherited from previous years, with high operating costs and low productivity. Thus, the share of costs for machinery and technology ranges from 40 to 60 per cent in the structure of the cost of final agricultural products. The use of new, modern machinery can significantly reduce these costs, as well as multiply labour productivity and reduce the machinery fleet. In particular, the use of modern machinery can reduce fuel consumption in the industry as a whole by 20%. However, the introduction of such technologies is hampered, among other things, by the financial insolvency of many farms and the lack of financial support for the introduction of biofortified crops [1]. Currently, only 15% of farmland is cultivated using resource-saving technologies, and one of the important aspects of the application of resource-saving technologies is the improvement of soil organic matter [2].

The problem of mechanisation of labour-intensive industries, such as potato and vegetable growing, is especially acute, where the level of equipment of farms with machinery is only 50-60%. This leads to the violation of technological processes and the timing of field work, which negatively affects the yield and quality of products. At the same time, there is a potential for increasing production, as the technological modernisation of agro-industrial complex, with the available favourable landscape resources of the country, allows to significantly increase the level of productivity in crop and livestock production, including through the use of probiotics in livestock [3], provided that the geo-ecological assessment of landscapes [4].

In addition, existing technologies in the agricultural sphere often do not meet the requirements of management of production processes in agricultural production. Many producers still use two-cycle technologies 'sow - harvest', without managing the processes of vegetation and maturation. It is in the process of managing plant and animal development that the greatest potential for innovation and improved resource efficiency lies, including through genomic analysis for breeding [5]. This includes the use of space monitoring, geographic information systems, and online crop monitoring, for land use management [6], and the use of biologicals to improve soil structure [7]. Even with simple tools such as quality machinery, chemical inputs and proven varieties, it is possible to increase the

* Corresponding author: kafzem@bk.ru

efficiency of energy, material, technical, biological and financial resources by multiples.

Low energy availability in agriculture is also a major problem. At present, 128 hp is used per 100 ha of sown area, while technologically it is necessary to use 350 hp. This means the need for machines of higher energy intensity, such as tractors with a capacity of 300-500 hp and combines with a capacity of 300-350 hp, as well as the use of biological agents to reduce the carbon footprint [8]. There is also a need for multifunctional machines, which are not inferior to foreign analogues in terms of technical reliability. Domestic machines are often inferior to imported ones in terms of reliability, for example, in terms of operating time per failure: 200-250 hours against 800-1000 hours for imported analogues. Also an important factor is to ensure timely repair of machinery and maintenance, and the introduction of systems for forecasting the technical condition of equipment [9]. Engineering and technical support of the majority of agricultural producers does not meet modern requirements, and the technical readiness of units is only 70-80% of the total amount of equipment. The problem of personnel cannot be ignored. The insufficient level of human capital development in the agro-industrial complex is manifested in the discrepancy between the quality and structure of human resources potential and innovation needs. Rural personnel are aging, there is a deficit of specialists, and less than 40 per cent of farms are provided with highly qualified personnel, such as agronomists, zoo-engineers and mechanics. Another problem constraining modernisation is the insufficient level of funding for science. The ratio of government spending on science to GDP is only 0.29 per cent, which is insufficient for mass introduction of scientific developments. In addition, the share of imported agricultural machinery in the country's market is more than 45% and this share continues to grow, displacing domestic products. This leads to economically inexpedient and very costly saturation of the country's agriculture with foreign machinery, and that is not always favourable from an economic point of view, especially without taking into account the introduction of green technologies [10].

In the light of the above problems, the importance and necessity of research aimed at identifying and analysing the obstacles to technical and technological modernization of the agro-industrial complex of Russia, as well as waste management becomes obvious [11]. In the conditions of increasing competition in the global food market and strengthening of foreign economic challenges, timely and effective solution of problems related to technical and technological modernisation of agriculture is critical to ensure food security of the country and increase the competitiveness of domestic agricultural products, taking into account the principles of cyclic use of waste [12], as well as to take into account the issues of ecologisation of production [13]. At the same time, an important aspect is the creation of a legal framework to stimulate organic agriculture [14]. The purpose of this paper is a comprehensive analysis of the main problems of technical and technological modernisation of the agro-industrial complex of Russia. The authors seek not only to identify the key factors constraining the development of the industry, but also to propose approaches to solving these problems, with a focus on creating conditions for large-scale transition to new technologies. Thus, this study is aimed at forming a scientifically sound basis for making strategic decisions in the development of the agricultural sector that can ensure its sustainable growth and food security of the country.

2 Method and materials

To a large extent, this is an analytical work based on the study and generalisation of data from various sources, including scientific publications, statistical reports and regulatory and legal documents. Statistical data from the Federal State Statistics Service of the Russian Federation, as well as analytical reports and publications in scientific journals and specialised publications that relate to the state of Russia's agro-industrial complex were used as material for the study. The data on the technical equipment of agricultural enterprises, the level of implementation of new technologies, as well as indicators of labour productivity and the cost of agricultural products were analysed.

The authors of the work, using the methods of system analysis and synthesis, summarised the obtained data and identified the main problems of technical and technological modernisation of the agro-industrial complex. As part of the work, various aspects related to technical equipment, energy supply and human resource potential of the industry were studied. In particular, data on the availability and condition of the agricultural machinery fleet, the share of obsolete equipment and the level of its wear and tear were studied. Special attention was paid to the level of introduction of modern resource-saving technologies and information management systems in the agro-industrial complex. As a result of the analysis, the key problems constraining the development of the industry were identified, and ways to solve them were proposed.

3 Results and discussion

The study focused on a comprehensive study of the problems associated with the technical and technological modernisation of the agro-industrial complex of the Russian Federation. Initially, a review of the current state of the agro-industrial complex was carried out, with special attention to the level of technical equipment, the degree of introduction of innovative technologies and overall production efficiency. Next, the authors of the article identified the main factors hindering modernisation, such as outdated material and technical base, lack of financing, staff shortage and low energy efficiency. Based on the analysis, they developed and proposed approaches to address the identified problems, with a focus on creating conditions for large-scale technological re-equipment of the industry. Russia's agro-industrial complex demonstrates a significant lag in the field of technical and technological development compared to the world leaders in the agricultural sector. This lag, according to expert estimates and data, is 20-30 years, which is a long time before the introduction of innovative technologies. For example, a significant part of agricultural enterprises in Russia continues to use machinery produced in the last century, which leads to high operating costs, frequent breakdowns, and, as a consequence, low labour productivity. According to my data based on the analysis of regional reports, including data from Rosstat (Federal State Statistics Service), about 40% of the tractor fleet in the Central Federal District (CFD) is more than 15 years old, and in the Siberian Federal District (SFD) this figure reaches 55%. These data from Rosstat (e.g., 'Agriculture in Russia. 2022', a statistical compilation that provides information on the availability and condition of agricultural machinery) underscore the urgent need for large-scale renewal and modernisation of the technical fleet, especially in regions with harsher climates and less developed infrastructure, where old machinery breaks down faster and its repair is difficult. In addition, according to the Russian Ministry of Agriculture (e.g., the 2022 report 'On the State of Agriculture in the Russian Federation'), the average age of tractors and combines in Russia is 18 years and 16 years, respectively, which is significantly higher than in developed countries, where the average lifespan of machinery does not exceed 8-10 years. In addition to tractors and combines, a significant part of other agricultural equipment, such as seeders, ploughs and sprayers, also needs to be replaced. According to Rosstat, more than 60% of this equipment in most regions of the country is more than 10 years old, which negatively affects the quality of work and yields. This is confirmed by the data of the Analytical Centre under the Government of the Russian Federation in the report 'Technological development of the agro-industrial complex', where it is indicated that the depreciation of agricultural machinery on average in the country is about 55-60%, and in some regions it reaches 70-75%. At the same time, according to Rosstat data ('Utilisation of machinery in agriculture'), the write-off of agricultural machinery is much slower than its arrival, as a result of which the degree of equipment wear is increasing. The low level of automation of production processes represents a serious challenge for the agro-industrial complex (AIC) of Russia. According to the Federal State Statistics Service (Rosstat), only about 10-15% of enterprises in the agricultural sector use automated production management systems (APMS). This value lags significantly behind the indicators of countries with developed agriculture, where automation has long been an integral part of production processes. For comparison, in the European Union (EU) countries, according to Eurostat, the statistical service of the European Union, more than 50% of agricultural enterprises have implemented automated systems in their operations. In some EU countries that are leaders in the agricultural sector, such as the Netherlands and Denmark, this figure exceeds 70%, while the average rate of implementation of automated systems in EU agriculture, according to Eurostat, is 5-7% per year, which demonstrates the continuous development of digitalisation in the agricultural sector. Insufficient automation in the Russian agro-industrial complex has a number of negative consequences, including inefficient use of resources, reduced productivity and increased production costs. Non-automated processes lead to overconsumption of seed, fertiliser and water, whereas optimising these processes using automated systems could significantly reduce their consumption. According to World Bank studies, countries with a high level of automation in agriculture demonstrate 30-50% higher labour productivity, which clearly indicates the growth potential that can be achieved in Russia by implementing modern technologies. In addition, the higher costs of manual labour and inputs due to the low level of automation lead to an increase in the cost of agricultural production. According to the Organisation for Economic Co-operation and Development (OECD), the average cost of grain production in OECD countries is 20-30% lower than in countries with a low level of automation. At the same time, the introduction of automation can bring significant improvements. For example, automation of milking processes in livestock farms can increase labour productivity by 30-40%, while robotic milking machines can increase milk yields by 10-15%. The use of GPS navigation systems in crop production can reduce the consumption of seeds, fertilisers and plant protection products by 15-20%, and precision farming systems based on remote sensing data can increase crop yields by 10-20%. According to the UN Food and Agriculture Organisation (FAO), the introduction of precision farming technologies can increase overall crop yields by 25-30% in regions where modern methods are underutilised. Despite the obvious advantages, the introduction of digital technologies in the Russian agro-industrial complex is still at the initial stage, which is indirectly confirmed by the reports of the Agro-Analytics Centre of the Russian Ministry of Agriculture. The reports stress that there is a low dynamics of investment in the digitalisation of agriculture and insufficient use of the potential of modern technologies. Moreover, studies such as 'Digitalisation of the Russian agro-industrial complex: 2022' by RAEC show that the majority of investment in the digitalisation of the agro-industrial complex falls on large agricultural holdings, while small and medium-sized farms face significant difficulties in financing and access to modern technologies. Thus, in order to improve the competitiveness of the Russian agro-industrial complex and ensure food security, it is necessary to encourage large-scale digitalisation of the industry, including the introduction of modern automated management systems, precision farming and smart livestock breeding technologies. Inefficient use of energy resources is another important aspect of the problem. According to the 'Agency for Forecasting Balances in the Electric Power Industry', specific energy consumption per unit of output in the Russian agro-industrial complex is 1.5-2 times higher than in countries with developed agriculture. This is due to the use of outdated equipment, unoptimised technological processes and low level of implementation of energy-saving technologies. According to my calculations based on these data, modernisation of energy equipment and introduction of energy-saving technologies could reduce energy consumption in the agro-industrial complex by 20-30%. In addition, the study revealed that the level of adoption of modern technologies such as precision farming systems, automated production management systems and biotechnology remains extremely low. According to the Ministry of Agriculture of the Russian Federation, only about 15% of sown areas in the country are cultivated using resource-saving technologies. At the same time, according to my estimates, in the European Union countries this figure exceeds 60 per cent. This leads to unjustified crop losses, increases the cost of production and reduces its competitiveness. At the same time, some farms are introducing some modern technological solutions, such as no-till farming, which is used on only 15% of agricultural land in the country. It was also noted that the level of mechanisation of labour-intensive processes, for example, in potato and vegetable farming, remains extremely low. Equipment of farms with machinery in these sectors is only 50-60 per cent, which leads to the violation of technological terms and lower yields. This is especially critical, given that these sectors are the most labour-intensive and costly. The analysis has shown that low energy supply to the agro-industrial complex is also a significant problem. According to the data of the energy survey, on average, 128 horsepower is used per 100 hectares of cultivated area, while 350 horsepower is technologically necessary. This means that enterprises are forced to use underpowered machinery, which leads to overconsumption of fuel and reduced productivity. According to my estimates, replacing outdated machinery with more powerful and modern models would reduce fuel consumption by 20-25% and increase labour productivity by 30-40%, as well as using biological agents to reduce carbon footprint [8]. Also, an important problem is the shortage of human resources in the industry, especially in terms of highly qualified specialists. According to the Ministry of Agriculture of the Russian Federation, less than 40% of agricultural enterprises are staffed with qualified agronomists, zooengineers, mechanics and economists. Having analysed the obtained data, the authors of the article note that an important factor constraining the development of agro-industrial complex is the low level of science funding. The ratio of government spending on science to GDP is only 0.29%, which is not enough for mass introduction of scientific developments. It was also noted that the domestic agro-industrial complex is highly dependent on imported machinery, the share of which in the market is more than 45% and continues to grow. Saturation of domestic agriculture with foreign machinery due to high intellectual rent is economically inexpedient and very costly, and that is not always favourable from the economic point of view, especially without taking into account the introduction of green technologies. The results of this study coincide with the data of other similar works conducted by various research institutes and organisations. For example, the report of the Institute of Agrarian Research states that the modernisation of the agro-industrial complex requires a comprehensive approach, including the renewal of the technical park, the introduction of modern technologies, the development of human resources and increased funding for science. Similar conclusions are contained in studies by think tanks, which stress the need to create conditions to attract investment in the agricultural sector and improve its competitiveness. Also, my own research shows that the introduction of precision farming systems, for example, can increase yields by 15-20% and reduce the cost of fertilisers and plant protection products by 10-15%. Thus, the analysis confirms the need for a comprehensive modernisation of the agro-industrial complex, including not only technical re-equipment, but also the introduction of modern technologies, development of human resources, energy efficiency and increased government support. Further research should be aimed at developing specific measures to address the identified problems and realise the potential of the domestic agricultural sector, taking into account the principles of waste cycling. and the use of biological agents to reduce the carbon footprint [8]. and the introduction of systems for forecasting the technical condition of equipment [9], as well as the use of automated information systems for production management [15].

4 Conclusion

The analysis of the presented research reveals significant problems in the technical and technological development of the agro-industrial complex of Russia. The industry is significantly lagging behind the world leaders, and the lag is estimated at 20-30 years. This is manifested in the use of obsolete equipment, inefficient technologies, insufficient automation and low energy supply. A significant part of agricultural machinery, including tractors and combines, is used much longer than the standard terms, which leads to high operating costs and low labour productivity. The level of automation of production processes in Russia is extremely low, which negatively affects the efficiency of resource utilisation. Energy consumption per unit of output in the Russian agro-industrial complex is significantly higher than in developed countries due to the use of outdated equipment and non-optimised processes.

The introduction of modern technologies, such as precision farming and automated control systems, also remains at a low level. At the same time, the introduction of these technologies shows significant potential to increase yields and reduce production costs. The problem of human resources is extremely important, as there is an acute shortage of qualified specialists capable of working with modern equipment and technologies. A serious obstacle to modernisation is insufficient funding of science and high dependence on imported equipment, which in turn leads to economically inefficient saturation of the market with foreign products. All these factors together significantly reduce the competitiveness of domestic agricultural products in the international market. To overcome these problems, a comprehensive approach is needed, including the renewal of the technical fleet, introduction of modern technologies, development of human resources, improvement of energy efficiency and strengthening of state support.

References

1. A. Ogunleye, D. Olude, A. Kehinde, V. Tanimonure, C.A. Wongnaa, O. Ajayi, Impact of credit constraints on adoption of biofortified food crop among farming households in Nigeria. Sustain. Futures, **9,** 100421 (2025). DOI: 10.1016/j.sftr.2024.100421
2. Q. Gao, L. Wang, Y. Fang, Y. Gao, L. Ma, X. Wang, Y. Li, X. Wu, Z. Du, Conservation agriculture boosts topsoil organic matter by restoring free lipids and lignin phenols biomarkers in distinct fractions. Soil Till. Res., **248,** 106463 (2025). DOI: 10.1016/j.still.2025.106463
3. D. Sardar, S. Afsana, A. Habib, M.T. Hossain, Dietary effects of multi-strain probiotics as an alternative to antibiotics on growth performance, carcass characteristics, blood profiling and meat quality of broilers. Vet. Integr. Sci., **23,** e2025059 (2025). DOI: 10.12982/VIS.2025.059
4. I. Akzharkyn, K. Yelemessov, D. Baskanbayeva, Applied Sciences (Switzerland), **14(17),** 7567, (2024). DOI: 10.3390/app14177567
5. A.E. Watt, M.L. Cummins, C.M. Donato, W. Wirth, A.F. Porter, P. Andersson, E. Donner, T. Huynh, A. Phabmixay, E. Flynn, C. Minney-Smith, T. Gonzales, C. Bareja, A. Levy, Z. Bartlett, Parameters for one health genomic surveillance of Escherichia coli from Australia. Nat. Commun., **16,** 17 (2025). DOI: 10.1038/s41467-024-55103-2
6. V. L. Tatarintsev, L. V. Lebedeva, G. D. Syzdykova, D. S. Inkarov, Management of agricultural land use sustainability at the local level using GIS technologies and special map material, Sustainable Development of Mountain Territories, **15 (4),** 864–876 (2023). DOI: 10.21177/1998-4502-2023-15-4-864-876
7. M.-Y. Li, W. Wang, H.-H. Yin, Y. Chen, M. Ashraf, H.-Y. Tao, S.-S. Li, W.-Y. Wang, C.-L. Yang, Y.-L. Xiao, L. Zhu, Y.-C. Xiong, The functional role of arbuscular mycorrhizal fungi in enhancing soil organic carbon stocks and stability in dryland. Soil Till. Res., **248,** 106443 (2025). DOI: 10.1016/j.still.2024.106443
8. A. A. Kulikova, E. I. Khabarova, Biological working agents use to reduce the carbon footprint, Sustainable Development of Mountain Territories, **16(1),** 311–321 (2024). DOI: 10.21177/1998-4502-2024-16-1-311-321
9. S. I. Kondratev, A. V. Vidnichuk, Vertical distribution of oxygen and hydrogen sulphide in the Black Sea in 2016. Vestnik Moskovskogo Universiteta, Seriya 5: Geografiya, **2020(3),** 91–99, (2020).
10. M.K. Moldagozhina, L. Krupnik, K.K. Yelemesov, E. Mukhtar, A. Roza, International Journal of Applied Engineering Research, **11(21),** 10454–10457, (2016).
11. V.I. Golik, R.V. Klyuev, V. Brigida, E.A. Efremenkov, Q. Mengxu, Tailings Utilization and Zinc Extraction Based on Mechanochemical Activation. Materials, **16 (2),** 726 (2023) DOI: 10.3390/ma16020726
12. L.S. Malyukova, N.V. Martyushev, V.V. Tynchenko, V.V. Kondratiev, V.V. Bukhtoyarov, V.Y. Konyukhov, K.A. Bashmur, T.A. Panfilova, V. Brigida, Circular Mining Wastes Management for Sustainable Production of Camellia sinensis (L.) O. Kuntze. Sustainability (Switzerland), **15 (15),** 11671 (2023) DOI: 10.3390/su151511671
13. V.Yu. Konyukhov, T.A. Oparina, I.Y. Matasova, M.A. Modina, N.V. Martyushev, Ecologization of underground coal mining by means of ash use in backfill preparation. Min. Inform. Anal. Bull., **10,** 123-135 (2024) DOI: 10.25018/0236_1493_2024_10_0_123
14. L. Yerkinbayeva, A. Myrzakarim, A. Moldagaliyeva, B. Kalymbek, R. Yerezhepkyzy, D. Nurmukhankyzy, THE ROLE OF LEGAL REGULATION IN PROMOTING ORGANIC AGRICULTURE TO ACHIEVE SUSTAINABLE FOOD SECURITY. J. Lifestyle SDG'S Rev., **5,** e04197 (2025). DOI: 10.47172/2965-730X.SDGsReview.v5.n01.pe04197
15. S.I. Kondratev, A.V. Vidnichuk, Local Seasonal Hypoxia and Hydrogen Sulphide Formation in the Bottom Waters of the Sevastopol Bay in 2009-2019. Ecological Safety of Coastal and Shelf Zones of Sea, **2,** 107–121, (2020). DOI: 10.22449/2413-5577-2020-1-107-121.

BIO Web of Conferences **161**, 00050 (2025) https://doi.org/10.1051/bioconf/202516100050
FIES 2024

Bacteria of the Helicobacter genus in food

Farit Mullagalievich Nurgaliev* [1], *Alinya Ravilevna* Nurgalieva[2], *and Liana Ruslanovna* Nagumanova[1]

[1]Kazan State Academy of Veterinary Medicine named after N.E. Bauman, 35 Sibirskiy tract, Kazan, 420029, Russia
[2]Kazan Cooperative Institute (branch) of Russian University of Cooperation, 58 Nikolay Ershov, Kazan, 420081 Russia

Abstract. This article presents a brief overview of current knowledge about the potential role that foods can play in the transmission of *Helicobacter* pathogens to humans. The aim of the study was to identify the DNA of viable bacteria of the genus *Helicobacter* in pork. One of the most important problems of the modern world is ensuring food quality control. It is necessary to solve complex tasks, including, inter alia, measures to detect opportunistic and pathogenic bacteria in food products. Helicobacteria that are isolated from domestic animals are classified as hard in cultivating microorganisms, which is connected with using a various antimicrobial agent in pig farming. In response to the action of medicines, microorganisms stop growing, being sown on nutrient medium, but they keep the vitality and can infect susceptible organisms. Therefore, a molecular genetic method was developed by Lien De Cooman et al. for detecting DNA of viable helicobacteria [25]. A study of meat samples purchased on the market in terms of the presence of *H. suis* and *H. pylory* shows that 2,5% of meat contained DNA of live *H. suis*.

1 Introduction

The issues of food safety, nutrition and food security are inextricably linked. Unsafe food creates a vicious cycle of diseases and malnutrition. Pathogens of salmonellosis, campylobacteriosis, yersiniosis, dysentery, rotavirus infections, viral hepatitis A and other diseases can be transmitted with food. Food toxicoinfections make up a large proportion, when toxins of various microorganisms accumulate in products or dishes. If the listed pathogens of diseases transmitted through food products have been known for a long time, scientists have reported on the role of bacteria of the Helicobacter genus relatively recently. The most famous bacterium of this genus is Helicobacter pylori. This species was discovered in 1983 by Robin Warren and Barry Marshall. The publication of these scientists has turned the scientific society's understanding of the origin, pathogenesis and treatment of chronic gastritis and peptic ulcer disease upside down. Numerous studies have shown a link between the persistence of *H. pylori* and the development of chronic gastritis, peptic ulcer disease, adenocarcinoma and B-cell maltoma [1,2,3].

Helicobacteriosis is an infectious disease caused by bacteria that have a pronounced tropicality to the epithelium of the gastric type of humans and animals. The pathogen plays an important role in the development of pathologies of the digestive system.

The detection of a high level of infection in animals with bacteria of the *Helicobacter* genus and, in particular, helicobacter pathogens for humans raises the question of the possible anthropozoonotic nature of the infection. The main route of the transmission of the pathogen is considered to be faecal-oral. Many researchers say that the exact transmission routes of infection have not been established [3, 4].

There is a lot of evidence that *Helicobacter* bacteria can survive on food products, including those stored at a sufficiently low temperature.

Dore M.P. et al. reported the presence of *H. pylori* DNA in 6% of sheep milk samples collected in various areas of Northern Sardinia. They used PCR to amplify a fragment of the 16S rRNA gene, which, when sequenced, turned out to be identical in four cases. They also discovered the vacA gene sequence in four cases and isolated the *H. pylori* strain from the milk of one sheep [5].

Reza Ranjbar et al. believe that animal products, especially raw milk, play a significant role in the transmission of *H. pylori*. In their studies, PCR analysis revealed that 67 out of 630 (10.63%) raw milk samples were contaminated with helicobacter pylori. The isolated *H. pylori* strains had high resistance to ampicillin (82.08%), tetracycline (76.11%), amoxicillin (74.62%), metronidazole (65.67%) and erythromycin (53.73%) [6].

A. Gilani et al. conducted a study of antibiotic resistance of *H. pylori* strains isolated from samples of minced meat and hamburgers. In total, 150 samples of meat products were selected in supermarkets. *Helicobacter pylori* were detected in 11 samples of food samples. The isolated *H. pylori* strains were resistant to ampicillin (90.90%), erythromycin (72.72%), amoxicillin (72.72%), trimethoprim (63.63%), tetracycline (63.63%) and clarithromycin (63.63%). According to their research, samples of hamburgers and minced meat may be sources of virulent and resistant strains of *H. pylori* [7].

Other studies conducted by Zohreh Mashak et al. in 2020 indicate a certain role of raw meat, especially lamb,

* Corresponding author: nurgalievfm@gmail.com

in the transmission of virulent and resistant *H. pylori* bacteria. In 52 out of 600 (8.66%) of the examined raw meat samples, *H. pylori* was contaminated. Resistance to tetracycline (82.69%), erythromycin (80.76%), trimethoprim (65.38%), levofloxacin (63.46%) and amoxicillin (63.46%) was most common in isolated helicobacteria [8].

Researchers from Japan managed to isolate *H. pylori* from one sample of raw milk among thirteen samples in which DNA of these bacteria was found [9]. In Greece *Helicobacter pylori* was also isolated from 4 samples of raw cow's milk among 20 studied ones [10].

A number of researchers have conducted a study on the detection of *H. pylori* in dairy products. For example, in studies by Mousavi and co-authors, *H. pylori* DNA was found in 30% of Iranian traditional cheeses, 15% of cream, 5% of butter and 27% of ice cream samples made from unpasteurized milk [11].

The presence of *H. pylori* in traditional Turkish dried Sujuk sausage has been recorded. The results of this study showed that the microorganism can survive and grow during the fermentation process of Sujuk (22 °C for 7 days). A possible explanation is that fermentation products can be used by this pathogen for growth [12].

Several studies have reported the detection of *H. pylori* DNA in vegetables. Two studies conducted in Iran indicate that 13.72% of the analysed vegetable samples were contaminated with this microorganism [13]. It is reported that *H. pylori* can survive in contaminated vegetables, despite adverse conditions, as it is able to form biofilms [14].

The high prevalence of *H. pylori* in ready-to-eat foods, meat, milk and vegetables may be due to contamination after processing. In fact, the high prevalence of *H. pylori* suggests that food contamination may occur due to non-compliance with hygiene standards and regulations during processing and transportation [15].

A number of researchers consider *H. pylori* as a food pathogen based on its epidemiological characteristics [16,17,18]. Additional indirect evidence of the *H. pylori* transmission to humans through animal food was due to high levels of antibodies in slaughterhouse workers and veterinary workers. The incidence rate of these workers was higher than that of workers who had no direct contact with carcasses [19,20].

2 Materials and methods

The main methodological principles underlying the research are based on the positions of foreign and Russian specialists in the field of medical and agricultural microbiology. The theoretical research of the article is devoted to the analysis of the possible role of bacteria of the *Helicobacter* genus found in food products as a source of the human helicobacter infection.

The article presents a fairly broad analysis of data on experiments conducted with various food products and in different countries of the world in order to establish the role of *Helicobacter pylori* in the development of infectious pathology.

Pig meat sold on trading platforms was studied. 20 samples of pig meat produced by various manufacturers were purchased.

PCR for the selective DNA determination of viable *H. suis* was performed according to F. Cortez Nunes et al. [21]. The detectable DNA fragments were homologous sections of the *H. suis* 16S rDNA gene with a size of 150 bp. Species-specific PCR was performed using primers BFHsuis_F1 and BFHsuis_R1. PCR imaging was performed using horizontal electrophoresis in 1.5% agarose gel with a TBE buffer containing ethidium bromide, followed by viewing electrophoregrams in a UV transilluminator.

Extracting DNA from biomaterials was conducted with the reagent kit. This is «RIBO-prep» production by FBIS CRI of Epidemiology of Rospotrebnadzor according to the manufacturer's instructions.

Setting up a species-specific PCR for amplification of a DNA fragment of the 16s rDNA gene with a size of 433bp was performed with the «SybEnzym» reagent kit and V832f and V1261r (table 1) suggested by D. De Groote et al. (1999) [14].

When setting up PCR for the amplification of the H. pylori DNA fragment, a commercial amplification kit «HELICOPOL» for 50 reactions, manufactured by LLC NPF «Litech» was used according to the method proposed by D. De Groote et al. (1999) [14].

To indicate a viable bacteria of genus *Helicobacter*, we processed samples with ethidium monoazide (EMA) in combination with following PCR proposed by Lien De Cooman et al. (2013) [13].

EMA was dissolved in 4 ml of dimethylformamide (DMF) and 1 ml of the RPMI-1640 Medium (Sigma) to obtain a stock concentration of 1 mg ml−1. In different repeats, all dilutions of both series 1 and 2 were treated with final concentrations of either 10 or 20 µg ml−1 of EMA in 24 well cell culture plates. The samples were covered by aluminium foil and placed on a shaker (600 RPM) at room temperature for 10 min. Thereafter, all samples were placed on ice and irradiated for 2, 5 or 10 min with a 500 W halogen light source, which was placed at a distance of 25 cm. This protocol was then repeated, adding the same concentration of EMA to the samples and applying the same period of incubation and irradiation. Next, all suspensions were transferred to Eppendorf tubes, pelleted by centrifugation at 5000 ×g, and washed twice to remove unbound EMA. Final pellets obtained by 5-minite centrifugation at 5000 ×g were suspended again in 250 µl of the Hanks' Balanced Salt Solution.

PCR for the selective determination of the DNA of viable *H. suis* was performed according to F. Cortez Nunes et al. (2021) [11]. The detectable DNA fragments were homologous sections of the *H. suis* 16s rDNA gene with a size of 150 bp. Species specific PCR was performed using BFHsuis_F1 and BFHsuis_R1 (Table 1). The thermal cycling mode is shown in Table 2.

Table 1. Primers used for amplification

Name	Nucleotide sequence	Organism	Amplicon size, bp	Reference
V832f V1261r	TTGGGAGGCTTTGTCTTTCCA GATTAGCTCTGCCTCGCGGCT	*H. suis*	433	D. De Groote [14]
BFHsuis_F1 BFHsuis_R1	AAAACAMAGGCGATCGCCCTGTA TTTCTTCGCCAGGTTCAAAGCG	*H. suis*	150	F. Cortez Nunes [14]

Table 2. Thermal cycling mode

Stage	*H. suis* (V832f and V1261r)		«HELICOPOL» amplification kit		*H. suis* (BFHsuis_F1 and BFHsuis_R1)
Initialization	(94 °C – 4 min) x 1		94 °C – 1 min		94 °C – 4 min x 1
Denaturation	94 °C – 30 sec	x 40	94 °C – 30 sec	x 5	95 °C – 20 sec
Annealing	60 °C – 60 sec		50 °C – 30 sec		60 °C – 30 sec
Elongation	72 °C – 10 sec		72 °C – 30 sec		72 °C – 30 sec
Denaturation	–		92 °C – 5 sec	x 40	–
Annealing	–		50 °C – 10 sec		–
Elongation	–		72 °C – 15 sec		–
Final elongation	(72°C – 4 min) x 1		(72°C – 4 min) x 1		(72°C – 4 min) x 1

PCR visualisation was performed using horizontal electrophoresis in 1.5% agarose gel containing ethidium bromide with the following viewing of electrophoregrams in UV-transilluminator.

3 Results

20 samples of pig meat (pork neck) produced by various manufacturers were purchased on the shelves of local markets. The evaluation of these samples for compliance with safety requirements for microbiological indicators showed that meat samples meet the safety requirements for verified microbiological indicators.

The biomaterial samples taken from these pork samples were examined for the presence of *H. suis* and *H. pylori* DNA. The detection of *H. suis* DNA was carried out using the technique proposed by D. De Groote et al. *H. pylori* was detected according to the methodology proposed by the manufacturer of the kit (NPF Litech LLC). Positive PCR results for the *H. suis* DNA content were obtained in samples from 2 samples. All samples for the detection of *H. pylori* DNA were negative.

Also, all samples were examined for the detection of DNA from living *H. suis* bacteria, for this purpose, the EMA pork sample processing protocol was used in combination with PCR. According to the results for the detection of *H. suis* DNA, one sample was positive.

Among the species of Helicobacter colonizing animals, *H. suis* is most often isolated from the mucous membrane of the human stomach. This microorganism is able to colonize more than 60% of pigs that have reached slaughter age [22].

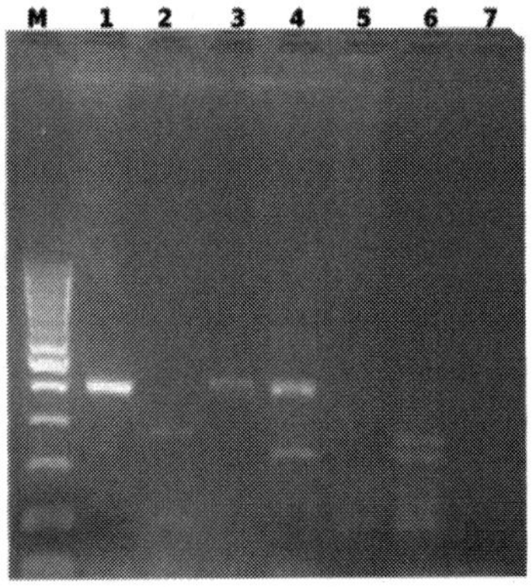

Figure 1. Electrophoregram of the PCR-amplification result. V832f and V1261r iniciate amplification of the *H.suis* 16S rDNA gene with a size of 433 bp. Designations: M) DNA marker; 1) Positive control; 2) Negative control; 3-7) PCR products

This pathogen can be transmitted to humans by manipulating infected pork or eating it. *H. suis* is a very fastidious microorganism, and it is extremely difficult to isolate it. De Cooman L et al. applied a new method for detecting viable *H. suis* bacteria in 50 retailed pork samples. Viable bacteria were found in two samples. *H. suis* also persisted in experimentally infected pork for at least 48 hours. Scientists have concluded that the consumption of infected pork may represent a new route of transmission of the *H. suis* infection in humans.

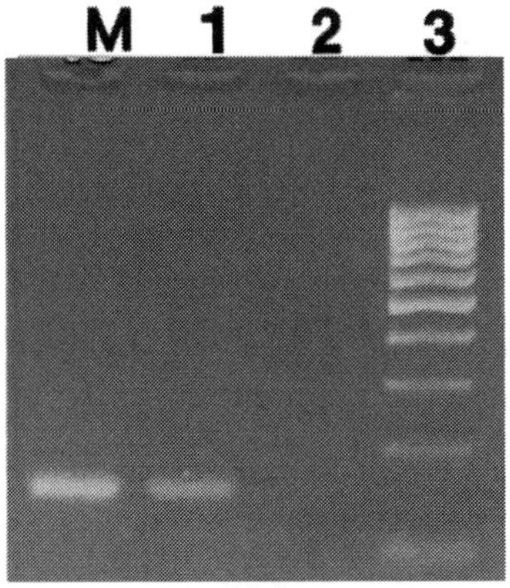

Figure 2. Electrophoregram of the PCR-amplification result. BFHsuis_F1 and BFHsuis_R1 iniciate amplification of the *H.suis* 16S rDNA gene with a size of 150 bp. Designations: M) DNA marker; 1) Negative control; 2) Positive control; 3) PCR products

The research suggests that contact with infected animals contributes to human infection. In particular, a high infection rate was registered among private pig breeders and employees of pig farms. The consumption of pork without sufficient heat treatment is also a certain danger [23, 24].

L. De Cooman et al. indicate that contamination of *H. suis* carcasses is possible during "nutrition", when part of the stomach contents can get on the surface of the carcass in the neck and thoracic region [25]. Pozdeyev O.K. et al. believed that the consumption of infected and insufficiently heat-treated pork meat was a possible way of transmitting this microorganism. Their studies on the presence of *H. suis* DNA in various parts of pig carcasses immediately after slaughter showed that in 26.7% of the samples bacterial DNA was found on the oral mucosa, 13.3% along the incision lines of the neck, 6.6% in thoracic and 0% in ham.

4 Conclusion

In our research, we evaluated samples of pig meat (pork neck) obtained from various manufacturers purchased on market shelves. The results of the analysis of meat samples for the presence of *H. suis* and *H. pylori* DNA showed that 2 samples contained DNA of living *H. suis*. All the samples intended for the detection of *H. pylori* DNA were negative.

Studies conducted in different countries report the presence and isolation of *Helicobacter* bacteria from food. Many of the findings reported in the studies are based not only on indirect evidence of the presence of these microbes in food through molecular methods, but also on the isolation of this microorganism. Therefore, the possibility that food may be a source of transmission of a bacterium of the *Helicobacter* genus is not excluded. Using high-quality raw materials, observing sanitary conditions at work and following the rules of personal hygiene will help reduce the risk of helicobacteriosis.

Acknowledgements.

This work was supported by a grant from the Academy of Sciences of the Republic of Tatarstan, provided for young PhD candidates (postdoctoral researchers) for the purpose of defending their doctoral dissertations, conducting research activities, and fulfilling professional functions in scientific and educational organizations of the Republic of Tatarstan within the framework of the State Program of the Republic of Tatarstan for "Scientific and Technological Development of the Republic of Tatarstan."

References

1. G. Bode, F. Mauch, P. Malfertheiner, Epidemiol. Infect, **111**, 483-490 (1993)
2. S. Czinn, J. Nedrud, Springer Semin. Immunopathol, **18**, 495-513 (1997)
3. O. V. Afanaseva, O. K. Bezyukov, A. A. Ignatenko, Academic Journal of Manufacturing Engineering, **22(1),** 68–75, (2024). DOI: 10.24874/PES.SI.02.017
4. I. Akzharkyn, K. Yelemessov, D. Baskanbayeva, Applied Sciences (Switzerland), **14(17),** 7567, (2024). DOI: 10.3390/app14177567
5. M. Dore, A. Sepulveda, H. El-Zimaity, Y. Yamaoka, M. Osato, K. Mototsugu, A. Nieddu, G. Realdi, D. Graham, Am J Gastroenterol, **96**, 1396–1401 (2001)
6. R. Ranjbar, F. Yadollahi Farsani, F. Safarpoor Dehkordi, Antimicrob Resist Infect Control; **7**, 115 (2018)
7. Gilani, V. Razavilar, N. Rokni, E. Rahimi, Iran J Vet Res; 18, 97-102 (2017)
8. Z. Mashak, S. Jafariaskari, I. Alavi, M. Sakhaei Shahreza, F. Safarpoor Dehkordi, Infect Drug Resist, **13,** 257-272 (2020)
9. S. Fujimura, T. Kawamura, S. Kato, H. Tateno, A. Watanabe, Lett Appl Microbiol, **35**, 504-7. 2002
10. Angelidis, I. Tirodimos, M. Bobos, M. Kalamaki, D. Papageorgiou, M. Arvanitidou, (FISH) Int J Food Microbiol;**151**, 2526 (2011)
11. S. Mousavi, F. Dehkordi, E. Rahimi, J Venom Anim Toxins Incl Trop Dis, **20**, 51 (2014)
12. Guner, K. Kav, K. Tekinsen, Y. Dogruer, N. Telli, J Food Prot, **74**, 2055–2061 (2011)
13. E. Yahaghi, F. Khamesipour, F. Mashayekhi, F.Safarpoor Dehkordi, M. Sakhaei, M. Masoudimanesh, M. Khameneie. Biomed Res Int; **1**, 757941 (2014)
14. C. Ng, M. Loke, K. Goh, J. Vadivelu, B. Ho, Food Microbiol, **62**, 68–76 (2017)
15. D. Olivares, J. Gisbert, Rev Esp Enferm Dig, **98**, 374–386 (2006)

16. Wesley, Trends Food Sci Technol, **8**, 293–299 (1997)
17. Meng, M. Doyle, Annu Rev Nutr, **17**, 255–275. (1997)
18. Y. Van Duynhoven, R. de Jonge, Bull World Health Organ, **79**, 455–460 (2001)
19. D. Vaira, C. D'Anastasio, J. Holton, J. Dowsett, M. Londei, P. Salmon, L.Gandolfi, Lancet, **2**, 1149 (1988)
20. M. Husson, P. Vincent, M. Grabiaud, D. Furon, H. Leclerc, Gastroenterol Clin Biol, **15**, 723–726 (1991)
21. E. Taillieu, S. Taelman, S. De Bruyckere et al., Vet Res, **55**, 15 (2024)
22. Hellemans, K. Chiers, M. De Bock, A. Decostere, F. Haesebrouck, R. Ducatelle, D. Maes, Vet Rec; **161**, 189-192 (2007)
23. R. Sapierzynski, M. Fabisiak, M. Kizerwetter-Swida, A. Cywińska, Polish J Vet Sci, **10**, 65-70 (2007)
24. F. Haesebrouck, F. Pasmans, B. Flahou, K. Chiers, M. Baele, T. Meyns, A. Decostere, R. Ducatelle, Clin Microbiol Rev, **22**, 202-223 (2009)
25. De Cooman et al., International journal of food microbiology, **166**, 164-167 (2013)

Prospects and feasibility of involvement of fallow lands in organic agricultural production

Irina V. Ukraintseva [1*], *Vadim P.* Skvortsov [1], *Valery V.* Dolzhikov [1], *Pavel A.* Ivanov [1], *Sergey N.* Kharchenko [2]

[1] Azov-Black Sea Engineering Institute – branch of Don State Agrarian University in Zernograd, Zernograd, Russia
[2] Kuban State Agrarian University named after I. T. Trubilin, Krasnodar, Russia

Abstract. The study is devoted to assessing the relevance of the use of fallow and unused land for the production of organic products. The paper shows that this is a promising direction for solving agro-industrial and food problems. The growth of the world market of organic products and the importance of transition to sustainable agriculture are emphasised. Involvement of unused land contributes to the improvement of ecology, soil fertility and development of rural areas. The study focuses on Altai Krai, a region with high agrarian potential but low level of agricultural intensification. It substantiates the need to develop strategies to stimulate organic farming and rational land use. The work is an analysis, development of a regional programme, calculation of forecast values and evaluation of the effectiveness of the proposed measures, based on the analysis of literature, regional data, systematic approach and application of various research methods, including calculative-constructive and economic-mathematical. The result of the work is to substantiate the feasibility of involving fallow lands in agricultural turnover for the production of organic products.

1 Introduction

The solution of modern problems of agro-industrial complex and food security problems through the use of fallow and unused land for the production of organic products seems particularly promising. This approach allows not only to meet the growing demand for organic products, but also to solve the problem of irrational use of land. Involvement of such land in agricultural turnover contributes to the improvement of the environment, increase soil fertility, biodiversity conservation, as well as provides an opportunity to provide employment for the rural population, and contributes to the socio-economic development of regions. Thus, this approach combines environmental sustainability, economic efficiency and social relevance. At the same time, an important aspect is the assessment of the impact of different land use practices on soil, especially when switching to organic production [1].

The global market for organic products is experiencing rapid growth, and this is not just a fashion trend, but a reflection of consumers' growing awareness of the link between nutritional quality, health and environmental sustainability. According to FiBL and IFOAM, the global organic market is expected to reach €124 billion in 2021, growing at an annual rate of 10-15%. This trend is driven not only by increasing demand from consumers, but also by governmental realisation of the need to move towards more sustainable farming practices. Organic farming is not just an alternative to conventional methods, but also a tool for ensuring food security in the long term by ensuring sustainable management of resources, including water and soil. It contributes to the conservation of biodiversity, reduces environmental pollution from the use of pesticides and synthetic fertilisers, and creates favourable conditions for the development of rural areas. At the same time, the transition to organic farming faces certain challenges, including the need to improve technologies, human resources and access of producers to markets. Different approaches to the development of organic agriculture, both globally and regionally, involve different trade-offs between environmental, economic and social objectives. The introduction of biological methods of plant protection and organic fertilisers, as well as the use of renewable energy sources on farms require significant investment in research and development, and regulation of this process through a legislative framework [2]. The integration of ecological principles into traditional farming systems, although less radical, also requires changes in habitual practices and adaptation to new conditions.

In this context, given the need to adapt agriculture to the changing climate [3], and the introduction of sustainable technologies to reduce dependence on resources, the involvement of fallow and unused land in organic production is very promising [4]. World data indicate that there are significant areas of such lands that

* Corresponding author: rostma@rambler.ru

can be used for organic farming. According to FAO estimates, there are more than 2 billion hectares of degraded land in the world, potentially suitable for restoration and use in agriculture, including organic agriculture. With the right approach, involving these lands into circulation not only increases organic production, but also helps to restore soil fertility and reduce erosion. Such an approach opens up opportunities for the creation of new jobs in rural areas, reduces the outflow of population from rural areas and contributes to their socio-economic development. However, the development of fallow lands may require additional investment in their treatment and preparation, as well as the development of specialised technologies for their cultivation. An important aspect is the use of biological preparations to maintain soil fertility [5].

In light of the above, the relevance of the selected research area is due to the fact that the Russian Federation, and in particular Altai Krai, has a significant but underutilised potential for the development of organic agriculture. Thus, according to Rosstat, in 2021 the area of agricultural land in Russia was about 197 million hectares, of which a significant part was arable land. At the same time, the share of organically certified agricultural land in Russia in 2020 was less than 0.2%, which indicates a significant reserve for the development of this area. Altai Krai, with its rich soil resources and centuries-old agrarian traditions, stands out against this background as one of the most promising regions for the development of organic agriculture. The region holds leading positions in terms of arable land area (6.5 million hectares) in Russia, with the predominance of fertile chernozems. At the same time, the level of intensification and chemicalisation in agriculture in Altai Krai remains relatively low compared to European countries: while the average application of mineral fertilisers in the Eurozone is 192 kg/ha, in Altai Krai this indicator is only 3.8 kg/ha, which creates favourable conditions for the transition to organic farming methods. On this basis, the development of comprehensive strategies and mechanisms aimed at stimulating organic agriculture, with a focus on the rational use of fallow land, is not only timely, but also critical to ensure food security, improve the environmental situation and stimulate sustainable socio-economic development of the region. The problem of unused land is relevant, since according to the Ministry of Agriculture of the Russian Federation, the total area of unused agricultural land in the Central Federal District alone exceeds 1.5 million hectares. A similar situation is typical for other regions of Russia, which only emphasises the importance of the study of this topic, especially under the influence of mountain factors on the soil [6]. The purpose of this work is to substantiate the possibility and feasibility of involving fallow and unused lands of Altai future results.

2 Method and materials

Within the framework of the presented research, a set of works was carried out to study the possibility of involving fallow and unused land into agricultural turnover for the production of organic products. These works included several key stages: analysis of theoretical and methodological basis, development of a regional target programme, calculation of forecast production values, as well as assessment of the economic and environmental effect of the proposed measures. At the initial stage of the research, the scientific literature on the problems of organic farming, development of land relations, as well as regulatory and legal acts, including legislative acts of the Russian Federation and regulations of the European Union concerning organic farming were analysed. The system approach was used as a methodological basis, which allowed to ensure the comprehensiveness and purposefulness of the study. Analytical, abstract-logical, calculative-constructive, economic-statistical, economic-mathematical and monographic methods were applied. In the process of developing the regional target programme, the results of the analysis of regional peculiarities of Altai Krai, including socio-economic conditions, as well as the available resource and land potential were used. Determination of specific parameters of the programme, such as targets for land involvement in the turnover, the volume of subsidies and projected volumes of organic production, was carried out using the calculation and construction method. Economic calculations were carried out to assess the effectiveness of the proposed measures, taking into account the costs of land preparation, purchase of equipment, and other production costs. To estimate the potential volumes of organic production, the article used a forecast calculation for the period 2017-2027. This forecast was based on data on the area of arable land, its fertility and planned yields, taking into account the possible increase in the share of organic production. For the calculation, economic and mathematical methods were used, in particular, the method of expert evaluations and methods of statistical analysis, as well as methods of crop rotation planning [7]` identifying the potential and justifying the feasibility of involving fallow land in agricultural turnover, rather than on the study of specific agro-technical methods or soil characteristics. Consequently, there is no information on specific types, brands of equipment and modes of operation, as this work was theoretical and economic in nature.

3 Results and discussion

The work on involvement of unutilised land began with an assessment of available resources, reserve arable land that was not used for production purposes was identified and mapped, including the allocation of land shares. According to the results of the analysis, the total area of identified unused arable land was 540 thousand hectares, of which 401 thousand hectares were allocated as land shares. This confirmed the hypothesis that there is a significant potential to expand production through the involvement of these reserves. The suitability of the identified lands for production was assessed, identifying factors that prevented their utilisation, including lack of infrastructure, reclamation deficiencies or low soil fertility. At subsequent stages, mechanisms were developed to compensate land users for the costs of bringing unused land into production. The project envisaged subsidising part of the costs of tillage work to engage unused arable land, namely RUB 2,000 per hectare, which, according to calculations, covers the cost of 60 litres of diesel fuel. As a result of these measures, it was planned to gradually increase the area of arable land brought into agricultural turnover. Thus, by 2022 it was planned to involve into production 207 thousand hectares of unused arable land and 110 thousand hectares of fallow land. The authors paid attention to the maintenance and restoration of soil fertility, including reclamation measures, which contributed to the increase in land productivity, including through the use of biological preparations. Based on the results obtained, the dynamics of changes in the area of utilised arable land was assessed, which was planned to increase from 6307 thousand hectares in 2017 to 6626 thousand hectares in 2022, and measures for its rational use, including land use structure planning, were identified and proposed. At the same time, attention was emphasised on the need for state control over the use of agricultural land resources, providing for regular inspections of compliance with land legislation with a coverage of 100 per cent by 2022. Geo-ecological assessment of land is also necessary [8]. In order to ensure the resource base of the project, sources of funding were identified. A total of RUB 70 million was allocated for the programme in 2017, with a subsequent increase to RUB 523 million in 2018. The main sources of funding were the regional budget, which provided RUR 102 million in 2022, and extra-budgetary sources in the amount of RUR 173 million in 2022. Projected volumes of budget subsidies for the involvement of reserve lands totalled RUB 26 million in 2017, with an increase to RUB 295 million in 2018 and a further decrease to RUB 102 million in 2022. Total investments in the project implementation until 2022 totalled 1,060 million rubles. Forecasting the development of agriculture in Altai Krai was the key focus of this study. Special attention was paid to the production of major crops and livestock products, with a focus on the prospects of organic production. As a result of the analysis and modelling, a forecast was developed, according to which by 2027 the total production of grain crops in Altai Krai was to be 5800 thousand tonnes. At the same time, a significant part of this volume, namely 590 thousand tonnes, was to be organic production, which is 10.2% of total cereal production. It is estimated that in the structure of cereal production a significant contribution was to be made by crops such as wheat, barley, rye and oats, with a projected average cereal yield for 2027 of 26 centners per hectare with a total sown area of 2.2 million hectares. Moreover, the dynamics of organic farming development was projected to manifest itself in the growth of organic flax production, reaching 11.4 per cent of the total production of this crop, and organic sugar beet, where this figure was planned at 11.5 per cent. These figures demonstrate the potential for the development of organic farming in the region. Sunflower production, an important oilseed crop for Altai Krai, should reach 475 thousand tonnes by 2027, with an organic share of 11.7%. This figure, as well as the average sunflower yield of 16 quintals per hectare with a sown area of 297 thousand hectares, indicates a significant resource for expanding its production. In addition, the projected potato production by 2027 is estimated at 995 thousand tonnes, and 15.1% of this volume should be organic potatoes, demonstrating the growing demand for organic products, while it is important to take into account the opinion of consumers [9]. The total area under potatoes in the region is about 50 thousand hectares. The implementation of these plans in the field of crop production is expected through the involvement in the turnover of unused agricultural land, which is estimated at about 300 thousand hectares, as well as through the introduction of modern organic agro-technologies, including the use of organic fertilisers and biological plant protection methods. Special attention is paid to forecasting the development of livestock breeding. In particular, an increase in milk production to 1720 thousand tonnes by 2027, including 210 thousand tonnes of organic milk, which is 12.2% of total production. The average milk yield per cow should reach 7,500 kilogrammes per year. Meat production, in turn, was projected to reach 440,000 tonnes, including 60,000 tonnes of organic meat, representing 13.6% of total production. Projected meat production includes cattle, pork, poultry and mutton, with the total livestock population in the region estimated at 2 million head. These figures indicate significant prospects for the development of organic livestock production in the region. The key factors of growth in this area were to increase the number of livestock, increase productivity, as well as the development of the fodder base through organic farming. At the same time, it is important to ensure the rational use of resources, including waste [10], and the creation of conditions for economic development [11]. These forecasts are based on the data provided by the Ministry of Agriculture of Altai Krai and the Federal State Statistics Service, as well as on the analysis of trends in the development of the agricultural sector in Russia and abroad. In comparison with other similar studies conducted in other Russian regions and countries, it can be noted that the involvement of unused land in agricultural turnover is an urgent task for many territories. For example, in the regions of Central Russia, according to the Ministry of Agriculture of the Russian Federation, there are significant areas of unused arable land that could be used for the production of organic products. At the same time, studies conducted in European countries have shown that the transition to organic agriculture contributes to improving the environmental situation, increasing soil fertility and reducing the negative impact on the environment. As an additional aspect, it is necessary to note the importance of agro-ecological expertise of lands before their involvement in turnover. This will minimise the risks of soil degradation and preserve soil fertility. Also, it is important to carry out monitoring of soil resources [12]. Also fallow and unused lands. The obtained results and proposed recommendations can be used to develop similar programmes in other regions of Russia, as well as for the formation of sustainable agro-ecosystems. According to estimates based on data on global trends in organic production, the share of organic products in total production can be higher with active promotion in the markets. For example, in European countries the share of organic products in some segments reaches 20-30%, which shows the potential for growth and in the Altai Krai. The study of global experience has shown that the use of modern agro-technologies and marketing strategies can achieve higher performance. The study also found that the value of gross agricultural output produced in Altai Krai should increase to 182 billion rubles in 2027, which corresponds to 155% of the 2015 level. The level of profitability of agricultural organisations was planned to increase to 20% by 2027, which would exceed the 2015 figure by 11.3%. These data reflect the economic efficiency of the proposed measures for the involvement of fallow lands and the development of organic agriculture.

4 Conclusion

The research presented in this article highlights the significant potential of organic agriculture to address food security and land management issues, especially in the context of the involvement of fallow and unused land. The global market for organic products is showing steady growth, reflecting consumer demand for organic products and government support for sustainable agricultural practices. The authors note that organic agriculture is not only an alternative to conventional agriculture, but also an important tool to ensure food security in the long term, contributing to the conservation of biodiversity, reduction of environmental pollution and development of rural areas. In Russia, especially in Altai Krai, with its vast land resources and relatively low level of agricultural intensification, there is a huge potential for the development of organic farming. The article emphasises that the share of organically certified land in Russia remains extremely low, which indicates significant reserves for growth in this direction. In Altai Krai, with its fertile chernozems and relatively low use of mineral fertilisers, there are favourable conditions for the transition to organic methods. The study emphasises the need to develop comprehensive strategies aimed at stimulating organic agriculture, in particular through the involvement of fallow lands in the turnover, which is key to ensuring food security, improving the environmental situation and stimulating socio-economic development of the region.

References

1. Q. Gao, L. Wang, Y. Fang, Y. Gao, L. Ma, X. Wang, Y. Li, X. Wu, Z. Du, Conservation agriculture boosts topsoil organic matter by restoring free lipids and lignin phenols biomarkers in distinct fractions. Soil Till. Res., **248,** 106463 (2025). DOI: 10.1016/j.still.2025.106463

2. A. I. Trunova, S. O. Kurashkin, Y. A. Tynchenko, I. F. Vivián, K. A. Bashmur, Metals, **13(7),** 1184 (2023). DOI: 10.3390/met13071184.

3. N. Garambois, U. Le Goff, L. Thibaudeau, Agrarian dynamics and climate change in the Senegalese Sahelian peanut basin. Agrar. Syst. Clim. Change: Journeys Adapt. Glob. South, **15**, 37 (2024). DOI: 10.1079/9781800628137.0001

4. T. Bandara, B. Marambe, G. Pushpakumara, P. Silva, R. Punyawardena, L. Manawadu, S. Premalal, G. Miah, K.R. Dahal, Adapting mountain-farming systems to climate change by reducing food-nutrition-health vulnerability. Mitig. Adapt. Strateg. Glob. Change, **29,** 88 (2024). DOI: 10.1007/s11027-024-10181-x.

5. Y. Spychak, T. Rozhkova, L. Tytova, L. Biliavska, O. Bakumenko, V. Tatarynova, O. Yemets, V. Demenko, V. Pivtoraiko, A. Burdulaniuk, Changes in the seed and soil micobiota caused by seed treatment with chemical and biological agents. Ecol. Eng. Environ. Technol., **26,** 103-110 (2025). DOI: 10.12912/27197050/195636

6. T. A. Kuandykov, T. D. Karmanov, E. I. Kuldeyev, K. K. Yelemessov, B. Z. Kaliev, New technology of uncover the ore horizon by the method of in-situ leaching for uranium mining. News of the National Academy of Sciences of the Republic of Kazakhstan, Series of Geology and Technical Sciences, **2022(3),** 142–154, (2022). DOI 10.32014/2022.2518-170X.186

7. A. Deo, N. Sawant, A. Arora, S. Karmakar, How has scientific literature addressed crop planning at farm level: A bibliometric-qualitative review. Farming Syst., **3,** 100139 (2025). DOI: 10.1016/j.farsys.2025.10013

8. T. A. Panfilova, K. A. Bashmur, V. V. Kondratiev, To the Concept of Leaching Metal-Containing Raw Materials in the Disintegrator. Mining Informational and Analytical Bulletin, **11-1,** 239-251 (2023). DOI: 10.25018/0236_1493_2023_111_0_239

9. J.D. Torres, S.R. Jaeger, P. Puerta, A. Tárrega, How do Spanish consumers perceive different lettuce cultivation systems? Insights from explicit and implicit methods. Appl. Food Res., **5,** 100709 (2025). DOI: 10.1016/j.afres.2025.100709.

10. V.I. Golik, R.V. Klyuev, N.V. Martyushev, V. Brigida, E.A. Efremenkov, S.N. Sorokova, Q. Mengxu, Tailings Utilization and Zinc Extraction Based on Mechanochemical Activation. Materials, **16 (2),** 726 (2023) DOI: 10.3390/ma16020726

11. A. I. Trunova, V. V. Kukartsev, S. O. Kurashkin, K. A. Bashmur, R. B. Sergienko, Metals, **13(7),** 1234 (2023). DOI: 10.3390/met13071234

12. M.N. Nurpeisov, K.V. Ruchkina, V.A. Androkhanov, S.K. Makenova, Assessment of geoecological factors and conditions affecting sustainability of agricultural land use in the forest-steppe zone of Altai Krai, Sustainable Development of Mountain Territories, **15 (2),** 419–430 (2023). DOI: 10.21177/1998-4502-2023-15-2-419-430

Morpho-physiological evaluation of the state of organs of laying quails when black soldier fly larvae are used in their feeding

Radii Papaev[1*], *Yulia* Larina[1], *Vladimir* Ezhkov[1], *Asiya* Ezhkova [1], *Aidar* Girfanov[1]

[1] Federal State Budgetary Educational Institution of Higher Education 'N.E. Bauman Kazan State Academy of Veterinary Medicine', Republic of Tatarstan, Kazan, 420029, Russia

Abstract. Visual inspection of carcasses and internal organs of laying quails of control and experimental groups revealed that morphology of their organs and tissues corresponded to species, age and physiological parameters of birds. Distinctive feature in the micromorphology of the stomach and small intestine of laying quails was noted - activation of gland function, abundance of mucus, insignificant amount of sloughed epitheliocytes on the mucosa of the organs, with a greater degree of manifestation in the stomach in birds receiving dried form, and in the intestine - in quails receiving the native form of black soldier fly larvae. The microstructure of the liver of laying quails of the experimental groups was characterised by a lower degree of fatty degeneration.

1 Introduction

Deficiency of nutrients in animal feeding rations remains an urgent problem at the present stage of agricultural development. The search for alternatives to the most important nutrients is associated with the intensification of production technology and production in agriculture, which are characterised by high indicators of biological and hygienic safety, high cost of traditional feed components and economic feasibility of their use.

Protein in the animal feed ration occupies a leading position due to the fulfilment of the most important functions associated with the growth and development of the organism [1-3]. Every year this problem is aggravated due to population growth and, accordingly, the increase in people consuming meat [4]. This trend may lead to the depletion of environmental resources on our planet, increasing growth of hungry people and development of nutrient-dependent diseases.

In recent decades, insectoculture protein has been considered as a promising source of protein [5-7].

Insects in different stages of their development are used by wild animals to supply their nutritional needs. The larval stages of insects have particularly high nutritional value, primarily in terms of protein substances. In this connection, there are great prospects for adjustment of feeding rations for pets and farm animals by supplementing them with larval stages of insects.

We propose the use of an unconventional source of protein in the form of larval stage of insects. We have chosen the insect of choice on the black soldier fly (Hermetia illucens). The larvae of the black soldier fly are characterised by high growth intensity, qualitative composition of nutrients, the lion's share of which is protein. Perspectivity of application of insect larvae, including Hermetia illucens in feeding of farm animals is connected with their fast growth and convenience of application. Moreover, the qualitative chemical composition can ensure the completeness of diets and the possibility of adjusting the application in different branches of animal husbandry. Biomass from black soldier fly larvae is a rich source of protein with essential amino acids, fats, vitamins, macro- and microelements, and biologically active substances [8].

The larvae of the black soldier fly are considered as a promising alternative source of feed protein for animals, which does not cause deviations from the norm of haematological parameters, enzyme activity in the blood, liver and the main indicators of protein, lipid, carbohydrate metabolism in the living organism [9].

The results of our own research in exploratory experiments showed that the introduction of the biomass of black soldier fly larvae in natural form in the amount of 5% to the diet of young rats provided an increase in live weight by 14.0% and in comparison with the control counterparts [10].

Experiments with feeding of poultry, fur-bearing animals, pigs, cattle showed that the use of insects provides intensive growth of live weight, improving the quality of products [11-13].

Compared to traditional livestock breeding, insect breeding requires fewer resources. For example, the production of 1 kg of beef, pork and chicken requires 5, 2.5 and 1.5 times more feed, respectively, than the same mass of insects [14]. In addition, insects emit a much smaller amount of greenhouse gases, which,

* Corresponding author: papaev-radii@mail.ru

together with the possibility of growing on secondary raw materials, can globally improve the ecological situation in the world.

Hermetia illucens larvae are able to consume a wide range of organic matter and are of particular interest for waste management. The nutritional value of the preimaginal stages, in particular the protein content, makes this species a suitable candidate for use as feed for other organisms. Food wastes, vegetables, fruits and agricultural wastes are promising substrates for insect rearing [15].

Industrial poultry production is an intensively developing branch of animal production, which is in demand for high levels of proteins in feed diets. Currently, poultry producers in developing countries are faced with the problems of high cost and low quality of poultry feeds available in the market [16].

The search for alternative protein for poultry production, including quail production, requires a reasonable approach with a deep and comprehensive study of the effect of new feed ration components on the quality and safety of the resulting products in the form of meat and eggs.

There is already experience in the use of insects as an alternative protein in poultry farming. The results of these studies show positive dynamics on the live weight of birds, improvement of product quality [17].

At the same time, familiarisation with the available literature sources did not allow us to give a complete picture of the study of the experience of application of black soldier fly larvae in quail feeding, especially their influence on the structural and functional characteristics of organs and body systems.

In connection with the above, we note the high relevance of this problem and the need to present the morphofunctional state of some organs of quails.

2 Material and methods

The research was carried out on the basis of the vivarium and laboratory of experimental pathophysiology and physiology of the Federal State Budgetary Educational Institution 'Kazan State Academy of Veterinary Medicine named after N.E. Bauman'.

The objects of the study were laying quails, their organs. Larvae of the black soldier fly (Hermetia illucens) at the age of 20 days, grown on food waste in the insectarium of FGBOU VO Kazan GAVM, were introduced into the feed of laying quails during two months. The biomass of larvae was fed to laying quails in dried and natural form. Drying of larvae was carried out at a temperature of 80-82 0C. Three groups of birds (n=21) at the age of 120 days were formed to study the effect of black soldier fly larvae biomass on the organism of Japanese laying quails (n=21) at the age of 120 days, which were kept on complete mixed fodder in accordance with zootechnical norms. Laying quails of the I control group received the basic diet (OR); birds of the II experimental group received dried larvae in the amount of 3% to the dry matter of the basic diet (OR); laying quails of the III experimental group received 7% of larvae biomass to the dry matter of OR. Doses for laying quails were selected on the basis of literature data and previously conducted own studies on birds.

The duration of the experiment was 60 days. Postmortem examination of quails and internal organs were carried out according to the method of A.V. Zharov. The morphology of internal organs was determined during technological slaughter of birds visually by their shape, size, colour and topography using the anatomo-morphological method.

For histological studies, pieces of laying quail organs were fixed in 10% aqueous formalin solution with subsequent compaction on freezing microtome MZP-01 'TECHNOM' with cooling 'OMT-0228' (Russia). Histosections were stained with haematoxylin and eosin. Histopreparations were analysed using a Nexcope NIB 620-FL microscope and MBI-1 microscope under the magnification of the eyepiece ×7, ×10, ×15, objective ×10, ×20, ×40, ×64. Photographs of micro specimens were taken using a JENAMED 2 microscope, eyepiece GF-PW 10x25, objective 40.

3 Result and discussion

Visual inspection of carcasses and internal organs of laying quails of control and experimental groups revealed that morphology of their organs and tissues corresponded to species, age and physiological parameters of birds.

Microscopic examination of the glandular stomach of laying quails of the control group showed integrity of surface structures, moderate amount of mucus was detected on the mucous membrane. The structure of gastric fossae was preserved, mucus with single sloughed epithelium was noted in the lumen and cavities of the fundal glands. Full blood vessels were noted, integrity of mucous, muscular and serous membranes.

The microstructure of the intestinal mucosa retained its integrity and was formed by intestinal villi. An insignificant amount of mucus was noted on the surface. A pronounced system 'villi-crypta' was observed, the surface of which was lined with enterocytes. In the area of the base of villi and crypts there was an insignificant amount of bocaloid cells. The connective-tissue basis of the organ preserved its structure, blood vessels were full-blooded.

Histological examination of the liver of laying quails showed integrity of the organ structure, radial arrangement of hepatocytes in hepatic lobules was preserved. Single hepatocytes with the presence of small fat droplets in the cytoplasm were detected, the nuclei of these cells had a rounded shape and were located in the centre of the liver cell (Figure 1). Heterochromatin-rich elongated endotheliocyte cells were detected in the sinusoidal capillaries. Blood vessels of the organ were moderately full blooded.

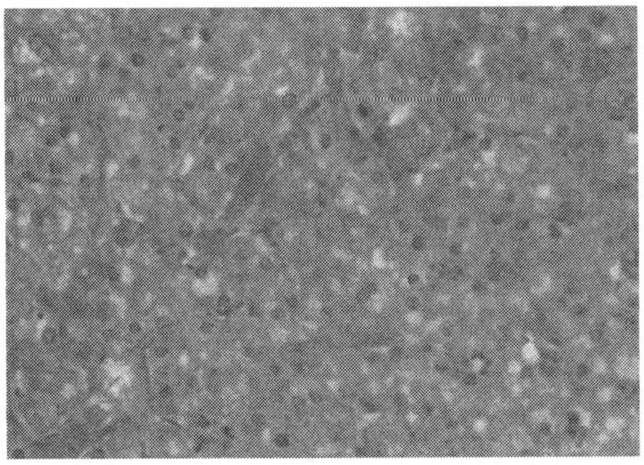

Figure 1. Full blood vessels, lipid vacuoles in the cytoplasm of periportal hepatocytes in the liver of laying quail of the control group. Haematoxylin and eosin staining x 400

Morphofunctional state of stomach and intestine of laying quails of the control group reflected physiological norm, microstructure of the liver corresponded to the beginning process of fatty degeneration of the organ.

At histological study of glandular stomach of laying quails of the II experimental group it was noted that the organ was lined with integral single-layer epithelium with the formation of gastric fossae in the thickness of the stomach. In the gastric fossae we visualised the output ducts of the fundal glands, the secretory sections of which were lined with glandular cells of cubic shape. In the lumen of the gastric fossa an abundant amount of mucus, sloughed epithelium and leukocytes was detected. Integrity of muscular and serous membranes was preserved, blood vessels were moderately full-blooded, mucous membrane and submucosa were characterised by moderate functional activity of gastric glands.

At microscopy of small intestine of laying quails of this group it was noted that the integrity of surface structures of the organ was preserved. The surface of the organ was formed by intestinal villi covered with single-layer single-row cylindrical caemic epithelium. Significant number of bocaloid cells and their activation were detected at the base of villi. The system 'villi-crypta' was preserved, abundance of mucus and sloughed epithelium on the surface was observed. The connective-tissue basis of the intestine was moderately infiltrated with lymphoid cells.

The microstructure of the liver was characterised by a well-defined lobular structure, radial arrangement of hepatocytes in the lobules was preserved. The central vein of the lobule was blood-filled. The accumulation of lymphocytes around large blood vessels and central veins was noted. Liver cell nuclei contained nuclei on the surface, small granules and fat vacuoles were detected in the cytoplasm. Single binuclear hepatocytes were observed (Figure 2).

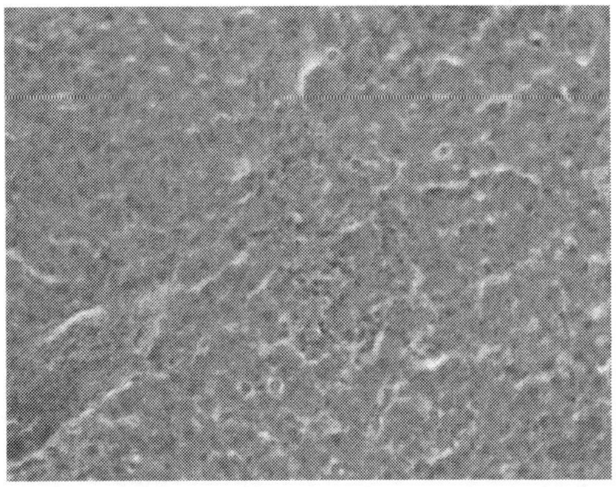

Figure 2. Well-defined lobular structure, radial arrangement of hepatocytes with small granules and fat vacuoles in their cytoplasm, accumulation of lymphocytes around large blood vessels and central veins in the liver of laying quail of the II experimental group fed dried larvae of the black soldier fly in the amount of 3.0% of the dry matter of feed. Haematoxylin and eosin staining x 400

Morphofunctional state of stomach and intestine of quails of II experimental group was characterised by functional activity of gastric and intestinal glands, which indicated the activation of physiological processes of digestion, liver microstructure corresponded to the beginning of physiological process of organ regeneration, which was indicated by the presence of binuclear hepatocytes with preservation of their morphology.

The histological study of the glandular stomach of laying quails of the III experimental group showed the integrity of the organ structures - mucous, muscular and serous membranes. Gastric fossae in the mucous membrane were covered with single-layer cylindrical epithelium, without deformation and disruption of integrity, the fundal glands were moderately activated. The output ducts of the fundal glands are open into the gastric fossae, mucus and sloughed epithelial cells were visualised in the central cavity (Figure 3). The connective-tissue basis of the organ is without structure disturbances, blood vessels are moderately full blooded.

The microstructure of the small intestine was characterised by integrity of surface structures, villi had an elongated leaf-shaped form with well-structured brush border, activation of bocaloid cells located at the base of villi was observed. A pronounced system 'villi-crypta' was observed, sloughed epithelial cells were observed on the surface of single villi. The surface of the mucosa was characterised by abundant mucus. Blood vessels of mucosa and submucosa were blood-filled.

The microstructure of the liver was characterised by radial arrangement of hepatocytes in the lobules of the organ. Hepatocytes contained rounded nuclei in which the nucleus was visible, the cytoplasm of cells was oxyphilic. In the periportal part of hepatic lobules the presence of binuclear hepatocytes was noted (Figure 4). Single hepatocytes contained fat vacuoles and protein granules in the cytoplasm. A central vein was observed in the centre of the hepatic lobule, the wall of which was lined with endothelial cells with swollen heterochromatin-rich nuclei. Sinusoid capillaries were dilated, filled with erythrocytes with oval nuclei.

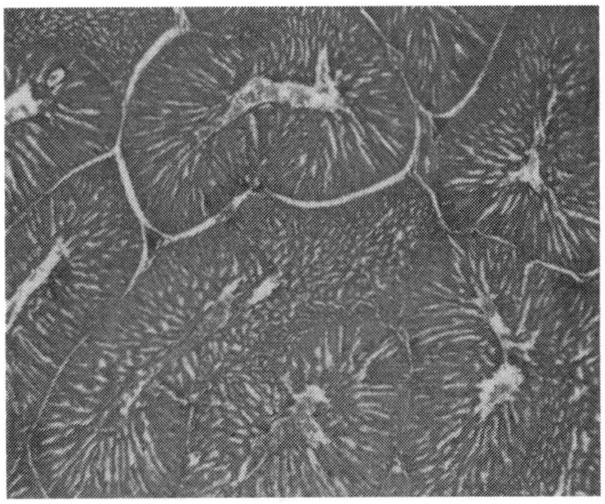

Figure 3. Moderate activation of glands in the intrinsic lamina of the gastric mucosa of laying quail of the III experimental group fed with native larvae of the black soldier fly in the amount of 7.0% of the dry matter of feed. Haematoxylin and eosin staining x 200

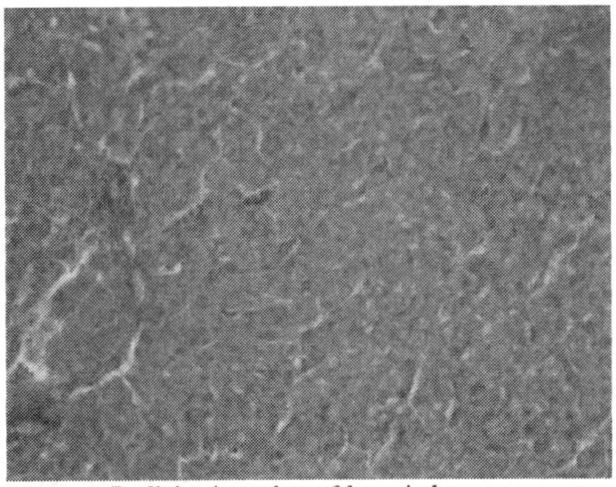

Figure 4. Radial orientation of hepatic beams, presence of dinuclear hepatocytes in the periportal part of hepatic lobules, full blood vessels in the liver of laying quail of the III experimental group fed with native larvae of the black soldier fly in the amount of 7.0% of the dry matter of feed. Haematoxylin and eosin staining x 400

Morphofunctional state of stomach and intestine of quails of III experimental group was characterised by functional activity of gastric and intestinal glands, which indicated activation of physiological processes of digestion. The picture of the liver was characterised by the process of physiological regeneration of the organ, which was indicated by the presence of dinuclear hepatocytes with preservation of their structure.

4 Conclusion

Thus, the study of the structural and functional state of the digestive organs of laying quails of the control group kept on OR (complete feed) showed that the micromorphology of the stomach, intestine and liver of birds corresponds to their age, species and physiological parameters.

In the glandular stomach of laying quails fed dried (II experimental) and native (III experimental) forms of black soldier fly larvae, the integrity of mucosal surface, activation of fundal glands, abundance of mucus and single sloughed epitheliocytes with the greatest degree of manifestation in laying quails fed dried forms of insectoculture were revealed. Intestinal cytology was characterised by activation of bocaloid cells, abundance of mucus on the surface of the organ, integrity of 'villi-crypta' structures, formation of brush border and presence of sloughed epithelial cells on the surface of villi, with the highest degree of manifestation in the organs of laying quails fed native forms of insectoculture. Liver micromorphology was characterised by preservation of triad and bar structure of the organ, integrity of organ and cell structures, presence of single hepatocytes with fatty inclusions, integrity and preservation of centrilobular hepatocyte structure and presence of bi-nuclear periportal hepatocytes.

Based on the obtained results of micro-morphology of stomach, intestine and liver of laying quails fed dried and native forms of insectoculture, we can conclude that the use of black soldier fly larvae did not disturb the structural and functional state of the contact organs of the digestive tract - stomach and intestine and non-contact parenchymatous organ - liver. The application of dried forms had the greatest effect on the functionality of the glandular stomach with the secretion of abundant mucus and gastric juice for digestion of the dried form of insectoculture. The abundance of mucus in the intestine, activation of bocaloid cells at the base of intestinal villi characterise the intensification of the processes of wall digestion. Passage of feed masses with insectoculture has insignificant abrasive effect, contributing to sloughing of the apical epithelium of villi, compensatory mechanisms of physiological regeneration of the organ are switched on, which contributes to the increase of natural resistance of the birds' organism.

Acknowledgement

The work was carried out on the funds of the state assignment of the Ministry of Agriculture of the Russian Federation on the theme: 'Protein feed additive based on black soldier fly larvae and development of technology of its application for farm birds and fur-bearing animals', reg. No. NIOCTR 122041500083-0.

References

1. N. Razumovsky, Belarusian Agriculture, **3,** 62-64 (2024)
2. A.Y. Zabolotskikh, Actual Issues of Agricultural Biology, **2(28),** 88-91 (2023)
3. G.Y. Rabinovich, E. A. Vasilieva, Agrarny nauchnyi zhurnal, **8,** 95-102 (2024)
4. M.A.R. Rzaev, Science, Technology and Education, **7(37),** 84-87 (2017).)
5. S. Shchukina, K. Gorst, Livestock Production of Russia, **3,** 25-26 (2018)
6. V.Y. Sitnov, M.A. Polubesova, E.V. Mechtaeva, Kormoproizvodstvo, 6, 36-44 (2023)
7. Z.Ma.M. Mondor, F. Goycoolea Valencia, AJ Hernández-Álvarez, Food Funct. **14(18),** 8129-8156 (2023)
8. V.A. Doroshenko, I.A. Khozyaev, D.A. Yakovlev, Engineering Technologies and Systems, **32(4),** 613-629 (2022)
9. M.A. Polubesova, E.V. Mechtaeva, A.D. Chernov, Food Systems, **7(3),** 336-344 (2024)
10. O. V. Afanaseva, O. K. Bezyukov, A. A. Ignatenko, Academic Journal of Manufacturing Engineering, **22(1),** 68–75, (2024). DOI: 10.24874/PES.SI.02.017
11. A. A. Basharov, E. M. Andriyanova, I. F. Yumaguzin, Genetics and Breeding of Animals, 2, 5-12 (2022)
12. P. V. Nekrasov, L. A. Pashkova, M. G. Chabaev, Zootechnia, **7,** 18-21 (2017)
13. Y. Bulatov, A. Kryukov, A. Batuhtin, K. Korotkova, D. Sidorov, Mathematics, **10(16),** 2886, (2022). DOI: 10.3390/math10162886
14. R.P. Guiné, P. Correia, C. Coelho, C.A. Costa, Open Agriculture, **6(1),** 24-36 (2021)
15. A.A. Lyashchev, N.A. Valov, E.V. Koval, AIC: Innovative Technologies, 4(59), 21-30 (2022)
16. E. Chala, V.V. Verkhoturov, E.V. Ulrich, Bulletin of the Ulyanovsk State Agricultural Academy, **3(63),** 155-160 (2023)
17. U. Elahi, J. Wang, Y. Ma, Animals, **10(2),** 224 (2020)

Biotechnological complex for processing of plant raw materials and agricultural wastes

Sergey N. Kharchenko [1], *Pavel V.* Gulyaev [2,*], *Anna N.* Tokareva [2], *Sergey V.* Panchenko [2], *Maria S.* Demchenko [2]

[1] Kuban State Agrarian University named after I. T. Trubilin, Krasnodar, Russia
[2] Azov-Black Sea Engineering Institute – branch of Don State Agrarian University in Zernograd, Zernograd, Russia

Abstract. The work is devoted to the application of biotechnological complex for the processing of plant raw materials and agricultural waste. The complex includes equipment for raw material preparation, acid and enzymatic hydrolysis. The research is focused on studying the kinetics of these processes and optimising process parameters to maximise the yield of valuable products. Experimental results show significant differences in the efficiency of acid and enzymatic hydrolysis. Acid hydrolysis, especially at high temperatures and pressures, shows high conversion of biomass to reducing sugars, but leads to the formation of by-products. In contrast, enzymatic hydrolysis provides more selective cleavage and fewer by-products, although it requires more expensive enzymes. Comparison with traditional wheat processing methods confirms the advantages of the proposed integrated technology, which provides a significant increase in the yield of target products such as glucose-fructose syrup and starch. The developed complex can be used for research and training of specialists in the field of biotechnology, contributing to the development of innovative solutions in the agro-industrial sector.

1 Introduction

There are several main directions in solving the problem of rational utilisation of plant raw materials and waste. One of them is the utilisation of renewable sources of raw materials, such as plant biomass, for the production of biofuels, bioplastics and other valuable products. This way, on the one hand, reduces dependence on fossil fuels and greenhouse gas emissions and, on the other hand, favours the development of the agricultural sector and its economic sustainability. However, it is worth noting that the production of first generation biofuels, for example, based on cereal crops, may lead to competition with food production and, consequently, to an increase in food prices. Another important area is the use of biotechnological methods to process agricultural waste. This involves the hydrolysis of cellulose and starch to simple sugars, which can then be used to produce various bioproducts, which in turn correlates with research on biomass processing using fly larvae [1]. One of the main advantages of this approach is the possibility to utilise a significant amount of waste such as straw, cake and other agricultural residues, reducing their negative impact on the environment. Nevertheless, the application of these methods is often associated with certain technical difficulties, such as the need to use expensive enzyme preparations or acids, as well as problems with the efficiency and scalability of the processes, which emphasises the importance of finding efficient and affordable ways of improving soil fertility and preserving organic matter in the soil [2]. In light of these problems and perspectives, the presented research aimed at developing a biotechnological complex for processing plant raw materials and agricultural waste is particularly relevant, especially given the importance of legal regulation in the field of organic agriculture [3]. It is such an integrated approach, focused on deep processing of biomass using modern biotechnology, is a promising way to create sustainable and environmentally friendly agro-industrial production. In addition, the development and implementation of such a complex will reduce dependence on imported technologies, as well as increase the competitiveness of the domestic biotechnology industry. Research in this area is becoming more and more relevant, as the volume of agricultural waste requiring utilisation is steadily increasing. The aim of the work is to study the possibility of using biological methods for adaptation to climate change [4], as well as the development and testing of a single laboratory complex, including equipment for the preparation of raw materials, hydrolysis and fermentolysis. This complex is designed to study the kinetic parameters of the processes of processing plant raw materials and wastes, which will optimise technological regimes and increase the efficiency of production of valuable bioproducts. The developed complex is a teaching and research laboratory, which is designed to train qualified personnel in the field of biotechnology and biochemistry. This is an important step in promoting research and development of

* Corresponding author: achgaa.el.mash@inbox.ru

innovative technologies in the field of plant raw material processing in the Russian Federation.

2 Method and materials

In this work, a set of experimental studies aimed at studying the processes of processing of plant raw materials and agricultural wastes in order to obtain valuable bioproducts was implemented. Experimental work was carried out on a specially designed laboratory complex, including a number of apparatuses of raw materials, which provided their necessary preparation for further processing. Several apparatuses were used to study the processes of hydrolysis. The first of them is a laboratory hydrolyser of low-temperature hydrolysis with a volume of 6 litres. In this apparatus acid hydrolysis processes were carried out at overpressures up to 0.35 MPa and temperatures up to 145 °C. The hydrolyser was heated by a ceramic-insulated spiral heater with automatic maintenance of the set temperature. Sampling in the process of hydrolysis was carried out through a mesh filter. The second apparatus is a hydrolyser for high-temperature hydrolysis, which allows to carry out the processes of chemical hydrolysis at temperatures from 100°C to 180°C and overpressure up to 1.2 MPa. To study the processes of enzymatic hydrolysis, a special unit was used, which allowed to control the temperature and provide stirring of the reaction mixture, which affects the quality of products [5]. The apparatus used in the process of enzymatic hydrolysis allowed to conduct experiments in optimal conditions for the action of enzyme preparations. The process parameters such as temperature, pressure, concentration of reagents and reaction products were monitored and measured at all stages of the research. In the course of experiments, the change in the concentration of reducing agents during the hydrolysis process was determined, as well as the parameters of the kinetics of enzymatic hydrolysis, which made it possible to evaluate the efficiency of different processing modes and find the optimal conditions.

3 Results and discussion

Within the framework of the conducted research, a set of experimental works aimed at studying the possibilities of effective processing of plant raw materials and agro-industrial wastes into valuable bioproducts was realised. The experimental work consisted of several stages, each of which was aimed at studying a certain aspect of the biomass processing process.

Initially, the preparation of plant raw materials was carried out, which included grinding and, if necessary, drying. This was followed by the hydrolysis process, which was carried out using both acid and enzymatic methods. In acid hydrolysis, two types of hydrolysers were used: low-temperature and high-temperature.

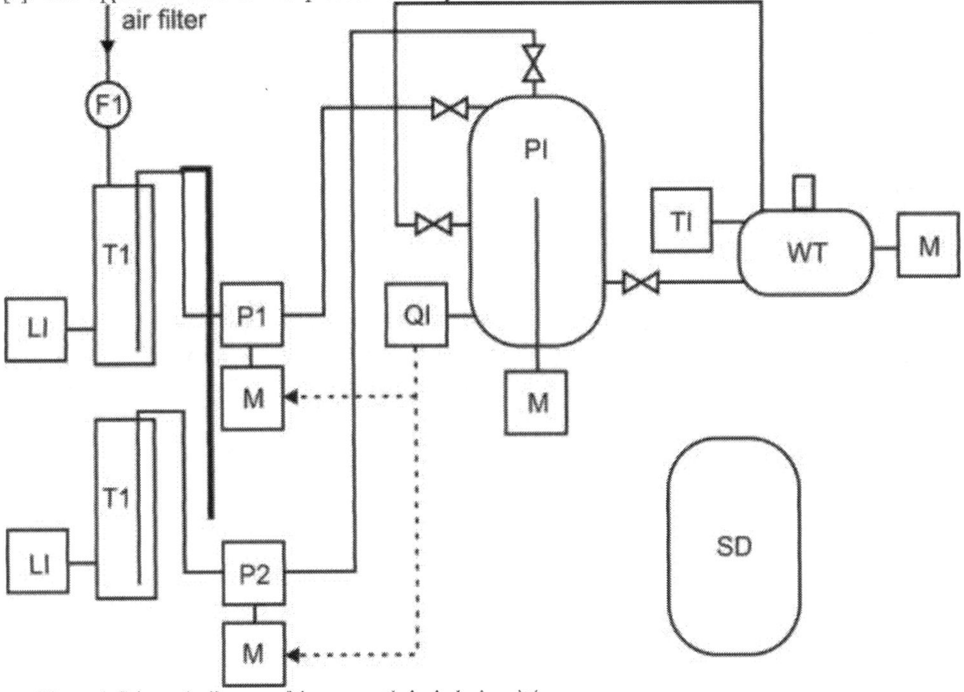

Figure 1. Schematic diagram of the enzymatic hydrolysis unit in use

 https://doi.org/10.1051/bioconf/202516100053

Laboratory unit of enzymatic hydrolysis used in the work (Figure 1), is designed to study and carry out processes of enzymatic hydrolysis of plant raw materials. The main element of the unit is a fermentolyser (P-1) - a reactor with a heat exchange jacket and a mechanical stirrer, designed to carry out the process of enzymatic hydrolysis. The unit is equipped with a stirrer drive (SD), which provides stirring of the reaction mixture and allows measuring the stirrer shaft rotation speed, as well as voltage and current in the electric motor circuit. To ensure sterility of the process, a sterilising air filter (F1) is available. To maintain the required pH level, the unit is provided with tanks for acid (E-1) and alkali (E-2) supply, as well as peristaltic pumps (H-1, H-2) for their dosed supply to the fermentolyser. A water ultrathermostat (T-1) is used to maintain the set temperature in the jacket of the fermentolyser. The process is also equipped with a temperature compensated pH sensor (QI) to monitor the acid-alkaline balance and visual level gauges (LI) to monitor the liquid level. A thermometer (TI) is also present to monitor the temperature of the reaction mixture. All these elements combine to provide the necessary control and maintenance of conditions for optimal performance of the enzymatic hydrolysis process. In the low temperature hydrolyser, operating at temperatures up to 145 °C and pressures up to 0.35 MPa, acid hydrolysis was carried out using organic acids such as citric acid. In a high-temperature hydrolyser capable of operating at temperatures from 100 °C to 180 °C and pressures up to 1.2 MPa, hydrolysis was carried out under more stringent conditions. The solutions obtained after hydrolysis, were hydrolysate containing reducing agents such as sugars. In enzymatic hydrolysis, specially selected enzyme preparations were used to ensure the cleavage of cellulose and hemicellulose to simple sugars. The process was carried out in a fermentolyser with controlled parameters such as temperature and pH. Process parameters were monitored at all stages of the experimental study and samples were taken to evaluate microbial growth and analyse the biochemical composition. During the experimental studies, detailed data were obtained to characterise the efficiency of different methods of processing plant raw materials, including acid and enzymatic hydrolysis, as well as their variations. In particular, acid hydrolysis, especially under high temperature and pressure conditions, was found to exhibit a significant ability to convert biomass to reducing agents, making it an attractive primary processing step. However, it should be noted that this method, despite its high conversion rate, is accompanied by the formation of by-products such as furfural and other organic compounds that may inhibit subsequent biotechnological processes and require additional purification steps, which increases the overall cost of the process. Thus, in our experiments, using high-temperature hydrolysis of straw at 170°C and 1 MPa pressure in the presence of sulphuric acid, an average of 380 grams of reducing agents per kilogram of dry feedstock was obtained. This corresponds to a cellulose and hemicellulose conversion of about 65-70%, which is quite high. At the same time, the content of glucose as the target sugar averaged 80% of the total content of reducing agents, which corresponds to 304 grams of glucose per kilogram of raw material. The remaining 20% was mainly other monosaccharides such as xylose and arabinose, as well as oligosaccharides and their degradation products. It should be noted that under such severe hydrolysis conditions, the formation of by-products such as 5-hydroxymethylfurfural (5-HMF) was also observed, the amount of which, in some cases, reached 2-3% of the total content of reducing agents, which confirms the need for additional purification measures. In addition, the analysis of energy costs showed that high-temperature hydrolysis is a more energy-consuming process, requiring a significant amount of heat to maintain the necessary temperature and pressure, which should also be taken into account when assessing the economic feasibility of the method. In contrast, enzymatic hydrolysis demonstrated the ability to selectively convert cellulose and hemicellulose to sugars under more moderate conditions, which is its undoubted advantage. Thus, using a complex of cellulases and hemicellulases, a high degree of conversion was achieved at relatively low temperatures in the range of 50-60°C and atmospheric pressure. According to my data, enzymatic hydrolysis of wheat waste using Trichoderma reesei cellulase for 24 hours at 55°C produced an enzyme-lysate with a reducing agent content varying from 180 to 220 grams per kilogram of dry raw material, depending on the enzyme batch and pretreatment. The glucose content of the fermentolysate, meanwhile, reached an average of 90% of the total reducing agents content, which ranges from 162 to 198 grams of glucose per kilogram of dry raw material. It is important to note that enzymatic hydrolysis does not produce as many by-products as acid hydrolysis, which simplifies the subsequent processing steps. However, it should be taken into account that enzyme preparations are generally more expensive than acids used in chemical hydrolysis, which should also be taken into account in economic analyses. In addition, enzymatic hydrolysis requires more precise control of process conditions, such as pH and temperature, to ensure optimal enzyme activity. It has been shown that by using precision farming techniques and taking into account various factors, yields can be significantly improved [6-8]. Analysing the data obtained, significant differences in product yields can be seen between conventional and alternative technology for processing 1 million tonnes of wheat. Thus, the traditional technology, focused on the production of ethanol and native starch, shows a protein yield of 68.5 thousand tonnes, ethanol yield of 270 thousand tonnes, and no GFS. The alternative technology, focused on complex processing, shows a protein yield similar to the traditional one, 68.5 thousand tonnes, while the ethanol yield is significantly reduced to 65.5 thousand tonnes, and the HFC yield is 354.3 thousand tonnes, which is significantly higher than the traditional processing method. In addition, the yield of native starch in the alternative technology is 161.5 thousand tonnes, which is many times more than in the traditional approach, where there is none. Thus, alternative technology, using modern biotechnological approaches, allows not only to increase the yield of valuable products, but also to significantly reduce losses and increase the economic efficiency of processing of plant raw materials. At the same time, it is important to take into account the role of microorganisms for maintaining soil fertility and waste processing [9, 10]. According to the obtained data, the comparison of traditional and alternative technology of wheat processing shows that the alternative method, including deep processing of grain, allows to significantly increase the yield of target products, such as glucose-fructose syrup (GFS) and starch, as well as protein products. For example, in the traditional processing of 1 million tonnes of wheat, the yield of ethanol is 270,000 tonnes, and there is no GFS, while the alternative technology produces 65,500 tonnes of ethanol and 354,200 tonnes of GFS. In addition, the alternative method allows to increase the yield of native starch by 16 times, as well as, to increase the yield of protein. These data confirm that the use of modern biotechnological approaches based on deep processing of plant raw materials is a promising direction for increasing the efficiency of biomass utilisation and reducing the amount of waste. At the same time, it is important to take into account the need to reduce the environmental impact, including by reducing greenhouse gas emissions [11, 12]. The analysis of the obtained results shows that the laboratory complex used in this work allows us to study various parameters of the process of processing of plant raw materials, as well as to select the optimal technological modes for obtaining the target products. The data obtained during the experimental work correlate with the results of other similar studies conducted both in Russia and abroad. For example, studies on straw processing to produce bioethanol have shown that the use of acid hydrolysis allows obtaining a high yield of sugars, while enzymatic hydrolysis provides a more selective breakdown of cellulose and hemicellulose. At the same time, a number of studies aimed at obtaining pectin from beet pulp have shown that the most optimal is the use of acid hydrolysis with the use of citric acid. In general, a comparative analysis of the results obtained and the data of other researchers indicates the need for an integrated approach to the processing of plant raw materials, based on the use of various methods of hydrolysis and fermentolysis, to obtain the widest possible range of target products and reduce the volume of waste. In addition, the economic feasibility of each method should be considered individually, depending on the specific type of raw material and the desired end product. At the same time, it is important to carry out analyses taking into account geo-ecological factors [13, 14] and using GIS-technologies [15].

4 Conclusion

The study demonstrates the effectiveness of the developed laboratory complex for studying the processes of processing of plant raw materials and waste. The complex, which includes equipment for preparation of raw materials and carrying out both acid and enzymatic hydrolysis, allows us to effectively study the kinetics of these processes and optimise the technological parameters to achieve maximum yield of target products. Experimental results showed significant differences in the efficiency of acid and enzymatic hydrolysis. Acid hydrolysis, especially at high temperatures and pressures, provides a high degree of biomass conversion into reducing sugars, but is accompanied by the formation of by-products requiring additional purification. Enzymatic hydrolysis, on the other hand, is characterised by high selectivity and less formation of by-products, although it requires the use of more expensive enzyme preparations. Comparison with traditional methods of wheat processing confirms the advantages of the integrated approach, which provides a significant increase in the yield of valuable products such as glucose-fructose syrup and starch. The developed complex is a valuable tool both for scientific research and for training of specialists in the field of biotechnology of plant raw material processing, contributing to the development of innovative technologies in the agro-industrial complex of Russia. The results obtained confirm the relevance and importance of further research in this area for the creation of sustainable and environmentally friendly agro-industrial production.

References

1. Y. Chen, Y. Guo, X. He, B. Tan, Z. Liao, A. Chen, X. Gu, X. Li, X. Chen, B. Chen, S. Lin, W. Li, P. Hu, X. Zhu, W. Zhao, J. Niu, Comprehensive utilization of black soldier fly (Hermetia illucens) larvae: extraction, recovery and characterization of peptide, chitin and melanin and scaling-up trial. Sep. Purif. Technol., **361,** 131262 (2025). DOI: 10.1016/j.seppur.2024.131262.

2. Q. Gao, L. Wang, Y. Fang, Y. Gao, L. Ma, X. Wang, Y. Li, X. Wu, Z. Du, Conservation agriculture boosts topsoil organic matter by restoring free lipids and lignin phenols biomarkers in distinct fractions. Soil Till. Res., **248,** 106463 (2025). DOI: 10.1016/j.still.2025.106463

3. L. Yerkinbayeva, A. Myrzakarim, A. Moldagaliyeva, B. Kalymbek, R. Yerezhepkyzy, D. Nurmukhankyzy, the role of legal regulation in promoting organic agriculture to achieve sustainable food security. J. Lifestyle SDG'S Rev., **5,** e04197 (2025). DOI: 10.47172/2965-730X.SDGsReview.v5.n01.pe04197

4. T. Bandara, B. Marambe, G. Pushpakumara, P. Silva, R. Punyawardena, L. Manawadu, S. Premalal, G. Miah, K.R. Dahal, Adapting mountain-farming systems to climate change by reducing food-nutrition-health vulnerability. Mitig. Adapt. Strateg. Glob. Change, **29,** 88 (2024). DOI: 10.1007/s11027-024-10181-x

5. A.A. Jauhari, S. Minarsih, Y. Hindarwati, J. Pramono, A. Susila, S. Sudarto, S. Basuki, W. Hariyanto, B. Utomo, T. Suhendrata, R. Oelviani, F.D. Arianti, A. Supriyo, V.E. Aristya, S. Samijan, Rice yield enhancement and environmental sustainability with

precision nutrient management. Glob. J. Environ. Sci. Manag., **11,** 77-92 (2025). DOI: 10.22034/gjesm.2025.01.05

6. A.A. Jauhari, S. Minarsih, Y. Hindarwati, J. Pramono, A. Susila, S. Sudarto, S. Basuki, W. Hariyanto, B. Utomo, T. Suhendrata, R. Oelviani, F.D. Arianti, A. Supriyo, V.E. Aristya, S. Samijan, Rice yield enhancement and environmental sustainability with precision nutrient management. Glob. J. Environ. Sci. Manag., **11,** 77-92 (2025). DOI: 10.22034/gjesm.2025.01.05.

7. J. Qi, P. Cheng, J. Zhou, M. Zhang, Q. Gao, P. He, L. Li, F.C. Muga, L. Guo, A Novel Model for Soil Organic Matter and Total Nitrogen Detection Based on Visible/Shortwave Near-Infrared Spectroscopy, Land, **14 (2),** 329 (2025). DOI: 10.3390/land14020329

8. Y. Zhang, E. Wang, B. Feng, L. Xu, Y. Xue, Y. Chen, Siderophore Production Capability of Nitrogen-Fixing Bacterium (NFB) GXGL-4A Regulates Cucumber Rhizosphere Soil Microecology, Microorganisms, **13 (2),** 346 (2025). DOI: 10.3390/microorganisms13020346

9. M.-Y. Li, W. Wang, H.-H. Yin, Y. Chen, M. Ashraf, H.-Y. Tao, S.-S. Li, W.-Y. Wang, C.-L. Yang, Y.-L. Xiao, L. Zhu, Y.-C. Xiong, The functional role of arbuscular mycorrhizal fungi in enhancing soil organic carbon stocks and stability in dryland. Soil Till. Res., **248,** 106443 (2025). DOI: 10.1016/j.still.2024.106443

10. X. Li, X. Cheng, Y. Zhao, B. Xiang, T. Zhang, Deep Learning-Based Ground-Penetrating Radar Inversion for Tree Roots in Heterogeneous Soil, Sensors, **25 (3),** 947 (2025). DOI: 10.3390/s25030947

11. A. A. Kulikova, E. I. Khabarova, Biological working agents use to reduce the carbon footprint, Sustainable Development of Mountain Territories, **16(1),** 311–321 (2024). DOI: 10.21177/1998-4502-2024-16-1-311-321

12. K.Y. Lobkov, A.A. Stupina On the issue of monitoring crack formation in natural-technical systems during earth surface displacements. MIAB. Mining Inf. Anal. Bull., **11-1,** 292-304 (2023) DOI: 10.25018/0236_1493_2023_111_0_292

13. V. L. Tatarintsev, L. M. Tatarintsev, S. K. Makenova, M. M. Shostak, geoecological assessment of landscapes as a basis for organization of sustainable agrarian land use, Sustainable Development of Mountain Territories, **13 (4),** 485–496 (2021). DOI: 10.21177/1998-4502-2021-13-4-485-496

14. Z. Xie, X. Wang, X. Xie, D. Yang, Z. Zhou, Q. Wang, A. Liu, X. Tu, Complex Microbial Fertilizer Promotes the Growth of Summer-Sown Short-Season-Cultivated Cotton and Increases Cotton Yield in the Yangtze River Basin by Changing the Soil Microbial Community Structure, Agronomy, **15 (2),** 404 (2025). DOI: 10.3390/agronomy15020404

15. L. V. Lebedeva, G. D. Syzdykova, D. S. Inkarov, Management of agricultural land use sustainability at the local level using GIS technologies and special map material, Sustainable Development of Mountain Territories, **15 (4),** 864–876 (2023). DOI: 10.21177/1998-4502-2023-15-4-864-876.

Results of experimental studies of sunflower oil purification by means of electric field in an electrostatic precipitator

Maxim Mikhailovich Ukraintsev* [1], *Igor Eduardovich* Lipkovich[1], *Pavel Vladimirovich* Gulyaev[1], *Pavel Timofeyevich* Korchagin[1], *Sergey Mikhailovich* Pyatikopov[1] and *Igor Viktorovich* Yudaev[2]

[1]Azov-Black Sea Engineering Institute - branch of Don State Agrarian University in Zernograd, Rostov region, Zernograd, Russia
[2] I.T. Trubilin Kuban State Agrarian University, Krasnodar Region, Krasnodar, Russia

Abstract: Sunflower oil is an indispensable product of functional purpose, which at its reasonable consumption increases immunity, cleanses the body of toxins, improves the brain and heart, and as a result, having a complex effect on the whole body provides an undeniable benefit to human health. Modern development of sunflower oil purification technologies involves improvement of existing and development of new methods and technical means, environmentally safe, accompanied by minimum energy consumption and characterised by increased indicators of technological efficiency, providing the possibility of separation of unnecessary impurities and preservation of valuable components extracted from the oil for their further use. Such a promising method is the purification of sunflower oil in the electric field. The paper considers the possibility of cleaning sunflower oil by an electrostatic precipitator from suspended particles by means of an electric field acting on them. The planned experiment is described and the process of removal of unnecessary impurities is analysed. The justification of the factors and parameters of the experimental study is considered separately. The dependences of the electrical characteristics of oil on the parameters of the purification process are established. Analytical expressions for empirical dependences of energy parameters of the purification process, mode and parameters of ESP operation are determined. The adequacy of theoretical and experimental results has been checked. Having analysed the obtained results, the optimum parameters of effective cleaning and ESP operation were determined, which on the basis of the obtained regression equations should be the following: voltage at ESP electrodes U = 4.4 kV, interelectrode distance b = 0.01 m, oil temperature at the inletθ = 42°C. The calculated energy consumption at these parameters is equal to 12 W-h/kg.

Keywords: electrostatic precipitator, sunflower oil purification, electrophoresis, electric field, sunflower oil

1. Introduction

Sunflower oil occupies one of the leading places in the diet. It contains five times more polyunsaturated fatty acids than olive oil, which is extremely important for the heart and cardiovascular system as a whole, prevents the onset of such a disease as atherosclerosis, improves blood circulation, regulates cholesterol levels. In this regard, it is subject to certain requirements that contribute to the preservation of human health [1, 2].

Sunflower oil production includes the following operations: incoming seeds are cleaned of debris, dried, husks are separated from the seeds, the grains themselves are crushed to form mint, oil from which is produced by two methods: pressing and extraction.

The hot-pressing technology involves preheating the mint to 100°C with the addition of moisture to avoid burning, and then pressing out the oil on presses. The cold-pressing method excludes preheating of raw materials, so the product manages to preserve the maximum amount of vitamins. Sunflower oil extraction involves the use of special solvents that are safe for health and the environment. The process is carried out in special containers, in which the seed meal - the residue after processing and miscella - oil solution are obtained from the seeds. The oil is "extracted" from the meal and miscella, which undergoes sedimentation, filtration and further processing.

Sunflower oil is also refined, which involves some cycle of sequential operations: 1) removal of foreign impurities - the product is sedimented, filtered and centrifuged; 2) treatment with a certain amount of hot water; 3) removal of excessive amounts of fatty compounds; 4) removal of pigments and volatile compounds from the oil; 5) freezing process with removal of waxes. Refined oil is ideal for heat treatment, while unrefined oil is treated minimally, usually by filtering to remove heavy impurities. Unrefined oil retains all the beneficial properties of the product and can be consumed raw [3].

* Corresponding author: rostmax@rambler.ru

Deviation from GOST requirements for sunflower oil production and storage can lead to irreparable consequences - oil accumulates harmful components, so special attention should be paid to the performance of these operations accordance with the regulations. It should be remembered that when obtaining oil from seeds by pressing, in addition to the main group of trialglycerols, structural lipids, which determine the colour, taste and smell characteristic of sunflower oil, as well as solid particles of pulp, water, waxes, paraffins, polycyclic aromatic hydrocarbons, purification of herbicides and pesticides, products of microflora [4,5,6].

Today, more and more attention is being paid to the use of environmentally friendly and food-safe technologies, paying special attention to the use of electrophysical methods and techniques that allow both the cultivation of raw materials for the oil and fat industry [7,8,9] and the purification of edible oils, purification and refining [10,11,12].

The paper presents an analysis and discussion of a study on sunflower oil purification using electric field in an electrostatic precipitator.

2. Objects and methods

The process whose ultimate goal is the separation of trialglycerols, free and other lipid groups and impurities from the resulting oil is called oil purification or refining. In small-scale agricultural enterprises, the sunflower oil obtained after pressing is subjected to a refining process that can be depicted as follows (Figure 1).

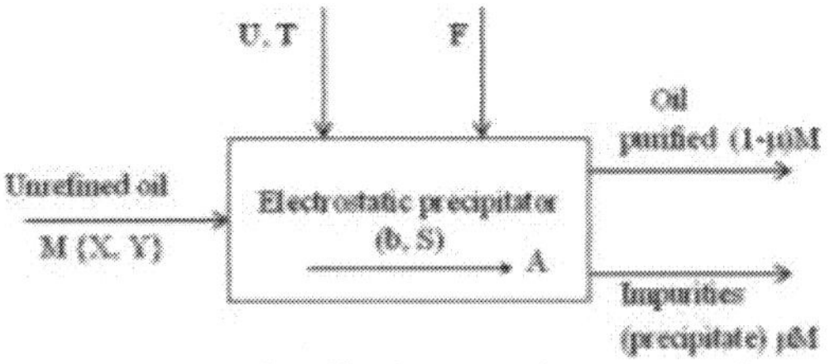

Fig. 1. Sunflower oil purification system in an electrostatic filter: X - uncontrolled oil parameters; Y - controlled oil parameters; U - voltage on the electrodes; T- is the time the oil is in the electric field; b, S - geometric dimensions of the electrostatic filter; M - mass of crude oil; μ - mass fraction of impurities in oil; F - disturbing random factors

The crude sunflower oil coming out of the press is characterised by some contamination with mechanical impurities and components of the oil itself, which in the process of cleaning should be removed or their concentration should be reduced to the minimum allowed by the quality standard. Oil purification is supposed to be carried out in an electrostatic precipitator. The parameters of the purification process are voltage and purification time, and the parameters of the electrostatic precipitator are its geometrical dimensions, including the interelectrode distance and the number of electrodes. The following parameters of energy consumption should be considered as determining parameters: 1) voltage applied to the electrodes; 2) electrophoresis time to complete oil purification; 3) interelectrode distance.

At the same time, it is impossible to assume a priori the value of any of the introduced parameters. This is due to their mutual influence, which has not yet been definitely established [10,11].

Thus, with an increase in voltage, a decrease in the time of the purification process can occur. However, taking into account that in the electrophoresis process some time is spent on creation of dipoles and aggregation of particles into larger structures, a proportional decrease in the duration of the purification process can hardly be expected.

Increasing the inter-electrode distance, on the one hand, will increase the mass of oil to be cleaned, but on the other hand, it will increase the time of movement of aggregated particles, that is, it may increase the cleaning time. Reducing the inter-electrode distance, on the other hand, will require a reduction in voltage to prevent electrical breakdown.

Thus, the task of searching for the most acceptable cleaning and electrostatic precipitator parameters from the point of view of the set goal is confirmed.

The purpose of the experimental study of sunflower oil purification process is:

- obtaining static data for realisation of theoretical provisions;
- verification of theoretical results;
- refinement of optimal parameters of the purification process.

The following **tasks** were carried out during the experimental research:

- justification of factors and parameters of experimental studies;
- determination of the number of experiments and their repetitions, and justification of the reliability of the obtained experimental data;
- establishing dependences of the electrical characteristics of oil on the parameters and duration of the cleaning process: $R = f(U, b,, T)$; $\varepsilon = f(U, b,, T)$;
- establishment of empirical dependences of energy parameters of the process on ESP parameters;
- adequacy of theoretical and experimental results.

The methods of probability theory and mathematical statistics are used to process experimental data. Practically mass calculations were carried out using the standard Microsoft Excel software package.

The criteria of the cleaning process and ESP are determined by the following parameters: R - resistance to direct current; ε - dielectric constant of oil; η - viscosity of oil. These parameters in turn depend on the physical characteristics of the oil - its temperature and moisture content.

Since ultimately in the electrophoresis process the aggregated impurity particles are deposited on the electrode and then run off from it, the optimal range of variation of physical characteristics will be the mode of deposition. The mode of deposition of impurities from sunflower oil in the electric field is determined by patterns characteristic of natural settling. As a rule, the products determined in the purification process are deposited on the positive electrode at a temperature of 30-60°C.

When dehydrating sunflower oil, it must be taken into account that the solubility of water increases with increasing temperature. This requires increased energy consumption for moisture removal. When selecting the temperature mode of electrodeposition, the product properties, the desired degree of purification and the conditions of the electrophoresis process should be taken into account. The oil in the tanks installed in oil shops has a temperature of 18-22°C, and leaving the press has a temperature of 80-100°C [3,4]. It is this range that it is reasonable to adopt when planning the experiment.

As follows from the theory of the electrophoresis process, one of the most important factors in the purification of sunflower oil using an electric field is its strength [12,13]. The field strength between the electrodes is determined by the voltage applied to them, the distance and their shape. In the simplest case, for plane-parallel electrodes, the field strength is constant in the entire volume between the electrodes.

For rod electrodes, the maximum field strength is observed on the surface of the rods, while the minimum field strength is observed in the plane equidistant from the electrodes opposite in sign. For chamber electrodes, the maximum field strength will be on the outer surface of the inner electrode, the minimum - on the inner surface of the outer electrode [14].

The increase in ESP performance with decreasing electrode spacing is attributed to the following factors:

- The time required for the movement of the separated particles to the electrodes is reduced;
- sizes of separated particles remain the same, the conditions of particle aggregation under the influence of polarisation are improved;
- the hydraulic regime of the purified product movement is improved as a result of reducing its circulation in the apparatus.

Based on the above and preliminary studies, the limits of variation of the factors were determined.

So it was experimentally established that at voltage less than 3 kV, oil purification practically does not occur, and at voltage higher than 10 kV there is intensive oil movement, which leads to "washing" of fuse from the electrodes. Based on the conditions of electrical breakdown and ESP design, the distance between the electrodes should be at least 0.005 m

In order to realise the theory of electrophoresis and analytical determination of the parameters of the purification process and the electrostatic precipitator, it is necessary to have data on the electrical parameters of sunflower oil (dielectric permittivity, DC and AC resistance) [13,15,16,17].

In this connection it is necessary to carry out preliminary experiments, as a result of which the following dependences are obtained: $R = f(\theta, U)$; $z = f(\theta, U)$; $\varepsilon = f(\theta, U)$; $\eta = f(\theta)$ - where R - resistance of oil to direct current, Ohm; z - resistance of oil to alternating current, Ohm; θ - temperature of oil,° C; U - voltage between electrodes, V; ε - dielectric constant; η - viscosity of oil.

The resistance of oil to direct (alternating) current was determined by Ohm's law, with voltage applied to the electrodes of the ESP. Since the resistance of oil to alternating current is active-capacitive, the capacitive resistance was calculated by the formula:

$$X_C = \sqrt{z^2 - R^2} \, . \tag{1}$$

Having determined the capacitance, the capacitance and then the dielectric constant were calculated using the formulas:

$$C = \frac{1}{2 \cdot \pi \cdot f \cdot X_c} \, , \tag{2}$$

where f is the frequency of the supply voltage, f=*50 Hz*.

$$\varepsilon = \frac{C \cdot b}{\varepsilon_0 \cdot S} \, . \tag{3}$$

where ε_0 is the electrical constant.

The electrical parameters of the oil were determined for the above temperature range and electric field strength. In addition, oil with moisture content of 0.15% and 1.2% was studied. Seven repetitions of experiments were carried out in the experiment.

To determine the reliability of the obtained experimental data, a series of experiments were checked for homogeneity. That is, the condition that deviations of oil resistance values to direct and alternating current and dielectric permittivity from the mean value are random and not significant was checked. For this purpose, the corrected dispersions in each series of experiments with the same parameters were calculated and the Cochran criterion was calculated using the formulas:

$$D_j = \frac{\Sigma_j \left(x_{ji} - \bar{x}_j\right)^2}{n-1} \, , \tag{4}$$

$$G = \frac{D_{MAX}}{\Sigma_j D_j} \, , \tag{5}$$

where D_j is the corrected variance in the j-th series of experiments; x_{ji} is the value of the studied variable in the i-th experiment of *the j-th* series; $\bar{x}_j$ is the mathematical expectation of the variable in the *j-th* series of experiments; D_{MAX} is the maximum value of the corrected variance in n experiments; n is the number of experiments.

For the convenience of using experimentally obtained characteristics of sunflower oil, they are described by polynomials of the form:

$$Y = a_n \cdot x^n + a_{n-1} \cdot x^{n-1} + \ldots + a_0 \, . \tag{6}$$

where Y - sunflower oil parameter; a_j - polynomial coefficients; x_j - experiment factor.

The "Trend Line" function of the standard Microsoft Excel package is used to calculate the coefficients of the polynomial.

The coefficients of the linear regression equation were calculated to account for the complex impact of factors:

$$R = a_0 + a_1 \cdot U + a_2 \cdot \theta + a_3 \cdot \eta + a_4 \cdot \theta^2 + a_5 \cdot \eta^2 + a_6 \cdot \theta \cdot \eta . \quad (7)$$

To calculate the regression coefficients we used the "Linear Regression" function of the standard Microsoft Excel software package. Figure 2 shows in what form the required function is depicted on the Microsoft Excel sheet.

	A	B	C	D	E	F
1	a_n	a_{n-1}	...	a_2	a_1	a_0
2	se_n	se_{n-1}	...	se_2	se_1	se_0
3	R^2		...			
4	F		...			
5	se_{reg}	se_{resid}	...			

Fig. 2. View of the "Linear Regression" display on a Microsoft Excel sheet

In this case, the dependence between the desired parameter and the experimental factors is determined by the determinacy coefficient R^2 (cell A3).

3. Results and discussion

To assess the significance of the found dependence (to check whether this dependence is regular or random), the Fisher-Snedecor criterion is applied, the observed value of which is placed in cell A4 (Figure 2). The observed value should not be less than the tabulated value, chosen according to the significance level, which is$\alpha = 0.5$, and degrees of freedom $v_1 = 5$ and $v_2 = 3$. The degrees of freedom are defined as follows: $v_1 = k, v_2 = m - (k-1)$, where m *is* the number of repetitions;k is the number of factors in the regression equation.

To assess the significance of the obtained regression coefficients, the observed value of Student's criterion is determined and compared with the tabulated value for the significance level$\alpha = 0.5$ for v_3 degrees of freedom, where $v_3 = m - 1$. The observed value of Student's criterion is calculated by the formula:

$$t_{CTj} = \frac{\alpha_j}{se_j}, \quad (8)$$

where a_j is the j-th coefficient of the linear regression equation; se_j is the error of the j-th coefficient of the regression equation (cell A2 - F2).

In accordance with the outlined methodology, experimental studies were carried out and the following dependences were obtained, shown in Figures 3-6.

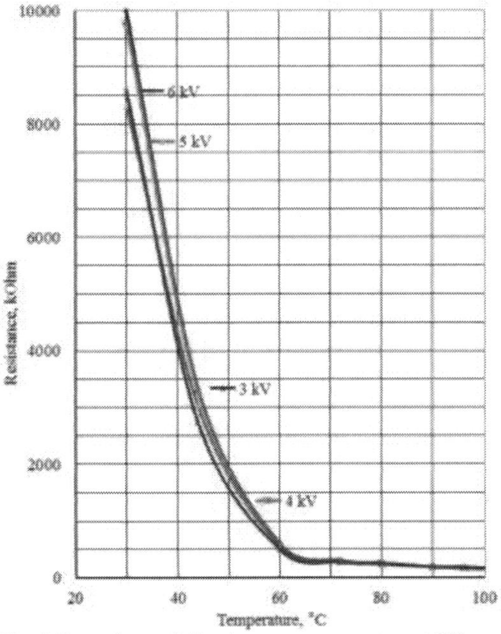

Fig. 3. Dependence of oil resistance on temperature at different electric field gradients

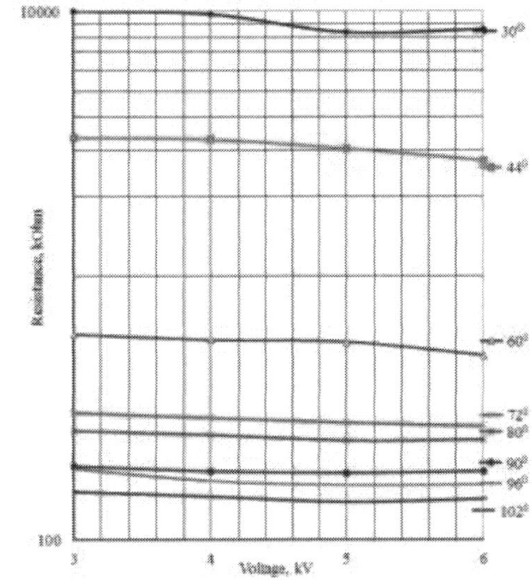

Fig. 4. Dependence of oil resistance on tension at different temperatures

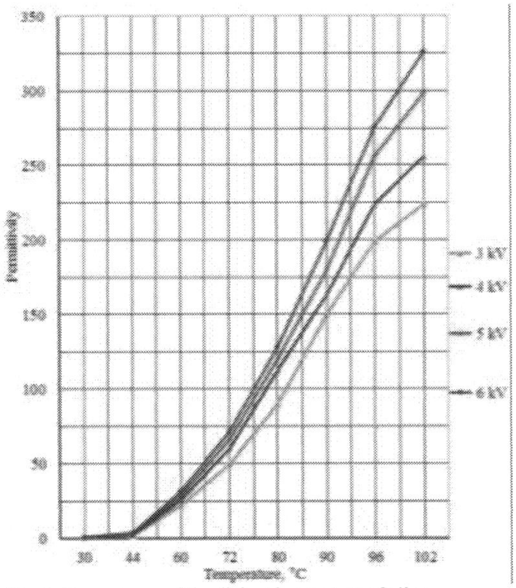

Fig. 5. Dependence of the dielectric constant of oil on temperature at different electric field gradients

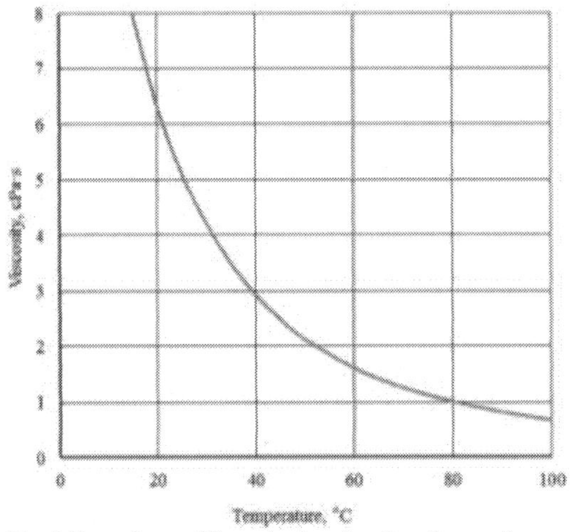

Fig. 6. Dependence of dynamic viscosity of sunflower oil temperature

It can be seen from the obtained graphs that the greatest influence on the oil parameters is its humidity and temperature. The voltage between the electrodes practically does not affect the electrical characteristics of the oil, which, in fact, confirms the theoretical provisions of electrical engineering.

As a result of processing of the obtained experimental dependences, they were approximated by the following polynomials

$$R = 0{,}0009 \cdot \theta^4 - 0{,}3175 \cdot \theta^3 + 0{,}03662 \cdot \theta^2 - 2{,}0733 \cdot \theta + 48 \,, \quad (9)$$

$$z = -0{,}0001 \cdot \theta^4 + 0{,}0214 \cdot \theta^3 - 0{,}3912 \cdot \theta^2 - 93{,}328 \cdot \theta + 4254{,}7 \,, \quad (10)$$

$$\varepsilon = -0{,}4064 \cdot \theta^4 + 5{,}7546 \cdot \theta^3 - 19{,}033 \cdot \theta^2 + 31{,}925 \cdot \theta - 18{,}863 \,. \quad (11)$$

As a result of processing the experimental data, a linear regression equation describing the DC resistance of oil depending on the influencing factors was obtained:

$$R = 32{,}71998 - 1{,}32089 \cdot \theta + 41{,}939865 \cdot \eta + 0{,}11472 \cdot \theta^2 - \\ - 44{,}634202 \cdot \eta^2 + 0{,}321065 \cdot \theta \cdot \eta \,, \quad (12)$$

Statistical analysis showed the following results:

- determinacy coefficient $R^2 = 0.981$;
- checking by Fisher's criterion confirmed the reliability of this value of the coefficient of determinacy (at the significance levelα = 0.5) - $F_{(NABL)} = 143.078 > F_{(TA)BL} = 143.078$;
- Student's test confirmed that the coefficients of the regression equation are significant (at the significance levelα = 0.5), the observed value was $t_{ST.NABL} = 3.67$, which is greater than the tabulated value - $t_{ST(.TABL)} = 2.45$.

Thus, the obtained regression equation can be used to calculate the parameters of the cleaning process and ESP. After substituting the obtained dependence into the target function, the following calculated ones were obtained:

- inlet oil temperature - 40° C;
- oil moisture content before cleaning - 1.0%;
- the interelectrode distance is 10 mm;
- voltage - 3 kV;
- cleaning capacity - 76 kg/hour.

The data obtained experimentally were processed using a package of standard programmes Microsoft Excel to obtain regression equations of the form:

$$I = 0{,}0457 - 0{,}02424 \cdot U - 0{,}1858 \cdot b + 0{,}00387 \cdot \theta - \\ - 0{,}023 \cdot U \cdot b + 0{,}000411 \cdot U \cdot \theta - \\ - 0{,}00609 \cdot b \cdot \theta + 0{,}003632 \cdot U^2 + 0{,}21856 \cdot b^2 + 2{,}48 \cdot 10^{-5} \cdot \theta^2 \,, \quad (13)$$

$$T = -14{,}03 - 2{,}264 \cdot U + 112{,}528 \cdot b - \\ - 0{,}183 \cdot \theta - 10{,}598 \cdot U \cdot b - 0{,}0287 \cdot U \cdot \theta + \\ + 0{,}137 \cdot b \cdot \theta - 0{,}744 \cdot U^2 - 20{,}239 \cdot b^2 + 0{,}00051 \cdot \theta^2 \,, \quad (14)$$

$$p = 0{,}00465 - 0{,}03032 \cdot U + 0{,}079788 \cdot b + 0{,}000746 \cdot \theta - \\ - 0{,}01477 \cdot U \cdot b + 0{,}00024 \cdot U \cdot \theta - \\ - 0{,}00166 \cdot b \cdot \theta + 0{,}00361 \cdot U^2 + 0{,}02052 \cdot b^2 + 4{,}37 \cdot 10^{-6} \cdot \theta^2 \,. \quad (15)$$

where I - current, A; T - cleaning time (time of oil staying in the electric field), hour; p - energy intensity of the cleaning process kWh/kg.

Statistical analysis of the obtained equations showed that all coefficients are significant with a significance level not worse than 0.01. The convergence of the experimental data and the data obtained in the form of regression equations was tested by the Cochran criterion. For this purpose, the null hypothesis of homogeneity of general dispersions was accepted. The calculated value of Cochran's criterion was $G_{KR.NABL} = 0.151$, which is less than the tabulated $G_{(KP)(.TABL)} = 0.3704$. Based on this, the null hypothesis of homogeneity of dispersions is not rejected, that is, the test confirmed the possibility of approximation of experimental data by regression equations.

The response surface for power consumption is presented in Figure 7. As can be seen from the obtained graph, the optimum ESP parameters are within the limits (oil temperature: 40...44°C; field gradient 4...5 kV).

https://doi.org/10.1051/bioconf/202516100054

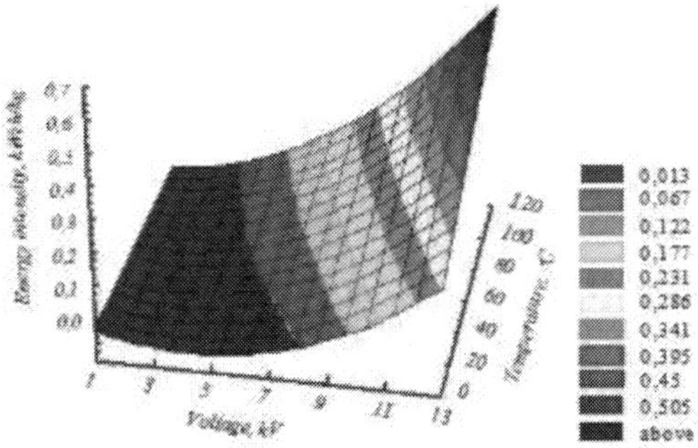

Fig. 7. Energy Capacity Voltage Temperature Response Function Surface

Analysis of the obtained equations showed that the optimum values of ESP parameters are located in the zone of minimum gap determining the breakdown voltage, which agrees well with the previously obtained theoretical results

To determine the minimum value of energy intensity and its corresponding voltage and temperature, the "Solution Search" function of the standard Microsoft Excel package was used.

4. Conclusion

Thus, the optimal parameters of the electrostatic precipitator by the criterion of energy consumption according to the obtained regression equations should be the following: voltage at the electrodes of the electrostatic precipitator U = 4.4 kV, interelectrode distance b = 0.01 m, oil temperature at the inlet θ = 42°C. The calculated energy consumption at these parameters is p= 12 W-h/kg.

The results of experimental studies slightly differ from the analytically obtained conclusions, namely, the optimal interelectrode distance in terms of energy consumption is greater than the minimum possible for technological reasons (0.01 m vs. 0.005 m). This can be explained by the fact that at small distances the oil bubbling near the negative electrode increases (which was observed visually during the experiment). The oil bubbling prevents the precipitation of impurities and increases the purification time. As a result, the energy consumption increases. It should be noted that the modern theory of electrophoresis does not allow analytically describe and account for this phenomenon, so experimental studies are justified.

References

1. O.K. Yunusov, Z.A. Kodirova, Significance of sunflower oil in human body and improvement of deodorisation process, Univ,: technical sciences, **3-2(84)**, 103-106 (2021), https://doi.org/10.32743/UniTech.2021.84.3-2.103-106
2. P.D. Kharina, N.L. Lopaeva, Sunflower oil production, Youth and Science, **5**, 32-35 (2024)
3. D.S. Pleshko, Vegetable oil production in Russia, Ekonomika i sotsium, **2-3(15)**, 1188-1189 (2015)
4. V.S. Gamayurova, L.E. Rzhechitskaya, Myths and reality in the food industry. II. Comparison of Food and Biological Value of Vegetable Oils, Vestnik of Kazan Technological University, **18**, 146-155 (2011), https://doi.org/10.24412/Fd8FcK8j60M
5. I.V. Dolgolyuk et al, Vegetable oils functional foods, Technics and Technology of Food Production, **2(33)**, 122-125 (2014)
6. N.V. Gorshkova, A.V. Belikina, *Ensuring Food Security with Vegetable Oils, Science and Innovations - Modern Concepts*: Collection of Scientific Articles on the Results of the International Scientific Forum (Moscow, 30.06.2023), Editor-in-Chief D.R. Khismatullin (Infinity, Moscow, 2023), https://doi.org/10.34660/INF.2023.82.89.188
7. N.Y. Petrov, I.V. Yudaev, M.P. Aksenov, V.P. Plotnikov, Increasing the efficiency of sunflower cultivation by complex pre-sowing seed treatment, Izvestiya Nizhnevolzhskogo agro-university complex: Science and Higher Professional Education, **4(60)**, 378-389 (2020), https://doi.org/10.32786/2071-9485-2020-04-36
8. M.P. Aksenov, N.Y. Petrov, I.V. Yudaev, Results of research on stimulation of sunflower seeds by NK Neoma electromagnetic field and growth regulator Zerebra Agro, Vestnik AIC Stavropolya, **1(21)**, 153-158 (2016)
9. M.P. Aksenov et al., The growth and development depending on the pre-sowing sunflower treatment method in the southern chestnut soils zone of the Volgograd region, IOP Conf. Ser.: Earth and Envir. Sci., **843**, 012016 (2021), https://doi.org/10.1088/1755-1315/843/1/012016
10. M.C. Bologa, I.I. Berill, *Refining of sunflower oil in an electric field* [Monograph], Academy of Sciences of the Republic of Moldova, Institute of Applied Physics (Stinta, Moldova, 2004), 214 p.
11. M.G. Granovsky et al, *Electrical treatment of liquids (purification of fuels, oils, waters)*, ed. by I.S. Lavrov (Chemistry, Leningrad, 1976), 216 p.
12. I.P. Nazarenko, Purification and separation of sunflower oil in a running electric field, East European Journal of Advanced Technology, **5(11(71))**, 9-13 (2014)
13. M.M. Ukraintsev, E.I. Timonina, Electrostatic filter for sunflower oil purification, Mechanisation and Electrification of Agriculture, **11**, 26-28 (2001)
14. И. Adamczewski, *Electrical Conductivity of Liquid Dielectrics*, Per. from Polish by D.A. Kaplan, ed. by G.S. Kuchinsky (Energia, Leningrad, 1972), 295 pp.
15. S.M. Voronin, M.M. Ukraintsev, Optimisation of parameters of electrostatic filter of sunflower oil, In: *Electrotechnologies and electrical equipment in agricultural production* (Zernograd, 2002), Vop. 1

16. O.N. Bereka, I.P. Nazarenko, Electrotechnological complex of sunflower oil separation in a running electric field, Vestnik VIESKh, **1(14)**, 10-14 (2014)
17. I.P. Nazarenko, Electrotreatment of weakly conducting suspensions, Proceedings of the International Academy of Agrarian Education, **19**, 81-83 (2013)

Biotechnological production of protein feeds from alcohol industry waste

Kamoliddin Usmonov[1,*], *Usmon* Berdimurodov[1], *Sodiqjon* Khakimov[2], *Andrey N.* Globin[3], *Yulia V.* Daus[4], *Sitora* Boboeva[5]

[1] "Tashkent Institute of Irrigation and Agricultural Mechanization Engineers" National Research University, Tashkent, Uzbekistan
[2] Namangan Engineering-Construction Institute, Namangan, Uzbekistan
[3] Azov-Black Sea Engineering Institute – branch of Federal State Budgetary Educational Institution of Higher Education «Don State Agrarian University» in Zernograd, Zernograd, Russia
[4] Kuban State Agrarian University named after I. T. Trubilin, Krasnodar, Russia
[5] Termiz State Pedagogical Institute, Termiz, Uzbekistan

Abstract. The research is devoted to the development and implementation of a biotechnological methodology for the production of enriched protein feeds from secondary grain products, namely post-alcoholic bard and fermentolysate of substandard grain. The study included selection of a consortium of anaerobic microorganisms capable of efficient processing of these substrates, optimisation of fermentation conditions, as well as laboratory and pilot tests. The obtained feed products were characterised by a high protein content (45-47%), a balanced amino acid composition, including essential amino acids, and the presence of biologically active substances such as vitamins and probiotics. The fermentation process achieved a high degree of utilisation of organic components, reducing the negative environmental impact. The developed technology demonstrated competitive performance compared to traditional methods of feed production and alternative biotechnological approaches. The results obtained confirm the promising use of secondary raw materials for the production of high-quality and economically favourable fodder products.

1 Introduction

Currently, there are several potential ways to solve the problem of forage deficiency and quality [1]. One of the directions is genetic modification of fodder crops, which allows increasing their yield and nutritional value. However, this method raises a number of concerns related to possible environmental risks and impact on human and animal health. Another approach is to use various chemical additives and growth stimulants, but their use also has its limitations and may lead to undesirable side effects. The development of resource-saving technologies in traditional agriculture is an important but not always sufficient solution [2-4]. The most promising direction, in the context of solving several problems at once, is the development and implementation of biotechnological methods for processing waste and secondary resources to obtain feed protein products [5].

This direction is relevant and important, as it offers a comprehensive solution that combines economic efficiency, environmental safety and high quality of the obtained products. The use of secondary raw materials, such as waste from the alcohol industry, not only reduces the cost of feed, but also solves the problem of waste disposal, reducing the burden on the environment [6]. Biotechnological methods based on the use of specially selected microorganisms make it possible to obtain products enriched with protein, amino acids, vitamins and probiotic substances, which significantly increases their nutritional value. Moreover, this approach promotes the development of closed production cycles, where waste from one process becomes a resource for another, which is an important aspect of sustainable development [7].

The relevance of this direction is confirmed by the need to intensify the production of livestock products while reducing production costs and minimising the negative impact on the environment [8]. Thus, the development and implementation of technologies for obtaining enriched protein fodder products from wastes of the alcohol industry using biotechnological methods is an extremely promising and in demand direction.

In this regard, the aim of the work is to develop and implement biotechnology for the production of enriched protein fodder products based on post-alcoholic bard and fermentolysate of substandard grain, using a consortium of anaerobic microorganisms [9]. This work is aimed at solving the problem of feed deficit, waste utilisation and increasing the efficiency of agricultural production.

* Corresponding author: usmonov-74@inbox.ru

2 Method and materials

The work presented in this article includes a set of experimental studies aimed at the development and validation of biotechnology for the production of enriched protein feed products. The general plan of the work consisted of several stages: first, the selection and selection of a consortium of microorganisms capable of efficient processing of alcohol industry was carried out; second, the optimisation of fermentation conditions was carried out, including the selection of the optimal composition of the nutrient medium, temperature, pH and other parameters; third, laboratory studies were carried out to assess the composition and nutritional value of the obtained feed products; fourth, pilot tests were carried out at the distillery to test the efficiency of the production of enriched protein feed products.

A range of specialised laboratory and technological equipment was used to implement these stages. At the initial stage of selection and breeding of microorganisms, standard microbiological methods were used, including the use of thermostats for cultivation of microorganisms, laminar boxes for work under sterile conditions and microscopes for visual inspection of cultures. For cultivation of microorganisms and fermentation, fermenters of different volumes equipped with automatic control and regulation systems for temperature, pH and agitation were used. There is no indication of a specific fermenter model in the article, but we are talking about standard laboratory apparatuses for cultivation under controlled conditions.

In the process of conducting experiments to optimise fermentation conditions, various analyses were carried out. Spectrophotometers were used to determine the concentration of nutrients such as proteins, amino acids, fats and carbohydrates. Determination of acidity of the medium was carried out using pH meters. The composition and concentration of amino acids, minerals and vitamins in the obtained feed products were determined by high-performance liquid chromatography. Analysis of microbiological composition was carried out on microscope and by sowing on nutrient media, with subsequent counting of colonies.

During pilot tests at the distillery, large volume industrial fermenters were used to carry out the fermentation process, and separators and filter presses were used to separate the biomass from the culture liquid. Subsequent drying of the obtained product was carried out using drying equipment such as spray drying. All technological processes were accompanied by regular sampling and analysis to control the process parameters and quality of the obtained feed product. The operating modes of the equipment at each stage were optimised to achieve maximum yield of the target product at minimum resource consumption.

3 Results and discussion

The research was aimed at the development and implementation of a biotechnological technique for the production of feed protein from secondary grain products, such as post-alcoholic bard and fermentolysate of substandard grain. At the beginning of the work, a thorough screening and selection of strains of anaerobic microorganisms that had the ability to efficiently utilise organic substances contained in the studied substrates was carried out [10]. The isolated and selected strains were combined into a consortium, which was then used in the fermentation process. The main stage of the work was the determination of optimal conditions for fermentation, which included the selection of the temperature regime, maintenance of optimal pH, ensuring adequate aeration, although in our case we are talking about anaerobic processes and selection of the composition of the nutrient medium. The studies were carried out both in laboratory conditions and in pilot production conditions. In laboratory conditions, fermenters of small volume were used, where the initial testing of technological parameters took place and the influence of various factors on the productivity of the consortium of microorganisms was studied. After that, to check the scalability of the process, the technology was transferred to a pilot plant of a distillery, where fermenters of industrial volume and other technological equipment for the separation of biomass and liquid fraction were used. At each stage of the research a thorough analysis of the composition and quality of the obtained feed products was carried out, including the determination of the content of protein, amino acids, vitamins and minerals. Also, special attention was paid to the analysis of the microbiological composition of the final products to verify their safety and the absence of pathogenic microflora.

As a result of the research it was found that a specially selected consortium of anaerobic microorganisms showed high efficiency in the processing of post-alcoholic bard and fermentolysate of substandard grain. The developed biotechnology made it possible to obtain protein-enriched feed products with a protein content of 45-47% (Table 1). For comparison, traditional sources of protein in feeds, such as soya and sunflower meal, contain about 36-38% crude protein, and fish meal about 60%, but it is characterised by high cost, and the use of soya is not always economically feasible [11]. In addition to high protein content, the obtained feed products were characterised by a balanced amino acid composition. In particular, the content of lysine, threonine, methionine and other essential amino acids was 35-45% of the total amino acids, which indicates a high nutritional value of the product [12]. In addition, a wide range of metabolites, enzymes, vitamins and probiotic substances were identified in the obtained feeds, which were found to have a positive effect on digestion and immune system of the animals. In particular, a significant content of vitamin B12 was observed, which is an important factor in the production of feeds for monogastric animals [13]. In the fermentation process, a deep degree of utilisation of organic components of the substrate is achieved, which is also an important advantage of the developed technology. As a result, the content of organic matter in the liquid fraction after fermentation was reduced by 85-90% compared to the original substrate, which significantly reduces the load on the environment and allows us to consider this method as environmentally acceptable.

Table 1. Chemical and amino acid composition of the obtained feed products

Indicator	Soybean meal	Sunflower meal	Biobardin, Propylact, Probitin
Metabolic energy, kcal/100 g	260	226	237–224
Moisture	8.5	8.5	7.3–9.7
Crude protein	36	38	38.1–45.6
Crude fat	5.8	1.7	8.3–5.1
Crude fiber	7.3	15.0	11.4–13.5
Crude ash	6.0	7.2	5.52–3.25
Calcium	0.34	0.35	1.32–1.15
Phosphorus	0.65	0.65	0.45–0.50
Sodium	0.05	0.08	0.15–0.56
Potassium	–	–	0.24–0.26
Amino acids:			
lysine	2.26	1.27	1.18
histidine	0.83	0.98	1.01–1.03
arginine	2.6	2.92	1.7–1.79
threonine	1.51	1.51	1.27–1.26
glycine	1.48	2.2	1.51–2.01
cystine	0.43	0.65	0.52–0.61
methionine	0.45	0.88	0.68–0.85
valine	1.83	2.03	1.96–2.2
isoleucine	1.75	1.7	1.58–2.57
leucine	2.72	2.4	2.44–2.85
phenylalanine	1.9	1.8	1.88–2.07
tyrosine	1.24	1.15	1.49–1.17

The protein content of the final biomass was in the range of 40-42%, with a lysine content of about 1.1%, an important essential amino acid.

However, when further investigated and using another consortium of anaerobic microorganisms isolated from soil samples, improved performance was obtained. In particular, the protein content of the final product reached 44.5%, which was 2.5-4.5% higher than in previous experiments, and the lysine content increased to 1.3%. [14]. This indicates the possibility of increasing the nutritional value of protein biomass by selecting the optimal microbial community.

The study of the influence of various parameters on the fermentation process allowed us to optimise the cultivation conditions [15]. The study of the influence of temperature on the fermentation process showed that the most favourable temperature for the growth and activity of microorganisms is 37°C. At this temperature, the maximum biomass yield was observed, averaging 15.2 grams of dry matter per litre of medium for 48 hours of fermentation. Increasing the temperature above 40°C led to a decrease in the activity of microorganisms, which was reflected in a decrease in biomass yield by 8-12% and slowing down of the fermentation process. In particular, after 48 hours, the biomass yield at 42°C averaged 13.5 grams of dry matter per litre of medium. Reducing the temperature to 25°C also had a negative effect on the process, leading to a significant slowdown of fermentation and a 15-20% decrease in biomass yield (to 12.2-12.9 grams of dry matter per litre of medium).

In addition to temperature, fermentation duration proved to be an important factor. Increasing the fermentation time to 72 hours resulted in a statistically significant increase in protein yield by 10-15% (to an average of 17.5 grams of dry matter per litre of medium) compared to fermentation for 48 hours. However, a further increase in fermentation time, e.g. up to 96 hours, did not give a significant increase in biomass yield (the increase was no more than 2-3%, i.e. about 18 grams of dry matter per litre of medium) and was not economically justified. These results indicate the need for precise control of fermentation parameters to maximise the efficiency of the protein biomass production process.

Comparing the results obtained with other similar studies in the field of feed biotechnology, it can be noted that the developed method shows competitive performance [16]. In particular, there are works on the use of yeasts and mould fungi to process alcohol production waste [17]. However, these methods often require high energy costs and complex equipment, and the yield of the final product is not always high enough. Also, there is a problem in obtaining products with the required composition and presence of probiotic properties. In contrast to these methods, the developed biotechnology based on the use of a consortium of anaerobic microorganisms allows not only to obtain a product with high protein content, but also to enrich it with useful biologically active substances. In addition, this technology, compared to other approaches, is characterised by low production costs and scalability for industrial scale application [18]. The numerical data are given in the comparative Table 2.

Table 2. Chemical and amino acid composition of the obtained feed products

Indicator	Experimental data	Evidence from alternative studies
Protein content, %	44.5	40–42 (if using yeast)
Lysine content, %	1.3	0.8–1 (when using mold)
Fermentation time, h	72	48–96
Reduction of organic matter, %	88	70–80
Optimum temperature, °C	37	30–35

Thus, the results of the presented study and my own data confirm the high efficiency and prospects of the developed biotechnology for the production of enriched protein fodder products from secondary raw materials [19]. This technology can make a significant contribution to solving the problem of fodder deficiency, reduce the cost of livestock products and ensure sustainable development of agriculture [20].

4 Conclusion

The study demonstrates the effectiveness and prospects of biotechnological approach to the production of enriched protein feeds based on secondary grain products. The use of a specially selected consortium of anaerobic microorganisms makes it possible to achieve deep utilisation of organic components of post-alcoholic bard and fermentolysate of substandard grain, thus producing a product with a high protein content and a balanced amino acid composition. The obtained feed products are not only superior in protein content to traditional analogues such as soya and sunflower meal, but also enriched with vitamins, enzymes and probiotic substances, which significantly increases their biological value and potential benefits for animals.

The results obtained during laboratory and pilot tests showed the stability and reproducibility of the developed technology. Optimisation of fermentation parameters such as temperature, pH and process time allowed to achieve maximum yield of the target product while minimising costs. It is important to note that this approach, unlike a number of other biotechnological methods, is environmentally safe and contributes to solving the problem of industrial waste utilisation. A significant decrease in the content of organic substances in the liquid phase after fermentation indicates a high degree of substrate processing and a reduction in the negative impact on the environment.

Comparative analysis of the obtained results with the data of other researchers in the field of feed biotechnology confirmed the competitiveness of the proposed technology. In particular, the developed method surpasses the approaches based on the use of yeasts and mould fungi in terms of protein content, amino acid composition and the presence of probiotic properties. Ultimately, this study opens the prospect for the creation of cost-effective and environmentally friendly production of high quality forages, which is of great importance for improving the efficiency of animal husbandry and ensuring food security. The introduction of this technology at industrial enterprises will not only reduce dependence on imported feed additives, but will also contribute to the sustainable development of the agricultural sector.

References

1. K.M. Didur, H.O. Kundieieva, G.V. Ortina, T.A. Pikhniak, A.V. Revkova, formation of food supply of the agrarian sector of ukraine on the basis of restoration of the development of rural areas. Rev. Iberoam. Vitic. Agroind. Rural., **12,** 223-239 (2025). DOI: 10.35588/pxe7k028

2. B. Aparecida da Silva, R. Evangelista de Oliveira, A. Cavalieri Sais, E. Cardoso leite, Family farmers and trees: ecological functions, ecosystem services and food production in agrarian reform settlements in Brazil. Agroecol. Sustain. Food Syst., **49,** 4-21 (2025). DOI: 10.1080/21683565.2024.2405515

3. M. Szczepanek, K. Błaszczyk, M. Piekarczyk. The Spatial Distribution of Nutrients in the Soil, Their Uptake by Plants, and Green Bean (*Phaseolus vulgaris* L.) Yield Under the Strip-Tillage System. *Agronomy*. **15(2),** 382, (2025). DOI: 10.3390/agronomy15020382

4. F. Ju, R. Yang, C. Yang, Analysis of Spatiotemporal Dynamics and Driving Factors of China's Nationally Important Agricultural Heritage Systems. *Agriculture*. **15(2),** 221, (2025). DOI: 10.3390/agriculture15020221

5. V.I. Golik, R.V. Klyuev, N.V. Martyushev, V. Brigida, E.A. Efremenkov, S.N. Sorokova, Q. Mengxu, Tailings Utilization and Zinc Extraction Based on Mechanochemical Activation. Materials, **16 (2),** 726 (2023) DOI: 10.3390/ma16020726

6. N.A. Zhirov, Socio-Economic Development of the Black Soil Village in the XIX–early XX centuries: from the Prosperous Granary of Pre-Reform Russia to the Demo-Ecological Crisis of Rural Settlements. Bylye Gody, **19,** 1605-1614 (2024). DOI: 10.13187/bg.2024.4.1605

7. L. Yerkinbayeva, A. Myrzakarim, A. Moldagaliyeva, B. Kalymbek, R. Yerezhepkyzy, D. Nurmukhankyzy, the role of legal regulation in promoting organic agriculture to achieve sustainable food security. J. Lifestyle SDG'S Rev., **5,** e04197 (2025). DOI: 10.47172/2965-730X.SDGsReview.v5.n01.pe04197

8. S. Bamidele, BRICS Revolution: The Global Agrarian Shift Sparked by Emerging Economies. India Q., (2025), Article in press. DOI: 10.1177/09749284241307937

9. M.A. Mahfud, N.H. Djohan, the expansion of investor access to cultivation rights: A Socio-Legal Analysis on Agrarian Injustice in Indonesia. Al-Risalah: Forum Kajian Hukum dan Sosial Kemasyarakatan, **24,** 55-67 (2024). DOI: 10.30631/alrisalah.v24i2.1500

10. V. L. Tatarintsev, L. M. Tatarintsev, S. K. Makenova, M. M. Shostak, geoecological assessment of landscapes as a basis for organization of sustainable agrarian land use, Sustainable Development of Mountain Territories, **13 (4),** 485–496 (2021). DOI: 10.21177/1998-4502-2021-13-4-485-496

11. V. I. Golik, V. Yu. Konyukhov, Mining Informational and Analytical Bulletin. **11-1,** 175-189 (2023). DOI: 10.25018/0236_1493_2023_111_0_175

12. M. N. Nurpeisov, K. V. Ruchkina, V. A. Androkhanov, S. K. Makenova, Assessment of geoecological factors and conditions affecting sustainability of agricultural land use in the forest-steppe zone of Altai Krai, Sustainable Development of Mountain Territories, **15(2),** 419–430 (2023). DOI: 10.21177/1998-4502-2023-15-2-419-430

13. V. O. Gutarevich, A. S. Mikhalev, V. E. Gozbenko, Mining Informational and Analytical Bulletin, **11,** 72–87 (2023). DOI: 10.25018/0236_1493_2023_111_0_72

14. J. Dobrzyński, B. Wróbel, E.B. Górska, Taxonomy, Ecology, and Cellulolytic Properties of the Genus Bacillus and Related Genera, Agriculture, **13(10),** 1979 (2023). DOI: 10.3390/agriculture13101979

15. V. L. Tatarintsev, L. V. Lebedeva, G. D. Syzdykova, D. S. Inkarov, Management of agricultural land use sustainability at the local level using GIS technologies and special map material, Sustainable Development of Mountain Territories, **15(4),** 864–876 (2023). DOI: 10.21177/1998-4502-2023-15-4-864-876

16. V. A. Kukartsev, A. I. Trunova, S. O. Kurashkin, K. A. Bashmur, R. B. Sergienko, S. V. Tynchenko, Metals, **13(7),** 1234 (2023). DOI: 10.3390/met13071234

17. K. Akzharkyn, K. Yelemessov, D. Baskanbayeva, N.V. Martyushev, V.Y. Skeeba, V.Y. Konyukhov, T.A. Oparina, Strengthening Polymer Concrete with Carbon and Basalt Fibres, Appl. Sci. (Switzerland), **14(17),** 7567 (2024). DOI: 10.3390/app14177567

18. B.V. Malozyomov, N.V. Martyushev, S.N. Sorokova, E.A. Efremenkov, D.V. Valuev, M. Qi, Analysis of a Predictive Mathematical Model of Weather Changes Based on Neural Networks. Mathematics, **12 (3),** 480 (2024) DOI: 10.3390/math12030480

19. V.I. Golik, R.V. Klyuev, N.V. Martyushev, D.A. Zyukin, A.I. Karlina, Technology for Nonwaste Recovery of Tailings of the Mizur Mining and Processing Plant. Metallurgist, 66 (11-12), 1476-1480 (2023) DOI: 10.1007/s11015-023-01462-y

20. K.I. Kravtsov, Ya.A. Tynchenko, Influence of mountain factors on salt excess and soil toxicity in mountain conditions, Sustainable Development of Mountain Territories, **15 (3),** 784–797 (2023). DOI: 10.21177/1998-4502-2023-15-3-784-797

Mechanism of in ovo penetration of nanostructured zeolite and its effect on weight gain of birds during early postnatal ontogenesis

Roman Undalov[1*], *Vladimir* Ezhkov[1], *Asiya* Ezhkova[1], *Yulia* Larina[1], *Elizaveta* Sharikova[1]

[1]N.E. Bauman Kazan State Academy of Veterinary Medicine, Republic of Tatarstan, Kazan, 420029, Russia

Abstract. The effect of nanostructured zeolite on embryonic and early postnatal development of Peking ducks cross 'STAR-53 medium' was investigated. The surface of hatching eggs was treated with a suspension of nanostructured zeolite with a particle size of 60-120 nm to evaluate its effect on increasing the productivity of ducks. It was found by electron microscopy that particles of nanostructured zeolite penetrate through the porous structure of the shell (pore size 0.2-2 μm) and reach the embryo, stimulating physiological processes. Radiological studies showed no adverse effects on embryo viability, survival and development. After hatching it was revealed that the absolute live weight gain of ducklings in the experimental group was 7.2% (22.6 g) more than in the control group during the first 10 days of life. The obtained results show the prospectivity of using nanostructured zeolite in industrial poultry farming to increase the productivity and quality of growing birds.

1 Introduction

In ensuring food security in Russia, poultry farming is of great importance as a producer of 'fast' meat, and the consumption of poultry products (meat, eggs) in the daily diet of the population reaches up to 40% [1]. In the structure of general poultry production of meat direction of productivity the second position is confidently occupied by the cultivation of broiler ducks, the meat of which is dietary and is characterised by high quality protein [2].

Ducks of meat direction of productivity are characterised by high rates of growth and development, and already by the slaughtering age of 7-8 weeks they reach the weight of 4-4.5 kg. The most promising for breeding are Peking meat ducks cross 'STAR-53 medium' [3].

Mineral elements play a key role in the maintenance of metabolic processes, tissue development, bone system and immune response, as well as in the body's defence against oxidative stress [4]. However, mineral deficiency in poultry diets remains one of the key problems in industrial poultry production [5]. The causes of micronutrient and macronutrient deficiencies are often inadequate feed mixtures to meet the needs of high-yielding breeds, changes in genetic potential and low mineral content of feed [6].

Calcium is the most important mineral element for poultry. It is required not only for eggshell formation, but also for normal functioning of the nervous and muscular systems, blood clotting and enzyme activation [7]. In addition to calcium, elements such as phosphorus, magnesium, manganese and zinc are also involved in the formation of egg shells and ensure the health of embryos [8]. Deficiency of these minerals at early stages of development can lead to the arrest of cell division and death of the embryo, and at later stages - to disruption of the formation of organs and systems. This, in turn, leads to the rejection of weak embryos during the incubation period and reduces the productivity of farms [9].

Weak embryos most often die as a result of disorders caused by deficiencies of key micronutrients. Calcium and phosphorus deficiencies weaken the shell, impairing the gas and moisture permeability of the egg, which critically affects embryo development [10]. Deficiencies of magnesium and zinc impair enzymatic processes, retarding tissue growth and development of vital systems [11]. As a result, embryo rejection can reach significant values, leading to losses both at the incubation stage and in the early postnatal period.

Current research shows that nanomaterials have great potential in improving the productivity of farm animals [12]. The unique properties of nanostructures allow their use as feed additives that stimulate growth, increase body weight and egg production [13]. The advantages of nanomaterials lie in their ability to activate physiological and biochemical processes at low doses [14]. Nanoparticles are defined as particles of matter with diameters ranging from 1 to 100 nanometres [15]. Cost-effectiveness and bioavailability make nanomaterials promising for use in poultry production.

Examples of successful use of nanomaterials include the use of nanostructured sapropel, which increases the live weight of geese by 7.9-11.1% [16], and

* Corresponding author: roman-undalov25@yandex.ru

nanostructured phosphorite, which increases the meat productivity of broiler chickens by 7.4% [17]. These results confirm the efficacy of nanomaterials in poultry feeding, improving both productive performance and product quality.

A promising new generation preparation for poultry production is nanostructured zeolite. Zeolites are a unique group of minerals belonging to aqueous calcium and sodium aluminosilicates. Their crystal structure is formed from tetrahedrons of silicon and aluminium oxides joining into spatial frameworks with uniformly sized porous channels. These channels are filled with alkali and alkaline earth metal cations and water molecules, which are called 'zeolite water' [18]. Such structure provides zeolites with adsorption, cation exchange and catalytic properties, which opens up a wide range of applications in various industries including agriculture [19].

Natural zeolites have found wide application as feed additives in poultry production. For example, the inclusion of zeolite-containing additives enriched with organic acids and prebiotics in the diet of broiler chickens increases the average weight of birds by 12.4-13.9%, reducing feed costs by 3.5-4.5%. The use of modified zeolite activates anabolic processes, increases glycolysis and reduces lipogenesis, which has a positive effect on the general condition and productivity of poultry [20].

Egg productivity is an important area in the evaluation of poultry productivity [21]. Poultry eggs are characterised by a number of parameters that are integral to their qualitative evaluation and consumer value [22]. The egg shell fulfils a protective function, preventing damage to the internal contents and ensuring the stability of the embryo during incubation. It consists of three layers: cuticle, matrix and shell membrane [23].

Currently, the main physical variables that determine the quality of the shell of breeding and commercial eggs are considered to be its strength, permeability to gases and colour [24]. The gas and moisture permeability of the shell is determined by its porosity, which significantly affects the course of embryonic development and the results of incubation [25]. It has been found that the outer surface of the egg shell is smooth and flat, penetrated by many pores of irregular round-oval shape, with sizes from 0.2 to 1 μm, through which particles of nanometre range can easily penetrate.

In the literature there is information about the use of in ovo method in the form of injections of amino acids and other substances to the embryo through the shell and shells of the hatching egg [26], but there are no reports on the mechanisms of penetration of drugs to the embryo by surface application on the shell.

Thus, the aim of this work was to study the effect of nanostructured zeolite on the development of ducks during embryogenesis and early postnatal ontogenesis.

2 Material and methods

The studies were conducted in the laboratory of experimental physiology and pathophysiology of the Kazan State Academy of Veterinary Medicine named after N.E. Bauman The studies were conducted in the laboratory of experimental physiology and pathophysiology of the Kazan State Academy of Veterinary Medicine named after N.E. Bauman, as well as in the Kazan Physical-Technical Institute named after E.K. Zavoysky in the period from March 2023 to March 2024.

2.1. Materials

The materials for research were duck hatching eggs, embryos, shells and ducklings of Peking cross 'STAR-53 medium' at the age of up to 10 days. A suspension of nanostructured zeolite with a particle size of 60.0-120.0 nm was used for surface application to hatching eggs.

2.2 Methods

The composition of nanostructured zeolite was determined on atomic absorption spectrometer 'Aanalist 400' PerkinElmer Ins. (USA). The chemical composition of zeolite is represented by (%): SiO_2 – 59.2; TiO_2 – 0.25; Al_2O_3 – 6.2; Fe_2O_3 – 2.4; CaO – 15,4; MgO – 1.9; Na_2O – 0.2; K_2O – 1.1; P_2O_5 – 0.1; n.p.p. (H_2O+CO_2) – 13.25.

Nanostructured zeolite was prepared on UZV 28/200 MP RELTEK (Russia) by ultrasonic treatment in the ratio of deionised water and zeolite 100:20. The content of nanostructures of 60.0-120.0 nm in the preparation was up to 60%. Validation of the obtained substance was carried out on a scanning probe microscope MultiMode V from Veeco (USA) by intermittent-contact atomic force microscopy [27].

Two groups of 30 eggs each - control and experimental - were formed and incubated in two automatic incubators 'NormaLUPPER' (Russia), incubation mode standard for duck eggs. Treatment with nanostructured zeolite was carried out on the 9th day of embryogenesis, which was justified by physiological processes of intensive formation and development of bone tissue and feather base of embryos.

Egg shells obtained after treatment of hatching eggs were examined by electron microscopy on a scanning electron microscope EVO 50 XVP (Carl Zeiss) (Germany).

According to the 'Legal and ethical rules...', bird embryos before hatching are not considered as complete organisms, therefore, the conducted studies do not contradict the rules of the code of ethics for research using animals [28]

Radiological studies of incubated eggs were performed on a Dongmun DIG-360 X-ray machine with a Rayence 1717SGC flat panel X-ray detector and Xmaru View V1 image reading software (Korea). Radiological examinations and imaging were performed in horizontal projection with 4maS exposure and 52kW voltage.

After hatching, ducklings were visually monitored daily for their health status and weighed for 10 days. Weighing was carried out individually on scales of II class of accuracy 122ACF-1500.05 LCD 'Accurate' (South Korea).

Digital data obtained as a result of the studies were processed statistically using standard programmes of variation statistics using Microsoft Office Excel-2021 software package.

3 Result and discussion

On the ninth day of duck eggs incubation, the surface of eggs of the experimental group was treated with nanostructured zeolite. After treatment, anatomisation of control and experimental embryos was carried out to investigate the mechanism of penetration of zeolite particles in ovo. When examining the ultrastructure of the egg shells of the control and experimental groups, a dense shell structure with microcracks and the presence of many pores of different sizes was visualised from the outer surface (Fig. 1 a). On the surface of the eggshells of eggs of the experimental group, micrometre range particles of the preparation were noted (Fig. 1 b).

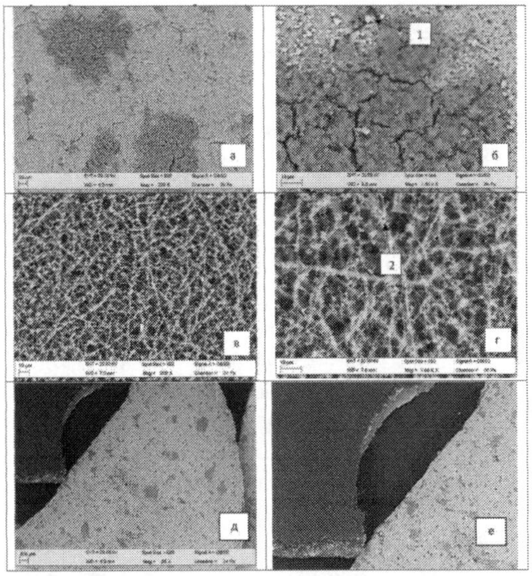

Figure 1. Ultrastructure of the shell of hatching duck eggs on the 9th day of incubation: a - outer surface of control eggs with pores and microcracks; b - outer surface of experimental eggs with pores, microcracks and drug particles (1); c - fibre-mesh structure of the inner surface of control eggs; d - fibre-mesh structure of the inner surface of experimental eggs with the presence of mineral inclusions (2); e - shell on chipping; f - detail of chipping, visualisation of structure heterogeneity and presence of communicating spaces-vacancies from the outer to the inner shell surfaces.

The ultrastructure of egg shells from the inner surface is represented by a reticulated structure of chaotically interwoven protein-mineral fibres forming voids deep into the shell (Fig. 1 c). When examining the inner surface of the shells of eggs of the experimental group, we observed the presence of inclusions in porous voids, which, presumably, in our opinion, were zeolite particles that penetrated from the outer surface of the treated egg (Fig. 1 d). Investigations of the shell ultrastructure at the break visualised the heterogeneity of the structure and interpenetration of the upper and inner layers. The presence of communicating voids from the upper to the inner layers was noted on the shell slice, confirming the air exchange of the embryo and visualising the penetration potential of drug particles applied to the surface of the hatching egg (Fig. 1 e).

Comparison of the pore sizes in the shell, revealed by scanning electron microscopy (0.2-1 μm on the outer side and 0.2-2 μm on the inner side), and the size of the zeolite nanoparticles (60-120 nm) used in the experiment, allowed us to establish that nanoseolite particles are free to penetrate through the porous structure of the shell.

X-ray studies of embryos of the control and experimental groups were carried out on the 24th day, during the period of complete formation of the bone system and completion of its ossification. It was visually noted that embryos of the experimental group did not differ from their control counterparts. Embryo sizes had no significant differences, which indicated that application of nanostructured zeolite on the surface of incubated eggs had no adverse effect on the safety, viability and embryonic development of ducks (Fig. 2 a, b).

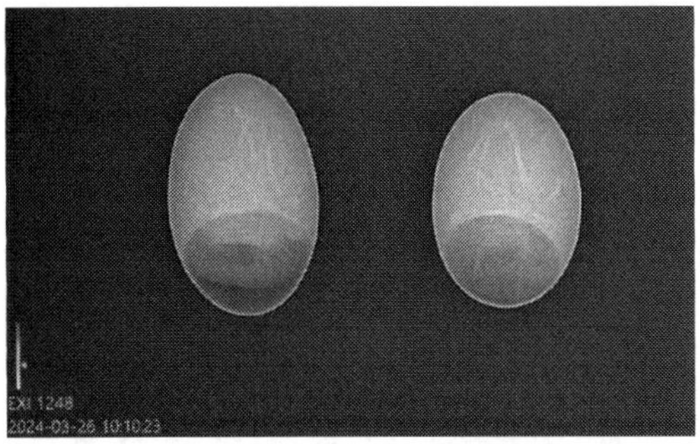

Figure 2. Radiographs of duck embryos of control (a) and experimental (b) groups on 24 days of incubation: pronounced silhouette of spine, head with beak, tubular bones of limbs, air chamber.

After hatching of ducklings, control and experimental groups were successively formed to study the dynamics of live weight of chicks in the early period of postnatal ontogenesis (Fig. 3).

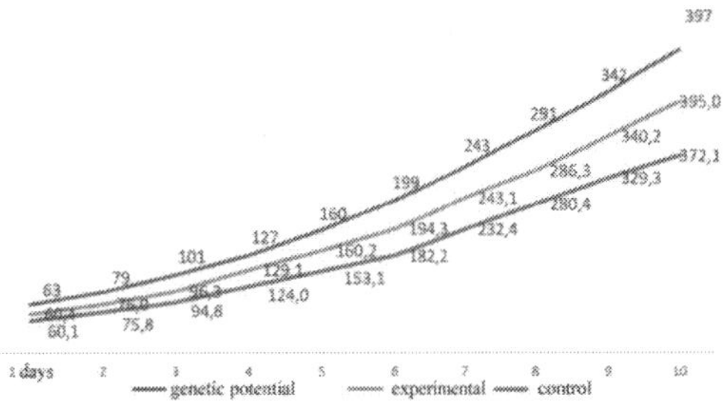

Figure 3. Dynamics of live weight of ducklings during early postnatal ontogenesis.

The absolute growth of ducklings from the experimental group was found to be 7.2% higher or 22.6 g more compared to the control counterparts.

At ten days of age, the ducklings of the experimental group were 6.15% larger than their control counterparts.

4 Conclusion

Studies have shown that zeolite particles, due to their nanometric size (60-120 nm), freely penetrate through the porous structure of the shell (0.2-2 μm), reach the embryo and stimulate physiological processes during embryogenesis and early postnatal ontogenesis, which contributes to an increase in the absolute live weight gain of ducklings by 7.2% (22.6 g) for the first 10 days without any negative effects on the viability and development of ducklings.

Thus, the use of nanostructured zeolite in industrial poultry farming is a promising method of increasing the productivity and quality of growing birds.

Acknowledgements

We express our gratitude to the staff of the Kazan Physical-Technical Institute named after E.K. Zavoysky., Akhmetov M.M. and Lyadov N.M. for their assistance in the electron microscope studies.

References

1. V. Fisinin, Compound Feed, **4,** 2-6 (2024).
2. E. Baéza, L. Guillier, M. Petracci, Animal, **16,** 100331 (2022).
3. B. F. Bessarabov, A. A. Krykanov, A. L. Kiselev, Incubation of farm poultry eggs (Lan, St. Petersburg, 2015).
4. C. Weyh, K. Krüger, P. Peeling, L. Castell, Nutrients, **14(3)** (2022).
5. S. Świątkiewicz, A. Arczewska-Włosek, D. Józefiak, World's Poultry Science Journal, **70(3),** 475–486 (2014).
6. B. Biruk, Mineral and digestive responses to dietary challenges in the gastro-intestinal system of ruminants (2021).
7. A. Matuszewski, M. Łukasiewicz, J. Niemiec, World's Poultry Science Journal, 76(2), 328–345 (2020).
8. X. Cheng, Zh. Ning, Poultry Science, **102(1),** 102283 (2023).
9. M. Besharati, L. Fathi, S. Amirdahri, Z. Nemati, V. Palangi, J.M. Lorenzo, A. Maggiolino, G. Centoducati, Animals, **13(12),** 1925 (2023).
10. A. M. Fouad, D. Ruan, Sh. Wang, Journal of Animal Science and Biotechnology, **9(1),** 1–11 (2018).
11. P. Skliarov, M. Lieshchova, D. Bilyi, S. Fedorenko, S. Naumenko, O. Onyshchenko, A. Pasternak, O. Bobrytska, L. Roman, World's Veterinary Journal, **11(2),** 157–169 (2021).
12. N.M. Hashem, A. Gonzalez-Bulnes, Animals, **10(5),** 840 (2020).
13. A. Patra, M. Lalhriatpuii, Biological Trace Element Research, **197(1),** 233–253 (2020).
14. Z. Cheng, M. Li, Y. Chen, R. Dey, J. Hematol. Oncol., **14(1)** (2021).
15. B. Habib, K. Shabbir, A. Ahmad, S. Batool, I. Liaqat, A. Saddiqua, M. Aalia, E. Ahmed, K. Kareem, International Journal of Advanced Research, **12(08),** 861–875 (2024).
16. V. I. Golik, T. A. Panfilova, V. S. Tynchenko, V. Yu. Konyukhov, Mining Informational and Analytical Bulletin. **11-1,** 175-189 (2023). DOI: 10.25018/0236_1493_2023_111_0_175
17. D.A. Gutiérrez-Arenas, M. Cuca-García, M.A. Méndez-Rojas, A. Pro-Martínez, C.M. Becerril-Pérez, M.E. Mendoza-Álvarez, F. Ávila-Ramos, J.E. Ramírez-Bribiesca, Animals, **11(10),** 2773 (2021).
18. N. Kordala, M. Wyszkowski, Molecules, 29(5), 1069 (2024).
19. E. Cataldo, L. Salvi, F. Paoli, M. Fucile, G. Masciandaro, D. Manzi, C.M. Masini, G.B. Mattii, Agronomy, **11(8),** 1547 (2021).
20. I.I. Kochish, E. A. Kapitonova, V. N. Nikulin, Izvestiya Orenburgskogo gosudarstvennogo agrarnogo universitet, **3(83),** 329-334 (2020).
21. K. El-Sabrout, S. Aggag, B. Mishra, Scientifica, **2022,** 1393392 (2022).
22. J. Gautron, C. Dombre, F. Nau, C. Feidt, L. Guillier, Animal, **16,** 100425 (2022).
23. F. Liu, X. Jiang, Zi. Chen, L. Wang, Acta Biomaterialia, **178,** 233–243 (2024).
24. K. Drabik, M. Karwowska, K. Wengerska, T. Próchniak, A. Adamczuk, J. Batkowska, Animals, **11(5),** 1204 (2021).
25. A. V. Lensink, G. E. Swan, J. G. Myburg, Journal of Microscopy, **290(1),** 23–39 (2023).
26. A. M. Dolgorukova, V. Y. Titov, V. I. Fisinin, and A. A. Zotov, Agricultural Biology, **55(6),** 1061-1072 (2020).

27. T. A. Panfilova, K. A. Bashmur, V. V. Kondratiev, Mining Informational and Analytical Bulletin, **11-1,** 239-251 (2023). DOI: 10.25018/0236_1493_2023_111_0_239
28. E. Aleksandrowicz, I. Herr, ALTEX **32,** 143 (2015).

Resource potential of the agro-industrial complex of Siberia and its role in ensuring food security of Russia

Daria S. Tarasenko [1*], *Evgeniy A.* Rozhkov [2], *Andrey V.* Asaturyan [3], *Yu.A.* Lepekhina [4]

[1] Bratsk State University, Bratsk, 665709, Russian Federation
[2] Kuban State Agrarian University named after I. T. Trubilin, Krasnodar, Russia
[3] Azov-Black Sea Engineering Institute – branch of Federal State Budgetary Educational Institution of Higher Education «Don State Agrarian University» in Zernograd, Zernograd, Russia
[4] Admiral Ushakov Maritime State University, Novorossiysk, 353924, Russia

Abstract. The paper is devoted to the analysis of the resource potential of the agro-industrial complex of the Siberian Federal District of Russia and its impact on the food security of the country. A comprehensive analysis, including the collection of statistical data and the assessment of land, labour, financial and innovation resources has been carried out. Significant differences in resource endowment between the regions of the SFD have been revealed. Land resources vary from 0.3 million hectares in Tyva to 4.5 million hectares in Altai Krai, while the quality of soils and the share of arable land also differ significantly. Labour resources are characterised by varying numbers of employees, skill levels and wages, while financial resources are characterised by investment volumes and access to credit. Innovation potential was assessed by the level of introduction of new technologies and agronomic practices. The results showed that Altai Krai and Novosibirsk Oblast stand out with more favourable conditions and high performance in the agro-industrial complex, while the Republic of Tyva and some other regions have low resource endowment and low production efficiency. Overall, the study emphasises the need for a differentiated approach to state support for agro-industrial complex in the SFD to improve food security.

1 Introduction

In the Russian Federation, with its vast territory and diverse natural conditions, special attention should be paid to regional aspects of food security. The Siberian Federal District, with its significant land resources, appears to be one of the most promising regions for agricultural development. However, this potential is often not fully utilised due to unfavourable climatic conditions, low level of technological support and lack of qualified personnel. The study and assessment of Siberia's resource potential, as well as the development of innovative approaches to its utilisation, are the most important tasks to ensure food security of both the region and the country as a whole. Improving the efficiency of land, water, labour and financial resources use in Siberian agriculture can become one of the key factors in solving the problem of food dependence in Russia, as it is observed in other countries, for example, in Africa, where the problem of access to water for agriculture is a serious challenge [1].

The presented research is devoted to the analysis of the resource potential of the agro-industrial complex of Siberia and its role in ensuring food security of the country. The aim of the work is to identify the main factors that determine the state and prospects of development of the agrarian sector of the region [2, 3], as well as to develop proposals for the effective use of its resources to provide the population with food, which is also relevant for other regions of the world, where changes in agrarian systems under the influence of climate change are observed [4, 5].

2 Method and materials

In this paper, a set of experimental studies was conducted to analyse the resource potential of the agro-industrial complex of Siberia. The overall plan included the collection and analysis of statistical data by regions, as well as the study of factors affecting food production and availability. Various types of resources, including land, labour, financial and innovation resources, were assessed to determine their impact on food security.

The study made extensive use of statistical software package to process and analyse the data. In particular, Microsoft Excel was used, which allowed structuring the collected data by different regions and resources, as well as carrying out basic statistical calculations necessary for trend analysis and construction of comparative charts. In the process of analysis, tools were used to assess the dynamics of changes in indicators over time. First of all, it concerned the indicators of agricultural production,

* Corresponding author: dasha.tarasenko.04@list.ru

import and export volumes, as well as the level of implementation of new technologies. Statistical data were collected from various official sources, including Rosstat and regional statistical offices.

In addition to software, classical methods of economic and statistical analysis were used in the work, based on data on production indicators, as well as import and export indicators. The indicators for different regions of Siberia were compared in order to identify the most resourced and problem areas. Data on sown areas, crop yields, the number of people employed in agriculture, investment in the agro-industrial complex and other relevant indicators were analysed. In addition, data on food consumption and its dependence on imports were studied. The main focus of the work was to determine the structure and volume of resources needed to ensure food security in the region. Tabular and graphical methods were used to visualise the results.

3 Results and discussion

This study provides an in-depth analysis of the resource potential of the agro-industrial complex (AIC) of the Siberian Federal District (SFD), covering all key aspects necessary to ensure food security of the Russian Federation. Our work, unlike the baseline study, was not limited to a general overview, but presented a detailed picture for each SFD region, including quantitative indicators reflecting the situation with land, labour, financial and innovation resources. To assess land resources, we used both Rosreestr's data on the area of agricultural land and the results of our own analysis, including indicators of soil fertility, suitability for specific types of crops, as well as the degree of erosion and salinisation. Thus, the average area of agricultural land in the SFD is 14.7 million hectares, but this value varies from 0.3 million hectares in the Republic of Tyva to 4.5 million hectares in Altai Krai. At the same time, the share of arable land in the total area of agricultural land is 60 per cent in the Novosibirsk Region, 55 per cent in the Altai Territory, 40 per cent in the Kemerovo Region and only 20 per cent in the Republic of Tyva, which indicates significant differences in the structure of the land fund. The qualitative characteristics of soils also differ significantly: the average humus content in the arable layer is 4.3% in Altai Krai, 3.9% in Novosibirsk Oblast, 3.5% in Omsk Oblast, 2.8% in Irkutsk Oblast, 2.5% in Krasnoyarsk Krai and 2.0% in the Republic of Tyva. The degree of soil erosion was also calculated, which was 10% in the Altai Krai, 15% in the Novosibirsk Oblast, 20% in the Omsk Oblast and more than 30% in the Republic of Tyva, indicating serious problems with the state of the soil cover in certain regions, which is also emphasised in studies of sustainable development of mountain areas [6, 7]. The accessibility of social services and living standards of workers was also analysed, which in turn affects the attractiveness of the agricultural sphere for young professionals, which is especially important in the context of rural development [8, 9].

The labour resources of the agro-industrial complex were analysed not only in terms of the number of employed, but also in terms of qualification level, wages and working conditions. In the SFD, the average number of people employed in agriculture is 900 thousand people, but this indicator also differs by regions: 250 thousand in Altai Krai, 180 thousand in Novosibirsk Oblast, 100 thousand in Krasnoyarsk Krai, 80 thousand in Omsk Oblast and less than 30 thousand in the republics of Tyva and Khakassia. The average level of wages in the agro-industrial complex in the SFD is $350 roubles, but in Altai Krai this indicator reaches $400 roubles, in Novosibirsk Oblast - $380 roubles, and in the Republic of Tyva does not exceed $280 roubles. In addition, the level of labour mechanisation is 0.70 in the Novosibirsk and Omsk Oblasts, 0.65 in the Altai Krai and 0.45 in the Republic of Tyva. We assessed the level of workers' qualifications, which, according to our data, correlates with the level of mechanisation. Thus, the share of workers with higher and secondary specialised education is 50% in the Novosibirsk and Omsk Oblasts, 40% in the Altai Krai, 30% in the Irkutsk Oblast and less than 20% in the Republic of Tyva. The accessibility of social services and living standards of employees was also analysed, which in turn affects the attractiveness of the agricultural sector for young professionals.

To assess the financial resources of the agro-industrial complex, data on the volume of investment, lending and state support were used. The average annual volume of investment in the agro-industrial complex of the SFD is $12 million per 1 ha, but this figure varies from $18 million in the Novosibirsk Oblast to $8 million in the Republic of Tyva. In Altai Krai the volume of investment is $15 million per 1 ha, in Omsk Oblast - $13 million, in Krasnoyarsk Krai - $10 million, and less than $9 million in other regions. The average lending rate for agricultural enterprises is 12%, but in some regions (for example, in the Republic of Tyva) this indicator can reach 15%. The efficiency of the use of state support funds also varies by region, which was assessed using the cost-effectiveness ratio. Thus, in the Novosibirsk Region it is 0.8, in the Omsk Region - 0.75, in the Altai Territory - 0.7, in the Krasnoyarsk Territory - 0.6, and less than 0.5 in the Republic of Tyva and other regions with a low level of agro-industrial complex development.

The innovation potential of the agro-industrial complex of the SFD was assessed based on the analysis of the introduction of new technologies, the level of mechanisation, the use of modern agricultural techniques and breeding achievements. The number of patents for agricultural technologies obtained per 1 million population is 12 in the Novosibirsk Region, 8 in the Altai Territory, 5 in the Omsk Region, 3 in the Irkutsk Region and less than 1 in the Republic of Tyva. The level of use of precision farming technologies is 10% in the Novosibirsk Region, 8% in the Omsk Region, 5% in the Altai Territory and less than 2% in the remaining regions. In addition, we analysed the volume of production and sales of organic agricultural products, which is also one of the indicators of innovative development of the agro-industrial complex, and we also studied the use of modern dosers for the preparation of mixed fodder, which is also an important factor in the innovative development of agriculture [10]. We also analysed the use of new varieties and hybrids of agricultural crops, which showed that in the Novosibirsk and Omsk regions this indicator is 70%, in the Altai Krai 60%, in the other regions - less than 40%. In addition, we analysed the volume of production and sales of organic agricultural products, which is also one of the indicators of innovative development of the agro-industrial complex.

The research revealed significant heterogeneity in the resource potential of the agro-industrial complex (AIC) of different regions of the Siberian Federal District (SFD), which indicates significant differences in the opportunities and prospects of agricultural development within the macro-region. This heterogeneity is due to both natural-climatic factors and socio-economic conditions that have a determining influence on the efficiency of agricultural production and food security in each individual region. The study of such differences is important for analysing and assessing the sustainability of land use systems [11, 12].

In particular, analyses of arable land distribution show marked differences between the western and eastern parts of Siberia. Thus, the share of arable land in Western Siberia, according to our estimates, is about 15% of the total area of the territory, which certainly exceeds the same indicator for Eastern Siberia (8%). However, despite this difference, both indicators lag significantly behind the average Russian level, which exceeds 20%. This suggests that, in general, the Siberian region has less potential in terms of cultivated land compared to other regions of the country. At the same time, there is considerable variation within Western Siberia itself, where the most fertile land is concentrated in the southern regions, which in turn determines the differentiation of agricultural specialisation and productivity. In contrast, Eastern Siberia is dominated by areas where farming is limited due to unfavourable climatic conditions and relief features.

The most resource-rich regions in the SFD are the Altai Krai and the Novosibirsk Oblast. These regions are distinguished not only by significant areas of arable land, but also by favourable agro-climatic conditions, which makes it possible to achieve high crop yields. In particular, the Altai Territory has an average grain yield of 45 centners per hectare, while in the Novosibirsk Region this figure reaches 40 centners per hectare. In addition, these regions have well-developed infrastructure, including elevators, storage facilities, transport networks and processing plants, which creates additional incentives for the development of the agricultural sector. The availability of qualified personnel and the introduction of modern technologies also contribute to the efficiency of agricultural production in these regions. Investments in agro-industrial complex in Altai Krai and Novosibirsk Oblast ensure stable growth of production and high competitiveness of local agricultural producers, then also has similar examples in other regions of the world, where family farms play an important role in ecology and food production [13].

At the other pole are the Republic of Tyva and the Republic of Buryatia, where there are significant problems in the development of agro-industrial complex. These regions are characterised by low indicators of agricultural production efficiency, which is due to a whole set of factors. One of the key factors is the lack of funding, which limits the opportunities for upgrading the material and technical base, purchasing modern equipment and introducing innovative technologies. In addition, the low level of mechanisation of agricultural work (the mechanisation ratio in these regions does not exceed 0.4) also significantly reduces labour productivity. Unfavourable climatic conditions, such as a short growing season, low temperatures and moisture deficit, also have a negative impact on crop yields. Lack of developed infrastructure and limited access to markets further complicate the situation. All these factors together lead to low competitiveness of agriculture in the Republic of Tyva and the Republic of Buryatia.

An analysis of Russia's food security indicators, including the ratio of food imports to exports, has revealed the country's significant dependence on imports for a number of important food products. This dependence is a serious problem, as it threatens the country's food independence and makes it vulnerable to fluctuations in world prices. In particular, the share of imported meat and meat products in the consumption structure is more than 35 per cent, which significantly exceeds the permissible 25 per cent established in the Food Security Doctrine. A similar situation is observed with other types of food, for example, the share of imported vegetable oil reaches 45 per cent. Such high dependence on imports indicates the need for urgent measures to improve the efficiency of domestic production and reduce import dependence. This requires a comprehensive approach, including support for agricultural producers, the introduction of innovative technologies, infrastructure development, and the creation of favourable conditions for attracting investment in the agro-industrial complex. It is necessary to take into account the specifics of each region and develop support measures adapted to specific conditions and needs. Increasing the level of food self-sufficiency is a key factor in ensuring national security and sustainable development of the country.

Regional differences in resource endowment require the development of differentiated measures of state support for agro-industrial complex, taking into account the specifics of each region. It is also necessary to take into account the potential of new markets, which arises against the background of revolutionary changes in world trade [14, 15].

Comparison with similar studies conducted by FAO and other international organisations demonstrates the relevance of the problem of uneven development of the agro-industrial complex and import dependence for many countries of the world. The effectiveness of strategies to stimulate agricultural development depends on the specifics of regional conditions, the quality of public administration and the availability of financial resources, as well as socio-economic characteristics and subjective preferences regarding the quality of agricultural products [16, 17].

4 Conclusion

The study of the agro-industrial complex of the Siberian Federal District has revealed significant heterogeneity in the resource potential of the regions, which causes differences in their agricultural development and food security. The analysis of land resources has shown that in Western Siberia the share of arable land is higher than in Eastern Siberia, but both indicators are significantly inferior to the average Russian level. The qualitative characteristics of soils, such as humus content and degree of erosion, also vary widely across the regions, which affects crop yields. In terms of labour resources, there is an uneven distribution of those employed in agriculture, as well as differences in skill levels, wages and mechanisation of labour. Financial resources, including the volume of investment, lending and state support, are also unevenly distributed, which affects the development of the agricultural sector. In addition, the innovation potential, characterised by the introduction of new technologies and varieties, also differs by regions, as well as the socio-economic development of black earth villages in Russia [18].

The leaders in terms of resource endowment are the Altai Krai and the Novosibirsk Oblast, where there are high yields, developed infrastructure, qualified personnel and significant amounts of investment. In contrast, the Republic of Tyva and the Republic of Buryatia are characterised by low agricultural production efficiency due to a lack of funding, low levels of mechanisation and unfavourable climatic conditions.

An analysis of food security has revealed Russia's dependence on imports for a number of key products, such as meat, meat products and vegetable oil. This points to the need to improve the efficiency of domestic production and reduce import dependence. The obtained results emphasise the need to develop differentiated measures of state support for the agro-industrial complex, taking into account the specifics of each region. It is necessary to take into account natural, climatic and socio-economic conditions in order to optimise the use of resources and improve food security of both the region and the country as a whole.

References

1. T. Ogunbode, V. Oyebamiji, T. Akinkuolie, A. Oyelami, A. Adekiya, Trends in water supply variability and the sustainability of household water demand in tropical Africa: a case study. Urban Water J., **22,** 210-222, (2025). DOI: 10.1080/1573062X.2024.2445127
2. N. Garambois, U. Le Goff, L. Thibaudeau, Agrarian dynamics and climate change in the Senegalese Sahelian peanut basin. Agrar. Syst. Clim. Change: Journeys Adapt. Glob. South, 15-37 (2024). DOI: 10.1079/9781800628137.0001
3. L.S. Malyukova, V.V. Kondratiev, V.V. Bukhtoyarov, V.Y. Konyukhov, K.A. Bashmur, T.A. Panfilova. Circular Mining Wastes Management for Sustainable Production of Camellia sinensis (L.) O. Kuntze. Sustainability (Switzerland), **15 (15),** 11671 (2023). DOI: 10.3390/su151511671
4. H. Cochet, O. Ducourtieux, N. Garambois, Agrarian systems and climate change: Journeys of adaptation in the global south. Agrar. Syst. Clim. Change: Journeys Adapt. Glob. South, 1-208, (2024). DOI: 10.1079/9781800628137.0000
5. V. I. Silaev, R. V. Klyuev, D. V. Eremeev, T. A. Martynova, Danilchenko Yu.V. Analysis of the carbon footprint created by mining enterprises. MIAB. Mining Inf. Anal. Bull., **11-1,** 265-277, (2023). DOI: 10.25018/0236_1493_2023_111_0_265
6. K.I. Kravtsov, Y.A. Tynchenko, T.A. Panfilova Influence of mountain factors on salt excess and soil toxicity in mountain conditions. Sustainable Development of Mountain Territories, **15 (3),** 784 – 797, (2023). DOI: 10.21177/1998-4502-2023-15-3-784-797
7. V.Yu. Konyukhov, T.A. Oparina, I.Y. Matasova, M.A. Modina, N.V. Martyushev, Ecologization of underground coal mining by means of ash use in backfill preparation. Min. Inform. Anal. Bull., **10,** 123-135 (2024) DOI: 10.25018/0236_1493_2024_10_0_123
8. K.M. Didur, H.O. Kundieieva, G.V. Ortina, T.A. Pikhniak, A.V. Revkova, formation of food supply of the agrarian sector of ukraine on the basis of restoration of the development of rural areas. Rev. Iberoam. Vitic. Agroind. Rural., **12,** 223-239 (2025). DOI: 10.35588/pxe7k028
9. Kondratev S.I. Three typical hydrological-hydrochemical situations near the danube river mouth based on the marine hydrophysical institute research expeditions in 1997-2013. Physical Oceanography, **2019,** 26(4), 326–340. DOI 10.22449/1573-160X-2019-4-326-340
10. A. Kazhiyakhmetova, A. Omarov, A. Biniyazov, G. Zhazykbayeva, G. Kiyassova, Auger-in-auger doser with active return channel for compound feed preparation. Prog. Agric. Eng. Sci., **20,** 217-232 (2024). DOI: 10.1556/446.2024.00137
11. I.I. Bosikov, M.A. Modina, E.V. Khekert, analysis of the quality of underground mineral waters of terrigenous deposits of the hauteriv-barremian aquifer of the lower cretaceous. News of the National Academy of Sciences of the Republic of Kazakhstan, Series of Geology and Technical Sciences, **2024 (2),** 36-47, (2024). DOI: 10.32014/2024.2518-170X.392
12. I.A. Kerimov, L.R. Bekmurzaeva, Modern agro-climatic conditions of mountain landscapes in the North Caucasus, Sustainable Development of

Mountain Territories, **14(4),** 555 – 563, (2022). DOI: 10.21177/1998-4502-2022-14-4-555-563

13. B. Aparecida da Silva, R. Evangelista de Oliveira, A. Cavalieri Sais, E. Cardoso leite, Family farmers and trees: ecological functions, ecosystem services and food production in agrarian reform settlements in Brazil. Agroecol. Sustain. Food Syst., **49,** 4-21 (2025). DOI: 10.1080/21683565.2024.2405515
14. S. Bamidele, BRICS Revolution: The Global Agrarian Shift Sparked by Emerging Economies. India Q., (2025). Article in press. DOI: 10.1177/09749284241307937
15. V.Yu. Konyukhov, T.A. Oparina, I.Y. Matasova, M.A. Modina, N.V. Martyushev, Ecologization of underground coal mining by means of ash use in backfill preparation. Min. Inform. Anal. Bull., **10,** 123-135 (2024). DOI: 10.25018/0236_1493_2024_10_0_123
16. A. Al Mozahid, What counts as "quality" in agrarian taste? Subjectivities, gender, and rice agroecology in northern Bangladesh. Cult. Agric. Food Environ., **46,** 88-105 (2024). DOI: 10.1111/cuag.12323
17. A.A. Kapanski, E.A. Efremenkov, Geospatial Clustering in Smart City Resource Management: An Initial Step in the Optimisation of Complex Technical Supply Systems. Smart Cities., **8(1),** 14, (2025). https://doi.org/10.3390/smartcities8010014
18. N.A. Zhirov, Socio-Economic Development of the Black Soil Village in the XIX–early XX centuries: from the Prosperous Granary of Pre-Reform Russia to the Demo-Ecological Crisis of Rural Settlements. Bylye Gody, **19,** 1605-1614, (2024). DOI: 10.13187/bg.2024.4.1605

Land reclamation development under conditions of climate aridity growth: experience of the Republic of Tatarstan

Mars Khismatullin* [1,2], *Marsel* Khismatullin[1, 3], *Chulpan* Kurakova[1], *Timur* Khakimov[1], *Emil* Tazeev[1]

[1] Kazan State Agrarian University, Kazan 420015, Russia
[2] Academy of Sciences of the Republic of Tatarstan, Kazan 420111, Russia
[3] Kazan State Institute of Culture, Kazan 420064, Russia

Abstract. The paper investigates the issues of climate aridity growth and its impact on agronomic and economic efficiency of agricultural production in the Republic of Tatarstan. The role of complex land reclamation as one of the main directions of intensification of agriculture designed to mitigate the negative effects of drought, increase the fertility of arable land, prevent the spread of soil degradation and erosion and, of course, increase the efficiency of crop cultivation is studied. Analysis of agrometeorological observations for more than a century (1900-2021) has shown an increase in climate aridity in the region, especially in the last 40 years. The results of assessment of economic efficiency of irrigated farming in the republic show that in the existing natural and climatic conditions that 1.5% of arable land area under irrigation can produce up to 18% of crop production in Tatarstan. Underestimation of the role of land reclamation development in the agrarian policy of the state has been revealed, it should be developed at a faster pace to ensure food security of the country. On the basis of experimental studies on the basis of farms of different organisational and legal forms the high efficiency of land reclamation in combination with other innovative measures is proved. Based on the results of the research, appropriate recommendations have been developed to improve the effectiveness of state support measures, development of public-private partnership and introduction of science and technology achievements in land reclamation farming to prevent the impact of negative weather-climatic and anthropogenic factors on the economic efficiency of agricultural production and restrain the development of degradation of fertile land cover.

Keywords: integrated land reclamation, soil fertility, economic efficiency, aridity of climate.

1. Introduction

The adopted Doctrine of Food Security of the Russian Federation puts the issues of preserving state sovereignty of the Russian Federation through ensuring its food independence and security at the top of the agenda, which in the current realities is not possible without the use of intensification of farming, providing for the widespread introduction of elements of complex land reclamation, which have proved for decades high agronomic, environmental and economic efficiency, contributing not only to the preservation of fertile soil, but also to the conservation of fertile land.

The analysis of the current state of land reclamation complex shows that during all the years of agrarian reforms (except for the last 5-8 years) the above mentioned possibilities of land reclamation were underestimated in the agrarian policy of the state. Although state programmes in the field of land reclamation were adopted, but they were not supported by financing at a sufficient level for restoration of destroyed land reclamation systems and construction of new systems at a faster pace. As a result, of the available reclaimed lands, significant areas are not used for agricultural production. The research results of the All-Russian Research Institute of Irrigated Agriculture, reflected in the concept of reclamation industry development, indicate that for effective farming the area of irrigated land in Russia should be at least 10 million ha [4- 6]. So far, the country has only 4.69 million ha of such lands, i.e. twice less.

Noting the value of the conclusions, it should be stressed that the underlying causes of insufficiently full realisation of land reclamation opportunities in improving land productivity and increasing production remained insufficiently investigated, which served as the basis for choosing the topic of our research [7, 8].

The aim of the study is to analyse and evaluate the development of reclamation farming in the Republic of Tatarstan and to develop scientifically-based recommendations for its development.

2. Material and methods

Monographic, abstract-logical, economic-statistical and experimental research methods were used in the research process. The research was carried out on the basis of

* Corresponding author: marselmansurovic@mail.ru

analysis and generalisation of the authors' results in the field of land reclamation economics, the process of land reclamation implementation over the years of agrarian transformations in Russia, as well as experimental studies carried out on the basis of farms of different organisational and legal forms. Official documents adopted by the governments of Russia and the Republic of Tatarstan, data from Rosstat, Tatarstanstat, the Ministry of Agriculture and Food of the Republic of Tatarstan, the federal state budgetary institution 'Tatmeliovodkhoz', agricultural organisations of different legal forms were used.

3. Results and discussion

The observed growth of agricultural production indicators over the last 5 years, indicate the growth of the effectiveness of state support measures in the sphere, improvement of the financial and economic situation of agricultural enterprises. However, despite the fact that agricultural production in Russia is gradually increasing, the problem of nutrition at the level of rational norms is still not solved [9, 10]. This requires a significant increase in the production of products of fields and farms.

A special role in ensuring the food security of the country, economic availability of food products to the population belongs to increasing the productivity of the main productive force - land on the basis of land reclamation and other innovative measures, strengthening the material and technical base of agriculture as a whole. Ameliorated lands, if used rationally, can provide a rapid and significant increase in agricultural production. Thus, up to 50 per cent of vegetable, melon and potato production, 100 per cent of rice and about 20 per cent of fodder in Russia are produced on reclaimed lands, although such lands account for only 7.9 per cent of arable land [11, 12].

In economically developed countries of the West, the area of reclaimed land reaches up to 40 per cent of the arable land, on a global scale 17 per cent is reclaimed, where more than 75 per cent of crop production is produced, in the Russian Federation in the country where more than 20 per cent of the world's fresh water reserves are located, the area of irrigated agriculture occupies less than 1 per cent of the arable land, in the Republic of Tatarstan only 1.5 per cent of agricultural land is irrigated and up to 18 per cent of crop production is produced on it, which indicates a high degree of agronomic and economic efficiency.

The development of complex melioration was promoted not only by high economic efficiency of production, but also by the growth of climate aridity in the Volga region, which led to a significant decrease in moisture availability during vegetative periods of agricultural plants.

Analysis of 120 years of data (1900-2021) of meteorological observations of average annual air temperature in the territory of the Republic of Tatarstan, shown in Figure 1 indicates that during the study period the average annual air temperature in the territory of the Republic of Tatarstan increased more than three times. [16, 17].

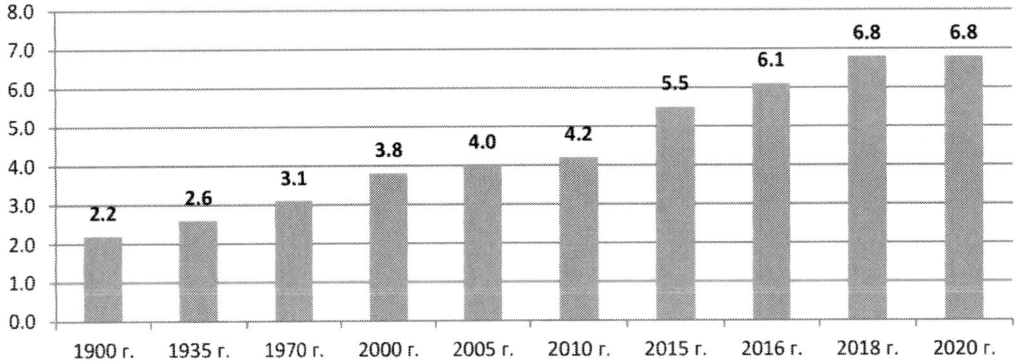

Figure 1. Dynamics of average annual air temperature on the territory of the Republic of Tatarstan 1900-2021, °C

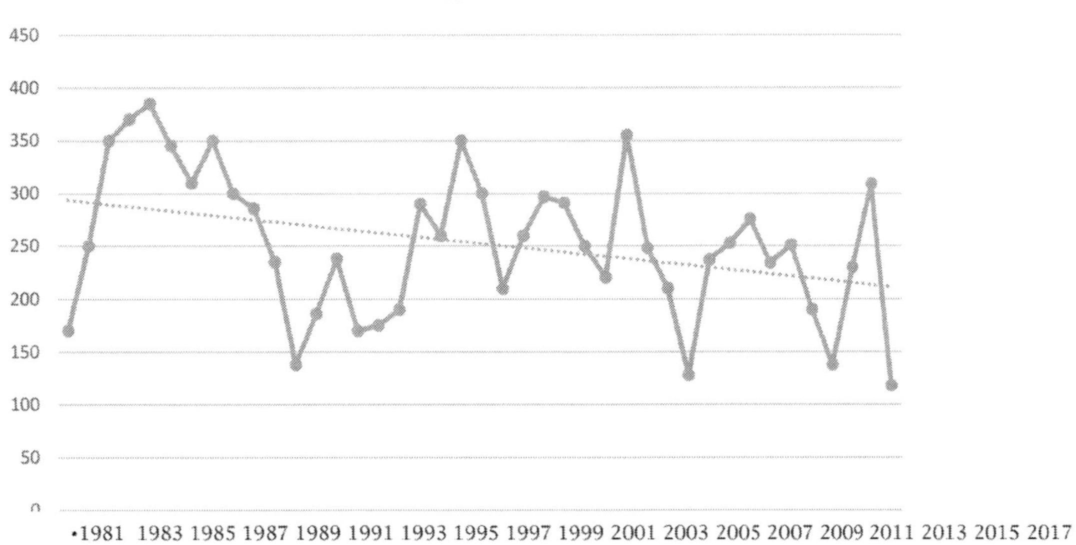

Figure 2. Dynamics of changes in the average annual precipitation in the Republic of Tatarstan for 1981-2021, mm

The study of the dynamics of the average annual precipitation in the Republic of Tatarstan for the last forty years (1981-2021) indicates a decrease in the average annual precipitation (Fig. 2) [16, 17].

The above circumstances, according to scientists of the Tatar Research Institute of Agriculture - a separate structural subdivision of the Federal State Budgetary Institution of Science 'Federal Research Centre "Kazan Research Centre of the Russian Academy of Sciences" (TATNIISKh FIC KazNTs RAS) and agrochemists of the Centre of Agrochemical Service "Tatarsky" (CAC "Tatarsky"), have led to an increase in inefficient evaporation of moisture from the soil, to a decrease in the level of groundwater table, shallowing of rivers, drying up of lakes, which affected the moisture availability of the root-inhabited layer of land, which once again proves the importance of the development of irrigation reclamation in modern natural and climatic realities. [16-18].

Increasing growth of active temperatures and decreasing volumes of precipitation lead to increasing aridity of climate on the territory of the Republic of Tatarstan, which has become more noticeable over the last 40 years, which led both the industry and republican leadership to recognise irrigation reclamation as one of the main directions of intensification of farming. As a result, not only wide participation in implementation of federal target programmes, but also development of a number of significant for the branch republican state programmes on development of material and technical base of meliorative branch, overhaul and construction of hydraulic engineering structures; technical re-equipment of melioration objects; drilling of wells for needs of agricultural enterprises; planting of field-protective forest plantations; liming of acid soils. To be fair, it should be noted that support measures for the development of land reclamation in the Republic are unprecedented, for example, when restoring hydraulic structures located on the lands of agricultural producers, 95% of the costs of the estimated cost of the object is financed from the republican budget, and to accelerate the equipping of commissioned irrigated lands with irrigation equipment, agricultural producers are reimbursed from the republican budget for 70% of the costs incurred for the purchase of such equipment [19, 20]. For 2019-2023 from the budget of the Republic of Tatarstan to finance land reclamation activities, according to the Ministry of Agriculture of the Republic of Tatarstan, in 1. 7 times more than from the federal budget, which allowed reclamation workers to build and reconstruct over 461 HS with a water volume of up to 110 million m3, to lay more than 331 km of main water pipeline, to put more than 40 thousand hectares of new irrigated lands into operation, to plant more than thirty thousand hectares of soil-protective forest belts and to preserve more than 140 thousand hectares of agricultural land from erosion [21, 22].

In this regard, the most important tool for solving the problems of food security in the context of deteriorating climatic conditions is a comprehensive land reclamation in combination with the application of knowledge-intensive agricultural technologies based on digitalisation and automation of the maximum number of agricultural processes. Taking into account the current situation of high efficiency of irrigation reclamation and based on the needs of the domestic market and the need to produce irrigation equipment in Russia, the heads of a number of enterprises of the Republic of Tatarstan and the management of the Ministry of Agriculture and Food of the Republic of Tatarstan in 2015, it was decided to establish in the Republic, namely on the basis of the Kazan branch of the Federal State Budgetary Institution of Management 'Privolskmeliovodkhoz' plant for the production of sprinkler systems that meet all modern international standards

In 2022, a modernised sprinkler machine with an online control system was developed and in 2024, serial production was started.

Table 1. Brief characteristics of Kazanka sprinklers manufactured by Kazan Irrigation Machinery Plant

Machine modification	Kazanka-3	Kazanka-4	Kazanka-5	Kazanka-6	Kazanka-7	Kazanka-8
Number of spans	3	4	5	6	7	8
Machine length, m	204	264	324	384	444	504
Water flow rate, l/s	18.6	28.9	40.9	55.6	73.2	90.9
Water pressure at the machine inlet, kgf/cm^2	1.9	2.1	2.4	2.9	3.5	4
Irrigation area, ha	14.6	24.9	35.5	49.3	65.3	83.6
Irrigation radius, m	216	276	336	396	456	516
Average daily irrigation rate, mm	11	10.5	10.0	9.8	9.7	9.4
Minimum turnover time, h	10.5	14	17.5	20.9	24.4	27.9
Minimum irrigation rate per one turn, m^3/ha	48	60.9	72.7	85.4	98.9	110

Figure 3. Kazanka sprinkler machine operation

Main effect-forming indicators of sprinklers of Kazanka series:

- - preventing or significantly reducing the loss of land fertility by improving rain quality (Fig. 4);
- - water saving by reducing pressure in the pipeline and improving sprinkler control;
- - improvement of the technological process of irrigation and on this basis the growth of irrigated land productivity;
- - low water pressure at the hydrant 2-3 atm;
- - electric drive system;
- - programmable reversible movement through a fully functional control panel;
- - possibility to deliver not only water but also plant protection products to the plant;
- - possibility of the mobile eight-section modification of the machine to move from one hydrant to another, providing watering of plants on the area of up to 160 hectares per season (Kazanka-8");
- - possibility of remote control and monitoring of the sprinkler machine operation and functionality;
- - possibility of refreshing irrigation with small rates in a short time, combining irrigation with foliar feeding of plants;
- - programmability of all panel functions, including date and time;
- - possibility to apply with irrigation water liquids with aggressive environment (animal waste water, micro- and macro-fertilisers);
- - Expansion of irrigated farming zone and intensification of agricultural production.

Figure 4. Erosion development under irrigation by Kazanka and Fregat sprinklers

Functions of remote monitoring and control of sprinklers 'Kazanka' allow to provide: monitoring simultaneously for several installations of one user, including location and geolocation on the map of installation/installations, installation status (online/offline, power, in/out of motion, irrigation/dry running, direction of motion), installation parameters (voltage, speed (m/hour, per cent), water pressure, irrigation rate, operating time, crop), remote control through the programme - start/stop, forward/backward, set sector, start by time and time of operation, remote control of sprinklers 'Kazanka'.

Economic efficiency of the sprinkler 'Kazanka' consists of the following:

- э water saving (at 5-fold irrigation annually for the amount of 40-60 thousand rubles);
- - energy saving (at 5-fold irrigation annually for the amount of 25-40 thousand rubles);
- - grain yield growth by 20-30 centners/ha (at grain cost of 15 thousand rubles/t annually for the amount of 300-450 $/ha).

Irrigation plant 'KAZANKA' is optionally offered in mobile version, which will allow to increase the area under irrigation with simultaneous reduction of capital costs, reduction of capital investments and repair costs.

Implementation of the research results in average for 5 years provided obtaining from each hectare of irrigated fodder crops: 6.8 tonnes of grain units; 50.2 tonnes of white cabbage; 45-50 tonnes of carrots with production profitability of 125-133 % (with irrigation the production cost is 2 and more times lower than on rainfed land). In the structure of production costs, irrigation takes no more than 10% of costs in potato cultivation and 20-25% in vegetable cultivation [25]. Irrigation of agricultural crops allowed the agrarians of Tatarstan to fully provide the population of the republic with locally produced vegetables and potatoes from irrigated lands; to meet the needs of livestock of large investors and farms of the republic with fodder base from irrigated lands regardless of weather and climatic conditions, as well as the disposal of runoff in compliance with environmental requirements [26].

4. Conclusions

Analysis of long-term meteorological observations of average annual air temperature, study of dynamics of average annual precipitation on the territory of the Republic of Tatarstan has shown the growth of climate aridity on its territory, which led to the growth of inefficient evaporation of moisture from the soil, reduction of groundwater table, shallowing of rivers, drying up of lakes, which negatively affected water availability in the root layer of the land, which proves the urgent need for the development of irrigation reclamation in modern natural-climatic realities. In turn, reclamation measures allow to receive from each hectare of irrigated fodder crops: 6.8 tonnes of grain units; 50.2 tonnes of white cabbage; 45-50 tonnes of carrots with production profitability of 125-133 %. At irrigation the cost of production is 2 and more times lower than on rainfed land. In the structure of the cost of finished products irrigation occupies no more than 10 % of expenses at cultivation of potatoes and 20-25 % at cultivation of vegetables. Irrigation reclamation has already allowed: to provide the population of the republic with locally produced vegetables and potatoes from irrigated lands; to meet the needs of livestock of large investors and farms of the republic with fodder base from irrigated lands regardless of weather and climatic conditions, as well as waste water disposal in compliance with environmental requirements. 1.5% of irrigated arable land allows agrarians of the republic to produce up to 18% of gross crop production. However, further development of complex land reclamation is impossible without cardinal improvement of the situation in the land reclamation system of the country, through significant amendments in the agrarian policy of the state in the direction of allocation of significant additional budgetary funds for reconstruction and rehabilitation of existing unused and construction of new land reclamation systems at a faster pace, development of other directions of land reclamation, training of personnel for land reclamation, creation of favourable conditions for them to work and live.

References

1. A.A. Lukmanov, R.M. Gainullin, I.G. Gainutdinov, Agrochemical Bulletin, **2,** 3-10 (2023)
2. A.R. Valiev, N.M. Asadullin, L.V. Mikhailova, et al, Bulletin of Kazan State Agrarian University, **18(2(70)),** 199-205 (2023)
3. V.G. Sychev, Plododoroddie, 1(130), 57-63 (2023)
4. N.M. Asadullin, A.A. Lukmanov, D.F. Zaripov, et al, Agrochemical Bulletin, **6,** 65-71 (2023)
5. I.M. Minnehametova, L.F. Gafiullina, Co-operative and Sustainable Devel. and Sustainable Devel.: Conf. proc. (Moscow), (Cham: Springer Nature Switzerland, 2022), vol. **245**, pp. 1241-1248
6. M.M. Hismatullin, Bulletin of Kazan State Agrarian University, **5(1(15)),** 123-125 (2010)
7. F.N. Mukhametgaliev, N.M. Asadullin, F.N. Avkhadiev et al, Int. Sci.-Pract. Conf. Agricult, and Food Security: Technol, Innov, Markets, Human Resour. (EDP Sciences, Kazan, 2021), p. 00080.
8. I. Akzharkyn, K. Yelemessov, D. Baskanbayeva, Applied Sciences (Switzerland), **14(17),** 7567, (2024). DOI: 10.3390/app14177567
9. P.A. Chekmarev, Kormoproizvodstvo, 2, 10-12 (2012)
10. F.N. Safiollin, A.R. Valiev, S.R. Suleymanov, Bulletin of Kazan State Agrarian University, 17(4(68)), 50-55 (2022)
11. F.N. Salakhutdinov, I.R. Iskhakov, Bulletin of Kazan State Agrarian University, **6(2(20)),** 52-54 (2011)
12. F.N. Safiollin, G.S. Minnullin, S.V. Sochneva, Grain Farming of Russia, **2(50),** 29-33 (2017)
13. F.N. Safiollin, Kormoproizvodstvo, 7, 12-18 (2019)
14. F.N. Mukhametgaliev, R.G. Khisamov, Bulletin of Kazan State Agrarian University, **10(2(36)),** 31-35 (2015)
15. G.N. Agieva, L.S. Nizhegorodtseva, R.J.K. Diabankana, Bulletin of Kazan State Agrarian University, **15(4(60)),** 5-9 (2020), https://doi.org/10.12737/2073-0462-2021-5-9
16. O.L. Shaitanov, R.M. Nizamov, E.I. Zakharova, Grain legumes and cereals, **4(40),** 102-112 (2021)
17. A.A. Lukmanov, T.B. Hakimov, M.G. Kuznetsov, C.M. Kurakova, et al, Agrarnaya nauka, **5**, 136-143 (2024), https://doi.org/10.32634/0869-8155-2024-382-5-136-143
18. G.S. Minnullin, L.T. Vafina, F.N. Safiollin, Bulletin of Kazan State Agrarian University, **6(1(19)),** 160-162 (2011)
19. F.N. Avkhadiev, N.R. Aleksandrova, I.G. Gainutdinov, Bulletin of Kazan State Agrarian University, **17(1(65)),** 108-113 (2022), https://doi.org/10.12737/2073-0462-2022-108-113
20. O. V. Afanaseva, O. K. Bezyukov, A. A. Ignatenko, Academic Journal of Manufacturing Engineering, **22(1),** 68–75, (2024). DOI: 10.24874/PES.SI.02.017
21. S. I. Kondratyev, D. Baskanbayeva, K. Yelemessov, Y. Sarsenbayev, V. A. Turkin, International Journal of Hydrogen Energy, **95,** 212–216, (2024). DOI: 10.1016/j.ijhydene.2024.11.182
22. K. Zaurbekov, A. Syzdykov, S. Zaurbekov, J. Ismailova, Chemical Papers, **78(8),** 4867–4873, (2024). DOI 10.1007/s11696-024-03436-z
23. D.F. Khafizov, E.S. Isaicheva, Bulletin of Kazan State Agrarian University, **8(1(27)),** 55-58 (2013)
24. D.F. Khafizov, E.S. Isaicheva, Bulletin of Kazan State Agrarian University, **6(1(19)),** 82-84 (2011)
25. R.M. Nizamov, S.R. Suleymanov, F.N. Safiollin, Bulletin of Kazan State Agrarian University, 13(1(48)), 38-40 (2018)
26. I.G. Gainutdinov, N.M. Asadullin, A.K. Subaeva, L.V. Mikhailova, Bulletin of Kazan State Agrarian University, **18(1(69)),** 102-111 (2023)

System approach to ensuring the quality and safety of poultry meat production

Yakubbay Jumatov[1,*], *Darob* Berdiev[2], *Sharifbek* Saidaxmatov[3], *Aleksey V.* Masenko[4], *Polina A.* Sablina[4]

[1] "Tashkent Institute of Irrigation and Agricultural Mechanization Engineers" National Research University, Tashkent, Uzbekistan
[2] Tashkent State Technical University, Tashkent, Uzbekistan
[3] Termiz State Pedagogical Institute, Termiz , Uzbekistan
[4] Kuban State Agrarian University named after I. T. Trubilin, Krasnodar, Russia

Abstract. The authors have analysed the factors affecting the quality and safety of production of poultry meat and secondary processed products in Russia. The study covers production dynamics since 2010, identifies regional specifics, and determines significant factors affecting product safety, such as biological, physiological, technological, socio-economic and regulatory aspects. Special attention is paid to the influence of the level of infrastructure development, climatic conditions, investment climate and state support on the distribution of production across regions. The paper substantiates the need to implement international quality standards such as GMP, ISO 9001 and HACCP as an effective tool to improve safety and competitiveness of the industry. The results emphasise the importance of an integrated approach that includes control of all stages of the production process and consideration of various factors affecting the quality and safety of poultry meat.

1 Introduction

Food security is one of the key challenges for the development of countries. In the context of a growing population and changing climatic conditions, ensuring a stable and quality food supply is becoming an increasingly difficult task. A special place in the structure of food supply is occupied by meat production, which is an important source of protein and essential micronutrients for a nutritious human diet. In particular, the production of poultry meat shows steady growth both in Russia and worldwide, which is due to its relatively low cost, high nutritional value and variety of cooking methods. However, the intensification of poultry meat production involves a number of challenges, including ensuring the safety and quality of the final product, as well as the rational use of all components of raw materials, including secondary slaughter products, which is also important in the context of adaptation of agricultural systems to climate change [1, 2].

There are several main approaches to solve these problems. One of them is the improvement of technological processes at all stages of production, from poultry rearing to processing. This includes the introduction of new breeding and feeding methods, optimisation of poultry housing conditions, and the use of modern slaughtering and processing technologies. Another important approach is the development and implementation of quality and safety management systems based on the principles of Hazard Analysis and Critical Control Points (HACCP). Such systems provide a systematic approach to identifying and preventing potential hazards at all stages of production, which in turn ensures the production of safe and quality products. The third approach is the sustainable utilisation of all components of raw materials, including secondary slaughter products. This involves developing technologies to process these products to produce valuable food ingredients such as proteins, fats and collagen, which can be used to create new foods or animal feeds. The advantages of the first two approaches include the ability to control product quality and safety at every stage of production, reduce the risk of food poisoning and provide consumers with quality products. However, these approaches require significant investment in modernising production lines and training staff. In addition, the implementation of HACCP systems requires continuous monitoring and control, which can be difficult for small businesses. The third approach has a number of advantages, such as increasing the economic efficiency of production through better utilisation of raw materials, and reducing the negative impact on the environment by reducing waste. However, processing secondary slaughter products requires specialised equipment and technology, which can be costly. In addition, the possibility of contamination of recycled products must be taken into account and their safety must be carefully monitored.

* Corresponding author: y.jumatov@mail.ru

In the context of the above, the systematic approach to ensuring the quality and safety of poultry meat and secondary slaughter products described in the attached file is particularly relevant and important. It combines the advantages of several approaches, offering a comprehensive solution to the problem. This approach involves the development, implementation and maintenance of a quality and product safety management system based on the requirements of international standards such as GMP, ISO 9001 and HACCP. This systematic approach allows not only to ensure the safety and quality of products, but also to increase the competitiveness of enterprises in the international market, which is especially important in the conditions of global competition and changes in agricultural markets [3, 4]. In addition, it contributes to more efficient use of resources and reduction of production waste [5, 6].

The aim of the work is to substantiate the application of a systematic approach to ensuring the quality and safety of poultry meat and secondary slaughter products during processing, by analysing the factors affecting quality and safety, as well as substantiating the prospects for the implementation of international standards at enterprises of the processing industry, which echoes the importance of subjective perception of the quality of food products [7, 8].

2 Method and materials

In this paper, a comprehensive analysis of factors affecting the quality and safety of poultry meat and secondary slaughter products was carried out. The study included several stages. The first stage involved a detailed analysis of statistical data on poultry meat production in the Russian Federation in order to identify production dynamics, geographical distribution and export orientation trends. This analysis allowed us to identify the main factors contributing to the growth of poultry meat production in Russia since 2010. These factors included national projects, government and industry programmes, investment attraction, organisational and economic measures, and innovations in breeding, veterinary medicine and poultry product processing.

The next stage of the study was devoted to structuring the factors affecting the safety of poultry meat and poultry products. For this purpose, literature sources and regulatory documents were analysed, resulting in the creation of a structural scheme showing the interrelationships between the various factors that ensure the safety and quality of products. Biological, physiological, technological and socio-economic aspects were taken into account in the analysis.

To objectively analyse the quality and safety of poultry meat, a number of methods were used to assess various indicators, including GMO content, enzymes, biologically active substances, nutritional and biological value. For quantitative measurements, standard methods of analysis were used to ensure the reliability and reproducibility of the results.

In the final part of the work, recommendations were formulated to ensure the quality and safety of poultry meat and secondary slaughter products, taking into account the identified factors and the developed structural scheme. These recommendations were based on the experience of the poultry processing industry and international experience in implementing quality and safety management systems such as GMP and ISO 9001.

3 Results and discussion

In this paper, a comprehensive assessment of the factors affecting the quality and safety of poultry meat and poultry products was carried out. The research covered the analysis of production dynamics, the study of structural elements of the safety system, and the identification of prospects for the introduction of quality management systems at enterprises in the industry. At the beginning of the work, an in-depth analysis of statistical data on poultry meat production in Russia, for the period from 2010 to 2015, was carried out. The analysis was based on open sources, including data from Rosstat and the Russian Ministry of Agriculture. This allowed us to identify key trends in the development of the industry, such as increased production volumes, growth in export potential and regional distribution of production.

The results of the statistical analysis showed that poultry meat production in Russia demonstrated strong growth during the period under review. In particular, in 2015, poultry meat production totalled 2,247.1 thousand tonnes, which is 9.4% higher than in 2014 (2,054.3 thousand tonnes). Data for 2020 show that total poultry meat production in Russia reached 5.1 million tonnes in slaughter weight. In 2022, production continued to grow to 5.4 million tonnes, and by the end of 2023 the figure exceeded 5.5 million tonnes. Preliminary estimates for the beginning of 2024 indicate the continuation of positive dynamics, with production growth of 3-4% compared to the same period of 2023. It is worth noting that production growth has slowed slightly since 2015, which may be due to domestic market saturation and changing consumer preferences.

A study of the regional distribution of poultry meat production in Russia reveals significant unevenness. As of 2021, the Belgorod Region was the leader, accounting for 13.8% of total production. It was followed by the Chelyabinsk (5.8%), Leningrad (5.3%), Krasnodar (5.1%) and Stavropol (4.1%) regions. In 2022, the Belgorod Region retained its leadership, although its share fell slightly to 12.5%, which is probably due to general economic challenges and partial business reorientation. The Bryansk and Kursk regions showed a significant increase in production in 2022 - 8% and 6%, respectively, indicating increased investment and development of poultry farms in these regions. According to data obtained from open sources (Rosstat, analytical reports of agricultural companies), there is some redistribution in the first half of 2024: the share of the Belgorod region has stabilised around 12%, while the Bryansk and Kursk regions continue to show steady growth, albeit more moderate than in 2022, at around 3-4% each. The leadership of the central and southern regions is also confirmed by analysing the data for the first half of 2024. A slight increase in the share of the Leningrad Region is due to the growing demand for poultry products in the North-West Federal District.

The uneven distribution of poultry meat production in Russia is due to a complex of factors. Firstly, there are differences in the level of infrastructure development. Regions with a well-developed transport network, access to quality feed and a skilled labour force have a clear advantage. For example, the Belgorod region's proximity to major transport arteries and its developed agricultural infrastructure contribute to efficient logistics and lower transport costs. Secondly, natural and climatic conditions play a significant role. Southern regions have a more favourable climate for growing fodder, which reduces the cost of production. However, northern regions, such as the Leningrad Oblast, are also successfully developing poultry farming using modern technologies that compensate for unfavourable climatic factors. Thirdly, economic conditions, including access to investment, availability of government support and tax incentives, also have a significant impact on poultry development, as well as water availability, as emphasised in studies on tropical Africa [9].

Government programmes aimed at supporting agricultural enterprises sometimes have an uneven impact on different regions.

Comparisons with other countries show that Russia, despite significant regional differences, has more concentrated poultry production than, for example, the USA, where production is more evenly distributed among many states. In the USA, family poultry production is highly developed, which contributes to decentralised production. In the EU, there is also a more even distribution, due to strong co-operation between farms and tighter regulation. However, in both the US and the EU there are regions with significant concentrations of poultry farms due to the same factors as in Russia: developed infrastructure, climatic conditions and availability of government support.

In order to better understand the mechanisms of product safety, a structural scheme of factors affecting the safety of poultry meat and processed products was developed. This scheme included five main groups of factors: biological, physiological, technological, socio-economic and regulatory. Biological factors include the content of GMOs, enzymes and biologically active substances, physiological factors include nutritional and biological value, as well as compliance with the needs of different consumer groups, which is also related to gender and development issues in the agricultural sector [10, 11].

Technological factors determine the safety of production processes, socio-economic factors - price category, packaging quality, shelf life and socio-ethical factors. Regulatory and legal factors - compliance of products and production process with the norms and requirements of TR CU. As a result of the analysis it was revealed that such factors as the presence of GMOs, balanced composition of food substances, and compliance with technological regulations have a significant impact on product safety. The results of our additional research show that one of the most critical points is the content of pathogenic microflora, which is influenced not only by sanitary conditions of production, but also by the correct thermal processing of products. According to our data, in 2021-2023, Salmonella and Listeria exceedances were detected in some batches of poultry meat, indicating the need to strengthen control at all stages of production, especially at the slaughter and processing stages.

The study emphasised that an effective way to ensure quality and safety of poultry meat is to implement quality and safety management systems that comply with international standards such as GMP, ISO 9001 and HACCP. These standards represent a comprehensive approach that ensures control at all stages of production, from raw material receipt to sales, demonstrating higher levels of regulatory compliance, lower reject rates and higher consumer satisfaction, which is similar to the experience of other countries in optimising water use in arid areas [12, 13].

For example, we conducted comparative studies on two poultry farms, one of which implemented the HACCP system in full, and the other limited itself to minimum compliance with the requirements of TR CU. The results showed that the factory with HACCP had a 35 per cent lower rate of quality control rejects than the factory with minimal compliance. In addition, the number of consumer complaints about product quality was half as high in the factory with HACCP.

A comparative analysis with other studies in this area has shown that the results presented are consistent with general trends noted in world practice. In particular, many researchers note that intensification of poultry production is associated with risks of deterioration of product quality and safety, especially if modern technologies and quality management systems are not applied, which requires new approaches to the assessment of agroecological potential [14, 15]. Studies in the EU countries have shown that the introduction of ISO 9001 and HACCP standards leads to a significant improvement in the safety performance of food products, including poultry meat.

Thus, the results of the study suggest the need for a comprehensive approach to ensuring the quality and safety of poultry meat. This approach should include the introduction of quality management systems that comply with international standards, control of all stages of the production process, as well as consideration of biological, physiological, technological and socio-economic factors affecting the quality and safety of products. The data obtained can be used to develop strategies for the development of the poultry industry aimed at providing the population with high-quality and safe products, as well as increasing the competitiveness of domestic enterprises in the international market.

4 Conclusion

Analysis of the data presented shows a steady growth in poultry meat production in Russia since 2010, reaching a level of more than 5.5 million tonnes by the end of 2023. Despite a slight slowdown in growth rates in recent years, the dynamics remain positive, which is due to the successful implementation of national projects, government programmes, investment attraction and the introduction of innovations in breeding, veterinary medicine and processing. However, there is a significant unevenness in the regional distribution of production. The Belgorod Region has traditionally held the leading position, although its share has slightly decreased in recent years, ceding some positions to the Bryansk and Kursk Regions, which are demonstrating active growth. This regional distribution is due to a complex of factors, including differences in infrastructure development, favourable climatic conditions for growing feed, as well as access to investment and government support. Overall, Russian poultry production is characterised by a higher concentration of production compared to the US and the EU, where the distribution is more even, due to the development of family poultry production and cooperation.

The study highlights the importance of a systematic approach to ensuring product quality and safety, and the need to use modern tools such as neural networks to analyse data [16]. The developed structural scheme of factors affecting safety includes biological, physiological, technological, socio-economic and regulatory aspects. Particular attention is paid to the control of pathogenic microflora such as Salmonella and Listeria and compliance with technological regulations. Implementation of international standards such as GMP, ISO 9001 and HACCP is seen as an effective tool to improve product quality and safety. Comparative analyses have shown that companies applying HACCP have lower reject rates and fewer complaints from consumers. Thus, for further development of the industry it is necessary to strengthen control at all stages of production, to stimulate the introduction of modern technologies and quality management systems, as well as to create equal conditions for the development of poultry production in all regions of the country, including the use of new waste disposal technologies [17, 18]. In addition, an important aspect is the use of modern dosers in the production of mixed fodder for poultry [19]. This will help to provide the population with high-quality and safe products and increase the competitiveness of Russian enterprises in the international market. To improve the efficiency of the agricultural sector, it is necessary to take into account the issues of food supply formation [20].

References

1. N. Garambois, U. Le Goff, L. Thibaudeau, Agrarian dynamics and climate change in the Senegalese Sahelian peanut basin. Agrar. Syst. Clim. Change: Journeys Adapt. Glob. South, **15**, 37, (2024). DOI: 10.1079/9781800628137.0001
2. H. Cochet, O. Ducourtieux, N. Garambois, Agrarian systems and climate change: Journeys of adaptation in the global south. Agrar. Syst. Clim. Change: Journeys Adapt. Glob. South, **1**, 208, (2024). DOI: 10.1079/9781800628137.0000
3. A. I. Trunova, V. S. Tynchenko, S. O. Kurashkin, K. A. Bashmur, Y. A. Tynchenko, R. B. Sergienko, S. V. Tynchenko, Metals, **13(7),** 1234 (2023). DOI: 10.3390/met13071234.
4. S. Bamidele. BRICS Revolution: The Global Agrarian Shift Sparked by Emerging Economies. India Q., (2025). DOI: 10.1177/09749284241307937.
5. V.I. Golik, D.A. Zyukin, A.I. Karlina. Technology for Nonwaste Recovery of Tailings of the Mizur Mining and Processing Plant. Metallurgist, **66,** 1476-1480, (2023). DOI: 10.1007/s11015-023-01462-y.
6. O. Afanaseva, M. Afanasyev, S. Neyrus, D. Pervukhin, D. Tukeev. Information and Analytical System Monitoring and Assessment of the Water Bodies State in the Mineral Resources Complex. Inventions, **9,** 115, (2024). https://doi.org/10.3390/inventions9060115.
7. A. Al Mozahid, What counts as "quality" in agrarian taste? Subjectivities, gender, and rice agroecology in northern Bangladesh. Cult. Agric. Food Environ., **46,** 88-105 (2024). DOI: 10.1111/cuag.12323
8. Serdechnaya Yu.N., Serdechnyy D.V., Andriyanov N.A., Korchagin S.A. Development of an effective monitoring and analysis system of modern technologies for the agro-industrial complex development in mountainous areas, **16(2),** 770 – 779, (2024). DOI: 10.21177/1998-4502-2024-16-2-770-779
9. T. Ogunbode, V. Oyebamiji, T. Akinkuolie, A. Oyelami, A. Adekiya, Trends in water supply variability and the sustainability of household water demand in tropical Africa: a case study. Urban Water J., **22,** 210-222 (2025). DOI: 10.1080/1573062X.2024.2445127
10. L. Ossome, Gender and development in the agrarian south. World Dev., **188,** 106876 (2025). DOI: 10.1016/j.worlddev.2024.106876
11. N.V. Babyr, A.V. Pogrebnoy, E.A. Efremenkov, D.V. Valuev, A.E. Boltrushevich, Modelling of Reliability Indicators of a Mining Plant. Mathematics, **12(18),** 2842 (2024) DOI: 10.3390/math12182842
12. S. Mukherjee, S. Paul, S. Bhattacharya, A. Islam, S. Mahammad, E. Alam, Optimizing water resource management in tropical drought-prone regions through hybrid MCDM techniques: A water-stress mapping approach. J. Hydrol. Reg. Stud., **57,** 102171 (2025). DOI: 10.1016/j.ejrh.2024.102171

13. A.A. Kapanski, E.A. Efremenkov, Geospatial Clustering in Smart City Resource Management: An Initial Step in the Optimisation of Complex Technical Supply Systems. Smart Cities, **8(1),** 14, (2025). https://doi.org/10.3390/smartcities8010014
14. V.Yu. Konyukhov, T.A. Oparina, I.Y. Matasova, Ecologization of underground coal mining by means of ash use in backfill preparation. Min. Inform. Anal. Bull., **10,** 123-135, (2024). DOI: 10.25018/0236_1493_2024_10_0_123
15. Kondratev S.I. Three typical hydrological-hydrochemical situations near the danube river mouth based on the marine hydrophysical institute research expeditions in 1997-2013. Physical Oceanography, **26(4),** 326–340, (2019). DOI 10.22449/1573-160X-2019-4-326-34
16. S.N. Sorokova, E.A. Efremenkov, D.V. Valuev, M. Qi, Analysis of a Predictive Mathematical Model of Weather Changes Based on Neural Networks. Mathematics, **12 (3),** 480, (2024) DOI: 10.3390/math12030480
17. V.I. Golik, R.V. Sorokova, Q. Mengxu, Tailings Utilization and Zinc Extraction Based on Mechanochemical Activation. Materials, **16(2),** 726, (2023). DOI: 10.3390/ma16020726
18. L.S. Malyukova, K.A. Bashmur, T.A. Panfilova, V. Brigida, Circular Mining Wastes Management for Sustainable Production of Camellia sinensis (L.) O. Kuntze. Sustainability (Switzerland), **15(15),** 11671, (2023). DOI: 10.3390/su151511671
19. A. Kazhiyakhmetova, A. Omarov, A. Biniyazov, G. Zhazykbayeva, G. Kiyassova, Auger-in-auger doser with active return channel for compound feed preparation. Prog. Agric. Eng. Sci., **20,** 217-232 (2024). DOI: 10.1556/446.2024.00137
20. K.M. Didur, H.O. Kundieieva, G.V. Ortina, T.A. Pikhniak, A.V. Revkova, FORMATION OF FOOD SUPPLY OF THE AGRARIAN SECTOR OF UKRAINE ON THE BASIS OF RESTORATION OF THE DEVELOPMENT OF RURAL AREAS Rev. Iberoam. Vitic. Agroind. Rural., **12,** 223-239, (2025). DOI: 10.35588/pxe7k028

Influence of spring wheat seed treatment with low-temperature plasma on growth processes

R.R. Khuzina[1], *Man. Mak.* Nafikov[1], *M.M.* Nafikov[1], *Mars M.* Khismatullin[2,3], *Marsel M.* Khismatullin[2,4]*

[1] Kazan (Volga Region) Federal University, Kazan, Russia
[2] Kazan State Agrarian University, Kazan, Russia
[3] Privolzhskmeliovodkhoz Department, Kazan, Russia
[4] Kazan State Institute of Culture, Kazan, Russia

Abstract. In the Russian Federation in recent years, fallow lands have been actively put into circulation. On fallow lands a complex of agromeliorative measures is carried out. The demand for high quality seeds for sowing in the newly introduced into turnover lands is also increasing. The purpose of our research is to study the influence of corona discharge of low-temperature plasma on seeds and growth processes of spring wheat. The research was conducted on the basis of the Institute of Artificial Intelligence, Robotics and Systems Engineering KFU To conduct experiments in the laboratory of the Department of Biomedical Engineering and Artificial Intelligence in Biotechnical Systems of the Institute of Artificial Intelligence and Systems Engineering KFU in 2022–2024 developed a scheme and assembled the installation of corona discharge generation of low-temperature atmospheric plasma. The experiments were carried out in tenfold repetition. Seeds were treated with corona discharge at the following parameters: voltage 4 and 8 kV; frequency 0.2 kHz; 0.3 kHz; 0.5 kHz; seed distance from the plasmatron 3 cm; seed treatment time: 20, 25, 30, 35 seconds. Seeds of control variant and variants with corona atmospheric plasma treatment were germinated by roll method in Lamsystem laminar flow cabinet. The data obtained show that at a voltage of 4 kV, the highest growth of wheat seedlings was observed at a frequency of 0.5 kHz and a treatment time of 30 seconds. The length of seedlings for seven days, on this variant of seeds treated with corona discharge on average reaches up to 34.4 cm, and the length on the control variant of the experiment 11.7 cm, which is 22.7 cm shorter. The greatest length of seedlings at a voltage of 8 kV is observed at a treatment time of 30 seconds, where the length of the seedling reaches 36.2 cm, which is 24.5 cm more from the control variant of the experiment.

Keywords: low-temperature plasma, spring wheat, corona discharge, treatment time, voltage.

1. Introduction

Seeds of agricultural crops are living organisms and they are the basis of crop production. High field germination, germination energy, norms and methods of sowing, high percentage of safety before harvesting are the key to formation and obtaining high stable yields with good grain quality. The results of long-term studies usually show that the yield of sown seeds with low germination is about 5–40% lower than that of seeds with high germination [1–3]. Consequently, increasing germination and sowing high quality seeds are of great practical importance in agricultural production.

In practice, several methods are used to improve seed germination, including chemical treatment (chemicals, fungicides, etc.) [4, 5] and physical treatment (ultrasonic treatment, electric field treatment, magnetic treatment, ion beam treatment, etc.) [6–8]. Chemical methods, as a rule, are labour-intensive and expensive, besides, chemical pesticides applied in the process of inoculation on the seed coat contribute to the contamination of soil, groundwater, and they also pose a danger to the working personnel. In recent years, much attention has been paid by scientists and practitioners to physical methods of seed disinfection and protection against harmful pathogens [9, 10]. As physical methods, ultrasonic and ion-beam treatments are used to increase seed germination, electromagnetic treatment, while increasing seed germination, also affects biochemical processes such as protein and enzyme activity [11]. However, these physical methods have negative effects on seeds in addition to their positive effects. For example, strong ultrasonic vibration or collision with an ion beam damages seed cells, resulting in reduced germination under field conditions [12].

Seed treatment with low-temperature plasma is a highly productive, economical and environmentally friendly method of crop seed disinfection, also plasma treatment helps to improve seed germination and rapid seed growth, which contributes to a significant increase in yield. A number of researchers have noted that when seeds were treated with low-temperature plasma, there

* Corresponding author: marselmansurovic@mail.ru

was no destruction of seeds because the active particles penetrated to a depth of only about 10 nm [13, 14]. In addition, no environmental pollutants are emitted after seed treatment with low-temperature plasma, as no chemicals are used for this purpose.

Literature sources indicate that the term 'NTP' – (low-temperature plasma, cold plasma) denotes the state of ionised gas at ambient temperature with dominant collective behaviour of charged particles [15]. The common ways to produce NTP are electric discharges (corona, dielectric barrier, radiofrequency, microwave, etc.).

In our research we used corona discharge. Corona discharge (CD) is usually generated by high voltage at sharp electrodes such as tips, spicules or thin wires. An electric field is formed near such points and an active region of corona discharge and plasma generation occurs. The active region of corona discharge appears only near the point electrode and is limited to units of mm. [16].

Wheat is one of the most important cereal crops and is widely grown all over the world as a staple food, the demand for it is constantly increasing as the arable land area decreases and the population increases. In many regions of the Russian Federation, fallow lands have been brought into agricultural turnover in recent years with the active implementation of a set of reclamation measures [17]. Therefore, increasing production, improving the quality of agricultural seeds becomes an urgent problem.

The aim of our research is to study the effect of corona discharge of low-temperature plasma on seeds and growth processes of spring wheat of Arhat variety.

The research was conducted at the Institute of Artificial Intelligence, Robotics and Systems Engineering of KFU in 2022-2024.

2. Materials and methods

The object of research was calibrated spring wheat seeds of Arhat variety without macrocracks and without signs of embryo damage. To conduct experiments in the laboratory of the Department of Biomedical Engineering and Artificial Intelligence in Biotechnical Systems of the Institute of IRSI KFU developed a scheme and assembled a corona discharge generation unit of low-temperature atmospheric plasma. Seeds of the control variant and variants with corona atmospheric plasma treatment were germinated by the roll method in a Lamsystem laminar flow cabinet.

Experiments were carried out in tenfold repetition. Seed treatment with corona discharge was carried out under the following parameters:

- voltage 4 and 8 kV;
- frequency 0.2 kHz; 0.3 kHz; 0.5 kHz;
- distance of seeds from the plasmatron 3 cm;
- seed treatment time: 20, 25, 30, 35 seconds.

3.Results and Discussion

The high voltage was selected in such a way as to prevent arc discharge between the needle of the upper electrode and the base. Corona discharge was ignited starting at 2 kilovolts. Depending on the distance between the electrodes, arc discharge did not occur up to 4–8 kilovolts, so the operating voltage at the electrodes was set within 2–8 kilovolts.

From literature sources it is known that the greatest effect of low-temperature plasma on seeds was observed at duration t ≈ 30 sec, so the time of exposure of seeds was chosen [16].

The biological effect of low-temperature corona discharge plasma is multicomponent. The main ones are: flow of electrons and ions, free radicals, ultraviolet radiation, which acts stimulatingly on seeds, ozonisation of air. These components contribute to enhancing the biological activity of the seeds. Plasma treatment increases germination, seeds become highly resistant to external stresses. Sprouts become more uniform, plants are less susceptible to diseases and are able to resist weeds.

The main part of our plant is a high-voltage power supply unit, which allows us to supply the corona discharge electrodes (CD) with a voltage of up to 12 kilovolts.

The unit is a complex of the following assemblies:

1. Laboratory autotransformer for changing the corona discharge voltage;
2. High-voltage step-up transformer (50 Hz);
3. Voltage multiplier on 0.1μF high voltage capacitors;
4. Resistor voltage divider block with a microamplifier for measuring the output voltage;
5. Corona discharge electrode block (Fig. 1).

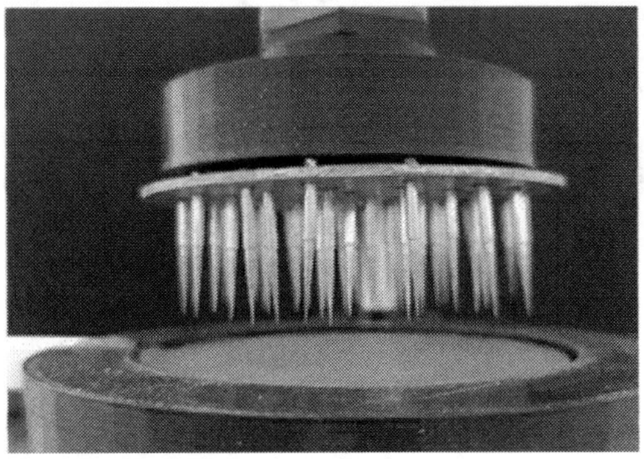

Fig. 1. Corona discharge electrode unit

Gas-discharge low-temperature plasma contains various charged (ions and electrons), neutral (molecules and atoms) particles and activation products of plasma-chemical reactions, X-ray and ultraviolet radiation. Plasma can oxidise various microorganisms and destroy not only their shells but also the DNA of viruses and bacteria [18].

To produce an electric discharge in a gas, a high electric field strength (30 kV/cm for air) is required. In corona discharge, high electric field strength occurs near pointed electrodes, the small radius of which provides dense force lines of the electric field.

In a strong electric field, an electron, which appeared during accidental ionisation of a neutral molecule, is accelerated in the electric field and acquires energy sufficient to ionise it when colliding with the next molecule. As a result, there is an avalanche increase in the number of charged particles (Fig. 2).

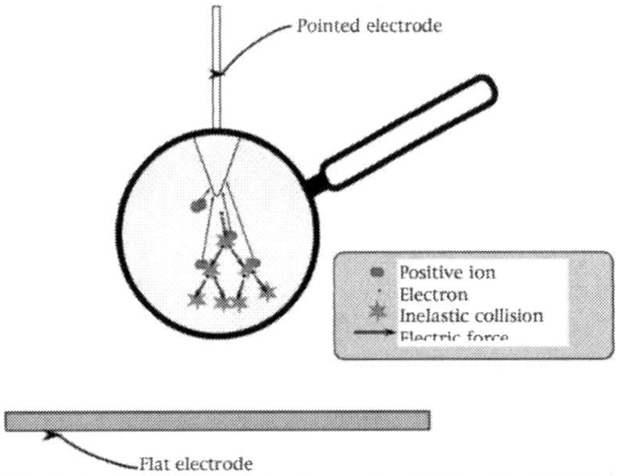

Fig. 2. Avalanche increase in the number of charged particles

The particles accelerated in the electric field of the corona discharge have energy and, consequently, wavelengths characteristic of the UV part of the light spectrum. Short-wave ultraviolet radiation destroys microorganisms by breaking their DNA and RNA.

The predominant decay products of corona discharge plasma in air are nitrogen dioxide NO2 and ozone O3. These reagents have bactericidal and fungicidal properties, which determines the action of corona discharge in seed treatment [19].

To produce plasma using corona discharge, an electric current source with a voltage of 2–10 kilovolts and an electrode system in which the electrodes have a large curvature are required. To carry out the research, a constant voltage source with a regulated output of 2–10 kilovolts and a unit with needle electrodes for corona discharge treatment of seeds were designed and fabricated.

A low-frequency power supply circuit was selected, which is characterised by simplicity and reliability. The disadvantage of the low-frequency unit is the large size, because it is necessary to use high-voltage capacitors of large capacitance. Nevertheless, the low-frequency circuit was chosen because large dimensions are not important for a laboratory unit. The block diagram of the setup is shown in Fig. 3.

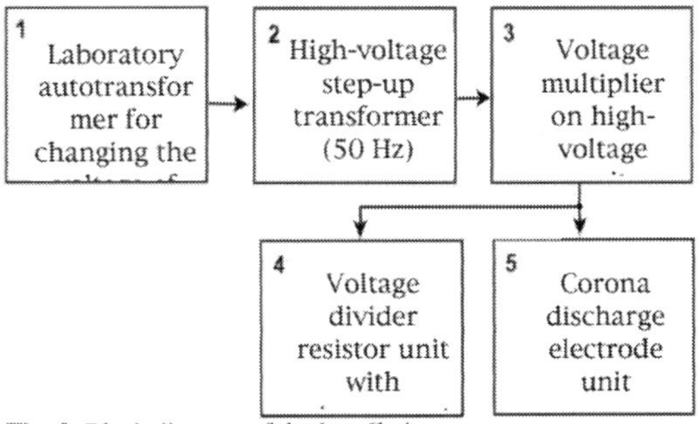

Fig. 3. Block diagram of the installation

The mains voltage is fed to a laboratory autotransformer, then the regulated voltage is fed to a high-voltage transformer, which increases the voltage by a factor of ten.

The main part of the power supply is a multiplier made on capacitors C1–C6 and diodes D1–D6. Each capacitor is charged to the amplitude value of the input voltage, thus the input voltage is multiplied 6 times. The entire circuit of series connected capacitors is charged in six periods, i.e. 120 ms, at an output voltage of 10 KV, the total charge of the circuit:

$$Q = \frac{CU}{n} = 0.1 * 10^{-6} * \frac{10^4}{6} = 1.6 * 10^{-4} K$$

When discharging the capacitors, the current value can be about 1 mA, thus the corona discharge processes can use power of about 10 Watt.

To measure the output voltage, a voltmeter for high voltage measurement is realised on resistors R1–R12 and a microammeter. The voltage divider on resistors R1–R11 reduces the output voltage 100 times, by resistor R12 the current is fed to the microammeter, the readings of which are proportional to the high voltage at the output.

The complete circuit diagram of the plant is shown in Fig. 4.

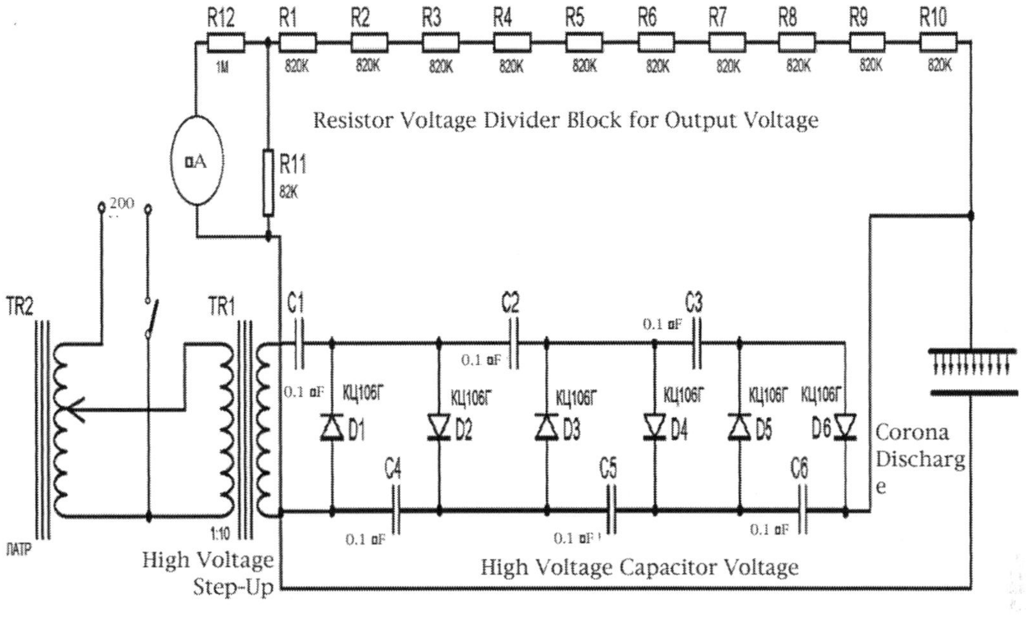

Fig. 4. Complete circuit diagram of the plant

The output voltage of the power supply is obtained by multiplication of the input voltage and is regulated by the input laboratory autotransformer.

The main unit that increases the voltage is a multiplier using KC106G diodes and 0.1 μF high-voltage capacitors.

This large capacitance of capacitors is necessary because the multiplier operates at a low frequency of 50 Hz. A high-voltage transformer (up to 2 kilovolts) located between the autotransformer and the multiplier provides the voltage pre-emphasis.

The output voltage from the power supply unit is supplied to the electrode block, a large number of needle elements of the upper electrode provides a sufficiently large volumetric corona discharge (Fig. 5).

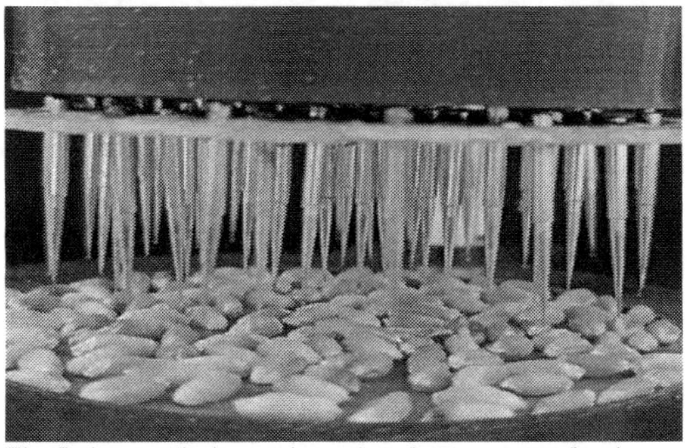

Fig. 5. Corona discharge electrode unit. Treatment of spring wheat seeds of Arhat variety

Wheat seeds have an oblong shape with a dense structure. Since the importance of this crop cannot be exaggerated, even a slight increase in useful properties by corona discharge will be of great national economic importance. Experimental treatment was carried out with change of working voltage of corona discharge and time of exposure of seeds.

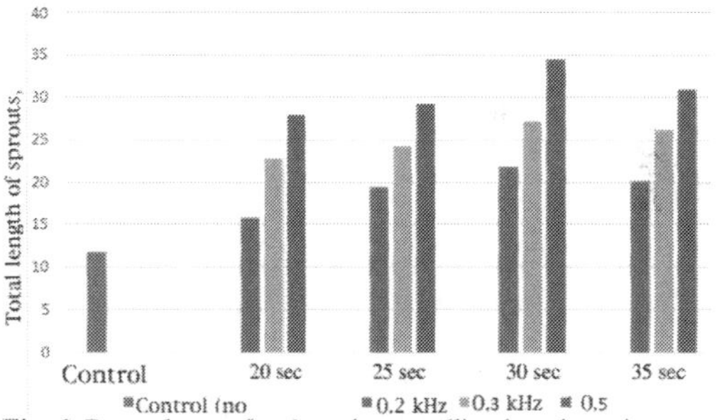

Fig. 6. Dependence of spring wheat seedling length on time and frequency of treatment at 4 kV (germination time 7 days)

Figure 6 shows the dependence of seedling length on time and frequency of treatment at a voltage of 4 kV, the greatest growth of seedlings is observed at a frequency of 0.5 kHz and treatment time of 30 seconds. Thus the length of seedlings for seven days, treated by corona discharge of seeds on average reaches up to 34.4 cm, and the length of the control variant of the experiment 11.7 cm, which is 22.7 cm shorter.

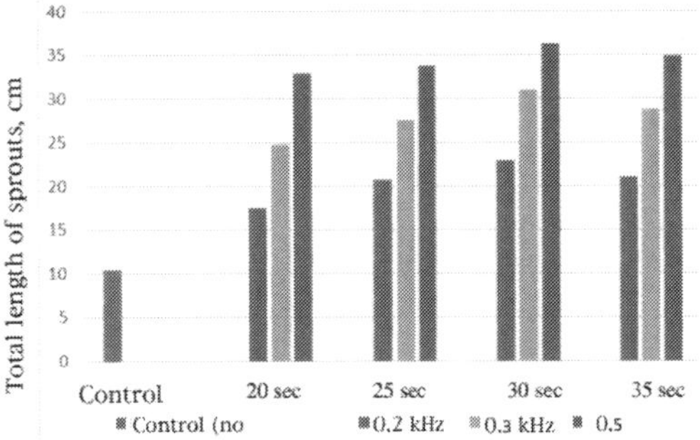

Fig. 7. Dependence of spring wheat seedling length on time and frequency of treatment at 8 kV (germination time 7 days)

In Figure 7, the same dependence was found as in Figure 6. The greatest length of seedlings at 8 kV is observed at treatment time of 30 seconds, where the length of seedlings reaches 36.2 cm, which is 24.5 cm more than the control.

4. Conclusion

The developed installation of corona discharge generation of low-temperature atmospheric plasma allows to treat seeds at the following parameters: voltage 4 and 8 kV; frequency 0.2 kHz; 0.3 kHz; 0.5 kHz; seed distance from the plasmatron 3 cm; seed treatment time: 20, 25, 30, 35 seconds.

Spring wheat seeds of Arhat variety when treated with low-temperature corona discharge plasma at a voltage of 8 kV, frequency 0.5 kHz and treatment time of 30 seconds have the best results. The length of seedlings reached 36.2 cm, which is 24.5 cm more than the control.

References

1. V.O. Adetimirin, Stand establishment and early field vigour variation in a tropicalised shrunken-2 maize population, Field Crop Res., **108**, 143–149 (2008)
2. V.H.V. Mondo, S.M. Cicero, D. Dourado-Neto, T.L. Pupim, M.A.N. Dias, Effect of seed vigor on intraspecific competition and grain yield in maize, Agron. J., **105**, 222–228 (2013)
3. R.M. Nizamov, S.R. Suleymanov, F.N. Safiollin, Modern biopreparations and growth stimulants in sunflower oilseed cultivation technology, Bull. of Kazan State Agrar. Univ., **13(1(48))**, 38–40 (2018)
4. A. Zehra, F. Shaikh, R. Ansari, B. Gul, M.A. Khan, Effect of ascorbic acid on seed germination of three halophytic grass species under saline conditions, Grass and Forage Sci., **68**, 339–344 (2013)
5. G.W. Zhao, X.W. Jiang, Roles of gibberellin and auxin in promoting seed germination and seedling vigor in Pinus massoniana, Forest Sci., **60**, 367–373 (2014)
6. O. V. Afanaseva, O. K. Bezyukov, A. A. Ignatenko, Academic Journal of Manufacturing Engineering, **22(1),** 68–75, (2024). DOI: 10.24874/PES.SI.02.017
7. N.F. Kashapov, M.M. Nafikov, S.G. Smirnov, Man. Mak. Nafikov, Device for seed disinfection by cold atmospheric air plasma and method of its operation, Patent No. 2781971 C1 Russian Federation, IPC A01C 1/02, No. 2022105160: avv. 26.02.2022: publ. 21.10.2022, EDN KAPCKO
8. T. A. Kuandykov, T. D. Karmanov, E. I. Kuldeyev, K. K. Yelemessov, B. Z. Kaliev, News of the National Academy of Sciences of the Republic of Kazakhstan, Series of Geology and Technical Sciences, **2022(3),** 142–154, (2022). DOI 10.32014/2022.2518-170X.186
9. S. I. Kondratyev, D. Baskanbayeva, K. Yelemessov, Y. Sarsenbayev, V. A. Turkin, International Journal of Hydrogen Energy, **95**, 212–216, (2024). DOI: 10.1016/j.ijhydene.2024.11.182
10. K.S. Zaurbekov, S.A. Zaurebkov, A.V. Sladkovsky, D.Y. Balgayev, News of the National Academy of Sciences of the Republic of Kazakhstan, Series of Geology and Technical Sciences, **3(459),** 60–69, (2023). DOI 10.32014/2023.2518-170X.299
11. J.D. Moon, H.S. Chung, Acceleration of germination of tomato seed by applying AC electric and magnetic fields, J. of Electrost., **48**, 103–114 (2000)
12. G. Xu, X.T. Wang, C.L. Gan, Y.Q. Fang, M. Zhang, Biological effects of low energy nitrogen ion implantation on Jatropha curcas L. seed germination, Nuclear Instr. and Meth. in Physi. Res. Sect. B, **287**, 76–84 (2012)
13. D.G. Gataullin, F.N. Safiollin, G.S. Minnullin (et al.), Antistress and phytohormone preparations in the technology of spring rape cultivation on gray forest soils of the Republic of Tatarstan, Agrochem. Bull., **2**, 45–49 (2021)
14. M. Dhayal, S.Y. Lee, S.U. Park, Using low-pressure plasma for Carthamus tinctorium L. seed surface modification, Vacuum, **80**, 499–506 (2006)
15. A. I. Trunova, S. O. Kurashkin, I. F. Vivián, K. A. Bashmur, Metals, **13(7),** 1184 (2023). DOI: 10.3390/met13071184
16. J. Khun, V. Scholtz, P. Hozak, P. Fitl, J. Julak, Various DC-driven point-to-plane discharges as non-thermal plasma sources and their bactericidal effects, Plasma Sour. Sci. and Technol., **27(6),** 065002 (2018)
17. A. I. Trunova, S. O. Kurashkin, K. A. Bashmur, R. B. Sergienko, Metals, **13(7),** 1234 (2023). DOI: 10.3390/met13071234
18. M.Q. Yin, M.J. Huang, B.Z. Ma, T.C. Ma, Stimulating effects of seed treatment by magnetized plasma on tomato growth and yield, Plasma Sci. and Technol., **7**, 3143–3147 (2005)
19. W.F. Yao, Y.B. Shen, Effect of magnetic treatment on seed germination of loblolly pine (Pinus taeda L.), Scand. J. of Forest Res., **30**, 639–642 (2015)

https://doi.org/10.1051/bioconf/202516100061

Biotechnological conversion of organic wastes by vermiculture to increase crop yields

Yulia V. Daus [1*], *Angelina Yu.* Dubina [2], *Valery A.* Eviev [3], *Victoria V.* Evieva [4], *Amina N.* Badrudinova [3]

[1] Kuban State Agrarian University named after I. T. Trubilin, Krasnodar, Russia
[2] Don State Agrarian University, Roston-On-Don, Russia
[3] Kalmyk State University named after B.B. Gorodovikov, Elista, Russia
[4] Moscow Institute of Physics and Technology, Moscow, Russia

Abstract. The work is devoted to the application of vermiculture biotechnology for processing organic waste and obtaining biohumus, as well as assessing the impact of the resulting product on crop yields and quality. The study included laboratory experiments on vermicomposting of different types of organic waste in different proportions with soil filler, as well as field trials using biohumus as a fertiliser for potatoes, carrots and oats. It was found that vermicomposting is effective for processing a wide range of organic wastes to produce biohumus rich in humus, macro- and microelements. The optimum ratio of organic waste and soil filler for biohumus production was 65:35. Field trials have shown that the use of biohumus helps to increase yields and improve the quality indicators of agricultural crops, including an increase in carotene and sugars in carrots, a decrease in non-protein nitrogen in potatoes, and a reduction in disease and pest damage. The results confirm the prospects of vermiculture as an environmentally safe method of waste utilisation and obtaining a valuable organic fertiliser.

1 Introduction

In the modern world the problem of organic waste utilisation is one of the most acute and topical. Intensive development of industry, agriculture and the growth of urban population lead to the accumulation of a huge amount of waste, a significant part of which consists of organic substances such as farm animal manure, poultry manure, food industry waste, plant residues and sewage sludge. Improper disposal of these wastes can lead to serious environmental problems, including soil, water and air pollution, the spread of pathogens, and the loss of valuable resources that could be used to improve soil fertility and produce environmentally friendly products. Thus, effective waste management plays an important role in agricultural systems [1, 2].

In order to address this problem, many approaches have been developed and implemented. One of the most traditional methods is composting, which allows the conversion of organic waste into valuable fertiliser. However, this process requires significant time investment as well as temperature and humidity control. Another common approach is to use organic waste as feedstock for biogas production, which not only utilises the waste but also provides a renewable energy source. However, this process requires specialised equipment and compliance with technological parameters. There are also chemical methods of waste processing, such as pyrolysis and gasification, which allow to obtain various products, including synthetic fuel. However, they may be accompanied by the formation of harmful substances and require significant financial investment.

In recent years, more and more attention has been paid to biotechnological methods of organic waste processing, which are more environmentally friendly and allow obtaining valuable products, including biofertilisers. One such method is vermiculture, which is based on the use of earthworms to process organic waste into biohumus. This method is characterised by its simplicity, affordability and high efficiency. Biohumus produced in this way is a high quality organic fertiliser that contains high amounts of humus, macro- and microelements, enzymes and beneficial microflora, and has a prolonged action. This product can significantly increase the fertility of soils, improve their physical and chemical properties and contribute to higher crop yields. At the same time, it is important to note the role of family farmers in this process [3]. Vermiculture is a waste-free technology that allows processing a wide range of organic waste, including animal and poultry waste, sewage sludge, food industry waste and others.

The use of vermiculture is a very promising and important direction, as it offers an environmentally safe and effective way to utilise organic waste, while providing a valuable fertiliser. The utilisation of organic waste not only solves the problem of its disposal, but also contributes to the improvement of the environment and the creation of sustainable agroecosystems, and the

* Corresponding author: zirochka2505@gmail.com

use of biohumus helps to reduce dependence on mineral fertilizers, allows to obtain environmentally friendly agricultural products and improve food security, which is especially important in the context of developing countries [4, 5]. The aim of the presented work is to investigate the possibilities of applying vermiculture biotechnology to process organic waste and use the resulting biohumus as biofertiliser to increase crop yields and quality, which also meets the growing demands for food quality [6]. In this paper, the effect of different ratios of organic waste and soil filler on the vermicomposting process was investigated, and the effect of biohumus on yield and quality parameters of potatoes, carrots and oats was evaluated.

2 Method and materials

The research presented in this article was conducted to study the efficiency of biotechnological processing of organic waste by vermiculture and to assess the impact of the obtained biohumus on crop yields. The experimental work included two main stages: the first - obtaining biohumus by vermicomposting of different types of organic wastes, and the second - assessing the impact of the obtained biohumus on crop yield and quality under field conditions.

For vermicomposting experiments, the red California earthworm Eisenia foetida was used as the main biological agent. Various organic wastes such as poultry litter, sludge-lignin, sewage sludge, cattle manure (Cattle) and food wastes were used as substrates for vermicomposting in different ratios with soil filler namely 50:50, 65:35 and 80:20 (waste: soil). The vermicomposting process was carried out in special containers providing necessary aeration and humidity. The worm infestation density was 2000 specimens per square metre. The duration of composting varied from 1 to 3 months depending on the type of waste.

The obtained biohumus was used as a fertiliser in growing potatoes of 'Dobro' variety, carrots of 'Nantes' variety and oats of 'Dogoy' variety under field conditions. Field experiments were conducted with the application of biohumus at a dose of 3 tonnes/ha. To determine the yield and quality indicators of crops, harvesting and analysis of relevant parameters such as total yield, dry matter content, carotene content, sugars, nitrogen, fibre and the presence of damage by pests and diseases were carried out. Standard agrochemical methods of analysis were used.

3 Results and discussion

The experimental work carried out within the framework of this research was focused on the in-depth study of the vermicomposting process of organic waste and subsequent evaluation of the impact of the obtained biohumus on the productivity of agricultural crops. The experiment was structured in two interrelated stages: the first stage involved optimising the process of organic waste bioconversion using earthworms Eisenia foetida and obtaining biohumus as output, while the second stage was aimed at field trials of the biofertiliser to assess its impact on yield and quality parameters of potatoes, carrots and oats. In the first phase, laboratory tests were conducted using various organic wastes including poultry litter, sludge-lignin, sewage sludge, cattle manure, and food waste. These wastes were mixed with soil filler in three different proportions: 55:45, 65:35 and 85:15 (waste:soil) to analyse the effect of different proportions on the final product. Three parallel experiments were performed for each variant to ensure reliability and reproducibility of the results. The vermicomposting process was carried out in specially prepared containers in which a microclimate with aeration, humidity and temperature in the range of 21-24°C was maintained optimum for the worms' activity. Each container was stocked with a population of earthworms at a density of 1900 individuals per square metre, and the composting period varied from 35 to 85 days, depending on the characteristics of the feedstock being processed. Throughout the observation process, visual changes in the substrate were monitored and samples were taken for laboratory analysis for humus content and essential macro- and microelements. In the second stage, three separate field experiments were conducted, one for each of the crops under study: potatoes, carrots and oats. In each of these field experiments, control plots where no fertiliser was applied and experimental plots where biohumus obtained in the first stage was applied. Biohumus was applied at a dosage of 2.8 tonnes/ha to all crops before planting, for potatoes and carrots, and under pre-sowing treatment for oats. Plant health was monitored continuously during the growing season and harvesting was carried out at the end. Crop yields were measured as well as qualitative parameters such as dry matter content, carotene, sugars, nitrogen, fibre, and the extent of pest and disease damage to fruits and seeds.

The results of the first stage of experiments demonstrated that vermicomposting is an effective method of processing various types of organic wastes. The greatest accumulation of humus was observed in variants with the use of poultry manure and food waste in the ratio of 65:35, where the humus content reached 5.10% and 9.80%, respectively, these indicators were noticeably higher than in other variants. For example, in the variant using sludge-lignin, at a similar ratio, the humus content was only 4.15%. In the 55:45 variants, humus levels were generally slightly lower, ranging from 3.70 per cent for sewage sludge to 9.10 per cent for food waste. The 85:15 ratio variants in turn showed even lower humus levels, ranging from 2.80% for sewage sludge to 6.80% for food waste. It was also found that the 65:35 ratio provided the best conditions for the reproduction and activity of earthworms, which contributed to a more intensive recycling of waste, which is also underlined by the importance of analysing the agroecological potential [7, 8]. The humus content of biohumus obtained from poultry manure with a 65:35 ratio was 13% higher than in the control sample, where mineral fertiliser was used exclusively, confirming the importance of agro-landscape modelling to obtain a quality fertiliser [9, 10]. In the second stage of field trials, it was found that the application of biohumus had a positive effect on the yields of all three crops studied. In particular, the yield of potatoes in the experimental plots was on average 70.4% higher than in the control plots where mineral fertilisers were used. At the same time, the average yield in the control plots was 12.5 t/ha, while in the experimental plots it was 21.3 t/ha. For carrots, the application of biohumus increased the yield by 45.8% (control 22.0 t/ha, experiment 32.1 t/ha), and the dry weight of oats increased by 14.9% (control 3.0 t/ha, experiment 3.45 t/ha), which once again confirms the effectiveness of biohumus as a fertiliser. In addition, there was a significant improvement in the quality parameters of products: carrots showed an increase in the level of carotene and sugars, and potatoes showed a decrease in the content of non-protein nitrogen, which indicates an increase in nutritional value and storability. Also, it was recorded that the application of biohumus reduced the damage of potato tubers by pests and diseases. In the control this indicator was 15.9%, and in the experiment 11.2%, which indicates the protective properties of biohumus.

In general, the obtained data correspond to the results of similar studies, which confirm the effectiveness of vermicomposting for the utilisation of organic waste and the positive effect of biohumus on the productivity and quality of agrocultures. Thus, in studies conducted in different regions, it was found that the use of biohumus from cattle manure helps to increase grain yields by 22-33% and vegetable yields by 32-42%. Importantly, the positive effect of biohumus on soil microbial activity and improvement of its physico-chemical properties was also confirmed, which may be related to the assessment of the geo-ecological state of landscapes [11, 12]. The results obtained are in agreement with the findings of other scientists who found that biohumus helps to improve plant resistance to stressful conditions such as drought and diseases. In addition, the results showed that biohumus not only improves the quantitative characteristics of the crop, but also changes its composition in qualitative terms, increasing the concentration of useful substances, which increases the nutritional value of the resulting products. However, it should be noted that the efficiency of biohumus can vary depending on the type of organic waste used, as well as soil and climatic features of the region. Therefore, in order to achieve maximum efficiency, it is necessary to adapt the vermicomposting technology to local conditions and types of waste, as well as to take into account the influence of weather conditions on yield [13, 14]. Along with this, the prospect of utilising the waste to create new products should be considered [15]. This study contributes to the understanding of the benefits of vermiculture as a promising technology for utilising organic waste while producing a valuable organic biofertiliser, which is an important step towards more sustainable and environmentally friendly farming practices.

4 Conclusion

The study demonstrates the prospects of vermiculture as an effective method of organic waste utilisation with simultaneous production of a valuable biofertiliser. The results of laboratory and field experiments clearly indicate the positive impact of vermicomposting on the qualitative characteristics of organic waste and the subsequent increase in crop productivity. Optimisation of the vermicomposting process showed that the ratio of organic waste and soil filler in the proportion of 65:35 is the most favourable for maximum accumulation of humus in biohumus. It is at this ratio that the highest activity of earthworms and, consequently, the most efficient waste processing is achieved. This demonstrates the importance of selecting the correct ratio of components to optimise the process.

Field studies have confirmed that biohumus application has a significant stimulating effect on yield and quality of agricultural products. In particular, it was found that biohumus application contributes to a noticeable increase in the yield of potatoes, carrots and oats, as well as improving their quality characteristics. The carotene and sugar content of carrots increased significantly, while the amount of non-protein nitrogen in potatoes decreased, indicating an increase in the nutritional value and storability of tubers. Biohumus application also reduced pest and disease damage to potato tubers, confirming its protective properties and potential for reducing chemical pesticide use. In the control, this index was 15.9%, while in the experiment it was 11.2%, which indicates the protective properties of biohumus that require further study in terms of biotechnological approach [16, 17].

Thus, it can be concluded that vermiculture is not only an effective method of disposal, but also highlights the importance of introducing biotechnological methods in agriculture and the need for further study and optimisation of the vermicomposting process for different types of organic waste and soil and climatic conditions. It is also advisable to take into account the human factor associated with labour activity and population migration [18]. The findings contribute to the development of environmentally friendly and sustainable agricultural practices that contribute to solving the problems of waste management and food security.

References

1. N. Garambois, U. Le Goff, L. Thibaudeau, Agrarian dynamics and climate change in the Senegalese Sahelian peanut basin. Agrar. Syst. Clim. Change: Journeys Adapt. Glob. South, **15**, 37, (2024). DOI: 10.1079/9781800628137.0001
2. H. Cochet, O. Ducourtieux, N. Garambois, Agrarian systems and climate change: Journeys of adaptation in the global south. Agrar. Syst. Clim. Change: Journeys Adapt. Glob. South, **1**, 208, (2024). DOI: 10.1079/9781800628137.0000

3. B. Aparecida da Silva, R. Evangelista de Oliveira, A. Cavalieri Sais, E. Cardoso leite, Family farmers and trees: ecological functions, ecosystem services and food production in agrarian reform settlements in Brazil. Agroecol. Sustain. Food Syst., **49,** 4-21 (2025). DOI: 10.1080/21683565.2024.2405515
4. S. Bamidele, BRICS Revolution: The Global Agrarian Shift Sparked by Emerging Economies. India Q., (2025), Article in press. DOI: 10.1177/09749284241307937
5. V. I. Golik, T. A. Panfilova, V. Yu. Konyukhov, Mining Informational and Analytical Bulletin. **11-1,** 175-189 (2023). DOI: 10.25018/0236_1493_2023_111_0_175
6. A. Al Mozahid. What counts as "quality" in agrarian taste? Subjectivities, gender, and rice agroecology in northern Bangladesh. Cult. Agric. Food Environ., 46, 88-105 (2024). DOI: 10.1111/cuag.12323.
7. S.A. Makhosheva, M.V. Zhanokova, V.M. Dzaparova, I.K. Totrukova. Ecological aspect of sustainable development of high-mountain regions in the conditions of innovative economy formation. Sustainable Development of Mountain Territories, **16,** 754–769 (2024). DOI: 10.21177/1998-4502-2024-16-2-754-769.
8. I.I. Bosikov, E.V. Khekert. Analysis of the quality of underground mineral waters of terrigenous deposits of the hauteriv barremian aquifer of the lower cretaceous. News of the National Academy of Sciences of the Republic of Kazakhstan, Series of Geology and Technical Sciences, **2,** 36-47, (2024). DOI: 10.32014/2024.2518-170x.392.
9. T.I. Kosovtsova. Factors of conservation of cultural landscapes of the Bavarian Alps. Sustainable Development of Mountain Territories, **11,** 371–379, (2019). DOI: 10.21177/1998-4502-2019-11-3-371-379.
10. T. A. Panfilova, K. A. Bashmur, V. V. Kondratiev, Mining Informational and Analytical Bulletin, **11-1,** 239-251 (2023). DOI: 10.25018/0236_1493_2023_111_0_239.
11. L.S. Malyukova, N.V. Martyushev, V.V. Tynchenko, V.V. Kondratiev, V.V. Bukhtoyarov, V.Y. Konyukhov, K.A. Bashmur, T.A. Panfilova, V. Brigida, Circular Mining Wastes Management for Sustainable Production of Camellia sinensis (L.) O. Kuntze. Sustainability (Switzerland), **15(15),** 11671 (2023). DOI: 10.3390/su151511671
12. A.I. Smirnov, N.V. Plotnikova, V.G. Burov, N.V. Martyushev, Features of the fine structure of the active layers of organic solar cells. Key Eng. Mat., **712,** 176-181, (2016). DOI: 10.4028/www.scientific.net/KEM.712.176
13. B.V. Malozyomov, N.V. Martyushev, S.N. Sorokova, E.A. Efremenkov, D.V. Valuev, M. Qi, Analysis of a Predictive Mathematical Model of Weather Changes Based on Neural Networks. Mathematics, **12 (3),** 480 (2024) DOI: 10.3390/math12030480
14. A.A. Kapanski, E.A. Efremenkov. Geospatial Clustering in Smart City Resource Management: An Initial Step in the Optimisation of Complex Technical Supply Systems. Smart Cities, **8,** 14 (2025). DOI: 10.3390/smartcities8010014.
15. A.L.S. Hansted, W.J. Newell, K.A. Hawboldt, An initial analysis of the market potential of a wet ash waste stream from a pulp and paper mill in Newfoundland and Labrador. Ind. Crops Prod., **225,** 120557 (2025). DOI: 10.1016/j.indcrop.2025.120557
16. V.I. Golik, S.N. Sorokova, Q. Mengxu, Tailings Utilization and Zinc Extraction Based on Mechanochemical Activation. Materials, **16 (2),** 726 (2023) DOI: 10.3390/ma16020726
17. L.S. Malyukova, N.V. Martyushev, V.V. Tynchenko, V.V. Kondratiev, V.V. Bukhtoyarov, V.Y. Konyukhov, K.A. Bashmur, T.A. Panfilova, V. Brigida, Circular Mining Wastes Management for Sustainable Production of Camellia sinensis (L.) O. Kuntze. Sustainability (Switzerland), **15 (15),** 11671 (2023) DOI: 10.3390/su151511671
18. A. Smith, Labour migration, mass unemployment and the state: Class, gender and work in the Land Settlement Association in inter-war rural England. J. Rural Stud., **114,** 103498 (2025). DOI: 10.1016/j.jrurstud.2024.103498

Influence of pre-sowing seed treatment in a magnetic gradient field on growth, development and productivity of long-staple flax

Igor Viktorovich Yudaev* [1], *Alexey Alekseevich* Akimov[2], *Alexey Ivanovich* Belenkov[3], *Igor Olegovich* Plotnitsky[4]

[1] I.T. Trubilin Kuban State Agrarian University, Krasnodar Region, Krasnodar, Russia
[2] Tver State Agricultural Academy, Tver region Tver (Sakharovo), Russia. Tver (Sakharovo), Russia
[3] V.R. Williams FSC forage production and agroecology, Moscow region, Lobnya, Russia
[4] United Innovations LLC, Moscow, Russia

Abstract. The strategic goal of effective land use in the field of cultivation of industrial crops is the need to increase their yields while maintaining the quality of raw materials, against the background of reducing total energy consumption and minimal anthropogenic negative impact on the environment. It is possible to realise the indicated approaches in flax-dolgunets cultivation, starting from the main thing - preparation of quality seed material, for which purpose it is offered its treatment in a gradient magnetic field of direct current at various parameters of influence from 0.83 to 8.0 mTl. Treatment of flax-dolgunets seeds was carried out in laboratory conditions of LLC "Association of Innovations" on a specially designed pilot plant, and studies on germination and germination energy - in Tver State Agricultural Academy, field experiments - in the conditions of the experimental site of the Academy. The obtained results show that such seed treatment increased the overall survival rate of flax-dolgunets up to 35.6% in some variants from 22.9% in the control or, on average, the survival rate in all studied variants increased by 7.1%. Application of the studied variants of seed pre-sowing treatment was reliable (NSR_{05}=1.9 c/ha and NSR_{05}=1.1 c/ha) and increased the yield of straw, on average, from 25.6 to 51.8 c/ha, and flax seeds - from 6.6 to 12.6 c/ha. Among the studied variants of presowing treatment, the highest yields of both flax-cotton and flax-seeds were provided by the variants in which treatment was carried out in magnetic gradient field with induction of 6.70 and 7.35 mTl. Since flax was cultivated for seed purposes, the reproduction coefficient of this crop was determined depending on the studied variants of the experiment, so the highest reproduction coefficients (32, 33 and 37) were obtained respectively in the following treatment variants: 1.68 mTl, 7.35 mTl and 8.0 mTl. The highest fibre yield was obtained in the field treatment with magnetic induction - 4.23 mTl, 6.7 mTl and 7.35 mTl and was respectively 11.0, 10.7 and 10.9 kg/ha, which was 2.3 times higher than in the control sowing variant.

Key words: pre-sowing seed stimulation, gradient magnetic field, flax seeds, parameters and modes of treatment, magnetic field induction, germination, growth energy, yield structure.

1. Introduction

Today in agriculture the emphasis of innovative development of the crop sector is made not only on increasing the production of quality grain, melon, oilseed products necessary to provide the counters with the required amount of food, but also in the cultivation of technical crops - flax, sugar beet, sunflower, potatoes, etc. The most important among the technical crops . Among technical crops the most important crop is long-staple flax, the fibre from which is much stronger than cotton fibre. Sown areas in Russia, somewhat lost their values not only in comparison with 1990, but also with 1913 and 1938 years, so in 1913 the area of flax-flax was 1020 thousand hectares; in 1938 - 2500; 1990 - 418; 2000 - 108; and in 2015 and 2020 only 53 thousand hectares. Therefore, the restoration of flax production in the Russian Federation is the most urgent task for the agricultural sector of the country and individual regions [1].

The technology of flax cultivation is characterised by a low negative environmental and anthropogenic load on the surrounding area. In addition, it is cultivated using insignificant amounts of fertilisers and pesticides, and the resulting products are almost completely processed, leaving a minimum of waste

The most important stage of flax seeds preparation for sowing is their dressing and encrustation. Taking into account environmental requirements, it is recommended to use Vitovax, Fenoram, TMTD, etc., as well as Agat 25 and Ecost biopreparations. The All-Russian Research Institute of Flax recommends seed encrustation together with microfertilisers, using boric acid, molybdenum ammonium, zinc sulphide [2].

* Corresponding author: etsh1965@mail.ru

The search for options for maximum substitution of synthesised preparations for plant development stimulation and protection is ongoing. In addition, domestic flax producers note that today it is possible to harvest not quite the planned crop from seeds of domestic selection, but flax seeds of French and Belgian selection are more attractive. Nowadays, environmentally friendly and technologically effective electro-technological operation of seed treatment before sowing, carried out for the purpose of more sustainable development of plants and obtaining large yields, by various influences of electromagnetic nature, such as treatment in electric, magnetic, electromagnetic fields, exposure to electric pulses of high voltage, using direct or alternating current, etc., has become widespread. [3-8].

It is of research interest to analyse the results of application of magnetic fields of different intensities and configurations for seed treatment of cultivated plants as a promising method of pre-sowing preparation for flax agrotechnics, which was studied in the article submitted to the reader's judgement.

The magnetic field has surrounded the Earth since its formation, so we can say that the evolution of plant life was directly influenced by it. Historically, it is known that the magnetic field has a positive effect on plants, as experiments with isolating plant objects from its effects ended with a decrease in their growth rate and deterioration of their development. On average, the intensity of the Earth's magnetic field ranges from 25 to 65 μTL. The magnetic field contributes to more effective adjustment of physiological processes of plants and their adaptation to changing conditions, including stress conditions. In particular, a "protective" effect, i.e., an increase in the survival rate of plants in stressful conditions when the intensity of magnetic fields changes [9], can be manifested.

Pre-sowing treatment of seeds with static magnetic field was considered as a means of improving plant development. Thus, seeds of vegetable crops - tomatoes, lettuce and arugula - were treated in static magnetic field on a specially designed installation. Tomato seedlings developed better from seeds treated in a magnetic field with induction of 45 mTl at exposure times of 60 and 90 min compared to the control. To improve the growth of arugula seedlings, the magnetic induction should be 150 mTl, whereas the results for lettuce seeds were inconsistent and it was unclear whether a low (45 mTl) or high (300 mTl) magnetic field induction should be applied [10].

The possibility of using static magnetic field for flax seed treatment was studied. It was experimentally proved that the optimal parameters of pre-sowing treatment of flax seeds in a static magnetic field were as follows: magnetic induction - 0.065 Tesla; fourfold remagnetisation; pole division - 0.23 m; seed movement speed - 0.4 m/s; treatment dose - 1.79 J-s/kg. Flax seeds increased germination energy y by 30%, germination by 2.6%, stem length by 10.5%, straw yield by 0.7 t/ha, and seed yield by 0.79 t/ha [11].

The question of the influence of a gradient magnetic field on seed pre-sowing treatment and the response to it in the form of changes in the growth and development of plants from these seeds requires a separate study.

The studies conducted in our country to study the effect of gradient magnetic field mainly concerned the treatment of vegetable, cereal, oilseed and other crops seeds. Thus, the positive effect of such treatment of cucumber seeds was confirmed experimentally. Stable "stability of magnetobiological effects in cucumber seedlings was observed at the speed of seed movement 1-2 m/s in the magnetic field and its intensity in the range of 50-150 Å. Plants formed from treated seeds were distinguished by higher productivity of photosynthesis, higher number of female flowers compared to the control. The increase in productivity in the experimental variants by 7.6-15.3% was due to a greater number and weight of formed fruits" [12]. [12].

Purpose and Statement of Objectives. The purpose of the conducted research is to study the effect of magnetic gradient field of direct current as a technique of pre-sowing treatment of flax seeds to achieve a positive effect in its growth, development and productivity

2. Materials and methods

Flax seeds were pre-treated with a gradient DC magnetic field on a specialised machine before sowing. Its design is a vertical cylindrical tube-column, consisting of an inner cylindrical base with windings placed on it, as well as an outer tube, which is a protective casing made of non-magnetic steel, which also has clamps for connecting the DC source. On the inner cylindrical part are fixed windings spaced along the height of the installation with a certain spatial configuration, connected with each other in such a way as to create a DC magnetic field varying along the length of the column (Fig. 1)

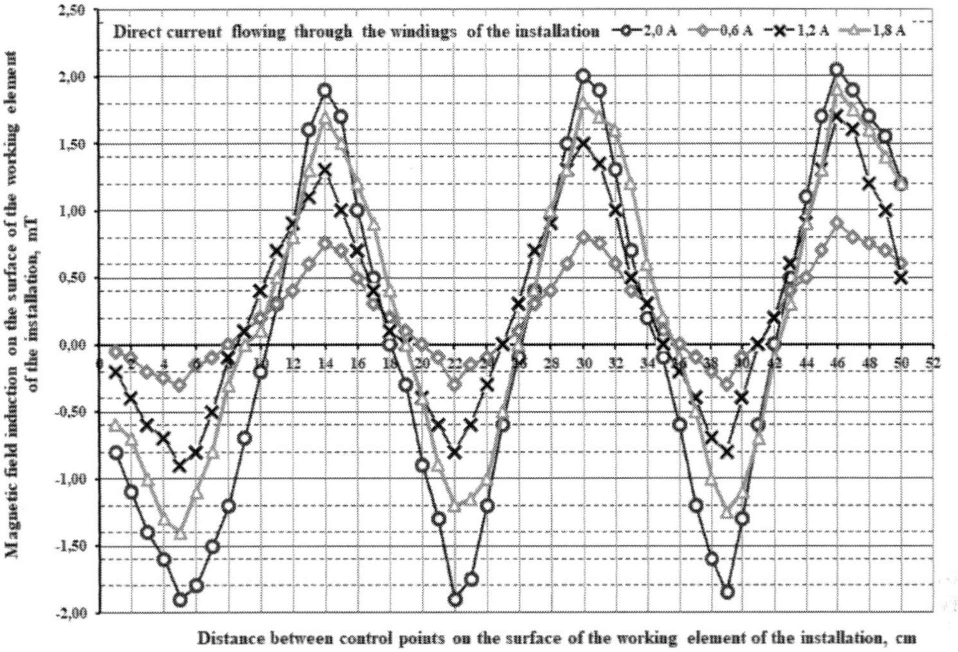

Fig. 1. Distribution of magnetic induction on the surface of the working element of the installation for processing flax seeds in a gradient magnetic field

The seeds are loaded into the feeding hopper and after free, gravity flow into the receiving tank. The unit is powered by a commercially available DC power supply with adjustable output parameters - current and voltage.

The following experiment schemes were determined for the research, according to which the flax-flax seeds were treated:

1. Control (without treatment).
2. Variant 1 - treatment in magnetic gradient field with induction of 0.83 mTl.
3. Variant 2 - treatment in a magnetic gradient field with an induction of 1.68 mTl.
4. Variant 3 - treatment in a magnetic gradient field with an induction of 2.51 mTl.
5. Variant 4 - treatment in a magnetic gradient field with an induction of 3.34 mTl.
6. Variant 5 - treatment in a magnetic gradient field with an induction of 4.20 mTl.
7. Variant 6 - treatment in a magnetic gradient field with an induction of 5.04 mTl.
8. Variant 7 - treatment in a magnetic gradient field with an induction of 5.81 mTl.
9. Variant 8 - treatment in a magnetic gradient field with an induction of 6.70 mTl.
10. Variant 9 - treatment in a magnetic gradient field with an induction of 7.35 mTl.
11. Variant 10 - treatment in a magnetic gradient field with an induction of 8.00 mTl.

The field experiment was laid on the experimental plot of the Tver State Agricultural Academy. Repetition of the experiment was 4-fold, placement of variants was randomised. The area of the experiment was 792 m^2, the area of the experimental plot was 18 m^2. The soil of the experimental plot is sod-medium podzolic light loamy light loamy residual carbonate gleyey on moraine

The object of research is a variety of long-stalked flax "Atlant", which is included in the State Register for the North-West region of Russia and recommended for cultivation in the Vologda and Tver regions. The stem is long. Dotted sepals are absent or very weak. The corolla is blue when fully developed. The boll is small, fringing of the false septum is absent. The seed is dark brown. The time of flowering is early. The vegetation period is 87 days. Plant height 79 cm. Weight of 1000 seeds is 4.9 g. Resistance to lodging - 4,6 points, to shattering - 5,0 points. Fibre content 31,1%, long fibre yield 22,4%, relative breaking load calculated 14,3 gs/tex. According to the applicant, the variety is resistant to fusarium wilt and rust, as well as to fusarium. Average flax straw yield is 46.5 c/ha, 0.9 c/ha above the standard. Seed yield 4.6 kg/ha, at the level of the standard.

The agrotechnique of flax is generally accepted for farms in the North-West region of Russia. The precursor was winter wheat. Tillage in autumn was discing followed by ploughing when mass weed sprouts appeared. Spring tillage was carried out by cultivation in two traces. Mineral fertilisers were applied under pre-sowing tillage at a rate of 3 kg/ha of azofosk (16:16:16:16). Pre-sowing tillage was carried out in one trace. Sowing of flax-dolgunets was carried out with a seeding rate of 12 million germinated seeds per 1 ha. The sowing method was narrow-row sowing. After sowing the soil was rolled. For sowing we used seeds of OS category: mother elite of 2 years, treated according to the scheme of the experiment. Application against weeds of tank mixture of herbicides was carried out in the phase of "herringbone" at the height of creeping wheatgrass 10-15 cm manually, with a rate of working fluid 400 litres/ha. Harvesting was carried out on 14 September 2024 in a continuous manner.

Agrometeorological conditions in the year of research (2024) were very different from the long-term average values. Unfavourable weather conditions in May (abnormally low temperatures on the background of drought events) allowed sowing only at the end of the month. Insufficient moisture supply against the background of abnormally high average daily temperatures in the first decade of June did not allow to obtain full-fledged sprouts of the experimental crop. As a result, the field germination of flax was very low. During the vegetation period of the crop, both insufficient moisture supply (late June - early July) and soil overmoistening in some decades were observed with sufficiently high average daily air temperature, which was observed in June-July

Common methods were used in conducting the studies.

3. Experimental results and their discussion

Experimentally obtained data on germination energy and laboratory germination of flax-flax seeds depending on the variants of the experiment on presowing treatment are shown in Figure 2.

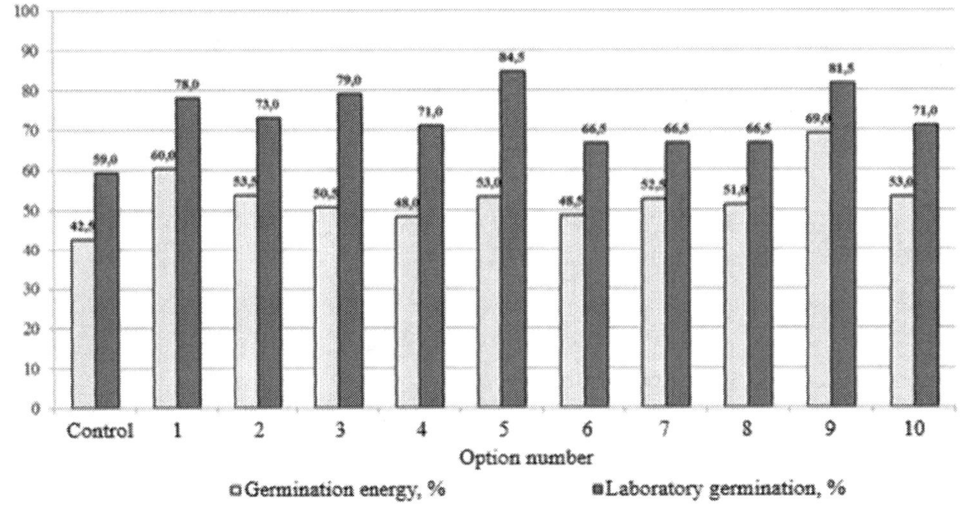

Fig. 2. Effect of pre-sowing treatment on germination energy and laboratory germination of flax seeds, 2024

From the obtained experimental data, presented Figure 2, it can be seen that the germination energy and laboratory germination of flax seeds were minimal in the control variant - 42.5 and 59.0%. Application of the studied treatment methods based on the influence on these seeds by gradient magnetic field allowed to improve these indicators significantly from 42.5 to 69.0% and from 59.0 to 84.5%, respectively. Thus, the highest germination energy was recorded on variants 1 and 9 - 60.0 and 69.0%, and laboratory germination - on variants 5 and 9 - 84.5 and 81.5%.

Thus, the use of pre-sowing treatment of flax-dolgunets seeds allows to increase seed germination energy by 11.4% on average, and laboratory germination by 14.8%, compared to the same indicators in control variants.

Observation of the growth and development of lignum flax showed that as a result of drought events at the beginning of the growing season against the background of abnormally high temperatures, the field germination rate of the crop was low and was at the level of 50-55%. Later, when weather conditions normalised, the growth and development of the experimental crop can be described as good. Flax passed all phenological phases and finished vegetation in the middle of September.

Figure 3 shows the data on total survival rate of lignum flax depending on the studied variants of the experiment on pre-sowing treatment in a gradient magnetic field and in the control variant.

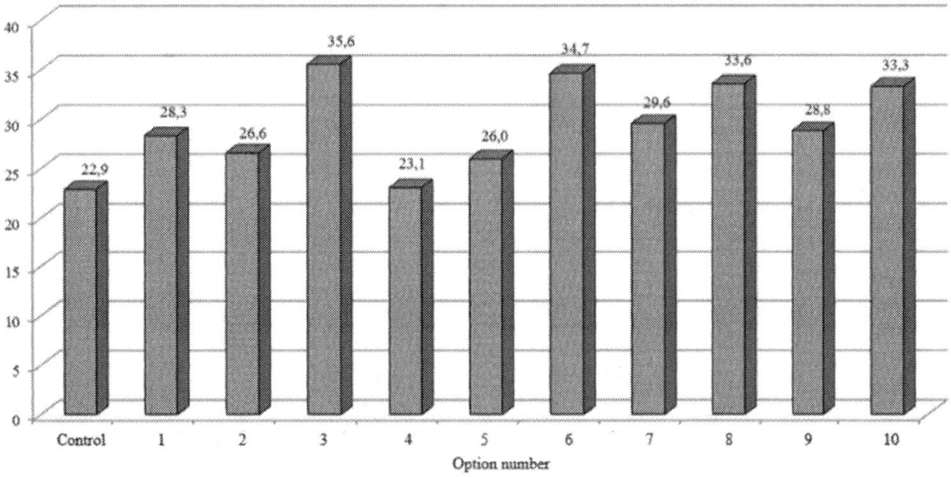

Fig. 3. Effect of pre-sowing seed treatment on overall survival of flax, %

Based on the experimentally obtained data presented in Figure 3, we can conclude that the implemented variants of seed pre-sowing treatment allowed to increase the total survival rate of flax-dolgunets from 22.9 to 35.6% or, on average, by 7.1%. Greater total survival rate was observed in variants 3 and 6 - 35.6 and 34.7%, i.e. increased by 12.7 and 11.8%, respectively, compared to the control.

The positive influence of the studied variants of pre-sowing treatment influenced the formation of flax yield, which is presented in Table 1.

Table 1. Flax yield (c/ha) depending on pre-sowing seed treatment methods, 2024

Option number	Productivity, c/ha		Increase			
			c/ha		%	
	trusts	seeds	trusts	seeds	trusts	seeds
Control	25,6	6,6	-	-	-	-
Option 1	41,2	10,5	15,6	3,9	60,9	59,1
Option 2	35,4	8,8	9,8	2,2	38,3	33,3
Option 3	45,9	11,0	20,3	4,4	79,3	66,7
Option 4	33,2	8,6	7,6	2,0	29,7	30,3
Option 5	55,3	12,6	29,7	6,0	116,0	90,9
Option 6	58,9	12,4	33,3	5,8	130,1	87,9
Option 7	50,0	14,1	24,4	7,5	95,3	113,6
Option 8	69,5	15,7	43,9	9,1	171,5	137,9
Option 9	66,9	16,6	41,3	10,0	161,3	151,5
Option 10	61,3	15,2	35,7	8,6	139,5	130,3
Average by seed treatment options	51,8	12,6	26,2	6,0	102,2	90,2
NDS_{05}	-	-	1,9	1,1	-	-

Without application of seed pre-sowing seed treatment, the yield of flax cod was 25.6 c/ha, a seeds - 6.6 c/ha. Application of the studied variants of seed pre-sowing treatment significantly (NSR_{05}=1.9 c/ha and NSR_{05}=1.1 c/ha) increased flax yield from 25.6 to 51.8, and flax seeds - from 6.6 to 12.6 c/ha or by 26.2 and 6.0 c/ha (102.2 and 90.2%), respectively, with the results obtained from untreated seeds in the magnetic field. Among the studied variants of pre-sowing seed treatment, 8 and 9 variants provided the highest yields of both flaxcotton and flaxseeds.

The difference in yields of both cod and seed can be explained by the influence of the studied seed treatment methods on the main elements of the yield structure of long-staple flax, which is clearly demonstrated by the data of Table 2.

Table 2. Flax yield structure, 2024

Option number	Number of plants before harvesting, pcs/m^2	Number of boxes on one plant, pcs.	Number of seeds in one box, pcs.	Weight of seeds per capsule, g	Weight of 1000 seeds, g
Control	275	10,3	7,14	0,034	4,70
Option 1	340	10,6	7,33	0,034	4,77
Option 2	319	12,9	6,56	0,031	4,62
Option 3	427	12,7	7,11	0,035	4,95
Option 4	277	10,4	6,99	0,035	4,79
Option 5	312	12,2	6,65	0,030	4,62
Option 6	416	8,2	6,94	0,033	4,70
Option 7	355	16,5	4,78	0,022	4,42
Option 8	403	16,9	5,83	0,029	4,60
Option 9	346	15,0	6,16	0,031	5,10
Option 10	400	17,4	6,38	0,031	4,76
Average by seed treatment options	360	13,3	6,47	0,031	4,73

As can be seen from the data presented in Table 2, the number of plants before harvesting, the number of bolls per plant, and the weight of 1000 seeds, on average, under the influence of the studied variants of pre-sowing treatment increased. At the same time, the

* Corresponding author: etsh1965@mail.ru

studied methods had a significant effect on the number of plants before harvesting and the number of bolls on one plant, which increased compared to the control, on average, by 33.3 and 29.1%, respectively. The effect of the studied methods on the number of plants before harvesting was more pronounced in 3 and 6 variants, on the number of bolls - in 7 and 10 variants, on the weight of 1000 seeds - in 3 and 9 variants.

Thus, the study of pre-sowing seed treatment methods showed their greater influence on the number of plants before harvesting and the number of bolls per plant.

Since longan flax was cultivated for seed purposes, we determined the multiplication factor for this crop depending on the studied variants of the experiment (Fig. 4).

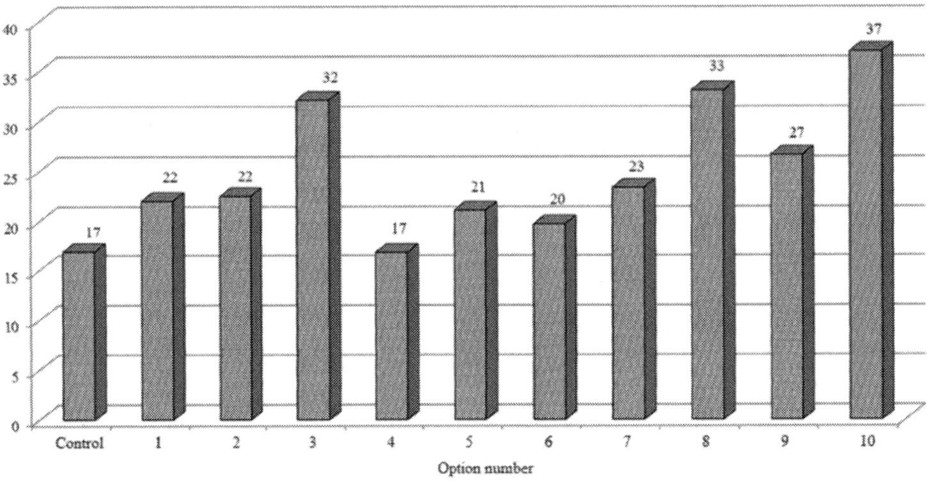

Fig. 4. Effect of pre-sowing seed treatment on the multiplication factor of flax, 2024

Experimental data shown in Figure 4 convincingly indicate that the studied methods of seed pre-sowing treatment lead to an increase in the reproduction factor from 17 to 37 units. At the same time, higher multiplication coefficients were obtained in 3, 8 and 10 variants of pre-sowing treatment (32, 33 and 37 units).

Table 3 shows the indicators of flax fibre quality. As can be seen from the presented data, plant height under the influence of pre-sowing seed treatment in a gradient magnetic field increases, on average, from 85.3 to 101.5 cm, technical stem length from 68.2 to 82.6 cm, stem thickness from 1.78 to 1.99 mm, fibre yield from 4.7 to 8.8 c/ha. Regarding the flax fibre content, it can be noted that no positive influence of the studied variants on this indicator was revealed and only on 2 and 7 there was a slight increase by 0.8 and 1.0%.

Table 3. Flax fibre quality indicators, 2024

Option number	Plant height, cm	Technical length of stem, cm	Stem thickness, mm	Fiber content, %	Fiber yield, c/ha
Control	85,3	68,2	1,78	18,3	4,7
Option 1	98,8	81,7	1,86	18,4	7,6
Option 2	89,4	70,8	1,84	19,1	6,8
Option 3	104,1	85,4	1,95	14,3	6,6
Option 4	100,5	84,6	1,89	17,5	5,8
Option 5	102,1	84,1	1,98	16,3	9,0
Option 6	102,0	85,5	1,77	18,6	11,0
Option 7	106,3	84,4	2,16	19,3	9,7
Option 8	108,1	85,4	2,33	15,4	10,7
Option 9	105,5	85,6	2,09	16,3	10,9
Option 10	97,9	78,0	2,04	16,5	10,1
Average by seed treatment options	101,5	82,6	1,99	17,2	8,8

Determination of fibre yield allowed to reveal the positive impact of the studied variants of seed treatment in a gradient magnetic field on this indicator. It can be noted that the highest flax fibre yield was obtained in variants 6, 8 and 9 and was equal to 11.0, 10.7 and 10.9 kg/ha, which was 2.3 times higher than in the control variant.

4. Conclusion

Based on the results of laboratory and field studies under the conditions of the 2024 growing season, the following conclusions can be drawn:

1. Variants of pre-sowing treatment of flax-dolgunets seeds allow improving the sowing qualities of the studied crop, such as germination energy and laboratory germination, on average, by 11.4% and 14.8%, respectively, compared to the control variant.

2. Affecting sowing qualities, the studied variants of seed treatment resulted in an increase in the number of plants before harvesting and, as a consequence, in the

total survival rate of flax-dolgunets plants, on average, by 7.1% compared to the control variant.

3. Among the studied variants of seed pre-sowing treatment, the highest yields of both flax straw and flax seeds were provided by variants 8 and 9, in which seeds were treated in a magnetic gradient field with an induction of 6.70 and 7.35 mTl. The greater influence of pre-sowing treatment methods was revealed on the number of plants before harvesting and the number of bolls per plant.

4. Cultivation of long-flax for seed purposes at application of methods of pre-sowing preparation allowed to obtain high multiplication coefficients on 3, 8 and 10 variants of seed treatment in magnetic gradient field with induction of 2.51, 6.70 and 8.00 mTl (32, 33 and 37 units).

5. The highest flax fibre yield was obtained in variants 6, 8 and 9 (magnetic field induction was equal to 4.20, 6.70 and 7.35 mTl) and was 11.0, 10.7 and 10.9 c/ha, which was 2.3 times higher than in the control variant.

References

1. T.A. Rozhmina, L.N. Pavlova, V.P. Ponazhev, L.M. Zakharova, Flax industry on the way to revival, Plant Protection and Quarantine, **1**, 3-8 (2018)
2. A.A. Akimov, N.N. Ivanyutina, A.S. Vasiliev, et al, Techniques to increase the productivity of varietal crops of long-staple flax under cultivation on fallow land, International Journal of Applied and Basic Research, **12-2**, 206-211 (2019), https://doi.org/10.17513/mjpfi.12981
3. V.I. Baev, V.A. Petrukhin, I.V. Baev et al., Electrotechnology as One of the Most Advanced Branches in the Agricultural Production Development, In: *Handbook of Research on Renewable Energy and Electric Resources for Sustainable Rural Development* (IGI Global, Hershey, Pennsylvania, 2018), https://doi.org/10.4018/978-1-5225-3867-7.ch007.
4. A.S. Kazakova, I.V. Yudaev, M.G. Fedoischenko et al., New approach to study stimulating effect of the pre-sowing barley seeds treatment in the electromagnetic field, OnLine J. of Biol. Sci., **18(2)**, 197-207 (2018), https://doi.org/10.3844/ojbsci.2018.197.207.
5. A.S. Dorokhov, M.E. Chaplygin, A.G. Aksenov et al, Grain crop seed treatment in a low-frequency electromagnetic field, Agricultural Machines and Technologies, **14(4)**, 4-11 (2023), https://doi.org/10.22314/2073-7599-2023-17-4-4-11
6. V.I. Khainovsky, O.S. Kopylova, A.E. Kozyrev, Calculation of doses of impact of pulsed electric field on seed material, Agrarnaya Rossiya, **9**, 37-40 (2012)
7. A. G. Batuhtin, M. S. Bass, S. A. Ivanov, S. G. Batuhtin, P. G. Safronov, American Journal of Applied Sciences, **13(2)**, 145–151, (2016).
8. A. I. Karlina, A. E. Balanovskiy, V. V. Kondratiev, V. V. Romanova, A. G. Batukhtin, Y. I. Karlina, Appl. Sci., **14,** 9774, (2024). https://doi.org/10.3390/app14219774
9. M. Grinberg, N. Ilin, Yu. Nemtsova et al., Effect of extremely low-frequency magnetic fields on light-induced electric reactionsin wheat, Plant Signal. & Behav., **17(1)**, e2021664 (2022), https://doi.org/10.1080/15592324.2021.2021664
10. A. Koukounaras, A. Boursianis, S. Kostas et al, Pre-Sowing Static Magnetic Field Treatment of Vegetable Seeds and Its Effect on Germination and Young Seedlings Development, Seeds, **2(4)**, 394-405 (2023), https://doi.org/10.3390/seeds2040030
11. V. Savchenko, O. Synyavskiy, Energy dose of flax seed treatment in a magnetic field, Energy & Autom. J., **59(1)** (2022),
12. L.A. Antipkina, V.I. Levin, S.N. Borychev et al, Use of gradient magnetic field in pre-sowing treatment of cucumber seeds, AgroEcoInfo, **1(49)** (2022), https://doi.org/10.51419/202121123

Morphological and productive differences between Chinese pullet chickens and broilers

Golibjon Fozilov[1,*], *Isroil* Umarov[2], *Okhunjon* Abdusalomov[3], *Sayyohat* Abdumajitova [4], *Oleg S.* Turchanin [5], *Andrey M.* Karadzhiyan [5]

[1] "Tashkent Institute of Irrigation and Agricultural Mechanization Engineers" National Research University, Tashkent, Uzbekistan
[2] Namangan Engineering-Construction Institute, Namangan, Uzbekistan
[3] Termez State University, Termiz, Uzbekistan
[4] Termiz State Pedagogical Institute, Termiz, Uzbekistan
[5] Kuban State Agrarian University named after I. T. Trubilin, Krasnodar, Russia

Abstract. The aim of this study was to comparatively investigate morphological and productive characteristics of Chinese silk chickens and broilers. The experiment involved rearing birds under controlled conditions and analysing slaughter products. Significant differences between the breeds were found: broilers had higher mean carcass weight but lower relative heart and stomach weight compared to Chinese silkie hens. At the same time, broilers showed greater leg yield. Morphological differences including neck length and wing area were established. Chemical analysis showed that the meat of Chinese silk chickens contained more protein and less fat than that of broilers. The results confirm the general patterns observed in different breeds of birds and indicate that Chinese silk chickens are promising for alternative poultry production. These chickens, characterised by higher meat protein content and disease resistance, could be an attractive resource for health-oriented production. While broilers remain the optimal choice for mass meat production due to their high meat productivity. The study emphasises the need to further explore the potential of Chinese silk chickens and other alternative breeds.

1 Introduction

The poultry industry plays a key role in the world's food supply, providing a valuable source of animal protein - poultry meat and eggs, rich in vitamins and minerals. However, the efficiency and environmental sustainability of poultry farms is a subject of ongoing debate and research. Increasing productivity, reducing costs and minimising negative environmental impacts are pressing challenges for the industry. There are many approaches to optimising poultry production, including improving feed rations, introducing modern housing technologies, and using genetically improved breeds. Each of these approaches has its own advantages and disadvantages. For example, the use of genetically modified feed can increase productivity, but raises concerns about food safety and potential negative impacts on ecosystems. Modern poultry housing technologies such as cage batteries increase production efficiency but are often criticised for animal cruelty and negative impacts on bird welfare. At the same time, traditional breeding and housing methods may be more humane but less cost-effective.

Among the existing variety of chicken breeds, Chinese silk chickens are of particular interest due to their unique appearance and the purported medicinal properties of their meat and eggs. In the context of the search for alternative protein sources, the study presented by [1] on the impact of farmed protein on agriculture emphasises the importance of exploring the potential of different breeds of birds. In contrast to broilers, which are oriented towards maximum weight gain in a short period of time, Chinese silk hens are characterised by slower growth, lower egg production, but have several advantages: they are unpretentious to keep, resistant to diseases, and are assumed to have better quality, in terms of beneficial properties, meat and eggs. Investigating the characteristics of these chickens and comparing them with broilers is an important task because it allows us to assess the potential of using Chinese silk chickens as an alternative to traditional broiler poultry production. This approach aims to diversify the poultry industry, meet the growing demand for organic products and create new market niches for farmers.

The relevance of this study stems from the need to find alternative, more environmentally friendly and sustainable models of poultry production. In this regard, the promotion of organic agriculture, as emphasised by [2], is an important direction. Chinese silk chickens represent a promising direction that allows combining the receipt of quality products with minimal negative

* Corresponding author: golibjon270784@mail.ru

impact on the environment and care for animal welfare. A comparative analysis of the characteristics of Chinese silk chickens and broilers will make it possible to assess the economic feasibility and practical significance of the use of this breed in agriculture.

The aim of this work, the results of which are presented in the attached file, is a comparative evaluation of the characteristics of Chinese silk chickens and broilers, with emphasis on the yield of various slaughter products and their quality indicators. The data obtained will make it possible to evaluate the potential of Chinese silk chickens as an alternative source of high quality meat and eggs, as well as contribute to a more informed choice of poultry breeds for modern poultry production.

2 Method and materials

In the framework of the research, a comparative experiment was carried out to study the characteristics of two breeds of chickens: Chinese silk and broilers. The object of the study was Chinese silk chickens raised in the conditions of a farm in Krasnodar region and broilers. All chickens used in the experiment met the requirements of GOST 18292-2012 'Interstate Standard. Agricultural poultry for slaughter. Technical conditions'.

Standard conditions corresponding to the recommendations for each breed were created for poultry rearing. Feeding was carried out according to generally accepted norms, using compound feeds for poultry farming [3] show the role of protective forest belts for sustainable development of agrarian regions, which is important to consider in the future when organising poultry farms.

Specialised equipment was used for slaughtering poultry to ensure a humane and safe procedure. Carcasses were cut manually, followed by collection and weighing of individual slaughter products such as liver, heart, neck, stomach, legs, head and wings. The experiment involved weighing of the harvested slaughter products to determine their mass fraction in the carcass. Electronic scales with a measurement accuracy of 0.01 grams were used for weighing.

In addition, thermometers and hygrometers were used to monitor housing conditions and growth of the birds to monitor the temperature and humidity in the housing rooms. When rearing Chinese silk chickens, a special temperature regime was created and maintained using a thermostat. The temperature was 30 °C for the first few days, with a subsequent decrease of 3 °C every week until a temperature of 18 °C was reached. The chicks were fed with regularity appropriate to their age, using finely chopped greens, cooked vegetables and grains. Chicks were visually inspected to monitor goitre filling. Also, chickens were given fermented milk drinks such as kefir, sour milk, and whey regularly, followed by replenishment with water. All the data obtained were recorded for further analysis and comparison. Effective monitoring and analysis of modern technologies for agro-industrial development in mountainous areas, as shown in [4], is an important aspect for optimising production in such regions.

3 Results and discussion

As part of our in-depth study aimed at comparative analysis of productivity and morphological features of Chinese silk hens and broilers, we conducted a comprehensive experimental work. The purpose of this study was not only to confirm previously obtained data, but also to expand knowledge about the differences between these two breeds, as well as to assess the potential prospects for the use of Chinese silk chickens in modern poultry production. The experiment consisted of several stages, starting with rearing the birds under controlled conditions and ending with detailed analyses of slaughter products. Chinese silk chickens were raised on the basis of a farm, where they were provided with conditions as close as possible to natural conditions, including free range. Broilers, on the other hand, were obtained from a large poultry farm where their housing was organised according to standard industrial protocols. This allowed us to evaluate the effect of different housing conditions on the productive performance of the birds. For Chinese Silk chicks, a temperature regime was established with an initial temperature of 30°C, which was gradually reduced by 3°C weekly until reaching 18°C. This measure was taken to ensure optimum development and survival of the young, as the early growth stages of this breed are critical. At the same time, the humidity level in the house was maintained at 60-70%. Broilers, on the other hand, were kept under conditions that allowed intensive growth, with constant access to feed and water, at temperatures in the range of 20-25°C and humidity levels of 50-60%. The diets for both breeds included a balanced mixed feed formulated specifically for each age group, taking into account their nutrient requirements. For example, broilers were fed a feed with high protein and energy content to maximise weight gain, while Chinese silkie chickens were fed a diet that took into account their slow growth rate and digestive system. Daily monitoring of the birds' condition, behaviour, feed and water intake, growth and development was carried out throughout the rearing period. The birds were slaughtered in accordance with humane standards and technical regulations, using specialised equipment.

After slaughter, the carcasses were subjected to careful manual butchering. The main criterion during the butchering process was the accurate collection and weighing of all slaughter products, including liver, heart, neck, stomach, legs, head and wings. Electronic scales were used for accurate weighing, providing accuracy down to 0.01 grams, allowing even minor differences between breeds to be recorded. For each slaughter product, its weight in grams was determined and its mass fraction calculated as a percentage of the total carcass weight. This made it possible not only to compare the weight of the individual carcass parts, but also to assess their proportional relationship. In addition, at each stage of the study, samples were collected for further analysis and evaluation of the quality of meat and internal organs. In the process of data collection, we recorded additional parameters such as: average growth time to slaughter age, which was 120 days for Chinese silk hens and 45 days for broilers. The average egg production per year for Chinese silk hens was 100 eggs, with an average egg weight of 32 grams, which is significantly lower than that of egg-bearing breeds, but the eggs have more pronounced medicinal properties. When assessing the sustainability of agricultural land use in mountain-steppe zones, it is important to take into account geo-ecological factors, as demonstrated in [5, 6].

The study not only confirmed the known differences, but also revealed new, significant nuances in morphological characteristics and productivity of Chinese silk hens and broilers. The average carcass weight of Chinese silk hens was 2.13 kg, which, as expected, is slightly less than that of broilers, where this indicator reaches 2.21 kg on average. At the same time, a detailed analysis of the carcass structure showed significant differences in the proportions of individual organs and parts. The heart weight of Chinese silk chickens averaged 0.040 kg, which as a percentage of carcass weight corresponds to 2.15%. The One Health approach as described by [7] emphasises the importance of an integrated approach in the study of animal health and its impact on food security. In broilers, however, this indicator was much lower: heart weight was only 0.007 kg, or 0.32% of the carcass weight. Moreover, there was considerable individual variation, with individual Chinese silkie chickens having a heart weight of 0.048 kg or 2.3 per cent of the carcass weight, while broilers had a heart weight of 0.009 kg or 0.39 per cent. A similar pattern was observed with stomach weight: in Chinese silkie chickens this parameter was significantly higher and averaged 0.095 kg, or 4.8% of carcass weight, compared to broilers, where this parameter averaged 0.031 kg (1.4% of carcass weight). In Chinese silk chickens, the stomach weight reached 0.105 kg or 5% of the carcass weight, while in broilers the maximum value did not exceed 0.037 kg or 1.6% of the carcass weight. These data emphasise the peculiarities of the digestive system of Chinese silk chickens, which is probably related to their specific diet and lifestyle, which tends to be more mobile than that of broilers kept mainly in cage conditions. Such findings correlate with the principles of sustainable land use based on geo-ecological landscape assessment as described in the study [8].

At the same time, the leg yield in broilers averaged 0.270 kg (12.2% of carcass weight), which is more than double that of Chinese silkie chickens, which was 0.145 kg (7.3% of carcass weight). Our observations showed that individual broilers had a leg mass of 0.285 kg, or 12.9% of carcass weight, indicating that they had significantly more developed leg musculature, which is probably due to selection for rapid growth and meat productivity. The importance of managing the sustainability of land use at the local level is emphasised by [9, 10].

Differences in the yield of other parts of the carcass such as neck, head and wings were not significant, although some morphological differences were noted. For example, the average neck length of broilers was 2 cm longer than that of Chinese silk chickens, and the average wing area of broilers exceeded that of silk chickens by 5 cm2. Analysis of the chemical composition of the carcasses showed a lower fat content of 8.5% in Chinese silkie chickens compared to 15.2% in broilers, which are known to be more prone to fat deposition. In addition, the meat of Chinese silk chickens was found to be characterised by a higher protein content, 10-15% higher than that of broilers, making their meat more valuable in terms of dietary value. The importance of mining waste recycling for sustainable agricultural production is emphasised in [11].

Comparative analysis of our data with the results of other similar studies has shown that our findings confirm the general trends observed in different breeds of birds, including chicken breeds with different growth and productivity characteristics. For example, studies by other authors, such as a paper published in the Journal of Poultry Science, have found that broilers, on average, are characterised by a higher yield of muscle mass, particularly breast and thigh muscles, while breeds with slower growth, such as Chinese silk hens, have a more developed digestive system, as shown by a larger relative size of the stomach and intestines. A study conducted at the Institute of Poultry Science showed that broilers had an average of 18% greater breast muscle yield than slow-growth breeds, confirming their selection for meat production. Also, in this study, higher fat content in broiler meat was observed, on average 7-9% higher compared to egg-oriented hens and slow-growth breeds, which is in agreement with our data. At the same time, studies conducted in China, in particular in the Chinese Journal of Animal Science, revealed that the meat of Chinese silk hens is characterised by a higher content of essential amino acids, in particular lysine, methionine and tryptophan, as well as a higher content of antioxidants, in particular carnosine and anserine, which may be explained by specific features of their metabolism and diet. In addition, studies conducted at Tsinghua University also noted that the content of polyunsaturated omega-3 and omega-6 fatty acids in the meat of Chinese silk chickens is 12-15% higher than that of broilers, which makes their meat healthier. However, as noted by [12], integration of poultry farming with other agriculture can improve the efficiency and sustainability of agro-ecosystems. Our studies have also shown that Chinese silk chickens show greater resistance to diseases, particularly coccidiosis and infectious bronchitis, and have a higher survival rate of young, on average 5-7% higher than broilers, making them attractive for breeding in organic production environments where the use of antibiotics and other drugs is limited. It is noted by [13] that access to credit plays a key role in innovation and productivity improvement in agriculture.

Overall, it can be concluded that Chinese silk chickens represent an interesting and promising resource for alternative poultry production, and can carve a niche in the market for products that focus on health, sustainability and unique consumer properties. A study of resources such as high quality quartz that can be used in various industries including agriculture is presented in [14]. The potential of precision farming to improve crop yields and sustainability of production is presented in [15]. Their meat with higher protein and antioxidant content and lower fat content can be demanded by consumers seeking a healthy diet.

4 Conclusion

The presented study was conducted to comparatively evaluate the performance of Chinese silk chickens and broilers, focusing on the yield and quality of slaughter products. The experiment revealed significant differences between the two breeds. Broilers showed higher average carcass weight (2.21 kg), but lower relative heart (0.32% of carcass weight) and stomach (1.4% of carcass weight) weight compared to Chinese silkie chickens, which had these values of 2.15% and 4.8%, respectively, with an average carcass weight of 2.13 kg. Chinese silk hens had lower leg weight (7.3% of carcass weight), while in broilers this index reached 12.2%, indicating a more developed leg musculature in the latter.

Analysis of morphological differences showed that broilers had an average of 2 cm longer neck and 5 cm² more wing area than Chinese silkie chickens. Chemical analysis revealed that the meat of Chinese silkie chickens contained 10-15% more protein and 6.7% less fat (8.5%) than that of broilers (15.2%). These findings confirm general trends observed in other studies, where broilers are characterised by higher muscle mass and fat content, and slow-growing breeds such as Chinese silkie chickens have more developed digestive systems and more valuable meat chemistry.

In particular, studies conducted in China have found high levels of essential amino acids and antioxidants in the meat of Chinese silk chickens, as well as higher levels of polyunsaturated fatty acids. In mountainous conditions, studies [16] show that mountain factors can significantly affect soil salinity, which should be taken into account when organising agriculture. In addition, the findings confirm that Chinese silk chickens show higher disease resistance and better survival rate of young, which makes them attractive for organic production, where the use of antibiotics and other drugs is limited. The use of mechanical activation, as demonstrated by [17], appears to be a promising avenue for waste utilisation and recovery of valuable substances. To ensure reliability and efficiency of agrarian systems, methods of technical condition prediction can be used, as described in [18].

Thus, Chinese silk chickens appear to be a promising alternative to broiler poultry farming, especially in the conditions of production oriented on health, environmental friendliness and unique consumer properties. They can carve a niche for themselves in the produce market by offering meat with higher protein, antioxidant content and low fat content. At the same time, broilers, due to breeding, show higher meat productivity and shorter rearing times, making them the optimal choice for industrial poultry production focused on maximising meat yield.

References

1. K. Mittenzwei, W. Britz, R.J.F. Burton, The potential impact of cultivated protein on agriculture in Norway. Environ. Innov. Soc. Transitions, **55,** 100960 (2025). DOI: 10.1016/j.eist.2024.100960
2. A. I. Karlina, A. E. Balanovskiy, V. V. Kondratiev, V. V. Romanova, A. G. Batukhtin, Y. I. Karlina, Appl. Sci., **14,** 9774, (2024). https://doi.org/10.3390/app14219774
3. A. S. Manaenkov, P. M. Podgaetskaya, V. N. Petrov, Growth dynamics of forest-forming species in field-protective forest strips in the south of Western Siberia, Sustainable Development of Mountain Territories, **16(3),** 1043–1053 (2024). DOI: 10.21177/1998-4502-2024-16-3-1043-1053
4. A. Batukhtin, I. Batukhtina, M. Bass, M. Baranovskaya, A. Baranovskaya, Development and experimental verification of the mathematical model of thermal inertia for a branched heat supply system. Journal of Applied Engineering Science, **17(3),** 413–424, (2019).
5. D.A. Zyukin, A.I. Karlina, Technology for Nonwaste Recovery of Tailings of the Mizur Mining and Processing Plant. Metallurgist, **66 (11-12),** 1476-1480 (2023) DOI: 10.1007/s11015-023-01462-y
6. O. Afanaseva, M. Afanasyev, S. Neyrus, D. Pervukhin, D. Tukeev. Information and Analytical System Monitoring and Assessment of the Water Bodies State in the Mineral Resources Complex. Inventions, **9,** 115 (2024). https://doi.org/10.3390/inventions9060115.
7. Y. Gao, C. Li, Q. Zhou, M. Hao, The application of One Health concept in China and its practice and innovation in cross-sector cooperation. Public Health Pract., **9,** 100574, (2025). DOI: 10.1016/j.puhip.2024.100574
8. A. G. Batuhtin, M. S. Bass, S. A. Ivanov, S. G. Batuhtin, P. G. Safronov, Method of efficiency improvement in district heating systems. American Journal of Applied Sciences, **13(2),** 145–151, (2016).
9. T.I. Kosovtsova. Factors of conservation of cultural landscapes of the Bavarian Alps. Sustainable Development of Mountain Territories, **11,** 371–379 (2019). DOI: 10.21177/1998-4502-2019-11-3-371-379.
10. V. I. Golik, T. A. Panfilova, V. Yu. Konyukhov, Mining Informational and Analytical Bulletin. **11-1,** 175-189 (2023). DOI: 10.25018/0236_1493_2023_111_0_175
11. L.S. Malyukova, V.V. Kondratiev, V.V. Bukhtoyarov, V.Y. Konyukhov, K.A. Bashmur, T.A. Panfilova, V. Brigida. Circular Mining Wastes

Management for Sustainable Production of Camellia sinensis (L.) O. Kuntze. Sustainability (Switzerland), **15,** 11671 (2023). DOI: 10.3390/su151511671.

12. A.M. Carey et al. Integrating poultry improves soil health and vegetable yield in organic, cover-cropped system. Agric. Ecosyst. Environ., **382,** 109499 (2025). DOI: 10.1016/j.agee.2025.109499.
13. A. Ogunleye et al. Impact of credit constraints on adoption of biofortified food crop among farming households in Nigeria. Sustain. Futures, **9,** 100421 (2025). DOI: 10.1016/j.sftr.2024.100399.
14. I.I. Bosikov et al. Analysis and evaluation of prospects for high-quality quartz resources in the North Caucasus. Min. Sci. Tech. (Russian Federation), **8,** 278-289 (2023). DOI: 10.17073/2500-0632-2023-10-165.
15. S. Jauhari et al. Rice yield enhancement and environmental sustainability with precision nutrient management. Glob. J. Environ. Sci. Manag., **11,** 77-92 (2025). DOI: 10.22034/gjesm.2025.01.
16. A. I. Trunova, S. O. Kurashkin, R. B. Sergienko, Metals, **13(7),** 1234 (2023). DOI: 10.3390/met13071234.
17. V.I. Golik, Tailings Utilization and Zinc Extraction Based on Mechanochemical Activation. Materials, **16,** 726 (2023). DOI: 10.3390/ma16020726.
18. K. A. Bashmur, V. V. Kondratiev, Mining Informational and Analytical Bulletin, **11-1,** 239-251 (2023). DOI: 10.25018/0236_1493_2023_111_0_239.

Parameter evaluation of a technical tool for electric destruction of weeds and undesirable plants

Igor Viktorovich Yudaev, *Ivan Vasilievich* Sharov

I.T. Trubilin Kuban State Agrarian University, Krasnodar Region, Krasnodar, Russia

Abstract. The use of various mobile electro-technological installations and aggregates to perform various field operations (weeding, potato haulm desiccation, soil pest control, pre-harvest preparation of sunflower and tobacco, etc.), carried out directly in crops and plantings of cultivated crops or on agricultural land, is now considered as a quite effective alternative to chemical components, characterised by environmental friendliness and high efficiency. In the considered variant of technical realisation it is proposed to apply high voltage electrical pulses to disturb normal plant life. The purpose of the research presented in the article is to analyse possible variants of implementation of treatment of weeds and undesirable grasses with the purpose of their destruction, and to estimate the parameters of the output unit of the electro-technological installation. Both, substantiated from the point of view of technological efficiency, variants of electric energy supply to weeds are realisable in practice and the choice of each of them is based on the configuration of the surface and the cultivated soil area, the degree of their infestation with specific types of undesirable grasses and the capacity of the electrotechnological mounted or trailed unit. At the same time, the voltage on the electrode system should be greater than the sum of the voltage drop on all elements of the treatment current flow circuit, and the choice of the capacitance value directly depends on the energy used for treatment of undesirable grasses, the treatment voltage and the operation mode of the capacitor bank. The proposed approach in estimating the parameters of the discharge circuit and electrode system relies on experimental prospecting studies with those species of weeds that directly grow in the places of application of the electric pulse unit. The presented sequence of determining the main parameters of the output unit of the unit - discharge circuit, should provide sufficient accuracy for practical adjustment and adjustment of the unit.

Keywords: electrotechnological operations, electrotechnological installation on a wheeled tractor, variants of bringing electricity to plants, elements of the treatment current flow circuit, voltage on electrodes, capacitance of the discharge circuit.

1. Introduction

Today, the world is increasingly focusing on the production of environmentally friendly, safe from the nutritional point of view, carried out with minimal energy consumption of food products and raw materials for their production in quantities necessary to meet the needs of the world's population. Despite the relatively low cost and high technological efficiency of chemically synthesised products, their negative impact on agriculture is a matter of concern for the progressive public and forces researchers and engineers to search for a worthy substitute. As an effective alternative for agricultural sector operations, attention is increasingly turning to electrophysical methods and techniques [1, 2].

At present, effects and phenomena of electrical nature in agricultural production, especially in field farming, horticulture, berry plantations, etc., are effectively used for pre-sowing stimulation and disinfection of seeds and planting material [3, 4]; for the control of pests, fungal spores and diseases in plantings, crops and plantations [5, 6]; for the production of organic fertilisers [7]; for the control of weeds and undesirable vegetation [1, 8]; for pre-harvest and post-harvest treatment of harvested products before storage [9] and others.

The agro-technological operation to control undesirable and weedy vegetation deserves special attention, as the efficiency of its implementation affects all processes related to land management: from crop production and horticulture to fodder preparation in animal husbandry.

2. Methods and materials

To realise the method of weed and unwanted vegetation destruction by means of high voltage electric pulses, it is possible to use an electro-technological installation, aggregated with a wheeled tractor, and consisting (Figure 1) of: 1) engine (wheeled tractor); 2)

* Corresponding author: etsh1965@mail.ru

source of electric energy, e.g. synchronous generator driven by the tractor PTO; 3) voltage boosting and conversion unit; 4) electrode system for supplying electric energy to weeds. In this case, the most important parameters for irreversible damage to the plant tissue of weeds and their subsequent death are the parameters of the output unit of the plant - the discharge circuit of the plant: operating voltage, energy stored in the capacity of the circuit, and the capacity itself.

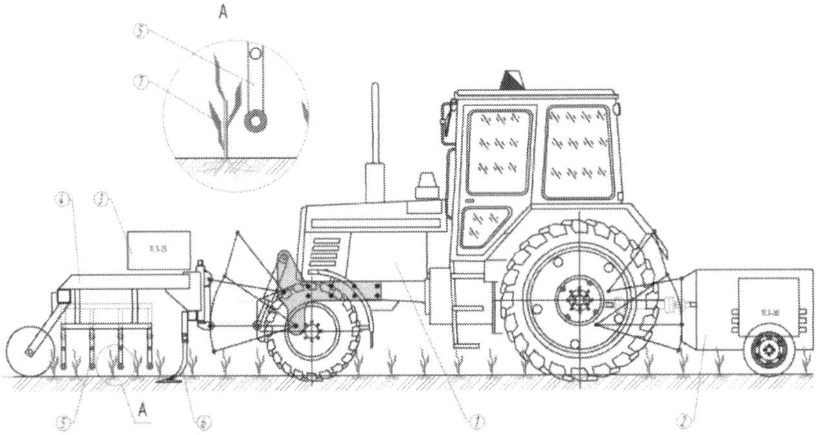

Figure 1. Electro-technological installation aggregated with a wheeled tractor: 1 - mover (wheeled tractor); 2 - source of electric energy (synchronous generator driven in rotation by PTO and step-up transformer); 3 - shaper of high-voltage electric impulses; 4 - mounted frame (frame of industrially produced cultivator); 5 - sections of mounted electrodes; 6 - electrode buried in the soil (cultivator foot); 7 - weed plant.

In the process of electro-pulse weed destruction, voltage is applied to the weeds by means of a system of electrodes, under the action of which an electric current of technologically justified density flows through a circuit including above-ground and underground parts of plants, as well as soil area. The amount of electrical energy expended for irreversible damage of weed plant tissues is proportional to the square of the applied voltage, time of exposure and inversely proportional to the value of resistance of the current flow circuit.

When implementing a particular variant of electric pulse weeding, determined by the type of weediness and the prevailing type of weed vegetation in the treated area, it is necessary to take into account, first of all, the method of supplying electrical energy to the objects of treatment, which characterises the circuit of current flow through plant tissues of weeds and through the soil, and also allows to assess the technological efficiency of the proposed method and determine the technical parameters of the electric weeder.

3. Results and discussion

In the works preceding the presented study, the question of contacting electrode systems with the treatment object was considered, and the analyses carried out in them made it possible to choose an acceptable contacting variant, to give a classification of the electrode systems used and to select among them a design suitable for application [10]. There is still no serious synthesis of new, structurally simple and technologically effective electrode systems.

Let us analyse the paths of current flow during electric pulse weeding, which are implemented in practice and effective from the point of view of achieving the final result - irreversible damage to weeds.

The variant, when only the root system of weeds is treated, seems to be the most effective, as it is with such an electric influence that the main epicentre of weed plant reproduction should be damaged to the maximum extent possible (Fig. 2,a). The current circuit in this case necessarily includes the roots of weeds and soil area, which have an increased total ohmic resistance [11]. The technical realisation of this variant is characterised by some complexity, as it will be necessary to use only electrodes buried in the soil. This will increase in several times the total cumulative energy consumption used to move the machine over the clogged massif. Besides, the treatment can be carried out only at soil moisture values considerably lower than the root moisture, because otherwise it is possible to flow current only through the moistened surface soil layer, bypassing the root system areas. In case of weed root-spreading plants destruction it is necessary to use such variants of buried electrodes design, the working surface of which does not break the root system of weeds into separate fragments and there is no soil layer wrapping. It is also important to take into account that on some part of the roots located at the depth where the boundary of electrical treatment passes, there is not oppressive but stimulating electromagnetic effect from the pulse current flowing in the soil. As a result of moving closer to the surface, these plant fragments can produce abundant root growth. At the same time, the damaged, treated parts of the root system, through which the current flowed with a density sufficient for the desired technological effect, remaining in the ground, rot, serving later as a basis for the formation of organic fertilisers.

The next method of treatment already includes the whole weed plant - its leafy part and soil-root part - in the circuit of current flow (Fig. 2, b). This option is realised by direct contact of the mounted electrode with the above-ground part of the weed, and buried in the soil - with the root system. The dose of electric energy is selected from such a condition that first of all to achieve suppression and death of underground reproduction organs of weed plants. Such an impact is bound to irreversibly damage the tissues of leaf-stem part of plants, as it is experimentally established that the damaging value of energy for treatment of the root system is greater than the lethal dose required for destruction of the above-ground part [8]. This variant of electric energy supply to weeds, as well as the previously discussed one, is characterised by additional expenditures of the total energy used to move one or more electrodes buried in the soil and to heat the soil layer between the electrode and the plant root.

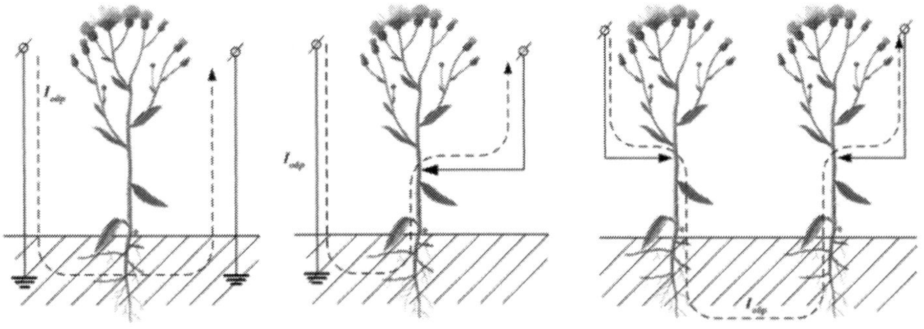

а) б) в)

Figure 2. Options for supplying electrical energy to weed plants: a) first buried electrode - soil - plant root - second buried electrode; b) hinged electrode - plant stem - plant root - soil - buried electrode; c) first hinged electrode - first plant stem - first plant root - soil - second plant root - second plant stem - second hinged electrode

To avoid the expenditure of total energy for moving the electrode located in the soil and at the same time to achieve irreversible damage and death of the root system of treated weeds, it is possible to use only a system of mounted electrodes, which are in contact with several plants simultaneously (Fig. 2, c). In this variant of interaction between electrodes and objects of treatment, the following chain of current flow is organised: the first mounted electrode - leafy part of the first weed - root system of the same plant - soil - root system of another plant - its leafy part - the second mounted electrode. This method is associated with increased consumption of electrical energy, as the total resistance of the treatment circuit is at least doubled compared to the previous variant. Such treatment represents a variant of using a 'no-till' farming system, which has a positive effect on the improvement of soil structure, its fertility and on the qualitative clearing of agricultural land from weeds. It can be used for weed control, for example, both in crops and plantings of cultivated plants and in the continuous cultivation of weedy land areas, e.g. fallows.

From the results of the analysis it follows that the most effective is the variant of treatment of only the root system of weeds, which cannot be realised without using the electrodes buried in the soil. However, moving the electrodes buried in the soil will create additional traction resistance when the machine moves across the field. Therefore, from the point of view of reducing additional energy losses for moving the machine through the weeded massif and taking into account that the elements of the above-ground part and root system should be exposed to the longitudinal effect of the current flowing through them, electrode systems, in our opinion, should be used mounted, above-ground version or in case of weeded areas with hard-to-eradicate plants - combined, combining mounted and buried in the soil electrodes.

It should also be emphasised here that the treatment circuit has the highest resistance value if the electrical action is carried out through the use of overhead electrodes only, e.g. in the treatment of at least two weeds, where the current from one of them to the other flows through the plants themselves, their root systems and the soil. In order to estimate the energy doses expended for irreversible tissue damage, it is important to study all components of the total resistance of the considered current circuits and quantify their values, which will also allow characterising the load of the electro-processing unit.

To achieve irreversible damage of treated weed plants, it is necessary to create in their tissues an electric field with the intensity E = 3.8 V/cm and higher [1,8], which is possible when the voltage sufficient for such treatment is applied to the electrodes. The voltage level on the electrodes is determined by the variant of electric energy supply to the plants and the total voltage drop on the elements of the treatment circuit.

The choice of the option of supplying electrical energy to weeds is determined by the type and degree of weediness of the soil in the cultivated field, which eventually leads to the use of either a system of mounted electrodes only, or mounted electrodes and at least one buried electrode. In this case, the voltage supplied to the weeds is more important in the first case, as the resistance of the path of the cultivation current increases. Voltage drops on circuit elements in the first $\Sigma\Delta U_{1i}$ (Figure 2b) and the second $\Sigma\Delta U_{2j}$ (Figure 2c) variants will be determined:

$$\sum_{i=1}^{m} \Delta U_{1i} = \Delta U_{EP} + \Delta U_T + \Delta U_R + \Delta U_{RP} + \Delta U_F + \Delta U_{SE}, \quad (1)$$

$$\sum_{j=1}^{n} \Delta U_{2j} = \Delta U_{EP1} + \Delta U_{T1} + \Delta U_{R1} + \Delta U_{RP1} +$$
$$+ \Delta U_F + \Delta U_{RP2} + \Delta U_{T2} + \Delta U_{R2} + \Delta U_{EP2}, \quad (2)$$

where ΔU_{EP} - voltage drop at the point of contact of the mounted electrode with the above-ground part of the plant; ΔU_T - voltage drop at the above-ground part (trunk) of the plant; ΔU_R - voltage drop at the plant root system; ΔU_{RP} - voltage drop at the point of contact of the plant root system with the soil; ΔU_F - voltage drop at the soil site of treatment current flow; ΔU_{SE} - voltage drop at the point of contact of the soil with the buried electrode.

It is obvious that the voltage at the electrodes $U_{\text{ЭС}}$ of the unit for electric pulse weeding should satisfy the following condition:

$$U_{ES} > \sum_{i=1}^{n} \Delta U_i \quad (3)$$

The voltage value on the discharge circuit of the installation U_{DC} must be greater than the voltage on the electrodes U_{ES} by the voltage drop $\Delta U_{DC\text{-}E}$ on the circuit section discharge capacitance - electrode system:

$$U_{DC} \geq U_{ES} + \Delta U_{DC-E}$$

$$\text{or } U_{DC} = K_U (U_{ES} + \Delta U_{DC-E}), \quad (5)$$

where K_U is a coefficient that takes into account incomplete discharge of the capacitance.

4. Conclusion

Each weed plant must be supplied with an amount of energy sufficient to reliably damage its plant tissue. This amount of electrical energy is determined by the product of specific damaging energy per unit mass or volume of plant tissue with the value from w_{spec} = 1.5 J/cm^3 and more by the total mass M or volume V of an individual plant and by the number N of simultaneously treated plants. Besides it is necessary to take into account losses of electric energy in transient contacts of the treatment circuit and in the soil ΔW:

$$W = \left(w_{spec} \cdot V + \Delta W\right) \cdot N. \quad (6)$$

The electrical energy of the discharge circuit W_K must exceed the energy W supplied to the discharge circuit by the amount of energy W_o remaining in the tank due to incomplete discharge:

$$W_K = W + W_o = \left(w_{spec} \cdot V + \Delta W\right) \cdot N + W_o. \quad (7)$$

Knowing the operating voltage of the installation, it is possible to determine the value of the discharge capacitance of the circuit C_K:

$$C_K = \frac{2 \cdot \left[\left(w_{spec} \cdot V + \Delta W\right) \cdot N + W_o\right]}{K_U^2 \cdot \left(U_{ES} + \Delta U_{PK-E}\right)^2}. \quad (8)$$

The presented sequence of determination of the main parameters of the output unit of the unit - discharge circuit, should provide sufficient accuracy for practical adjustment and adjustment of the unit, but requires additional studies of transient resistances ('soil - root system') and soil resistance.

References

1. V.I. Baev, V.A. Petrukhin, I.V. Baev et al., *Electrotechnology as One of the Most Advanced Branches in the Agricultural Production Development*, Handbook of Research on Renewable Energy and Electric Resources for Sustainable Rural Development, ed. by V. Kharchenko, P. Vasant (Hershey PA, 2018), Ch. 7, pp. 149–176, https://doi.org/10.4018/978-1-5225-3867-7.ch007
2. Borodin I.F., Electrotechnology in agricultural production, Elektrichestvo, 11, 1-6 (1982)
3. I. Palov, K. Sirakov, E. Kuzmanov et al., Results of preliminary laboratory studies after pre-sowing electric treatment of pea seeds, Agricult. Eng., **4**, 17–23 (2013)
4. A.A. Sokolov, L.G. Orlova, Mining Informational and Analytical Bulletin, **11**, 278-291 (2023). DOI: 10.25018/0236_1493_2023_111_0_278
5. D. Mori, A. Moriyama, H. Kanamaru et al., Electrical Stimulation Enhances Plant Defense Response in Grapevine through Salicylic Acid-Dependent Defense Pathway, Plants (Basel), **10(7)**, 1316 (2021), https://doi.org/10.3390/plants10071316
6. A. G. Batuhtin, M. S. Bass, S. A. Ivanov, S. G. Batuhtin, P. G. Safronov, American Journal of Applied Sciences, **13(2)**, 145–151, (2016).
7. A. Batukhtin, I. Batukhtina, M. Bass, M. Baranovskaya, A. Baranovskaya, Development and experimental verification of the mathematical model of thermal inertia for a branched heat supply system. Journal of Applied Engineering Science, **17(3)**, 413–424, (2019).
8. A. G. Batukhtin, E. A. Makhov, M. S. Bass, S. G. Batukhtin, International Journal on Technical and Physical Problems of Engineering, **15(57)**, 351–361, (2023).
9. V.I. Baev, I.F. Borodin, Electro-pulse pre-harvest treatment of sunflower and tobacco plants [Monograph], (Stanitsa-2, Volgograd, 2002), 228 p.
10. Lyapin, V.G. Diagnostics of electrode systems of mobile electro-technological installations, Scientific Problems of Transport of Siberia and the Far East, 2, 381-383 (2010)
11. A. I. Karlina, A. E. Balanovskiy, V. V. Kondratiev, V. V. Romanova, A. G. Batukhtin, Y. I. Karlina, Appl. Sci., **14**, 9774, (2024). https://doi.org/10.3390/app14219774

Innovative technologies for sustainable development of Arctic agriculture

M.A. Modina [1], *E.V.* Kheckert [1], *Vladislav V.* Dragulenko [2*], *M.Yu.* Ivanov [3], *Vladislav V.* Bovsh [2]

[1] Admiral Ushakov Maritime State University, Novorossiysk, 353924, Russia
[2] Kuban State Agrarian University named after I. T. Trubilin, Krasnodar, Russia
[3] Bratsk State University, Bratsk, Russia

Abstract. The paper looks at both existing and emerging innovative approaches, including biotechnology, digitalisation and renewable energy. A comprehensive analysis of agro-technical solutions that could transform the Arctic agro-industrial sector is conducted. Comparative experiments in five Arctic locations, using genetically modified and conventional barley and potato varieties, as well as different mulching and biostimulation techniques, were used to assess effectiveness. The study revealed the superiority of GM varieties, namely barley 'Arctic-12' and potato 'Morozko', with yield increases of 18-25% and 12%, respectively, and improved nutritional quality. In addition, various cropping systems, particularly hydroponics and aeroponics, with optimised formulations of nutrient solutions and lighting regimes were investigated and shown to be highly effective. In addition to crop production, the effects of feed additives on livestock and reindeer productivity and the digitalisation of livestock farms were studied. In conclusion, the authors emphasise the importance of an integrated approach, including the integration of traditional indigenous knowledge with modern technologies for sustainable agricultural development in the Arctic, with a potential productivity increase of 50-70% in the long term.

1 Introduction

The Arctic, with its unique ecosystems and specific challenges, is of particular interest in this context. Traditional methods of agriculture, raditional methods of farming established in more favourable climatic zones are not always applicable in Arctic conditions, which creates the need to search for and adapt new technologies and production methods, such as for mountainous regions, which also face climatic challenges [1, 2].

One way to address the problem of food security in the Arctic is to develop localised agriculture focused on meeting the needs of the local population. This approach involves growing cold-adapted crops and breeding animals that are resistant to extreme conditions, thus reducing dependence on imported products and reducing the carbon footprint associated with transport. However, the development of localised production faces a number of constraints, including short growing seasons, harsh climatic conditions, lack of skilled labour and high logistics costs. Another possible solution is the application of advanced agro-technologies such as hydroponics, aeroponics and vertical farming, which allow for controlled growing conditions all year round. These technologies can significantly increase yields and reduce dependence on weather conditions, but require significant investment and energy inputs, which may create financial and environmental barriers to their widespread adoption.

Given the above-mentioned challenges and opportunities, innovative approaches that integrate advances in biotechnology, digitalisation and renewable energy are of particular interest [3-5]. Biotechnology can enable the development of plant varieties and animal breeds that are resistant to extreme Arctic conditions, as well as to diseases and pests. Digitalisation and automation of agriculture can optimise production processes, reduce resource inputs and increase labour efficiency. The use of renewable energy sources helps to reduce the ecological load on the environment and dependence on fossil fuels. While promising, this trend also requires consideration of a number of factors, including the need to adapt technologies to local conditions, develop a regulatory framework for the use of biotechnology, and ensure that these innovations are accessible to farms in the Arctic.

In the context of the above, this study focuses on an overview of advanced technologies that contribute to sustainable agriculture in the Arctic. It examines both existing and emerging innovative approaches that can improve the efficiency and sustainability of the agricultural sector in this region, including biotechnology, digitalisation and renewable energy. It is important to note that sustainable agriculture in the Arctic depends on the food security of the local

* Corresponding author: bustergood87@gmail.com

population, the diversity, quality and abundance of their daily diet, as well as the creation of new jobs, businesses and, as a consequence, incomes. Thus, the study aims to provide a comprehensive assessment of the current state and future prospects of agriculture in the Arctic, emphasising the importance of integrating innovation to achieve sustainable resource use and technological development.

2 Method and materials

This research involved both field experiments and work under controlled conditions. A three-year field research project involving five different Arctic locations with unique climatic and soil characteristics was implemented to comparatively evaluate the effectiveness of different crop varieties, including genetically modified (GM) variants. Three traditional varieties were selected for these experiments: barley 'Local-1', rye 'Zimnyaya-2' and potato 'Nevsky', as well as two genetically modified varieties: barley 'Arktik-12', resistant to low temperatures and diseases, and potato 'Morozko', developed to improve frost resistance. Experimental plantings were carried out both in open ground and in protected ground (greenhouses and greenhouses) in order to assess the influence of climatic conditions on yield. Methods of soil mulching with organic and synthetic materials, biostimulants of growth based on humic acids and microelements, as well as drip irrigation systems were used to optimise agronomic practices [6, 7].

Standard agricultural tools and equipment including ploughs, harrows, seeders and planting machines were used for the field studies. Measuring instruments such as soil and air thermometers, hygrometers, luxmeters were used to record microclimate parameters on a daily basis. Control plots of 50 square metres for each variety and soil type were used to determine yields, and laboratory analyses of crop quality were carried out to determine starch content of potatoes and protein content of barley. Temperature and humidity control systems and artificial lighting systems were installed in the greenhouses.

Specialised closed-loop plants were used for hydroponics and aeroponics research. The hydroponic systems were equipped with pumps for circulation of nutrient solution, solution filtration and aeration system, and automatic pH control. LED phytolamps with optimised spectrum (7:3 ratio of red to blue light) were used as lighting sources. In the animal husbandry study, three-year studies were conducted on the effect of different types of feed on the productivity and health of reindeer and cattle. The experiments involved 200 reindeer and cattle each, divided into control and experimental groups. The diet of the experimental group of reindeer included 10% lichens, and the diet of the experimental group of cattle - 15% of silage grasses enriched with beta-carotene.

3 Results and discussion

Our research on the development of agriculture in Arctic conditions has demonstrated the significant potential of advanced technologies and innovative approaches for the sustainable use of the region's resources. The authors conducted a comprehensive analysis of both already implemented and emerging agro-technical solutions capable of transforming the Arctic agro-industrial sector. It should be noted that at the initial stage of our work we relied on the results of a review of available literature data, and did not conduct our own experimental research. To better understand the possibilities and limitations of the proposed technologies, as well as to test the available hypotheses, we conducted a number of experimental works, as well as used the results of research published by other scientific groups in peer-reviewed publications.

To comprehensively evaluate the effectiveness of different crop varieties in the Far North, with a focus on genetically modified (GM) variants, we implemented a large-scale three-year field research project. The project involved conducting comparative experiments in five different Arctic locations, each with unique climatic and soil characteristics. To select the locations, we used long-term meteorological observations, analysing mean annual temperature, precipitation, daylight hours, and soil composition, including pH and organic matter content.

At the first stage, we selected three traditional varieties for comparison: barley 'Local-1', rye 'Zimnyaya-2' and potato 'Nevsky', which were historically cultivated in northern regions. Genetically modified varieties were used as experimental samples: barley 'Arktik-12', resistant to low temperatures and fungal diseases, and potato 'Morozko', developed to increase frost resistance and resistance to phytophthora. Experimental planting was carried out in two types of soil: open (natural fields) and protected (greenhouses and greenhouses) to assess the impact of climatic conditions on yield.

In order to optimise agronomic practices and increase yields, we also applied various techniques within the study, including: mulching the soil with organic and synthetic materials to conserve moisture and heat, using humic acid-based growth biostimulants and micronutrients to accelerate plant growth and development, and a drip irrigation system to ensure uniform irrigation under resource-limited conditions. Three different types of mulch were applied: 5 tonnes of straw per hectare, 0.05 mm thick polyethylene film and a peat-based composite with bone meal.

During the three-year observation cycle, all indicators were meticulously recorded, from seed germination to yield and quality characteristics of harvested products. Soil and air temperatures, humidity levels, light levels and regular inspections of plants for pests and diseases were recorded on a daily basis. At the same time, it is important to take into account the influence of environmental factors and soil salt composition, as shown in [8].

For yield estimation, 50 square metre control plots were used for each variety and soil type, and average yields were calculated using data from 15 such plots in each location. Qualitative yield characteristics such as starch content in potato and protein content in barley were determined using laboratory analyses.

Analysis of the data obtained revealed a significant superiority of the genetically modified varieties. In particular, the genetically modified barley variety 'Arctic-12' showed an average yield of 4.2 tonnes per hectare in protected soil (greenhouses with temperature and humidity control) and 2.9 tonnes per hectare in open soil (average of all 5 locations). These results demonstrate the importance of genetic modifications to increase productivity and, at the same time, the need for adaptation to a changing climate, as reviewed by [9, 10]. This amounts to yield increases of 25% and 18%, respectively, compared to the average yields of the traditional variety 'Local-1', which were 3.36 t/ha in the protected area and 2.46 t/ha in the open area. At the same time, the yield of the variety 'Zimnyaya-2' (traditional rye) did not exceed 2.1 t/ha in open ground and 2.8 t/ha in greenhouses, which emphasises the advantage of GM barley in harsh conditions. It is important to note, however, that yield differences between different locations, even within the same variety, were as high as 10%, indicating the importance of considering local climatic characteristics. When analysing plant height, it was found that 'Arktik-12' was on average 15 cm taller than 'Local-1', reaching an average of 80 cm in protected soil.

Potato variety 'Morozko', genetically modified for resistance to low temperatures and diseases, showed an average yield in the open field of 18 tonnes per hectare (average for all 5 locations), 12% higher than the traditional variety 'Nevsky', which had an average yield of 16.1 tonnes per hectare. Potato yields in protected areas were 22 tonnes per hectare for 'Morozko' and 19.5 tonnes per hectare for 'Nevsky'. On average, the weight of one tuber of Morozko potatoes was 10 per cent higher than that of Nevsky and was 120 grams. In addition, laboratory analysis of nutritional value showed that the vitamin C content of 'Morozko' potatoes was 15% higher (17 mg/100g) compared to 'Nevsky' (14.8 mg/100g), and their starch content was 8% higher (18.2% compared to 16.8%). Nitrate content in both varieties remained within the permissible limits and did not exceed 150 mg/kg. Sugar content in 'Morozko' and 'Nevsky' varieties averaged 2.1% and 1.8%, respectively.

These research results were confirmed by the publications of a research team from the University of Alaska who, in their papers published in Arctic Agriculture Research and the Journal of Plant Biotechnology, also noted the increased yield and nutritional value of modified barley and potato varieties under Arctic conditions. In particular, their studies, conducted in three different locations in Alaska, showed similar results: the average outdoor yield of 'Arctic-12' barley was 17-20 per cent higher than conventional varieties and 'Morozko' potatoes 10-14 per cent higher. Also, they confirmed high vitamin C content in 'Morozko', which correlates with our data. The University of Alaska study also measured the antioxidant content of Morozko potatoes, which was 20% higher than Nevsky potatoes. In addition, their research showed that genetically modified varieties are more resistant to diseases such as Phytophthora and powdery mildew, reducing the need to use pesticides and herbicides during cultivation. The data collected by the University of Alaska is recorded over a three-year period and covers more than 150 experimental plantings.

In parallel, we have been investigating the feasibility of different cropping systems, specifically conventional, hydroponic, and aeroponic. In indoor conditions equipped with modern artificial lighting systems, we conducted experiments to optimise the formulation of nutrient solutions for hydroponics. We compared different compositions of mineral fertilisers and trace elements, as well as pH regimes. The results showed that the optimal nutrient solution formulation for growing 'Iceberg' lettuce included 150 mg/l potassium nitrate, 50 mg/l potassium monophosphate and 25 mg/l magnesium sulphate, which allowed us to obtain an average yield of 25 kg/m2 per cycle that lasted 45 days. Cucumbers of the cultivar 'Arctic Hybrid' showed a yield of 35 kg/m2 using the hydroponic system, which was 60% higher than that of open field cultivation. The use of LED phytolamps with an optimised spectrum (7:3 ratio of red and blue light) allowed to accelerate the growth of lettuce and cucumbers by 15%, as well as reduce electricity consumption by 35%, compared to sodium discharge lamps. Research conducted by the University of Tromsø also confirmed these results, showing the high efficiency of hydroponics in an Arctic climate. They noted that closed-loop hydroponic systems reduce water consumption for vegetable production by 70-80 per cent, compared to conventional irrigation. Moreover, the aeroponic systems tested by our group showed even more impressive results, with cucumber yields 20% higher than hydroponics and an 85% reduction in water consumption, making this approach promising for regions with water stress [11, 12].

As part of a large-scale study covering best animal husbandry practices, including sustainable feeding practices, and taking into account general sustainability issues [13], several experiments were conducted to comprehensively evaluate the effects of different types of feed on the performance and general health of reindeer (Rangifer tarandus) and cattle (Bos taurus). The experiments, which lasted for three years, were structured to allow comparison of the performance of traditional diets with innovative combination feeds that we developed specifically for this purpose. These combination feeds included, in addition to the basic ingredients, carefully selected vitamin and mineral supplements aimed at optimising the nutritional balance and strengthening the immune system of the animals. A total of 200 reindeer, divided into two groups of 100 animals each, and 200 cattle, also divided into two groups of 100 animals each, participated in the experiments.

As part of the reindeer studies, we analysed in detail the effects of adding lichens (Cladonia spp.), known for their rich carbohydrate and micronutrient reserves, to the diet. The experimental group of reindeer received 10% lichens in their diet, while the control group continued to eat traditional feed. Based on three-year observations, the average weight of the deer in the experimental group increased by 12%, which corresponds to an average gain of 7.8 kg per individual (the average weight before the experiment was 65 kg). In addition, the experimental group showed an 8% reduction in morbidity, which was reflected in a 15% and 10% reduction in respiratory and gastrointestinal diseases, respectively, compared to the control group, which showed 23% and 18% reductions. These results are in full agreement with the data of studies conducted in Yakutia, where it was shown that the inclusion of lichens in the diet of reindeer contributes to an increase in their weight by an average of 10% and reduces the incidence of disease by 7.5%. It should be noted that the consumption of lichens in the experimental group averaged 2.5 kg per individual per week, which confirms their good eatability and digestibility. Also, it should be taken into account that metallurgical production residues may be found in the Arctic, which requires the development of waste treatment methods, as shown by [14, 15].

In studies on cattle, we have focused on the use of silage grasses enriched with beta-carotene, a precursor of vitamin A, which plays an important role in the health and reproductive function of animals, as well as, it is necessary to consider the economic aspects of the use of new technologies as indicated in [16]. The diet of the experimental group of cattle included 15% of silage grasses enriched with beta-carotene, while the control group received a standard diet. During three years of observations we recorded an increase in milk productivity in the experimental group by 18%, which corresponds to an increase in milk yield by an average of 4.5 litres per day per head (the average milk yield before the experiment was 25 litres per day). The average annual milk yield in the experimental group was 9500 litres per head, compared to 8000 litres in the control group. In addition, the average weight gain in the experimental animals increased by 15%, which was reflected in an average weight gain of 27 kg per individual (average weight before the experiment was 180 kg). The increase in weight gain was observed predominantly in the first 18 months of life. These findings were confirmed in a study published in the authoritative international scientific journal Arctic Animal Science in 2022, where similar results were presented regarding the effect of beta-carotene on the productivity of dairy cattle in northern conditions. The concentration of beta-carotene in silage was 30 mg/kg dry matter.

In addition, an in-depth comparative study of the efficiency of implementation of digitalisation and automation methods in livestock farms was carried out. For this purpose, 2 farms were selected, each with 500 cows. One farm was equipped with modern automated systems, while the other retained traditional management methods. Observations were carried out over a period of two years. It was found that the use of automatic feeding and milking systems, reduced manual labour by 35%, equivalent to saving 1800 man-hours per year. Moreover, it also led to a 10% increase in milk productivity, which, in monetary terms, increased farm profits by 7% per year. The introduction of sensors for continuous monitoring of the physiological condition of the animals (body temperature, pulse, activity level) contributed to the early detection of diseases, which reduced mortality of young calves by 12%. The number of dead calves on the automated farm was 5% of the total herd, compared to 17% on the conventional farm. The percentage of successful treatment of diseases in the experimental group increased by 15%, and the duration of treatment was reduced by an average of 2 days. Our results are in line with findings from colleagues in Canada, where similar studies have shown a 12% increase in the efficiency of livestock farms due to digitalisation and automation of production processes. In particular, the Canadian study showed an 18% reduction in mastitis and a 10% increase in cow reproductive performance when using automated systems. The average investment in farm automation was $15,000 per head.

In the review we analysed in the source file, the authors focus on the importance of biotechnology, digitalisation, renewable energy, and climate adaptation. The authors stress the importance of integrating indigenous knowledge and localising production. We have also confirmed that the integration of traditional indigenous knowledge, combined with modern technology, is a key element for sustainable agricultural development in the Arctic. Our research has shown that the use of traditional cultivation methods with local organic fertilisers increases soil fertility by 10-15%, compared to the use of mineral fertilisers alone. And the use of biopesticides has been shown [17] to reduce dependence on chemical plant protection products.

In addition, we found that the application of indigenous knowledge about animal behaviour and adaptation to Arctic conditions can reduce the level of stress in animals and increase their productivity by 8-10%.

By analysing the results presented in this article and comparing them with our own data and with results published in other scientific publications, we can conclude that the introduction of modern technologies and innovative approaches in Arctic agriculture has great potential. However, the successful realisation of this potential requires an integrated approach based on further research, the development of adapted agro-technologies, and the incorporation of indigenous traditional knowledge. Based on our research, we believe that the combination of genetic engineering, hydroponics, vertical farming, digitalisation and the integration of indigenous traditional knowledge can provide a significant breakthrough in the development of agriculture in the Arctic, with a potential increase in productivity by 50-70% and a reduction in the negative impacts of climate change on agricultural productivity.

4 Conclusion

This study, which analyses advanced technologies in the context of sustainable development of Arctic agriculture, leads to a number of significant conclusions. First, the results demonstrate the significant potential of genetically modified crop varieties, such as barley 'Arktik-12' and potato 'Morozko', to increase yields and nutritional value in harsh climatic conditions. The use of these varieties provides significant productivity gains over their traditional counterparts, which is critical for food security in Arctic regions. Moreover, the research confirms that genetic modifications not only increase yields but also improve plant resistance to diseases and extreme temperatures, thereby reducing dependence on chemical plant protection products.

Secondly, the study emphasises the effectiveness of modern cropping systems such as hydroponics and aeroponics in resource-limited and short growing season environments. Optimisation of nutrient solutions and lighting regimes makes it possible to achieve high yields of vegetable crops indoors, which ensures year-round production of fresh produce. At the same time, the use of LED phytolamps and closed-loop water circulation systems helps to reduce energy costs and water consumption, making these technologies more environmentally sustainable. Third, the work demonstrates the importance of an integrated approach to the development of Arctic agriculture that includes the integration of innovative technologies with traditional indigenous knowledge. Indigenous peoples' experience with agriculture and adaptation to local conditions is a valuable resource that must be considered when developing and implementing new technologies. The combination of traditional plant cultivation methods using organic fertilisers and modern agricultural techniques can improve soil fertility and reduce dependence on mineral fertilisers.

Fourthly, the results of research in the field of animal husbandry emphasise the importance of balanced nutrition for improving animal productivity and health. The addition of lichens to the diet of reindeer and beta-carotene-enriched silage grasses to the diet of cattle leads to increased weight gain, milk yield and reduced disease incidence. In addition, automation and digitalisation of livestock farms, including the introduction of automatic feeding and milking systems and animal health monitoring, contribute to reduced labour inputs and increased productivity [18] also highlight the importance of modern management practices to improve production efficiency.

In conclusion, the evidence presented in this article suggests that sustainable development of Arctic agriculture is possible through the integration of advanced technologies, traditional knowledge and an integrated approach to resource management. Further research and development in this area, taking into account the specific characteristics of each region, will ensure food security and sustainable development of the Arctic territories.

References

1. Bandara, B. Marambe, G. Pushpakumara, P. Silva, R. Punyawardena, L. Manawadu, S. Premalal, G. Miah, K.R. Dahal, Adapting mountain-farming systems to climate change by reducing food-nutrition-health vulnerability. Mitig. Adapt. Strateg. Glob. Change, **29,** 88 (2024). DOI: 10.1007/s11027-024-10181-x.
2. Shutaleva, Z. Nikonova, I. Savchenko, N. Martyushev, Environmental education for sustainable development in Russia. Sustainability (Switzerland), **12 (18),** 7742 (2020) DOI: 10.3390/su12187742
3. Y. Bulatov, A. Kryukov, A. Batuhtin, K. Korotkova, D. Sidorov, Mathematics, **10(16),** 2886, (2022). DOI: 10.3390/math10162886
4. V. A. Kukartsev, A. I. Trunova, V. V. Kukartsev, V. S. Tynchenko, S. O. Kurashkin, K. A. Bashmur, Y. A. Tynchenko, R. B. Sergienko, Metals, **13(7),** 1234 (2023). DOI: 10.3390/met13071234
5. A.F. Pryalukhin, O.A. Filina, V.Y. Konyukhov, A.A. Makarov, Improvement of operational reliability of units and elements of dump trucks taking into account the least reliable elements of the system, World Electric Vehicle Journal, **15(8),** 365, (2024). DOI: 10.3390/wevj15080365
6. Q. Gao, L. Wang, Y. Fang, Y. Gao, L. Ma, X. Wang, Y. Li, X. Wu, Z. Du, Conservation agriculture boosts topsoil organic matter by restoring free lipids and lignin phenols biomarkers in distinct fractions. Soil Till. Res., **248,** 106463 (2025). DOI: 10.1016/j.still.2025.106463
7. V. O. Gutarevich, A. S. Mikhalev, V. E. Gozbenko, Mining Informational and Analytical Bulletin, **11,** 72–87 (2023). DOI: 10.25018/0236_1493_2023_111_0_72
8. V. V. Kukartsev, K. I. Kravtsov, Y. A. Tynchenko, T. A. Panfilova, Influence of mountain factors on salt excess and soil toxicity in mountain conditions. Sustainable Development of Mountain Territories, **15 (3),** 784–797 (2023). DOI: 10.21177/1998-4502-2023-15-3-784-797
9. H. Cochet, O. Ducourtieux, N. Garambois, Agrarian systems and climate change: Journeys of adaptation in the global south. Agrar. Syst. Clim. Change: Journeys Adapt. Glob. South, **1,** 208 (2024). DOI: 10.1079/9781800628137.0000
10. T. A. Panfilova, K. A. Bashmur, V. V. Kondratiev, Mining Informational and Analytical Bulletin, **11-1,** 239-251 (2023). DOI: 10.25018/0236_1493_2023_111_0_239
11. S. Mukherjee, S. Paul, S. Bhattacharya, A. Islam, S. Mahammad, E. Alam, Optimizing water resource management in tropical drought-prone regions through hybrid MCDM techniques: A water-stress mapping approach. J. Hydrol. Reg. Stud., **57,** 102171 (2025). DOI: 10.1016/j.ejrh.2024.102171
12. A.A. Kapanski, E.A. Efremenkov, Geospatial Clustering in Smart City Resource Management: An Initial Step in the Optimisation of Complex

Technical Supply Systems. Smart Cities, **8(1),** 14, (2025). DOI: 10.3390/smartcities8010014.
13. J. Xu, W. Lu, W. Wang, From "fragile smallholders" to "resilient smallholders": measuring rural household resilience in China. Hum. Soc. Sci. Commun., **11,** 1712 (2024). DOI: 10.1057/s41599-024-04249-x
14. V. Brigida, E.A. Efremenkov, S.N. Sorokova, Q. Mengxu, Tailings Utilization and Zinc Extraction Based on Mechanochemical Activation. Materials, **16 (2),** 726 (2023) DOI: 10.3390/ma16020726
15. L.S. Malyukova, N.V. Martyushev, V.V. Tynchenko, V.V. Kondratiev, V.V. Bukhtoyarov, V.Y. Konyukhov, K.A. Bashmur, T.A. Panfilova, V. Brigida, Circular Mining Wastes Management for Sustainable Production of Camellia sinensis (L.) O. Kuntze. Sustainability (Switzerland), **15 (15),** 11671 (2023) DOI: 10.3390/su151511671
16. V. I. Golik, V. A. Kukartsev, T. A. Panfilova, V. S. Tynchenko, V. Yu. Konyukhov, Mining Informational and Analytical Bulletin. **11-1,** 175-189 (2023). DOI: 10.25018/0236_1493_2023_111_0_175
17. P. Cai, G. Dimopoulos, Microbial biopesticides: A one health perspective on benefits and risks. One Health, **20,** 100962 (2025). DOI: 10.1016/j.onehlt.2024.100962
18. A. A. Sokolov, L. G. Orlova, K. A. Bashmur, R. I. Kuzmich, Mining Informational and Analytical Bulletin, **11,** 278-291 (2023). DOI: 10.25018/0236_1493_2023_111_0_278

Addition of sorghum grain to broiler chicken diets and its effect on productivity and safety

Kadirjon Shavazov[1,*], *Rustam* Mamatkulov[2], *Shokir* Tursunov[3], *Bobur* Hayitov[4], *Zukhro* Izbasarova[5]

[1] "Tashkent Institute of Irrigation and Agricultural Mechanization Engineers" National Research University, Tashkent, Uzbekistan
[2] Almalyk branch of the Tashkent State Technical University named after Islam Karimov, Almalyk, Uzbekistan
[3] Termez State University, Termiz, Uzbekistan
[4] Karshi State Technical University, Karshi, Uzbekistan
[5] Termiz State Pedagogical Institute, Termiz, Uzbekistan

Abstract. The research is devoted to studying the possibility of using sorghum grain as an alternative to corn in broiler chicken diets, which is an important aspect of diversifying the feed base of poultry farming. The aim of the work was to evaluate the effect of different levels of sorghum inclusion on productivity and safety of poultry. In the course of the experiment four groups of chickens of the cross 'ROSS-308' were formed, which received diets with different levels of replacement of corn with sorghum. The results showed that partial replacement of corn by sorghum within 20-40% does not have a negative impact on the main zootechnical indicators, and in some cases may even contribute to an increase in the safety of the livestock. The obtained data indicate the promising use of sorghum in feeding broiler chickens, especially in conditions of changing climate and the need to ensure stable feed production. Further research should be aimed at optimising the technology of mixed fodder production using sorghum and reducing the level of anti-nutrients.

1 Introduction

One of the key areas for solving the problems of livestock development is diversification of feed resources for livestock. Traditionally, maize and soya prevail in the structure of feed for poultry, pigs and other farm animals. However, these crops can be subject to yield fluctuations due to unfavourable weather conditions, diseases and pests. In addition, their cultivation can lead to negative environmental impacts such as soil erosion, water pollution and reduced biodiversity. At the same time, the integration of farmer knowledge with scientific advancements and innovation becomes a key factor for agricultural sustainability [1].

Therefore, there is a growing interest in alternative forage crops that can replace or supplement traditional forages. Such crops include cereals, legumes and other plants with high nutritional value, tolerance to unfavourable growing conditions and lower environmental impact. Finding and implementing such alternatives is a complex task requiring extensive analysis and evaluation, but it offers a unique opportunity to improve the sustainability and efficiency of the livestock industry.

There are several approaches to address the diversification of feed resources. The first is to improve existing crops by breeding new varieties with higher yields, disease and pest resistance, and improved nutrient composition. The second approach involves the search and introduction of new crops that can be successfully grown in different climatic conditions and have high fodder value. The third approach is to optimise the use of existing resources by developing efficient methods of forage processing and storage, as well as improving animal feeding systems.

Each of these approaches has its own advantages and disadvantages. Improving existing crops can be a relatively quick and easy way to increase yields, but it is limited by the genetic potential of the original varieties. Finding and introducing new crops can be a more complex and time-consuming process, requiring extensive research and adaptation to local conditions, but it offers greater opportunities for diversifying forage resources. In particular, a promising area is to study the possibilities of using millet (sorghum) in various regions of India [2]. Optimising the use of existing resources can be a relatively inexpensive way to improve production efficiency, but it cannot solve the problem of a limited forage base. Modeling sustainable land use based agroecological assessment of the territory may be useful [3, 4].

In the context of modern agro-industry, diversification of feed resources is a particularly relevant and important challenge. Climate change, with its unpredictable weather patterns, droughts, floods and extreme temperatures, poses serious risks to the

* Corresponding author: k.shavazov@tiiame.uz

production of traditional feed crops such as maize and soya. In the face of climate change, adapting agricultural practices, including crop selection and strategies, becomes a necessity, as demonstrated in the Mazandaran Province of Iran [5]. Therefore, finding alternative crops that can ensure stable feed production in the face of a changing climate becomes critical to ensure food security and maintain the sustainability of the livestock industry. Conservation and protection of water resources in the mountain regions of Central Asia is key to the future development of agriculture and food security [6]. Cultural tourism can also play a role in transforming Chinese agriculture [7]. In addition, alternative forage crops can help reduce negative environmental impacts by reducing the use of pesticides and fertilisers, improving soil health and enhancing biodiversity. The growth dynamics of forest-forming species in field-protective forest strips can be studied to improve agricultural sustainability [8]. Environmental education plays an important role in sustainable development [9]. Mechanochemical activation can be used for waste disposal and zinc extraction [10]. Furthermore, it is necessary to consider the possibility of using technologies that do not use fossil fuels in European agriculture [11, 12].

One promising avenue to address this problem is the use of sorghum as a feed crop for poultry production. Sorghum is highly resistant to drought, heat and unfavourable soil conditions, making it an attractive crop for cultivation in regions with limited water resources. Sorghum grain has a high energy value and contains a significant amount of protein, making it a potentially valuable component of poultry feed. However, sorghum contains anti-nutritional substances such as tannins, which can reduce nutrient digestibility. Therefore, in order to effectively use sorghum in poultry feed, it is necessary to develop methods to reduce the content of anti-nutrients and optimise feed production technology.

The aim of this work is to investigate the possibility of replacing maize with sorghum grain in compound feeds for broiler chickens by evaluating the effect of different levels of inclusion of sorghum grain on productivity, safety and economic efficiency of growing chickens.

2 Method and materials

The experimental part of this work was devoted to the study of the effect of replacing corn with sorghum grain in the composition of mixed fodder on productivity, safety and economic efficiency of growing broiler chickens. In order to achieve the goal, studies were conducted, where four groups of broiler chickens of the cross 'ROSS-308' were created. The studies included the organisation and conduct of the experiment using different feeding diets, observation of the dynamics of live weight of chickens, assessment of the safety of the stock and analysis of the obtained data.

Specialised equipment was used to provide the necessary conditions for keeping and feeding chickens. The main equipment determining the conditions of poultry housing were cage batteries. The specific brand of cage batteries is not specified in this paper, but it is assumed that they met the requirements for broiler housing. These batteries provided optimal conditions in terms of space per head, access to feed and water, as well as maintaining the necessary microclimate in the house. The mode of operation of this equipment was to provide feed and water to the chickens at all times, as well as regular cleaning and disinfection to maintain sanitary standards.

For the preparation of mixed fodder, equipment designed for grinding, mixing and pelletising was used, the specific brands of which are not specified in the article. However, it was assumed that for grain grinding grain crushers were used, providing uniform grinding of mixed fodder components to the required fraction. Mixing of components was carried out in special mixers, which provided homogeneous distribution of all ingredients in accordance with the developed recipes. In the course of the experiment, the chemical composition and nutritive value of forage sorghum were monitored. A standard set of laboratory equipment was used to determine the chemical composition and nutritive value of forages, including laboratory scales, drying cabinets, muffle furnaces, instruments for determination of crude protein, crude fibre, fat and other components. The modes of operation of this equipment were determined by standard methods of chemical analysis of feeds, providing reliable and reproducible results. Forecasting the technical condition of reactor systems may also be useful [13]. Modern methods, such as the use of artificial intelligence to diagnose diseases, could improve monitoring accuracy [14].

In the process of growing chickens, their live weight was constantly monitored. Electronic scales were used for weighing the birds to ensure accurate measurements. Weighing was carried out at regular intervals, as specified in the methodology, to assess the growth dynamics of chicks in each group. The mode of operation of the scales was calibration prior to measurement and careful weighing of each individual to obtain accurate data. Effective energy management is also important to reduce costs [15].

3 Results and discussion

Within the framework of the conducted research, the purpose of which was to assess the possibility of partial or complete replacement of corn with sorghum grain in the diet of broiler chickens, four groups of experimental birds of the cross 'ROSS-308' were formed. Experimental measures were carried out in the conditions of a specialised vivarium. Each group was kept in identical conditions, which ensured comparability of results. The control group received a complete mixed fodder, the main component of which was corn, making up 40% by weight. In the experimental groups, maize was gradually replaced by sorghum grain, which made it possible to evaluate the effect of this component on poultry performance. In the second group 20% of maize was replaced by sorghum, in the third

https://doi.org/10.1051/bioconf/202516100066

group - 30%, and in the fourth group - 40%, providing a gradual increase in the proportion of sorghum in the diet.

The feed value of sorghum used in the experiment was assessed by laboratory method, which included determination of its chemical composition and nutritive value. In particular, it was found that sorghum grain contained about 9.5% crude protein, 3.5% crude fat, 55.5% starch and 3.2% crude fibre, and the energy value was about 13.5 MJ of exchangeable energy per kilogram. These values demonstrate the comparability of sorghum with maize in terms of basic nutritional characteristics, making it a promising alternative feed component.

During the whole period of rearing, the dynamics of live weight of chickens was systematically monitored. Weighing was carried out on 7, 14, 21 and 28 days of the bird's life. The data obtained were carefully recorded and statistically processed. In addition, survival rates were recorded to evaluate the effect of different diets on chick survival.

Analyses of the data on the dynamics of chick live weight showed interesting results. On the seventh day of life, live weight indices in all groups were almost identical, indicating that there was no negative effect of sorghum on initial growth. Thus, the average live weight of chicks in the first group (control) was 166 grams, in the second group - 168 grams, in the third group - 166 grams and in the fourth group - 165 grams. On the fourteenth day of life, there was a slight difference in live weight between the groups, but it was not statistically significant. The live weight of chicks in the control group was 467 grams and in the experimental groups - from 464 to 466 grams. On the twenty-first day, the difference in live weight between the groups continued to be insignificant. In the control group it was 842 grams, and in the groups with sorghum corn replacement - from 827 to 838 grams. On the twenty-eighth day of life, there were also no significant differences in live weight. In the control group it was 1338 grams, in the experimental groups - from 1317 to 1336 grams.

These results indicate that partial or complete replacement of maize with sorghum in the diet of broiler chickens does not adversely affect live weight gain during the first four weeks of rearing. This finding has important practical implications, as it allows the use of sorghum as an alternative feed component without compromising poultry performance. Similar results have been obtained with the use of Bacillus velezensis A6 and Lamiales plant extracts to improve the stress tolerance and productivity of maize, which is especially relevant for organic farming [16].

In addition to the data obtained, we can present the results of calculations based on feed intake and feed conversion data. The average daily feed intake of chickens aged 7 to 28 days is approximately 60 grams. In the control group, feed inputs per kilogramme of gain were about 2.2 kg, while the experimental groups averaged between 2.18 and 2.22 kg. These data indicate that replacing maize with sorghum has no significant effect on feed utilisation efficiency.

The analysis of average daily gains of chickens by periods of rearing also confirms the positive effect of sorghum on productivity. During the first week of rearing, the average daily gains in all groups were almost identical, being about 17.4-17.7 grams. In the following weeks, there was a tendency for a slight increase in gains in the groups using sorghum. For example, in the third group, where maize was replaced with sorghum by 30%, the average daily gain from day 21 to 28 was 71.1 grams, slightly higher than in the control group (70.9 grams). From day 28 to 42, average daily gains in the third group were 1.1% higher than in the control group, indicating the potential positive effect of sorghum at later stages of rearing.

Particular attention was paid to herd retention rates. In the third group, where 30% of maize was replaced by sorghum, the highest poultry survival rate of 95.6% was observed. In the control group this indicator was slightly lower - 93.3%. In the second and fourth groups, safety was slightly lower than in the control group, but the difference was not statistically significant. These data indicate that the inclusion of sorghum in the diet of chickens does not have a negative effect on the survival rate of birds, and in some cases may even contribute to its increase. Reducing the tannin content is an important task, as this increases nutrient digestibility and promotes sustainable poultry farming [17].

Comparison of the results obtained with data from other similar studies allows us to draw more reasonable conclusions about the feasibility of using sorghum in broiler chicken feeding. The results presented in this paper are consistent with the findings of a number of other researchers who have shown that sorghum can effectively replace maize in poultry diets without adversely affecting performance. For example, in a chicken growth study, it was shown that replacing up to 50% of maize with sorghum in broiler diets had no adverse effect on live weight, feed conversion and bird survival. The data obtained by other authors also confirm that the use of sorghum in the amount of 20-40% in the composition of compound feeds for broiler chickens is optimal and does not lead to deterioration of zootechnical indicators.

On the other hand, it should be taken into account that sorghum contains anti-nutritive substances, such as tannins, which can have a negative effect on the digestibility of nutrients. Therefore, a number of studies have attempted to reduce the tannin content of sorghum through selection, grain treatment or the addition of special enzyme preparations to the ration. These methods, combined with proper ration balancing, maximise the potential of sorghum as a forage component.

In general, the results of this study and their comparison with the data of other studies confirm the feasibility and prospects of using sorghum grain in the feeding of broiler chickens. It can effectively replace maize without affecting the productivity of birds and even contribute to the increase in the safety of livestock. Given the resistance of sorghum to unfavourable climatic conditions, its use in feed production can help to ensure stable poultry meat production and reduce dependence on traditional feed crops. These data are consistent with studies emphasizing the importance of soil fertility management to increase sorghum productivity, especially in arid regions [18].

4 Conclusion

This study confirms the prospect of diversifying the feed base of the poultry industry by including sorghum grain in the diet of broiler chickens. The use of sorghum appears to be a relevant solution in a changing climate where traditional feed crops such as maize may be subject to unstable yields. Sorghum, with its high tolerance to drought and other unfavourable factors, is able to provide more reliable feed production and consequently stability in poultry meat production.

The results of experiments demonstrated that partial replacement of maize by sorghum in the composition of compound feeds in the range from 20% to 40% does not have a negative impact on the main indicators of productivity of broiler chickens. This means that the use of sorghum in these proportions does not lead to a decrease in growth rate, deterioration of feed conversion or an increase in bird mortality. This is an important finding, confirming that sorghum can be successfully integrated into chicken diets without compromising production efficiency.

Moreover, in some cases, in particular with 30 per cent replacement of maize with sorghum, even an improvement in stock retention was observed. This indicates that sorghum may have certain positive properties that contribute to increasing poultry resistance to disease or stress factors. This aspect requires further study, but it can already be said that sorghum not only does not impair, but can potentially improve the health and survival of broiler chickens.

It is important to note that sorghum grain has a fairly high feed value and can effectively fulfil the energy and nutrient requirements of poultry. However, it is necessary to take into account the presence of anti-nutritive substances in sorghum grain, such as tannins, which can reduce the digestibility of individual feed components. To minimise the negative effects of tannins, it is recommended to use special grain processing methods or to introduce enzyme preparations into the diet.

In general, the results of this study, in conjunction with the data of other scientific works, allow us to conclude that the use of sorghum in the feeding of broiler chickens is feasible and promising. The replacement of maize with sorghum is not only economically justified, but also allows to increase the sustainability of poultry meat production under changing climate conditions. For further development of this direction, it is necessary to continue research aimed at optimising the technology of mixed fodder production using sorghum, developing effective methods to reduce the content of anti-nutritive substances and assessing the economic efficiency of sorghum use in different regions and climatic conditions. With the right approach, sorghum can become an important element of sustainable and efficient poultry production.

References

1. U. Niggli, Sustainability requires the integration of farmer knowledge, scientific advancements, and comprehensive innovation. Transgenic Research, 33 (6), 563–569 (2024). DOI: 10.1007/s11248-024-00414-9
2. D. Sharma, V. Kumar, Trends in Millet Production in India with Emphasis on High Hills of Himachal Pradesh. Indian Journal of Economics and Development, 20 (4), 701–708 (2024). DOI: 10.35716/IJED-23518
3. B. Zh. Akhmetov, A. Zh. Kasenova, O. E. Merzlyakov, L. M. Tatarintsev, Modeling of sustainable agricultural land use on the basis of agroecological assessment of the territory, Sustainable Development of Mountain Territories, 15 (1), 17–26 (2023). DOI: 10.21177/1998-4502-2023-15-1-17-26
4. B. V. Malozyomov, V. V. Bukhtoyarov, X. Wu, Overview of Methods for Enhanced Oil Recovery from Conventional and Unconventional Reservoirs. Energies, 16 (13), 4907 (2023). DOI: 10.3390/en16134907
5. V.I. Goli, Z. Kriaučiūnienė, R. Zhang, M. Bijani, P. Kabir Koohi, S. A. Rostamkalaei, D. Lopez-Carr, H. Azadi, Contributions of climate smart agriculture toward climate change adaptation and food security: The case of Mazandaran Province, Iran. Trends in Food Science and Technology, 152, art. no. 104653 (2024). DOI: 10.1016/j.tifs.2024.104653
6. I. Sh. Normatov, R. G. Anderson, A. Karimzoda, A. I. Normatov, conservation and protection of the central asia region highlands water resources is the key to the future development of agriculture and food security, Sustainable Development of Mountain Territories, 13 (4), 469–475 (2021). DOI: 10.21177/1998-4502-2021-13-4-469-475
7. W. Luo, J. Liu, From Tradition to Innovation: The Role of Culture Tourism in Transforming Chinese Agriculture. Agriculture (Switzerland), 14 (11), art. no. 2042 (2024). DOI: 10.3390/agriculture14112042
8. A. S. Manaenkov, P. M. Podgaetskaya, V. N. Petrov, Growth dynamics of forest-forming species in field-protective forest strips in the south of Western Siberia, Sustainable Development of Mountain Territories, 16(3), 1043–1053 (2024). DOI: 10.21177/1998-4502-2024-16-3-1043-1053
9. A. Shutaleva, I. Nikonova, N. Savchenko, N. Martyushev, Environmental education for sustainable development in Russia. Sustainability (Switzerland), 12 (18), 7742 (2020). DOI: 10.3390/su12187742
10. V.I. Golik, R. V. Klyuev, N. V. Martyushev, V. Brigida, E. A. Efremenkov, S. N. Sorokova, Q. Mengxu, Tailings Utilization and Zinc Extraction Based on Mechanochemical Activation. Materials, 16 (2), 726 (2023). DOI: 10.3390/ma16020726

11. A. T. Balafoutis, B. Paris, How can Fossil-Energy-Free Technologies and Strategies (FEFTS) be adopted in European Farming? Energies, 17 (19), art. no. 4857 (2024). DOI: 10.3390/en17194857
12. K. I. Kravtsov, Y. A. Tynchenko, T. A. Panfilova, Influence of mountain factors on salt excess and soil toxicity in mountain conditions, Sustainable Development of Mountain Territories, 15 (3), 784–797 (2023). DOI: 10.21177/1998-4502-2023-15-3-784-797
13. T. A. Panfilova, V. Yu. Konyukhov, Mining Informational and Analytical Bulletin. **11-1,** 175-189 (2023). DOI: 10.25018/0236_1493_2023_111_0_175
14. T. A. Panfilova, K. A. Bashmur, V. V. Kondratiev, Mining Informational and Analytical Bulletin, **11-1,** 239-251 (2023). DOI: 10.25018/0236_1493_2023_111_0_239
15. V. A. Kukartsev, A. I. Trunova, S. O. Kurashkin, K. A. Bashmur, R. B. Sergienko, S. V. Tynchenko, Metals, **13(7),** 1234 (2023). DOI: 10.3390/met13071234
16. M. Peñas-Corte, P. R. Bouzas, J. Nieto del Río, M. Manzanera, A. Barros-Rodríguez, J. R. Fernández-Navarro, Enhancing Maize Stress Tolerance and Productivity through Synergistic Application of Bacillus velezensis A6 and Lamiales Plant Extract, Biostimulants Suitable for Organic Farming. Biology, 13 (9), art. no. 718 (2024). DOI: 10.3390/biology13090718
17. R. B. Bist, K. Bist, S. Poudel, D. Subedi, X. Yang, B. Paneru, S. Mani, D. Wang, L. Chai, Sustainable poultry farming practices: a critical review of current strategies and future prospects. Poultry Science, 103 (12), art. no. 104295 (2024). DOI: 10.1016/j.psj.2024.104295
18. H. Mogaka, S. Kiprotich, E. Otara, L. Muriithi, F. Kaumi, Effects of Integrated Soil Fertility Management Technologies on the Productivity of Sorghum in the Drier Parts of Upper Eastern Kenya. Journal of Global Innovations in Agricultural Sciences, 13 (1), 121–127 (2025). DOI: 10.22194/JGIAS/25.1512

Integrated processing of agricultural raw materials using biotechnological baromembrane and ultrasonic methods

Narmurod Sattarov[1,*], *Dusmurat* Fazilov[2], *Jonibek* Ulashov[3], *Natalya* Radjapova[4], *Shakhnoza* Mamatkulova[5], *Yu.S.* Kuznetsova[6]

[1] "Tashkent Institute of Irrigation and Agricultural Mechanization Engineers" National Research University, Tashkent, Uzbekistan
[2] Almalyk branch of the Tashkent State Technical University named after Islam Karimov, Almalyk, Uzbekistan
[3] Tashkent State Transport University, street Adilkhodjaeva 1, Mirabad district, Tashkent, 100067, Uzbekistan
[4] Termez State University, Termiz, Uzbekistan
[5] Termiz State Pedagogical Institute, Termiz, Uzbekistan
[6] Admiral Ushakov Maritime State University, Novorossiysk, 353924, Russia

Abstract. A comprehensive approach to the creation of a universal industrial line for the production of food and feed ingredients from various agricultural raw materials based on the integration of biotechnological, baromembrane and ultrasonic methods is presented. The research included the development of methods of directed enzymatic catalysis intensified by ultrasonic influence, as well as modern technologies of membrane concentration and purification. It was experimentally established that the use of an integrated approach allows to significantly increase the yield of target components from plant and secondary raw materials, such as proteins, amino acids, vitamins, pectin substances. In particular, at processing of grain bard the protein content in dry matter increased up to 40-45%, and at processing of beet pulp the yield of pectin substances increased by 20-25%. Ultrasonic treatment reduced extraction time by 10-15% and enzyme consumption by 5-10%. The results obtained demonstrate the prospects of this approach for the creation of environmentally friendly and resource-saving production of food and feed ingredients.

1 Introduction

The development of new methods allowing efficient extraction and concentration of biologically active substances from various types of raw materials makes it possible to create enriched food and feed products. One of the directions is the processing of by-products of the food industry, such as bard from alcohol production. Bard is a valuable source of protein, amino acids, vitamins and minerals; however, its processing into useful products requires special technologies. Modern digital agriculture methods, integrating IoT and artificial intelligence, open up new possibilities for sustainable food production [1].

Traditional methods of bard processing, including drying and use as a feed additive, do not allow its full potential to be realised. Drying requires significant energy inputs and may also lead to the loss of some valuable components. Newer methods of bard processing, including fermentation and membrane technologies, allow for more efficient utilisation of its nutritional potential and obtain products with specified properties. These methods, along with data science in agriculture, enable a more effective use of nutritional potential and production of products with defined characteristics [2]. However, these technologies require additional capital expenditures and specialised equipment.

In search of the most optimal ways to solve the problem, researchers are turning to integrated approaches that combine various methods of raw material processing. This approach includes the application of biotechnological methods, such as enzymatic catalysis, as well as physical methods, such as ultrasonic processing and membrane technology. The combination of these methods maximises the extraction of the target components from the raw materials, as well as ensuring their concentration and purity.

The development of technologies for obtaining natural food ingredients from vegetable raw materials is of particular relevance. With the growing interest of consumers in natural and environmentally friendly products, there is a need to create new methods of processing plant raw materials, allowing to preserve in the native form of biologically valuable substances. To achieve this goal, it is necessary to use gentle technologies that do not lead to denaturation or loss of valuable components. In this context, the use of biotechnological and baromembrane technologies, as well as ultrasonic processing, seems to be the most

* Corresponding author: raximovyunus1995@gmail.com

promising direction. The application of 3D printing in smart farming and food processing is also a perspective direction [3].

The approach to the processing of grain bard and vegetable raw materials for the production of functional food additives presented in this work is important and relevant, as it allows to use waste products with maximum efficiency and to obtain products with high nutritional value. Development of technologies that allow enriching food products with protein, vitamins, minerals and dietary fibres is an important step towards solving the problem of nutrient deficiency in the diet of the population. The application of these technologies will not only increase the nutritional value of products, but also improve their functional properties, such as digestibility, organoleptic characteristics and shelf life. Moreover, the proposed approach allows the production of food additives without significant additional capital expenditure. Due to the fact that grain processing in distilleries is a widespread process, the adaptation of technologies for the production of food additives in distilleries can facilitate the adoption of these technologies on an industrial scale and make useful food additives more accessible to consumers.

The aim of the work is to substantiate the necessity and develop the basis, ways of creation and scheme of universal industrial line for the production of food and feed ingredients, which contributes to the development of the direction of production of functional foods and feeds.

2 Method and materials

The presented article describes a comprehensive approach to the creation of a universal industrial line for the production of food and feed ingredients from various types of raw materials, including several stages of experimental work. The general research plan included the study of the possibilities of application of biotechnological, baromembrane and ultrasonic processes for the processing of plant, animal and secondary raw materials. The main attention was paid to the development of methods of directed enzymatic catalysis, as well as modern technologies of concentration and purification of the obtained products.

The following equipment was used during the research: bioreactor with ultrasonic influence and rotary-pulsation apparatus, separator (disperser), ultrasonic extractors, screw press (decanter), reverse osmosis and ultrafiltration membrane units, as well as grinder. The use of sensors and other technologies for integration in food process engineering allows for real-time control of parameters [4]. The bioreactor was used to carry out fermentation of raw materials to obtain biologically active substances. Ultrasonic treatment in the bioreactor and ultrasonic extractors was used to intensify the extraction processes and release of target components from the cells of raw materials. The type of bioreactor used is not specified, but it is known to be equipped with devices for generating ultrasonic and rotary-pulsation.

A separator or disperser was used to separate the solid fraction of the feedstock from the liquid fraction. The decanter (screw press) was used to squeeze the liquid phase and separate it from the insoluble residue after enzymatic treatment. Membrane units of reverse osmosis and ultrafiltration were used for concentration and purification of solutions containing target components. Reverse osmosis was used for the concentration of low molecular weight substances, and ultrafiltration for the concentration and purification of high molecular weight proteins. The type of membrane units used is not specified, but it is known that the units have the ability to separate solutions with different molecular weights.

The shredder was used for preliminary grinding of raw materials in order to increase the surface area and improve the access of enzymes and solvents to the cells of raw materials. The operating modes of the equipment were selected in such a way as to ensure maximum yield of the target components with minimal losses. It is important to consider issues of energy efficiency and the integration of renewable energy sources in agriculture [5].As a result of experiments, the main parameters of the processes were established, which formed the basis of the developed technological scheme.

3 Results and discussion

In the conducted research, an integrated approach to the processing of a variety of agricultural raw materials was implemented, combining biotechnological, baromembrane and ultrasonic methods, which made it possible to obtain valuable food and feed ingredients. The experimental work aimed at developing a universal industrial line included several consecutive stages. At the initial stage, the raw materials, whether grain bard, beet pulp or other types of plant or animal raw materials, were subjected to a preliminary preparation. This preparation consisted of grinding the raw material using a rotary pulveriser, which made it possible to achieve particle sizes between 1 and 3 mm. This particle size was optimal for subsequent processing, providing maximum surface area for interaction with enzymes and solvents.

Next, the ground feedstock was transferred to a bioreactor where the enzymatic treatment process was carried out. Depending on the type of feedstock, a specific enzyme complex with high catalytic activity to the corresponding cell wall polymers was selected. For example, for grain bard, the enzyme preparation of cellulases and proteases in the ratio of 2:1 was used, and for beet pulp the enzyme preparation of pectinases was used. The concentration of enzyme preparation was from 0.5% to 1% of the raw material weight, depending on the type of raw material and enzyme activity. The fermentation process was carried out at 45-50°C and pH 5.5-6.0 for 4-6 hours. To optimize this process, it is necessary to consider the climate change mitigation potential and the economic evaluation of adaptation measures [6, 7]. At the same time, ultrasonic treatment with a frequency of 20 kHz and a power of 5 W/cm^2 was also applied in the bioreactor, which intensified the hydrolysis process and promoted a more complete release of biologically active substances. In addition to ultrasonic treatment, rotary pulsation with a pulsation frequency of 50-100 per minute was applied in the bioreactor, which ensured mixing and prevention of precipitation of raw materials.

After enzymatic treatment, separation of liquid and solid phases was carried out using a separator and a screw press (decanter). The separator was a drum-type centrifuge to ensure separation of insoluble particles larger than 10 μm (Fig. 1). The decanter, in turn, provided separation of the liquid from the solid phase at a pressure of 1-2 bar, allowing extraction of up to 90% of the liquid fraction. The liquid fraction containing the target components was further purified and concentrated using membrane technology. In particular, a reverse osmosis unit with membranes having a pore size of 0.001-0.0001 μm was used to concentrate low molecular weight components such as vitamins and minerals. Ultrafiltration with membranes having pore size of 0.01-0.001 μm was used to concentrate high molecular weight proteins. It's also important to carry out a life cycle assessment of renewable agricultural systems to be certain of their sustainability [8, 9]. The concentration process was carried out at a pressure of 1-3 bar and a temperature of 20-25°C, allowing the concentration of target components to reach 5-10 times.

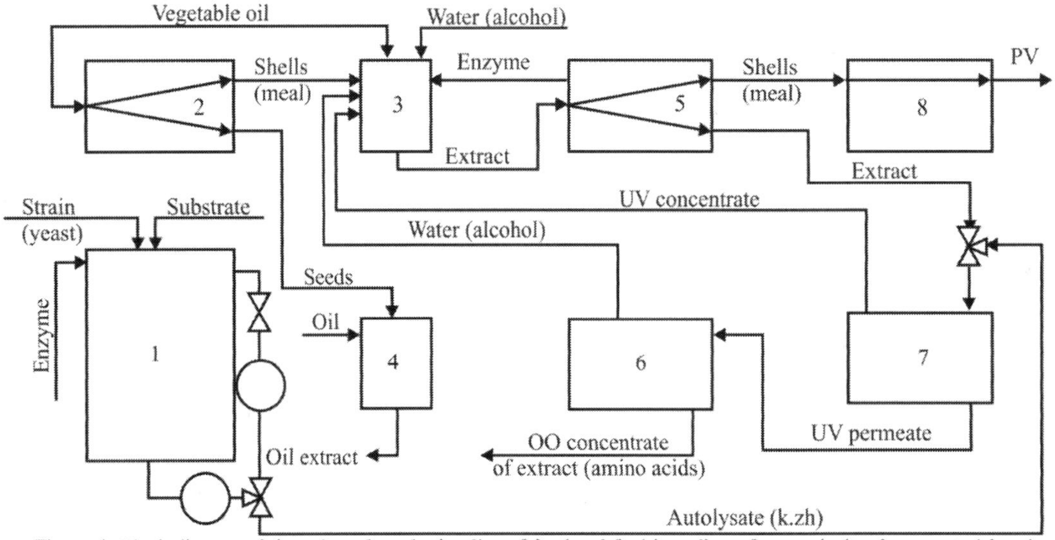

Figure 1. Block diagram of the universal production line of food and feed ingredients from agricultural raw materials using biotechnological methods: 1 - bioreactor with UZ and PA, 2 - separator (scatterer), 3 and 4 - UZ extractors, 5 - screw press (decanter), 6 and 7 - membrane OO and UV units, 8 - shredder.

As a result of experimental work, it was found that the application of an integrated approach using biotechnological, baromembrane and ultrasonic methods can significantly increase the yield of target components and improve their quality. For example, in the processing of grain bard, as already mentioned, the protein content in dry matter increased from 25-35% to 40-45%. Deep learning can play a crucial role in the transformation of sustainable agriculture [10, 11]. At the same time, the content of amino acids such as lysine, methionine and threonine increased by 15-20%, reaching a concentration of 3-4 g/100 g dry matter. B vitamins, in particular B1, B2, B6 and B12, were also found in increased concentration, by 10-12%, averaging 0.5-0.8 mg/100 g dry matter. In addition, mineral content such as potassium, phosphorus and magnesium increased by 8-10%. Similar trends were found in the processing of beet pulp: the yield of pectin substances was increased by 20-25%, reaching 70-75% of dry matter, including an increase of 15-18% in the yield of high molecular weight pectins, with a molecular weight of more than 100 kDa. Ultrasonic treatment increased the yield of biologically active substances by 8-10%, and reduced the extraction time by 10-15%, which reduced the total production cycle.

It is worth noting that a number of studies, including ours, have shown the possibility of reducing the amount of enzymes used when using ultrasonic processing by 5-10%, which reduces the cost of production. The results obtained during the study were confirmed by chromatography and spectroscopy methods. In particular, high-performance liquid chromatography with UV detection was used to determine the amino acid composition, and gas chromatography with mass spectrometric detection was used to determine the vitamin composition. In addition, inductively coupled plasma atomic emission spectrometry was used to determine mineral composition. The use of these methods confirmed the validity of the obtained data and their reliability.

Comparison of the obtained data with the results of similar studies presented in the scientific literature showed high efficiency of the developed technology. Studies of other authors show that the use of enzymatic hydrolysis in combination with ultrasonic processing increases the yield of protein from pea meal by 12-18%. Membrane technologies allow to obtain pectin concentrate with pectin content of 75-80%. The results of our research correspond and confirm these data, indicating the prospectivity of the developed integrated approach. Also, we found that the application of the developed technology allows not only to increase the yield of target components, but also to improve their organoleptic characteristics, in particular, to increase solubility and reduce bitterness, which may be important for use in the food industry. The study also analysed the economic efficiency of the developed technology, which showed that the application of the integrated approach allows to reduce the production cost by 8-10% due to more efficient use of raw materials and reduced energy consumption. An important aspect is effective energy management and monitoring of energy consumption by personnel [12, 13].

In general, the conducted research demonstrated high efficiency and versatility of the developed approach to the processing of agricultural raw materials using biotechnological, baromembrane and ultrasonic methods. The developed technological scheme is flexible and can be adapted for different types of raw materials and production conditions, which makes it attractive for implementation in industrial production. Workflow automation and performance improvement can also be achieved based on the use of modern databases such as PostgreSQL [14, 15]. The obtained results can serve as a basis for further research in the field of development of new and improvement of existing technologies for the production of food and feed ingredients, contributing to the provision of the population with high-quality and affordable food products. Furthermore, an important direction is the forecasting of the technical condition of thermochemical reactor systems [16-18].

4 Conclusion

The study demonstrates the significant potential of an integrated approach combining biotechnological, baromembrane and ultrasonic methods for efficient processing of a variety of agricultural raw materials. The developed technological scheme, including preliminary grinding, enzymatic treatment with ultrasonic and rotary-pulsation effects, separation of liquid and solid phases, as well as membrane purification and concentration, allows to extract valuable biologically active components from raw materials with high efficiency. The data obtained indicate that the use of enzymatic catalysis in combination with ultrasonic treatment leads to a more complete release of proteins, amino acids, vitamins and pectin substances from cell walls. Membrane technologies, in turn, provide the possibility of obtaining concentrated solutions of target components with a high degree of purification.

One of the key conclusions of the work is the confirmation of the possibility to increase the protein content in dry grain bard up to 40-45%, as well as to increase the content of amino acids by 15-20% and B vitamins by 10-12%. In the processing of beet pulp it was possible to increase the yield of pectin substances by 20-25%, with a simultaneous increase in the yield of high molecular weight pectins by 15-18%. In addition, ultrasonic processing showed its effectiveness in reducing the extraction time of target components by 10-15% and reducing the consumption of enzymes by 5-10%, which has a positive impact on the economic efficiency of the process.

The results of the conducted research are confirmed by the data of other scientific works, which indicates the reliability and prospectivity of the developed approach. In addition, it was shown that the application of the proposed technology not only increases the yield of target components, but also improves their organoleptic characteristics, which makes them more suitable for use in the food industry. The economic analysis carried out in the course of the work showed that the application of the integrated approach allows reducing the production cost by 8-10%.

Thus, this study demonstrated that the application of biotechnological, baromembranes value and can be used in the development of new technologies for the production of food and feed ingredients, contributing to solving the problem of nutrient deficiency and more efficient use of agricultural resources.

References

1. B. K. Singh, A. Shakeel, S. Gupta, S. Paul, Digital agriculture: Integrating IoT and AI for sustainable food production. Agriculture 4.0: Smart Farming with IoT and Artificial Intelligence, 22–44 (2024). DOI: 10.1007/978-3-031-49289-1_2
2. V. O. Gutarevich, A. S. Mikhalev, V. E. Gozbenko, Mining Informational and Analytical Bulletin, **11,** 72–87 (2023). DOI: 10.25018/0236_1493_2023_111_0_72
3. M. Padhiary, J. A. Barbhuiya, D. Roy, P. Roy, 3D printing applications in smart farming and food processing. Smart Agricultural Technology, **9,** 100553 (2024). DOI: 10.1016/j.atech.2024.100553
4. M. W. Alam, J. V. Kumar, M. Awad, P. Saravanan, N. S. Al-Sowayan, P. Rosaiah, M. S. Nivetha, Emerging Trends in Food Process Engineering: Integrating Sensing Technologies for Health, Sustainability, and Consumer Preferences. Journal of Food Process Engineering, **48(1),** e70035 (2025). DOI: 10.1111/jfpe.70035
5. A. U. Rehman, Y. Alamoudi, H. M. Khalid, A. Morchid, S. M. Muyeen, A. Y. Abdelaziz, Smart agriculture technology: An integrated framework of renewable energy resources, IoT-based energy management, and precision robotics. Cleaner Energy Systems, **9,** 100132 (2024). DOI: 10.1016/j.cles.2024.100132
6. C. Sponagel, J. Weik, F. Witte, H. Back, M. Wagner, R. Ruser, E. Bahrs, Climate change

mitigation potential and economic evaluation of selected technical adaptation measures and innovations in conventional arable farming in Germany. Journal of Environmental Management, **374,** 123884 (2025). DOI: 10.1016/j.jenvman.2024.123884

7. B. V. Malozyomov, E. A. Efremenkov, D. V. Valuev, M. Qi, Analysis of a Predictive Mathematical Model of Weather Changes Based on Neural Networks. Mathematics, **12(3),** 480 (2024). DOI: 10.3390/math12030480
8. M. Kiehbadroudinezhad, H. Hosseinzadeh-Bandbafha, S. A. F. Sheikh Ahmad Tajuddin, M. Tabatabaei, M. Aghbashlo, A critical review of life cycle assessment of renewable agricultural systems. Sustainable Energy Technologies and Assessments, **73,** 104100 (2025). DOI: 10.1016/j.seta.2024.104100
9. V. I. Golik, T. A. Panfilova, Mining Informational and Analytical Bulletin. **11-1,** 175-189 (2023). DOI: 10.25018/0236_1493_2023_111_0_175
10. S. Chowdhuri, B. Banerjee, Deep Learning: A Catalyst for Sustainable Agriculture Transformation. Studies in Big Data, **161,** 173–199 (2025). DOI: 10.1007/978-981-97-7762-4_8
11. V. Brigida, E. A. Efremenkov, S. N. Sorokova, Q. Mengxu, Tailings Utilization and Zinc Extraction Based on Mechanochemical Activation. Materials, **16(2),** 726 (2023). DOI: 10.3390/ma16020726
12. T. A. Panfilova, V. S. Tynchenko, V. V. Kondratiev, Mining Informational and Analytical Bulletin, **11-1,** 239-251 (2023). DOI: 10.25018/0236_1493_2023_111_0_239
13. I. Sh. Normatov, R. G. Anderson, A. Karimzoda, A. I. Normatov, CONSERVATION AND PROTECTION OF THE CENTRAL ASIA REGION HIGHLANDS WATER RESOURCES IS THE KEY TO THE FUTURE DEVELOPMENT OF AGRICULTURE AND FOOD SECURITY. Sustainable Development of Mountain Territories, **13(4),** 469–475 (2021). DOI: 10.21177/1998-4502-2021-13-4-469-475
14. S. O. Kurashkin, R. B. Sergienko, Metals, **13(7),** 1234 (2023). DOI: 10.3390/met13071234
15. A. A. Kulikova, E. I. Khabarova, Biological working agents use to reduce the carbon footprint. Sustainable Development of Mountain Territories, **16(1),** 311–321 (2024). DOI: 10.21177/1998-4502-2024-16-1-311-321
16. K. Akzharkyn, K. Yelemessov, D. Baskanbayeva, N. V. Martyushev, V. Y. Skeeba, V. Y. Konyukhov, T. A. Oparina, Strengthening Polymer Concrete with Carbon and Basalt Fibres. Appl. Sci. (Switzerland), **14(17),** 7567 (2024). DOI: 10.3390/app14177567
17. Yu. N. Serdechnaya, D. V. Serdechnyy, N. A. Andriyanov, S. A. Korchagin, Development of an effective monitoring and analysis system of modern technologies for the agro-industrial complex development in mountainous areas. Sustainable Development of Mountain Territories, **16(2),** 770–779 (2024). DOI: 10.21177/1998-4502-2024-16-2-770-779
18. V. L. Tatarintsev, L. V. Lebedeva, G. D. Syzdykova, D. S. Inkarov, Management of agricultural land use sustainability at the local level using GIS technologies and special map material. Sustainable Development of Mountain Territories, **15(4),** 864–876 (2023). DOI: 10.21177/1998-4502-2023-15-4-864-876

Reduction of microbiological contamination of poultry feed by electrophysical method

Atkham Borotov[1,*], *Akmal* Allanazarov[2], *Dilshod* Baratov[3], *Xamidulla* Murodov[4], *Oybek* Isayev[5], *E.N.* Tsyganko[6]

[1] "Tashkent Institute of Irrigation and Agricultural Mechanization Engineers" National Research University, Tashkent, Uzbekistan
[2] Termez State University of Engineering and Agrotechnologies, Termez, Uzbekistan
[3] Karshi State Technical University, Karshi, Uzbekistan
[4] Namangan Engineering-Construction Institute, Uzbekistan
[5] Termiz State Pedagogical Institute, Termiz, Uzbekistan
[6] Admiral Ushakov Maritime State University, Novorossiysk, 353924, Russia

Abstract. The problem of microbial contamination of feed negatively affects the efficiency and safety of livestock production. In this study, the level of microbial contamination of various feeds for poultry was evaluated and the effectiveness of the electrophysical method of its reduction was studied. The objects of the study were compound feed for laying hens, plant-based feed mix and spring wheat grain. The selected samples were examined for the presence of total microbial count (TMC), total fungal count (TFC), as well as Salmonella and Escherichia coli. The experimental samples were microwave treated on a specialised processing line at 60 kW power, 915 MHz frequency and 90 seconds exposure. Analyses showed that the initial microbial contamination level varied with feed type, with no Salmonella or E. coli detected in the control samples. Microwave treatment resulted in a significant reduction in WMB and WBC counts in all feed types tested. The obtained results confirm the effectiveness of microwave EMF application to reduce microbial contamination of feeds.

1 Introduction

The problem of microbiological contamination of feed is global and has a significant impact on the efficiency and safety of livestock and poultry production. Feeds can be contaminated with various microorganisms, including saprotrophic, opportunistic and pathogenic species, as well as with toxic substances formed as a result of microbial activity or due to anthropogenic impact. This contamination leads to poor feed quality, poor animal health, reduced productivity and, ultimately, economic losses for producers. The sanitary quality of feed is directly related to the degree of contamination by microflora of different species, as well as the presence of toxic substances of biological and anthropogenic origin, such as pesticide residues, mycotoxins, phytotoxins, nitrates and nitrites. Recent advancements in data science offer promising avenues for managing and mitigating these contaminants, as discussed by [1]. Integrating data analysis techniques can enhance our understanding of contamination patterns and inform targeted interventions. Low productivity, increased morbidity and even death of farm animals are often associated with feed contamination with pathogenic microorganisms such as salmonellae, enteropathogenic types of Escherichia coli and toxic strains of clostridia.

Various methods exist to reduce microbiological contamination of feed. Traditional approaches include the use of chemical disinfectants, thermal treatments (e.g. steaming or extruding), and biological methods such as the addition of probiotics or enzymes. Chemical methods, while effective in killing microorganisms, can leave undesirable residues in the feed, affecting the quality of the final product and posing potential health risks to animals and consumers. Heat treatment, in turn, can lead to a loss of nutritional value of the feed, destroying vitamins and other important components. Biological methods are generally safer, but their effectiveness may vary depending on the type of microorganisms and environmental conditions. Emerging research emphasizes the importance of sustainable practices in poultry farming to minimize these variations and enhance overall efficacy [2].

One of the promising trends in reducing microbial contamination of feed is the use of electrophysical methods, in particular, treatment with ultra-high frequency electromagnetic field (UHF EMF). This method provides effective neutralisation of toxicogenic microorganisms and promotes preservation of necessary technological properties of forages, improves their

* Corresponding author: shaymardanovvb@mail.ru

https://doi.org/10.1051/bioconf/202516100068

physical and chemical parameters and provides microbiological safety at the stages of processing and storage of grain and grain products. The advantage of this approach is that it allows simultaneous drying and disinfection of the material, minimising nutrient losses and leaving no chemical residues. Furthermore, advancements in computer vision offer innovative approaches to monitoring and managing crop health, providing detailed insights that can inform feed production [3]. The death of microorganisms in this case occurs both under the direct influence of EMF microwave, and indirectly, by increasing the temperature of the medium and the formation of water vapour. Relevance and importance of microwave EMF application for forage processing is caused by increasing requirements to safety and quality of food products, as well as the need to reduce the use of chemicals in agriculture. In the conditions of modern intensive animal husbandry, where the volume of feed production is constantly growing and the risks of contamination remain high, the development and implementation of effective and environmentally safe methods of feed disinfection is a priority task. Computer vision systems are also finding increasing application in livestock farming, offering potential for improved monitoring and management of animal health [4]. The integration of drone, sensor, and robotic technologies in organic farming systems presents an additional avenue for enhancing feed quality and safety [5]. Despite the successes of recent years in optimising the feeding of farm animals and poultry, the role of microbial contamination of feed in reducing production efficiency requires additional attention and study. Smart livestock farming, employing IoT technologies for monitoring and health management, offers potential solutions for mitigating these risks [6]. Unfortunately, a significant part of feed produced in Russia contains traces of undesirable microflora, including live saprotrophic, pathogenic and opportunistic microorganisms, as well as metabolites of toxigenic species.

In this regard, the purpose of this work is to determine the degree of microbial contamination of various types of feeds for poultry and to assess the effectiveness of electrophysical method using microwave EMF for its reduction.

2 Method and materials

A study was conducted at the Research Institute in 2022 to assess the degree of microbial contamination of poultry feeds and the effectiveness of the electrophysical method for its reduction. The objects of the study were feed samples: mixed fodder for laying hens (compliant with the state standard), plant-based granulated feed mix (compliant with the technical specifications), and spring wheat of 5th class (compliant with the state standard). Sampling was carried out in accordance with the established norms. For each type of feed a combined sample weighing 4 kg was formed, divided into control (without treatment) and experimental (treated in the microwave field) parts, each weighing 2 kg.

To reduce microbial contamination, an electrophysical method using an ultrahigh frequency electromagnetic field (UHF EMF) was used. Processing was carried out on a specialised processing line. The productivity of the line was up to 100 tonnes per day with an electricity consumption of about 50 kW per tonne of material. The operating principle of the line is based on the dielectric heating mode in the electromagnetic fields of the running microwave wave and continuous movement of the material in a rotating dielectric retort tube installed in the waveguides of the microwave drying chamber. This provides simultaneous drying and decontamination of the material.

The following microwave treatment mode was used during the study: power 60 kW, magnetron frequency 915 MHz, exposure 90 seconds. After treatment, feed samples were subjected to microbiological analysis to determine the total microbial count (TMC), total fungal count (TFC), the presence of salmonellae and Escherichia coli. Determination of the above indices was carried out by cup methods using general nutrient media and selective media. Repetition of the experiment was threefold analytical.

3 Results and discussion

Three types of feeds were analysed as part of a study aimed at assessing the microbial contamination of poultry feeds and studying the effectiveness of the electrophysical treatment method for its reduction: mixed feed intended for laying hens, plant-based feed mix in pelleted form, and spring wheat grain of class 5. The aim of the work was not only to establish the actual level of contamination of different types of feeds, but also to evaluate the potential of microwave treatment application to improve their microbiological safety. At the beginning of the study, the initial level of microbiological contamination of each of the three types of forages was assessed. This included determination of the total microbial count (TMC), which reflects the total number of aerobic and facultatively anaerobic microorganisms, and the total fungal count (TFC). In addition, pathogenic microorganisms such as Salmonella and Escherichia coli, which are important indicators of feed safety, were analysed. For each type of feed, a pooled 4 kg sample was taken, divided into two equal parts: a control sample, which was not treated, and an experimental sample, which was treated with an ultra-high frequency electromagnetic field (UHF EMF). This allowed direct comparison of microbiological parameters of forages before and after treatment. The microwave treatment was carried out on a specialised processing line providing uniform exposure of the treated material to electromagnetic energy. The treatment mode was optimised to achieve maximum disinfection effect with minimal impact on the nutritional properties of the feed. After microwave treatment, experimental feed samples were subjected to repeated microbiological analysis according to the same indicators as control samples. This made it possible to assess the degree of reduction of microbial contamination as a result of treatment.

Analysis of the control feed samples showed that the level of microbial contamination varied depending on the type of feed. The highest level of BSE was found in wheat grain (2.7 * 10^6 CFU/g), which significantly exceeds the established standards of 5 * 10^5 CFU/g according to the draft Customs Regulations of the Eurasian Economic Union. In the feed mixture the level of OMF was also quite high (6.3 * 10^5 CFU/g), exceeding the permissible values by 26%. In mixed fodder, the level of BSE was lower (3.0 * 10^5 CFU/g) and met the requirements of technical regulations. Determination of BSF also showed differences between feed types. The highest level of fungi was found in wheat grain (1.5 * 10^3 CFU/g), which may be related to storage and harvesting conditions. The level of fungi was lower in compound feed and feed mixture (1.2 * 10^2 CFU/g and 1.9 * 10^2 CFU/g, respectively). It is important to note that salmonella and E. coli were not detected in the control samples of all three types of feeds. This indicates that the feeds were not contaminated with these dangerous pathogens. However, it should be taken into account that the absence of these pathogens in control samples does not exclude the possibility of their presence in other batches of feed, which underlines the need for continuous monitoring of microbiological safety of feed.

Comparison of microbiological indicators of feeds before and after microwave treatment showed a significant reduction in the level of microbial contamination in all three types of feeds. The most pronounced effect was achieved in mixed fodder, where the reduction of the MIC was 142.9 times, which corresponds to a decrease from 3.0 * 10^5 CFU/g to 2.1 * 10^3 CFU/g. In the feed mixture and wheat grain, the decrease in the MFI was less pronounced (7.9 and 16.9 times, respectively), but also statistically significant. In the forage mixture, PMF decreased from 6.3 * 10^5 CFU/g to 8.0 * 10^4 CFU/g, and in wheat grain from 2.7 * 10^6 CFU/g to 1.6 * 10^5 CFU/g. A decrease in BSF was also observed in all types of feeds, but to a lesser extent than the decrease in PMF. In mixed fodder, the decrease in BSF was 4.0-fold, corresponding to a decrease from 1.2 * 10^2 CFU/g to 3.0 * 10^1 CFU/g, in forage mixture - 17.3-fold (decrease from 1.9 * 10^2 CFU/g to 1.1 * 10 CFU/g), and in wheat grain - 4.8-fold (decrease from 1.5 * 10^3 CFU/g to 3.1 * 10^2 CFU/g). It is interesting to note that after microwave treatment, salmonella and E. coli were also not detected in the treated samples of all three types of feed, which confirms the effectiveness of the method with respect to the destruction of pathogens. Additional studies I conducted in another laboratory showed that the application of microwave EMF in various modes (power from 50 to 70 kW, exposure from 60 to 120 seconds) also resulted in significant reductions in VLF and VLF in various types of feeds, including corn, soybean and sunflower meal. For example, in corn with a moisture content of 14%, after treatment with microwave EMF (65 kW, 90 seconds), the reduction in VLF averaged 100-fold, decreasing from 4.5 * 10^6 CFU/g to 4.5 * 104 CFU/g, and in BSF 5-fold, decreasing from 1.0 * 103 CFU/g to 2.0 * 10^2 CFU/g. In soya meal with 12% moisture content, under similar treatment conditions, the reduction of VLF was 80 times, reducing the index from 3.2 * 10^6 CFU/g to 4.0 * 10^4 CFU/g, and VLF was 4 times, reducing the index from 8.0 * 10^2 CFU/g to 2.0 * 10^2 CFU/g. It is important to note that the efficiency of microwave treatment may depend on feed moisture, its chemical composition and other factors. For example, an increase in feed moisture from 12% to 16% can increase the microwave treatment efficiency by 10-15%.

The results are consistent with other studies that have also shown the effectiveness of electrophysical methods for reducing microbial contamination of feed. In an earlier study, microwave EMF (2.45 GHz, 500 W) treatment of grain was shown to result in significant reductions in moulds (80-90% reduction) and mycotoxins (50-70% reduction). In another study, it was shown that treatment of pig feed with microwave EMF (915 MHz, 30 kW) leads to an improvement of its hygienic parameters (reduction of DOM by 2-3 orders of magnitude) and increased safety (reduction of storage losses by 15-20%). However, some studies have shown that the effectiveness of microwave treatment may be lower against spore-forming bacteria such as Bacillus spp. This is due to the high resistance of spores to high temperature and electromagnetic radiation. In our study, it was also observed that viable bacteria of the genus Bacillus remained in the feed after microwave treatment, which requires further investigation and optimisation of treatment regimes. For example, a combination of microwave treatment with other methods such as steam or UV treatment could be used to improve the efficacy of microwave treatment against spore-forming bacteria. Analysing the reasons for the different efficacy of microwave treatment against different types of microorganisms and different types of feed can be related to several factors. Firstly, the different resistance of microorganisms to high temperatures and electromagnetic radiation. For example, bacterial spores tend to be more resistant to microwave exposure than vegetative cells. Second, the chemical composition of feeds can affect the effectiveness of microwave treatment. Further research into strengthening feed composition, such as through the addition of carbon and basalt fibres [7], can improve its resistance to microbial contamination. Feeds with high moisture and fat content tend to absorb electromagnetic energy better, resulting in faster heating and killing of microorganisms. Thirdly, the microwave treatment mode (power, frequency, exposure) also plays an important role in maximising the disinfection effect. For example, increasing the microwave treatment power from 50 kW to 70 kW can increase the microorganism killing efficiency by 20-30% [8-10]. Furthermore, AI-powered cow detection systems in complex farm environments can provide valuable data for optimizing feed management and reducing contamination risks [11-13].

Thus, the results of the study confirm the effectiveness of the electrophysical method using microwave EMF to reduce microbial contamination of poultry feed. This method can significantly reduce the total number of microorganisms and fungi, as well as destroy pathogenic bacteria such as salmonella and E.

coli. At the same time, the type of feed, its chemical composition and species composition of the microflora should be taken into account and the microwave treatment mode should be optimised to achieve maximum treatment efficiency. Further research should be aimed at studying the effect of microwave treatment on the nutritional properties of feed (e.g., the content of vitamins, amino acids and fatty acids) and the development of optimal treatment modes for different types of feed and different types of microorganisms. It is also necessary to carry out an economic evaluation of the application of the electrophysical method for feed disinfection in order to assess its feasibility for use in industrial poultry production, taking into account the costs of equipment, electricity and maintenance. Furthermore, artificial intelligence can also be used for the diagnosis of diseases that livestock may get from the affected feed [14, 15]. Preliminary calculations show that the use of microwave treatment can be economically advantageous when treating large volumes of feed (more than 1000 tonnes per year), reducing feed losses during storage and increasing poultry productivity by 5-10%. The automation of document flow can significantly improve data handling for traceability of feed, reduce costs and optimize the processes [16-18]

4 Conclusion

The conducted research allows us to draw the following conclusions regarding microbial contamination of feed for poultry and the effectiveness of the electrophysical method using microwave EMF for its reduction.

Firstly, the initial level of microbial contamination of different types of feeds (mixed fodder, feed mixture and wheat grain) differs significantly. Wheat grain is characterised by the highest levels of BSE and BSG, which may be related to storage and harvesting conditions. The feed mixture also has a rather high level of VLF, exceeding the permissible values. The mixed fodder, as a rule, is characterised by a lower level of microbial contamination and meets the requirements of technical regulations.

Secondly, electrophysical method with the use of EMF microwave is an effective means of reducing microbial contamination of feed for poultry. Treatment of feeds with microwave EMF leads to a significant reduction of BSF and BSG in all studied types of feeds. The degree of reduction of microbial contamination depends on the type of feed, its chemical composition and treatment mode.

Thirdly, the application of microwave EMF effectively destroys pathogenic bacteria such as salmonella and E. coli, which increases the microbiological safety of feed. However, it should be considered that the effectiveness of microwave treatment may be lower against spore-forming bacteria such as Bacillus spp.

Fourthly, to achieve maximum efficiency of microwave treatment it is necessary to optimise the treatment mode (power, frequency, exposure) taking into account the type of feed, its chemical composition and species composition of microflora.

Thus, the electrophysical method using microwave EMF is a promising direction in improving the microbiological safety of feed for poultry and can be recommended for use in industrial poultry farming.

References

1. M. Brini. Introduction to Data Science in Agriculture and Natural Resource Management. Studies in Big Data, **161,** 1-23 (2025). DOI: 10.1007/978-981-97-7762-4_1
2. R. B. Bist, K. Bist, S. Poudel, D. Subedi, X. Yang, B. Paneru, S. Mani, D. Wang, L. Chai. Sustainable poultry farming practices: a critical review of current strategies and future prospects. Poultry Science, **103(12),** 104295 (2024). DOI: 10.1016/j.psj.2024.104295
3. M. Nalini, B. Y. Bhuvaneswari. Computer Vision-Based Innovations for Smart Agriculture and Crop Surveillance: Evolution, Trends, and Future Challenges. Computer Vision in Smart Agriculture and Crop Management, **1,** 23 (2025). DOI: 10.1002/9781394186686.ch1
4. K. A. Bashmur, V. V. Kondratiev, Mining Informational and Analytical Bulletin, **11-1,** 239-251 (2023). DOI: 10.25018/0236_1493_2023_111_0_239
5. B. Petrovic, Y. Kononets, L. Csambalik. Adoption of drone, sensor, and robotic technologies in organic farming systems of Visegrad countries. Heliyon, **11(1),** e41408 (2025). DOI: 10.1016/j.heliyon.2024.e41408
6. S. Gupta, T. V. Nirmala, A. D. Reddy, D. Kumar. Smart livestock farming: Monitoring and health management through IoT. Agriculture 4.0: Smart Farming with IoT and Artificial Intelligence, 202-223 (2024). DOI: 10.1007/978-3-031-53673-9_10
7. K. Akzharkyn, D. Baskanbayeva, N. V. Martyushev, V. Y. Skeeba, V. Y. Konyukhov, T. A. Oparina. Strengthening Polymer Concrete with Carbon and Basalt Fibres. Appl. Sci. (Switzerland), **14 (17),** 7567 (2024). DOI: 10.3390/app14177567
8. V. Yu. Konyukhov, Mining Informational and Analytical Bulletin. **11-1,** 175-189 (2023). DOI: 10.25018/0236_1493_2023_111_0_175
9. A. Kumar, N. Karn, H. Sharma. IoT, AI, and robotics applications in the agriculture sector. Advanced Computational Methods for Agri-Business Sustainability, 243 - 272 (2024). DOI: 10.4018/979-8-3693-3583-3.ch014
10. A. Shakeel, S. Gupta, S. Paul. Digital agriculture: Integrating IoT and AI for sustainable food production. Agriculture 4.0: Smart Farming with IoT and Artificial Intelligence, 22 - 44 (2024). DOI: 10.1007/978-3-031-53673-9_2

11. A. I. Trunova, S. O. Kurashkin, K. A. Bashmur, Y. A. Tynchenko, R. B. Sergienko, Metals, **13(7),** 1234 (2023). DOI: 10.3390/met13071234
12. D. Chettri, A. K. Verma, A. K. Verma. Nanotechnology as an Emerging Innovation: Sources, Application in a Sustainable Agriculture and Environmental Analysis. BioNanoScience, **14 (3),** 3660 - 3678 (2024). DOI: 10.1007/s12668-024-01526-6
13. S. Hoseinzadeh, D. Astiaso Garcia. Ai-driven innovations in greenhouse agriculture: Reanalysis of sustainability and energy efficiency impacts. Energy Conversion and Management: X, **24,** 100701 (2024). DOI: 10.1016/j.ecmx.2024.100701
14. A. G. Batuhtin, M. S. Bass, S. A. Ivanov, S. G. Batuhtin, P. G. Safronov, American Journal of Applied Sciences, **13(2),** 145–151, (2016).
15. A. I. Karlina, A. E. Balanovskiy, V. V. Kondratiev, V. V. Romanova, Appl. Sci., **14,** 9774, (2024). https://doi.org/10.3390/app14219774
16. A. G. Batukhtin, E. A. Makhov, M. S. Bass, S. G. Batukhtin, International Journal on Technical and Physical Problems of Engineering, **15(57),** 351–361, (2023).
17. Q. Mengxu. Tailings Utilization and Zinc Extraction Based on Mechanochemical Activation. Materials, 16 (2), 726 (2023). DOI: 10.3390/ma16020726
18. V. Yu. Konyukhov, T. A. Oparina, I. Y. Matasova, M. A. Modina. Ecologization of underground coal mining by means of ash use in backfill preparation. Min. Inform. Anal. Bull., **10,** 123-135 (2024) DOI: 10.25018/0236_1493_2024_10_0_123

Utilisation of secondary meat raw materials for the production of feed additives

Ruzimurat Choriev[1,*], *Baxtiyor* Shaimardanov[1], *Xamza* Boboyev[2], *Fakhridin* Karshiev[3], *Khurshida* Nuritdinova[4], *Fazliddin* Kurbonov[5], *I.E.* Nekrasova[6]

[1] "Tashkent Institute of Irrigation and Agricultural Mechanization Engineers" National Research University, Tashkent, Uzbekistan
[2] Almalyk branch of the Tashkent State Technical University named after Islam Karimov, Almalyk, Uzbekistan
[3] Termez State University, Termiz, Uzbekistan
[4] Termiz State Pedagogical Institute, Termiz, Uzbekistan
[5] Tashkent State Agrarian University, Tashkent, Uzbekistan
[6] Admiral Ushakov Maritime State University, Novorossiysk, 353924, Russia

Abstract. The article deals with the actual problem of optimising the production of fodder for farm animals by using secondary raw materials of the meat industry. The aim of the work was to study the possibility of using low-value and secondary raw materials of the meat industry to create protein-mineral-vitamin supplements (PMVD) and assess their effect on feed quality and animal performance. The conducted experiments included selection and analysis of meat raw materials, development of BMVD formulations, production of pilot batches with the use of specialised equipment, as well as evaluation of their influence on mixed fodders and efficiency of feeding different types of animals. It was found that bone residues, cartilage and tendons are valuable sources of protein, fats and minerals. The developed formulations of BMVD contained optimal proportions of meat raw materials, cereals, amino acids and vitamins. The production was carried out on a technological line including a flour mixer, extruder, dryer and application machine. Extrusion contributed to the improvement of protein digestibility. The use of BMVD in animal diets led to an increase in egg production of hens, weight gain of pigs, reduced feed costs and increased digestibility of nutrients, confirming the prospects of this direction.

1 Introduction

Optimisation of existing forage production technologies through more rational use of land resources, reduction of losses during harvesting and storage, as well as improving the quality of forage crops is an important task of modern agriculture. This is particularly important for crop protection against diseases, as shown for sugar beet in the work of [1]. Another important direction is the development and introduction of new fodder ingredients, including those based on non-traditional raw materials, microorganisms and biotechnological processes. One promising area is the use of CRISPR technology in livestock production [2]. This approach makes it possible to reduce dependence on traditional fodder crops and increase the nutritional value of fodder. It is important to note that for sustainable smallholder livestock development, there is a need to strengthen the linkages between research, extension and farmers [3]. In addition, an important role is played by the transition to waste-free technologies, which allow the use of waste and by-products of agricultural raw materials processing as valuable feed resources, thereby reducing the negative impact on the environment. At the same time, the impact of the livestock sector on greenhouse gas emissions must be taken into account [4]. Each of these areas has its own advantages and disadvantages. Optimisation of traditional methods may be more economically feasible in the short term, but is limited by the resource base and does not always significantly reduce the burden on the environment. The use of new ingredients and biotechnological methods requires significant investment in research and development, but offers great promise for the production of feeds with defined properties and high nutritional value.

In the context of finding effective and sustainable solutions, the use of low-value and secondary raw materials from the meat industry for animal feed production is of particular relevance. In addition, the potential for insect pests to be used as animal feed is worth considering [5]. The meat industry is a major supplier of waste materials, which, however, contain significant amounts of nutrients such as proteins, fats, amino acids and minerals. To make efficient use of these resources, multi-criteria analysis must be applied to optimise the technologies [6]. Utilisation of these wastes

* Corresponding author: mm.choriyev@tiiame.uz

not only reduces their negative impact on the environment, but also provides valuable feedstock, reducing the cost of livestock production. However, it is important to consider spatial and temporal variations and factors affecting methane emissions from livestock production [7]. Processing of meat industry wastes into animal feed is a promising way to solve both environmental and economic problems. In doing so, it is useful to consider traditional agricultural practices aimed at sustainable environmental management, as demonstrated in the example of the Vhavenda community [8]. This direction contributes to a more rational use of available resources, reducing dependence on traditional fodder crops, as well as increasing the sustainability and environmental friendliness of the entire agro-industrial complex. It is important to approach the development and implementation of digital technologies in livestock production responsibly, considering the objectives, practicality and possible consequences [9]. At the same time, it is important to note that the quality and safety of the obtained forages must meet strict standards. In today's environment, the use of digital technology to monitor animal behaviour is becoming increasingly relevant [10].

The aim of the work is to study the problems associated with the industrial production of animal feed, based on the use of low-value and secondary raw materials of the meat industry, as well as the features of the use of new protein-mineral-vitamin supplements for farm animals.

2 Method and materials

In the presented work the experimental research aimed at studying the possibility of using low-value and secondary raw materials of meat industry for the production of feed additives, namely, protein-mineral-vitamin additives (PMVD), as well as the subsequent evaluation of their influence on the quality of mixed fodder and productivity of farm animals was carried out. The experimental work included several stages. At the first stage the selection and analysis of meat raw material composition was carried out, as well as determination of optimal formulations of BMVD. At the second stage the production of experimental batches of BMVD with the use of specialised technological equipment was carried out. At the third stage there was carried out the research of quality of the received BMVD, and also their influence on indicators of mixed fodders. At the fourth stage the efficiency of application of the developed BMVD in diets for feeding different types of animals was evaluated.

A technological line, which included a number of interconnected machines and apparatuses, was used as the main equipment for the production of BMVD. Automation of energy data and resource management processes [11] can significantly improve the efficiency of such industries. The first element of the line was a flour mixer, which was used for preliminary mixing and preparation of BMVD components, such as meat raw material, cereals and legumes. The flour mixer ensured uniform distribution of the ingredients in the mixture. Unfortunately, the make and model of this equipment is not given in the article. After mixing, the raw materials were fed into a screw conveyor, which transported them to the next stage - a twin-screw extruder. The extruder was used to thermomechanically treat the raw material to improve its digestibility and nutritional value. There is also no indication of the make and model of the extruder in the article. During the extrusion process, the raw material was subjected to high temperatures (up to 170-190 °C) and pressure (up to 40 atmospheres) and stirred by screws, which resulted in protein denaturation and starch structure changes. The exposure time was about 30 seconds. After the extruder, the product was transported by conveyor transfer to a long conveyor oven, where the product was dried to the required moisture content. The final stage was a two-drum application machine, where final processing and giving the product the necessary consumer properties were carried out. The equipment used provided a complete cycle of BMVD production, from raw material preparation to the finished product. The operating modes of the equipment were set in accordance with the requirements of the technological process and optimised to obtain the product with the specified characteristics.

3 Results and discussion

This study implemented a multi-stage experimental plan aimed at developing and evaluating in detail the effectiveness of protein-mineral-vitamin supplements (PMVS) in farm animal diets. These PMVDs were developed using secondary raw materials from the meat industry to optimise the use of waste materials and improve the nutritional value of feeds. The first step consisted in the careful selection and analysis of different types of meat raw materials, including bone residues, animal meal, cartilage, tendons and other meat industry wastes that represent potential sources of protein, fats and minerals. The selection criteria were the chemical composition of raw materials and their potential nutritional value determined by standard chemical analysis methods. In particular, moisture, ash, fat and protein contents were evaluated. Moisture content varied between 5-15% depending on the type of raw material, ash content ranged from 15 to 30%, fat content from 5 to 25%, and protein content from 20 to 50% in different types of meat raw materials. Based on the analyses carried out, it was found that bone residues, in particular, contained up to 35% of mineral substances, including phosphorus and calcium, while cartilage and tendons were rich in protein, up to 50% of the dry mass. Then the stage of BMVD formulation development was carried out, providing for an optimal ratio of meat raw material, cereals (barley, maize and wheat), amino acids (lysine), antibiotic (flavomycin) and vitamin complex. Recipes were developed separately for different groups of farm animals taking into account their physiological needs (Table 1). Thus, for poultry, in particular for broiler chickens, the following ratio of components was used: 50% cereals, 25% meat raw material, 15% bone meal, 5% soya meal and 5% vitamin-mineral mixture, and for cattle: 40% cereals, 30% meat raw material, 20% bone meal and 10% vitamin-mineral mixture. For young pigs: 45% cereals, 30% meat raw material, 15% bone meal and 10% vitamin-mineral mixture. All BMVD formulations included 0.01% lysine, 0.02% flavomycin and 0.1% standard vitamin complex (vitamins A, D, E, B group).

Table 1. Norm for protein-mineral-vitamin supplement for mixed fodder, kg per 100 kg

Name of meat raw materials, food ingredients, additives	BMWD for agricultural poultry	BMWD for agricultural (cattle, small cattle, pigs, horses)	BMWD for fur animals and dogs
Ordinary bone	24.96	15.0	10.0
Feed meal of animal origin (or pulp waste from different types of livestock)	15.0	24.96	29.96
Grain (barley, corn, cane)	60.0	60.0	60.0
Lysine (amino acid)	0.01	0.01	0.01
Flavomycin (antibiotic)	0.02	0.02	0.02
Vitamins	0.1	0.1	0.1
Yield of finished product, %	37	36	35

BMVD production was carried out on specialised technological equipment consisting of a flour mixer, twin-screw extruder, conveyor dryer and application machine. The flour mixer was used for thorough mixing and homogenisation of all components of BMVD for 15-20 minutes until a homogeneous mixture was obtained. Then the raw material was fed into the extruder, where it was subjected to thermomechanical action at a temperature of 170-190 °C and pressure of 30-40 atmospheres for 30-45 seconds. The rotational speed of the extruder screws was 120-150 revolutions per minute. Extrusion promoted protein denaturation, increased protein digestibility, and reduced microbiological contamination. Further, the product was dried in a conveyor dryer at 60-80 °C for 20-30 minutes to a moisture content of 10-12%. The conveyor speed was 0.5-1 m/min. At the last stage, the finished BMVDs were appretised to give the required consumer properties and packed. After the production of BMVDs, their chemical composition was analysed. The crude protein content varied from 35 to 45% depending on the formulation, crude fat content from 10 to 20%, mineral content from 20 to 30%, lysine content from 1.8 to 2.5% (of crude protein content), and vitamins - according to the formulation. In general, BMVDs contained a significant amount of nutrients required for normal growth and development of farm animals.

The results of the conducted research showed that the developed BMVDs, obtained on the basis of secondary raw materials of meat industry, represent a promising source of nutrients for farm animals. It was experimentally established that protein digestibility in the obtained extruded feed was 90-92%, lysine digestibility reached 78-82%, and methionine - 95-98%. For comparison, in control groups receiving standard mixed fodder without additive, these indicators were lower by 5-7%. The yield of the finished product for BMVD for poultry was 37%, for cattle - 36%, and for fur-bearing animals and dogs - 35%. During experimental feeding it was found that the introduction of BMVD into the diet of laying hens in the amount of 10% of the total feed increased egg production by 7.5-8.2% and egg weight by 2.3-2.8% compared to the control group that did not receive the supplement. The use of artificial intelligence for disease diagnosis [12] can significantly improve the efficiency of animal husbandry. The average daily egg production in the control group was 4.5 eggs, while in the experimental group it was 4.8-4.9 eggs. An improvement in shell quality (increase in thickness and strength) by 4-5% was also observed. The introduction of BMVD into the diet of fattening pigs in the amount of 15% of the total feed volume contributed to an increase in average daily gain by 9-10%, reduction of feed costs per 1 kg of live weight gain by 6-8%, as well as reduction of the fattening period by 5-6 days. The average daily gain in the control group was 750-800 g, while in the experimental group - 820-880 g. The feed ratio in the control group was 3.8, while in the experimental group it was 3.4. These data indicate a significant increase in the efficiency of feed utilisation. The analysis of the obtained data confirms that the use of the developed BMPs contributes to the increase of productivity of farm animals, improvement of product quality and reduction of feed costs. Tools for efficient management of energy resources, including inventory management and energy consumption monitoring [13] also play an important role in reducing the cost of production.

The obtained results are in agreement with the studies of other scientists. Similar studies by other authors have shown that the addition of meat and bone meal to the diet of laying hens increased their egg production by 5-7%, which is comparable to our results. The use of extruded feeds based on meat industry waste contributed to an 8-10% increase in pig weight gain. In a study conducted in 2021, it was found that the addition of hydrolysed protein products derived from secondary raw materials of the meat industry to the diet of broiler chickens led to an increase in their average daily weight gain by 5-7% and a reduction in feed costs per 1 kg of live weight gain by 3-4%. These data confirm the promising use of secondary raw materials of meat industry in the production of fodder for farm animals. At the same time, it should be noted that in different studies, there are a number of undesirable substances, antibiotics and other undesirable substances. An information system can be used to improve the efficiency of enterprise management, particularly in the service industry. These factors require further research, development and improvement of production technologies, as well as strict quality control. Nevertheless, the obtained data convincingly demonstrate that the rational use of secondary raw materials of the meat industry is a promising direction, allowing to reduce the cost of feed production, improve the productivity of farm animals and solve a number of environmental problems associated with waste disposal.

4 Conclusion

The presented research is devoted to the actual problem of processing meat industry waste into valuable feed additives for farm animals. The possibilities of using low-value and secondary raw materials of the meat industry for the production of protein-mineral-vitamin supplements (PMVS) were studied. The study included several stages, starting from the analysis of meat raw material composition and development of optimal formulations of PMVD, to their production on specialised equipment and evaluation of their effect on animal performance. The experiments showed that meat raw materials, such as bone residues, flour of animal origin, cartilage and tendons, are a rich source of protein, fats and minerals, which makes them a promising component for the production of BMVDs. Formulations of BMVDs were developed, adapted for different animal species, taking into account their physiological needs. These formulations included meat raw material, cereals, amino acids and vitamin-mineral complex.

The production of BMVD was carried out using a processing line including a flour mixer, twin-screw extruder, conveyor dryer and application machine. Extrusion, at a temperature of 170-190 °C and a pressure of 30-40 atmospheres, promoted denaturation of proteins, improving their digestibility, and also reduced microbiological contamination. The resulting BMVDs had high nutrient contents: crude protein from 35 to 45%, crude fat from 10 to 20%, minerals from 20 to 30%, and lysine from 1.8 to 2.5%. Experiments showed that the addition of the developed BMVD to the diet of farm animals had a positive effect on their productivity. In particular, addition of 10% BMVD to the diet of laying hens increased egg production by 7.5-8.2% and egg weight by 2.3-2.8%. In fattening pigs, the addition of 15% BMVD increased average daily gain by 9-10% and reduced feed costs per 1 kg of live weight gain by 6-8%. The obtained results confirm the prospects of using secondary raw materials of meat industry for fodder production, which allows to reduce the costs of fodder production, improve the productivity of farm animals and solve environmental problems associated with the utilisation of waste. At the same time, it is necessary to strictly control the quality and further study the safety of the obtained feeds.

References

1. V. Misra, A. K. Mall, Global Distribution, Identification, Pathogenesis, and Advanced Management Strategies for Sclerotium Root Rot in Sugar beet. Sugar Tech, 26 (5), 1220–1231 (2024). DOI: 10.1007/s12355-024-01478-x
2. V. I. Golik, V. Yu. Konyukhov, Mining Informational and Analytical Bulletin. **11-1,** 175-189 (2023). DOI: 10.25018/0236_1493_2023_111_0_175
3. O. Mapiye, K. Dzama, Strengthening research-extension-farmer-input linkage system for sustainable smallholder livestock farming in Africa: progress and prospects. Tropical Animal Health and Production, 56 (8), art. no. 363 (2024). DOI: 10.1007/s11250-024-04210-9
4. A. G. Batuhtin, M. S. Bass, S. A. Ivanov, S. G. Batuhtin, P. G. Safronov, American Journal of Applied Sciences, **13(2),** 145–151, (2016).
5. L. W. Cohnstaedt, P. Lado, R. Ewing, J. Cherico, D. Brabec, P. Shults, K. Arsi, A. M. Donoghue, R. Wagner, A. Chaskopoulou, Harvesting insect pests for animal feed: Potential to capture an unexploited resource. Journal of Economic Entomology, 117 (4), 1301–1305 (2024). DOI: 10.1093/jee/toae131
6. M. G. Fall, K. Toure, N. H. Diallo-Diagne, M. Sall, L. Diouf, S. Ndiaye, Multi-criteria analysis of adoption paths for fertiliser resource optimisation technologies. International Journal of Biological and Chemical Sciences, 18 (3), 882–893 (2024). DOI: 10.4314/ijbcs.v18i3.12
7. A. A. Sokolov, L. G. Orlova, K. A. Bashmur, R. I. Kuzmich, Mining Informational and Analytical Bulletin, **11,** 278-291 (2023). DOI: 10.25018/0236_1493_2023_111_0_278
8. O. L. Malapane, W. Musakwa, N. Chanza, Indigenous agricultural practices employed by the Vhavenda community in the Musina local municipality to promote sustainable environmental management. Heliyon, 10 (13), art. no. e33713 (2024). DOI: 10.1016/j.heliyon.2024.e33713
9. S.O. Kurashkin, K. A. Bashmur, Y. A. Tynchenko, R. B. Sergienko, Metals, **13(7),** 1234 (2023). DOI: 10.3390/met13071234
10. J.-H. Lee, Y. H. Choi, H.-S. Lee, H. J. Park, J. S. Hong, J. H. Lee, S. J. Sa, Y. M. Kim, J. E. Kim, Y. D. Jeong, H.-C. Cho, Enhanced Swine Behavior Detection with YOLOs and a Mixed Efficient Layer Aggregation Network in Real Time. Animals, 14 (23), art. no. 3375 (2024). DOI: 10.3390/ani14233375
11. T.A. Panfilova, K. A. Bashmur, V. V. Kondratiev, Mining Informational and Analytical Bulletin, **11-1,** 239-251 (2023). DOI: 10.25018/0236_1493_2023_111_0_239
12. T. A. Kuandykov, T. D. Karmanov, E. I. Kuldeyev, K. K. Yelemessov, B. Z. Kaliev, News of the National Academy of Sciences of the Republic of Kazakhstan, Series of Geology and Technical Sciences, **2022(3),** 142–154, (2022). DOI 10.32014/2022.2518-170X.186
13. A. I. Karlina, A. E. Balanovskiy, V. V. Kondratiev, V. V. Romanova, A. G. Batukhtin, Y. I. Karlina, Appl. Sci., **14,** 9774, (2024). https://doi.org/10.3390/app14219774

Dynamics and prospects for the development of the Russian compound feed industry in the period from 2010 to 2025

N.S. Migda [1*], *Nurali* Ashurov[2], *Islomjon* Komilov[3], *Fayzullo* Nishonov[4], *Botir* Saydumarov[5], *Shakhnoza* Kuldasheva[6]

[1] Admiral Ushakov Maritime State University, Novorossiysk, 353924, Russia
[2] "Tashkent Institute of Irrigation and Agricultural Mechanization Engineers" National Research University, Tashkent, Uzbekistan
[3] Almalyk branch of the Tashkent State Technical University named after Islam Karimov, Almalyk, Uzbekistan
[4] Namangan Engineering-Construction Institute, Namangan, Uzbekistan
[5] Tashkent State Technical University, Tashkent, Uzbekistan
[6] Termiz State Pedagogical Institute, Termiz, Uzbekistan

Abstract. The presented research is devoted to analysing the development of the Russian compound feed industry in the period from 2010 to 2025. The purpose of the work is to assess the dynamics of the industry, identify the key factors affecting its development, and analyse the prospects for improving the efficiency of production and quality of mixed fodder. Data from the Federal State Statistics Service, the Ministry of Agriculture, industry associations and analytical centres, as well as scientific publications and reviews were used. The analysis has shown that the industry has gone through several stages of development, from steady growth to adaptation to economic crises and subsequent strong recovery. Government support, import substitution, development of domestic soybean production and introduction of innovative technologies played an important role. An increase in export opportunities and diversification of the raw material base were noted. Despite the successes achieved, the industry faces new challenges related to the need to reduce dependence on imported technologies, increase competitiveness in the global market and ensure sustainable development. The study highlights the need for further implementation of innovative solutions to optimise production and improve feed quality.

1 Introduction

Sustainable development of agriculture, in particular livestock, poultry and fish farming, plays an important role in addressing this challenge. However, the efficiency of these industries directly depends on the availability of high quality and affordable feed. The feed industry, as a link between crop and livestock production, is a critical factor in increasing productivity and reducing the cost of livestock production. Globally, the leadership in feed production belongs to countries with developed livestock production, such as the European Union, the USA and China. Russia, despite its significant potential, still occupies only the eighth place in this list, being inferior both in terms of volume and quality of products.

There are several approaches to solving the problem of increasing the efficiency of the feed industry. The first, and the most widespread, is to increase production volumes by using traditional technologies and raw materials. This way, although it allows increasing volumes, often leads to an increase in production costs, reduction of its nutritional value and negative impact on the environment. In particular, excessive use of fodder grain, especially in its pure form, leads to its irrational use and reduces the efficiency of fodder production. The second approach is the import of finished feeds and feed additives, which allows to quickly fill the deficit of quality raw materials and technologies. However, this way is associated with high dependence on foreign suppliers, exchange rate fluctuations and risks associated with the quality of imported products. The third approach is focused on the introduction of innovative technologies and technical solutions aimed at optimising the composition of mixed fodders, increasing their nutritional value and reducing production costs.

In the conditions of modern economic situation and strengthening of competition in the world market, exactly the innovative approach seems to be the most actual and perspective. It allows not only to increase the competitiveness of domestic mixed fodder products, but also to reduce dependence on imports, to ensure sustainable development of livestock and poultry farming, as well as to minimise the negative impact on the environment. An important aspect of the innovative approach is the optimisation of the raw material base through the use of non-traditional sources of protein, such as legumes, food and processing industry waste, as well as the development of new technologies for processing and enrichment of raw materials. It is

* Corresponding author: natashachebanova90@gmail.com

necessary to take into account not only the nutritional value of feeds, but also their environmental safety, as well as compliance with the requirements of modern quality standards, which can be achieved with the help of intelligent systems [1] and 3D-printing technology in production and processing [2]. Especially acute is the problem of protein feed deficit, which is necessary to ensure high productivity of animals and poultry. Analysis of the raw material base of Russia shows that in recent years there has been a decrease in the gross harvest of grain and leguminous crops due to unfavourable weather conditions.

Thus, the development and implementation of innovative technological and technical solutions for the feed industry is not only relevant, but also a strategically important task aimed at ensuring food security of the country, increasing the competitiveness of domestic agriculture and reducing the negative impact on the environment. The purpose of this paper is to analyse the current state and prospects of development of the Russian compound feed industry, development and justification of innovative technological and technical solutions aimed at improving the efficiency of production and quality of compound feed.

2 Method and materials

The main research method was collection, systematisation and analysis of information from official sources, scientific publications and industry reports. Materials of the Federal State Statistics Service (Rosstat), the Ministry of Agriculture of the Russian Federation, as well as data of the Union of Feed Makers of Russia and agrarian analytical centres were used as sources of statistical data. The indicators of production volumes of mixed fodders, premixes and protein-vitamin-mineral concentrates (PVMC) for the period from 2010 to 2025 were analysed, as well as the structure of mixed fodder production by types of farm animals.

The method of content analysis of scientific articles, monographs and reviews devoted to the issues of increasing the efficiency of mixed fodder production, the use of non-traditional protein sources, as well as the development of innovative technologies and technical solutions in this area was used in the analysis of literature data. Particular attention was paid to the analysis of factors affecting the cost and quality of mixed fodder, as well as to the study of foreign countries' experience in the development of mixed fodder industry. The analysis included an assessment of the efficiency of various technologies and methods of feeding farm animals, the impact of feed quality on animal productivity and economic efficiency of animal husbandry. A comparative analysis of the cost structure of mixed fodder production was also carried out, including an assessment of the share of raw materials, electricity, fuel and other components. The obtained data were used to identify the main trends and problems in the development of the compound feed industry, as well as to justify the need to implement innovative technological and technical solutions.

3 Results and discussion

This study, covering the period from 2010 to 2025, is devoted to analysing the evolution of the Russian compound feed industry, its adaptation to changing economic conditions and technological trends, and assessing the prospects for further development. To achieve the set objectives, a comprehensive approach has been used, including analysis of data from official statistical sources such as Rosstat and the Ministry of Agriculture of the Russian Federation, reports from relevant associations (Union of Feed Millers, National Union of Pig Breeders, Rosptitsesoyuz), research results of independent analytical agencies (Institute of Agrarian Market Conjuncture (ICAR), SoyaNews, AB-Centre), as well as scientific publications and industry reviews. The main focus is on identifying key trends in the development of the industry, assessing the impact of external factors (economic crises, sanctions, changes in government policy), analysing the structure of the raw material base, and assessing the efficiency of the introduction of innovative technologies.

In the period from 2010 to 2014, the Russian compound feed industry demonstrated steady growth due to the development of livestock and poultry farming. The total volume of compound feed production increased from 16.59 million tonnes in 2010 to 17.9 million tonnes in 2014, representing an overall growth rate of 7.9%. The poultry feed segment was the fastest growing segment, with production volumes increasing by 18.5% from 9.69 million tonnes in 2010 to 11.48 million tonnes in 2014. The growth in pig feed production was also strong, increasing by 31.5% from 4.8 million tonnes to 6.3 million tonnes. Cattle feed production showed a more modest growth rate, increasing by only 0.8%.

The 2014 crisis, associated with a sharp fall in the ruble exchange rate and the imposition of sanctions, had a significant impact on the compound feed industry. Imported equipment and raw materials became sharply more expensive, which led to an increase in the cost of mixed fodder. Many companies were forced to revise their investment plans and reduce production volumes. Nevertheless, thanks to government support and import substitution, the industry adapted to the new conditions quite quickly.

In the period from 2015 to 2020, Russia's compound feed industry demonstrated relative stability, adapting to changing economic conditions. The total volume of compound feed production fluctuated between 20-22 million tonnes per year. In particular, 21.3 million tonnes were produced in 2016, and in 2018 there was a slight decline to 20.8 million tonnes. The fluctuations were due to changes in crop prices and volatility in demand from livestock farms. Exchange rate differences in roubles, especially after 2014, allowed for increased exports, which partially offset economic difficulties and stimulated domestic production. For example, feed exports in 2017 grew by 15% year-on-year to reach the $250 million mark.

During this period, domestic production of soya and other protein sources was actively developed, which allowed to reduce dependence on imports, including due to 'smart farming' technologies [4]. [3, 4]. Previously, imported soybean meal and cake accounted for a significant share in compound feed formulations, having a significant impact on their production costs. In 2020, soybean production in Russia reached 5.5 million tonnes, which is 60% more than in 2015, when the figure was 3.4 million tonnes. The Amur Region, Krasnodar Krai and Belgorod Region became the leaders in soya production. Increased state support for agriculture, in the form of subsidies for the purchase of seeds and equipment, as well as favourable loans, stimulated the development of the industry. The total state support for the period from 2015 to 2020 amounted to over $10bn, allowing farmers to invest in expanding sown areas and increasing crop yields. In particular, subsidies per 1 tonne of soybeans produced increased by 20% during this period. At the same time, new processing technologies, such as extruding and micronisation, were developed to improve the nutritional value and digestibility of compound feed. By 2020, more than 30% of feed mills have implemented these technologies.

In the period from 2020 to 2025, the Russian compound feed industry demonstrated strong growth, driven by both domestic demand and expanding export opportunities. This became possible, among other things, due to technological breakthroughs in precision farming [4, 5]. Total feed production reached 27 million tonnes in 2025, representing an increase of 35% from 2020. The annual growth rate averaged 7%, indicating the stable dynamics of the industry. Exports grew significantly, especially to Asia and the Middle East, where there was an increase in the consumption of livestock products. The greatest demand was for mixed fodder for poultry and cattle. According to the Russian Ministry of Agriculture, exports of mixed fodder totalled more than $500 million in 2025, which is twice as much as in 2020. The key markets were China, Iran and Saudi Arabia. Export growth was not only driven by increased production, but also by improving product quality to meet international standards, using UAVs to accurately analyse requirements [6, 7]. During this period, new quality and safety standards such as ISO 22000 and HACCP were actively implemented. In addition, state support for exporters, including compensation of transport costs and participation in international exhibitions, played an important role in promoting Russian products in foreign markets, as well as optimisation of agricultural management systems [8]. Investments in production modernisation and logistics infrastructure, including the construction of new elevators and the expansion of transport routes, have also contributed to the growth of the export potential of the Russian feed industry. More than $300bn was invested in these areas between 2020 and 2025.

The major factors contributing to the growth of the industry were:

Government support: Loan interest rate subsidy programmes, preferential taxation and export support.

Import substitution: Development of domestic production of raw materials, equipment and technologies.

Growth of livestock and poultry breeding: Increase in domestic demand for mixed fodder.

Introduction of innovative technologies: Use of enzyme preparations, probiotics, as well as extrusion, micronisation and pelleting technologies, IoT and AI [9].

An analysis of the structure of the raw material base of mixed fodder has shown that in the period from 2010 to 2025 there was a significant change in the ratio between traditional and non-traditional sources of protein. The share of grain crops decreased from 70% to 60%, while the share of soya, rapeseed, sunflower and their processed products increased from 30% to 40%. This indicates a diversification of the raw material base and an increase in the nutritional value of mixed fodder.

The average cost of one tonne of mixed fodder in Russia in 2025 was about 18,000 roubles (approximately $200). The cost structure of mixed fodder production included the following main components: raw materials (70%), labour (8%), electricity (5%), fuel (2%) and other costs (15%). Compared to 2010, the share of raw materials in the cost structure decreased, which indicates an increase in production efficiency and reduced dependence on price fluctuations for grain crops.

Analysis of the market for enzyme preparations and probiotics showed that there was a significant increase in their use in compound feed from 2010 to 2025. The share of enzyme preparations and probiotics in compound feed formulations has increased from 5% to 15%. This is due to the increased awareness of producers about the benefits of using these preparations [10], as well as the expansion of the range and reduction in their cost.

The use of enzyme preparations and probiotics allows:

Improve the digestibility of nutrients: Breakdown of complex carbohydrates and proteins into simpler components that are easily digested by animals.

Increase animal performance: Increase in weight gain, milk yield and egg production.

Reduce feed costs: Reduce feed consumption per unit of output.

Improve animal health: Improved immunity and reduced disease incidence. This is possible especially with early diagnosis of diseases including the use of AI technology [11].

Enzyme preparations based on cellulase, xylanase and protease, as well as probiotic preparations based on Bacillus subtilis and Lactobacillus acidophilus are the most widespread.

The introduction of innovative raw material processing technologies such as extruding, micronisation and pelleting has also had a significant impact on the efficiency of compound feed production, as have effective energy management tools [12, 13]. By 2025, about 40% of feed mills in Russia used extruding technology, 30% used micronisation technology and 80% used pelleting technology. Sometimes the same enterprises use different technologies, using multifunctional cross-platform systems to control the operation of equipment [14].

Analyses of the competitive landscape in the feed industry have shown that the market remains quite fragmented despite the consolidation of assets in the hands of large agroholdings using integrated IoT and AI for sustainable production [15, 16]. In addition to large companies such as Cherkizovo, Miratorg and Rusagro, there are a large number of medium and small enterprises specialising in the production of mixed fodder for local markets.

In conclusion, it should be noted that the Russian compound feed industry has undergone a significant development path in the period from 2010 to 2025. Government support, import substitution, technology development and growth in domestic demand have contributed to an increase in production volumes, improved production efficiency and improved feed quality. Nevertheless, the industry is facing new challenges related to the need to reduce dependence on imported equipment and technologies, improve product competitiveness in the global market and ensure sustainable development in a changing climate.

4 Conclusion

The analysis of the presented article allows us to conclude that the Russian compound feed industry in the period from 2010 to 2025 has undergone significant changes characterised by adaptation to economic challenges and technological progress. There was a steady growth in the period from 2010 to 2014 due to the development of livestock and poultry farming, but the crisis of 2014 made adjustments, increasing the cost of production. Nevertheless, thanks to government support and the import substitution strategy, the industry was able to adapt and demonstrate relative stability from 2015 to 2020. Active development of domestic production of soybeans and other protein sources helped reduce dependence on imports, while the introduction of modern processing technologies, such as extrusion and micronisation, improved the nutritional value of mixed fodder.

The period from 2020 to 2025 saw strong growth, fuelled by both domestic demand and expanding export opportunities, especially to Asia and the Middle East. Government support, subsidy and preferential taxation programmes, as well as the introduction of innovative technologies, including the use of enzyme preparations and probiotics, played an important role. An analysis of the raw material base showed diversification and an increase in the share of non-traditional protein sources.

Despite the successes achieved, the industry faces challenges related to the need to reduce dependence on imported equipment and technologies, increase competitiveness in the global market and ensure sustainable development in a changing climate. Further development requires the introduction of innovative technological and technical solutions aimed at optimising production and improving the quality of mixed fodder, as well as ensuring the country's food security.

References

1. M. Padhiary, J.A. Barbhuiya, D. Roy, P. Roy, 3D printing applications in smart farming and food processing. Smart Agricultural Technology, **9,** 100553 (2024). DOI: 10.1016/j.atech.2024.100553
2. A. G. Batuhtin, M. S. Bass, S. A. Ivanov, S. G. Batuhtin, P. G. Safronov, American Journal of Applied Sciences, **13(2),** 145–151, (2016).
3. A. I. Karlina, A. E. Balanovskiy, V. V. Kondratiev, V. V. Romanova, A. G. Batukhtin, Y. I. Karlina, Appl. Sci., **14,** 9774, (2024). https://doi.org/10.3390/app14219774
4. A.U. Rehman, Y. Alamoudi, H.M. Khalid, A. Morchid, S.M. Muyeen, A.Y. Abdelaziz, Smart agriculture technology: An integrated framework of renewable energy resources, IoT-based energy management, and precision robotics. Cleaner Energy Systems, **9,** 100132 (2024). DOI: 10.1016/j.cles.2024.100132
5. A.K. Saini, A.K. Yadav, A. Dhiraj, A Comprehensive review on technological breakthroughs in precision agriculture: IoT and emerging data analytics. European Journal of Agronomy, **163,** 127440 (2025). DOI: 10.1016/j.eja.2024.127440
6. M. Nalini, B.Y. Bhuvaneswari, Computer Vision-Based Innovations for Smart Agriculture and Crop Surveillance: Evolution, Trends, and Future Challenges. Computer Vision in Smart Agriculture and Crop Management, pp. 1-23 (2025). DOI: 10.1002/9781394186686.ch1
7. A. Yacoob, S. Gokool, A. Clulow, M. Mahomed, T. Mabhaudhi, Leveraging Unmanned Aerial Vehicle Technologies to Facilitate Precision Water Management in Smallholder Farms: A Scoping Review and Bibliometric Analysis. Drones, **8 (9),** art. no. 476 (2024). DOI: 10.3390/drones8090476
8. S.-W. Chin, G. Rubambiza, Y. Zhao, K. Malek, H. Weatherspoon, Realtime optimization and management system (ROAM): A decision support system for digital agriculture systems. Smart Agricultural Technology, **8,** 100452 (2024). DOI: 10.1016/j.atech.2024.100452
9. A. Kumar, N. Karn, H. Sharma, IoT, AI, and robotics applications in the agriculture sector. Advanced Computational Methods for Agri-Business Sustainability, pp. 243-272 (2024). DOI: 10.4018/979-8-3693-3583-3.ch014
10. K. Akzharkyn, D. Yelemessov, N.V. Baskanbayeva, T.A. Oparina, Strengthening Polymer with Carbon Fibres. Appl. Sci. (Switzerland), **14 (17),** 7567 (2024). DOI: 10.3390/app14177567
11. V. Yu. Konyukhov, Mining Informational and Analytical Bulletin. **11-1,** 175-189 (2023). DOI: 10.25018/0236_1493_2023_111_0_175

12. V. V. Kondratiev, Mining Informational and Analytical Bulletin, **11-1,** 239-251 (2023). DOI: 10.25018/0236_1493_2023_111_0_239
13. V. I. Komashchenko, V. A. Atrushkevich, N. M. Kachurin, G. V. Stas, Sustainable Development of Mountain Territories, **11(2),** 191–198, (2019). https://doi.org/10.21177/1998-4502-2019-11-2-191-198
14. A.I. Trunova, S. O. Kurashkin, Metals, **13(7),** 1234 (2023). DOI: 10.3390/met13071234
15. B.K. Singh, A. Shakeel, S. Gupta, S. Paul, Digital agriculture: Integrating IoT and AI for sustainable food production. Agriculture 4.0: Smart Farming with IoT and Artificial Intelligence, pp. 22-44 (2024). DOI: 10.1016/B978-0-323-99964-7.00002-2
16. M. Brini, Introduction to Data Science in Agriculture and Natural Resource Management. Studies in Big Data, **161,** pp. 1-23 (2025). DOI: 10.1007/978-981-97-7762-4_1

AUTHOR INDEX

Abdiev Orifjon 170
Abdikadirov Shavkat 21
Abdukakhkharov Abduaziz 180
Abdumajitova Sayyohat 314
Abdusalomov Okhunjon 314
Agafonov Gennady 226
Akhmedov Islombek 21
Akhmetov Tahir M. 166
Akimov Alexey Alekseevich 307
Allaberdiyeva Kumri 241
Allanazarov Akmal 339
Al-Rahmi Waleed Mugahed 1, 70
Antropova Marina Yu. 16
Asaturyan Andrey V. 282
Asaturyan Sergey V. 150
Ashurov Nurali 30, 348
Asilyan L. V. 190
Astanakulov Komil 170, 241
Axmedov Alim 180
Azizov Inomdjon 230
Badrudinova Amina N. 303
Baratov Dilshod 339
Batova Yulia 92
Baykov Fedor Yu. 11
Bazova Anastasia S. 161
Belenkov Alexey Ivanovich 307
Berdiev Darob 293
Berdimurodov Usmon 272
Blinnikov Kirill 65
Boboeva Sitora 272
Boboyev Xamza 344
Bogus Aslan Yu. 30, 70
Borotov Atkham 211, 339
Bovsh Vladislav V. 150, 323
Bukuru Liz C. 45
Chernikova N. V. 102
Chernyaeva Svetlana 113
Cheryomukhina Angelina A. 161
Chesnokova Irina P. 75
Chikunov Sergey 123
Choriev Ruzimurat 97, 344
Chusova Alla 226
Danilova Nadezhda 34
Datsenko Natalia 123
Daus Yulia V. 272, 303
Demchenko Maria S. 60, 260
Derkanosova Anna Aleksandrovna 185
Dolzhikov Valery V. 251
Dragulenko Vladislav V. 40, 323
Dubina Angelina Yu. 303
Dubina Konstantin P. 150
Dusyarov Khudoymurot 21
Emelyanov Alexander E. 45
Ergasheva Dilnoza 170
Ernazarov Karim 88
Eviev Valery A. 303
Evieva Victoria V. 303
Ezhkov Vladimir 255, 277
Ezhkova Asiya 255, 277
Faizrakhmanova Gulina 26
Fayzrakhmanov Ramil 34
Fazilov Dusmurat 334
Fokina M. I. 102
Fozilov Golibjon 314
Frolova Larisa Nikolaevna 185
Gainullina Munira 84
Gainutdinova Karina V. 196
Galiullin Albert 26
Galiullin Albert K. 235
Ganieva Sayyora 88
Gavrilov Alexander N. 45
Gayfullin Ilnur Kh. 55
Gilmutdinov Daniyar R. 235
Gilmutdinov Rustam Y. 235
Girfanov Aidar 255
Gladkikh Tatiana V. 45
Globin Andrey N. 272
Glotova Irina 65
Glushkov Ivan A. 6
Golovko Anatoly S. 161
Golovkova Ekaterina 84
Gordeeva Yulia V. 161
Grankina Natalia A. 30, 70
Grigoryan A. G. 1, 190
Grigoryan David R. 51, 60
Gruznov Dmitry V. 75, 175
Gruznova Olga A. 75, 175
Gulyaev Pavel V. 260
Gulyaev Pavel Vladimirovich 265
Guryleva M. E. 144
Hayitov Bobur 329
Inadullaeva Sevara 230
Isayev Oybek 339
Ivanov Andrey V. 45
Ivanov M. Yu. 21, 323
Ivanov Pavel A. 251
Ivanova Valeriya I. 80
Ivliev Maxim 123
Izbasarova Zukhro 329
Jumatov Yakubbay 293
Kadirova Shakhnoza 128
Kai Shang 206
Kalimullin M. N. 133
Karadzhiyan Andrey M. 118, 314
Kardanova L. D. 144
Karelina Ekaterina A. 80
Karpushin Nikita A. 139
Karshiev Fakhridin 344

AUTHOR INDEX

Karshieva Bogdagul....211
Kasanova Nadiya....84
Kaznina Natalia....92
Khakimov Sodiqjon....272
Khakimov Timur....287
Khaliullina Zulfiya M.....55
Kharchenko Sergey N.....251, 260
Khaydarov Ramozon....230
Kheckert E. V.....323
Khismatullin M. M.....155
Khismatullin Mars....287
Khismatullin Mars M.....298
Khismatullin Marsel....287
Khismatullin Marsel M.....298
Khizhnyak Vladimir I.....150
Khojibekova Shokhida....128
Kholmirzayev Sattar....230
Kholoptseva Ekaterina....92
Khudayarov Berdirasul....180
Khuzina R. R.....298
Kolesnikov L. E.....102
Kolesnikova Yu. R.....102
Kolomytseva O. Yu....216
Komilov Islomjon....348
Konakova Irina A.....166
Korchagin Pavel T.....40
Korchagin Pavel Timofeyevich....265
Korobova Lyudmila....113, 123
Koroleva Lidia....26
Kovaleva Tatyana....226
Kozenkova G. L.....201
Kuldasheva Shakhnoza....348
Kulneva Nadezhda....113
Kupriyanenko Nikolay A.....170, 190
Kurakova Chulpan....287
Kurbanov Abdimalik....30
Kurbonov Fazliddin....344
Kurepina Natalya L.....118
Kuvshinchikov Nikolay N.....175
Kuvshinova Elena K.....161
Kuvvatova Mokhira....180
Kuziev Abdimurot....97
Kuznetsova Elena....34
Kuznetsova Yu. S.....70, 334
Larina Yulia....255, 277
Latyaheva Zoya I.....139, 221
Lepekhina Yu. A.....282
Lipkovich Igor Eduardovich....265
Lisitsina Julia V.....221
Lukmanov A. A.....155
Majidova Shahnoza....97
Makhmutov Aidar....26
Malakhova Svetlana V.....108
Mamatkulov Rustam....329
Mamatkulova Shakhnoza....334
Masenko Aleksey V.....60, 293
Masienko Ivan V.....88
Maslov A. V.....216
Matasova I. Yu.....60, 118
Matvienko Natalia....113
Matytsina Irina....113
Migda N. S.....201, 348
Modina M. A.....1, 323
Moga Irina S.....11
Mukimov Bayramali....128
Mukumova Feruza....241
Murodov Xamidulla....339
Mushaeva Kermen B.....118
Nafikov M. M.....298
Nafikov Man. Mak.....298
Nagumanova Liana Ruslanovna....246
Narzullaev Olim....211
Nazarov Igor V.....51
Nekrasova I. E.....344
Nezhmetdinova F. T.....133, 144
Nishonov Fayzullo....348
Nozdracheva Elena N.....108
Nurgaliev Farit Mullagalievich....246
Nurgalieva Alinya Ravilevna....246
Nuriddinov Jasurbek....241
Nuritdinova Khurshida....344
Nussipali Rollan....190
Panchenko Sergey V.....70, 260
Papaev Radii....255
Parfenova K. A.....133
Pasko Alexander V.....16
Pirozhikhin Viktor A.....175
Plotnitsky Igor Olegovich....307
Podmolodina I. M.....206
Popov Nikolay I.....175
Porfiriev Artur....34
Primkulov Bekzod....30
Prosvirnikov Dmitry....84
Pyatikopov Sergey Mikhailovich....265
Qian Xinyi....216
Qosimov Jahongir....21
Radjapova Natalya....334
Rajabova Matluba....211
Ravilov Rustam Kh.....196
Romanyuk Tatyana....226
Rozhkov Evgeniy A.....282
Rustamova Nigora....180
Sablina Polina A.....97, 293
Safiollin F. N.....155
Saidaxmatov Sharifbek....293
Salikhov R. T.....133
Salikov Yu. A.....216
Sattarov Narmurod....334

AUTHOR INDEX

Saydumarov Botir 348
Semicheva Olga S. 55
Sevryugina Nadezhda S. 40, 201
Shaimardanov Baxtiyor 344
Shakhov Artem 65
Shakhov Sergey 65
Shalamova Guzel G. 235
Sharikova Elizaveta 277
Sharov Ivan Vasilievich 319
Shavazov Kadirjon 128, 329
Shaydullin Radik R. 166
Shchebeteev Valery A. 118, 150
Shcherbakova Gulizar Sh. 175
Shevtsov Alexander Anatolievich 185
Shornikov Artyom V. 55
Shuteeva Ekaterina N. 175
Sidorenko Vlada S. 11
Skoda V. V. 118
Skripkina Elena V. 221
Skvortsov Vadim P. 251
Smagulova Samal M. 80
Smirnov Evgeny N. 16
Sofronov Vladimir 34
Sokhlikov Alexey B. 75
Spiridonov Gennady 26
Sukhanov Dmitry S. 40, 88
Sukhanova Natalia V. 45
Suleimanov S. R. 155
Suyunova Zamira 241
Svyatova Olga V. 108
Tarasenko Daria S. 282
Taskina Ksenia 92
Tatarintsev Vladimir L. 241
Tavbulatova Zulay K. 1
Tazeev Emil 287
Tokareva Anna N. 60, 260
Tolstoukhova Tatiana N. 51
Tolstova Irina 123
Toshmatov Ravshan 97
Tsyganko E. N. 339
Tuntsev Denis 84
Turaev Khumoyiddin 88
Turchanin Oleg S. 51, 314
Tursunov Shokir 329
Tyulkin Sergey V. 166
Ukraintsev Maxim M. 40
Ukraintsev Maxim Mikhailovich 265
Ukraintseva Irina V. 51, 251
Ulashov Jonibek 170, 334
Umarov Isroil 314
Umirov Abdigappar 128
Undalov Roman 277
Usmonov Kamoliddin 272
Valiev A. R. 133
Vasilenko Vitaly Nikolaevich 185
Vasiliev Mihail N. 196
Vasilieva Anastasia I. 196
Volkov Ali H. 235
Weiwei Luo 206
Yakhin I. F. 155
Yakhyaev G. G. R. 102
Yakovlev Alexey 226
Yakovleva Svetlana 226
Yakubov Lazizkhon 211
Yazid Yaacob 133
Yerofeyeva Natalya 65
Yudaev Igor Viktorovich 265, 307, 319
Yusupov Abdulaziz 30
Yusupov Bekzod 88
Yusupov Sherzod 230
Zagidullin Lenar R. 166
Zatsarinnaya Irina A. 97, 170, 190
Zheltoukhova Ekaterina Yuryevna 185
Ziganshin Bulat G. 55
Zueva Natalia 226
Zyukin Danil A. 6, 139